Fuzzy Logic and Mathematics:
A Historical Perspective

Fuzzy Logic and Mathematics:
A Historical Perspective

Radim Bělohlávek, Joseph W. Dauben, and George J. Klir

OXFORD

UNIVERSITY PRESS

OXFORD
UNIVERSITY PRESS

Oxford University Press is a department of the University of Oxford. It furthers
the University's objective of excellence in research, scholarship, and education
by publishing worldwide. Oxford is a registered trade mark of Oxford University
Press in the UK and certain other countries.

Published in the United States of America by Oxford University Press
198 Madison Avenue, New York, NY 10016, United States of America.

© Oxford University Press 2017

Library of Congress Cataloging-in-Publication Data
Names: Bělohlávek, Radim. | Dauben, Joseph Warren, 1944- | Klir, George J., 1932-
Title: Fuzzy logic and mathematics : a historical perspective / Radim
Bělohlávek, Joseph W. Dauben, George J. Klir.
Description: Oxford ; New York, NY : Oxford University Press, [2017] |
Includes bibliographical references.
Identifiers: LCCN 2016024541| ISBN 9780190200015 (hardcover) |
ISBN 9780190200039 (online content)
Subjects: LCSH: Fuzzy logic. | Logic, Symbolic and mathematical.
Classification: LCC QA9.64 .B4525 2017 | DDC 511.3/13–dc23 LC record available
at https://lccn.loc.gov/2016024541

Contents

Preface

THE INITIAL IDEA for writing a book of this kind emerged from discussions at the Eighth World Congress of the International Fuzzy Systems Association (IFSA) in Taipei in August of 1999. With the approaching new millennium, the focus of the congress was on reexamining the past and pondering the future of theoretical developments as well as applications of fuzzy logic and mathematics based on fuzzy logic. It was repeatedly suggested in the many discussions at the congress that the time was ripe for writing a book describing comprehensively and in sufficient detail the relatively short history of fuzzy logic and assessing its significance and likely prospects in the new millennium. One of us (Klir) participated in these discussions and the idea of writing such a book was appealing to him. He realized, however, that the scope of this kind of book was too large for a single person to write, and so the idea of writing such a book remained idle in his mind for more than a decade. In 2011, while working jointly with a former colleague (Bělohlávek) on another book involving fuzzy logic (*Concepts and Fuzzy Logic*, MIT Press, 2011), he mentioned the idea and they eventually agreed to work on this larger book together. However, due to the historical nature of the book, they felt it would be useful to enlist someone with expertise in history, especially the history of mathematics, and so they extended an invitation to Joseph Dauben to join them in working on this challenging book, which he accepted.

In 2012, the three of us began by converting the initial, somewhat haphazard idea into a well-defined writing project. That is, we formulated our vision of the aims and scope of the book and described how we intended to implement it. We also defined a general structure for the book and determined roughly the envisioned contents of individual chapters. This overall description of the book ultimately served as the basis for our introductory chapter 1. We then explored several prospective publishers and settled eventually on Oxford University Press, which we all considered the ideal publisher for a book like this.

Although we intended to work on the book as a team, it was reasonable and practical that each of us would assume a particular role in the actual writing process. Since two of us—Bělohlávek and Klir—had been active participants in research, education, and other activities regarding theory and applications of fuzzy logic, we took the responsibility for most of the writing and divided it according to our interests: Bělohlávek assumed responsibility for chapters 4 and 5; Klir for chapters 2, 3, and 6. Even though several parts—particularly of chapters 2 and 6—represent joint work, the individual chapters differ slightly in their styles of writing. Because these distinctions in style also reflect distinct backgrounds and experiences—indeed, Klir and Bělohlávek represent two distinct generations of fuzzy-logic researchers—we considered it desirable to preserve them. Although the primary role of Dauben— a relative newcomer to fuzzy logic—was to provide his expertise in the history of mathematics, he also critically evaluated the content of chapters 2–6 and participated actively in writing chapter 7.

Our cooperation with Oxford University Press, primarily via Peter Ohlin—the editor responsible for this book in the New York office of OUP—has been efficient and cordial from the beginning, for which we are very grateful. Initially, we underestimated the enormous scope of information to be covered in a book like this, but we were determined to include all of the relevant and valuable historical information that we could find, some of which had nowhere been previously documented. Nevertheless, we have had to be very selective and, in some cases, we have had to shorten some parts of our original text. However, it is our intent to publish at least some of the deleted material in some form elsewhere in the near future.

We are aware that the publication of this book coincides with the 50th anniversary of the genesis of fuzzy logic. This is fortunate, even though initially it was not planned to happen this way. However, we believe that the book can now be succinctly described as covering in considerable detail the history of fuzzy logic during the first fifty years of its existence, as well as offering a carefully argued assessment of the significance of fuzzy logic at the end of its first fifty years. We hope our work may also serve as a benchmark in so far as our vision of possible future developments of fuzzy logic may be compared at some future time whenever another historical assessment of fuzzy logic and its significance may be made.

Radim Bělohlávek, Joseph W. Dauben, and George J. Klir

Acknowledgments

We would like to express our gratitude to the following persons: Eduard Bartl, a colleague of Radim Bělohlávek at Palacký University in Olomouc, Czech Republic, who played a major role in typesetting the entire book using a LaTeX style that he created. Similarly, Ellen Tilden, editorial assistant to George Klir, read and copyedited virtually the entire manuscript from a stylistic point of view, and we are grateful to her for improving the overall readability of this book.

Notes for the Reader

THESE NOTES provide information about how the material covered in this book is organized. The aims and scope of the book as well as the content of its main chapters, chapters 2–7, are presented in chapter 1. In addition to these chapters, the book contains three appendices: appendix A concerns Cox's theorem, which is discussed in chapter 2; appendix B provides an overview of classical logic; appendix C consists of photographs with short biographical sketches of the major contributors to the early development of fuzzy logic. The book also contains a glossary of mathematical symbols widely used in the text, a list of references, a name index, and a subject index.

The list of references at the end of the book contains all of the major publications on which we have drawn in writing this book. Additional references not directly concerned with its main subject are presented in footnotes, which are numbered consecutively within individual chapters, and also contain further information related to the main text. For works published in languages other than English, the original title is given with an English translation in brackets; whenever an English translation of such works is also available, the respective bibliographic item contains a complete reference to the translated version as well.

For stylistic matters, the 16th edition of the *Chicago Manual of Style* is generally followed. In cross-references, "p." stands for "page" and "n." for "footnote." In mathematical theorems, we use "iff" for "if and only if" and "w.r.t." for "with respect to." The following abbreviations are used throughout:

ACM	Association for Computing Machinery
AMS	American Mathematical Society
Ann	*Annals*
Bull	*Bulletin*
IEEE	Institute of Electrical and Electronics Engineers
Int	*International*
J	*Journal*
Proc	*Proceedings*
Symp	*Symposium*
Trans	*Transactions*
Univ.	University

Aeq Math	*Aequationes Mathematicae*
American J Math	*American Journal of Mathematics*
American J Phys	*American Journal of Physics*
Ann Math	*Annals of Mathematics*
Ann Pure Appl Logic	*Annals of Pure and Applied Logic*
Arch Math Logic	*Archive for Mathematical Logic*
Artif Intell Rev	*Artificial Intelligence Review*
Bull AMS	*Bulletin of the American Mathematical Society*
Civil Eng Syst	*Civil Engineering Systems*
Comput Intell	*Computational Intelligence*
Commun ACM	*Communications of the ACM*
Europ J Operational Research	*European Journal of Operational Research*
Fund Inform	*Fundamenta Informaticae*
Fund Math	*Fundamenta Mathematicae*
Fuzzy Sets Syst	*Fuzzy Sets and Systems*
IEEE Trans Automat Contr	*IEEE Transactions on Automatic Control*
IEEE Trans Fuzzy Syst	*IEEE Transactions on Fuzzy Systems*
IEEE Trans Syst Sci Cyb	*IEEE Transactions on Systems Science and Cybernetics*
IEEE Trans Syst Man Cyb	*IEEE Transactions on Systems, Man and Cybernetics*
Inf Control	*Information and Control*
Inf Sci	*Information Sciences*
Int J Approx Reason	*International Journal of Approximate Reasoning*
Int J Gen Syst	*International Journal of General Systems*
Int J Intell Syst	*International Journal of Intelligent Systems*
Int J Man Mach Stud	*International Journal of Man-Machine Studies*
Int J UFKBS	*International Journal of Uncertainty, Fuzziness and Knowledge-Based Systems*
J Applied Non-Classical Logics	*Journal of Applied Non-Classical Logics*
J American Chem Soc	*Journal of the American Chemical Society*
J ACM	*Journal of the ACM*
J Comput Syst Sci	*Journal of Computer and System Sciences*
J Logic Comput	*Journal of Logic and Computation*
J Math Anal Appl	*Journal of Mathematical Analysis and Applications*
J Pure Appl Algebra	*Journal of Pure and Applied Algebra*
J Symb Logic	*Journal of Symbolic Logic*
Lect Notes Comput Sci	*Lecture Notes in Computer Science*
Lect Notes Logic	*Lecture Notes in Logic*
Lect Notes Math	*Lecture Notes in Mathematics*
Logic J IGPL	*Logic Journal of the Interest Group of Pure and Applied Logic*
Math Logic Quart	*Mathematical Logic Quarterly*
Notices AMS	*Notices of the American Mathematical Society*
Philos Sci	*Philosophy of Science*
Proc London Math Soc	*Proceedings of the London Mathematical Society*
Proc Natl Acad Sci USA	*Proceedings of the National Academy of Sciences of the United States of America*
Tatra Mt Math Publ	*Tatra Mountains Mathematical Publications*
Theor Comput Sci	*Theoretical Computer Science*
Trans AMS	*Transactions of the American Mathematical Society*
Z Math Logik Grundlagen Math	*Zeitschrift für mathematische Logik und Grundlagen der Mathematik*

Chapter **I**

Aims and Scope of This Book

THE SUBJECT OF THIS BOOK—fuzzy logic and its role in mathematics—has a relatively short history of some fifty years. The overall aim of this book is to cover this short history as comprehensively as possible. This means that we intend to cover not only theoretical and practical results emanating from fuzzy logic, but also motivations and creative processes that led to these results. Such a retrospective reflection is in our opinion essential for properly assessing the overall significance and impact of fuzzy logic, and we feel that the time is ripe for it.

It seems reasonable to expect that the aims and scope of any scholarly book be expressed in a nutshell by its title. Our choice of "Fuzzy Logic and Mathematics: A Historical Perspective" was indeed intended to do so. First, although "fuzzy logic" has multiple connotations, their common ground is the rejection of a fundamental principle of classical logic—the principle of bivalence. This is basically an assumption, inherent in classical logic, that any declarative sentence has only two possible truth values, *true* and *false*.

By rejecting the principle of bivalence, fuzzy logic does not abandon the classical truth values—true and false—but allows for additional ones. These truth values, which are interpreted as *degrees of truth*, may be construed in various ways. Most commonly, they are represented by numbers in the unit interval $[0, 1]$. In this interpretation, 1 and 0 represent the boundary degrees of truth that correspond, respectively, to the classical truth values true and false. The numbers between 0 and 1, with their natural ordering, represent intermediate degrees of truth. Either all real numbers from $[0, 1]$ or their various subsets, each containing 0 and 1, may be employed as truth values. Other sets of truth values are possible as well, provided that they are at least partially ordered and bounded by the classical truth values.

Embracing additional truth values and interpreting them as degrees of truth has an appealing motivation, which may briefly be described as follows. Classical logic is appropriate for reasoning that involves bivalent propositions such as "7 is a prime number," that is, propositions that may in principle be only true or false. Most propositions that people use to communicate information about the physical world are, however, not bivalent—their *truth is a matter of degree.*

As an example, consider the proposition "the outside humidity is high." In classical logic, one is forced to assign to this proposition the truth value of either false or true. To do so, one needs to choose a threshold value h such that the proposition is considered true if and only if the outside humidity is larger than h (or, alternatively, larger than or equal to h). Then, for an arbitrarily small deviation, ϵ, in the humidity from h, the proposition is considered true when the actual humidity is $h + \epsilon$ and false when the humidity is $h - \epsilon$. This abrupt change from truth to falsity, which clearly does not agree with our common-sense understanding of the proposition "the outside humidity is high," can be avoided when the principle of bivalence is rejected.

In fuzzy logic, one may assign to any proposition not only 1 (true) or 0 (false) but also other, intermediate truth degrees, such as 0.8. For a (slightly) higher humidity, one naturally assigns the proposition a (slightly) higher truth degree, for example 0.85. Such employment of degrees of truth meets our common-sense understanding of this proposition (and other propositions expressed in natural language) quite well; it is one of the attractive capabilities of fuzzy logic. How to assign truth degrees in the context of each particular application of fuzzy logic is an important issue that we discuss later in this book.

What is the effect of degrees of truth on predicates? As is well known, the extension of each predicate in classical logic is a unique set defined within some predetermined universe of discourse. If the proposition "x is P" is true, then object x is a member of P's extension; if it is false, then x is not a member of P's extension. In fuzzy logic, where the requirement that "x is P" be either true or false is rejected, a broader class of predicates is admitted. It includes predicates, such as "high humidity," "tall man," or "low inflation rate," for which the truth of the proposition "x is P" is, in general, a matter of degree. Such predicates are called fuzzy predicates. It is natural to view the degree of truth of each particular proposition "x is P" as a degree of membership of object x in P's extension. Such an extension is then a nonclassical set whose membership is a matter of degree. These nonclassical sets, which are referred to as fuzzy sets, are thus essential components of predicate fuzzy logic. However, their role in fuzzy logic extends beyond their connection with fuzzy predicates, as is explained throughout this book.

So far, we have used the term "fuzzy logic" in its common-sense meaning. It is in this way that the term is most often used, referring to all aspects of representing and manipulating knowledge that employ degrees of truth. In addition, two other meanings are recognized, namely fuzzy logic in the narrow sense and fuzzy logic in the broad sense. Fuzzy logic in the narrow sense, also called mathematical fuzzy logic, deals with formal logical systems in which truths of propositions are a matter of degree. These systems provide foundations for fuzzy logic in the broad sense, which has a much wider agenda. Its primary aim is to emulate common-sense human reasoning in natural language and other human capabilities and utilize them in various applications.

These few remarks characterize fuzzy logic in a generic way, as a variety-embracing concept. In this sense, the term "fuzzy logic" stands for a wide number of special systems, which are distinguished from one another by various properties such as the set of truth degrees employed, their algebraic structure, truth functions chosen for logic connectives, and other properties. Since it is virtually impossible to cover in detail the entirety of fuzzy logic in a single book, we have been selective and focus primarily on the principal aims of this book. Whatever we consider essential for achieving our aims, we cover in detail. The rest we try to cover as satisfactorily as possible via short notes and accompanying references to the literature.

The literature dealing with fuzzy logic in its various forms and applications has been rapidly growing for several decades and is now very extensive. It covers, on the one hand, many important theoretical results concerning foundations of fuzzy logic and, on the other hand, a broad spectrum of impressive applications of fuzzy logic in engineering, science, and other areas of human affairs. It seems reasonable to view these established theoretical results, successful applications, and the ongoing rapid growth of the literature as indicators of the growing acceptance and viability of fuzzy logic.

As the title of this book also suggests, our aim is not only to examine fuzzy logic alone, but also to examine its role in mathematics. The proper way to examine the role of fuzzy logic in mathematics and to assess its overall impact is to approach this problem from a historical perspective. Let us elaborate a little on why we consider this important.

As is well known, most technical publications in virtually any area focus largely on presenting results and usually pay little or no attention to motivations and creative processes leading to those results. This means that valuable information regarding the emergence, evolution, and propagation of ideas within a given subject is habitually suppressed. Such details often remain only in the memories of the people who created them, and then pertinent historical details are lost forever when these people pass away.

In fuzzy logic, fortunately, many of the pioneers and key contributors are still alive, making it to a large degree feasible to reconstruct important but ephemeral information. This fleeting opportunity motivates us to embark on this project at this time. Our ultimate goal is to assess the overall significance of fuzzy logic. To this end, we believe, it is essential to characterize fuzzy logic not only from the standpoint of the current situation—by existing results—but also from the historical perspective or rather retrospective—by documenting processes leading to these results. To achieve the above-mentioned aims, the book is organized as follows.

In chapter 2, we first identify instances throughout the long history of classical logic when the principle of bivalence was challenged. This may loosely be viewed as a sort of "prehistory" to fuzzy logic. We then describe in some detail circumstances associated with the emergence of fuzzy logic. Furthermore, we survey and present chronologically events that contributed in a major way to the development of fuzzy logic from its very beginnings to its present state. We also survey evolving attitudes toward fuzzy logic, especially within the academic community, and summarize various debates between members of the emerging fuzzy logic community and opponents of fuzzy logic.

In chapter 3, which is devoted to fuzzy logic in the broad sense, we trace the emergence of a broad spectrum of ideas motivated by the ultimate goal of developing sufficiently expressive means for emulating common-sense human reasoning in natural language and some other unique capabilities of human beings.

In chapter 4, we first survey the development of many-valued logic before the emergence of fuzzy sets. We then cover in a fairly comprehensive way the evolution of fuzzy logic in the narrow sense since the late 1960s and examine the various propositional, predicate, and other fuzzy logic calculi that have been advanced.

In chapter 5, we explore foundations of mathematics based on fuzzy logic and examine some of its areas. We show that the various concepts of mathematics based on bivalent logic may be viewed as special cases of their extended counterparts in mathematics based on fuzzy logic, and also that some additional useful concepts emerge in the latter that have no counterparts in the former. In addition, we examine various semantic issues such as interpretation of truth degrees, the phenomenon of vagueness, and resolution of certain paradoxes by fuzzy logic.

In spite of its relatively short history, fuzzy logic has already achieved high visibility, primarily due to its impressive applications in many areas. These applications are the subject of chapter 6. There, we cover fairly comprehensively the many applications of fuzzy logic that have so far been developed in various branches of engineering, science, medicine, management, and other areas.

An appraisal of the overall significance of fuzzy logic is the subject of chapter 7. We regard fuzzy logic as a new paradigm, provide a retrospective overview of its foundations and applications, and assess its significance from a variety of viewpoints. We conclude by assessing the prospects of fuzzy logic as we see them on its 50th anniversary.

Chapter **2**

Prehistory, Emergence, and Evolution of Fuzzy Logic

IN THIS CHAPTER, various challenges to the principle of bivalence of classical logic throughout the whole history of logic are briefly surveyed. Circumstances that led eventually to the emergence of fuzzy logic are then described in greater detail. We examine how the initial, strongly negative attitudes toward fuzzy logic by the academic community have gradually become more positive. In this context, we summarize various documented debates between researchers criticizing fuzzy logic and researchers supporting it, which usually have also contributed to its development. Finally, we outline several visible stages in the evolution of supporting infrastructure for fuzzy logic.

2.1 Prehistory of fuzzy logic

Our aim in this section is to focus solely on one aspect of the long and complex history of classical, bivalent logic and mathematics based on this logic. Our only interest here is to identify episodes when the principle of bivalence of classical logic was challenged. Such episodes were actually extremely rare and of minor influence prior to the 19th century.

The mainstream view of the history of logic (Bocheński 1961) associates the beginnings of logic with ancient Greece, and Aristotle (384–322 BC) is routinely considered the founder of classical logic. This is primarily due to his development of a system of inference schemes (syllogisms) for deductive reasoning, which influenced the development of classical logic for some two thousand years. This system, usually referred to as *syllogistic,* is described in one part of Aristotle's major work entitled *Organon* (McKeon 1941, 62–107).

In developing his syllogistic, Aristotle fully subscribed to the principle of bivalence. However, he questioned the applicability of this principle to propositions concerning future contingencies. This issue is discussed in one part of *Organon,* which is entitled *De interpretatione* (*On Interpretation*) (McKeon 1941, 40–61). The following is the essence of his arguments (p. 48):

> Everything must either be or not be, whether in the present or in the future, but it is not always possible to distinguish and state determinately which of these alternatives must necessarily come about.
>
> Let me illustrate. A sea-fight must either take place to-morrow or not, but it is not necessary that it should take place to-morrow, neither it is necessary that it should not take place, yet it is necessary that it either should or should not take place to-morrow. Since propositions correspond with facts, it is evident that when in future events there is a real alternative, and a potentiality in contrary directions, the corresponding affirmation and denial have the same character.
>
> This is the case with regard to that which is not always existent or not always nonexistent. One of the two propositions in such instances must be true and the other false, but we cannot say determinately that this or that is false, but must leave the alternative undecided. One may indeed be more likely to be true than the other, but it cannot be either actually true or false. It is therefore plain that it is not necessary that of an affirmation and a denial one should be true and the other false. For in the case of that which exists potentially, but not actually, the rule which applies to that which exists actually does not hold good. The case is rather as we have indicated.

Even though Aristotle certainly questioned the applicability of the principle of bivalence for propositions involving future occurrences, he actually did not abandon the principle. He dealt with the problem regarding propositions involving future events via the modalities of possibility and necessity.

While the above thoughts of Aristotle are well known, it seems virtually unknown that he was also interested in a completely different phenomenon which challenges the principle of bivalence and which is—unlike the problem of future contingencies—in fact directly related to the idea of fuzzy logic, namely vagueness of human concepts. As shown by the following excerpt from another part of *Organon,* called *Categoriae* (*Categories*) (McKeon 1941, 7–37), Aristotle clearly recognized that certain human categories apply to objects to various degrees and hence do not have sharp boundaries. He was also aware of borderline cases, which he called intermediates, and alluded to the violation of the law of excluded middle by such categories (pp. 27, 29–30):

> Qualities admit of variation of degree. Whiteness is predicated of one thing in a greater or less degree than of another. . . .
>
> The qualities expressed by the terms 'triangular' and 'quadrangular' do not appear to admit of variation of degree. . . . Thus it is not all qualities which admit of variation of degree. . . .
>
> Those contraries which are such that the subjects in which they are naturally present, or of which they are predicated, must necessarily contain either the one or the other of them, have no intermediate, but those in the case of which no such necessity obtains, always have an intermediate. . . . Odd and even, again, are predicated of number, and it is necessary that the one or the other should be present in numbers. . . . On the other hand, in those contraries with regard to which no such necessity obtains, we find an intermediate. Blackness and whiteness are naturally present in the body, but it is not necessary that either the one or the other should be present in

> the body, inasmuch as it is not true to say that everybody must be white or black. . . . These pairs
> of contraries have intermediates: the intermediates between white and black are grey, sallow, and
> all the other colours that come between; the intermediate between good and bad is that which
> is neither the one nor the other.

In his monumental book entitled *Metaphysica* (*Metaphysics*) (McKeon 1941, 689–926), Aristotle even exercises the related idea that some claims may be more true than others (p. 743):

> Again, however much all things may be 'so and not so', still there is a more and a less in the nature
> of things; for we should not say that two and three are equally even, nor is he who thinks four
> things are five equally wrong with him who thinks they are a thousand. If they are not equally
> wrong, obviously one is less wrong and therefore more right. If then that which has more of any
> quality is nearer the norm, there must be some truth to which the more true is nearer.

However, neither of these considerations led him to abandon bivalence—then a deeply rooted pillar of logical investigations.

Aristotle was not the only philosopher in ancient Greece who questioned the universal applicability of the principle of bivalence in logic. The principle was questioned even more emphatically by one of his contemporaries, Epicurus (341–270 BC), and his followers—Epicureans. These philosophers basically rejected the principle of bivalence on the basis of their strong belief in free will and their associated commitment to the doctrine of indeterminism.

Aristotle's hesitation regarding the general applicability of the principle of bivalence was criticized by Stoics, a group of Greek philosophers with strong interests in logic, who were fully committed to this principle. Otherwise, it has not received much attention, contrary to the great and lasting influence of his syllogistic. There are only a few rare episodes in the history of logic before the end of the 19th century when it reappeared as a subject of discussion and further development.

The first of these rare episodes occurred in the 14th century and is primarily attributable to the well-known English philosopher William of Ockham (c. 1287–1347). Educated in theology at Oxford University, he was a complex figure who made important contributions to some areas of philosophy as well as theology. In this book, we are only interested in his careful logical analysis of Aristotle's discussion about the truth status of propositions regarding future contingencies.[1] Our principal source in this regard is the book edited by Philotheus Boehner (1945), which contains Latin originals of Ockham's *Treatise on Predestination, God's Foreknowledge, and Future Contingents*, and an extensive English commentary on Ockham's works by Boehner. Details of Ockham's logical analysis of chapter 9 of Aristotle's *De interpretatione* and of parts of Aristotle's *Topics* are covered in his *Commentary to Perihermenias* and in chapter 30 of the third section of part III of his *Summa*

---

[1] There were apparently other philosophers who were attracted to this issue after Aristotle's ideas became available in Latin due to the efforts of Albert the Great (1193–1280) during the second half of the 13th century. However, Ockham's analysis seems to be the only one, according to currently available historical facts, which showed that Aristotle's arguments lead to the violation of the principle of bivalence and the need for a three-valued logic.

Logicae, whose Latin originals are included in Boehner 1945 as appendices II and III, respectively.[2]

Ockham's analysis shows that the arguments made by Aristotle about future contingencies lead inevitably to propositions that are neither true nor false. Boehner (1945) introduces the name *neuter* (N) as the third value—neither true nor false—for such propositions. After examining Ockham's arguments in greater detail, he shows (Boehner 1945, 58–66) that in fact they contain a description of a truth function of three-valued implication, $p \rightarrow q$, which is shown in the following table:[3]

$$
\begin{array}{c|ccc}
\rightarrow & F & N & T \\
\hline
F & T & T & T \\
N & N & T & T \\
T & F & N & T
\end{array}
\tag{2.1}
$$

Although Ockham explained by his own analysis that Aristotle's argument about the truth status of propositions concerning future contingencies led inevitably to a violation of the principle of bivalence, he did not accept these arguments, partially for theological reasons and partially due to his strong belief in the principle. Therefore, he did not take his prospective three-valued logic seriously, as is also confirmed by Boehner (1945, 66–69). As a consequence, his idea of three-valued logic had virtually no influence on the development of logic.

In 1465, Aristotle's arguments regarding the truth status of propositions of future contingencies became a subject of extensive and heated debates at the University of Louvain (now in Belgium). The debates involved members of the Faculty of Arts, who supported Aristotle's arguments, and members of one fraction in the Faculty of Theology, who rejected them on strictly theological grounds. These debates, which lasted for ten years (1465–75) and extended far beyond the university, are captured in considerable detail in English translations of transcripts of the various debates that have been collected by Baudry (1989).[4]

This book offers fascinating and insightful reading. It contains an extensive introduction, in which Baudry characterizes in detail the historical context within which the debates took place, explains how they evolved, and introduces the key debaters—Peter de Rivo, representing the Faculty of Arts, and Henry de Zomeren, representing the Faculty of Theology. It is clear that the principal issue in these debates was the principle of bivalence. In the end, however, neither side in these ten-year debates was an obvious winner.

---

[2] English translations by Marilyn McCord Adams and Norman Kretzmann of the *Treatise* (Ockham 1321–24) and appendices II and III of Boehner 1945 are included in Adams and Kretzmann 1983; the two appendices from Boehner 1945 are included in Adams and Kretzmann 1983 as appendices II and III. Appendix II contains, in addition, English translations (from medieval Latin versions) of relevant passages from Aristotle's *De interpretatione*.

[3] Adapted from p. 63 of Boehner 1945.

[4] Translated into English by Rita Guerlac from the French original, L. Baudry, *La querelle des futurs contingents: (Louvain 1465–1475)* (Paris: Vrin, 1950).

Since the end of the 17th century, the considerations on vagueness of human concepts reappeared, with no apparent influence of Aristotle's early analyses, first in the works of John Locke and then continued to appear in textbooks on philosophy and logic, such as in the established textbook of Watts (1724) and that of Bain (1870). Vagueness of concepts and in particular the existence of their borderline cases and unsharp boundaries naturally fitted Locke's empiricism and became clearly recognized in these works. Interestingly, in order to describe such concepts Bain even deemed it necessary to use a "numerical scale for stating the amount or degree of each property." The respective developments along with relevant quotations are covered in section 5.4.3 below. Even though the principle of bivalence is clearly challenged by such considerations, these works did not question it explicitly.

In the late 19th century, the principle was challenged again by logicians and mathematicians. They argued that the principle is unnecessarily restrictive and that it is possible to develop various alternative logics by abandoning it. Some alternative logics, in each of which more than two truth values were recognized, started to emerge in the late 1910s and early 1920s. However, some relevant initial ideas regarding such logics had already appeared in the literature in the late 19th century and early 20th century. Authors of these earlier ideas, which we cover only briefly, were Hugh MacColl (1837–1909), a Scottish mathematician, Charles Sanders Peirce (1839–1914), the well-known American philosopher with strong interests in logic and mathematics, and Nikolai A. Vasil'ev (1880–1940), a Russian philosopher. It is nevertheless interesting to note that already in his analysis of classical logic, George Boole (1815–1864) briefly considered the idea that some naturally derived algebraic expressions allow for trichotomy in logical values but concluded that such expressions are "not interpretable in the system of logic" (Boole 1854, chapter III).[5]

Within the period from 1877 through 1908, MacColl published numerous papers and a book (MacColl 1906), in which he presented many thought-provoking ideas addressing a broad spectrum of issues pertaining to symbolic logic.[6] He is considered a forerunner of many-valued logics due to his work in modal logic, which was apparently motivated by his distrust of material implication and his search for strict implication. He outlined a system of propositional logic in which propositions are qualified as *true, false* (both "in a particular case or instance," in MacColl's terms), *certain* ("always true," "true in every case," "its probability is 1"), *impossible*

---

[5] Several interesting remarks in this respect are contained in a thorough analysis of Boole's works by T. Hailperin. In section 3.6 of his *Boole's Logic and Probability*, 2nd ed. (Amsterdam: North-Holland, 1986), he shows that certain expressions, which Boole used in addition to the classical truth values 0 and 1, may be thought of as truth values of a particular four-valued logic. In "Boole's algebra isn't Boolean algebra," *Mathematics Magazine* 54 (1981): 172–84, Hailperin argues that what Boole described was actually an algebra more general than a Boolean algebra, namely a nonidempotent commutative ring with unit and without nilpotents which may be seen as an algebra of nonclassical logic from today's perspective.

[6] This book and MacColl's papers are reprinted, together with some of his correspondence and published debates regarding his ideas, in S. Rahman and J. Redmond, eds., *Hugh MacColl: An Overview of His Logical Work with Anthology* (London: College Publications, 2007). This book also contains biographical notes and a critical overview of MacColl's logical work.

("its probability is 0"), and *variable* ("neither impossible nor uncertain," "probability is . . . some proper fraction" between 0 and 1). MacColl, however, treated these qualifications as modalities of propositions and seems not to have been considering explicit assignments of intermediate truth degrees such as probability degrees to propositions.

In 1902, Peirce wrote an essay entitled "Minute logic" (published only posthumously in his *Collected Papers*) in which he envisioned a new mathematics based on three-valued logic.[7] He referred to this mathematics as a *trichotomic mathematics* and to the underlying three-valued logic as a *triadic logic*. Unfortunately, his work on triadic logic is documented only by occasional remarks in his publications or letters and, more importantly, in his handwritten working notes. Analyzing this fragmentary evidence, Fisch and Turquette (1966) came to a reasonable conclusion that Peirce had succeeded by 1909 in developing basic ideas of some three-valued logics. These logics are among those that were introduced and studied in the 1920s by the pioneers of many-valued logics, as we describe later in this section. Why Peirce did not publish these ideas remains, according to Fisch and Turquette, an unsolved mystery.

In the early 1910s, Nikolai A. Vasil'ev (1880–1940), who was at that time with the Kazan University in Western Russia, explored the possibility of a new logic, distinct from classical logic, which he called *imaginary logic*. He presented his ideas for the first time in a lecture at the Kazan University on May 18, 1910, and published them in Vasil'ev 1910. During 1910–13, he developed his logic further and wrote several papers, including the important *Воображаемая (неаристотелва) логика* [Imaginary (non-Aristotelian) logic]. Except for a small note written in 1925,[8] his works are written in Russian.[9] Vasil'ev chose the term "imaginary logic" because he saw a parallel between his ideas and those of N. A. Lobachevsky who in the early 19th century had also been at Kazan University. Lobachevsky developed his *imaginary geometry*—a kind of non-Euclidean geometry describing in a sense a different world than that described by classical, Euclidean geometry. Vasil'ev's main idea was that in addition to our world, which he saw as being described by the Aristotelian logic, there are other, imaginary worlds for which one needs logics different from the classical one. Each such logic, which he therefore called imaginary, differs from classical logic in the laws it satisfies. In particular, Vasil'ev considered logics that violate the law of noncontradiction and the law of excluded middle. His ideas are, however, somewhat ambiguously described in his writings because he never attempted to for-

---

[7] C. Hartshorne and P. Weiss, eds., *Collected Papers of Charles Sanders Peirce*, vol. 4 (Cambridge, MA: Harvard Univ. Press, 1933), paragraphs 4.307–4.323, 248–62.

[8] "Imaginary (non-Aristotelian) logic," in *Atti del Quinto Congresso Internazionale di Filosofia*, ed. G. della Valle (Naples, 1925), 107–9.

[9] They are available in N. A. Vasil'ev, *Избранные труды* [Selected works] (Moscow: Nauka, 1989). Vasil'ev life and work are covered in V. A. Bazhanov, *Николай Александрович Васильев (1880–1940)* [Nikolai Alexadrovich Vasil'ev (1880–1940)] (Moscow: Nauka, 1988); see also the short account in English, V. A. Bazhanov, "The fate of one forgotten idea: N. A. Vasiliev and his imaginary logic," *Studies in Soviet Thought* 39 (1990): 333–41.

malize them. As a consequence, the numerous existing works assessing his contributions contain mutually incompatible views—while some regard Vasil'ev as a pioneer
of many-valued logic, paraconsistent logic, and sometimes even intuitionistic logic,
some regard such conclusions as unwarranted and claim that he never propounded
the idea of multiple truth values. His idea of alternative logics nevertheless clearly
makes him a forerunner of nonclassical logics.

The first formal systems of many-valued logics were developed by Jan Łukasiewicz (1878–1956), Paul Bernays (1888–1977), and Emil Post (1897–1954). Although
they invented their logics independently and at about the same time, Łukasiewicz
not only has a slight priority in being first but his logics were also most influential
in subsequent developments. We therefore cover his ideas in some detail.

Łukasiewicz first developed a three-valued logic and presented it in a lecture to
the Polish Philosophical Society that he delivered in June of 1920 in Lvov. The
essence of this lecture was summarized in a short, two-page paper (Łukasiewicz
1920). In this paper, Łukasiewicz allows each proposition to have three values. Two
of them are the classical values 1 and 0, which represent truth and falsity, respectively.
The third value represents neither truth nor falsity and is interpreted as "possibility." This value is denoted in this early paper by 2, but later on, the convenient symbol $1/2$ was adopted. Łukasiewicz defines the truth function, $\rightarrow$, of his three-valued
implication, as shown by the following table:

$$
\begin{array}{c|ccc}
\rightarrow & 0 & 1/2 & 1 \\
\hline
0 & 1 & 1 & 1 \\
1/2 & 1/2 & 1 & 1 \\
1 & 0 & 1/2 & 1
\end{array}
\tag{2.2}
$$

Furthermore, he defines the truth function, $\neg$, of negation for any truth value a as

$$\neg a = a \rightarrow 0,$$

which yields

$$
\begin{array}{c|ccc}
a & 0 & 1/2 & 1 \\
\hline
\neg a & 1 & 1/2 & 0
\end{array},
$$

and defines the truth functions of disjunction and conjunction in terms of $\rightarrow$ and
$\neg$ via the formulas

$$a \vee b = (a \rightarrow b) \rightarrow b,$$
$$a \wedge b = \neg(\neg a \vee \neg b),$$

respectively. When defining his implication, Łukasiewicz remarks that this can be
done in different ways in three-valued logics, but he does not explain in this paper his
rationale for choosing his particular definition. He addressed this issue in another
paper ten years later (Łukasiewicz 1930). We describe his motivations in detail in
chapter 4. It is significant that the three-valued implication defined by Łukasiewicz
(2.2) is the same except for notation as the one defined by Ockham (2.1).

Although Łukasiewicz's first written presentation of his three-valued logic appeared in 1920, it is clear from the Farewell Lecture he delivered at Warsaw University on March 7, 1918 (Borkowski 1970, 84–86) that he had already developed it in the summer of 1917. The following are a few excerpts from this interesting lecture:

> In this farewell lecture, I wish to offer a synthesis of my research, based on autobiographical confession. I wish to describe the emotional background against which my views have developed.
>
> I have declared a spiritual war upon all coercion that restricts man's free creative activity.
>
> There are two kinds of coercion. One of them is physical, which occurs either as an external force that fetters the freedom of movement, or as internal impotence that incapacitates all action. . . .
>
> The other kind of coercion is *logical*. We must accept self-evident principles and the theorems resulting therefrom. . . .
>
> That coercion originated with the rise of Aristotelian logic and Euclidean geometry. The concept was born of *science* as a system of principles and theorems connected by logical relationship. The concept came from Greece and has reined supreme. The universe was conceived after the pattern of a scientific system: all events and phenomena are connected by causal links and follow from one another as theorems in scientific theory. All that exists is subject to necessary *laws*.
>
> In the universe conceived in this way there is no place for a creative act resulting not from a law but from a spontaneous impulse. . . .
>
> The creative mind revolts against this concept of science, the universe, and life. A brave individual, conscious of his value, does not want to be just a link in the chain of cause and effect, but wants himself to affect the course of events. . . .
>
> And what does [*sic*] a scientist to do? . . .
>
> He has two paths to choose from: either to submerge himself in skepticism and abandon research, or to come *to grips with the concept of science based on Aristotelian logic*. . . .
>
> Logical coercion is most strongly manifested in a priori sciences. In 1910 I published a book on the principle of contradiction in Aristotle's work, in which I strove to demonstrate that the principle is not so self-evident as it is believed to be.[10] I strove to construct non-Aristotelian logic, but in vain.
>
> Now I believe to have succeeded in this. My path was indicated to me by *antinomies*, which prove that there is a gap in Aristotle's logic. Filling the gap led me to a transformation of the traditional principles of logic.
>
> This gave rise to system of *three-valued logic*, which I worked out in detail last summer. That system is coherent and self-consistent as Aristotle's logic, and is much richer in laws and formulae.
>
> The possibility of constructing different logical systems shows that logic is not restricted to reproduction of facts but is a free product of man, like a work of art. Logical coercion vanishes at its very source.

Later in the 1920s, Łukasiewicz also worked on generalizing his three-valued logic to n-valued logics and decided to label, for each particular n, the truth values by a set $Ł_n$ of equidistant rational numbers in $[0, 1]$ defined in the following way:

$$Ł_n = \{0 = \tfrac{0}{n-1}, \tfrac{1}{n-1}, \ldots, \tfrac{n-1}{n-1} = 1\}.$$

Under this more natural notation the logic operations of his three-valued logic can be conveniently expressed for all $a, b \in \{0, \tfrac{1}{2}, 1\}$ by the following formulas:

$$a \rightarrow b = \min(1, 1 - a + b),$$

$$\neg a = 1 - a,$$

$$a \vee b = \min(1, a + b),$$

---

[10] See Łukasiewicz 1910.

$$a \wedge b = \max(0, a + b - 1).$$

These formulas also describe every n-valued generalization of Łukasiewicz's three-valued logic, as presented together with other results in an important paper by Łukasiewicz and Tarski (1930). They also describe the case when the set of truth values are all real numbers in the unit interval $[0, 1]$, as shown earlier by Łukasiewicz (1923).

Łukasiewicz's finitely- and infinitely-valued logics play an important role in fuzzy logic and we therefore discuss them in detail in chapter 4. In that chapter, we also thoroughly examine the above-mentioned logics of Bernays and Post as well as additional many-valued logics which started to emerge since the early 1930s. They include the logic developed in the context of intuitionistic logic by Kurt Gödel (1906–1978), which plays an important role in fuzzy logic in the narrow sense, the three-valued logic of Dmitri A. Bochvar (1903–1990) for analyzing some paradoxes of classical logic and set theory, in which the third value is interpreted as "meaningless," the three-valued logic of Stephen C. Kleene (1909–1994), in which the third value is interpreted as "not defined," as well as some probabilistic logics and logics developed in the context of quantum mechanics. This information and much more on many-valued logics is covered in chapter 4.

The literature on many-valued logics has been rapidly growing since the early 1920s. Early major sources in this area include the excellent and comprehensive survey of the literature on many-valued logics from the very beginnings until about the mid-1960s prepared by Rescher (1969). His book covers almost 600 relevant references. More references, covering the period from 1966 to 1974, are given in Wolf 1977. We should also mention the early classic monographs written by Rosser and Turquette (1952), Zinov'ev (1963), and Ackermann (1967), as well as the edited books by Dunn and Epstein (1977) and Rine (1984).

The late 19th century witnessed a revival of interest in vagueness. Peirce, whom we mentioned in connection with his three-valued logic, elaborated on distinctions between various facets of linguistic imprecision. Moreover, he recognized that it is primarily the vagueness of general terms, ubiquitous in natural language, which makes classical logic incapable of formalizing common-sense reasoning with statements described in natural language. His definition of vagueness is clear and convincing:[11]

> A proposition is vague when there are possible states of things concerning which it is *intrinsically uncertain* whether, had they been contemplated by the speaker, he would have regarded them as excluded or allowed by the proposition. By intrinsically uncertain we mean not uncertain in consequence of any ignorance of the interpreter, but because the speaker's habits of language were indeterminate.

Peirce repeatedly returned to vagueness in his many writings. He considered it unavoidable and thus important to study, and he captured his strong views about

---

[11] J. M. Baldwin, ed., *Dictionary of Philosophy and Psychology*, vol. 2 (New York: Macmillan, 1902), 748.

vagueness in his short and remarkably clear statement: "Vagueness is no more to be done away with in the world of logic than friction in mechanics."[12]

Shortly after the death of Peirce, the famous Engllish philosopher, logician, and mathematician Bertrand Russell (1872–1970) wrote a paper fully devoted to vagueness (Russell 1923). In this paper, Russell straightforwardly exposed the limitations of classical two-valued logic, to which he himself had immensely contributed. The following excerpts from the paper capture quite well the essence of Russell's criticism (italics added):

> Vagueness and precision alike are characteristics which can only belong to a representation, of which language is an example. They have to do with the relation between a representation and that which it represents. Apart from representation, whether cognitive or mechanical, there can be no such thing as vagueness or precision; things are what they are, and there is an end of it . . .
>
> *The law of excluded middle is true when precise symbols are employed, but it is not true when symbols are vague, as, in fact, all symbols are. . . .*
>
> The notion of "true" and "false" can only have a precise meaning when the symbols employed—words, perceptions, images or what not—are themselves precise. Since propositions containing nonlogical words are the substructure on which logical propositions are built, it follows that logical propositions also, so far as we know them, become vague through the vagueness of "truth" and "falsehood." We can see an ideal of precision, to which we can approximate indefinitely; but we cannot obtain this ideal. . . . *All traditional logic habitually assumes that precise symbols are being employed. It is therefore not applicable to this terrestrial life, but only to imagined celestial existence.*

It is interesting to see, at the beginning of this quotation, that Russell viewed vagueness and precision as characteristics of a representation rather than as characteristics of the things being represented. This is compatible with Peirce's view, as well as with our own view. Notice also Russell's casual and rather radical remark in one of these excerpts that all symbols are vague. He elaborates on this point as follows:

> Let us consider the various ways in which common words are vague, and let us begin with such a word as "red." It is perfectly obvious, since colours form a continuum, that there are shades of colour concerning which we shall be in doubt whether to call them red or not, not because we are ignorant of the meaning of the word "red," but because it is a word the extent of whose application is essentially doubtful. . . . All words describing sensible qualities have the same kind of vagueness which belongs to the word "red." This vagueness exists also, though in a lesser degree, in the quantitative words which science has tried hardest to make precise, such as a metre or a second. . . . The metre, for example, is defined as the distance between two marks in a certain rod in Paris, when the rod is at a certain temperature. Now the marks are not points, but patches of a finite size, so that the distance between them is not a precise conception. Moreover, temperature cannot be measured with more than a certain degree of accuracy, and the temperature of the rod is never quite uniform. For all these reasons the conception of a metre is lacking in precision. The same applies to a second. The second is defined by a relation to the rotation of the earth, but the earth is not a rigid body, and two parts of the earth's surface do not take exactly the same time to rotate; moreover all observations have a margin of error. There are some occurrences of which we can say that they take less than a second to happen, and others of which we can say that they take more, but between the two there will be a number of occurrences of which we believe that they do not last equally long, but none of which we can say whether they last more or less than a second. Therefore, when we say an occurrence lasts a second, all that it is worth while to mean is that no possible accuracy of observation will show whether it lasts more or less than a second.

---

[12] C. Hartshorne and P. Weiss, eds., *Collected Papers of Charles Sanders Peirce*, vol. 5 (Belkamp Press of Harvard University, 1934), paragraph 5.512, 361.

This thought-provoking paper by Russell is now considered a classic in the literature on vagueness, but it was Russell's rather singular paper on the subject. After its publication, strangely, Russell showed no interest in developing further his ideas about vagueness.

Vagueness was discussed again by Cohen (1927) who spoke of concepts' "twilight zones," and, more thoroughly than ever before, in a paper by the American philosopher Max Black (1909–1988). After publication of his paper (Black 1937), which we examine in some detail in the next section, interest in vagueness began to grow. We return to the various views and aspects of vagueness throughout this book, in particular in section 5.4.3.

In addition to arguments based on vagueness of natural language, other arguments against the principle of bivalence can be found in philosophical literature in the late 19th century. One type of argument emerged from metaphysical views of a major English philosopher, Francis H. Bradley (1846–1924); another type of argument, based on epistemological grounds, was put forward by the influential French physicist, mathematician, and philosopher, Pierre Duhem (1861–1916).

Bradley developed his ideas primarily in his book *Appearance and Reality* (Bradley 1893). This book, which is based on Bradley's metaphysical view that true reality is immune to self-contradiction, is divided into two parts. In the first part, which is rather short, Bradley basically argues that any human experiential domain, regardless of how it is organized, is subject to unavoidable inconsistencies. This means that, according to Bradley's metaphysical view, it is only an appearance of reality, but not true reality. In the much longer second part of the book, Bradley's principal aim is to reconcile the inconsistent and fragmentary appearance of reality with the consistent, harmonious, and all-inclusive true reality. Through a long sequence of thoroughly discussed arguments, each including a response to potential criticism, Bradley concludes that the only sensible way to reconcile reality with appearance of reality is by employing degrees of truth, which means in turn rejecting the principle of bivalence. His argumentation leading to this conclusion is so extensive that even any meaningful summary of it would be long; instead, we include here only a few representative short quotations from his book (Bradley 1893, 364–65, 396):

> [To] be more or less true, and to be more or less real, is to be separated by an interval, smaller or greater, from all-inclusiveness or self-consistency. Of two given appearances the one more wide, or more harmonious, is more real. It approaches nearer to a single, all-containing, individuality. To remedy its imperfection, in other words, we should have to make a smaller alteration. The truth and the fact, which, to be converted into the Absolute, would require less re-arrangement and addition, is more real and truer. And this is what we mean by degrees of reality and truth. To possess more the character of reality, and to contain within oneself a greater amount of the real, are two expressions for the same thing. . . . There cannot for metaphysics be, in short, any hard and absolute distinction between truths and falsehoods. With each assertion the question is, how much will be left of that assertion, if we suppose it to have been converted into ultimate truth? . . .
>
> A predicate, we may say, in no case is, as such, really true. All will be subject to addition, to qualification and re-arrangement. And truth will be the degree up to which any predicate, when made real, preserves its own character.

It is worth mentioning that Bradley's metaphysical arguments for rejecting the principle of bivalence were further advanced almost half a century later by the American philosopher, Brand Blanshard (1892–1987), in his large, two-volume book Blanshard 1939. In fact, almost half of the second volume is devoted to this issue. Blanshard is basically reformulating Bradley's metaphysical ideas in the framework of a coherence theory of truth. It suffices for our purposes to present here his summary in paragraph 12 of chapter XXII on "Coherence and Degrees of Truth" (306–7):

> We have been setting out the logical ground for the doctrine of degrees of truth. Is its connection with the doctrine now clear? To make sure, let us briefly repeat: To think of any object whatever is to think of it in the relation to what is beyond it. There are always some of these relations that are so vital to the thing's nature, and therefore to our concepts of its nature, that neither could be what it is if cut off from them. Thus our concept can never be adequate till we have embraced these in our thought. And since we never do grasp them all, our thought remains inadequate. Not that as it stands it is worthless; to say that would be self-contradictory. It is plainly not wholly adequate; it is equally plainly not wholly worthless. Its adequacy is a matter of degree.

Switching now to Duhem, we feel that he should be mentioned in this section due to his original, and in many ways unorthodox, views about the role of mathematics in science. These views, expressed primarily in Duhem 1906, envisioned in some sense mathematics based on fuzzy logic—the subject of our chapter 5. Duhem's ideas are characterized in a nutshell through the following excerpts:[13]

> Between the concrete facts, as the physicist observes them, and the numerical symbols by which these facts are represented in the calculations of the theorist, there is an extremely great difference.
>
> It is impossible to describe a practical fact without attenuating by the use of the word "approximately" or "nearly"; on the other hand, all the elements constituting the theoretical fact are defined with rigorous exactness.
>
> A practical fact is not translated therefore by a single theoretical fact but by a kind of bundle including an infinity of different theoretical facts. Each of the mathematical elements brought together in order to constitute one of these facts may vary from one fact to another; but the variation to which it is susceptible cannot exceed a certain limit, namely, the limit of error within which the measurement of this element is blotted. The more perfect the methods of measurement are, the closer is the approximation and the narrower the limits but they never become so narrow that they vanish. . . .
>
> Thus every physical law is an approximate law. Consequently, it cannot be, for the strict logician, either true or false; any other law representing the same experiments with the same approximation may lay as just a claim as the first to the title of a true law or, to speak more precisely, of an acceptable law. . . .
>
> A mathematical deduction is of no use to the physicist so long as it is limited to asserting that a given *rigorously* true proposition has for its consequence the *rigorous* accuracy of some such other proposition. To be useful to the physicists, it must still be proved that the second proposition remains *approximately* exact when the first one is *approximately* true. And even that does not suffice. The range of these two approximations must be delimited; it is necessary to fix the limits of error which can be made in the result when the degree of precision of the methods of measuring the data is known; it is necessary to define the probable error that can be granted the data when we wish to know the result within a definite degree of approximation.
>
> Such are the rigorous conditions that we are bound to impose on mathematical deduction if we wish this absolutely precise language to be able to translate without betraying the physicist's idiom, for the terms of this latter idiom are and always will be vague and inexact like the perceptions which they are to express. On these conditions, but only on these conditions, shall we have a mathematical representation of the *approximate*.

---

[13] From Duhem 1954 (pp. 133–35, 171–72, 143)—an English translation of the second edition of Duhem 1906.

> But let us not be deceived about it; this "mathematics of approximation" is not a simpler and cruder form of mathematics. On the contrary, it is a more thorough and more refined form of mathematics, requiring the solution of problems at times enormously difficult, sometimes transcending the methods at the disposal of algebra today.

We consider it suggestive that mathematics based on fuzzy logic, especially some of its branches, may play the role of the "mathematics of approximation" envisioned by Duhem more than a century ago.

We began this section with Aristotle's syllogistic, his considerations about future contingencies and those about vague properties. It is a happy coincidence that we can close it by returning to Aristotle and show a compatibility of his broader views with the more specific views of Duhem that we just quoted. We have in mind Aristotle's statements such as "It is the mark of an educated man to look for precision in each class of things just so far as the subject admits" (McKeon 1941, 936). Or "We must also remember what has been said before, and not look for precision in all things alike, but in each class of things such precision as accords with the subject matter, and so much as is appropriate to the inquiry" (McKeon 1941, 943). Such statements appear throughout his writings in various forms and in various contexts.

Our reason for returning to Aristotle at the end of this section is to point to the little-known fact that Aristotle was tolerant toward imprecision. This is contrary to some writings that portray him as the "high priest" of logical precision and, unfortunately, therefore refer to any logic that deviates from classical logic as non-Aristotelian logic. The guardians of the principle of bivalence in ancient Greece were the Stoics, not Aristotle. They were actually the precursors of modern two-valued mathematical logic, as convincingly argued by Łukasiewicz (1930). As he correctly remarked, the nonclassical logics that do not accept the principle of bivalence should—after the great Stoic philosopher Chrysippus (279–c. 206)—be termed "non-Chrysippean" rather than "non-Aristotelian."

2.2 Emergence of fuzzy logic

There is general agreement that the emergence of fuzzy logic is uniquely associated with one particular event in 1965: the publication of a paper by Lotfi A. Zadeh (1965a), in which he introduced the concept of a fuzzy set. Because of this paper, Zadeh is generally acknowledged as the founder of fuzzy logic. However, he has also played a central role in its development, as is amply documented throughout this book.

Lotfi A. Zadeh was born in 1921 in Baku, the former Soviet Azerbaijan, to an Iranian father, Rahim Ali-Asker Zadeh, who worked in Baku as a foreign correspondent for a daily newspaper, *Iran*, and to a Russian-Jewish mother, Fanny Zadeh, who was a pediatrician. He went to a Russian school until 1931, when the family moved to Tehran. There he studied at the American College, an American Presbyterian Missionary School later renamed as Alborz College, where he learned English and Persian. However, Russian was still his primary language and he spoke it at home. He continued his studies at the University of Tehran, where he received a BS degree in electrical engineering in 1942. The time he spent at the American College apparently influenced him deeply and, as he recalls, "it instilled in me a deep desire to live in the United States."[14] He decided to immigrate to the United States in 1943 and, due to his strong interests in science and engineering, he applied for admission to the graduate program in electrical engineering at MIT. One year later, when he was admitted into the program, he actually did immigrate. Before starting his studies at MIT, he decided to spend some time in New York City and visited a young woman, Fay, whom he knew from Tehran and who happened to live in New York. It was a happy reunion. They stayed in contact during his studies at MIT and were married in 1946, shortly before he received his MS degree in electrical engineering from MIT that same year. Although he could have continued his studies at MIT for a PhD in electrical engineering, he decided to move to New York City to be close to his parents. During his studies at MIT, they had also immigrated to the United States and settled in New York. He was offered a position at Columbia University later that year as an instructor in electrical engineering, and was also admitted to their doctoral program. After he received his PhD from Columbia in 1949, he served on the Electrical Engineering Faculty of Columbia University for another 13 years. During these years, he made several important contributions to electrical engineering, initiating a new direction in frequency analysis of time-varying systems, an extension of Wiener's theory of prediction, and a new way of analyzing sampled-data systems based on a modified z-transformation. These contributions earned him a promotion to the rank of full professor in 1957. He also did some work and taught courses in circuit analysis, system theory, and information theory. He also became increasingly interested in the emerging computer technology, whose significance he quickly recognized.

In 1959, Zadeh left Columbia University and joined the Electrical Engineering Department of the University of California at Berkeley, with which he has been associated ever since. From 1963 to 1967, he served as department chairman. In this position, he emphasized the growing importance of computer technology and succeeded in transforming it into the Department of Electrical Engineering and Computer Science.

During his early years at Berkeley, Zadeh worked on various problems emerging from system theory, including problems pertaining to adaptive and time-varying systems, optimal control, and system identification. In the early 1960s, he began to question the adequacy of conventional mathematics for dealing with highly complex systems, as exemplified by the following quotation from one of his papers on systems theory (Zadeh 1962, 857):

---

[14] Interview with Lotfi Zadeh by Betty Blair, from "Famous people: Then and now" (*Azerbaijan International,* Winter 1999, 7.4).

Among the scientists dealing with animate systems, it was a biologist—Ludwig von Bertalanffy—who long ago perceived the essential unity of systems concepts and techniques in various fields of science and who in writings and lectures sought to attain recognition for "general systems theory" as a distinct scientific discipline.[15] It is pertinent to note, however, that the work of Bertalanffy and his school, being motivated primarily by problems arising in the study of biological systems, is much more empirical and qualitative in spirit than the work of those system theorists who received their training in exact sciences. In fact, there is a fairly wide gap between what might be regarded as "animate" system theorists and "inanimate" system theorists at the present time, and it is not certain that this gap will be narrowed, much less closed, in the near future. There are some who feel that this gap reflects the fundamental inadequacy of conventional mathematics—the mathematics of precisely-defined points, functions, sets, probability measures, etc.—for coping with the analysis of biological systems, and that to deal effectively with such systems, which are generally orders of magnitude more complex than man-made systems, we need a radically different kind of mathematics, the mathematics of fuzzy or cloudy quantities that are not describable in terms of probability distributions. Indeed, the need for such mathematics is becoming increasingly apparent even in the realm of inanimate systems, for in most practical cases the *a priori* data as well as criteria by which the performance of a man-made system is judged are far from being precisely specified or having accurately-known probability distributions.

Although this quotation appears to be quite suggestive of Zadeh's evolving views about classical mathematics during the early 1960s, Zadeh's own recollections are even more illuminating:[16]

In the process of writing the book, I became conscious of the fact that there are certain things that did not seem to lend themselves to precise formulation.[17] However, I was still a strong believer of classical mathematics, so I tried to define these things the way they are defined in mathematics. I slowly began to move from this very strong feeling that classical mathematics could solve all these problems. And then, in 1964, I got the idea of creating a membership function, which is a very simple and very basic idea. Once the idea occurred to me, I felt I had to do something with it. In a very short time I wrote a paper on fuzzy sets, which first appeared as a report in 1964 and then as a published paper in 1965. This was my first paper on fuzzy sets. As a result, the course of my thinking changed; in other words, I sort of put aside the more traditional approaches and directed my attention towards fuzzy systems. I was still very much into systems, but I began to develop an interest specifically in fuzzy systems.

The paper to which he refers in this excerpt is, of course, his seminal paper (Zadeh 1965a), in which he introduced the concept of a fuzzy set. After that, his research has been almost exclusively oriented to the development of fuzzy set theory and, later, to fuzzy logic in the broad sense. We cover this research later in the book, primarily in chapter 3. Here, we want to describe in detail the content of his seminal paper and assess its significance.

The key idea in Zadeh's paper of 1965 is of course the concept of a *fuzzy set*, which is a generalization of the classical concept of a set. Intuitively, a *classical set* is any collection of definite and distinct objects that are conceived as a whole. Objects

---

[15] Ludwig von Bertalanffy (1901–1972) was a distinguished Austrian biologist. The genesis and development of his general systems theory (or systemology) have been thoroughly researched by David Pouvreau, a French historian. It is covered in a three-part paper published in the *Int J Gen Syst* I: 36, no. 3 (2007): 281–337, II: 43, no. 1–2 (2014): 172–245; and III: 44, no. 5–6 (2015): 523–71. Parts I and III were coauthored with Manfred Drack, an Austrian theoretical biologist.

[16] Excerpted from an unpublished interview with Lotfi Zadeh by George Klir, which was conducted on the occasion of the Second IEEE International Conference on Fuzzy Systems in San Francisco in 1993.

[17] He refers to the book by L. A. Zadeh and C. A. Desoer, *Linear System Theory: The State Space Approach* (New York: McGraw-Hill, 1963).

that are included in a set are usually called its members. Each classical set must satisfy two requirements. First, members of each set must be distinguishable from one another; and second, any given object either is or is not a member of the set. That is, the proposition "*a* is a member of *A*" for any given object *a* and any given set *A* is either true or false. We say that each classical set has a sharp boundary, which separates objects that are members of the set from those that are not its members. Fuzzy sets differ from classical sets by rejecting the second requirement. As a consequence, their boundaries are not necessarily sharp.

In each particular application of set theory, as is well known, all objects of concern are members of a set called a *universe of discourse* (a *universe* for short) or a *universal set*. A common way of defining an arbitrary classical set, *A*, within a given universe of discourse *U*, is to assign the number 1 to each member of *U* that is supposed to also be a member of set *A*, and to assign the number 0 to the remaining members of *U*. In classical set theory, this assignment is called a *characteristic function* of *A*, and it is usually denoted by χ_A. It is a function of the form $U \to \{0, 1\}$.

Zadeh begins his paper with a clear and convincing motivation for his groundbreaking concept (Zadeh 1965a, 338–339):

> More often than not, the classes of objects encountered in the real physical world do not have precisely defined criteria of membership. For example, the class of animals clearly includes dogs, horses, birds, etc. as its members, and clearly excludes such objects as rocks, fluids, plants, etc. However, such objects as starfish, bacteria, etc. have an ambiguous status with respect to the class of animals. The same kind of ambiguity arises in the case of a number such as 10 in relation to the "class" of all real numbers which are much greater than 1.
>
> Clearly, the "class of all real numbers which are much greater than 1," or "the class of beautiful women," or "the class of tall men," do not constitute classes or sets in the usual mathematical sense of these terms. Yet, the fact remains that such imprecisely defined "classes" play an important role in human thinking, particularly in the domains of pattern recognition, communication of information, and abstraction.
>
> The purpose of this note is to explore in a preliminary way some of the basic properties and implications of a concept which may be of use in dealing with "classes" of the type cited above. The concept in question is that of a *fuzzy* set, that is, a "class" with a continuum of grades of membership. . . . The notion of a fuzzy set provides a convenient point of departure for the construction of a conceptual framework which parallels in many respects the framework used in the case of ordinary sets, but is more general than the latter and, potentially, may prove to have a much wider scope of applicability. . . . Essentially, such a framework provides a natural way of dealing with problems in which the source of imprecision is the absence of sharply defined criteria of class membership.

He then defines a fuzzy set *A* within a given universe of discourse *U* as a function that assigns to each member of *U* a real number in the unit interval $[0, 1]$. He denotes this function as f_A and calls it a membership function of the fuzzy set *A*. Moreover, he refers to the number $f_A(x)$ as the *grade of membership* of object *x* in fuzzy set *A*. Fuzzy sets defined in this way are now commonly referred to as *standard fuzzy sets*. Other types of fuzzy sets, which were introduced later, are surveyed in chapter 3. Zadeh was apparently aware of other possible types of fuzzy sets when he inserted a footnote, right after his definition of the membership function f_A, with the following remark: "In a more general setting, the range of the membership function can be taken to be a suitable partially ordered set."

A remark regarding notation should be made at this point. The symbol f_A introduced by Zadeh for the membership function of fuzzy set A was soon replaced in the literature on fuzzy sets (by Zadeh as well as others) with the symbol μ_A. The intent of this replacement was to designate the Greek letter μ for membership functions of fuzzy sets and reserve the letter f for other purposes. Later, the notation was simplified by defining standard fuzzy sets as functions

$$A : U \to [0,1].$$

This notation is predominant in current literature and we use it in this book as well. In particular, we use this notation (and not Zadeh's original notation) in our overview of Zadeh's seminal paper.

Given a fuzzy set A defined within some universe U, the proposition "x is a member of fuzzy set A" is certainly a meaningful claim regarding the membership status of a specific object from U in A. Clearly, the proposition may naturally be regarded as true when $A(x) = 1$ and false when $A(x) = 0$. In all other (borderline) cases, the proposition is regarded as true to a degree that is expressed by the grade of membership of x in A, i.e. by the value $A(x)$. All real numbers in the unit interval $[0,1]$—the range of function A—can thus be interpreted for each $x \in U$ either as grades of membership of x in A or as degrees of truth of the proposition "x is a member of A." Zadeh does not explicitly discuss this important connection between fuzzy sets and fuzzy propositions in his first paper on fuzzy sets, but he suggests it in a footnote. However, he does discuss some other ideas that were important at the time the paper was published for guiding early developments of fuzzy set theory.

One of these ideas is the introduction of the concept a *fuzzy relation* as a fuzzy set defined within a universe of discourse that is a Cartesian product of two or more sets. Also introduced in the paper is the notion of a *composition* of binary fuzzy relations as well as the notion of various *projections* of n-ary fuzzy relations ($n \geq 2$). Particular attention is given in the paper to fuzzy relations defined on n-dimensional Euclidean spaces $\mathbb{R}^n$ ($n > 1$).

Zadeh also recognized, somewhat indirectly, that each standard fuzzy set is associated with a special family of classical sets. These classical sets are now commonly known as level-cuts, but Zadeh did not introduce any name for them in the paper.[18] We denote these sets (level-cuts) for each given fuzzy set A by $^\alpha A$, where $\alpha \in [0,1]$. Given a standard fuzzy set A, the full family of its level-cuts is defined by

$$^\alpha A = \{x \in U \mid A(x) \geq \alpha\}$$

for all $\alpha \in [0,1]$. In his paper, Zadeh does not examine any properties of these families of classical sets. He only employs them in the following ways for defining the concepts of convexity and boundedness for fuzzy sets defined on n-dimensional Euclidean spaces $\mathbb{R}^n$ ($n \geq 1$): Fuzzy set A defined on $\mathbb{R}^n$ for some $n \geq 1$ is convex (or bounded) if and only if its level-cuts are convex (or bounded) in the classical sense

---

[18] Often these also are called alpha-cuts, α-cuts, or simply cuts.

for all $\alpha \in [0, 1]$. This means that he had already introduced in this seminal paper one particular principle for extending various concepts of classical mathematics to their counterparts in mathematics based on fuzzy sets. Any concept obtained by this principle is now usually called a *cutworthy concept*—a concept extended from classical mathematics via the level-cut representation of fuzzy sets. Zadeh did not address the issue of whether the level-cuts represent the associated fuzzy set A completely and uniquely when he wrote the paper (he addressed this issue later), but he was apparently confident at that time that they do.

Also discussed in the paper was the idea of fuzzy sets induced by mappings. From this simple idea, Zadeh developed and published ten years later his extension principle (Zadeh 1975a,c). This is an important principle by which any function $f : X \rightarrow Y$, where X and Y are classical sets, can be extended to act on fuzzy sets defined on X and Y (section 3.2 below).

Zadeh devotes almost one third of his 1965 paper to discussing operations on fuzzy sets. For each standard fuzzy set A, he defines its complement, $\overline{A}$, by the formula

$$\overline{A}(x) = 1 - A(x) \text{ for all } x \in U.$$

Furthermore, he defines the intersection and union of any two standard fuzzy sets A and B by the formulas

$$(A \cap B)(x) = \min\{A(x), B(x)\} \text{ for all } x \in U,$$
$$(A \cup B)(x) = \max\{A(x), B(x)\} \text{ for all } x \in U,$$

respectively. These three operations are now usually referred to in the literature as standard operations on fuzzy sets. Zadeh also defines the meaning of containment of one fuzzy set in another in the following way: fuzzy set A is contained in fuzzy set B, denoted by $A \subseteq B$, if and only if $A(x) \leq B(x)$ for all $x \in U$. Moreover, he derives various properties of the standard operations. He is well aware that the operations can be defined in many other ways and gives a few illustrative examples. A few other ideas regarding primarily convex fuzzy sets are also presented in this paper.

That same year, Zadeh published another paper on fuzzy sets (Zadeh 1965b), not as well-known as his seminal paper, but also significant. In this paper, which is based on his presentation at the Symposium on System Theory held at the Polytechnic Institute of Brooklyn in April 1965, he discusses the potential role of fuzzy sets in system theory. He defines discrete-time fuzzy dynamic systems, discusses some basic issues involved in the optimization of systems under fuzzy constraints, and introduces the interesting notion of fuzzy classes of systems, such as the class of systems that are "approximately equivalent" to a given system, the class of systems that are "approximately linear," the class of systems that are "adaptive," and the like. He argues that "most of the adjectives used in system theory to describe various types of systems, such as: linear, nonlinear, adaptive, time-invariant, stable, etc., are in reality names for fuzzy classes of systems." In fact, he discussed one of these classes—the class of adaptive systems—in a short note published earlier in the 1960s.[19]

Why is the publication of Zadeh 1965a generally considered a turning point that marks the emergence of fuzzy logic? To address this question adequately, other relevant developments must be considered.

Somewhat surprisingly, the book on holism by Jan Smuts (1926) seems to be the first publication that characterizes vagueness of concepts in a holistic way that is highly suggestive of fuzzy sets. The following excerpt taken from pp. 16–18 of the book clearly shows the connection between Smuts's holistic view of concepts and Zadeh's fuzzy sets:

> The science of the nineteenth century was like its philosophy, its morals and its civilisation in general, distinguished by a certain hardness, primness and precise limitation and demarcation of ideas. Vagueness, indefinite and blurred outlines, anything savouring of mysticism, was abhorrent to that great age of limited exactitude. The rigid categories of physics were applied to the indefinite and hazy phenomena of life and mind. Concepts were in logic as well as in science narrowed down to their most luminous points, and the rest of their contents treated as nonexistent. Situations were not envisaged as a whole of clear and vague obscure elements alike, but were analysed merely into their clear, outstanding, luminous points. A "cause," for instance, was not taken as a whole situation, which at a certain stage insensibly passes into another situation, called the effect. No, the most outstanding feature in the first situation was isolated and abstracted and treated as the cause of the most outstanding and striking feature of the next situation, which was called the effect. Everything between this cause and this effect was blotted out, the two sharp ideas or rather situations of cause and effect were made to confront each other in every case of causation like two opposite forces. This logical precision immediately had the effect of making it impossible to understand how the one passed into the other in actual causation. . . . And all this is due to the initial mistake of enclosing things or ideas or persons in hard contours which are purely artificial and are not in accordance with the natural shading-off continuities which are or should be well known in science and philosophy alike. . . . We have to return to the fluidity and plasticity of nature and experience in order to find the concepts of reality. When we do this we find that round every luminous point in experience there is a gradual shading off into haziness and obscurity. A "concept" is not merely its clear luminous centre, but embraces a surrounding sphere of meaning or influence of smaller or larger dimensions, in which the luminosity tails off and grows fainter until it disappears.

Throughout the whole book, Smuts does not use any mathematics. However, we consider his verbal description of concepts an excellent verbal description of fuzzy sets, especially of their level-cut representations.

Ideas, even more closely connected with fuzzy sets, were presented by the well-known American philosopher Max Black (1909–1988). His paper (Black 1937) is a penetrating study of the gap between a mathematical model and experience and the need to bridge it. When introducing the paper, Black describes its purpose clearly and concisely (427–29):

> It is a paradox, whose important familiarity fails to diminish, that the most highly developed and useful theories are ostensibly expressed in terms of objects never encountered in experience. The line traced by a draftsman, no matter how accurate, is seen beneath the microscope as a kind of corrugated trench, far removed from the ideal line of pure geometry. And the "point-planet" of astronomy, the "perfect gas" of thermodynamics, or the "pure species" of genetics are equally remote from exact realization. Indeed the unintelligibility at the atomic or sub-atomic level of

---

[19] L. A. Zadeh, "On the definition of adaptivity," *Proc IEEE* 51 (1963): 469–70.

a rigidly demarcated boundary shows that such objects not merely are not but could not be encountered. While the mathematician constructs a theory in terms of "perfect" objects, the experimental scientist observes objects of which the properties demanded by theory are and can, in the very nature of measurement, be only approximately true. As Duhem remarks,[20] mathematical deduction is not useful to the physicist if interpreted rigorously. It is necessary to know that its validity is unaltered when the premise and conclusion are only "approximately true." But the indeterminacy thus introduced, it is necessary to add in criticism, will invalidate the deduction unless the permissible limits of variation are specified. To do so, however, replaces the original mathematical deduction by a more complicated mathematical theory in respect of whose interpretation the same problem arises, and whose exact nature is in any case unknown.

    This lack of exact correlation between a scientific theory and its empirical interpretation can be blamed either upon the world or upon the theory. . . . On either view there remains a gap between scientific theory and its application which ought to be, but is not, bridged. To say that all language (symbolism, or thought) is vague is a favorite method for evading the problems involved and lack of analysis has the disadvantage of tempting even the most eminent thinkers into the appearance of absurdity. . . .

    We shall not assume that "laws" of logic or mathematics prescribe modes of existence to which intelligible discourse must necessarily conform. It will be argued, on the contrary, that deviations from the logical or mathematical standards of precision are all pervasive in symbolism; that to label them as subjective aberrations sets an impassable gulf between formal laws and experience and leaves the *usefulness* of the formal sciences an insoluble mystery. And it is the purpose of the constructive part of the essay to indicate in outline an appropriate symbolism for vagueness by means of which deviations from a standard can be absorbed by a reinterpretation of the same standard in such a way that the laws of logic in their usual absolutistic interpretation appear as a point of departure for more elaborate laws of which they now appear as special or limiting cases.

At the end of his paper, Black shows that vagueness of each linguistic term, L, in some language can be characterized by a function, C, that he calls a *consistency profile* of the given term. This is basically a function that assigns to each considered object, x, in some domain of applicability of L its degree of consistency, $C(x, L)$, with a given linguistic term. Black also develops a simple method by which the consistency for each considered x with respect to a given L can be operationally determined. The method employs a group of users of the given language. For each particular x, each of the users is asked for his or her opinion of whether or not x is consistent with L. Given n positive answers and m negative answers, where $n + m$ is the number of users employed, the degree of compatibility of the considered object x with respect to the given linguistic term can be expressed as $n/n + m$.[21] The values of $C(x, L)$ defined in this way are always in the interval $[0, 1]$ and express for each x the relative frequency of positive answers. Clearly, this concept of the consistency profile is strikingly similar to the concept of a fuzzy set. In fact, the simple method suggested by Black for determining consistency profiles for various linguistic terms is also one of the many methods that have been developed for constructing fuzzy sets.

    Generalized characteristic functions were considered by Szpilrajn (1936) in his study of point sets under the term "fonction caractéristique." However, they only

---

[20] Duhem 1954 (p. 143).

[21] In fact, Black uses the ratio n/m in his paper, but agreed later with a critical remark made by Hempel (1939) that the ratio $n/n + m$ is preferable. His agreement is expressed in a note on the reprinted version of his 1937 paper (Black 1949, 249).

played an auxiliary role and Szpilrajn was not interested in this notion per se. Shirai (1937) introduced certain three-valued sets, which he called pseudo-sets. He defined the operations of intersection, union, and complementation in such a way that the pseudo-sets form a ring, but except for a simple arithmetical example he did not provide any motivation for his notion.

In 1940, Hermann Weyl (1885–1955), an American mathematician of German origin, observed that "Predicates or properties of a point in a continuum are often of the *'more or less'* type, so that the question is not whether an individual has this property, but *to what degree*" (Weyl 1940). He suggested describing such predicates "by a function $f(x)$ whose argument x varies over the points of the given space while the value f is a real number in the interval $[0, 1]$." This function is clearly the same as the one representing standard fuzzy sets. Weyl, furthermore, suggested for his functions f operations that are equivalent to the standard operations on fuzzy sets. However, as far as we know, Weyl never returned to this idea to develop it further.

The concept of a many-valued relation, and thus in particular also of a many-valued set, is automatically present whenever one considers many-valued predicate logics because many-valued relations interpret symbols of relations in these logics. The first many-valued predicate logics appeared in the late 1930s and their development before the mid 1960s, which includes many contributions particularly since the 1950s, is covered in section 4.2.4. However, apart from a few exceptions, such as the first such logic developed by Bochvar (1938), no particular attention was paid to the meaning of such many-valued relations. Worth noting is the fact that standard fuzzy sets and relations played an important role in these developments because they were the many-valued relations in the infinitely-valued predicate Łukasiewicz logic.

Interesting considerations are found in the works of the Austrian mathematician, physicist, and philosopher Friedrich Waismann (1896–1959)—a key member of the Vienna Circle and a leading figure in logical positivism. In his paper, "Are there alternative logics?" (Waismann 1945–46), he convincingly argues in favor of a new logic:

> [W]e should be blind if we did not admit that the ideas "true" and "false" are often employed in ways running counter to orthodox logic. Thus we say, . . . "This is not entirely true", . . .— phrases which strongly suggest that we regard the ideas "true" and "false" as capable of gradation. . . . the matter can be looked upon as showing the rudiments of a novel type of logic.

Waismann then outlines this new logic, a three-valued one, and says that

> [W]e might as well introduce a logic with a larger number of truth-gradations, for instance a four-valued logic ("true", "nearly true", "not quite false", "entirely false"). Let no one say that these logics are a mere play with symbols. For there are sub-domains in our language in which a logic with graduated truth-values is quite natural. . . . in doing so one *accomplishes the transition to a logic with a graduated scale of truth values*. The same holds good of most statements describing properties capable of gradations.

In 1951, Kaplan and Schott, both then with the Philosophy Department at UCLA, published a paper (Kaplan and Schott 1951) whose aim they described as:

> To construct a calculus which will provide a more adequate explication of classes in their scientific use than is afforded by the conventional "calculus of classes," which we prefer to call a calculus of sets. The procedure of this paper is to construct, with the help of the theory of probability and the set calculus, entities which have a degree of vagueness characteristic of actual classes, but which, when the degree is minimal, correspond to the precise sets of *Principia Mathematica*.[22]

Kaplan and Schott refer to the classes of concern as *empirical classes* and define them as follows. They assume a framework, called *articulation*, consisting of a set U of individuals representing a field of inquiry and a set of *categories*, such as color and length, each of which consists of a collection of subsets of U, called *qualities* of the given category. Thus, qualities of the category color may be "red," "green," and so on. Any selection of a single quality from each category is called a *profile*. It represents the individuals in U which have the chosen qualities. An arbitrary function assigning to every profile a number in the real unit interval $[0, 1]$ is called an *indicator*. Clearly, indicators—which are thought of as models of empirical classes—may be identified with standard fuzzy sets. On the one hand, an indicator is a fuzzy set because it is a function to $[0, 1]$. On the other hand, every standard fuzzy set in U may be conceived of as an indicator in the articulation consisting of a single category which contains for each $u \in U$ the singleton quality $\{u\}$. Interestingly, the authors introduce standard operations with fuzzy sets, inclusion of fuzzy sets, and derive some of their properties. Even though the very concept of indicator does not rely on probabilistic concepts, the authors' probabilistic view of indicators is apparent not only from their formula for union of classes but, more importantly, from a probabilistic way of constructing indicators which they provide. A large part of Kaplan and Scott's paper is devoted to a discussion of the rationale for developing this calculus and for discussing its properties.

When exploring the Poincaré paradox of a physical continuum (p. 334), Menger (1951b) introduced *probabilistic relations*, which are standard fuzzy relations with a probabilistic interpretation of membership degrees. In his paper, written in French, Menger (1951a) also considered unary probabilistic relations and called them "ensembles flous"—a French term for "fuzzy sets." We examine Menger's investigations in greater detail in section 5.4.2 and on pp. 274 and 278.

Another area, rather important and mathematically advanced, in which fuzzy sets and relations appeared is the axiomatic theory of many-valued sets. It started with T. Skolem (1957), who explored set theories within the finitely- and infinitely-valued Łukasiewicz logics and hence considered standard as well as finitely-valued fuzzy sets. Early subsequent contributions were made by C. C. Chang (1963), J. E. Fenstad (1964), and later also D. Klaua (1965, 1966) who examined cumulative hierarchies of fuzzy sets within finitely-valued Łukasiewicz logics. Worth noting is also the work of Rasiowa (1964) who studied a many-valued generalization of a formal theory of fields of sets. Rasiowa represents the Polish school whose many works on

---

[22] They mean, of course, the classic book Whitehead and Russell 1910–13.

predicate logic involve various kinds of many-valued sets.[23] These contributions, which play an important role in later developments of many-valued and fuzzy set theories, are examined in section 5.2.3. However, none of these early studies attached any significance for applications to the many-valued sets.

The above list of works in which the idea of a set with graded membership or some closely related idea appeared could continue and include further works, such as Kubiński 1958, 1960, which examine "nazwy nieostre" [unsharp names] and make an attempt to "bring logic nearer to colloquial language"; Körner 1951, 1966 with the ostensive predicates modeled in a three-valued logic; and Salij 1965, which develops sets and relations with values in an arbitrary lattice and studies e.g. relational compositions. Nevertheless, we stop here because now our question of why the publication of Zadeh 1965a represents the turning point that marks the emergence of fuzzy logic can be properly addressed.

The following list of important features addresses this question directly:

1. The paper contains a clearly stated motivation for the notion of a fuzzy set as a fundamental, broadly applicable, and much needed concept.
2. The concept of a fuzzy set introduced in the paper is simple, easy to understand, and general, not restricted to any special interpretation.
3. The paper was published in a highly respectable and visible journal with a broad orientation to science and engineering and written for a broad spectrum of readers.
4. The paper not only introduces fuzzy sets, but also examines some of their fundamental properties and indicates directions for further research.

These features, we believe, made Zadeh's paper qualitatively different from all of the other related publications. Its significance was ultimately demonstrated by the fact that it incited, after a few-years incubation period, a strong response from the academic community, both positive and negative. A characterization of this response is the subject of the next section.

2.3   Evolving attitudes toward fuzzy logic

The response of the academic community to the publication of Zadeh's seminal paper was initially (in the 1960s) lukewarm at best, which seems to indicate that the ideas presented in the paper were not taken seriously and were, by and large, ignored. However, a handful of researchers, mostly young, considered this new area so interesting and potentially important that they opted to work on its further development. One of these young researchers, Joseph A. Goguen (1941–2006), stands

---

[23] The first seems to be A. Mostowski's "Proofs of non-deducibility in intuitionistic functional calculus," *J Symb Logic* 13 (1948): 204–7.

out for his important contributions to fuzzy logic, but above all for his two remarkable early papers (Goguen 1967, 1968–69).[24] In the first paper, he developed an important generalization of standard fuzzy sets to the so-called *L*-fuzzy sets by replacing the unit interval of membership grades with a more general and well-conceived algebraic structure of a complete residuated lattice. In the second paper, he developed basic ideas for developing logic for reasoning with inexact concepts in which fuzzy sets play the role of inexact predicates. In addition to these important papers, he made other contributions to fuzzy logic in the 1970s. Later, he made significant contributions to several other areas, including theoretical computer science, artificial intelligence, and systems science. Unfortunately, he died in 2006, at age 65, much too young. In an obituary for him, Zadeh (2007) recalled Goguen as follows:

> My first paper on fuzzy sets was published in June 1965, while I held the position of the Chair of the Department of Electrical Engineering at UC Berkeley. Shortly after the publication of my paper, a student walked into my office and identified himself as Joe Goguen, a graduate student in mathematics. He told me that he had read my paper and was interested in developing the concept of a fuzzy set within the framework of category theory. We discussed his ideas for a while. At the end of our discussion, he asked me to become his research supervisor. I responded affirmatively, since it was quite obvious that Joe Goguen was not an average graduate student—he was a superior intellect. This meeting was the beginning of my lifelong relationship with Goguen. We met frequently to discuss various issues of fuzzy set theory. Nobody in the mathematics department took interest in his work.

Two additional doctoral students at the University of California at Berkeley were among the young contributors to fuzzy logic in the 1960s, C. L. Chang, who investigated the applicability of fuzzy sets to pattern recognition in his dissertation (Chang 1967) and developed also the first formulation of fuzzy topological spaces (Chang 1968), and E. T. Lee, who investigated the notion of fuzzy languages (Lee and Zadeh 1969).

It was especially fortunate for fuzzy logic that three established and well-known researchers not only endorsed it, but also contributed to its development at this early stage. Just one year after the publication of Zadeh's paper, Richard Bellman (1920–1984) and Robert Kalaba (1926–2004) joined with Zadeh in writing a significant early paper on the role of fuzzy sets in abstraction, generalization, and pattern classification (Bellman, Kalaba, and Zadeh 1966). This was possible since they had been exposed to Zadeh's ideas regarding fuzzy sets even before they were published, which happened through discussions during their occasional meetings at the RAND Corporation. Bellman, in particular, was quick to recognize the significance of these ideas and encouraged Zadeh to work on their development. He himself developed some of them later in the 1970s. The third important early contributor to fuzzy set theory was King-Sun Fu (1930–1985) of Purdue University, a pioneer in the field of pattern recognition and machine intelligence, who coauthored a significant early paper on fuzzy automata and learning systems (Wee and Fu 1969), and contin-

---

[24] Goguen's contributions are examined in detail in chapters 4 and 5.

ued contributing to the emerging area of fuzzy set theory until his untimely death in 1985 at age 54.[25]

We cover here these details about the very early history of fuzzy logic, focusing only on the period from 1965 to 1969, to show how quickly some of the ideas sketched in Zadeh's seminal paper were further developed in various ways by a tiny number of talented researchers, young and old, to whom these ideas appealed. However, the rest of the academic community did not seem to notice these promising initial developments or, perhaps, did not consider them seriously. This changed abruptly in the 1970s, when the few initial movers were joined by many others. This resulted in an impressive growth of the literature on fuzzy logic, through which many new ideas were introduced. This literature included, for example, a series of four books on fuzzy set theory by Arnold Kaufmann that were published in French between 1973 and 1977.[26] These early books, which covered fuzzy set theory quite extensively and in a human-friendly way, were historically important in the sense that they made the French academic community aware of the existence of fuzzy logic at a very early date. Lotfi Zadeh wrote a foreword to the first volume, which was the only one that was later published in English translation (Kaufmann 1975). Another important publication in the 1970s was a book by two Romanian authors, Constantin Negoiță and Dan Ralescu (1975a), who were at that time unknown, but who are now widely recognized for their pioneering work on foundations as well as applications of fuzzy logic.[27] Published so early in the history of fuzzy logic, this was a surprisingly advanced book of remarkable mathematical sophistication. It was a very timely publication, as it surveyed in a comprehensive way all the best results that were available in fuzzy logic at that time. Each of the chapters was supplemented by highly valuable historical and bibliographical remarks, and the book's bibliography contained references to all publications on fuzzy logic that were at that time worth mentioning. This was undoubtedly an important book published at the right time, whose historical value is enormous.

For some, emergence of all of these ideas meant that the time was ripe to examine them and assess their value. To many, however, these ideas were fundamentally repulsive and unacceptable, and they felt that a strong stand had to be taken to discredit them. This aroused in the 1970s some hostile, irrational, and often emotional and wholesale criticisms of fuzzy logic. This is well documented in a historically valuable paper by Zadeh (1999a)—his personal reflections on the evolution of fuzzy logic from its emergence in 1965 to the end of the 20th century.

In order to illustrate the hostility toward fuzzy logic when it was still in its infancy and thus vulnerable to criticism, we reproduce here a fully recorded discussion

---

[25] He was the first president of the International Association of Pattern Recognition.

[26] *Introduction a la théorie des sous-ensembles flous: A l'usage des ingénieurs*: vol. 1 (1973); vol. 2 (1975); vol. 3 (1975); vol. 4 (1977) (Paris: Masson et Cie Editeurs).

[27] This book is an English translation of a revised version of the Romanian original, *Mulțimi vagi și aplicațiile lor*, which was published in 1974 by Editura Technică in Bucharest, Romania.

that took place at the International Conference on Man and Computer in Bordeaux, France, on September 11, 1972, after Zadeh's plenary presentation, in which he introduced and discussed systems based on linguistic variables (section 3.6). This presentation along with the discussion was published two years later in the proceedings of the conference and includes the following record of discussion between Rudolf E. Kalman, a key figure in mathematical system theory, and Lotfi Zadeh (Zadeh 1974b, 93–94):

> KALMAN: I would like to comment briefly on Professor Zadeh's presentation. His proposal could be severely, ferociously, even brutally criticized from a technical point of view. This would be out of place here. But a blunt question remains: Is Professor Zadeh presenting important ideas or is he indulging in wishful thinking?
>
> The most serious objection of "fuzzification" of systems analysis is that lack of methods of system analysis is *not* the principal scientific problem in the "systems" field. *That* problem is one of developing basic concepts and deep insight into the nature of "systems," perhaps trying to find something akin to the "laws" of Newton. In my opinion, Professor Zadeh's suggestions have no chance to contribute to the solution of this basic problem.
>
> To take a concrete example, modern experimental research has shown that the brain, far from fuzzy, has in many areas a highly specific structure. Progress in brain research is now most rapid in anatomy where the electron microscope is the new tool clarifying regularities of structure which previously were seen only in a fuzzy way.
>
> No doubt Professor Zadeh's enthusiasm for fuzziness has been reinforced by the prevailing political climate in the US—one of unprecedented permissiveness. "Fuzzification" is a kind of scientific permissiveness; it tends to result in socially appealing slogans unaccompanied by the discipline of hard scientific work and patient observation. I must confess that I cannot conceive of "fuzzification" as a viable alternative for the scientific method; I even believe that it is healthier to adhere to Hilbert's naïve optimism, "Wir wollen wissen: wir werden wissen."
>
> It is very unfair for Professor Zadeh to present trivial examples (where fuzziness is tolerable or even comfortable and in any case irrelevant) and then imply (though not formally claim) that this vaguely outlined methodology can have an impact on deep scientific problems. In any case, if the "fuzzification" approach is going to solve any difficult problem, this is yet to be seen.
>
> ZADEH: To view Professor Kalman's rather emotional reaction to my presentation in a proper perspective, I should like to observe that, up to a certain point in time, Professor Kalman and I have been traveling along the same road, by which I mean that both of us believed in the power of mathematics, in the eventual triumph of logic and precision over vagueness. But then I made a right turn, or maybe even started turning backwards, whereas Professor Kalman has stayed on the same road. Thus, today, I no longer believe, as Professor Kalman does, that the solution to the kind of problems referred to in my talk lies within the conceptual framework of classical mathematics. In taking this position, I realize, of course, that I am challenging scientific dogma—the dogma that was alluded to by Professor Caianiello in his remarks.[28]
>
> Now, when one attacks dogma, one must be prepared to become the object of counterattack on the part of those who believe in the status quo. Thus, I am not surprised when in reaction to my views I encounter not only enthusiasm and approbation, but also criticism and

---

[28] Presented by Prof. Eduardo R. Caianiello (1921–1993), a well-known Italian physicist, under the title "Intelligence in man and machine" in another plenary lecture that took place just before Zadeh's presentation. These are some of his introductory remarks to which Zadeh refers: "The hardest thing of all is perhaps to dispose of prejudices that come with our very education as scientists; von Neuman was quite clear about it when he stated that our present forms of logic and mathematics are not necessities, but historical accidents such as the fact that we happen to speak English, Chinese, or some other language. The assumption that the extant mathematical methods, which were invented to describe some aspects of physical reality (as the infinitesimal calculus for gravitation and mechanics) are also adequate for the task at hand may well be a presumption; it is to be hoped, rather than feared, that unbiased experimental research may originate new branches of mathematics, in terms of which that goal may be better approached, which Poincaré enunciated when he said that what matters most is not 'whether a problem is solved' but 'how much solved' it is"; see the conference proceedings, M. Marois, ed., *Man and Computer* (Amsterdam: North-Holland, 1972), 33–54.

derision. Nevertheless, I believe that, in time, the concepts that I have presented will be accepted and employed in a wide variety of areas. Their acceptance, however, may be a grudging one by those who have been conditioned to believe that human affairs can be analyzed in precise mathematical terms. The pill, though a bitter one, will be swallowed eventually. Indeed, in retrospect, the somewhat unconventional ideas suggested by me may well be viewed as self-evident to the point of triviality. Thank you.

KALMAN: Professor Zadeh misrepresents my position if he means to say that I view scientific research solely in terms [of] rigidly precise or even "classical" mathematical models. The desire to have useful insights into vaguely defined phenomena is very old; for instance, it has led to the development of probability theory and topology. René Thom has recently reintroduced and greatly expanded this old aspect of topology by his highly imaginative theory of "catastrophes" governing discontinuous changes in the evolution of natural systems. All of these theories use precise reasoning to reach conclusions about imprecisely defined situations. They also lead to striking insights.

The question, then, is whether Professor Zadeh can do better by throwing away precise reasoning and relying on fuzzy concepts and algorithms. There is no evidence that he can solve any nontrivial problems.

This is not to argue that only rigorously rational methods (of conventional science) should be used. But if one proposes to deprecate this tool (which, when properly understood and used, has given us many striking successes), he should at least provide some hard evidence of what can be gained thereby. Professor Zadeh's fears of unjust criticism can be mitigated by recalling that the alchemists were not prosecuted for their beliefs but because they failed to produce gold.

I, too, share the hope that science will eventually penetrate into areas which are too complex or too ill defined at present. But Professor Zadeh should refrain from claiming credit before this has been accomplished. He will surely receive credit *afterwards*, if he was in fact one of those who really had the "ideas."

ZADEH: I will rest on what I have said, since only time can tell whether or not my ideas may develop into an effective tool for the analysis of systems which are too complex or too ill-defined to be susceptible of analysis by conventional techniques. Conceding that I am not an unbiased arbiter, my belief is that, eventually, the answer will turn out to be in the affirmative.

Another example showing the hostility toward fuzzy logic in the 1970s is the following statement that was made in 1975 by a colleague of Zadeh at UCB, William Kahan, a professor of computer science and mathematics (Zadeh 1999a, 893):

> Fuzzy theory is wrong, wrong, and pernicious. I cannot think of any problem that could not be solved by ordinary logic. . . . What Zadeh is saying is the same sort of thing as, "Technology got us into this mess and now it can't get us out." Well, technology did not get us into this mess. Greed and weakness and ambivalence got us into this mess. What we need is more logical thinking, not less. The danger of fuzzy theory is that it will encourage the sort of imprecise thinking that has brought us so much trouble.

Zadeh was able to withstand this kind of harsh and emotional criticism of fuzzy logic, which was rather common in the 1970s, and he always responded to it in a rational and friendly way. In spite of the criticism, the number of researchers choosing to work in this area steadily grew during the decade and this, in turn, resulted in an impressive growth of literature dealing with fuzzy logic. Zadeh himself published 36 papers on fuzzy logic in the 1970s, almost all of which opened new directions of research.

In the face of these developments, it was gradually more difficult for opponents of fuzzy logic to ignore it or dismiss it on emotional grounds. Fuzzy logic was still criticized in the 1980s, but the criticism gradually became less emotional and more focused. Most commonly, it involved some mixture of the following three arguments, usually expressed more or less like this:

1. We have observed that the number of publications under the names of "fuzzy logic" (or "fuzzy set theory") has been growing, but the problem with all these publications is that whatever they cover is not based on any sound foundations.

2. We have been told that fuzzy logic is essential for developing some innovative applications for which classical logic is not adequate. So far, we have not seen any such applications, so we are waiting for the supporters of fuzzy logic to show us at least one nontrivial application.

3. Fuzzy logic advocates argue that fuzzy logic is essential for dealing with some aspects of uncertainty. We are not able to envision any problem involving uncertainty that cannot be correctly dealt with by probability theory.

In the 1980s, the first two arguments were perfectly justifiable and advocates of fuzzy logic were not in a position to refute them. To develop sound foundations of fuzzy logic still required not only more time and hard work, but also a conviction that this work was important. To develop some significant and successful real-world applications based on fuzzy logic required, in addition, obtaining some industrial or governmental sponsors willing to support the actual development of the envisioned applications. Through the work of a growing number of researchers working on the theory or applications of fuzzy logic, sound foundations as well as convincing applications were eventually established, so the two above-mentioned arguments critical of fuzzy logic were no longer valid. This is documented in detail in chapters 3–6. In the meantime, fuzzy logic continued to be a subject of many debates throughout the 1980s and even the early 1990s. Fortunately, some of these debates are well documented. We summarize these debates and emphasize some of their peculiarities in the next section.

2.4 Documented debates

One of the early debates, partially documented, took place at the Conference on the Calculus of Uncertainty in Artificial Intelligence and Expert Systems that was held at the Institute for Reliability and Risk Analysis at George Washington University in Washington, DC, on December 28–29, 1984. There were four invited speakers, among them Lotfi Zadeh and Dennis Lindley, a strong supporter of probability theory. An attempt was made three years later to document the debate via written versions of the conference presentations in *Statistical Science* 2, no. 1 (1987): 1–44. Unfortunately, Zadeh was not able to deliver a written version of his presentation in time for inclusion in this debate issue of *Statistical Science*; so fuzzy logic plays only a minor role in this incomplete record of the actual debate. We mention this debate issue here only to focus on an overly strong and somewhat peculiar position taken by Lindley (1987), as expressed in the following excerpts (pp. 17 and 24):

> Our thesis is simply stated: *the only satisfactory description of uncertainty is probability*. By this is meant that every uncertainty statement must be in the form of probability; that several uncertainties must be combined using the rules of probability; and that the calculus of probabilities is adequate to handle *all* situations involving uncertainty. In particular, alternative descriptions of uncertainty are unnecessary. . . . We speak of "the inevitability of probability." . . . Our argument may be summarized by saying that probability is the only sensible description of uncertainty. All other methods are inadequate. . . . My challenge that anything that can be done with fuzzy logic, belief functions, upper and lower probabilities, or any other alternative to probability, can better be done with probability, remains.

We describe here this rather extreme position by Lindley because it plays a role in some of the other debates we are going to summarize next.

An extensive and well-documented debate regarding various ways of dealing with uncertainty, mainly within the context of artificial intelligence, was published in McLeish 1988. A position paper was written by Peter Cheeseman (1988a), then a researcher at NASA Ames Research Center in California, and 23 respondents were chosen to comment on it. Fuzzy logic was represented by only five respondents in this debate, which included Lotfi Zadeh and four other major contributors to fuzzy logic at that time—Didier Dubois, Henri Prade, Enrique Ruspini, and Ronald Yager. In the position paper, Cheeseman addressed fuzzy logic only indirectly by referring to his earlier paper (Cheeseman 1986), in which he directly and strongly challenged fuzzy logic by arguing that "Everything that can be done with fuzzy logic can better be done with probability." It was thus natural that the participating supporters of fuzzy logic addressed both papers in their commentaries.

In his response to the commentaries, Cheeseman (1988b) expressed basically the same views as Lindley in the previous debate. When summarizing the response, he writes (p. 142):

> The message [of this position paper] is that if you are constructing an AI system that must represent and reason about the world, then you should use probability to do so. The system's knowledge of the world is necessarily uncertain because it is induced from sensory data. As a result, you should not use (categorical) logic or any of the proposed alternative uncertainty calculi (fuzzy logic, Dempster-Shafer, Endorsements, etc.). Let a single flower bloom—there is a reason why monoculture dominates in agriculture.

In arguing that probability is the only sensible way to deal with uncertainty of any kind, Cheeseman invoked, as Lindley and other strong advocates of probability theory routinely do, the so-called proof by Cox. Dubois and Prade in their joint commentary, as well as Ruspini in his commentary, convincingly criticized the validity of this proof. Since Cox's proof plays a central role in other debates in which strong advocates of probability theory participate, we describe it in appendix A.

Another debate about uncertainty in the context of artificial intelligence was published in the *Knowledge Engineering Review* 3, no. 1 (1988): 59–91. This included responses by ten respondents to a previously published paper by Saffiotti (1987) and an overview of the responses by the organizer of the debate, Dominic Clark. Six of the respondents were active contributors to fuzzy logic. The paper by Saffiotti is basically a review of various formalisms for representing and dealing with uncertainty and a discussion of their suitability for artificial intelligence. His presentation is well balanced and the responses to it are constructive and informative.

The next debate, which was held during the Eighth Workshop on Maximum Entropy and Bayesian Methods at St. John's College, Cambridge, England, August 1–5, 1988, was in many respects very different from all of the other debates. It focused on one issue—the inevitability of probability theory as the only mathematical theory to correctly represent and deal correctly with any kind of uncertainty. It involved only two debaters—Peter Cheeseman, who represented the strong probabilistic position expressed in previous debates by Lindley and Cheeseman himself, and George Klir, who challenged this position. It was decided one year earlier at the preceding Seventh Workshop in Seattle, Washington, August 4–7, 1987, where both debaters were recruited by organizers of the forthcoming Eighth Workshop and agreed to participate in this one-to-one debate. This means that the debaters had one full year to prepare for the debate.

Since the challenger in this interesting debate is one of the authors of this book (Klir), his personal recollections serve to shed considerable light on this debate:

> I came to the Seventh Workshop on "Maximum-Entropy and Bayesian Methods in Science and Engineering" (these Workshops were often abbreviated as "MaxEnt Workshops"), which was held in Seattle (WA) August 4–7, 1987, to present a paper on the methodological use of principles of maximum and minimum entropy for constructing probabilistic systems from data, in which I also explored similar principles for constructing systems formalized in terms of possibility theory and in some other uncertainty theories that were of interest at that time (Klir 1988). In my overview of possibility theory, I surveyed various interpretations of the theory and emphasized especially its fuzzy-set interpretation. This reference to fuzzy sets apparently caught the attention of some people in the audience and it was suggested that an informal debate on probability theory versus fuzzy set theory be arranged during the Workshop, and Peter Cheeseman and I were nominated as the debaters. When I was approached with an official request to represent fuzzy set theory at this somewhat improvised debate, I hesitated but eventually agreed to participate. The debate did go well until a point when Peter started to claim that probability is all we need for representing and dealing properly with uncertainty. Of course, I do not remember the exact words he used, but I do remember that my response was a strong disagreement with these claims. Then, according to my best recollection, Peter's response was roughly like this: "George, you have every right to disagree with me on this point, but disagreeing with something that was mathematically proven is another matter. How do you respond to the Cox's proof?" At that point I had to admit that I did not know anything about this proof and, hence, I was not able to respond to Peter's question. After this impasse, there was no need to continue debating, but someone suggested that a serious, well-prepared debate on the same theme and by the same debaters be arranged at the next MaxEnt Workshop to be held at St. John's College in Cambridge, England, August 1–5, 1988. The advantages for this proposed arrangement were that there was about one year for preparing the debate and that St. John's College offered extraordinary facilities designed specifically for such debates. Both Peter and I enthusiastically endorsed this suggestion and we also agreed to a few simple rules whose purpose was to ensure a high quality debate. According to one of these rules, we were supposed to prepare a written version of our position, which would be given to all participants of the 8th MaxEnt Workshop prior to the debate. Another rule was that in order to avoid surprises such as the reference to the Cox's proof in the improvised debate, we should inform each other well before the debate about any specific examples, mathematical results, and the like that we intended to use as arguments in the debate. According to these rules, I prepared a written version of my opening statement, and I sent to Peter a few specific examples that I planned to use in my argumentation. Peter did not send me any examples or other material, which was OK since he did not use any unexpected material in his argumentation as well. Unfortunately, he did not prepare a written version of his opening statement either. The debate proceeded smoothly and the response of the audience was

the most interesting part of the whole debate for me. It was perfectly understandable that within the context of a conference devoted to probability theory my point of view would be criticized by most participants, and it actually was criticized, in some cases rather brutally, but there was some encouragement for me: a few participants were actually supporting in different ways my point of view. When the debate ended and the vote was taken, most participants abstained and I lost only by two votes. It was only disappointing for me that Peter did not prepare a written version of his opening statement. It seems that due to this omission my statement was not included in the Proceedings of the Workshop either. In order to preserve it for historical purposes, such as this book, I decided to publish it with proper explanation as a separate paper (Klir 1989).

A few years after the Cambridge debate, another debate, perhaps the most peculiar one of all of the debates involving fuzzy logic, was published in *IEEE Expert* 9, no. 4 (August 1994): 2–49, under the title "The paradoxical success of fuzzy logic." This debate was not organized like the other debates, but emerged rather naturally from a sequence of events described in some detail as follows.

In July of 1993, Charles Elkan presented a paper entitled "The paradoxical success of fuzzy logic" at the Eleventh National Conference on Artificial Intelligence in Washington, DC, sponsored by the American Association for Artificial Intelligence (AAAI). The paper was also included in the conference proceedings (Elkan 1993). In this paper, Elkan claims that the apparent success of fuzzy logic in many practical applications is paradoxical since fuzzy logic collapses upon closer scrutiny to classical, bivalent logic. To support this claim, he uses one definition and one theorem expressed in terms of the following notation: A, B denote assertions; $t(A)$, $t(B) \in [0, 1]$ denote the degrees of truth of A, B, respectively; and $\wedge, \vee, \neg$ denote the logical connectives of conjunction, disjunction, and negation, respectively. The following is the definition (exactly as it appears in the paper), which is supposed to define a particular system of fuzzy logic:

Definition 2.1. $t(A \wedge B) = \min\{t(A), t(B)\}$,

$\qquad t(A \vee B) = \max\{t(A), t(B)\}$,

$\qquad t(\neg A) = 1 - t(A)$,

$\qquad t(A) = t(B)$ if A and B are logically equivalent.

The definition is followed by a clarification of the term "logically equivalent": "In the last case of this definition, let 'logically equivalent' mean equivalent according to the rules of classical two-valued propositional calculus." Next is the statement of Elkan's theorem, preceded by his short remark: "Fuzzy logic is intended to allow an indefinite variety of numerical truth values. The result proved here is that only two different truth values are in fact possible in the formal system of Definition 2.1."

Theorem 2.1. *For any two assertions A and B, either $t(A) = t(B)$ or $t(A) = 1 - t(B)$.*

This statement of the theorem is then followed by a fairly long proof. This proof, whose length may be seen as demonstrating a kind of ingenuity of Elkan's finding, is correct but a big problem is Elkan's assumption regarding logically equivalent formulas. Namely, such an assumption has never been proposed by fuzzy logicians and is in fact foreign to fuzzy logic.[29] Therefore, Elkan in fact defined a system of nonfuzzy, bivalent logic by imposing a hidden condition of bivalence and then proved that such logic indeed is bivalent. It is not surprising that information about this peculiar paper spread quickly throughout the fuzzy logic community, especially after the announcement that it was a winner of the Best Paper Award at the conference. Many comments on the paper were circulated via email, virtually all expressing surprise that a paper of this kind was not only accepted for publication, but was also chosen as the best paper of the conference.

It is worth mentioning that some casual remarks about Elkan's paper in popular literature added to the confusion by unwittingly inflating the significance of the paper. A typical example is a short article entitled "The logic that dares not speak its name" that appeared in the *Economist* (April 16, 1994): 89–91. In this note, the author (not given) reviews in a highly positive way some of the impressive applications of fuzzy logic developed in Japan in the late 1980s and early 1990s. However, the author includes the following remark about Elkan's paper:

> Nevertheless there is still resistance from some quarters. A theorem devised by Charles Elkan, of the University of California, San Diego, suggests that fuzzy logic may not be all that it claims— that it collapses to a more traditional sort of two-valued logic under close scrutiny. Such arguments might be the stuff of Nobel prizes, were they to be awarded for mathematics. What has caused nonacademics to sit up and take notice of fuzzy logic is that, in the physical world, systems that incorporate it seem to work rather well.

This remark (and similar ones in popular literature) introduced a suspicion about fuzzy logic not only among the general public, but among some members of the academic community not familiar with fuzzy logic as well.

Under these circumstances, it was increasingly recognized that some more formal debate of Elkan's paper was needed, and the Editor of *IEEE Expert*, B. Chandrasekaran, felt that this journal was the right one for just such a debate. Jointly with one member of the editorial board, Lakendra Shastri, they describe how they proceeded: "We invited Elkan to contribute an updated version of his paper, which we circulated among experts in fuzzy logic and reasoning under uncertainty. We invited these experts to write commentaries on the paper, and then sent the commentaries to Elkan for rebuttal." The debate issue of *IEEE Expert*, which contains 15 commentaries written by 22 contributors, was published almost exactly one year after Elkan's conference presentation.[30] It is a valuable document that captures well the state of fuzzy logic at that time.

In the updated version of his paper, Elkan did not change his definition 2.1, but rewrote the clarification of the term "logically equivalent" in the following way: "Depending how the phrase 'logically equivalent' is understood, definition 2.1 yields

---

[29] This has convincingly been shown by Belohlavek and Klir (2007) who also presented a very short proof of Elkan's theorem as well as of Elkan's revised theorem which we present below.

[30] *IEEE Expert* 9, no. 4 (1994): 3–49.

different formal systems." He also rewrote the short remark preceding theorem 2.1 as "A fuzzy logic system is intended to allow an indefinite variety of truth values. However, for many notions of logical equivalence, only two different truth values are possible given the postulates of definition 2.1." A major change in the updated version is the following revised version of theorem 2.1.

Theorem 2.2 (revised version of theorem 2.1). Given the formal system of defini-tion 2.1, if $\neg(A \wedge \neg B)$ and $B \vee (\neg A \wedge \neg B)$ are logically equivalent, then for any two assertions A and B, either $t(A) = t(B)$ or $t(B) = 1 - t(A)$.

This revised version is somewhat more meaningful than theorem 2.1 as it shows that adding just this one particular logical equivalence from classical logic to the sys-tem of standard fuzzy logic suffices to reduce the latter to a bivalent logic. In this case, the logical equivalence involves two distinct formulas of classical logic that both express in classical logic material implication in terms of the three given logic oper-ations, conjunction, disjunction, and negation. What is the point of adding this equivalence of classical logic to the given system of fuzzy logic except to show that, if we do so, the fuzzy logic ceases to be fuzzy? In fuzzy logic, as defined by Elkan, the two formulas of classical logic lead actually to fundamentally different classes of implications. The assumption that they are logically equivalent in the revised theorem 2.1 constrains again the given system of fuzzy logic to a special system of bivalent logic. Clearly, many other equivalences of classical logic can be employed as assumptions in other revisions of Elkan's theorem, with the same result.

Before leaving this debate, let us mention that it stimulated some researchers working in the area of fuzzy logic to write more extensive articles trying to clarify the confusion emanating from the two versions of Elkan's article. Among them are articles by Nguyen, Kosheleva, and Kreinovich (1996); Pacheco, Martins, and Kan-del (1996); Klawonn and Novák (1996); Trillas and Alsina (2001); and Belohlavek and Klir (2007). In addition, French sociologist Claude Rosental took advantage of the various incarnations of Elkan's theorem as an example in his extensive study of social aspects in the development of logic. This study was first published in French as *La trame de l'évidence* in 2003 by Presses Universitaires de France in Paris; its English translation by Catherine Porter is Rosental 2008.

Almost at the same time when the debate issue on Elkan's articles was pub-lished in *IEEE Expert*, another debate regarding fuzzy logic was organized by James Bezdek, who edited at that time a relatively new journal entitled *IEEE Transactions on Fuzzy Systems*. This debate was published in issue 1 of volume 2 of this jour-nal in February 1994, pp. 1–45. The subject is captured well by the title of Bezdek's introduction—"Fuzziness vs. probability—again (! ?)." The inspiration for Bezdek to organize the debate was a paper by Laviolette and Seaman (1992), in which the au-thors recognized the growing interest in the theory as well as applications of fuzzy logic and argued that the time was ripe for evaluating methods based on fuzzy logic and comparing them with well-established probabilistic methods. Bezdek invited

Laviolette and Seaman to write a position paper focusing on this issue and they did so under the title "The efficacy of fuzzy representations of uncertainty" (Laviolette and Seaman 1994). The debate involves six commentaries that are helpful in clarifying some of the misunderstandings between probability theory and fuzzy logic.

In Laviolette and Seaman 1994, the authors announced their forthcoming paper Laviolette et al. 1995, which became a position paper in yet another debate regarding probability and fuzzy logic. The debate is covered in *Technometrics* (37, no. 3, August 1995, 249–92) and involves six commentaries. As in the previous debates, almost all debaters from the fuzzy logic camp viewed probability theory and fuzzy logic as complementary, while debaters from the probability camp viewed them routinely as competing. A few new and interesting points were made in the commentaries, but one can sense an overall state of fatigue among the commentators—the same issues had been debated for too long and had been gradually losing their relevance. It thus comes as no surprise that no more debates of this sort were organized.

The challenges to fuzzy logic in all of the debates described here were prompted either by probability theorists, especially those practicing the Bayesian methodology, or by researchers working in the area of mainstream artificial intelligence. In the early debates, opponents of fuzzy logic largely dismissed it as useless or even wrong. In later debates, the success of fuzzy logic in some areas of application was recognized. However, it was either implicitly suggested that such success was a result of good marketing (and that probability theory could deal with each of the applications better), or that fuzzy logic was not responsible for the success at all (as Charles Elkan attempted to prove mathematically). Since the mid-1990s, criticism of fuzzy logic has gradually subsided. Some other criticisms are considered at pertinent places throughout the remainder of this book.

2.5 Evolution of supporting infrastructure for fuzzy logic

The evolution of fuzzy logic can be characterized basically in three ways: by theoretical advances, by developments of successful applications, and by advancements in infrastructure supporting fuzzy logic, such as relevant professional societies, dedicated conferences, specialized journals, educational programs, and the like. Evolution of fuzzy logic in terms of its theoretical advances or in terms of its successful applications are characterized in chapters 3–6. On the other hand, its characterization in terms of the supporting infrastructure seems well suited to be covered here.

The supporting infrastructure for fuzzy logic was nonexistent prior to 1978, i.e. for 13 years after the publication of Zadeh's seminal paper. When researchers working on fuzzy logic felt the need to meet during this period, they tried to organize various kinds of special sessions at well-established conferences and were sometimes able to publish their results in conference proceedings. It was due to a few open-minded editors that some results of this research were also published in established journals during this period. The only exception was the US-Japan Seminar on Fuzzy Sets, sponsored by the US-Japan Cooperative Science Program and supported jointly by the US National Science Foundation and the Japan Society for the Promotion of Science, which was held at the University of California at Berkeley on July 1–4, 1974. Papers presented at this seminar were published in Zadeh et al. 1975, an early book of historical significance.

The situation changed in the most favorable way for fuzzy logic in the late 1970s, when North-Holland accepted a proposal from Hans-Jürgen Zimmermann, who had decided to launch a new international journal entitled *Fuzzy Sets and Systems*. The first volume of the journal was published as a quarterly in 1978. *Fuzzy Sets and Systems* was the very first of what we call the supporting infrastructure of fuzzy logic. The following is a summary of editorial policies set by the journal's three founding editors, Negoiţă, Zadeh, and Zimmermann:[31]

> The primary purpose of the international journal *Fuzzy Sets and Systems* is to improve professional communication between scientists and practitioners who are interested in, doing research on, or applying Fuzzy Sets and Systems. The journal seeks to publish articles (survey papers, original research and application papers and notes) of real significance, broad interest, and of high quality. They will normally be refereed by at least two competent referees and the editors will secure additional refereeing when warranted.

The journal also had other features of great importance for the growing community of researchers and practitioners working in the area, such as a "Bulletin" (reporting regularly on relevant events), special sections on "Current Literature," "Book Reviews" and "Who's Who in Fuzzy Sets." In addition to its historical significance, this journal has over the years substantially increased its size and it is now generally recognized as the principal journal devoted to fuzzy sets, fuzzy logic, and fuzzy systems.

Shortly after the first issue of *Fuzzy Sets and Systems* was published, three additional events occurred that greatly enhanced the evolving infrastructure supporting fuzzy logic. The first was publication of the book by Dubois and Prade (1980a), which was the first encyclopedic book on fuzzy set theory. It was an important resource for many researchers throughout the 1980s.

The second event was the International Seminar on Fuzzy Set Theory organized by Peter Klement, which was held at Johannes Kepler University in Linz, Austria, on September 24–29, 1979. The characterization of the seminar, which was supported by the Austrian Ministry of Science and Research and the Johannes Kepler University, is described in the proceedings of the seminar:[32]

> The main goal of this seminar was to provide a whole week for discussions and common work for a small group of mathematicians working on closely related problems in fuzzy set theory. To give

---

[31] From "Editorial Policies" on the inside cover of *Fuzzy Sets Syst* 1, no. 1 (1978).

[32] E. P. Klement, ed., *Proc Int Seminar on Fuzzy Set Theory* (Linz: Johannes Kepler Univ., 1979), 1–2.

a brief survey of the state-of-the-art, we started with lectures of the participants on their most recent work. These lectures, which are collected in this volume, focused mainly on fuzzy measures, fuzzy topology, and fuzzy Boolean algebras. In the following days, many aspects of these fields were discussed. There were a lot of additional remarks, examples and counterexamples. An additional part of the discussions was the problem (of) how to introduce fuzzy real numbers in a natural way.

The 1979 International Seminar on Fuzzy Set Theory was quite successful and this provided an incentive to organize it again as an annual event, which later became known as the Linz Seminar. The 30th Linz Seminar in 2009 was somewhat different; it was devoted to the theme "The Legacy of 30 Seminars—Where Do We Stand and Where Do We Go?" Its aim was to determine the state-of-the-art of fuzzy set theory and to assess its impact on other fields, especially on mathematics. The significance of the Linz Seminar, which is still ongoing, has been threefold. First, the seminar has focused each year primarily on key mathematical issues concerning fuzzy set theory at the time of the seminar. Second, without running any parallel sessions, it has allowed all participants (usually restricted to 40) to concentrate fully on extensive presentations of key issues by invited speakers. Third, it has provided ample time for thorough discussions of the major issues of interest to the participants. The seminar has thus provided over the years a working environment from which many mathematical ideas pertaining to fuzzy set theory have emerged and have been subjected to serious critical examination.

The third event that substantially enhanced the supporting infrastructure for fuzzy logic was the publication of a quarterly bulletin known as *BUSEFAL*, which was initiated by Didier Dubois and Henri Prade at the University of Paul Sabatier in Toulouse, France, in 1980.[33] The aim of *BUSEFAL* was to serve as a medium for quick communication of new ideas and other relevant information within the growing fuzzy-set community. This is how its founders characterized *BUSEFAL*:[34]

> BUSEFAL is a communication medium conveying exchanges on current research. Opinions, controversies, correspondences, short survey papers, research summaries, technical notes, research team presentations, congress or workshop announcements, etc. are welcome. BUSEFAL is not a regular international journal, and should not be considered as such. There is no refereeing system. Papers are published under authors' responsibility without copyright. They are not to be viewed as publications in their final form.

The important role of *BUSEFAL* as an effective communication medium within the fuzzy-set community in the 1980s and 1990s is now generally recognized. It is fortunate that Dubois and Prade chronicled this period by preparing a booklet that contains a complete table of contents for issues 1–76 of *BUSEFAL*, published between 1980 and 1998.[35] This booklet is an invaluable resource for historians of science

---

[33] "BUSEFAL" is an acronym for the French *Bulletin pour les Sous-Esembles Flous et leurs AppLications*, as well as for its English translation, *BUlletin for Studies and Exchanges on Fuzziness and its AppLications*.

[34] From the announcement of *BUSEFAL* in *Fuzzy Sets Syst* 5, no. 3 (1981): 330, which also contains the editorial address, French editorial board, and a list of foreign correspondents.

[35] Published as *BUSEFAL: Table of Contents, Issues 1–76* by the Institut de Recherche en Informatique de Toulouse—C.N.R.S. (UMR 5505), Université Paul Sabatier, 31062 Toulouse Cedex 4—France (ISSN 0296–3698).

and mathematics as it shows the beginnings of some profound ideas that were fully developed later as well as new ideas that have remained, for various reasons, undeveloped. Since 1999, *BUSEFAL* has not been published by the University of Paul Sabatier (details are covered in the booklet) and gradually it has been losing its significance, primarily due to emerging new media.

The first professional organization supporting fuzzy logic was formed in 1981, the North American Fuzzy Information Processing Society (NAFIPS). The initial program of NAFIPS was to organize annual meetings. The first NAFIPS Meeting was held in Logan, Utah, in 1982. In 1987, NAFIPS launched its own journal, *International Journal of Approximate Reasoning*, under the initial editorship of James Bezdek.

In the meantime, another professional organization was formed in 1984, the International Fuzzy Systems Association (IFSA), with a plan to organize biennial IFSA World Congresses on fuzzy systems and related areas. The first congress was held one year later in Palma de Mallorca. In 1995, IFSA began to operate as a federation of national and regional organizations supporting fuzzy logic, and adopted *Fuzzy Sets and Systems* as its official journal.

An important addition to the fuzzy-logic infrastructure was an endorsement of the area of fuzzy systems by a major engineering society, the Institute of Electrical and Electronics Engineers (IEEE). The endorsement was expressed by two specific decisions. The first decision was to initiate annual IEEE International Conferences on Fuzzy Systems (generally known as FUZZ-IEEE Conferences) with the first one held in San Diego, California, in 1992. The second decision was to start publishing *IEEE Transactions on Fuzzy Systems* in 1993. In both of these decisions, Bezdek, an important contributor to fuzzy systems, again played a major role, and he served as the general chairman of the first FUZZ-IEEE Conference, as well as the first editor of the *IEEE Transactions on Fuzzy Systems*.

In the late 1980s and early 1990s, many national or regional organizations supporting fuzzy logic and related areas (professional societies, associations, institutes, etc.) were formed, such as the Japan Society for Fuzzy Theory and Systems, the Korea Fuzzy Mathematics and Systems Society, the Indian Society for Fuzzy Mathematics and Information Processing, the Spanish Association of Fuzzy Logic and Technologies, which was later integrated into the European Society for Fuzzy Logic and Technology (EUSFLAT), and many more. Unfortunately, they organized their activities (conferences, workshops, summer schools, and the like) more or less independently of one another, which resulted in scheduling conflicts, duplications, and other undesirable side effects. The IFSA Council (the governing body of IFSA) recognized this problem at the Fifth IFSA World Congress in Seoul, South Korea, in 1993, and it was felt that IFSA, as the only international organization in this area, was in a unique position to alleviate it. The question was how to do that. This led to an extensive debate among members of the IFSA Council and other involved parties, especially representatives of the national and other societies mentioned above, prior to the Sixth IFSA Congress in São Paulo, Brazil, in 1995. To make a long story short, a unanimous decision was reached at this congress that IFSA would function as a federation of institutional members, each defined as "a national or regional not-for-profit organization whose purpose is related to and compatible with the purpose of IFSA." As a result, most of the national and regional organizations supporting fuzzy logic and related areas have become institutional members of IFSA and partners in coordinating worldwide activities in these areas.

In addition to the three primary journals mentioned earlier, many new journals devoted to various aspects of fuzzy logic have been founded since the late 1980s, too many in fact to list them here. The following are just a few examples:

- *Japanese Journal of Fuzzy Theory and Systems* (published by the Japan Society for Fuzzy Theory since 1989);
- *International Journal of Uncertainty, Fuzziness and Knowledge-Based Systems* (published by World Scientific since 1993);
- *Fuzzy Optimization and Decision Making* (published by Kluwer since 2002).

The supporting infrastructure in any area must also provide relevant encyclopedic resources and textbooks. In fuzzy logic, the former are available in the form of three handbooks: an extensive early *Handbook of Fuzzy Computation* (Ruspini, Bonissone, and Pedrycz 1998), a multiple-volume *Handbooks of Fuzzy Sets Series* organized by Didier Dubois and Henri Prade and published since 1998,[36] and a more recent, two-volume *Handbook of Mathematical Fuzzy Logic* (Cintula, Hájek, and Noguera 2011). As far as textbooks on fuzzy logic are concerned, they are only a tiny fraction of all books published in this area, and the earliest of them appeared in the late 1980s. They included two textbooks by Zimmermann (1985, 1987) and one by Klir and Folger (1988). Since the 1990s, textbooks on fuzzy logic have been published more frequently. Some were written as general textbooks, such as those by G. Bojadziev and M. Bojadziev (1995), Klir and Yuan (1995), Nguyen and Walker (1997), and Pedrycz and Gomide (1998), while others were written for courses in various application areas, such as those by Ross (1995), Passino and Yurkovich (1998), and Sakawa (1993). Of special significance is a recent textbook by Merrie Bergmann (2008), which is the first textbook devoted to fuzzy logic in the narrow sense.

From this short overview of how the infrastructure supporting fuzzy logic has evolved, it is clear that fuzzy logic is now recognized as a well-established and strong academic area. This is one criterion, even though perhaps a secondary one, for assessing the overall evolution of fuzzy logic. The primary criteria—the evolution of theoretical aspects of fuzzy logic and of its applications—can be properly assessed only in the context of the developments described in the following chapters.

---

[36] Originally published by Kluwer and later taken over by Springer. The following are the individual volumes that have already been published in this series: Słowiński 1998; Nguyen and Sugeno 1998; Höhle and Rodabaugh 1999; Bezdek, Dubois, and Prade 1999; Bezdek et al. 1999; Zimmermann 1999; and Dubois and Prade 2000.

Chapter 3

Fuzzy Logic in the Broad Sense

3.1 Introduction

ALTHOUGH THE DISTINCTION between fuzzy logic in the narrow sense and fuzzy logic in the broad sense was already briefly mentioned in chapter 1, we now need to elaborate on it further. Zadeh introduced this distinction in the mid-1990s through his short remarks on various occasions (conference presentations, forewords to various books, etc.). He described it more completely on p. 2 in Zadeh 1996a:[1]

> From its inception, fuzzy logic has been an object of controversy. By now, the controversies have abated, largely because the successful applications of fuzzy logic are too visible to be ignored. Nevertheless, there are still many misconceptions regarding the structure of fuzzy logic and its relationships with other methodologies which—like fuzzy logic—address some of the basic issues that center on uncertainty, imprecision and approximate reasoning.
>
> Many of the misconceptions about fuzzy logic stem from differing interpretations of what "fuzzy logic" means. In the narrow sense, fuzzy logic is a logical system that focuses on modes of reasoning that are approximate rather than exact. In this sense, fuzzy logic, or FLn for short, is an extension of classical multivalued logical systems, but with an agenda that is quite different in spirit and in substance.
>
> In the wide sense, which is in predominant use today, fuzzy logic, or FLw for short, is almost synonymous with the theory of fuzzy sets. The agenda of FLw is much broader than that of FLn, and logical reasoning—in its traditional sense—is an important but not a major part of FLw. In this perspective, FLn may be viewed as one of the branches of FLw. What is important to note is that, today, most of the practical applications of fuzzy logic involve FLw and not FLn.

This distinction between FLn and FLw is a useful one, and it is now generally accepted in the literature on fuzzy logic.

---

[1] Observe that Zadeh uses the term "fuzzy logic in the wide sense" rather than "fuzzy logic in the broad sense." However, we (and some other authors) consider the adjective "broad" more natural in this case and we prefer to use it in this book.

Fuzzy logic in the narrow sense, which is covered in detail in chapter 4, is basically the study of formal logical systems in which truth is a matter of degree. Its beginnings are usually associated with the second classical paper by Goguen (1968–69). Its primary agenda is to develop logical systems for truth-preserving deductive reasoning in which the principle of bivalence of classical logic is rejected.

The agenda of fuzzy logic in the broad sense, the subject of this chapter, can be loosely characterized as the research program that has been pursued under the leadership of Zadeh since the publication of Zadeh 1965a. This program is centered on fuzzy sets in all their theoretical as well as applied manifestations. On the theoretical side, its overall aim is to generalize intuitive set theory[2] to intuitive fuzzy set theory by replacing classical sets with fuzzy sets. On the applied side, its aim is to utilize the various new capabilities of intuitive fuzzy set theory for expanding the applicability of mathematics based on classical set theory.[3]

Fuzzy logic in the broad sense is a huge undertaking that has been shaped over the years by many contributors. Among them, however, Zadeh has played a leading role. Since the initial ideas he presented in his seminal paper, he has been consistently introducing important new ideas pertaining to fuzzy logic in the broad sense, which have gradually expanded the agenda of this research program. He usually developed each idea to some point and then proceeded to the next one. From 1965 until the mid-1990s, the emergence of these new ideas is well documented in two large volumes of his selected papers that were edited by Yager et al. (1987) and Klir and Yuan (1996), but a few additional prime ideas have emerged since the mid-1990s, including the ideas of soft computing (Zadeh 1994a,b), computing with words and perceptions (Zadeh 1996b, 1999b), and a general theory of uncertainty (Zadeh 2005, 2006). Our aim in this chapter is to present a comprehensive overview of the development of key ideas of fuzzy logic in the broad sense by Zadeh and others.

3.2 Basic concepts of fuzzy sets

Fuzzy logic in the broad sense has developed predominantly within the framework of standard fuzzy sets. It is thus reasonable to assume that fuzzy sets considered in this chapter are standard fuzzy sets. An exception is section 3.10, in which we survey some of the nonstandard fuzzy sets whose roles in fuzzy logic in the broad sense have only lately become subjects of serious investigations. That is, the term "fuzzy set" in sections 3.2–3.9 refers to a function of the form

$$A : U \to [0,1], \tag{3.1}$$

where U denotes a universal set—a classical (nonfuzzy) set of all objects that are of interest in a given application context. For each given object $x \in U$, the value of

---

[2] See e.g. R. R. Stoll, *Set Theory and Logic* (San Francisco: Freeman, 1961).

[3] Axiomatization of fuzzy set theory is discussed in section 5.2.3.

$A(x)$ is interpreted as the degree (or grade) of membership of object x in fuzzy set A or, alternatively, as the degree of truth of the proposition "x is a member of set A." We denote the set of all fuzzy sets of the form (3.1) by the symbol $[0,1]^U$.

For any fuzzy set A, the set of all objects of U for which $A(x) > 0$ is called a *support* of A, and the set of all objects of U for which $A(x) = 1$ is called a *core* of A.

$$h(A) = \sup_{x \in U} A(x)$$

is called the *height* of fuzzy set A. When $h(A) = 1$, fuzzy set A is called *normal*; otherwise, it is called *subnormal*. If $h(A) = 0$, A is called *empty* and is denoted by the usual symbol $\emptyset$.

Given two fuzzy sets $A, B \in [0,1]^U$, A is said to be a subset of B, written as $A \subseteq B$, if and only if $A(x) \le B(x)$ for all $x \in U$. According to this concept of subsethood, which was already introduced in Zadeh's 1965 paper, the associated proposition "A is a subset of B" is either true or false—it is a bivalent proposition.

Given fuzzy set $A \in [0,1]^U$ and a particular number $\alpha \in [0,1]$, the classical set

$$^\alpha A = \{x \,|\, A(x) \ge \alpha\}$$

is called a *level-cut* of A (often also called an α-cut of A). Clearly, if $\alpha_1 \le \alpha_2$, where $\alpha_1, \alpha_2 \in [0,1]$, then $^{\alpha_1}A \supseteq {}^{\alpha_2}A$. This means that all distinct level-cuts of any fuzzy set A form a family

$$\mathscr{A} = \{^\alpha A \,|\, \alpha \in [0,1]\} \tag{3.2}$$

of nested classical sets.[4] This family is important since it is an alternative representation of A:[5] for any given family defined by (3.2), the associated fuzzy set A is uniquely determined for all $x \in U$ by the formula

$$A(x) = \sup_{\alpha \in [0,1]} \{\alpha \cdot {}^\alpha A(x)\}.$$

Another way of representing fuzzy sets is to view them as special random sets.[6] Let

$$f_\alpha(x) = \begin{cases} 1 & \text{if } x \in {}^\alpha A(x), \\ 0 & \text{if } x \notin {}^\alpha A(x), \end{cases}$$

for all $x \in U$. Then,

$$A(x) = \int_0^1 f_\alpha(x)\,d\alpha$$

for all $x \in U$.

The significance of the level-cut and random-set representations of fuzzy sets is that they connect fuzzy sets with classical sets. They allow us to extend various properties defined on classical sets to their counterparts defined on fuzzy sets. A property defined on fuzzy sets is a level-cut extension of some property defined on classical sets if the latter property is preserved in all level-cuts of the former. Properties of classical sets that are extended to fuzzy sets via the level-cut representation are commonly

---

[4] The set of all values of α that represent distinct level-cuts of a given set A, C_A, is usually called a *level set* of A. Formally, $C_A = \{\alpha \,|\, A(x) = \alpha \text{ for some } x \in U\}$.

[5] See, for example, section 2.2 in Klir and Yuan 1995. This representation is further discussed on p. 271 below.

[6] See, for example, pp. 46–47 in Dubois and Prade 2000. This representation is further discussed on p. 273 below.

referred to in the literature as *cutworthy properties*. We identify some examples of cutworthy properties throughout this chapter.

An important question is that of how to fuzzify classical functions. Given a function $f : X \rightarrow Y$, where X and Y denote classical sets, we say that the function is *fuzzified* when f is extended to a function F which maps fuzzy sets defined on X to fuzzy sets defined on Y. Formally, F has the form

$$F : [0,1]^X \rightarrow [0,1]^Y.$$

To qualify as a fuzzified version of f, function F must conform to f within its domain and range. This is guaranteed when the so-called *extension principle* is employed. According to this principle, $B = F(A)$ is determined for any given fuzzy set $A \in [0,1]^X$ via the formula

$$B(y) = \max_{x|y=f(x)} A(x) \tag{3.3}$$

for all $y \in Y$. Clearly, when the maximum in (3.3) does not exist, it is replaced by the supremum. The inverse function, $F^{-1} : [0,1]^Y \rightarrow [0,1]^X$, of F is defined, according to the extension principle, for any given $B \in [0,1]^Y$, by the formula

$$[F^{-1}(B)](x) = B(y)$$

for all $x \in X$, where $y = f(x)$. Clearly, $F^{-1}[F(A)] \supseteq A$ for all $A \in [0,1]^X$, where the equality is obtained when f is a one-to-one function. This extension principle was introduced by Zadeh (1975c) but it was outlined already in Zadeh 1965a.

For any given fuzzy set A defined on a finite universe of discourse U, it is reasonable to express its *size*, $s(A)$, by the formula

$$s(A) = \sum_{x \in U} A(x),$$

which can be viewed as a fuzzy-set counterpart of the concept of cardinality of finite classical sets. In the literature, $s(A)$ is called a *scalar cardinality* or *sigma count* of A.

3.3 Operations on fuzzy sets

Historical background

In his seminal paper, Zadeh already recognized that the classical operations of set intersection, union, and complementation could be extended to their fuzzy-set counterparts in different ways. However, he did not address the issue of how to characterize these extensions. In two early papers, Bellman and Giertz (1973) and Fung and Fu (1975) explored axioms for characterizing these extensions, but they did not address the issue either. Bellman and Giertz's primary aim was to show that the standard operations on fuzzy sets introduced by Zadeh are unique under quite reasonable axiomatic assumptions. Fung and Fu were interested in characterizing rational aggregates of fuzzy sets in the context of decision making and under the assumption that the range of fuzzy sets involved is not the unit interval [0, 1], but a simpler space.

The issue was finally addressed in the late 1970s in a rather unusual way: a natural connection was suggested between prospective classes of intersections and unions on fuzzy sets and classes of binary operations on $[0, 1]$ that were known as triangular norms (or t-norms) and triangular conorms (or t-conorms), respectively.[7] This turned out to be a good suggestion, which quickly attracted the attention of a fair number of researchers. Since 1980, the study of t-norms and t-conorms in the context of fuzzy set theory has been a rapidly growing new area of research.

When the connection of intersections and unions of fuzzy sets with t-norms and t-conorms, respectively, was recognized in the late 1970s, the latter had already been extensively researched for many years in the context of probabilistic metric spaces.[8] A timely summary of results emanating from this research was the book by Schweizer and Sklar (1983), which was of great help to researchers investigating at that time t-norms and t-conorms in the context of fuzzy sets. This book also contains an extensive historical overview, whose updated version was later prepared by Schweizer (2005). We refer the reader to these sources for historical background regarding t-norms and t-conorms; for general overviews of the rather extensive theoretical results regarding these operations we recommend the monographs by Klement, Mesiar, and Pap (2000); Alsina, Frank, and Schweizer (2006); and the book edited by Klement and Mesiar (2005).

In the following subsections, we examine each of the three basic operations on fuzzy sets—intersections, unions, and complements—separately and in combinations. Then, we examine two additional types of operations on fuzzy sets—averaging operations and modifiers—which have no counterparts in classical set theory.

Intersections

It should be clear from our discussion in this section that the remarkable operations known as t-norms emerged somewhat unexpectedly at the right time as the natural operations of intersections on fuzzy sets. Since we are not interested in t-norms as purely mathematical objects, as e.g. in Alsina, Frank, and Schweizer 2006, but only in their role for the operations of intersection of fuzzy sets in this section, we use the suggestive symbol i for t-norms in the context of fuzzy sets. In this sense, an intersection operation on fuzzy sets associated with a binary operation i on $[0, 1]$ (a t-norm) is for convenience denoted also by i and is defined as

$$[i(A, B)](x) = i[A(x), B(x)]$$

for all $x \in U$. We say that an intersection of fuzzy sets defined in this way is based on the associated t-norm operation i.

---

[7] For details, see p. 178.
[8] Inspired by the idea of statistical metrics introduced by Menger (1942).

Each t-norm i is a function of the form $i : [0,1]^2 \rightarrow [0,1]$ that satisfies at least the following requirements for all $a,b,c,d \in [0,1]$:

(i1) $i(a, i(b,d)) = i(i(a,b),d)$ (associativity),

(i2) $b \leq d$ implies $i(a,b) \leq i(a,d)$ (monotonicity),

(i3) $i(a,b) = i(b,a)$ (commutativity),

(i4) $i(a,1) = a$ (boundary requirement).

The first three conditions are simple generalizations of properties of classical conjunction on which the classical set-intersection is based. The boundary condition (i4) guarantees that i collapses to the classical operation of set intersection when $a,b,c,d \in \{0,1\}$. Various other requirements, such as continuity or strict monotonicity of i, may be added as needed.

The following are examples, with their usual names, of some well-known t-norms (each defined for all $a,b \in [0,1]$) that play important roles as intersection operations on fuzzy sets:[9]

$i_s(a,b) = \min(a,b)$ (minimum),

$i_p(a,b) = a \cdot b$ (product or algebraic product),

$i_b(a,b) = \max(0, a+b-1)$ (Łukasiewicz),

$i_d(a,b) = \begin{cases} \min(a,b) & \text{if } \max(a,b) = 1, \\ 0 & \text{otherwise} \end{cases}$ (drastic t-norm).

One can easily prove that all intersection operations on fuzzy sets satisfy for all $a,b \in [0,1]$ the following inequalities:

$$i_d(a,b) \leq i(a,b) \leq i_s(a,b). \tag{3.4}$$

Moreover, the four important examples of intersections of fuzzy sets satisfy for all $a,b \in [0,1]$ the following inequalities:

$$i_d(a,b) \leq i_b(a,b) \leq i_p(a,b) \leq i_s(a,b). \tag{3.5}$$

In addition to individual intersection operations, such as those involved in these inequalities, various parameterized classes of intersections based on t-norms have been introduced. Some of these classes capture a whole range of intersection operations given by (3.4) within the designated range of the parameter involved. As an example, we present a class that was introduced by Schweizer and Sklar (1983), which has some interesting properties.[10] This class employs parameter λ whose range is $(-\infty, \infty)$. For each $\lambda \in (-\infty, 0) \cup (0, \infty)$, the intersection operation $i_\lambda(a,b)$ is defined for all $a,b \in [0,1]$ by the formula

$$i_\lambda(a,b) = (\max(0, a^\lambda + b^\lambda - 1))^{\frac{1}{\lambda}}.$$

The limiting t-norms for $\lambda \rightarrow -\infty$, $\lambda \rightarrow 0$, and $\lambda \rightarrow \infty$ are i_s, i_p, and i_d, respectively.

---

[9] The role of t-norms as truth functions of conjunctions in fuzzy logic is examined in chapter 4.

[10] This parameterized class of t-norms is usually referred to as a *Schweizer and Sklar class*.

For $\lambda = 1$, clearly, the t-norm is i_b, and for $\lambda = -1$, we obtain the t-norm defined by

$$i_{\lambda=-1}(a,b) = \frac{a \cdot b}{a+b-a \cdot b}.$$

Many more properties, especially the theory for producing t-norms and classes of t-norms from simple functions called generators (n. 100 on p. 275 below), can be found in Klement, Mesiar, and Pap 2000 and in Alsina, Frank, and Schweizer 2006.

Unions

While intersections of fuzzy sets are naturally based on t-norms, unions of fuzzy sets are based on t-conorms, u, which are functions that are dual to t-norms in a sense explained later. Each t-conorm u is a function of the form $u : [0,1]^2 \rightarrow [0,1]$ that satisfies at least the following requirements for all $a, b, c, d \in [0,1]$:

(u1) $u(a, u(b,d)) = u(u(a,b),d)$ (associativity),

(u2) $b \leq d$ implies $u(a, b) \leq u(a,d)$ (monotonicity),

(u3) $u(a,b) = u(b,a)$ (commutativity),

(u4) $u(a,0) = a$ (boundary condition).

Clearly, t-norms and t-conorms differ solely by their distinct boundary conditions. Condition (u4) guarantees, together with monotonicity and commutativity, that u collapses to the classical operation of set union when $a, b, c, d \in \{0, 1\}$.

It is well known and easy to prove that a function $u : [0,1]^2 \rightarrow [0,1]$ is a t-conorm if and only if there exists a t-norm i such that for all $(a, b) \in [0, 1]^2$,

$$u(a, b) = 1 - i(1-a, 1-b). \tag{3.6}$$

Functions u and i satisfying equation (3.6) are called mutually dual.

For each given t-norm, we obtain the corresponding (dual) t-conorm by (3.6). Hence, applying (3.6) to the introduced t-norms i_s, i_p, i_b, and i_d, we obtain, respectively, the following dual t-conorms:

$$u_s(a,b) = \max(a, b),$$

$$u_p(a,b) = a+b-a \cdot b,$$

$$u_b(a,b) = \min(1, a+b),$$

$$u_d(a,b) = \begin{cases} \max(a, b) & \text{when } \min(a, b) = 0, \\ 1 & \text{otherwise.} \end{cases}$$

Clearly, the following inequalities follow from (3.5) and (3.6) for all $a, b \in [0, 1]$:

$$u_s(a,b) \leq u_p(a,b) \leq u_b(a,b) \leq u_d(a,b).$$

Moreover, the range of t-conorms is defined for all $a, b \in [0, 1]$ by the inequalities

$$u_s(a,b) \leq u(a,b) \leq u_d(a,b),$$

which are counterparts of those in (3.4).

Applying the duality equation to the Schweizer and Sklar class of t-norms, we obtain the corresponding class of t-conorms, which is defined for all $\lambda \in (-\infty, 0) \cup (0, \infty)$ by the formula

$$u_\lambda(a, b) = 1 - (\max(0, (1-a)^\lambda + (1-b)^\lambda - 1))^{\frac{1}{\lambda}}. \tag{3.7}$$

The limiting t-conorms for $\lambda \to -\infty$, $\lambda \to 0$, $\lambda \to \infty$ are, respectively, u_s, u_p, u_d. Moreover, $u_{\lambda=1} = u_b$ and $u_{\lambda=-1}(a, b) = \frac{a+b-2ab}{1-ab}$.

To illustrate the great variety of possible parameterized classes of dual t-norms/t-conorms, we examine another example (Yager 1980a). In this case, it is easier to define a class of t-conorms first and use the duality equation (3.6) to derive the dual class of t-norms. The range of parameter λ is $(0, \infty)$. For each $\lambda \in (0, \infty)$ a particular t-conorm is defined for all $a, b \in [0, 1]$ by the formula

$$u_\lambda(a, b) = \min(1, (a^\lambda + b^\lambda)^{\frac{1}{\lambda}}). \tag{3.8}$$

For $\lambda \to 0$ and $\lambda \to \infty$ the limiting t-conorms are u_d and u_s, respectively. For $\lambda = 1$, clearly, $u_{\lambda=1}(a, b) = \min(1, a + b) = u_b$. Dual t-norms, derived by applying (3.6) to (3.8), are expressed for all $\lambda \in (0, \infty)$ by the formula

$$i_\lambda(a, b) = 1 - \min(1, ((1-a)^\lambda + (1-b)^\lambda)^{\frac{1}{\lambda}}).$$

The limiting t-norms for $\lambda \to 0$ and $\lambda \to \infty$ are i_d and i_s, respectively. Clearly, $i_{\lambda=1} = i_b$.

Complements

Like intersections and unions of fuzzy sets, complements of fuzzy sets are not unique operations. In general, each complement is based on a function $c : [0, 1] \to [0, 1]$ that satisfies at least the following two requirements:

     (c1)   if $a \leq b$, then $c(a) \geq c(b)$ (antitonicity),

     (c2) $c(0) = 1$ and $c(1) = 0$ (boundary condition).

A complement of fuzzy set A, $c(A)$, that is based on function c is defined for all $x \in U$ by $[c(A)](x) = c[A(x)]$. Additional requirements may be imposed on function c in some application contexts. Among them, the most common are the following two:

     (c3) c is a continuous function   (continuity),

     (c4) $c(c(a)) = a$, for all $a \in [0, 1]$ (involution).

It is known that the set of all involutive complements (each satisfying conditions (c1), (c2), and (c4)) is a proper subset of all continuous complements (each satisfying (c1), (c2), and (c3)), and the latter is a proper subset of general complements (each satisfying only (c1) and (c2)). The standard complement introduced by Zadeh (1965a), which is based on the function $c(a) = 1 - a$ of standard negation, is clearly an involutive complement. The complement based on function

$$c(a) = \tfrac{1}{2}(1 + \cos \pi a),$$

$a \in [0, 1]$, is an example of a continuous complement that is not involutive. An important class of general complements is threshold-type complements, each defined for one particular value of parameter $t \in [0, 1)$, called a threshold of c, and all $a \in [0, 1]$ by the equation

$$c(a) = \begin{cases} 1 & \text{for } a \leq t, \\ 0 & \text{for } a > t. \end{cases}$$

Various other classes of complements have been proposed for fuzzy sets. The first parameterized class of involutive complements was discovered by Sugeno (1977) as a byproduct in his study of nonadditive measures. These complements are based on a class of functions defined for all $a \in [0, 1]$ by the formula

$$c_\lambda(a) = \frac{1-a}{1+\lambda a}, \tag{3.9}$$

where $\lambda \in (-1, \infty)$. Each value of the parameter λ defines one particular involutive complement. According to a well-established theory, various classes of involutive complements can now be conveniently derived from simple generators, see e.g. Klir and Yuan 1995.

Combinations of intersections, unions, and complements of fuzzy sets

In classical set theory, the set inclusion is unique and forms a partial ordering on the power set 2^U of any given universal set U. The operations of set intersection, union, and complement are also unique. The subsets of U along with these operations thus form a unique algebra on 2^U, usually referred to as *algebra of sets*.

As already explained in this section, none of the three basic operations on fuzzy sets is unique. As explained earlier, each of them can be chosen from a suitable parameterized class of operations, such as the Schweizer and Sklar class of intersections defined by (3.7) or the Sugeno class of complements defined by (3.9), each containing an infinite number of operations. Clearly, there is an infinite number of possible combinations $\langle i, u, c \rangle$. However, regardless of which combination is chosen, some of the properties of classical set operations are necessarily violated as a consequence of abandoning the principle of bivalence in fuzzy set theory. Which of the properties are violated depends on the chosen combination. For example, the combination consisting of the standard operations on fuzzy sets violates the law of excluded middle and the law of contradiction, while the combination $\langle i_b, u_b, c_s \rangle$ violates the properties of distributivity, idempotency, and absorption. It is interesting that the standard operations of intersection and union of fuzzy sets introduced by Zadeh in his 1965 paper are the only ones that are idempotent. For the standard intersection of fuzzy sets, we can prove it easily as follows: Clearly, $\min(a, a) = a$ for all $a \in [0, 1]$, which means that the standard intersection is idempotent. Assume now that there exists an intersection operation i such that $i(a, a) = a$ for all $a \in [0, 1]$. Then, for any $a, b \in [0, 1]$, if $a \leq b$, then $a = i(a, a) \leq i(a, b) \leq i(a, 1) = a$ by monotonicity and the boundary condition. Hence, $i(a, b) = a = \min(a, b)$. Similarly, if $a \geq b$, then $b = i(b, b) \leq i(a, b) \leq i(1, b) = b$ and, consequently, $i(a, b) = b = \min(a, b)$. Hence, $i(a, b) = \min(a, b)$ for all $a, b \in [0, 1]$. The proof for the standard union is analogous.

Although any of the possible combinations can be used in principle, it may be desirable to use those that preserve as many properties of classical set operations as possible. In this sense, it is desirable to work with combinations that preserve the duality between intersections and unions in classical set theory, as expressed by the De Morgan laws. In fuzzy set theory, the De Morgan laws are expressed by

$$c(i(a, b)) = u(c(a), c(b)) \quad \text{and} \quad c(u(a, b)) = i(c(a), c(b)).$$

Any triplet $\langle i, u, c \rangle$ that satisfies these equations is usually called a *dual triplet* (or *De Morgan triplet*) of basic operations on fuzzy sets. Many procedures for constructing dual triples or even dual triples that are required to satisfy some specific properties are now available, see e.g. Klir and Yuan 1995 (pp. 83–88).

Averaging operations

Intersections and unions of fuzzy sets are special types of operations for aggregating fuzzy sets: given two or more fuzzy sets, they produce single fuzzy sets—aggregates of the given ones—that are viewed as generalizations of intersections and unions of classical sets. In addition to intersections and unions, fuzzy sets can be aggregated in other ways. Among all possible aggregation operations (Grabisch et al. 2009), we describe only a class of *averaging operations*, as an example of useful operations on fuzzy sets that are not applicable within classical set theory. Indeed, an average of characteristic functions of classical sets is not in general a characteristic function.

Contrary to intersections and unions, averaging operations are not associative. Hence, they must be defined as functions of n arguments for any $n \geq 2$. Averaging operations for fuzzy sets, which we denote by h, are thus based on functions of the form $h : [0, 1]^n \to [0, 1]$ in that $A(u) = h(A_1(u), A_2(u), \ldots, A_n(u))$ for all $u \in U$. To qualify as an intuitively meaningful averaging function, h must satisfy the following requirements:

(h1)   for any $\mathbf{u}, \mathbf{u}' \in [0, 1]^n$, if $\mathbf{u} \leq \mathbf{u}'$ then $h(\mathbf{u}) \leq h(\mathbf{u}')$ (monotonicity),

(h2) $h(u, u, \ldots, u) = u$ for all $u \in [0, 1]$ (idempotency),

(h3) h is a continuous function   (continuity),

(h4)   for any permutation p of $\mathbb{N}_n$,

$$h(u_1, u_2, \ldots, u_n) = h(u_{p(1)}, u_{p(2)}, \ldots, u_{p(n)}) \qquad \text{(symmetry)}.$$

It is significant that requirements (h1) and (h2) imply the following inequalities:

$$\min(u_1, u_2, \ldots, u_n) \leq h(u_1, u_2, \ldots, u_n) \leq \max(u_1, u_2, \ldots, u_n) \qquad (3.10)$$

for all n-tuples $\langle u_1, u_2, \ldots, u_n \rangle \in [0, 1]^n$. To see this, let

$$u_* = \min(u_1, u_2, \ldots, u_n) \quad \text{and} \quad u^* = \max(u_1, u_2, \ldots, u_n).$$

If h satisfies (h1) and (h2), then the inequalities (3.10) follow from

$$u_* = h(u_*, u_*, \ldots, u_*) \leq h(u_1, u_2, \ldots, u_n) \leq h(u^*, u^*, \ldots, u^*) = u^*.$$

That is, the averaging operations cover a range between the operations of standard intersection and the standard union. The latter thus play a pivotal role in the three types of aggregating operations on fuzzy sets. Owing to their associativity and idempotency, they qualify not only as extensions of the classical set intersection and union in fuzzy set theory, but also as extreme averaging operations of fuzzy sets.

Various parameterized classes for averaging operations have been suggested. We describe here as an example only one class of averaging operations which forms an interval between min and max operations. This class is defined by the formula

$$h_\lambda(u_1, u_2, \ldots, u_n) = \left(\frac{u_1^\lambda + u_2^\lambda + \cdots + u_n^\lambda}{n} \right)^{\frac{1}{\lambda}}, \tag{3.11}$$

where λ is a parameter whose range is the set of all real numbers except 0. One particular averaging operation is obtained for each value of the parameter. For $\lambda = 0$, function h_λ is not defined by the formula but by the limit

$$\lim_{\lambda \to 0} h(u_1, u_2, \ldots, u_n) = (u_1 \cdot u_2 \cdots u_n)^{\frac{1}{n}},$$

which is the well-known geometric average. Moreover,

$$\lim_{\lambda \to -\infty} h_\lambda(u_1, u_2, \ldots, u_n) = \min(u_1, u_2, \ldots, u_n),$$

$$\lim_{\lambda \to \infty} h_\lambda(u_1, u_2, \ldots, u_n) = \max(u_1, u_2, \ldots, u_n).$$

For $\lambda = 1$, equation (3.11) clearly yields the arithmetic average

$$h_1(u_1, u_2, \ldots, u_n) = \frac{u_1 + u_2 + \cdots + u_n}{n},$$

and for $\lambda = -1$, it yields the harmonic average

$$h_{-1}(u_1, u_2, \ldots, u_n) = \frac{n}{\frac{1}{u_1} + \frac{1}{u_2} + \cdots + \frac{1}{u_n}}.$$

Another class of averaging operations, which was introduced by Yager (1988), has turned out to be very useful in many applications. Operations in this class are called *ordered weighted averaging operations*, often known in the literature as OWA operations. As the name suggests, these operations are weighted averages, but they still satisfy conditions (h1)–(h4).

Each OWA operation, $h_\mathbf{w}$, is based on a chosen *weighting vector* $\mathbf{w} = \langle w_1, w_2, \ldots, w_n \rangle$ such that $w_i \in [0, 1]$ for all $i \in \mathbb{N}_n$ and $\sum_{i \in \mathbb{N}_n} w_i = 1$; it is defined by

$$h_\mathbf{w}(u_1, u_2, \ldots, u_n) = \sum_{i \in \mathbb{N}_n} w_i \cdot v_i,$$

where v_i denotes the i-th largest value of $u_1, u_2, \ldots, u_n$. That is, the vector $\langle v_1, v_2, \ldots, v_n \rangle$ is a permutation of $\langle u_1, u_2, \ldots, u_n \rangle$ in which $v_i \geq v_j$ when $i < j$ ($i, j \in \mathbb{N}_n$). Clearly, the lower and upper bounds of $h_\mathbf{w}$ are obtained, respectively, for the weighting vectors $\underline{\mathbf{w}} = \langle 0, \ldots, 0, 1 \rangle$ and $\overline{\mathbf{w}} = \langle 1, 0, \ldots, 0 \rangle$, i.e.

$$h_{\underline{\mathbf{w}}}(u_1, u_2, \ldots, u_n) = \min(u_1, u_2, \ldots, u_n),$$

$$h_{\overline{\mathbf{w}}}(u_1, u_2, \ldots, u_n) = \max(u_1, u_2, \ldots, u_n).$$

For $\mathbf{w} = \langle \frac{1}{n}, \frac{1}{n}, \ldots, \frac{1}{n} \rangle$, $h_{\mathbf{w}}$ is the arithmetic mean. In general, the whole range of OWA operations between min and max is obtained by varying the assignments of weights from $\underline{\mathbf{w}}$ to $\overline{\mathbf{w}}$.

Modifiers

Another class of operations on fuzzy sets that are not applicable to classical sets consists of unary operations that are called *modifiers*. The primary purpose of these operations is to properly modify fuzzy sets to account for *linguistic hedges*, such as *very*, *fairly*, *extremely*, *moderately*, and the like, in representing expressions of natural language. They were first introduced by Zadeh (1971a, 1972a) and stimulated an extensive discussion by Lakoff (1973) about their role in linguistics and the psychology of concepts. While Zadeh continued to pay attention to modifiers representing linguistic hedges in the context of his later developments, especially in connection with linguistic variables (Zadeh 1975c), mathematical aspects of modifiers of fuzzy sets attracted attention of other researchers, as is well described in an overview paper by Kerre and De Cock (1999); logical aspects are examined on p. 205 in chapter 4.

As with the other operations on fuzzy sets overviewed in this section, we can associate each modifier with a function

$$m : [0, 1] \to [0, 1],$$

which assigns to each membership grade $A(x)$ of a given fuzzy set A a modified grade $m(A(x))$. Denoting the fuzzy set thus defined by HA, where H denotes the linguistic hedge represented by modifier m, we have

$$HA(x) = m(A(x))$$

for all $x \in U$. For example, if fuzzy set A represents (in a given context) the concept of a *low interest rate* and H represents the linguistic hedge *very*, then fuzzy set HA represents the concept of a *very low interest rate*.

Modifiers are naturally defined as order-preserving operations. That is, if $a \leq b$ then $m(a) \leq m(b)$ for all $a, b \in [0, 1]$. While this is the only requirement for general modifiers, it is often also required that m be continuous and that $m(0) = 0$ and $m(1) = 1$.

Among the great variety of possible modifiers, the most common modifiers either increase or decrease all membership grades of a given fuzzy set. A convenient class of functions, m_λ, for representing modifiers is defined for all $a \in [0, 1]$ and each $\lambda \in (0, 1) \cup (1, \infty)$ by the simple formula

$$m_\lambda(a) = a^\lambda. \tag{3.12}$$

The value of the parameter λ distinguishes modifiers of two types:

1. When $\lambda \in (0, 1)$ the modifiers defined by (3.12) are increasing their arguments and are suitable for representing linguistic hedges such as *fairly, somewhat, more or less,* and the like. The smaller the value of parameter λ, the stronger the modifier. Modifiers of this type are sometimes called *dilators.*
2. When $\lambda \in (1, \infty)$ the modifiers defined by (3.12) are decreasing their arguments. They are suitable for representing linguistic hedges such as *very, highly, extremely,* and the like. The larger the value of parameter λ, the stronger the modifier. Modifiers of this type are sometimes called *concentrators.*

These two types of modifiers can also be combined in such a way that a dilator is applied to membership grades that are smaller than some threshold value (usually 0.5) and a concentrator is applied to the remaining membership grades, or vice versa. And, of course, there are many other ways of defining modifiers or classes of modifiers, some even employing more than one parameter.

Although the primary purpose of modifiers is to represent linguistic hedges, they are also needed for other purposes. For example, any operation by which a subnormal fuzzy set is converted to a normal one under specified requirements (usually called a normalization) is in fact a fuzzy-set modifier.

3.4 Fuzzy intervals, fuzzy numbers, and fuzzy arithmetic

Historical background

Fuzzy sets defined on the set of real numbers $\mathbb{R}$ (i.e. $U = \mathbb{R}$) are of special significance as they can be viewed, when properly defined, as fuzzy numbers. Fuzzy numbers are of great importance in fuzzy logic and its applications.[11]

There is no doubt that Zadeh recognized the significance of fuzzy sets defined on $\mathbb{R}$ already in Zadeh 1965a. Although he does not explicitly define fuzzy numbers in this paper, he introduces basic ingredients needed for such definition, such as convexity and boundedness of fuzzy sets defined on $\mathbb{R}$. Moreover, he devotes almost half of the paper to the discussion of fuzzy sets defined on $\mathbb{R}$. Ten years later, he showed that convex and bounded sets defined on $\mathbb{R}$ can be combined not only by set-theoretic operations, but also by arithmetic operations (Zadeh 1975c). This stimulated some researchers to formulate more precisely the concept of a fuzzy number and investigate arithmetic operations on fuzzy numbers. Some important early contributions to these issues were made by Mizumoto and Tanaka (1976a), Jain (1976b), Dubois and Prade (1987b), Nahmias (1978), and Nguyen (1978). Ideas presented in

---

[11] Fuzzy numbers and their role in mathematics based on fuzzy logic are examined in detail in section 5.3.4.

these early papers stimulated other researchers to work in this area, which became known as *fuzzy arithmetic*. In the mid-1980s, the time was already right for publishing the first book specializing on fuzzy arithmetic (Kaufmann and Gupta 1985).

Fuzzy intervals and fuzzy numbers

It is appropriate to define the concept of a *fuzzy number* within a more general concept of a *fuzzy interval*, and we define the latter as any fuzzy set defined by a function of the form $A : \mathbb{R} \to [0, 1]$ that satisfies the following requirements:[12]

(fi1)   level sets of A, $^{\alpha}A$, are closed intervals of real numbers for all $\alpha \in (0, 1]$,

(fi2)   the support of A is bounded,

(fi3) A is a normal fuzzy set,

(fi4) A is a continuous function.

The requirement of continuity in the last condition is sometimes replaced with a weaker requirement of piecewise continuity. The piecewise-continuous fuzzy intervals are clearly more general than the continuous fuzzy intervals, but this generalization also has some drawbacks. They are computationally more demanding than continuous fuzzy intervals and their practicability is somewhat questionable.

We can now define the concept of a fuzzy number as any special fuzzy interval whose core consists of exactly one real number. To explain the rationale for this requirement, let us consider all fuzzy intervals whose cores consist of the same real number, say number r. These fuzzy intervals may be interpreted as representations of various statements in natural language that refer in some way to this particular real number r, such as "numbers that are very close to r," "numbers that are approximately equal to r," "numbers that are around r," and the like. These special fuzzy intervals—fuzzy numbers—thus play in fuzzy set theory the role of fuzzy approximators of real numbers.

Continuous fuzzy intervals can be conveniently defined for all $x \in \mathbb{R}$ in the form

$$A(x) = \begin{cases} 0 & \text{when } x < a, \\ f_A(x) & \text{when } x \in [a, b], \\ 1 & \text{when } x \in [b, c], \\ g_A(x) & \text{when } x \in [c, d], \\ 0 & \text{when } x > d, \end{cases} \tag{3.13}$$

where a, b, c, d are real numbers such that $a \leq b \leq c \leq d$, f_A is a strictly increasing function from 0 to 1, and g_A is a strictly decreasing function from 1 to 0. This form, which was introduced by Dubois and Prade (1978), is sometimes called a *canonical form of fuzzy intervals*. Observe that equation (3.13) defines a fuzzy number whenever $b = c$, it defines a real number when $a = b = c = d$, and it defines a closed interval of real numbers when $a = b$ and $c = d$.

---

[12] A more general definition is given on p. 299.

When A is a fuzzy interval defined by (3.13), then the level-cut representation of A is given for all $\alpha \in (0, 1]$ by the simple formula

$$^{\alpha}A = [f_A^{-1}(\alpha), g_A^{-1}(\alpha)], \qquad (3.14)$$

where symbols f_A^{-1} and g_A^{-1} denote the inverses of functions f_A and g_A.

As an example, let us consider a specific fuzzy interval A expressed in the form (3.13) for which $a = 0$, $b = 1$, $c = 2$, $d = 4$, $f_A(x) = x^2$ and $g_A(x) = \frac{(4-x)^2}{4}$. Function f_A is clearly strictly increasing from 0 to 1 on interval $[0, 1]$ and function g_A is strictly decreasing from 1 to 0 on interval $[c, d]$. After calculating the inverse functions of f_A and g_A, $f_A^{-1}(\alpha) = \sqrt{\alpha}$ and $g_A^{-1}(\alpha) = 4 - 2\sqrt{\alpha}$ we obtain by (3.14) the following level-cut representation of the fuzzy interval A: $^{\alpha}A = [\sqrt{\alpha}, 4 - 2\sqrt{\alpha}]$ for all $\alpha \in (0, 1]$.

Special fuzzy intervals with linear functions $f_A(x) = \frac{x-a}{b-a}$ and $g_A(x) = \frac{d-x}{d-c}$, which are called *trapezoidal fuzzy intervals* (due to the shape of function A), are frequently employed in applications. They are easy to handle and are uniquely represented by the quadruple $\langle a, b, c, d \rangle$, so their level-cut representations are expressed for all $\alpha \in (0, 1]$ by the convenient formula

$$^{\alpha}A = [a + (b-a)\alpha, d - (d-c)\alpha]. \qquad (3.15)$$

When $b = c$, they are called *triangular fuzzy numbers*.

Standard fuzzy arithmetic

Employing the level-cut representation of fuzzy intervals, we can perform arithmetic operations on fuzzy intervals in terms of arithmetic operations on classical closed intervals of real numbers. The latter are the subject of *interval computation* (also called *interval analysis*), a relatively new area of mathematics whose beginnings are usually associated with the publication of an influential book by Moore (1966); see section 5.3.4, p. 298 below for more information. Interest in this area has been strong and steadily growing. Literature on interval computation, which is now quite extensive, is well captured by several monographs, especially those written by Moore (1966, 1979), Alefeld and Herzberger (1983), Neumaier (1990), and Hansen (1992).

Given any pair of classical closed intervals of real numbers, $A = [a_1, a_2]$ and $B = [b_1, b_2]$, the four basic arithmetic operations on these intervals are defined in general as

$$(A * B) = \{x * y \mid x \in A, y \in B\}, \qquad (3.16)$$

where $*$ denotes any of the four basic arithmetic operations. However, when the operation is the division A/B, it is required that $0 \notin B$. The individual operations can also be defined, more efficiently, in terms of the endpoints:

$$A + B = [a_1 + b_1, a_2 + b_2],$$
$$A - B = [a_1 - b_2, a_2 - b_1],$$
$$A \cdot B = [\min\{a_1 b_1, a_1 b_2, a_2 b_1, a_2 b_2\}, \max\{a_1 b_1, a_1 b_2, a_2 b_1, a_2 b_2\}],$$
$$A/B = A \cdot 1/B, \text{ where } 1/B = [1/b_2, 1/b_1] \text{ if } b_1 > 0 \text{ or } b_2 < 0.$$

All of these definitions are consistent with (3.16) and easy to understand on intuitive grounds. When we want to apply them to fuzzy intervals (denoted for convenience by the same symbols A and B), we just need to replace A and B in the above definitions with $^{\alpha}A$ and $^{\alpha}B$ defined by (3.14). Clearly, the endpoints a_1, a_2, b_1, b_2 are now functions of α.

As an example, consider two fuzzy intervals, A and B. For convenience, let A be the fuzzy interval defined earlier in this section by which we illustrated the computation of level-cut representations, and we showed that $^{\alpha}A = [\sqrt{\alpha}, 4 - 2\sqrt{\alpha}]$, $\alpha \in (0, 1]$. Let B be another fuzzy interval that is defined by $a = 0$, $b = c = 1$, $d = 2$ and $f_A(x) = g_A(x) = 2x - x^2$. Clearly, $^{\alpha}B = [1 - \sqrt{1 - \alpha}, 1 + \sqrt{1 - \alpha}]$, $\alpha \in (0, 1]$. Now, we can use the endpoints of these intervals for each $\alpha \in (0, 1]$ and calculate, according to the above rules, the four basic arithmetic operations of A and B. We easily obtain, for example, $^{\alpha}A + {}^{\alpha}B = [\sqrt{\alpha} + 1 - \sqrt{1 - \alpha}, 5 - 2\sqrt{\alpha} + \sqrt{1 - \alpha}]$ for $\alpha \in (0, 1]$ and, similarly, the level-cut representations of $A - B$, $A \cdot B$, and A/B.

Another way of defining arithmetic operations on fuzzy intervals is to employ the extension principle for fuzzy sets (outlined in section 3.2) to functions of two variables. According to this principle, the four basic arithmetic operations on any two fuzzy intervals, A and B, are defined via their counterparts on real numbers by the common formula

$$(A * B)(c) = \sup_{c = a * b} \min\{A(a), B(b)\}, \qquad (3.17)$$

where $a, b, c \in \mathbb{R}$ and $*$ denotes again any of the four arithmetic operations. In important cases, the two ways above coincide (see section 5.3.4 for a detailed account). Extensive information on this particular use of the extension principle can be found in any of the books that specialize on fuzzy arithmetic: Kaufmann and Gupta 1985, Mareš 1994, and Hanss 2005.

Constrained fuzzy arithmetic

Each arithmetic expression containing symbols of real-valued variables is evaluated by an appropriate sequence of basic arithmetic operations. Each of these operations takes into account only numerical values of the symbols involved, not their meanings. This is not sufficient, in general, for evaluating arithmetic expressions that contain symbols of interval-valued variables. In this case, the evaluation depends, in general, not only on the intervals represented by the symbols, but also on the symbols themselves. This feature can be demonstrated by any arithmetic expression of interval-valued variables in which the same symbol of some interval-valued variable appears more than once. This means that such a symbol appears in more than one operation needed to evaluate the expression. In each appearance, the uncertainty of the associated interval is fully accounted for. As a result, this uncertainty is counted more than once, even though it should be counted only once as it represents the same variable. Due to this overestimation of uncertainty, the resulting interval is almost always wider than the one whose endpoints would be obtained globally by determining the minimum and maximum of the function associated with the evaluated expression within a restricted domain defined by the intervals of all distinct variables in the given expression.

This fundamental distinction between real-valued arithmetic and interval-valued arithmetic was already recognized in the pioneering book by Moore (1966), and it is routinely discussed in the more recent books on interval computation. For example, the discussion of this issue in the book by Neumaier (1990, 16–20) contains the following two interesting propositions, easy to understand on intuitive grounds:

1. Two arithmetic expressions that are equivalent in real arithmetic are equivalent in interval arithmetic when every variable occurs only once on each side.
2. If f, g are two arithmetic expressions that are equivalent in real arithmetic then the inclusion $f(x_1, x_2, \ldots, x_n) \subseteq g(x_1, x_2, \ldots, x_n)$ holds in interval arithmetic if every variable occurs only once in f.

Due to the recognized limitations of simple interval arithmetic based on (3.16), research in interval computations has increasingly concentrated on the investigation of methods for global optimization, as exemplified by the work of Hansen (1992).

The focus of early research work on fuzzy arithmetic was on investigating various forms of fuzzy intervals (referred to in the early publications as fuzzy numbers) and on basic arithmetic operations defined either by level-cut extension of (3.16) or via the extension principle (3.17). The limitations of simple interval arithmetic defined by (3.16), well recognized in the area of interval computation, were not mentioned in early publications on fuzzy arithmetic. It seems that it was tacitly assumed during this early research that arithmetic operations on fuzzy intervals could be applied in the same way as classical arithmetic operations on real numbers. At the same time, some practitioners began to use the simple fuzzy arithmetic without being aware of its limitations. When they repeatedly obtained fuzzy intervals that were unreasonably wide, their confidence in fuzzy arithmetic was shaken. As a result of this disappointment, some of them abandoned not only fuzzy arithmetic, but also lost interest in everything else offered at that time by fuzzy set theory.

To alleviate this unfortunate situation, a new concept of fuzzy arithmetic was introduced under the name *constrained fuzzy arithmetic* in a series of papers by Klir and Cooper (1996), Klir (1997a,b), and Klir and Pan (1998).[13] The principle aim of these papers was to show that the dubious results of standard fuzzy arithmetic have one common cause—a deficient generalization from the arithmetic on real numbers to the arithmetic on fuzzy intervals. It is argued that standard fuzzy arithmetic is deficient because it ignores requisite constraints among the fuzzy intervals involved. This means, in turn, that available information is ignored and, consequently, the obtained results are often information-deficient or, in other words, overly imprecise.

---

[13] Constraints in the extension principle were considered for the first time in Zadeh 1975c; see p. 302 below.

After the introduction of constrained fuzzy arithmetic, it still took some time before the deficiencies of standard fuzzy arithmetic were fully recognized. In this respect, it is pertinent to insert here a short quotation from the concluding section of Klir 1997a (p. 174):

> Although some authors have recognized the need to revise standard fuzzy arithmetic to account for the various dubious results when dealing with linguistic variables, others still maintain that there is no reason for any such revision. For example, when an extended abstract of a previous paper discussing constrained fuzzy arithmetic (Klir and Cooper 1996) was reviewed by two anonymous referees, they both expressed their opposition to any revision of standard fuzzy arithmetic, as can be seen from their comments: "There is no way (and no reason) to try to change $V - V \neq 0$; it is due to the definitions of the operations with sets. It is known that $V - V = 0$ iff V is a singleton. Also if probabilities add to exactly 1, then they should be crisp" (first referee); and "The sole motivation (for the constrained fuzzy arithmetic), it seems, is that classical unconstrained fuzzy arithmetic does not satisfy several desirable (?) crisp properties. However the example $V - V = 0$ shows that the addition of the equality constraint has the sole effect of producing unintuitive results" (second referee).[14]
>
> Comments like these indicate that the difficulties in using standard fuzzy arithmetic for dealing with linguistic variables are not generally recognized as yet. This paper is a modest attempt to clarify some of the basic issues involved.

The deficiencies of standard fuzzy arithmetic are now generally recognized and considerable research work within the fuzzy-set community has lately been devoted to developing computationally efficient procedures for sound fuzzy-interval computation.

3.5 Fuzzy relations

n-ary relations

A fuzzy set that is defined on universe $X = X_1 \times X_2 \times \cdots \times X_n$, where $n \geq 2$ and X_i ($i \in \mathbb{N}_n$) are nonempty classical sets, is called an *n-ary fuzzy relation*. That is, each *n*-ary standard fuzzy relation, R, is a function of the general form

$$R : X_1 \times X_2 \times \cdots \times X_n \rightarrow [0, 1]. \tag{3.18}$$

For each particular *n*-tuple $\langle x_1, x_2, \ldots, x_n \rangle \in X_1 \times X_2 \times \cdots \times X_n$, $R(x_1, x_2, \ldots, x_n)$ is interpreted as the strength of relationship among elements of the *n*-tuple.

All concepts and operations that are applicable to fuzzy sets are applicable to fuzzy relations as well. However, some additional concepts and operations are applicable to fuzzy relations due to the special type of functions defined by the general form (3.18). The aim of this section is to survey these additional concepts and operations.

Important operations that are applicable to arbitrary fuzzy relations are projections. In order to define them, let $Y = \times_{i \in J} X_i$, where $J \subset \mathbb{N}_n$. For each pair of tuples, $\mathbf{x} = \langle x_i \mid i \in \mathbb{N}_n \rangle$ and $\mathbf{y} = \langle x_i \mid i \in J \rangle$, we say that $\mathbf{x}$ contains $\mathbf{y}$ if and only

---

[14] Fortunately, the program committee of the conference overruled these strange comments and accepted the paper.

if $x_i = y_i$ for all $i \in J$, and we write $\mathbf{x} \succ \mathbf{y}$. Then, given a fuzzy relation R on X, a *projection* of R on Y, denoted as $[R \downarrow Y]$, is defined for all $\mathbf{y} \in Y$ by the formula

$$[R \downarrow Y](\mathbf{y}) = \max_{\mathbf{x} \succ \mathbf{y}} R(\mathbf{x}).$$

As can be seen from this formula, the operation of projection converts a given n-ary fuzzy relation R to the largest $|J|$-ary fuzzy relation $[R \downarrow Y]$ consistent with R.

An operation that is inverse to a projection is called a *cylindric extension*. By this operation, a given fuzzy relation on Y, R_Y, which may be viewed as projection $[R \downarrow Y]$, is converted to the largest fuzzy relation on X that is consistent with R_Y. The cylindric extension of R_Y to X, which we denote by $[R_Y \uparrow X]$, is defined for each $\mathbf{x}$ such as $\mathbf{x} \succ \mathbf{y}$ by the formula

$$[R_Y \uparrow X](\mathbf{x}) = R_Y(\mathbf{y}).$$

Given a set $\{[R \downarrow Y_k], k \in \mathbb{N}_p\}$ of p projections R_{Y_k} of an unknown relation R defined on X, where $Y_k = \times_{i \in J_k \subset \mathbb{N}_n} X_i$, a *cylindric closure* of these projections, $C(R_{Y_k} | k \in \mathbb{N}_p)$, is defined as the standard intersection of their cylindric extensions. Hence, for all $\mathbf{x} \in X$,

$$C(R_{Y_k} | k \in \mathbb{N}_p)(\mathbf{x}) = \min_{k \in \mathbb{N}_p} \{[R_{Y_k} \uparrow X](\mathbf{x})\}. \tag{3.19}$$

Cylindric closure is clearly the smallest relation on X that is consistent with all the given projections and always contains the unknown relation. That is, $C(R_{Y_k} | k \in \mathbb{N}_p)(\mathbf{x}) \supseteq R$ for any given set of projections.

Special n-ary fuzzy relations defined on the n-dimensional Euclidean space $\mathbb{R}^n$ ($n \geq 1$) were introduced and discussed in three early papers by Zadeh (1965a,b, 1966) with a particular focus on convex n-ary relations. All of the above concepts were introduced in Zadeh 1966, but no name was used for C in (3.19). The name "shadow" was defined, in general, as the orthogonal projection of the given relation on any specified hyperplane. Zadeh has employed these concepts of n-ary fuzzy relations frequently in his later writings for various purposes, but it is rather rare to find them in writings of other authors.

Consider now distinct fuzzy relations P and Q defined on sets $X_p = \times_{i \in I_p \subset \mathbb{N}_n} X_i$ and $X_Q = \times_{i \in I_Q \subset \mathbb{N}_n} X_i$ $(I_p \neq I_Q)$, respectively, and let $\mathbf{p} \in X_p$ and $\mathbf{q} \in X_Q$. Then, the *relational join* of P and Q, denoted by $P * Q$, is a fuzzy relation defined for all $\mathbf{r} \in \times_{i \in I_p \cup I_Q} X_i$ (i.e. concatenations of $\mathbf{p}$ and $\mathbf{q}$) by the equation

$$[P * Q](\mathbf{r}) = \min(P(\mathbf{p}), Q(\mathbf{q})). \tag{3.20}$$

Observe that $P * Q = Q * P$ due to commutativity of set union and, more importantly, that the operation of relational join can be applied recursively on a given set of projections of some n-ary fuzzy relation R. For example, let the fuzzy relation obtained by the specific operation of relational join defined by (3.20) be joined with another fuzzy relation, say S, defined on $X_S = \times_{i \in I_S \subset \mathbb{N}_n} X_i$ and let $\mathbf{s} \in X_S$. Then,

$$[(P * Q) * S](\mathbf{r}) = \min(\min(P(\mathbf{p}), Q(\mathbf{q})), S(\mathbf{s})) = \min(P(\mathbf{p}), Q(\mathbf{q}), S(\mathbf{s}))$$

for all $\mathbf{r} \in \times_{i \in I_P \cup I_Q \cup I_S} X_i$. Since the set union is associative, we have

$$(P * Q) * S = (P * S) * Q = (Q * S) * P.$$

It is thus convenient to write $P * Q * S$ regardless of the order in which the final relation was obtained.

Given a set of projections $P_1, P_2, \ldots, P_m$ of some unknown n-ary fuzzy relation R, it is established that the relation $P_1 * P_2 * \cdots * P_m$, which can be obtained recursively in any order, is the same as the cylindric closure of the given relations (Cavallo and Klir 1982). However, the use of the relational join reduces the computational complexity of obtaining the cylindric closure.

The concepts of projections, cylindric extensions, relational joins, and cylindric closures are principal tools for dealing with general n-ary fuzzy relations. They play an important role in many applications, as clearly recognized by Zadeh, and we point to some of these applications throughout this book.

Binary fuzzy relations

When fuzzy relations are binary, i.e. $n = 2$ in (3.18), the concepts of projections, cylindric extensions, cylindric closures, and relational joins are still applicable if we view fuzzy sets as degenerate, 1-ary fuzzy relations. In addition to these general concepts, however, two additional concepts are applicable to binary fuzzy relations: inverses and relational compositions.

The *inverse* of a binary fuzzy relation R on $X \times Y$, denoted by R^{-1}, is a fuzzy relation on $Y \times X$ such that $R^{-1}(y, x) = R(x, y)$ for all pairs $\langle y, x \rangle \in Y \times X$. Clearly, $(R^{-1})^{-1} = R$ holds for any binary fuzzy relation.

Consider now two binary fuzzy relations P and Q that are defined on sets $X \times Y$ and $Y \times Z$, respectively. Observe that these relations are connected via the common set Y. A composition of such relations always produces another binary fuzzy relation on set $X \times Z$, but this resulting relation (or composite relation) depends on how the composition is defined. There are numerous ways of defining the operation of composition for binary fuzzy relations. The first definition, which was already introduced in Zadeh 1965a, is the so-called sup-min composition, $P \circ Q$, defined for all pairs $\langle x, z \rangle \in X \times Z$ by the formula[15]

$$[P \circ Q](x, z) = \sup_{y \in Y} \min(P(x, y), Q(y, z)). \tag{3.21}$$

A few years later he suggested in Zadeh 1971b that it is meaningful to extend the sup-min composition (3.21) to a class of compositions

$$[P \circ_i Q](x, z) = \sup_{y \in Y} i(P(x, y), Q(y, z)), \tag{3.22}$$

where i denotes more general operations such as a t-norm. Compositions in this

---

[15]Sometimes, when the supremum in (3.21) can be replaced with the maximum, this composition is referred to as *max-min composition*.

class are usually called *sup-t-norm compositions*. In a more general setting, sup-t-norm compositions were studied in a remarkable paper by Goguen (1967).

Another class of compositions of binary fuzzy relations was introduced by Bandler and Kohout (1980b,c) in the context of their extensive study of operations of implication in fuzzy logic (Bandler and Kohout 1980a). This class is defined by

$$[P \triangleleft_{r_i} Q](x,z) = \inf_{y \in Y} r_i(P(x,y), Q(y,z)), \qquad (3.23)$$

where r_i denotes the operation of residuum for a given t-norm i; it is defined for all $a, b \in [0,1]$ by $r_i(a,b) = \sup\{x \in [0,1] \,|\, i(a,x) \leq b\}$. This operation, which may be interpreted as an operation of implication based on t-norm i, plays an important role in fuzzy logic in the narrow sense. Compositions defined by (3.23) are usually referred to as *inf-residuum compositions*.

As with classical relations, binary fuzzy relations can be conveniently represented and manipulated by matrices. In particular, the various types of compositions of binary fuzzy relations can be efficiently performed in terms of matrix operations that resemble standard matrix multiplication, see e.g. Klir and Yuan 1995.

Fuzzy relational equations

Fuzzy relational equations are closely associated with compositions of binary fuzzy relations. Each fuzzy relational equation involves two binary fuzzy relations P and Q defined on sets $X \times Y$ and $Y \times Z$, respectively, which are composed via one of the possible compositions to form a composite fuzzy relation R on $X \times Z$. Thus, for example, in each fuzzy relational equation based on the sup-min composition (called a sup-min fuzzy relational equation) the three relations are connected in a way that can be expressed symbolically as $R = P \circ Q$. This symbolic form represents in fact a set of simultaneous equations of the form

$$R(x,z) = \sup_{y \in Y} \min(P(x,y), Q(y,z)),$$

one for each pair $\langle x,z \rangle \in X \times Z$. Fuzzy relational equations for the sup-t-norm and inf-residuum compositions are defined in similar ways.

The problem of solving a given sup-min fuzzy relational equation $R = P \circ Q$ (or, alternatively, a sup-t-nom equation $R = P \circ_i Q$ or an inf-residuum equation $R = P \triangleleft_{r_i} Q$) has the following two possible formulations: Given relation R and relation P (resp. Q), determine all possible relations Q (resp. P) for which the given fuzzy relational equation is satisfied. Observe that either of these formulations describes inverse problems for the various compositions of binary fuzzy relations. It is thus suggestive to refer to these problems as *decomposition problems*.

Solving decomposition problems via fuzzy relational equations was suggested for the first time and investigated by Sanchez (1976). His study was restricted to sup-min fuzzy relational equations, for which he derived the following important results (written here in our notation):

1. Given R and Q, the set of solutions of $R = P \circ Q$ for P is not empty iff $\hat{P} = (Q \triangleleft_{r_{\min}} R^{-1})^{-1}$ is a solution. If it is, which can be easily verified by checking if the equality $R = \hat{P} \circ Q$ holds, then $\hat{P}$ is the largest solution.

2. Given R and P, the set of solutions of $R = P \circ Q$ for Q is not empty iff $\hat{Q} = P^{-1} \triangleleft_{r_{\min}} R$ is a solution. If it is, then $\hat{Q}$ is the largest solution.

For obtaining these results, Sanchez introduced the inf-residuum composition, but only for $i = \min$, as indicated in the above formulas.

     Sanchez began to investigate fuzzy relational equations during his studies at the Medical School of the Aix-Marseille University in Marseille, France. Recognizing their potential for formalizing medical diagnosis and the processes of constructing medical knowledge, he chose fuzzy relational equations as the subject of his thesis.[16] His basic ideas about the use of fuzzy relational equations for medical diagnosis and for constructing medical knowledge are outlined in Sanchez 1977. Within a few years, fuzzy relational equations attracted the attention of a fair number of other researchers who produced some early results in this subject area. For example, Pedrycz (1983) introduced and investigated the general sup-t-norm fuzzy relational equations defined by (3.22); Miyakoshi and Shimbo (1985) introduced and investigated the general inf-residuum fuzzy relational equation defined by (3.23); and some of the first attempts to compute minimal solutions of finite max-min fuzzy relational equations, most notably Czogala, Drewniak, and Pedrycz 1982, were surveyed and unified by Higashi and Klir (1984). Many of the existing results were soon incorporated into a book by Di Nola et al. (1989). The book also contains a short exploratory chapter on approximate solutions of fuzzy relational equations and covers fairly extensively the role of fuzzy relational equations in knowledge-based fuzzy systems.

     Since the publication of this book, theoretical research on fuzzy relational equations continued primarily in two main directions: research on developing efficient solution methods (in some cases methods for obtaining solutions in some required special forms), and research on investigating the notion of approximate solutions. While the former is too specific and detail-oriented to be covered in this book, the latter deserves to be covered here as it illustrates par excellence the spirit of fuzzy logic—thinking in degrees rather then in absolutes.

     Investigating approximate solutions of fuzzy relational equations is important since it is quite common that no exact solution exists. Were this to happen in any of the many applications of fuzzy relational equations (to be covered in chapter 6) the application would fail completely. However, it can be made applicable via an approximate solution. In the following we present a formulation of this approximation problem and an overview of key results pertaining to its solution. For details, we refer the reader to Gottwald 1993 and Klir and Yuan 1994, 1995.

---

[16] *Equations de Relations Floues* [Equations of fuzzy relations], Thèse Biologie Humaine, Faculté de Médecine de Marseille, 1974.

Consider a sup-t-norm fuzzy relational equation

$$P \circ_i Q = R, \tag{3.24}$$

in which relations Q and R are given, and assume that this equation is not solvable for P. In order to convert it into a solvable equation, we need to increase Q or decrease R. Assume that the equation

$$P' \circ_i Q' = R', \tag{3.25}$$

where $Q' \supseteq Q$ and $R' \subseteq R$, is solvable and P' is a solution. In order to accept P' as an approximate solution of (3.24), it is reasonable to require that Q' and R' in (3.25) be the closest relations to Q and R, respectively, for which this equation is solvable. This requirement can be formally stated as follows: For any P'', Q'', R'' such that $Q \subseteq Q'' \subseteq Q'$, $R' \subseteq R'' \subseteq R$, and $P'' \circ_i Q'' = R''$, we have $Q'' = Q'$ and $R'' = R'$. When Q' and R' in (3.25) satisfy this requirement, then P' is acceptable as an approximate solution of (3.24), and we denote it by $\tilde{P}$.[17] We are now ready to state one of the key results regarding approximate solutions of fuzzy relational equations: $\tilde{P} = (Q \triangleleft_{r_i} R^{-1})^{-1}$ is the greatest approximate solution of (3.24).

Although this result provides us with a simple calculation of the greatest approximate solution of (3.24), we still do not know how good this approximation is. To this end, we need to define what we mean by goodness of approximation. A proper way to do this in fuzzy set theory is to view the equality of fuzzy sets as a fuzzy concept. Denoting the degree of equality of fuzzy sets A and B by $A \approx B$, it is reasonable to define it in terms of the degree of inclusion of A in B, $S(A,B)$, and the degree of inclusion of B in A, $S(B,A)$, by the formula[18]

$$A \approx B = \min(S(A,B), S(B,A)).$$

It is then quite natural to define the goodness of approximate solution $\tilde{P}$ as the degree of equality $(\tilde{P} \circ_i Q) \approx R$.[19] Another key result regarding approximate solutions of sup-t-norm fuzzy relational equations is the following: Fuzzy relation $\tilde{P} = (Q \triangleleft_{r_i} R^{-1})^{-1}$ is the best approximate solution of (3.24).

Other issues regarding approximate solutions of fuzzy relational equations have been addressed by Gottwald (1993, 1994, 1995, 2008), Perfilieva and Gottwald (2003), and Belohlavek (2002c). See also a survey of solution methods for fuzzy relational equations by De Baets (2000).

Regarding the problem of computing minimal solutions of sup-min and sup-t-norm fuzzy relational equations, many papers that appeared after Czogala, Drewniak, and Pedrycz 1982 up until recently claimed to have proposed efficient algorithms for this problem. Only lately, the relationship between this problem and the well-known set cover problem— a basic problem in discrete optimization—has

---

[17] This concept of approximate solution appeared in Wu 1986.

[18] Degrees of inclusion and equality are examined in detail in chapter 5. The present results are valid for $S(A,B) = \inf_{u \in U} r_i(A(u), B(u))$.

[19] This concept of goodness of solution is due to Gottwald (1986).

been recognized (Markovskii 2005). The question of algorithmic feasibility of the problem of computing minimal solutions of fuzzy relational equations has only recently been solved by Bartl and Belohlavek (2015). They proved that there is no efficient algorithm for solving this problem.[20] Hence, the problem is inherently complex and the above mentioned claims of efficiency were mistaken.

Binary fuzzy relations on a single set

The purpose of binary fuzzy relations defined on $X^2 (= X \times X)$ is to characterize various ways in which objects of some set X can be related to themselves. Among the most important relations of this type are equivalence relations, compatibility relations, and various types of ordering relations.

There are several well-known basic properties of classical relations on $X \times X$ by which relations can be usefully classified. The most important of them are called reflexivity, symmetry, transitivity, and antisymmetry. Their counterparts for fuzzy relations were first introduced by Zadeh (1971b) as follows.

Let R denote a fuzzy relation on X^2. When $R(x,x) = 1$ for all $x \in X$, R is said to be *reflexive*. When $R(x,y) = R(y,x)$ for all $x,y \in X$, it is called *symmetric*. When

$$\min(R(x,y), R(y,z)) \leq R(x,z) \qquad (3.26)$$

for all $\langle x,z \rangle \in X^2$, R is called *transitive*, more specifically *max-min transitive*. Observe that (3.26) can be written as $R \supseteq R \circ R$. This indicates that other ways of defining transitivity are possible, such as $R \supseteq R \circ_i R$ or $R \subseteq R \triangleleft_{r_i} R$. However, definition (3.26) is the only definition of transitivity that is cutworthy. Antisymmetry, which Zadeh (1971b) defined by the requirement that $R(x,y) > 0$ and $R(y,x) > 0$ implies $x = y$ for all $x,y \in X$, is a much more delicate property and we examine it along with the concept of fuzzy ordering in chapter 5.[21]

Following Zadeh 1971b, a *similarity relation* is defined as any fuzzy relation S on X^2 that is reflexive, symmetric, and transitive.[22] When transitivity is defined by (3.26), S may be viewed as a fuzzy counterpart of a classical equivalence relation in that each level-cut of S is a classical equivalence relation for all $\alpha \in (0, 1]$. Moreover, each of these equivalence relations is associated with a particular partition of set X. These partitions are nested and increasingly refined when the value of α increases.

Zadeh's conception of fuzzy similarity is broader than the above-mentioned concept of cutworthy extension of classical equivalence relations. For each similarity relation (not necessarily based on max-min transitivity), he defines a similarity class, $S[x]$, associated with a particular element $x \in X$ as

$$(S[x])(y) = S(x,y)$$

---

[20] Here efficient means to have a polynomial time complexity.

[21] In chapter 5, fuzzy relations and various of their properties are examined in detail. In particular, antisymmetry is discussed in section 5.2.1, pp. 248 and 278.

[22] Such relations are nowadays called fuzzy equivalence relations; see section 5.3.1, p. 274 for a detailed account.

for each $y \in X$. He argues that similarity classes of S are not, in general, disjoint and shows, more specifically, that the height of the standard intersection of any two similarity classes, $S[x]$ and $S[y]$, satisfies for all $x, y \in X$ the inequality

$$h(S[x] \cap S[y]) \leq S(x, y).$$

Another important class of fuzzy relations on X^2 consists of fuzzy counterparts of classical *compatibility relations*.[23] These are relations that are reflexive and symmetric. Since the above definitions of reflexivity and symmetry for fuzzy relations are clearly cutworthy, the concept of a fuzzy compatibility relation is cutworthy as well. That is, level cuts of every fuzzy compatibility relation are classical compatibility relations for all $\alpha \in (0, 1]$. However, since compatibility relations are not transitive, they do not induce partitions of set X at the various level cuts. Although fuzzy compatibility relations are not mentioned in Zadeh 1971b, they can be viewed as generalized similarity relations that need not be transitive.

Given a binary fuzzy relation R on X^2 that is not transitive (in a given sense), its *transitive closure*, R_T, is the smallest fuzzy relation that contains R and is transitive (in the given sense). Transitive closure is an important concept in some applications of fuzzy relations, in which a given relation is a fuzzy compatibility relation (i.e. reflexive and symmetric relation), but it should be on intuitive grounds an equivalence relation (i.e., also transitive in the given sense of the definition of transitivity). The lack of transitivity may be caused by deficiency of data from which the relation was derived, inconsistent opinions of experts, or some other shortcomings. The problem of computing a transitive closure of a given fuzzy relation is solvable by efficient algorithms and has been thoroughly examined in various contexts, see e.g. De Baets and De Meyer 2003a,b, Garmendia and Recasens 2009, and Garmendia et al. 2009.

Particularly important are compatibility and similarity relations on fuzzy sets. These are usually conceived as functions $s : [0, 1]^U \times [0, 1]^U \to [0, 1]$, usually called *similarity measures of fuzzy sets*, assigning to any two fuzzy sets A and B in the universe U a degree $s(A, B) \in [0, 1]$ and are inspired by the pioneering paper by Tversky (1977). This degree, often called a comparison index, is interpreted as the degree to which A and B are similar. For details and references, see the influential writings Dubois and Prade 1982a; Zwick, Carlstein, and Budescu 1987; Shiina 1988; Bouchon-Meunier, Rifqi, and Bothorel 1996; Fonck, Fodor, and Roubens 1998; Cross and Sudkamp 1998; and De Baets and De Meyer 2005.

Among the most important fuzzy relations on X^2 are fuzzy orderings of various types, some of which were first introduced by Zadeh (1971b). All fuzzy orderings are transitive relations that also satisfy some sort of asymmetry. Fuzzy orderings are examined in detail in chapter 5 (p. 278).

Since the publication of Zadeh 1971b, literature dealing with fuzzy relations has grown quite rapidly, but concentrating primarily on fuzzy binary relations. Fairly

---

[23]Sometimes also called *tolerance relations* or *proximity relations*.

comprehensive overviews of this literature are Belohlavek 2002c, Peeva and Kyosev 2004, and Ovchinnikov 2000. Unfortunately, general n-ary fuzzy relations, which are important in some applications, have been rather neglected. In particular, apart from Zadeh's papers, we are not aware of any systematic study of projections, cylindric extensions, or cylindric closures of general n-ary fuzzy relations.

Fuzzy partitions

The first definition of a fuzzy partition appeared in Ruspini 1969. Accordingly, a *fuzzy partition* is a family of nonempty fuzzy sets $F_1, F_2, \ldots, F_n$ on a universe U such that $\sum_{k \in \mathbb{N}_n} F_k(u) = 1$ for all $u \in U$.[24] This was an ad hoc notion, introduced in Ruspini 1969 only implicitly in the context of fuzzy clustering (section 3.8). Although it was later found useful for other purposes as well, such as granulating base variables of linguistic variables (section 3.6) and others, it is not based on generalizing properly the concept of a partition in classical set theory to fuzzy set theory.

It is well known that in classical set theory there is a bijective correspondence between the set of all partitions that can be defined on any given universe U and the set of all equivalence relations that can be defined on U^2. The classical definition of a partition of set U—a family of nonempty subsets of U that are pair-wise disjoint and whose union is set U—can be generalized to fuzzy set theory in numerous ways, as it involves the broad classes of intersections and unions on fuzzy sets. By the same token, the classical definition of equivalence relation can be generalized to fuzzy set theory in numerous ways, as it involves a broad variety of possible definitions of transitivity. This means that the correspondence between fuzzy partitions and fuzzy equivalence relations is considerably more complex.

Although the connection between fuzzy equivalence relations and fuzzy partitions was briefly mentioned in Zadeh 1971b, it has been thoroughly investigated only since the mid-1990s. Important contributions to this topic include Gottwald 1993; Kruse, Gebhardt, and Klawonn 1994; Thiele and Schmechel 1995; Schmechel 1996; De Baets and Mesiar 1998; Höhle 1998; and Belohlavek 2002c. We examine them in detail in chapter 5 (p. 276).

Among numerous other papers addressing the correspondence between fuzzy partitions and fuzzy equivalence relations, two papers are worth mentioning. In the first paper by Mesiar, Reusch, and Thiele (2006), the authors introduce for this purpose operations of fuzzy conjunctions that are more general than the usual fuzzy conjunctions that are based on t-norms. This operation, which they call a *conjunctor*, is a function of the form $C : [0, 1]^2 \to [0, 1]$ whose only requirement is that it coincides with the classical conjunction when its domain is restricted to $\{0, 1\}^2$. This very general framework opens numerous new, and often surprising ways of looking at the relationship between fuzzy partitions and fuzzy equivalence relations.

---

[24] The term "fuzzy partition" itself did, however, not appear in Ruspini 1969.

In the second paper by Jayaram and Mesiar (2009), the authors investigate the relationship under a very general operation of fuzzy implication, called *implicator*, which they define as a function of the form $I : [0,1]^2 \rightarrow [0,1]$ whose only requirement is that it coincides with the classical operation of implication when its domain is restricted to $\{0,1\}^2$. Again, they obtain many interesting and often unexpected mathematical results within this very general framework and compare them with results obtained in Mesiar, Reusch, and Thiele 2006. Covering these interesting but highly technical results is beyond the scope of this chapter.

3.6 Approximate reasoning

Historical background

Virtually all papers Zadeh published in the period 1973–77 were devoted to building foundations for approximate reasoning in natural language. Some initial ideas germane to approximate reasoning were introduced in Zadeh 1973. They are well captured in the following quotation from section 1:

> Essentially, our contention is that the conventional quantitative techniques of system analysis are intrinsically unsuited for dealing with humanistic systems or, for that matter, any system whose complexity is comparable to that of humanistic systems. The basis of this contention rests on what might be called the *principle of incompatibility*. Stated informally, the essence of this principle is that as complexity of a system increases, our ability to make precise and yet significant statements about its behavior diminishes until a threshold is reached beyond which precision and significance (or relevance) become almost mutually exclusive characteristics....
>
> An alternative approach outlined in this paper is based on the premise that the key elements in human thinking are not numbers, but labels of fuzzy sets, that is, classes of objects in which the transition from membership to nonmembership is gradual rather than abrupt....
>
> The approach in question has three main distinguishing features: 1) use of so-called "linguistic" variables in place of or in addition to numerical variables; 2) characterization of simple relations by conditional fuzzy statements; and 3) characterization of complex relations by fuzzy algorithms.

The three distinguishing features of the suggested new approach, which are mentioned in this quotation, are explained in the paper, but mostly by examples. They are more extensively covered in three follow-up papers. In Zadeh 1974a, the focus is on discussing the role of linguistic variables as tools for dealing with some approximation problems; the term "approximate reasoning" is introduced in this paper for the first time to refer to one of these approximation problems. In Zadeh 1975a, the focus is solely on approximate reasoning. Zadeh describes the content of this paper concisely as follows (p. 2):

> The *calculus of restrictions* is essentially a body of concepts and techniques for dealing with fuzzy restrictions in a systematic fashion. As such, it may be viewed as a branch of the theory of fuzzy relations. However, a more specific aim of the calculus of fuzzy restrictions is to furnish a conceptual basis for fuzzy logic and what might be called *approximate reasoning*, that is, a type of reasoning which is neither very exact nor very inexact. Such reasoning plays a basic role in human decision-making because it provides a way of dealing with problems which are too complex for precise solution.

In Zadeh 1975b, the term "approximate reasoning" is given a special prominence by being included in the title of this paper, which offers a fairly comprehensive study of the meaning and applicability of fuzzy truth values in approximate reasoning and a clearly explained rationale for the compositional rule of inference (introduced in the previous paper) as a special case of the general procedure for computing a cylindric closure (described in section 3.5).

The climax of Zadeh's efforts to develop foundations for approximate reasoning based on fuzzy logic was reached when he published a series of three connected papers (Zadeh 1975c) on the role of linguistic variables in approximate reasoning. In these three papers, all of the previously developed ideas were integrated with some relevant new ideas into a comprehensive framework eminently suitable for further research on approximate reasoning. As a result, approximate reasoning became, in the late 1970s and throughout the 1980s, the most active research area within fuzzy logic in the broad sense. Some of the reasoning procedures suggested by Zadeh were further developed in greater detail by many contributors, new procedures were suggested, and numerous efforts were devoted to computer implementation of the various procedures. So many specific results were produced in the area of approximate reasoning during this period that it would not be sensible to try to cover them here. We consider it sufficient to refer the reader to a survey of these many developments from the late 1970s until the end of the 1990s, the most creative period of research on approximate reasoning, that was prepared by Bouchon-Meunier et al. (1999). However, in the rest of this section, we present an overview of basic ideas of approximate reasoning, focusing primarily on a broad conceptual coverage of these ideas.

Linguistic variables

The concept of a linguistic variable is central to approximate reasoning, and so we begin by introducing this concept first. Each linguistic variable consists of three interrelated components—base variable, a set of linguistic terms, and a set of fuzzy sets. These components are defined as follows, together with their exemplifications in terms of a particular linguistic variable expressed visually in figure 3.1.

A base variable, $\mathbf{V}$, is a traditional variable, which is usually (but not necessarily) numerical. We denote the set of states (or values) of $\mathbf{V}$ by V and a particular state in V by v. The base variable is exemplified in figure 3.1 by the physical variable of relative humidity expressed as a percentage. In this example, clearly, $V = [0, 100]$.

A set of linguistic terms, $L = \{L_i \mid i \in \mathbb{N}_n\}$, consists of simple expressions in natural language that approximately characterize states of the base variable. In our illustrative example in figure 3.1, the linguistic terms are *low*, *medium*, and *high*, and it is understood that they all refer to states of the base variable—humidity.

A set of fuzzy sets, $F = \{F_i \mid i \in \mathbb{N}_n\}$, which are all defined on V, is such that fuzzy sets in F are paired with linguistic terms in L via the common index i. For each pair $\langle L_i, F_i \rangle$, fuzzy set F_i is supposed to represent the meaning of linguistic term L_i.

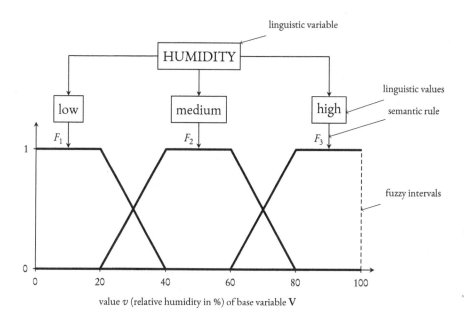

Figure 3.1: Example of a linguistic variable.

In this context, it is suggestive to interpret membership degrees $F_i(v)$ for all $v \in V$ as degrees of compatibility of F_i with L_i. In our example, F consists of trapezoidal fuzzy intervals, F_1, F_2, and F_3, which are visually depicted in figure 3.1. They can be defined either directly by (3.13) or, more conveniently, via their level-cuts by (3.15): $^{\alpha}F_1 = [0, 40 - 20\alpha]$, $^{\alpha}F_2 = [20 + 20\alpha, 80 - 20\alpha]$, and $^{\alpha}F_3 = [60 + 20\alpha, 100]$. Observe that these fuzzy sets form a fuzzy partition of the interval $[0, 100]$.

At this point, we should make a remark regarding the use of linguistic hedges. The linguistic variable in our example does not involve any linguistic expressions with linguistic hedges. Such expressions, for example *very low* or *very high*, can be handled within the framework of this linguistic variable only by applying appropriate modifiers to fuzzy sets F_1, F_2, and F_3. However, an alternative linguistic variable may be defined in which set F contains five linguistic terms—*very low, low, moderate, high,* and *very high*, each of which is represented by a fuzzy set in such a way that these five fuzzy sets form a fuzzy partition of the interval $[0, 100]$. If desirable, these five sets may be appropriately modified to account for additional linguistic expressions, such as *very very high, extremely high, extremely low,* or *more or less moderate*. It is clear that the meaning of linguistic terms *very low* and *very high* is represented quite differently by fuzzy sets within the frameworks of the two linguistic variables. The representation by the alternative linguistic variable is in some sense a refinement of the one by the linguistic variable of our example illustrated in figure 3.1. Both representations are meaningful; which one to choose depends, as always, on the context in which the linguistic variable is applied.

Granulation versus quantization

Each linguistic variable basically groups states of its base variable into fuzzy sets. These fuzzy sets are usually called *granules*, and the process of grouping is called *granulation*. The latter is a generalization of the classical process of grouping states of variables, which is called *quantization*. In quantization, states of a given variable are grouped into classical sets that form a partition. In granulation, the states are grouped into fuzzy sets that usually (but not necessarily) form a fuzzy partition. Quantization is often unavoidable due to limited resolution of measuring instruments. Assume for example a numerical variable, v, whose values range over an interval $[a, b]$, where a and b denote values that are measured in some particular units of measurement. Assume further that the measuring instrument employed is capable of measuring the variable with a precision of up to one of these measurement units. Then each of the measurements $a, a+1, a+2$, etc., really means *around a, around a + 1, around a + 2*, etc., which in classical mathematics must be represented by appropriate half-open intervals such as $[a-0.5, a+0.5), [a+0.5, a+1.5), [a+1.5, a+2.5)$, etc. This is not a satisfactory mathematical representation due to the obvious discontinuities at values $a+0.5$, $a+1.5$, $a+2.5$, and so on. Granulation, on the other hand, does not involve any such discontinuities, as it allows for smooth transitions from one granule to the next. This is illustrated for three quanta and the corresponding granules in figure 3.2, where we use for convenience triangular fuzzy numbers to represent granules. The three quanta in the top part of figure 3.2 are represented by the following half-open intervals of real numbers: $Q_a = [a-0.5, a+0.5), Q_{a+1} = [a+0.5, a+1.5), Q_{a+2} = [a+1.5, a+2.5)$. The three granules in the bottom part of figure 3.2 are represented by the following triangular fuzzy numbers (expressed via their level-cut forms): ${}^\alpha G_a = [a-1+\alpha, a+1-\alpha]$, ${}^\alpha G_{a+1} = [a+\alpha, a+2-\alpha]$, ${}^\alpha G_{a+2} = [a+1+\alpha, a+3-\alpha]$.

Forced quantizations or granulations such as those imposed by the limited resolution of measuring instruments must be distinguished from optional quantizations or granulations. The latter are employed for various pragmatic reasons, such as simplification or reduction of computational complexity. An example of optional granulation is the representation of *low, medium,* and *high* humidity by the three granules (trapezoidal fuzzy intervals) in figure 3.1.

Basic types of linguistic terms

Linguistic variables may involve any of the following basic types of linguistic terms:

1. Terms like *young, tall, intelligent, honest, heavy, expensive, fast, low, high, medium, normal, beautiful,* etc., possibly with some applicable linguistic hedges, such as *very young, extremely intelligent, highly honest, below normal,* etc. These linguistic terms, which are in each particular linguistic variable applied

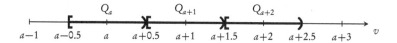

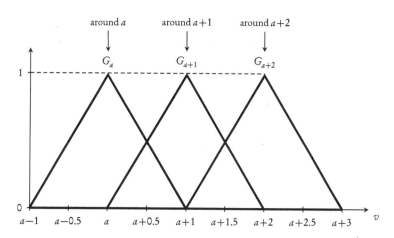

Figure 3.2: Quantization (top) versus granulation (bottom).

to states of the base variable and are represented by specific fuzzy sets, are called *fuzzy predicates*.

2. Terms like *true, false, fairly true, very false, more or less true*, etc., which are always applied to a special base variable whose values are in the unit interval $[0, 1]$ and represent degrees of truth. These terms, which are represented in each particular linguistic variable of this sort by appropriate fuzzy sets defined on $[0, 1]$, are called *fuzzy truth values*.

3. Terms like *usually, likely, very likely, extremely unlikely, around v* (for some $v \in [0, 1]$), etc., which are always applied to a special base variable whose values are in $[0, 1]$ and represent probability values. These terms, which are represented in each particular linguistic variable of this kind by appropriate fuzzy sets defined on $[0, 1]$, are called *fuzzy probabilities*.

4. Terms like *many, most, almost all, virtually none, several, about one half, around* 20%, etc., which are applied to states of its base variable whose values are either natural numbers or real numbers in $[0, 1]$. These terms, which are represented by appropriate fuzzy sets defined either on a given set of natural numbers or on $[0, 1]$, are called *fuzzy quantifiers*. They are of two kinds: *absolute fuzzy quantifiers* (*several, about twenty, much larger than ten, not very many*, etc.), which refer to a set of natural numbers; and *relative* (or *proportional*) *fuzzy quantifiers* (*about one half, almost all, around* 20%, etc.), which refer to real numbers in $[0, 1]$.

Types of fuzzy propositions

In this section, we introduce four basic types of fuzzy propositional forms and fuzzy propositions based on them. We also show how each of these types can be quantified via either an absolute or relative fuzzy quantifier. We use the notation introduced in the context of linguistic variables.

Unconditional and unqualified fuzzy propositions are based on the canonical form

$$\mathbf{V} \text{ is } F_i,$$

in which fuzzy set F_i represents a fuzzy predicate L_i ($i \in \mathbb{N}_n$). For each particular state $v \in V$ of variable $\mathbf{V}$, we obtain a fuzzy proposition

$$P_{i,v} : \mathbf{V} = v \text{ is } F_i.$$

The degree of truth of $P_{i,v}$, which we denote by $T(P_{i,v})$, is in this case the same as the degree of membership of v in fuzzy set F_i. That is,

$$T(P_{i,v}) = F_i(v). \tag{3.27}$$

Consider again the linguistic variable depicted in figure 3.1, and assume that the actual humidity (in a given context) is $v = 75\%$. Then, according to (3.27), we obtain $T(P_{1,75}) = 0$, $T(P_{2,75}) = 0.25$, and $T(P_{3,75}) = 0.75$. These results can be expressed in natural language as follows:

The degree of truth of proposition "humidity is low" is 0 (i.e. false).

The degree of truth of proposition "humidity is moderate" is 0.25.

The degree of truth of proposition "humidity is high" is 0.75.

Unconditional and qualified fuzzy propositions are characterized by involving some additional conditions that are called qualifiers. In the following, we exemplify qualified fuzzy propositions by describing truth-qualified fuzzy propositions and probability-qualified fuzzy propositions.

Truth-qualified fuzzy propositions are based on the canonical form

$$\mathbf{V} \text{ is } F_i \text{ is } Q_t,$$

where Q_t denotes a fuzzy truth qualifier—a fuzzy set defined on $[0, 1]$ that represents linguistic truth claims such as *true, false, fairly true, very false*, and the like. Denoting individual fuzzy propositions based on this form by $P_{i,v,t}$ ($i \in \mathbb{N}_n, v \in V$), we have

$$P_{i,v,t} : \mathbf{V} = v \text{ is } Q_t.$$

The degree of truth of $P_{i,v,t}$ is obtained in this case by composing function F_i with function Q_t:

$$T(P_{i,v,t}) = Q_t(F_i(v)).$$

When Q_t represents the linguistic claim *true*, then it is the identity function: $Q_t(a) = a$ for all $a \in [0, 1]$. In this case, clearly, $T(P_{i,v,t}) = F_i(v)$, which is the same as the truth degree for a comparable but unqualified fuzzy proposition. This means that unqualified fuzzy propositions are, in fact, special truth-qualified propositions in which the truth qualifier is tacitly assumed to represent the linguistic claim *true*. When Q_t represents the linguistic truth claim *false*, $Q_t = 1-a$ for all $a \in [0,1]$. Most other linguistic truth claims involve these two basic truth claims, *true* and *false*, with various linguistic hedges, and can be represented by appropriately modified fuzzy set $Q_t(a) = a$ or fuzzy set $Q_t(a) = 1-a$.

Probability-qualified fuzzy propositions are based on the canonical form

$$\text{Pro}(\mathbf{V} \text{ is } F_i) \text{ is } Q_p,$$

where $\text{Pro}(\mathbf{V} \text{ is } F_i)$ denotes a probability associated with fuzzy sets F_i (fuzzy events) and Q_p stands for a probability qualifier that represents a linguistic probability claim such as *likely, very likely, extremely unlikely,* close to 0, around 0.5, and so on. For any given probability of fuzzy event F_i, say $\text{Pro}(F_i) = p_i$, we obtain the fuzzy proposition

$$P_{i,p} : \text{Pro}(\mathbf{V} \text{ is } F_i) = p_i \text{ is } Q_p,$$

whose truth degree is

$$T(P_{i,p}) = Q_p(\text{Pro}(F_i)).$$

If this proposition were also truth-qualified, its truth degree would be calculated by the following composition of three functions Pro, Q_p and Q_t:

$$T(P_{i,p,t}) = Q_t(Q_p(\text{Pro}(F_i))).$$

The notion of *probabilities of fuzzy events* was introduced in one of the very early papers on fuzzy sets by Zadeh (1968a). When variable $\mathbf{V}$ has a finite number of states (V is finite) and a probability distribution function, f, on set V is given, the probabilities $\text{Pro}(F_i)$ are calculated for each $i \in \mathbb{N}_n$ by the formula

$$\text{Pro}(F_i) = \sum_{v \in V} F_i(v)f(v).$$

When states of variable $\mathbf{V}$ are real numbers (V is usually a closed interval of real numbers), sets F_i are fuzzy intervals defined on V (as in the example in figure 3.1), and a probability density function, g, on set V is given, the probabilities $\text{Pro}(F_i)$ are calculated for each $i \in \mathbb{N}_n$ by the formula

$$\text{Pro}(F_i) = \int_{\mathbb{R}} F_i(v)g(v)dv.$$

Conditional fuzzy propositions involve two linguistic variables, whose base variables we denote by $\mathbf{V}, \mathbf{W}$ and their state sets by V, W, respectively. They connect fuzzy sets defined on V to those defined on W by a chosen operation of fuzzy implication. When unqualified, they are based on the canonical form

$$\text{If } \mathbf{V} \text{ is } F_{i,V}, \text{ then } \mathbf{W} \text{ is } F_{j,W},$$

where fuzzy sets $F_{i,V}$ ($i \in \mathbb{N}_n$) and $F_{i,W}$ ($i \in \mathbb{N}_n$) are defined on sets V and W, respectively. A convenient alternative form is

$$(\mathbf{V}, \mathbf{W}) \text{ is } R, \tag{3.28}$$

where R denotes a fuzzy relation on $V \times W$.[25] This relation is determined for all $v \in V$ and all $w \in W$ by the formula

$$R(v, w) = r[F_{i,V}(v), F_{j,W}(w)],$$

where r denotes a binary operation on $[0, 1]$ that represents a suitable fuzzy implication in the context of each given application.[26] A conditional fuzzy proposition based on the alternative form (3.28), $P_{v,w,r}$, is obtained when $\mathbf{V} = v$ and $\mathbf{W} = w$. That is,

$$P_{v,w,r} : (\mathbf{V} = v, \mathbf{W} = w) \text{ is } R.$$

The truth degree of this proposition is

$$T(P_{v,w,r}) = r[F_{i,V}(v), F_{j,W}(w)].$$

When the proposition is qualified, its truth degree is obtained by composing function r with one or both qualifying functions Q_t and Q_p. Conditional fuzzy propositional forms are fundamental to various fuzzy rules of inference in approximate reasoning. The most common fuzzy inference rule, used here as an example, is a fuzzy rule that is analogous to but different from the rule of *modus ponens* in classical logic. The following inference schema expresses its simplest version, which employs unqualified propositional forms:

$$\text{If } \mathbf{V} \text{ is } F, \text{ then } \mathbf{W} \text{ is } G$$
$$\underline{\mathbf{V} \text{ is } A}$$
$$\mathbf{W} \text{ is } B$$

As is explained above, this schema can be reformulated as:

$$(\mathbf{V}, \mathbf{W}) \text{ is } R$$
$$\underline{\mathbf{V} \text{ is } A}$$
$$\mathbf{W} \text{ is } B$$

This formulation of the inference rule can be conveniently implemented via the sup-min composition by putting

$$B(w) = \sup_{v \in V} \min(A(v), R(v, w))$$

for each $w \in W$. This rule of inference is usually referred to as the *compositional rule of inference*.[27] When desirable in some application contexts, the sup-min composition may be replaced with a more fitting sup-t-norm composition.

---

[25] This captures the essence of conditional fuzzy propositions and avoids the many nuances resulting from the great variety of possible fuzzy implications, as described, for example, in Klir and Yuan 1995 (pp. 304–17) or in Baczyński and Jayaram 2008.

[26] This relation can be obtained from data, for example, by solving appropriate fuzzy relational equations, as shown by Wu (1986).

[27] For details, see n. 200 on p. 222.

Quantifiers in fuzzy logic

In classical logic, quantifiers are essential ingredients of predicate calculus, and they play a similar role in FLn, as we discuss in chapter 4. In FLb, they are of primary interest within the area of approximate reasoning. The challenge is to identify quantifiers of various types in natural language, develop their formal representations via resources of fuzzy set theory, and employ these representations for dealing with natural language statements that contain linguistic quantifiers of all kinds. In other words, the challenge is to add appropriate fuzzy quantifiers into all of the fuzzy propositional forms introduced above—that is to quantify them, and to employ the quantified propositional forms for implementing various inferential schemata.

The variety of possible statements in natural language that involve quantifiers is very high and, inevitably, the variety of the corresponding quantified fuzzy propositional forms is of comparable order. Consequently, this section focuses on tracing the development of fuzzy quantifiers and providing the reader with a few key references for the many details that cannot be considered here. A detailed examination of some fuzzy quantifiers within FLn is covered in chapter 4.

The notion of a fuzzy quantifier was for the first time introduced informally by Goguen (1968–69, 369–72). In particular, he showed possible ways of representing linguistic quantifiers *most, some, a few,* and *many* by fuzzy sets, as well as some inferences involving these quantifiers.

Zadeh began to discuss fuzzy quantifiers some ten years after the publication of his seminal paper. In some of his publications during the period from 1975 to 1992, none of which was fully devoted to fuzzy quantifiers, he gradually explored possible ways of representing the various quantifiers that are routinely employed in natural language. In Zadeh 1975a, he made only a small remark about the quantifier *most,* which he defined as a fuzzy subset of $[0,1]$. In Bellman and Zadeh 1977 and Zadeh 1979a, 1981a,b the discussion of linguistic quantifiers and their fuzzy-set representations is gradually expanded. The first paper fully devoted to fuzzy quantifiers is Zadeh 1983a.[28]

In general, Zadeh views fuzzy quantifiers as fuzzy intervals defined either on $\mathbb{R}$ or on $[0,1]$. As explained on p. 73, the former are called absolute and the latter relative quantifiers. Normally, each quantifier is applied within a finite universe U. The focus is on a certain relevant attribute of objects in U (*height, weight, age, price, performance,* etc.), which is expressed by a function $f : U \to A$. A propositional form involving an absolute fuzzy quantifier Q and fuzzy set F on A representing a predicate can be expressed quite naturally as

$$\text{There are } Q \ u\text{'s such that } f(u) \text{ is } F, \tag{3.29}$$

where F denotes a fuzzy set defined on set A, which represents a fuzzy predicate.

---

[28]Ideas covered in this paper were actually presented a year earlier at the Fourth Biennial Conference of the Canadian Society for Computational Studies of Intelligence at the University of Saskatchewan in Saskatoon in May 1982. An abridged version of the paper is included in the proceedings of the conference.

The degree of truth of a particular proposition p based on this form is then usually defined by the formula

$$T(p) = Q\left(\sum_{u \in U} F(f(u))\right).$$

Observe that the argument of Q is a particular scalar cardinality of fuzzy set $F(f(u))$ expressed as the sigma count of the fuzzy set. When desirable in some application contexts, other cardinalities may be explored (Wygralak 2003).

The counterpart of (3.29) for a proportional fuzzy quantifier Q is the form:

Among u's such that $f(u)$ is 1_A, there are Q u's such that $f(u)$ is F,

where 1_A denotes here the function $1_A : A \rightarrow \{1\}$. The degree of truth of a particular proposition p based on this form is defined as

$$T(p) = Q[S(1_A, F(f))],$$

where S in the argument of Q denotes the degree of subsethood of 1_A in $F(f)$. Again, S can be defined in several alternative ways.

In addition to propositional forms involving quantifiers, Zadeh also gradually introduced in some of the above-mentioned papers, and especially in Zadeh 1983b, the various possible inference rules involving quantified propositional forms. In several papers published in the 1980s, he also investigated the so-called dispositions involved in linguistic propositions. His primary paper in this area is Zadeh 1987, in which he explains (p. 39):

> A disposition may be interpreted as a proposition which is preponderantly, but not necessarily always, true. In this sense, *birds can fly* is a disposition, as are the propositions *Swedes are blond*, *snow is white*, and *slimness is attractive*. An idea which underlies the theory described in this article is that a disposition may be viewed as a proposition with implicit fuzzy quantifiers which are approximations to *all* and *always*, e.g., *almost all, almost always, most, frequently, usually*, etc. For example, *birds can fly* may be interpreted as the result of suppressing the fuzzy quantifier *most* in the proposition *most birds can fly*. . . . The process of transforming a disposition into a proposition with explicit fuzzy quantifiers is referred to as *explicitation* or *restoration*. Explicitation sets the stage for representing the meaning of a disposition through the use of test-score semantics [see Zadeh 1978a, 1981b]. In this approach to semantics, a proposition, p, is viewed as a collection of interrelated elastic constraints . . . which are induced by p.

Zadeh also addressed the issue of how to represent the meaning of the linguistic term *usually*. This term sometimes appears explicitly in linguistic propositions. However, it is more common that it does not appear and it is only implicitly assumed. In the latter case, the linguistic proposition is a disposition. Zadeh discusses some issues regarding this important kind of disposition in Zadeh 1986, 1987.

We should also mention an important early contribution to quantified fuzzy propositions by Yager (1980b), which he presented at the Second International Seminar on Fuzzy Set Theory in Linz, Austria, in September 1980. The same paper was also published two years later as Yager 1983. In this paper, Yager approaches fuzzy quantifiers in an original way, focusing on their role in summarizing large collections of data. He described this role particularly well on pp. 69–70 in Yager 1982a as follows:

> The ability to summarize data provides an important method for getting a grasp on the meaning of the content of a large collection of data. It enables humans to help understand their environment in a manner amenable to future useful manipulations.
>
> At one extreme lies the large mass of undigested data while at the other extreme lies the usual summarization in terms of the mean or average value of the data. The mean does help in understanding the content of data, but, in some respect, it may provide too terse a summarization. The variance provides a means of judging the validity of the mean as the summary. In many instances, especially in situations involving presentations to non-quantitatively oriented people, an alternative form of summarization may be useful. This form of summarization may be especially useful if it can provide us with a summary that is not as terse as the mean, and also if it can be used for summarization of non-numerical data. . . . The facility to summarize data or observations plays an important role in the ability to allow a person to communicate his observations about the world in a useful and understandable manner. It also provides a starting point for the ability to make useful inferences from large collections of data. The statement that "many Chinese like rice," which is a summarization of some observations, allows us to make inferences about the viability of a rice shop in China.

Yager published more papers on fuzzy quantifiers in which he further explored their role in summarizing data (Yager 1985, 1986a, 1988), and there were also many other authors who contributed in different ways to this subject area. Instead of covering further details here, we refer only to a comprehensive overview of fuzzy quantifiers in a book by Glöckner (2006), as well as the more recent overview of the literature on fuzzy quantifiers by Delgado et al. (2014).

Before closing this section on approximate reasoning, we should mention that it is this area of FLb that is most closely connected with FLn. The goal of approximate reasoning is to emulate mathematically, as much as possible, human reasoning in natural language. This goal has been pursued within fuzzy set theory largely intuitively and without the rigor of FLn. However, some of the ideas emerging from research on approximate reasoning have inspired new directions of research in FLn. On the other hand, FLn serves as a safeguard against possible ill-conceived developments in approximate reasoning.

3.7 Possibility theory

Possibility theory is a mathematical theory that has various interpretations. Our primary interest is its fuzzy-set interpretation. However, in order to fully understand this interpretation, we first introduce a common axiomatic characterization of possibility theory.

Axiomatic characterization of possibility theory

In a strictly formal sense, possibility theory is a special branch of generalized measure theory (Wang and Klir 2009). It is based on two dual semicontinuous generalized measures that are monotone, but not additive—possibility measures (semicontinuous from below) and necessity measures (semicontinuous from above). The theory can be axiomatically characterized in terms of either of these two classes of generalized measures. In the following definitions, symbols U and $\mathscr{A}$ denote, respectively, a universe of discourse and some given family of nonfuzzy subsets of U that is closed under arbitrary unions and intersections, and under complementation in U—usually referred to as an *ample field* on U (De Cooman and Kerre 1993).

Given a universe of discourse U, a *possibility measure*, Pos, is a set function[29]

$$\text{Pos} : \mathscr{A} \rightarrow [0,1]$$

that satisfies the following requirements:

(Pos1) $\text{Pos}(\emptyset) = 0$ (vanishing at $\emptyset$),

(Pos2) $\text{Pos}(U) = 1$ (normalization),

(Pos3)   for any family of sets $\{A_i \mid A_i \in \mathscr{A}, i \in I\}$,

$$\text{Pos}\left(\bigcup_{i \in I} A_i\right) = \sup_{i \in I} \text{Pos}(A_i). \tag{3.30}$$

Given a possibility measure, Pos, the dual measure, Nec, defined by

$$\text{Nec}(A) = 1 - \text{Pos}(\overline{A}) \tag{3.31}$$

for all $A \in \mathscr{A}$, is called the necessity measure corresponding to Pos. However, it may also be defined independently.

Given a universe of discourse U, a *necessity measure*, Nec, is a function

$$\text{Nec} : \mathscr{A} \rightarrow [0,1]$$

that satisfies the following requirements:

(Nec1) $\text{Nec}(\emptyset) = 0$,

(Nec2) $\text{Nec}(U) = 1$,

(Nec3)   for any family of sets $\{A_i \mid A_i \in \mathscr{A}, i \in I\}$,

$$\text{Nec}\left(\bigcap_{i \in I} A_i\right) = \inf_{i \in I} \text{Nec}(A_i). \tag{3.32}$$

The following are some well-known properties that can be derived fairly easily from either axiomatic characterization: for any $A, B \in \mathscr{A}$,

$$\text{Nec}(A) \leq \text{Pos}(A), \tag{3.33}$$

$$\text{Pos}(A \cap B) \leq \min\left(\text{Pos}(A), \text{Pos}(B)\right),$$

$$\text{Nec}(A \cup B) \geq \max\left(\text{Nec}(A), \text{Nec}(B)\right),$$

$$\text{Pos}(A) + \text{Pos}(\overline{A}) \geq 1,$$

$$\text{Nec}(A) + \text{Nec}(\overline{A}) \leq 1, \tag{3.34}$$

$$\max\left(\text{Pos}(A), \text{Pos}(\overline{A})\right) = 1,$$

$$\min\left(\text{Nec}(A), \text{Nec}(\overline{A})\right) = 0, \tag{3.35}$$

$$\text{Pos}(A) < 1 \text{ implies } \text{Nec}(A) = 0,$$

$$\text{Nec}(A) > 0 \text{ implies } \text{Pos}(A) = 1.$$

These properties capture well some of the nuances of possibility theory.

---

[29] In a more general formulation, which is not followed in this book, Pos is a function from $\mathscr{A}$ to a given complete lattice (De Cooman 1997).

Fuzzy-set interpretation of possibility theory

As an attempt to develop a natural mathematical framework for representing information expressed by fuzzy propositions, Zadeh (1978b) introduced a meaningful connection between fuzzy propositions and possibility measures. This connection is usually referred to as the *standard fuzzy-set interpretation of possibility theory*.

In order to explain the meaning of this interpretation of possibility theory, let **V** denote a variable that takes values from set U. Assume that information about the actual value of the variable is expressed by a fuzzy proposition **V** is F, where F is a fuzzy set defined on U. To express this information in measure-theoretic terms, it is natural, as Zadeh argued, to interpret the membership degree $F(u)$ for each $u \in U$ as the degree of possibility that **V** $= u$. This, in effect, defines possibility values for all singletons of U. That is, $\text{Pos}_F(\{u\}) = F(u)$ for all $u \in U$. The set of possibility values for all singletons of U is usually called a *possibility distribution*.[30] To simplify the notation for possibility distributions, let $r_F(u) = \text{Pos}_F(\{u\})$. The standard fuzzy-set interpretation of possibility theory is then expressed as

$$r_F(u) = F(u) \tag{3.36}$$

for all $u \in U$. Given a possibility distribution (3.36) for some fuzzy set F, the possibility value for any set $A \in \mathcal{A}$ is determined by the simple formula $\text{Pos}_F(A) = \sup_{u \in A} r_F(u)$ that is based on equation (3.30) of requirement (Pos3). This value has the following meaning: Given information that **V** is F, the degree of possibility that the actual value of variable **V** is in set A is $\text{Pos}_F(A)$. This can also be expressed as: It is possible to degree $\text{Pos}_F(A)$ that the actual value of **V** is in set A. Using the duality equation (3.31), we obtain $\text{Nec}_F(A) = 1 - \text{Pos}_F(\overline{A})$, whose meaning can be expressed verbally as: It is necessary to degree $\text{Nec}_F(A)$ that the actual value of **V** is in set A.

Let us consider now $\text{Pos}_F(U) = \sup_{u \in U} r_F(u)$ and $\text{Nec}_F(U) = 1 - \text{Pos}_F(\emptyset)$. Clearly, we always get $\text{Pos}_F(\emptyset) = 0$, so (Pos1) is satisfied, and $\text{Nec}_F(U) = 0$, so (Nec1) is also satisfied. Moreover, if F is a normal fuzzy set, then $\text{Pos}_F(U) = 1$, so (Pos2) is satisfied, and $\text{Nec}_F(U) = 1$ so (Nec2) is also satisfied. That is, the interpretation (3.36) is coherent.

If F is a subnormal fuzzy set whose height is h_F, then the situation is very different. Clearly, $\text{Pos}_F(U) = h_F < 1$, which violates condition (Pos2). This means that the interpretation (3.36) is not coherent and all kinds of weird results can be obtained from it. Consider, for example, the support of F, which we denote by S_F and its complement by $\overline{S}_F$; then, $\text{Pos}_F(S_F) = h_F$ and $\text{Pos}_F(\overline{S}_F) = 0$. Hence, $\text{Nec}_F(S_F) = 1$ and $\text{Nec}_F(\overline{S}_F) = 1 - h_F$ by the duality equation (3.31). The resulting inequalities, $\text{Nec}_F(S_F) > \text{Pos}_F(S_F)$ and $\text{Nec}_F(\overline{S}_F) > \text{Pos}_F(\overline{S}_F)$ violate not only a basic property of possibility theory, but also our common-sense understanding of the linguistic

---

[30] The suitability of the term *possibility distribution* has lately been questioned. The term is somewhat misleading since, in fact, $\{\text{Pos}_F(\{u\}) \mid u \in U\}$ for any given F is not a distribution of some fixed value. In Wang and Klir 2009, for example, the more fitting term *possibility profile* is introduced and consistently used. However, due to the historical orientation of this book, we have decided to use the traditional term *possibility distribution*.

terms *possibility* and *necessity*. Indeed, we cannot imagine anything that is necessary to a larger degree than the degree to which it is possible.

Zadeh (1978b) did not address the connection of possibility measures with necessity measures and it seems that he tacitly assumed that F in (3.36) is a normal fuzzy set. He introduced necessity measures in Zadeh 1979b, but did not recognize the difficulty with subnormal fuzzy sets. This difficulty was recognized for the first time by Yager (1986b). He attempted to overcome it by replacing the necessity measure with a new function, Cert, defined for all $A \in \mathscr{A}$ by the formula $\mathrm{Cert}_F(A) = \min\{\mathrm{Pos}_F(A), \mathrm{Nec}_F(A)\}$, and called a measure of certainty. Although the new measure satisfies the essential inequality (3.33), it has other severe drawbacks, as was argued by Dubois and Prade (1987a): Not only is the interpretation of the function Cert not clear, the function also does not satisfy (3.32), which is one of the basic requirements of possibility theory. Dubois and Prade suggested a way to resolve the problem by keeping the necessity measure, but replacing the duality equation (3.31) with a generalized equation $\mathrm{Nec}_F(A) = h_F - \mathrm{Pos}_F(\overline{A})$. Although the function Nec_F defined by this equation satisfies both (3.32) and (3.33), it has other severe drawbacks: It does not satisfy $\mathrm{Pos}_F(U) = \mathrm{Nec}_F(U) = 1$, and, as a consequence, it also violates properties (3.34) and (3.35).

For any given subnormal fuzzy set F that is not empty, it is of course always possible to obtain a coherent fuzzy-set interpretation of possibility theory by normalizing it and applying equation (3.36) to its normalized form, say $\hat{F}$. Among several possible normalizations, the most common is expressed by the equation $\hat{F}(u) = \frac{F(u)}{h_F}$ for all $u \in U$. The issue of which normalization is appropriate in this context was investigated by Klir (1999) on epistemological grounds. He showed that the only normalization that captures faithfully the information conveyed by fuzzy set F is obtained by the equation $\hat{F}(u) = F(u) + 1 - h_F$ for all $u \in U$. Given this result, the standard fuzzy-set interpretation of possibility theory (3.36) can be replaced with a generalized interpretation expressed for all $u \in U$ by the equation

$$r_F(u) = F(u) + 1 - h_F,$$

which is applicable to all standard fuzzy sets.

Using this generalized fuzzy-set interpretation or any other interpretation (justified in some other way) for which $\mathrm{Pos}(U) = \sup_{u \in U} r_F(u) = 1$, we obtain

$$\mathrm{Pos}_F(A) = \sup_{u \in U} \min(A(u), r_F(u))$$

for any fuzzy set A defined on universe U.

Zadeh's 1978 paper on a fuzzy-set interpretation of possibility theory, which was published as the first paper in the first issue of *Fuzzy Sets and Systems*, attracted considerable attention within the fuzzy-set community. It was quickly recognized that the new, largely unexplored theory of graded possibilities was a fertile area for research. Some in the community soon joined Zadeh in researching the many nuances of the theory, such as those emerging from possibility measures derived from fuzzy relations (e.g. the issues of joint, marginal, and conditional possibilities), or those involving information-theoretic aspects of possibility theory (e.g. the key issue of finding proper ways of measuring possibilistic uncertainty and information—fully justifiable on intuitive as well as mathematical grounds, the issue of developing and implementing sound methodological principles of possibilistic uncertainty or information, and various other related issues). Due to the growing number of researchers working in this area, the theory was already fairly well developed just ten years after the publication of Zadeh 1978b, when the first book on the theory of graded possibilities was published (Dubois and Prade 1988). The most comprehensive and rigorous exposition of the theory to date was written by De Cooman (1997).

Possibilistic uncertainty and information

A simple possibility theory of uncertainty and information was introduced within the framework of classical set theory by Hartley (1928). This theory is a special case of the previously introduced possibility theory such that $Pos : \mathcal{A} \rightarrow \{0, 1\}$ and $Nec : \mathcal{A} \rightarrow \{0, 1\}$. Otherwise, the requirements for possibility and necessity measures remain the same. Assume again that variable $\mathbf{V}$ takes values from set U, assumed here to be nonempty and finite. Given information that the actual value of $\mathbf{V}$ is in some subset of U, say E, the set interpretation of possibility theory is defined in this case via the characteristic function of E. That is,

$$r_E(u) = \begin{cases} 1 & \text{when } u \in E, \\ 0 & \text{when } u \notin E. \end{cases}$$

Hartley's contribution was that he showed convincingly, even though rather informally, that the only sensible way to measure the amount of possibilistic uncertainty regarding the actual value of variable $\mathbf{V}$ when we only know that it is not possible that the value is outside set E is by employing function $H(E) = \log_2 |E|$, provided that we choose to measure the uncertainty in bits.[31] This function is usually referred to as the Hartley uncertainty. Its uniqueness was later established formally in various ways, most interesting among them being a proof by Rényi (1970, chapter 9).

Hartley uncertainty has also been modified for measuring uncertainty associated with closed intervals of real numbers or, more generally, convex subsets of the n-dimensional Euclidean space $\mathbb{R}^n$ ($n \geq 1$); see e.g. pp. 45–57 in Klir 2006. This modified form is called a Hartley-like uncertainty and it is usually denoted HL. For each particular interval $[a, b]$, $HL([a, b]) = \log_2(1 + b - a)$. For convex subsets A of $\mathbb{R}^n$ ($n \geq 2$), the formula for $HL(A)$ is more complex.

It is well established that the Hartley uncertainty is also applicable to the theory of graded possibilities via the following formula based on the fuzzy-set interpretation of the theory

$$H(F) = \int_0^1 H(^\alpha F) \, d\alpha.$$

---

[31]Uncertainty of 1 bit may be understood as uncertainty regarding the truth of a bivalent proposition about whose truth status we do not have any evidence.

According to this formula, possibilistic uncertainty associated with a fuzzy set F is defined as the expected value of uncertainty of the level cuts of F for all $\alpha \in [0, 1]$. An analogous formula with HL replacing H applies to Hartley-like uncertainty.

Other interpretations of graded possibility theory

Although the introduction of fuzzy-set interpretation of possibility theory by Zadeh (1978b) brought the attention of the academic community to the uncertainty theory of graded possibilities, it was not the first time graded possibilities were considered. In the late 1940s, the British economist George L. S. Shackle (1903–1992) began to question the use of probability theory in economics and, at the same time, started thinking about graded possibilities as a viable alternative (Shackle 1949). These ideas were certainly quite radical in economics at that time. However, they seem to have been a natural product of Shackle's unorthodox views about economics, apparent from his early papers on uncertainty in economics (Shackle 1955, 228–29, 244–45):

> Economics is not a natural science, but a human science. Why then does the economist say that psychology is none of his business? Do we say that economics is concerned with men's preferences, choices, hopes, fears, beliefs, decisions, imagining, motives, behaviour, and yet say that the inner truth about the workings of their minds are none of our business? . . .
>
> Economics is a field of study enclosed within arbitrary boundaries. These boundaries are not part of the nature of things. No doubt after 175 years we can say that they correspond to a consensus of powerful minds. . . . Nevertheless, these boundaries are artificial. Is there any natural discontinuity between a man's economic motives and conduct on the one hand and, on the other, his conduct as politician, scientist, philosopher, poet, lover, or mystic? . . .
>
> Some of us may remember a verse by Edmund Clerihew Bentley, which explains the difference between geography and biography:
>
> > Geography is about maps
> > But biography is about chaps.
>
> I doubt whether a geographer would accept the suggestion here that geography is not a human science; it is very largely concerned with human activities. But far more so is economics. Economics is entirely concerned with men's doings and arrangements, their wants and their means of satisfying them, their hopes and fears, beliefs, ambitions, conflicts of interest, their valuation and decisions, their governments and their material well-being. Economics emphatically is about chaps.
>
> Many men have attempted to say in a sentence or a paragraph what economics is about; it is better, in my view, to begin by saying that economics is about human nature, human conduct and human institutions, and then to say which part of this huge field the economist is willing to leave to others.

The central tenet of Shackle's approach is that the essence of economics is the study of choice and decision making in human affairs. For him, choice is a decision of the chooser at present to commit to a particular imagined course of action in the future. The chooser may be an individual, a group of individuals, an organization, or even the whole nation.

Shackle recognized early in the course of developing his approach that decisions are made on the basis of expectations and these, in turn, are based on imagination constrained by what seems possible to some degree. He interpreted possibility as the absence of discernible obstacles or, in other words, as the chooser's ease in conceiving a particular sequel of some imagined course of action as feasible. However, he formalized degrees of possibility in terms of their inverses, which he called degrees of potential surprise. He argued that the concept of *potential surprise* has a more natural interpretation in economics than that of possibility.[32] He characterizes it well (Shackle 1961, 68):

> It is the degree of surprise to which we expose ourselves when we examine an imagined happening as to its possibility, in general or in prevailing circumstances, and assess the obstacles, tensions and difficulties which arise in our minds when we try to imagine it occurring, that provides the indicator of degree of possibility. This is the surprise we *should* feel, if the given thing *did* happen; it is *potential* surprise.

Shackle developed his initial ideas regarding uncertainty in economics in a series of articles published during the period 1937–49, and presented them as a coherent whole in his pioneering book Shackle 1949.[33] After publishing this book, he elaborated and further developed various aspects of his unorthodox approach to economics in three additional major books. In the first one (Shackle 1958), which is based on lectures he presented at the Institute of Social Studies in The Hague, he elaborates on the complex concept of time in economics in comparison with the usual concept of time in natural sciences. In the second book (Shackle 1961), which is perhaps the most comprehensive statement of his unorthodox views on economics, he further refines some of his many ideas. In the third book (Shackle 1979), which was also his last, Shackle manages to summarize his approach to economics in a brilliant way in just 152 pages.

After describing the context within which Shackle developed his interpretation of the theory of graded possibilities (albeit indirectly, via the concept of potential surprise), it remains to explain how his theory of graded potential surprise is formally connected with the theory of graded possibilities. For convenience and clarity, we describe this connection by using the notation employed in this book, not the one employed by Shackle.

Shackle defines *potential surprise* as a function of the form $s : U \rightarrow [0, s_m]$, where s_m denotes the largest conceivable potential surprise. In each application of this function, set U consists of all conceived sequels of an imagined course of action (e.g. all conceived monetary gains and losses). For any given sequel $u \in U$, the value $s(u)$ stands for the assessed degree of potential surprise that the chooser would experience if the sequel did actually happen. Since Shackle considered the choice of s_m solely a matter of measurement unit, we may assume that $s_m = 1$ and, hence, work with the following more specific form of function s:

$$s : U \rightarrow [0, 1].$$

---

[32] The concept of potential surprise was first introduced by the well-known logician John Venn in his book *The Logic of Chance* (London: Macmillan, 1866). It is possible, even likely, that Shackle's choice of the term was inspired by this book, to which he occasionally refers in his writings, but we have not been able to find any statement by Shackle that would confirm this.

[33] According to a remark by Shackle in his preface, he actually completed the book in August of 1946.

It is reasonable to call this function a *potential-surprise distribution function*. It follows directly from Shackle's conceptual framework that this function is connected for all $u \in U$ to possibility distribution function r via the equation

$$r(u) = 1 - s(u).$$

Now, for each nonempty set A in a given ample field $\mathscr{A}$ on U, let $\mathrm{Sur}(A)$ denote the potential surprise on A. Then, we have

$$\mathrm{Pos}(A) = \sup_{u \in A} r(u) = \sup_{u \in A}(1 - s(u)) = 1 - \inf_{u \in A} s(u) = 1 - \mathrm{Sur}(A).$$

Similarly,

$$\mathrm{Nec}(\overline{A}) = 1 - \mathrm{Pos}(A) = 1 - (1 - \inf_{u \in A} s(u)) = \inf_{u \in A} s(u) = \mathrm{Sur}(A).$$

Let Doc denote a measure that is defined for all $A \in \mathscr{A}$ via the duality equation

$$\mathrm{Doc}(A) = 1 - \mathrm{Sur}(\overline{A}).$$

Then, we can conclude from all the above facts that each of the four considered measures, Pos, Nec, Sur, and Doc, can be expressed for all sets in a given ample field $\mathscr{A}$ via the possibility distribution function r or, alternatively, via the potential-surprise distribution function s:

$$\mathrm{Pos}(A) = \sup_{u \in A} r(u) \qquad \text{or} \qquad \mathrm{Pos}(A) = 1 - \inf_{u \in A} s(u),$$
$$\mathrm{Nec}(A) = 1 - \sup_{u \in \overline{A}} r(u) \qquad \text{or} \qquad \mathrm{Nec}(A) = \inf_{u \in \overline{A}} s(u),$$
$$\mathrm{Sur}(A) = 1 - \sup_{u \in A} r(u) \qquad \text{or} \qquad \mathrm{Sur}(A) = \inf_{u \in A} s(u),$$
$$\mathrm{Doc}(A) = \sup_{u \in \overline{A}} r(u) \qquad \text{or} \qquad \mathrm{Doc}(A) = 1 - \inf_{u \in \overline{A}} s(u).$$

Shackle's potential surprise may thus be viewed as a particular interpretation of the theory of graded possibilities and, according to our best knowledge, it was the first one proposed and studied. It preceded Zadeh's interpretation by nearly 40 years.

Apparently, Zadeh was not aware of Shackle's work on the theory of graded possibilities when he introduced his fuzzy-set interpretation of possibility theory in 1978. Shackle only learned about fuzzy set theory and fuzzy-set interpretation of possibility theory in the early 1980s from a letter he received from Dubois and Prade.[34] They wrote to him after they discovered his work in the late 1970s and informed him about the close connection of his and Zadeh's interpretations of graded possibilities.[35]

The theory of graded possibilities also played an essential role in the inductive logic developed by L. Jonathan Cohen in his book (Cohen 1970).[36] In this book, he argues at length for the following principle, which he calls a *conjunction principle*. In any particular field of inquiry, the conjunction of two relevant propositions, P_1

---

[34] According to our personal communication with Didier Dubois and Henri Prade.

[35] The book by Shackle (1961) was already mentioned in the first encyclopedic book on fuzzy set theory and applications by Dubois and Prade (1980a).

[36] L. J. Cohen (1923–2006) was a highly original British philosopher, a fellow of Queen's College, Oxford University, with a strong interest in natural language.

and P_2, always has the same grade of inductive support on the evidence of a third proposition, E, as the one whose support is smaller. More formally,

$$\text{if } s(P_1, E) \geq s(P_2, E), \text{ then } s(P_1 \& P_2, E) = s(P_2, E),$$

where P_1 and P_2 are either universally quantified conditionals of the same category or substitution instances of them, $P_1 \& P_2$ denotes the conjunction of P_1 and P_2, and s is the support function appropriate for propositions of that category. We can see that this principle is compatible with the theory of graded necessities, as it basically describes the principal requirement of necessity theory expressed by equation (3.32).

Besides the conjunction principle, another key feature of Cohen's approach to induction is the use of precisification (Cohen's term) of inductive hypotheses expressed in natural language, which he developed in an earlier book (Cohen 1962). The following quotation captures in a nutshell the highly unorthodox role of precisification in his approach to inductive reasoning in science (Cohen 1962, 301–3):

> We must view scientific argument as being designed, ideally, neither to verify nor falsify hypotheses, but rather to precisify hypothesis-sentences. Hypotheses must be viewed neither as empirically verifiable in principle nor as empirically falsifiable, but as *a priori* truths that are empirically precisifiable, Hence the process of testing a hypothesis by experimental variation of circumstances is to be viewed as the process of discovering exactly what meanings we can assign a certain culture-sentence, viz. the sentence to the truth of which, in some meaning, we are committed so long as we maintain the hypothesis. The wider the meaning we can successfully assign this sentence, the better confirmed is the hypothesis. The narrower the meaning, the better we should do by turning our attention to the precisification of some quite different hypothesis. . . .
> Empirical precisifiability is the *sine qua non* of hypotheses in natural science.

Cohen supported his unorthodox approach to inductive logic not only by vigorous philosophical arguments (Cohen 1962, 1970, 1977, 1989), but he also argued for its superiority in scientific inquiries (Cohen 1973; Cohen 1977, chapter 13) and in judicial practice (Cohen 1977).

It is interesting that Cohen's idea of precisification of meaning in natural language is quite similar to Zadeh's idea of precisiation of meaning, which is described in the following short quotation (Zadeh 1984, 373):[37]

> The traditional approach to precisiation of meaning of utterances in a natural language is to translate them into an unambiguous synthetic language—which is usually a programming language, a query language, or a logical language such as predicate calculus. The main limitation of this approach is that the available synthetic languages are nowhere nearly as expressive as natural languages. Thus, if the target language [were] the first order predicate calculus, for example, then only a small fragment of natural language would be amenable to translation, since the expressive power of first order predicate calculus is extremely limited in relation to that of a natural language.
> To overcome this limitation, what is needed is a synthetic language whose expressive power is comparable to that of natural languages. A candidate for such language is PRUF (Zadeh

---

[37] Both Cohen and Zadeh apparently tried to find a single word for describing the translation of meanings of sentences in natural language into an unambiguous synthetic language. Both of them were interested in a synthetic language whose expressive power is comparable to that of natural languages and, in fact, they both came to the conclusion, via very different arguments, that the proper language is the one of graded possibilities. Since they did not find in current dictionaries any English word suitable for this purpose, they invented their own words. Cohen's choice was "precisification" while Zadeh chose "precisiation." It is likely that one of these words or possibly both of them will eventually be added to common dictionaries of the English language.

1978a)—which is a meaning representation language for natural languages based on the concept of possibility distribution (Zadeh 1978b).

In essence, a basic assumption underlying PRUF is that the imprecision which is intrinsic in natural languages is possibilistic rather than probabilistic in nature.

Another interpretation of graded possibility and necessity was introduced in the book by Shafer (1976). In this book, Shafer develops an uncertainty theory based on a pair of dual nonadditive measures—belief measure and plausibility measure (these are also called Dempster-Shafer measures). One way of defining these measures is to employ a set function $m : 2^U \to [0, 1]$ for which $m(\emptyset) = 0$ and $\sum_{A \in 2^U} m(A) = 1$.[38] For each particular function m with these properties, a belief measure, Bel, is defined by $\mathrm{Bel}(B) = \sum_{C \subseteq B} m(C)$ for all $B \in 2^U$ and a plausibility measure, Pl, is defined by $\mathrm{Pl}(B) = \sum_{C \cap B \neq \emptyset} m(C)$ for all $B \in 2^U$. Belief measures are totally monotone, super-additive, and continuous from above, and are connected with plausibility measures via the duality equation $\mathrm{Pl}(B) = 1 - \mathrm{Bel}(\overline{B})$. Plausibility measures are subadditive and continuous from below. For more information, see Wang and Klir 2009.

In chapter 10 of his book, Shafer introduces special classes of belief and plausibility measure, which he calls consonant. These special classes are characterized by the following restriction on the underlying function m: the family $\{A \in 2^U \mid m(A) \neq 0\}$ is nested. He shows that under this restriction belief measures defined on a finite universe U become necessity measures (his theorem 10.4). He also makes several remarks regarding the connection of these special belief and plausibility measures with Cohen's inductive logic and Shackle's conception of potential surprise.

Three distinct interpretations of graded possibility theory had thus been introduced and investigated in very different contexts prior to the fuzzy-set interpretation proposed by Zadeh (1978b). However, it was mostly the latter that attracted considerable attention to the theory of graded possibilities. Various other interpretations of graded possibilities are now recognized (Dubois and Prade 2000), such as a *similarity interpretation* (in which $r(u)$ reflects the degree of similarity between u and an ideal prototype u_i whose possibility degree is 1), a *feasibility interpretation* (in which $r(u)$ expresses the ease of achieving u), or the *modal logic interpretation* developed by Klir and Harmanec (1994). However, these interpretations have been explored by and large after the publication of Zadeh's fuzzy-set interpretation.

3.8 Fuzzy clustering

Fuzzy versus classical clustering: A historical overview

In general, *clustering* (or cluster analysis) refers to the task of classifying a given set of objects, each of which is characterized by a set of features, into *clusters* (a generic name for groups, classes, clumps, types, etc.). The aim of clustering is to obtain a set

---

[38] 2^U denotes the power set of a given universe U.

of clusters in such a way that objects in the same cluster are more similar to each other, according to some specified measure of similarity among their features, than to those in other classes.

Clustering also appears under other names, such as classification, partitioning, grouping, categorization, typology, numerical taxonomy, and others. In classical clustering, clusters are required to be classical sets of the objects that are classified. This requirement is abandoned in fuzzy clustering, where clusters are, in general, fuzzy sets.

It can be argued that classification has always been one of the most fundamental activities of human beings, but the first attempts to deal with it in a scientific way were only made in the 1930s. These early attempts were oriented primarily to the fields of biology, zoology, and anthropology. However, Robert Tryon formulated some basic ideas of classical clustering as early as 1939.[39] Later, some actual methods for classical clustering began to emerge in the early 1950s, but became widely recognized only some ten years later, when computer technology became sufficiently powerful to be able to accommodate computational demands of classification problems, which are typically quite extensive. Classical cluster analysis emerged in the early 1960s as a unifying methodological area for dealing with a broad class of problems of classification, discrimination, categorization, grouping, pattern recognition, and the like. It has rapidly advanced, side by side with advances in computer technology, ever since.

The perceived need for fuzzy pattern classification, fuzzy clustering, fuzzy partitioning, or whatever name we choose to use, was apparently one of the principal motivations for introducing the concept of a fuzzy set by Zadeh (1965a). This is best captured by Zadeh's own words in the following quotation from his foreword to an anthology of principal early contributions to fuzzy clustering, fuzzy pattern classification and other related areas, edited by James Bezdek and Sankar Pal (1992):

> It is of historical interest to note that the initial development of the theory of fuzzy sets was motivated in large measure by problems in pattern classification and cluster analysis.
>
> In the spring of 1964, I was invited by Richard Bellman to spend a part of the summer at Rand in Santa Monica to work on problems in pattern classification and systems analysis. At the time, we both felt that conventional approaches to the analysis of large-scale systems were ineffective in dealing with systems that are complex and mathematically ill-defined, as are most of the real-world systems in which human perception and intuitive judgment play important roles. The question was: What can be done to capture the concept of imprecision in a way that would differentiate imprecision from uncertainty?
>
> A very simple idea that occurred to us at that time was to generalize the concept of the characteristic function of a set to allow for intermediate grades of membership. This, in effect, was the genesis of the concept of a fuzzy set. Although, as it was noted in my 1965 paper on fuzzy sets, the concept of a grade of membership bears a close relation to the truth value of a predicate, the agenda of fuzzy set theory is quite different from that of multivalued logic. In particular, the concept of a fuzzy set fits very naturally into the framework of pattern recognition. Indeed, the papers collected in this volume provide a convincing demonstration of the effectiveness of fuzzy-set theoretic techniques in both formulation and solution of problems in these fields.
>
> The starting point in the application of fuzzy set theory to pattern classification was the

---

[39] R. C. Tryon, *Cluster Analysis* (Ann Arbor, MI: Edwards Bros., 1939).

1966 paper by Bellman, Kalaba, and myself. In this paper, the problem of pattern classification was formulated as the problem of interpolation of the membership function of a fuzzy set, and thereby a link with the basic problem of system identification was established.

The keen interest of Zadeh in fuzzy clustering and related areas is demonstrated not only by his insightful remarks on developments in these areas on various occasions (at conferences, forewords to relevant books, interviews, and the like), but also by his own early paper devoted fully to fuzzy cluster analysis and pattern classification (Zadeh 1977).

The anthology edited by Bezdek and Pal (1992) traces in some sense the history of fuzzy clustering and related areas. Chapter 2, in particular, is devoted to general aspects of fuzzy cluster analysis. It contains 20 papers, which played important roles in developing foundations of fuzzy cluster analysis, and an editors' introduction that describes how principal ideas in this area evolved via these papers. Among others, three early contributors whose work was instrumental in initiating research on fuzzy clustering were: Ruspini (1969), who introduced the concept of a fuzzy partition that became essential in fuzzy cluster analysis; Dunn (1973), who generalized a widely accepted method of classical cluster analysis, known as the k-means method and its extension, ISODATA (Iterative Self-Organizing Data Analysis) method, to a fuzzy ISODATA method; and Bezdek (1973), who further developed the fuzzy ISODATA method in a form which is now usually referred to as the *fuzzy c-means clustering method*.[40] Bezdek also resolved some important open theoretical issues pertaining to the soundness of the method, such as the issue of its convergence, in Bezdek 1980 and others. His extensive work on fuzzy clustering over the period of 1973–81 culminated in his book Bezdek 1981, which was the first book devoted to the role of fuzzy cluster analysis in pattern recognition.

After the publication of Bezdek's book, literature on fuzzy cluster analysis and related areas virtually exploded, including the publication of more books devoted to this area. Most of these publications focus on further development of algorithms or on exploration of applications of fuzzy cluster analysis. Due to the enormous scope of these rather narrowly oriented developments, we do not intend to trace them here. However, a fairly comprehensive book on fuzzy cluster analysis by Höppner et al. (1999) is frequently cited in the literature, and a book of fuzzy clustering by Pedrycz (2005) builds upon his previous work on fuzzy clustering and develops further the important connection between fuzzy clustering and granular computing.

Fuzzy c-means clustering algorithm

In order to illustrate the computational aspects of fuzzy cluster analysis, we describe the standard fuzzy c-means clustering algorithm developed by Bezdek (1981). Let $X = \{\mathbf{x}_k \mid k \in \mathbb{N}_n\}$ denote a given set of data to be analyzed, where each $\mathbf{x}_k$ is assumed

---

[40]The classical ISODATA method was developed by G. H. Ball and D. J. Hall, "A clustering technique for summarizing multivariate data," *Behavioral Science* 12, no. 2 (1967): 153–55.

to be, in general, a vector: $\mathbf{x}_k = \langle x_{k1}, x_{k2}, \ldots, x_{kr} \rangle \in \mathbb{R}^r$. The aim of fuzzy c-means clustering is to represent the data by a family of c fuzzy subsets of X, $\mathscr{F} = \{ F_i \mid i \in \mathbb{N}_c \}$, that form a fuzzy partition of X in the sense of Ruspini's definition (p. 68), i.e. $\sum_{i \in \mathbb{N}_c} F_i(\mathbf{x}_k) = 1$ for all $k \in \mathbb{N}_n$. Moreover, $0 < \sum_{k \in \mathbb{N}_n} F_i(\mathbf{x}_k) < n$ for all $i \in \mathbb{N}_c$ is required. The resulting fuzzy partition $\mathscr{F}$ is expected to be optimal according to some specified performance index. The algorithm assumes that c is a fixed (desired) number of clusters chosen from the set $\{2, 3, \ldots, n-1\}$.

The performance index of a c-fuzzy partition $\mathscr{F} = \{ F_i \mid i \in \mathbb{N}_c \}$ is based on the set of c cluster centers, $\mathbf{v}_i$, which are calculated by the formula

$$\mathbf{v}_i = \frac{\sum_{k \in \mathbb{N}_n} (F_i(\mathbf{x}_k))^m \mathbf{x}_k}{\sum_{k \in \mathbb{N}_n} (F_i(\mathbf{x}_k))^m}$$

for all $i \in \mathbb{N}_c$, where $m > 1$ is a real number that governs the influence of individual fuzzy sets in fuzzy partition $\mathscr{F}$. When $m \to 1$ the fuzzy c-means algorithm converges to the classical c-means algorithm; when $m \to \infty$, all cluster centers tend to converge to the centroid of the data set X. That is, the fuzzy partition becomes fuzzier with increasing value of m.

The performance index, $J_m(\mathscr{F})$, for each fuzzy partition $\mathscr{F}$ is then defined in terms of these cluster centers by the formula

$$J_m(\mathscr{F}) = \sum_{k \in \mathbb{N}_n} \sum_{i \in \mathbb{N}_c} (F_i(\mathbf{x}_k))^m \| \mathbf{x}_k - \mathbf{v}_i \|^2,$$

where $\| \mathbf{x} \|$ denotes the norm (or length) of any vector $\mathbf{x} \in \mathbb{R}^r$, so $\| \mathbf{x}_k - \mathbf{v}_i \|^2$ is the quadratic Euclidean distance between vectors $\mathbf{x}_k$ and $\mathbf{v}_i$ in $\mathbb{R}^r$. It was shown by Bezdek (1981, theorem 11.1, 66–69) that the goal of fuzzy clustering can be expressed as the problem of minimizing $J_m(\mathscr{F})$, and he developed the fuzzy c-means algorithm as an algorithm for solving this optimization problem.

The algorithm is based on the assumption that the following is given: a data set X, a desired number of clusters c, a real number $m > 1$, and a small positive number ε that serves as a stopping criterion. The algorithm is defined via the following steps:

Step 1. Let $t = 0$. Select randomly an initial fuzzy partition $\mathscr{F}^{(0)}$.

Step 2. Calculate c cluster centers $\mathbf{v}_i$, $i \in \mathbb{N}_c$ for $\mathscr{F}^{(t)}$ and for the given m.

Step 3. Update $\mathscr{F}^{(t)}$ to $\mathscr{F}^{(t+1)}$ as follows. For each $\mathbf{x}_k \in X$, if $\| \mathbf{x}_k - \mathbf{v}_i^{(t)} \| > 0$ for all $i \in \mathbb{N}_c$, then define

$$F_i^{(t+1)}(\mathbf{x}_k) = \left[\sum_{j=1}^{c} \left(\frac{\| \mathbf{x}_k - \mathbf{v}_i^{(t)} \|}{\| \mathbf{x}_k - \mathbf{v}_j^{(t)} \|} \right)^{\frac{2}{m-1}} \right]^{-1};$$

if $\| \mathbf{x}_k - \mathbf{v}_i^{(t)} \| = 0$ for some $i \in I \subseteq \mathbb{N}_c$, then define $F_i^{(t+1)}(\mathbf{x}_k)$ for all $i \in I$ by any nonnegative real numbers such that $\sum_{i \in I} F_i^{(t+1)} = 1$, and define $F_i^{(t+1)}(\mathbf{x}_k) = 0$ for all $i \in \mathbb{N}_c - I$.

Step 4. Compare $\mathscr{F}^{(t+1)}$ and $\mathscr{F}^{(t)}$. If $\max_{i \in \mathbb{N}_c, k \in \mathbb{N}_k} |F_i^{(t+1)}(\mathbf{x}_k) - F_i^{(t)}(\mathbf{x}_k)| \leq \varepsilon$, then stop; otherwise, increase t by one and return to step 2.

Additional remarks on fuzzy clustering

Notwithstanding the importance of fuzzy cluster analysis for a great variety of applications (see chapter 6), including applications to fuzzy set theory itself—as one way of constructing fuzzy sets and fuzzy rules (section 3.9)—we do not consider it suitable to survey in this book the extensive literature in this area, which is predominantly oriented to the development of ever-improving algorithms for analyzing various types of data. Thus, for example, we do not cover here the problem of cluster validity, primarily for reasons outlined in the following short quotation from the book by Höppner et al. (1999, 185):

> The cluster validity problem is the general question whether the underlying assumptions (cluster shapes, number of clusters, etc.) of a clustering algorithm are satisfied at all for the considered data set. In order to solve this problem, several cluster quality (validity) measures have been proposed. It is impossible to answer all questions without any knowledge about the data. It has to be found out for each problem individually which cluster shape has to be searched for, what distinguishes good clusters from bad ones, and whether there are data without any structure.

Nevertheless, we feel that we should make a few general remarks about fuzzy cluster analysis. We begin by examining two assumptions upon which the fuzzy c-means clustering algorithm described above is based.

The first assumption, which is stated explicitly in the algorithm, is that the number of desired clusters is fixed. This is a strong assumption. If the chosen number is not based on any knowledge about the data, it may lead to very bad clusters. This difficulty can be partially alleviated by replacing the requirement that the number of clusters be fixed with a weaker requirement that only the maximum acceptable number of clusters be fixed, say $c_{\max}$. The global optimum is then obtained via some search through local analyses for $c \in \{2, \ldots, c_{\max}\}$. The search may be constrained by some additional requirements, possibly fuzzy, such as "c should be close to some number c_i" (ideal number of clusters), and the like, and c as well as the additional requirements may be included in the definition of a global performance index. Various algorithms of this kind have been developed, but their use adds a relatively small improvement when no knowledge regarding the natural structure in the analyzed data is known.

The second assumption, which is not explicitly stated in the described fuzzy c-means algorithm, is that the algorithm is based on the Euclidean distance in $\mathbb{R}^r$. This assumption is even stronger than the first one. Again, with no knowledge about the structure of the analyzed data, any algorithm based on this rigid conception of a distance does not have enough flexibility to unravel the structure. This limitation can be somewhat alleviated by employing some well-known classes of distance functions defined on $\mathbb{R}^r$, such as the Minkowski class and especially the Mahalanobis class.

The Minkowski class of distance functions, d_p, based on parameter $p \geq 1$, is defined for each pair of data points $\mathbf{x}, \mathbf{y} \in \mathbb{R}^r$ by the formula

$$d(\mathbf{x}, \mathbf{y}) = \left(\sum_{i \in 1}^{r} |x_i - y_i|^p \right)^{1/p}.$$

For $p = 1$, $d(\mathbf{x}, \mathbf{y})$ is clearly the well-known city-block distance (sometimes called the Hamming distance); for $p = 2$, it is the Euclidean distance; and for $p \to \infty$, it converges to

$$d_\infty(\mathbf{x}, \mathbf{y}) = \max_{i \in \mathbb{N}_r} |x_i - y_i|.$$

The Minkowski class adds some flexibility to fuzzy cluster analysis, but not enough.

The Mahalanobis class of distance functions[41] is much more flexible. It is defined for each pair of data points $\mathbf{x}, \mathbf{y} \in \mathbb{R}^r$ by the formula

$$d_A(\mathbf{x}, \mathbf{y}) = \left([x_j - y_j \mid j \in \mathbb{N}_r]^\mathrm{T} \mathbf{A}[x_j - y_j \mid j \in \mathbb{N}_r] \right)^{\frac{1}{2}},$$

where $\mathbf{A}$ denotes a symmetric $r \times r$ matrix that is positive definite, by which distances within this class are distinguished, and the superscript T indicates that the vector is transposed.

The capabilities of the various distance functions can be roughly summarized as follows. The Euclidean distance makes it possible to recognize only such natural clusters in data whose shapes are close to any hyperspherical shapes; the Minkowski class of distances allows one to recognize clusters whose shapes are anywhere between hyperspherical and hypercubic shapes; and the Mahalanobis class has the capability of recognizing clusters that are close to arbitrary hyperellipsoidal shapes (Mao and Jain 1996). In spite of these added capabilities, cluster analysis (fuzzy as well as classical) still depends heavily on understanding the nature of data by the investigator in each particular application context. We close this section by a relevant quotation from the book by Xu and Wunsch (2009, 30), in which the term "proximity measure" is used instead of our term "distance function":

> Different proximity measures will affect the formation of the resulting clusters, so the selection of an appropriate proximity function is important. How can effective measures be selected? How can data objects with features that have quite different physical meanings be compared? Furthermore, what kinds of standards can be used to weight the features or to select the features that are important to the clustering? Unfortunately, we cannot answer these questions in a conclusive and absolute way. Cluster analysis is a subjective and problem-dependent process. As an indispensable part of this exploration, a proximity measure, together with the features selected, ultimately has to be put in the framework of the entire data set and determined based on the understanding and judgment of the investigators.

---

[41] Introduced by Prasanta C. Mahalanobis (1893–1972), an Indian statistician, in "On the generalised distance in statistics," *Proc of the National Institute of Sciences of India* 2, no. 1 (1936): 49–55.

3.9 Methods for constructing fuzzy sets

General discussion

Fuzzy set theory, viewed either in the intuitive sense introduced by Zadeh (1965a) or in the axiomatic sense that we discuss in section 5.2.3, is a mathematical theory. As such, it deals with abstract phenomena that are not connected with any particular experimental domain, but are potentially applicable to any of them. However, the emergence of a new mathematical theory is often driven by some perceived needs for it. This is how the idea of fuzzy set theory emerged, as is apparent from the introduction to Zadeh's seminal paper (see the quotation on p. 20). It is clear that Zadeh introduced the concept of a fuzzy set with a clear interpretation: For each object u in some universe of discourse U and a particular fuzzy set A defined on U, $A(u)$ is interpreted as the grade (or degree) of membership of object u in a class described by a statement in natural language. Fuzzy sets are thus connected with meanings of statements expressed in natural language. In other words, the aim of fuzzy sets is to represent meanings of statements in natural language. To achieve this aim for any given natural-language statement, L, we need somehow to construct the corresponding fuzzy set, A_L. That is, we need to determine $A_L(u)$ for all $u \in U$.

Although the connection of fuzzy sets with natural-language statements clearly delineates the application domain of fuzzy set theory, this connection depends on various factors, among which the application context within which these statements are made is predominant. The meaning of *high temperature*, for example, is very different when applied in the broad context of weather forecasting, or in the broad context of nuclear reactors. Moreover, neither of these broad contexts is sufficiently specific and both subsume a multitude of contexts and, consequently, a multitude of meanings of the term *high temperature*. In weather forecasting, for example, these different contexts are distinguished from one another by different geographical locations, different times of the year, and possibly by other distinguishing factors. Similarly, meanings of terms such as *young* or *old* depend on the objects to which they are applied. Thus, for example, they significantly vary when they are applied to different animal species, and they are drastically different when applied to geological formations. These examples, and there are seemingly an endless number of examples like these, illustrate that the construction of fuzzy sets for representing a given statement expressed in natural language is meaningful only in some sufficiently specific application context.

The problem of constructing fuzzy sets is not a problem of fuzzy set theory per se, but it is a problem that is crucial for applications of fuzzy sets. The need for developing methods for constructing fuzzy sets in the context of various application contexts has been recognized since the early 1970s. This recognized need incited a growing number of researchers in the community to embark on research dealing with this problem. The outcome of this research for over 40 years has been a great variety of methods of many diverse types for constructing fuzzy sets. A fairly extensive survey of the whole range of these methods was prepared by Bilgiç and Türkşen (2000). In this section, we complement this survey with a historical overview of the development of the main ideas bearing upon the problem of fuzzy-set construction. We also return to this problem in chapter 6, where we describe in greater detail some construction methods that were developed for specific applications.

Historical overview

Zadeh did not raise the issue of constructing fuzzy sets in either of his first two papers on fuzzy sets (Zadeh 1965a,b). However, the issue was raised in Bellman, Kalaba, and Zadeh 1966, in Zadeh 1966, and then again in Zadeh 1969. In these three early papers, a simple approach was suggested to address this issue. The approach is based on the following two assumptions: (i) the fuzzy set to be constructed (say fuzzy set A on universe U) is exemplified by a finite set of samples, say set $\{A(u_1), A(u_2), \ldots, A(u_n)\}$; and (ii) a family of functions from U to $[0, 1]$ is postulated whose members are considered (on intuitive or some other grounds) reasonable candidates for A. Under these assumptions, the construction of fuzzy set A is converted to an identification problem: identify a function in the postulated family of functions that fits best (in some specified sense) the given samples. The suggested approach to constructing fuzzy sets in these early papers was not sufficiently specific as it did not address the issue of how to satisfy the two rather strong assumptions. Nevertheless, it offered a meaningful direction for further research.

After this suggestion, Zadeh never returned again to the specific question of how to construct fuzzy sets in the context of various applications. Instead, he embarked in the early 1970s on related, but more fundamental research whose aim was to develop a quantitative semantics of natural languages based on fuzzy set theory. This became one of his primary research interests. In his first paper devoted to this research, Zadeh (1971a) discusses the issues involved and introduces a formal framework for dealing with them. In his own words (pp. 159–60):

> Few concepts are as basic to human thinking and yet as elusive of precise definition as the concept of "meaning." Innumerable papers and books in the fields of philosophy, psychology, and linguistics have dealt at length with the question of what is the meaning of "meaning" without any definitive answers. In recent years, however, a number of fairly successful attempts at the formalization of semantics—the study of meaning—have been made by theoretical linguists on the one side, and workers in the field of programming languages and compilers on the other.[42] These attempts reflect, above all, the acute need for a better understanding of the semantics of both natural and artificial languages—a need brought about by the rapidly growing availability of large-scale computers for automated information processing.
>
> One of the basic aspects of the notion of "meaning," which has received considerable attention in the literature of linguistics, but does not appear to have been dealt with from a quantitative point of view, is that of fuzziness of meaning. . . . In fact, it may be argued that in the case of natural languages, most of the words occurring in a sentence are names of fuzzy rather than nonfuzzy sets, with the sentence as a whole constituting a composite name for a fuzzy subset of the universe of discourse.

---

[42] He supports this claim with 32 references.

> Can the fuzziness of meaning be treated quantitatively, at least in principle? The purpose of the present paper is to suggest a possible approach to this problem based on the theory of fuzzy sets. . . .
>
> Our initial goal is to formalize the notion of "meaning" by equating it with a fuzzy subset of a "universe of discourse."

In this paper, Zadeh basically formulates a framework for further research regarding the computation of the meaning of composite linguistic terms from knowing the meanings of the simple terms involved. He assumes that each fuzzy set representing a simple term can be defined or at least partially defined (exemplified) on a finite set of sample points. Although he does not address the issue of how to actually do that, this unresolved issue does not affect his development of quantitative semantics of natural languages, as is exemplified by his large paper Zadeh 1976a. This research led eventually to a powerful meaning representation language for natural languages based on possibility theory, PRUF (Zadeh 1978a), and its applications (Zadeh 1982a,b, 1984).[43]

The question of how to construct fuzzy sets remained by and large open until the early 1980s, when some researchers began to investigate it in the context of various emerging applications of fuzzy set theory. One exception was a paper by Saaty (1974), in which he made a preliminary suggestion of a method for constructing fuzzy sets based on some early ideas of his analytic hierarchy process (Saaty 1980), on which he was still working at that time. He returned to the construction problem 12 years later in Saaty 1986, where he described a fully developed method with its theoretical justification. In this method, membership grades are derived indirectly from assessments of their pairwise comparisons by an expert. The gist of the method is its ability to deal with assessments that are not fully consistent.

Another exception in the 1970s was the following suggestion made in a paper on pattern recognition by Bremermann (1976, 134):

> Since 1965 a rich literature on fuzzy sets has developed (compare [Zadeh 1973]). There are, however, few methods (outside probabilistic interpretations of fuzzy sets) to actually compute fuzzy set membership functions. The method of deformable prototypes is such a method.[44] If we consider the similarity function Sim of a prototype, then Sim(prototype) = 0, since the prototype is a perfect match for itself and there is no distortion. When the match is good and distortion is low, then Sim(object) has small values and outside the prototype class it has large values. Thus Sim itself does not qualify as a membership function μ, however
>
> $$\mu(\text{object}) = 1 - \frac{\text{Sim(object)}}{\max \text{Sim}}$$
>
> will do. Here max Sim is the least upper bound of Sim in the universe U. Thus $\mu(\text{object}) = 1$ for the prototype itself and it will be small for objects that are very different from the prototype. Of course, our definition of μ is not the only possible one. Any function that maps the interval $[0, 1]$ onto itself, leaving 0 and 1 fixed when composed with μ will be a possible membership function for defining a fuzzy set associated with the prototype.

This method has been found useful in certain specific applications of fuzzy set the-

---

[43] PRUF is an acronym for "Possibilistic Relational Universal Fuzzy."

[44] The method is described on pp. 129–33 in Bremermann 1976.

ory, such as handwritten character recognition, face recognition, or image processing, for which the method of deformable prototypes has been developed in detail.

Numerous methods for constructing fuzzy sets representing the meaning of specified linguistic terms of a natural language (for example terms involved in linguistic variables) were suggested in the 1980s. They were by and large two-stage methods, which were of course always applied in a specific context of a particular application. In the first stage, the constructed fuzzy set defined on some specific universe was exemplified for a finite number of elements of this universe by polling a group of selected subjects (experts in the given application area or just users of the given natural language). Various methods were distinguished from one another by the type of questions the subjects were asked and how their answers were processed. One such method was proposed already by Black (1937). In the second stage, the exemplified values were employed for constructing the fuzzy sets of concern for all elements of the given universe. This of course can be done in many different ways, which further multiplies the number of possible methods. In most cases, the construction was formulated as an optimization problem within an intuitively reasonable parameterized class of functions: identify a function from the chosen class (via the parameters) that fits best (in some specified sense) the exemplified values. Observe that all of these methods were actually based on the approach suggested by Zadeh in the mid-1960s and mentioned earlier in this section. However, Zadeh only assumed that the fuzzy set to be constructed is exemplified by a finite set of samples, while each of the methods developed in the 1980s employs an operational procedure to obtain these samples. Greater details of these methods are described in an overview paper by Türkşen (1991), who himself was one of the early contributors to this area.

The 1990s was a decade of remarkable advancement of fuzzy-set applications, in which improved methods for constructing fuzzy sets played an important role. Most of these applications involved control of complex industrial or other processes, where knowledge-based fuzzy systems performed surprisingly well.[45] The critical issue in developing these systems was the elicitation of relevant knowledge from domain experts in these various applications and representing it in terms of fuzzy inference rules. Knowledge elicitation is the subject of a relatively new engineering area, known as *knowledge engineering*, which emerged in the mid-1980s. This was very good timing for the rapidly expanding applications of fuzzy logic in the 1990s, when some of the early knowledge-engineering techniques turned out to be quite useful. During the 1990s, these techniques were combined efficiently with learning capabilities of artificial neural networks, as described in Wang 1994 or in greater detail in Klir and Yuan 1995 (pp. 295–300, 467–75). In this combination, knowledge engineering techniques are employed for the exemplification stage, often in combination with appropriate fuzzy clustering methods, while the role of neural networks

---

[45] We discuss some of these systems in chapter 6.

is to construct the full membership function by learning from given examples (usually via the well-established backpropagation algorithm). These tools have recently been further enhanced by incorporating the capabilities of genetic and evolutionary computation, as is thoroughly explained in the book by Cordón et al. (2001).

The outlined methods for constructing fuzzy sets have worked surprisingly well in applications that involve concepts represented by linguistic variables whose base variables are traditional numerical variables. Thus far, most applications of fuzzy logic have dealt with concepts like these. However, many, perhaps most, concepts in social sciences are not of this kind. These are concepts that often elude, for various reasons, the traditional nominal, ordinal, interval, log-interval, and ratio scales of measurement. This requires that new, nontraditional theories of measurement be developed, drawing on expertise in areas such as psycholinguistics and psychometrics. Some initial attempts addressing these matters are discussed in papers by Verkuilen (2005) and Verkuilen, Kievit, and Zand Scholten (2011), as well as in books by Smithson and Verkuilen (2006) and Ragin (2000, 2008).

Sensitivity of constructed fuzzy sets

The following question, whose answer has important implications for the construction of fuzzy sets, has often been raised on various occasions and stimulated relevant discussion: To what extent do variations in input fuzzy sets affect output fuzzy sets in various types of fuzzy systems? This question was usually motivated by the perceived discrepancy between, on one hand, the linguistic imprecision that each fuzzy set is supposed to represent and, on the other hand, the required precision of its definition.[46] A typical response of practitioners experienced in applications of fuzzy sets was that relatively small variations in membership grades of input fuzzy sets do not actually have a substantial impact on the output fuzzy sets or, in other words, that the performance of fuzzy systems employed in practical applications is little affected by such variations. The practitioners employed this insight, gained by experience and expressed only qualitatively, as a "rule of thumb." In an important paper[47] entitled "Do exact shapes of fuzzy sets matter?", Belohlavek (2007) confirmed this by a rigorous mathematical analysis performed for a broad class of mathematical structures in which an output fuzzy set is determined by several input fuzzy sets via a logical formula. He specifically analyzed the extension principle for fuzzy sets, various compositions of fuzzy relations, including the compositional rule of inference, fuzzy automata, and various properties of fuzzy relations defined on a single set. The paper basically answered qualitatively the question posed in its title in the negative: Exact shapes of fuzzy sets by and large do not matter in most applications. In addition, the question is also answered quantitatively for each of the mathematical

---

[46] In some nonstandard fuzzy sets that we discuss in section 3.10, membership degrees are not required to be precise, but the description of their impressions is required to be precise. This means that the issue of sensitivity of constructed fuzzy sets cannot be completely avoided; it can only be moved to a higher level.

[47] Recipient of the 2007 Best Paper Award of the *International Journal of General Systems*.

structures analyzed by deriving a formula showing to what extent the exact shapes of fuzzy sets do not matter.

Constructing fuzzy inference rules

In addition to the problem of constructing fuzzy sets that adequately capture the meanings of linguistic terms in various applications, we should also mention the related problem of constructing operations on fuzzy sets or, in other words, the problem of determining operations on fuzzy sets that adequately capture the meanings of their linguistic descriptions. The issue of constructing operations on fuzzy sets was largely neglected prior to the 1990s.

A breakthrough in this neglected domain was made by Takagi and Hayashi (1991). In this paper, they described a sophisticated approach to constructing not only fuzzy sets and operations on fuzzy sets, but the whole sets of fuzzy inference rules needed for specific fuzzy-control applications or other applications of fuzzy rule-based systems.[48] The approach is based on learning the rules by neural networks from data obtained by observing the process to be controlled. It also involves fuzzy clustering of data to a specified number of clusters, each associated with a particular inference rule. That is, the user specifies the number of clusters according to the number of inference rules desired for the designed fuzzy controller.

Takagi and Hayashi's approach has provided a broad framework for dealing with the construction problem in fuzzy-rule-based systems within which further algorithmic and other developments have been made in various directions, resulting in a fairly extensive literature of practical construction methods. Two books by Rutkowski (2004, 2008) offer comprehensive treatments of automatic construction of inference rules, in which the great flexibility of fuzzy logic is fully utilized.

3.10 Nonstandard fuzzy sets

In order to keep the size of this book within reasonable limits, we decided not to cover various nonstandard fuzzy sets in detail. However, they substantially enlarge the inventory of methodological tools of fuzzy logic, so we at least introduce in this section their definitions and their historical backgrounds. In addition, we also examine in chapter 6 some of their established applications.

Interval-valued fuzzy sets

Interval-valued fuzzy sets are functions of the form $A : U \to I$, where I denotes the set of all closed intervals of real numbers contained in $[0, 1]$. That is, $A(u) = [\underline{a}, \overline{a}]$

---

[48] Although this paper is the first full description of their approach, it seems that the approach was already fairly well developed in 1988 when they presented it at two conferences.

for each $u \in U$. An alternative way is to view interval-valued fuzzy sets as pairs, $A = \langle \underline{A}, \overline{A} \rangle$, of standard fuzzy sets $\underline{A}$ and $\overline{A}$ such that $\underline{A}(u) \leq \overline{A}(u)$ for all $u \in U$.

Interval-valued fuzzy sets were introduced by Zadeh (1975c) and also by the British historian of mathematics Grattan-Guinness (1976). Since they are relatively easy to handle, especially when defined as pairs of standard fuzzy sets, papers exploring their utility in various applications have appeared in the literature fairly regularly since about the mid-1980s.

Fuzzy sets of type 2 and higher types

Fuzzy sets of type 2 are functions of the form $A : U \rightarrow \tilde{I}$, where $\tilde{I}$ denotes the set of all fuzzy intervals defined on $[0, 1]$. That means that $A(u)$ is a fuzzy interval on $[0, 1]$ for each $u \in U$. These sets were introduced by Zadeh (1975c) and were quite thoroughly investigated in early papers by Mizumoto and Tanaka (1976b, 1981) and Nieminen (1977). Clearly, it is computationally much harder to deal with type-2 fuzzy sets than with interval-valued fuzzy sets. The latter are in fact just special, more tractable cases of the former.

In a paper surveying research on type-2 fuzzy sets in the 1980s and 1990s, John (1998) appraises their theoretical significance and practical utility. Karnik, Mendel, and Liang (1999) described a complete design of a fuzzy logic system based on type-2 fuzzy sets. This paper initiated more thorough and practical research on type-2 fuzzy sets. It resulted eventually in a book by Mendel (2001) describing in detail logic systems based on type-2 fuzzy sets and providing the reader with relevant computer software (in MATLAB). Interest in fuzzy sets of type 2 continues to be strong, as is exemplified by a recent book edited by Sadeghian, Mendel, and Tahayori (2013).

It is quite natural to introduce fuzzy sets of higher types than 2. This generalization was first suggested by Zadeh (1975c, definition 3.1 in part I) and further discussed by Hisdal (1981). Let A^k denote fuzzy sets of type $k \geq 1$. It is convenient to interpret A^1 as standard fuzzy sets, so that $A^1 : U \rightarrow [0, 1]$. Then, A^k is defined for all $k \geq 1$ as a function of the form $A^k : U \rightarrow \tilde{I}^{k-1}$, where $\tilde{I}^{k-1}$ denotes the set of all fuzzy intervals of type $k - 1$ defined on $[0, 1]$, and $\tilde{I}^0$ is interpreted as the set $\{[a, a] \mid a \in [0, 1]\}$ of degenerate fuzzy intervals $[0, 1]$ or, equivalently, the interval $[0, 1]$.

Fuzzy sets of order k are certainly theoretically and philosophically significant. However, their practical utility is severely limited for $k \geq 3$ primarily due to prohibitive computational complexity for dealing with them.

Fuzzy sets of level 2 and higher levels

Fuzzy sets of level 2 are functions of the form $A^2 : \tilde{\mathbf{A}}_U \rightarrow [0, 1]$, where $\tilde{\mathbf{A}}_U$ denotes a set of standard fuzzy sets defined on U. This definition can be easily generalized to higher levels. Let A^l denote a set of fuzzy sets of level $l \geq 1$ with the understanding that A^1 denotes a standard fuzzy set on U. Then A^l are for all $l \geq 1$ functions of the form $A^l : \mathbf{A}_U^{l-1} \rightarrow [0, 1]$, where $\mathbf{A}_U^0$ is interpreted as U.

Fuzzy sets of level 2 as well as higher levels were introduced by Zadeh (1971a, definition 9) as a part of his formulation of a comprehensive framework for the development of meaning representation language for natural languages. However, fuzzy sets of higher levels appeared in the late 1960s in the work on cumulative hierarchies in many-valued set theories by Klaua, which was continued in the 1970s by Gottwald. This work is examined in detail in section 5.2.3. Clearly, fuzzy sets of higher levels are crucial for representing meanings of complex concepts.

Rough fuzzy sets

The notion of a fuzzy set may be augmented by ideas from *rough sets* (Pawlak 1991) in several ways. One leads to what is sometimes called rough fuzzy sets. The idea is to equip the universe set U with an equivalence relation R representing a given indiscernibility. For a given fuzzy set $A : U \rightarrow [0,1]$, its level cuts ^{a}A, $a \in [0,1]$, may be approximated as usual in rough set theory, namely by two sets,

$$\underline{^{a}A_{R}} = \bigcup \{[u]_{R} \mid [u]_{R} \subseteq {}^{a}A, u \in U\} \text{ and}$$

$$\overline{^{a}A_{R}} = \bigcup \{[u]_{R} \mid [u]_{R} \cap {}^{a}A \neq \emptyset, u \in U\},$$

called, respectively, the lower and upper approximation of ^{a}A. That is, $\underline{^{a}A_{R}}$ is the union of all classes of R included in ^{a}A and $\overline{^{a}A_{R}}$ is the union of all classes of R that have a nonempty intersection with ^{a}A. The sets $\underline{^{a}A_{R}}$ and $\overline{^{a}A_{R}}$ thus represent, respectively, the largest subset and the smallest superset of ^{a}A that may be observed given the indiscernibility constraint represented by R. Instead of A, one may therefore only consider its approximation which is represented by the family of the approximations $\langle \underline{^{a}A_{R}}, \overline{^{a}A_{R}} \rangle$ of its level cuts.

Various other combinations of fuzzy sets and rough sets habe been explored, including the one in which rough sets are generalized by employing a fuzzy equivalence relation in the definition of a rough set. Dubois and Prade 1990, 1992 represent influential early contributions to this topic. The literature in this area has lately been growing fairly rapidly.

Intuitionistic fuzzy sets

Intuitionistic fuzzy sets A defined on a universe U are pairs $A = \langle A_{m}, A_{n} \rangle$ of functions of the forms $A_{m} : U \rightarrow [0,1]$ and $A_{n} : U \rightarrow [0,1]$, such that $A_{m}(u) + A_{n}(u) \in [0,1]$ for each $u \in U$. The values $A_{m}(u)$ and $A_{n}(u)$ are interpreted, respectively, as the degree of membership of u in A and the degree of nonmembership of u in A. They are supposed to be determined independently of one another, but their sum is required to be a value in $[0,1]$.

The conception of intuitionistic fuzzy sets (IFS) was conceived by Krassimir Atanassov, a Bulgarian mathematician, in 1983. He announced it first at a conference in Poland (Atanassov and Stoeva 1983) and published his initial ideas regarding this class of nonstandard fuzzy sets more thoroughly in Atanassov 1986. In spite of some initial skepticism within the fuzzy-set community about these ideas, it was soon recognized that IFS have a significant potential for applications in many areas. In the 1990s, many researchers were attracted to IFS, which resulted in a rapidly growing literature on this subject. In 1995, Atanassov founded a specialized journal entitled *Notes on Intuitionistic Fuzzy Sets*, fully devoted to IFS.[49] In 1997, the Annual International Conference on IFS was inaugurated in Bulgaria and, moreover, the Annual International Workshop on IFS was initiated in Slovakia in 2005. Papers from these annual meetings have been mostly published in *Notes in IFS*. In addition to the many papers published in *Notes in IFS*, a large number of papers on IFS have also been published in various other journals, edited volumes, or conference proceedings since the early 1990s. The most important among all these publications was a comprehensive monograph on IFS by Atanassov (1999). This publication record deserves further examination.

Let us begin with a foreword to Atanassov 1999, in which Atanassov writes:

> In the beginning of 1983, I came across A. Kaufmann's book "Introduction to the theory of fuzzy sets" (Academic Press, New York, 1975). This was my first acquaintance with the fuzzy set theory. Then I tried to introduce a new component (which determines the degree of non-membership) in the definition of these sets and to study the properties of the new objects so defined. The late George Gargov (1947–1996) is the "godfather" of the sets I introduced—in fact, he has invented the name "intuitionistic fuzzy", motivated by the fact that the law of the excluded middle does not hold for them.

Although the utility of IFS was already well recognized when Atanassov's 1999 book was published, its appearance has significantly stimulated further research into the theory as well as applications of his nonstandard fuzzy sets. However, Atanassov's somewhat haphazard choice of the term "intuitionistic" has been contested on various occasions, most notably in debate papers by Dubois et al. (2005) and Grzegorzewski and Mrówka (2005), and even more thoroughly in a comprehensive analysis made by Cattaneo and Ciucci (2006) of the connection, or rather lack of it, between IFS and the concept of fuzzy sets within the scope of intuitionistic logic, as derived by Takeuti and Titani (1984). The following assessment appears on pp. 3198–99 in Cattaneo and Ciucci 2006:

> The mathematical structure introduced [by Atanassov and Stoeva (1983) and Atanassov (1986, 1999)] on the basis of ortho-pairs of fuzzy sets, and called "intuitionistic fuzzy sets", recently raised a terminological debate [(see Cattaneo and Ciucci 2003a,b)] based on a first comment [that] appeared several years before in [Cattaneo and Nisticò 1989, p. 183]. Successively an explicit discussion about this terminological controversy [was] published in [Dubois et al. 2005], with the consequent answers in [Grzegorzewski and Mrówka 2005 and Atanassov 2005].
>
> The main point of discussion is that in the above quoted seminal Atanassov's papers it is constructed a particular distributive lattice equipped with a complementation $\neg p$ which does not satisfy the algebraic version of excluded middle $\forall p : p \vee \neg p = 1$. Since this law is not accepted by intuitionistic logic, Atanassov claimed that the structure can be characterized as intuitionistic.

---

[49] The journal *Notes on IFS* is published by Marin Drinov Academic Publishing House in Sofia, Bulgaria. It is a major Bulgarian academic publisher with a long history reaching back to the 19th century. With the exception of three volumes, the journal has been published regularly as a quarterly.

The crucial point with respect to this claim is that Atanassov's complementation satisfies the algebraic version of the "strong" double negation law ($\forall p : \neg\neg p = p$, algebraically called property of involution), which is rejected by intuitionism (precisely, it is rejected that $\forall p : \neg\neg p \leq p$), and does not satisfy the contradiction law (($\forall p : p \wedge \neg p = 0$), which on the contrary is assumed to hold in intuitionistic logic. . . .

Our point of view in this debate is that it is not at all correct to assume a term (precisely, intuitionistic), very articulated in its assumptions, from the fact that only some of its asserted principles are satisfied. It is as if in the ancient Greek mathematical tradition, after the widely accepted definition of a circle as a plane figure whose points have the same distance from a fixed point called center, someone asserts that a square is a circle from the only fact that its vertices have the same distance from the square center.

Let us elaborate a little more on the conflict between the well-defined and useful mathematical structures of these new nonstandard fuzzy sets introduced by Atanassov and his unfortunate choice of calling these structures "intuitionistic fuzzy sets." This name has inevitably suggested their association with intuitionistic mathematics, in the same way as the well established names "intuitionistic arithmetic," "intuitionistic continuum," "intuitionistic set theory," and the like. Since this connection, and hence the appropriateness of the name, was challenged by Dubois et al. (2005), Atanassov has tried to defend the name by arguing that his nonstandard fuzzy sets are actually genuine fuzzy sets of intuitionistic mathematics. We are in agreement with Cattaneo and Ciucci (2006) that this is a futile effort. We feel that a good way to resolve this controversy would be to call these nonstandard fuzzy sets "Atanassov fuzzy sets." After all, they are solely the product of Atanassov's vision.

Chapter 4

Fuzzy Logic in the Narrow Sense

I N THIS CHAPTER we describe the development of fuzzy logic in the narrow sense. A natural milestone in this regard was the inception of fuzzy sets in the 1960s and we indeed emphasize the period since the 1960s until the present. Nevertheless, we cover the developments in many-valued logic up to the 1960s as well. A number of concepts and methods fundamental for fuzzy logic in the narrow sense appeared in that period and these provide relevant historical context. Moreover, two significant fuzzy logics—the Łukasiewicz and Gödel logics—were actually advanced well before the 1960s. Although the history of many-valued logic may be traced back to ancient times, we start in our coverage at the end of the 1910s because this period marks the emergence of modern many-valued logic.[1] Note that we take the 1960s as a milestone not only because fuzzy sets were first defined then but—perhaps more importantly—also because the emergence of fuzzy sets indeed represents a significant point in the development of contemporary logic.[2]

We assume familiarity with classical logic, whose main concepts are provided in appendix B. This serves as a reference point to illustrate differences between fuzzy logic and classical logic as well as to explain conceptually new issues arising in fuzzy logic. To facilitate our presentation, we survey in section 4.1 the principal concepts and issues that concern many-valued and fuzzy logic and explain our notation.

---

[1] Earlier contributions are surveyed in section 2.1.
[2] We return to this claim several times in this chapter.

4.1 From classical to fuzzy logic: Principal issues

The acceptance of additional truth values implies that certain issues in fuzzy logic and in many-valued logics in general are radically different from those in classical logic. In this section we present core issues of this kind. We also introduce notions to which we refer in the subsequent sections and define notation.

Truth values, truth degrees

The primary difference between many-valued logics and classical logic is that the set of truth values in many-valued logics may contain truth values distinct from 0 and 1. We denote the set of all truth values by L and assume that it always contains 0 and 1. In fuzzy logic, L naturally comes with a partial order $\leq$ with 0 and 1 being its lower and upper bounds. If $a \leq b$, then a proposition with truth value b is considered to be at least as true as that with truth value a. The truth values are hence also called *truth degrees* or *degrees of truth* in fuzzy logic. It is conceptually convenient not to exclude the two-element set $L = \{0, 1\}$—thus classical logic may be considered as a particular, limit case. Important sets of truth values include the three-, four-, and in general the n-element set

$$L = \{0, 1/2, 1\}, \quad L = \{0, 1/3, 2/3, 1\}, \quad \text{and} \quad L = \{0, 1/n-1, \ldots, n-2/n-1, 1\}.$$

Another important set is the real unit interval

$$L = [0, 1] = \{a \mid a \text{ is a real number} \geq 0 \text{ and} \leq 1\}.$$

Language and formulas

We first consider the propositional case. A *language* of a propositional many-valued logic usually does not differ substantially from its classical counterpart. It contains propositional symbols, symbols of logical connectives, and auxiliary symbols. Sometimes, it also contains symbols of certain truth values, so-called truth constants.[3] If, for instance, 1/2 is a truth value in L, the language may contain the symbol $\overline{1/2}$ representing 1/2.[4] *Formulas* are defined in the usual way, i.e. following the pattern

(a) every propositional symbol p is a formula;
(b) if c is a symbol of an n-ary connective and $\varphi_1, \ldots, \varphi_n$ are formulas, then $c(\varphi_1, \ldots, \varphi_n)$ is a formula, with the usual convention of writing $c\varphi$ if c is unary, such as a negation, and $\varphi_1 c \varphi_2$ if c is binary, such as a conjunction.

As symbols of a many-valued negation, conjunction, disjunction, implication, and equivalence, we usually employ $\neg$, $\otimes$, $\oplus$, $\rightarrow$, and $\leftrightarrow$, respectively. Since there may be

---

[3] Truth constants may be considered as 0-ary connectives. This view brings a slight technical simplification.

[4] Having truth constants in the language is not a substantial conceptual difference because some formulations of classical logic involve truth constants; see e.g. Church 1956.

several reasonable candidates for some connectives, we also use additional symbols, such as $\wedge$ for conjunction, and thus have formulas such as $(\varphi \otimes \psi) \rightarrow (\varphi \wedge \psi)$.

Truth functions of connectives, truth functionality

Most many-valued and fuzzy logics are *truth functional*.[5] This means that truth values are assigned to formulas in a way generalizing the situation in classical logic. Namely, one associates to every n-ary connective c the *truth function of c*,

$$c^{\mathbf{L}} : L^n \rightarrow L,$$

which assigns to every n-tuple consisting of truth values $a_1, \dots, a_n$ in L a truth value $c^{\mathbf{L}}(a_1, \dots, a_n)$ in L. We also use just c if $\mathbf{L}$ is obvious or if we just need to make clear that we refer to a truth function of c. Sometimes we even drop the superscript and denote both c and $c^{\mathbf{L}}$ simply by c. For constants representing truth and falsity, we denote the corresponding truth values as 1 and 0 rather than use the somewhat awkward $\overline{1}^{\mathbf{L}}$ and $\overline{0}^{\mathbf{L}}$. The boldface superscript in $c^{\mathbf{L}}$ refers to the fact that the set L of truth values equipped with the truth functions forms an algebra

$$\mathbf{L} = \langle L, \{c^{\mathbf{L}} \mid c \text{ is a symbol of connective}\}\rangle,$$

called a *structure of truth values*.[6] We adopt the custom of writing just $\langle L, c_1^{\mathbf{L}}, \dots, c_m^{\mathbf{L}}\rangle$ instead of $\langle L, \{c_1^{\mathbf{L}}, \dots, c_m^{\mathbf{L}}\}\rangle$. As an example, consider a language with $\neg, \otimes, \oplus, \overline{0}$, and $\overline{1}$. If $L = \{0, 1\}$ and if $\neg^{\mathbf{L}}, \otimes^{\mathbf{L}}$, and $\oplus^{\mathbf{L}}$ are the classical truth functions (appendix B), then $\mathbf{L} = \langle L, \otimes^{\mathbf{L}}, \oplus^{\mathbf{L}}, \neg^{\mathbf{L}}, 0, 1\rangle$ is just the two-element Boolean algebra, i.e. the structure of truth values of classical logic. However, if $L = [0, 1]$ and the truth functions are defined by $a \otimes^{\mathbf{L}} b = \min(a, b)$, $a \oplus^{\mathbf{L}} b = \max(a, b)$, and $\neg^{\mathbf{L}} a = 1 - a$, then

$$\mathbf{L} = \langle L, \otimes^{\mathbf{L}}, \oplus^{\mathbf{L}}, \neg^{\mathbf{L}}, 0, 1\rangle \text{ becomes } \mathbf{L} = \langle [0, 1], \min, \max, 1 - x, 0, 1\rangle,$$

a different structure which is not a Boolean algebra. Yet another structure is obtained with $L = [0, 1]$ by letting $a \otimes^{\mathbf{L}} b = \max(0, a + b - 1)$, $a \oplus^{\mathbf{L}} b = \min(1, a + b)$, and $\neg^{\mathbf{L}} a = 1 - a$. This indicates that neither the set L nor the truth functions need to be fixed for a particular many-valued logic. That is, one may consider different sets of truth values for a particular language and thus have the possibility of interpreting formulas in various structures of truth values.[7] In such cases, one usually specifies a certain *class* $\mathcal{L}$ of structures one works with, such as the class of all Boolean algebras, the class of all MV-algebras, or the class of all structures with $L = [0, 1]$ satisfying certain (logically relevant) conditions. Note also that with a small set of truth values, truth

---

[5] We do not consider non-truth-functional logics in this section. These are discussed in section 4.5.

[6] This notation follows the custom in algebra of denoting the algebraic structure with a support set A by $\mathbf{A}$ and denoting the operations of this structure by symbols with superscript $^{\mathbf{A}}$. This way, we may denote the addition of real numbers by $+^{\mathbb{R}}$, the addition of integers by $+^{\mathbb{Z}}$, the addition of $n \times n$ matrices by $+^{\mathbf{M}_n}$, etc.

[7] The reason may be purely mathematical but may also have an extramathematical motivation. As an example, one might want to be able to use for L the set $\{0, 1/n-1, \dots, n-2/n-1, 1\}$ instead of the real unit interval $[0, 1]$ as a kind of discretization and perform this discretization for computational or other reasons.

functions are conveniently described by *truth tables* such as

$\wedge$	0	$1/3$	$2/3$	1
0	0	0	0	0
$1/3$	0	$1/3$	$1/3$	$1/3$
$2/3$	0	$1/3$	$2/3$	$2/3$
1	0	$1/3$	$2/3$	1

which describes a truth function $\wedge^{\mathbf{L}}$ of $\wedge$ defined on the set $L = \{0, 1/3, 2/3, 1\}$.

Now, let $\mathbf{L}$ be a structure of truth values for a given propositional language and let e be a truth evaluation in L, also called an *L-evaluation*, i.e. a mapping assigning to every propositional symbol p its truth value $e(p)$ in L. The *truth value* $\|\varphi\|_e^{\mathbf{L}}$ *of a formula* φ in e is defined inductively in a truth-functional manner as follows:

(a) atomic formulas: $\|p\|_e^{\mathbf{L}} = e(p)$,
(b) compound formulas: $\|c(\varphi_1, \ldots, \varphi_n)\|_e^{\mathbf{L}} = c^{\mathbf{L}}(\|\varphi_1\|_e^{\mathbf{L}}, \ldots, \|\varphi_n\|_e^{\mathbf{L}})$.

Property (b) is called *truth functionality*: a truth degree of a c-compound formula is obtained from the truth degrees of the constituent formulas by the truth function of c. If there is no danger of ambiguity, we also write just $\|\varphi\|_e$ instead of $\|\varphi\|_e^{\mathbf{L}}$. The pattern of this definition is clearly the same as in classical logic which becomes a special case where $\mathbf{L}$ is the two-element Boolean algebra.

As in the classical setting, logical connectives may be defined in terms of other connectives, but there are important differences. For example, while classical disjunction is definable in terms of classical negation and implication, i.e. $\varphi \vee \psi$ may be defined as $\neg\varphi \to \psi$ or, equivalently, as $(\varphi \to \psi) \to \psi$, these two definitions need not be equivalent in a many-valued setting. Many such examples are encountered later.

Designated truth values, tautologies, entailment

In classical logic, a formula is a tautology if it is always true, i.e. its value is 1 in any truth evaluation. In a many-valued setting, we may adopt exactly this notion of tautology because 1 is always one of the truth values involved. Similarly, we may adapt the notions of satisfiability, contradiction, entailment, and other notions to a many-valued setting. However, one may proceed in a more general way by declaring as *designated* some subset D of the set L of truth values that contains 1, does not contain 0, and satisfies possibly further conditions.[8] Thus, in the case of $L = \{0, 1/2, 1\}$, we may have $D = \{1/2, 1\}$ or, classically, $D = \{1\}$. Then, an *L-tautology with respect to D* is any formula φ whose truth value is always designated, i.e. $\|\varphi\|_e^{\mathbf{L}} \in D$ for every truth evaluation e. We omit $\mathbf{L}$ and D if these are clear from context. We denote the

---

[8] The terminology "designated value" is due to Bernays (1926), who used the German term "ausgezeichnete Werte." The idea of using designated truth values appears in Bernays 1918 and Post 1921.

A pair $\langle \mathbf{L}, D \rangle$, where $\mathbf{L} = \langle L, \ldots \rangle$ is an algebra of truth values and D a subset of its support set, L, is called a *logical matrix* in the Polish school of logic; see p. 158.

set of all tautologies w.r.t. D of a logical calculus C by

$$\mathrm{Taut}_D^L(\mathsf{C}), \quad \text{or} \quad \mathrm{Taut}^L(\mathsf{C}), \quad \text{or just} \quad \mathrm{Taut}(\mathsf{C}),$$

in particular if 1 is the only designated truth value. As we shall see later, different sets of designated truth values may yield different sets of tautologies. A formula φ *follows (w.r.t. a set $D \subseteq L$) from a set T* of formulas, in symbols

$$T \models_D^L \varphi, \quad \text{or} \quad T \models^L \varphi, \quad \text{or just} \quad T \models \varphi,$$

if, when using $\mathbf{L}$, the truth value of φ is designated in every evaluation for which the truth values of all the formulas in T are designated. In particular, if T is empty, $T \models \varphi$ means that φ is a tautology. Obviously, with $L = \{0, 1\}$ and $D = \{1\}$ we obtain the ordinary notions of classical logic. Instead of relating the above notions to a single structure $\mathbf{L}$, one may relate them to a class $\mathscr{L}$ of such structures with their designated truth values. Thus for instance, a formula is an $\mathscr{L}$-tautology if it is an L-tautology for every $\mathbf{L} \in \mathscr{L}$; φ follows from T with respect to $\mathscr{L}$, if $T \models_D^L \varphi$ for each $\mathbf{L} \in \mathscr{L}$. One then uses

$$\mathrm{Taut}_D^{\mathscr{L}}(\mathsf{C}) \quad \text{or just} \quad \mathrm{Taut}^{\mathscr{L}}(\mathsf{C}) \quad \text{and} \quad T \models_D^{\mathscr{L}} \varphi \quad \text{or just} \quad T \models^{\mathscr{L}} \varphi.$$

Proofs, axiomatization

In most systems of many-valued logic, the concepts of *deduction rule, axiom, theory,* and *proof* are defined essentially as in classical logic (appendix B). Accordingly, we denote that formula φ is provable from theory T by

$$T \vdash \varphi, \quad \text{or just} \quad \vdash \varphi \quad \text{if } T \text{ is empty.}$$

One may then examine the problem of *axiomatization* with respect to a particular semantics, or the problem of finding a *syntactico-semantically complete* set of axioms, i.e. axioms such that the formulas provable from them are just the tautologies, or, more generally such that the formulas provable from the axioms and T are just those entailed by T.

Completeness: Syntactico-semantical, functional, and w.r.t. consequence

In addition, two other notions of completeness are considered. The first is a property of a set F of functions on a set L, such as the set $F = \{\neg, \rightarrow\}$ of truth functions of classical negation and implication defined on $L = \{0, 1\}$. A set F is called *functionally complete* if every n-ary function $f : L^n \rightarrow L$, n being an arbitrary nonnegative integer, may be obtained as a composition of some functions in F. A set of propositional connectives is functionally complete if the set of the truth functions of these connectives is complete. Every propositional formula φ with propositional symbols $p_1, \ldots, p_n$ induces a function $\varphi^\cdot : L^n \rightarrow L$, defined by

$$\varphi^\cdot(a_1, \ldots, a_n) = \|\varphi\|_e, \quad \text{where } e \text{ is any evaluation with } e(p_i) = a_i, \qquad (4.1)$$

for $i = 1,\ldots,n$. Therefore, a set of propositional connectives is functionally complete if for every function $f : L^n \to L$ there exists a formula φ involving only the connectives in the set such that $f = \varphi$.

The second is completeness relative to a consequence operator C. A set T of formulas is called *complete w.r.t.* C if adding to T any formula φ that is not a C-consequence of T renders a C-inconsistent set of formulas, i.e. if for any $\varphi \notin C(T)$, the set $C(T \cup \{\varphi\})$ contains all formulas. Two important cases obtain if C is the syntactic consequence $C_\vdash$, i.e. $C_\vdash(T) = \{\varphi \mid \varphi \text{ is provable from } T\}$, or the semantic consequence $C_\models$, i.e. $C_\models(T) = \{\varphi \mid \varphi \text{ semantically follows from } T\}$, for the given notions of provability and semantic entailment.

From propositional to predicate logic

An extension of a propositional to a predicate many-valued logic may, in principle, follow the same pattern as in classical logic. In this basic extension, the language contains symbols of connectives as in the propositional case, symbols $x,\ldots$ of individual variables, symbols $r,\ldots$ of relations, also called predicates, symbols $\forall$ and $\exists$ of quantifiers, and possibly also symbols of truth values and symbols of functions. More general conceptions are possible and are discussed later. In the basic extension, there are no significant differences as far as the syntactic notions are concerned—the notions of term, formula, theory, proof and the like are basically defined as in the classical case. On the semantic side, however, the differences from classical logic are considerably more substantial. The presence of additional truth values entails more involved notions as well as conceptually new problems compared to classical logic.

To start with, consider the basic semantic notion of predicate logic—the notion of structure for a given language. Let $\mathbf{L}$ be a structure of truth values with L as its set of truth values. A structure $\mathbf{M}$ with truth values in $\mathbf{L}$, or an $\mathbf{L}$-*structure* for short, consists of a nonempty universe set M and interpretations of symbols of relations, functions, and truth values of the given language. While symbols of functions and truth values are interpreted in a standard manner, i.e. as ordinary functions on M and as truth values in L, respectively, symbols r of relations are interpreted differently. An interpretation $r^\mathbf{M}$ of a k-ary symbol r is a k-ary L-*valued relation*, or just L-*relation*, in M, i.e. a mapping

$$r^\mathbf{M} : M^k \to L$$

assigning to every tuple $\langle m_1,\ldots,m_k \rangle$ of elements in M a truth value $r^\mathbf{M}(m_1,\ldots,m_k)$ in L. This value is interpreted as a truth value of the proposition "$m_1,\ldots,m_k$ are related by r."

The concept of an L-relation clearly generalizes that of an ordinary relation in that if $L = \{0,1\}$, then L-relations are just the characteristic functions of ordinary relations. Hence, the concept of $\mathbf{L}$-structure generalizes that of structure in classical logic.[9] Notice that (standard) fuzzy sets and fuzzy relations are particular L-valued relations. Namely, for $L = [0, 1]$, a fuzzy set in M is just a unary L-relation in M and similarly for fuzzy relations. This means that the concept of a fuzzy set is automatically present and dealt with when considering predicate many-valued logic.

Given an L-structure M and a valuation v of individual variables, values $\|t\|_{M,v}$ of terms are defined as in classical logic. The truth value $\|\varphi\|^L_{M,v}$, or simply $\|\varphi\|_{M,v}$, of a formula φ in M and v is defined as follows. For atomic formulas, one puts

$$\|r(t_1, \ldots, t_k)\|_{M,v} = r^M(\|t_1\|_{M,v}, \ldots, \|t_k\|_{M,v}).$$

For compound formulas of the form $\varphi \to \psi$—and analogously for other propositional connectives—one puts $\|\varphi \to \psi\|_{M,v} = \|\varphi\|_{M,v} \to^L \|\psi\|_{M,v}$, as in the propositional case. For quantified formulas, a natural option is available if there exists a partial order $\leq$ on L. Namely, one then defines $\|(\forall x)\varphi\|_{M,v}$ to be the largest truth value a in L such that $a \leq \|\varphi\|_{M,v'}$ for every valuation $v' =_x v$, i.e. a valuation that differs from v only in the values assigned to x. That is, one puts

$$\|(\forall x)\varphi\|_{M,v} = \bigwedge\{\|\varphi\|_{M,v'} \mid v' =_x v\}. \tag{4.2}$$

Similarly, one may put

$$\|(\exists x)\varphi\|_{M,v} = \bigvee\{\|\varphi\|_{M,v'} \mid v' =_x v\}. \tag{4.3}$$

Note that the infima $\bigwedge \ldots$ and suprema $\bigvee \ldots$ in the above definitions always exist if L forms a complete lattice, such as the unit interval $[0, 1]$ of reals standardly used in fuzzy logic. For $L = \{0, 1\}$, this yields the classical definitions.

The notions of designated truth values, tautologies, entailment, proof, and the like then may be defined analogously to those of the propositional case.

Further issues

As was mentioned at the beginning, the overview in this section is not exhaustive. For one, we do not discuss certain important notions, such as nonstandard quantifiers. Secondly, we do not go beyond mentioning that fuzzy logics surpass classical logic also in that they involve certain language constructs that are rather important in the presence of intermediate truth degrees but are degenerate in classical logic and thus not recognized in it. A case in point is represented by linguistic hedges such as "very," "rather," and "more or less," which may be treated as unary connectives. Thirdly, we only point out that the treatment of some notions in a many-valued setting may be different, sometimes radically, from what has been discussed above. As an example, entailment, provability, and other notions need not be considered bivalent. Instead, one may naturally introduce the notion of degree of entailment and that of degree of provability for fuzzy logics.

---

[9] Note that in earlier texts on logic, including Church 1956 and Kleene 1967, symbols of relations are interpreted as characteristic functions of relations, rather then relations themselves, which is the reason for using the term "functional calculus" instead of "predicate calculus" in the past.

4.2 Many-valued logics until the 1960s

4.2.1 Łukasiewicz finitely- and infinitely-valued logics

Łukasiewicz's work is generally considered the first systematic attempt to develop a many-valued logic. Even though Łukasiewicz's original motivation was different from that of fuzzy logic and was later criticized, we cover the development of Łukasiewicz logics in detail because they represent very important fuzzy logics.[10]

The emergence of Łukasiewicz logic

To understand properly the emergence of Łukasiewicz logic, it is appropriate to start at the end of the 19th century when the so-called Lvov-Warsaw school of logic appeared.[11] The school was founded by Kazimierz Twardowski (1866–1938), a professor of philosophy in Lvov. Twardowski, a former student of Brentano, was a charismatic teacher who intended to introduce Brentano's rigorous style in Poland.[12] He demanded clarity of language and thinking and gave the first course in mathematical logic in Poland. Among Twardowski's students were Jan Łukasiewicz and Stanisław Leśniewski (1886–1939)—two founders of the Warsaw school of logic, which grew out of the Lvov-Warsaw school. Łukasiewicz, who studied philosophy and mathematics in Lvov, graduated with a degree in philosophy in 1902 and was appointed professor of philosophy at the University of Warsaw in 1915.[13]

Both Leśniewski and Łukasiewicz were very active and attracted many young people. Several of them, including Stanisław Jaśkowski (1906–1965), Adolf Lindenbaum (1904–1941), Andrzej Mostowski (1913–1975), Mojżesz Presburger (1904–1943), Jerzy Słupecki (1904–1984), Bolesław Sobociński (1904–1980), Alfred Tarski (1901–1983), and Mordchaj Wajsberg (1902–unknown),[14] later became famous logicians who also made important contributions to many-valued logic. An important aspect of the Warsaw school was the passion of its members for doing logic, convinced that by researching and teaching logic, they were making an important contribution to society. This view is in keeping with the school having its roots not only in mathematics but, perhaps more importantly, in philosophy—probably the most decisive reason for the great success of the school. According to Łukasiewicz (1929b, 606):[15]

---

[10] For the criticism see p. 323 below.

[11] See e.g. J. Woleński, *Logic and Philosophy in the Lvov-Warsaw School* (Dordrecht: Kluwer, 1989); Woleński 2004; and U. Wybraniec-Skardowska, "Polish logic, some lines from a personal perspective" (notes for Logic Seminar at the Department of Philosophy, Stanford University, May 29, 2009).

[12] Franz Brentano (1838–1917), a German philosopher and psychologist. Besides Twardowski, among his students were Edmund Husserl, Sigmund Freud, Alexius Meinong, and Tomáš G. Masaryk.

[13] See Łukasiewicz 1994.

[14] According to S. J. Surma, ed., *Mordchaj Wajsberg: Logical Works* (Wrocław: Wydawnictwo Polskiej Akademii Nauk, 1977), ii, "Since the outbreak of the Second World War we have no reliable information concerning Wajsberg. Sure is only the fact that he perished during the war."

[15] Translation of the quoted text is from p. 403 in Woleński 2004.

> It was a happy coincidence that mathematicians and philosophers worked together in creating Polish mathematical logic. . . . mathematicians will not permit the transformation of mathematical logic into fruitless speculation, while philosophers will defend it against the servile application of mathematical methods and the restriction of its role to an auxiliary mathematical discipline.

Philosophical motivations were particularly important for Łukasiewicz's work in logic. He thought that logic could help solve philosophical problems. Thanks to his knowledge of Greek, acquired in a philological gymnasium, he read the original works of the great Greek philosophers, particularly Aristotle, the Stoics, and the Epicureans, and had a good knowledge of their philosophical and logical thoughts. His views were influenced by Meinong's theory of contradictory objects which had been decisively shaped by Brentano and influenced by Twardowski. The theory claimed the existence of objects with contradictory properties, such as a squared circle. Łukasiewicz shared these views. He published a book on the principle of contradiction by Aristotle, a synopsis of which appeared in Łukasiewicz 1910, wherein he argued that the principle is not self-evident and attacked it.[16] Łukasiewicz highly valued this book and planned its English edition in 1955.[17] He was deeply interested in the problems of determinism, causality, and modality, which go back to Aristotle and the debates mentioned in chapter 2. Łukasiewicz himself was an advocate of indeterminism. Important stimuli for him were the debates in the Lvov-Warsaw school regarding freedom and creativity and Kotarbiński's suggestion that two-valued logic should be revised because it seemed to interfere with the freedom of human thinking.[18] According to his Farewell Lecture (Łukasiewicz 1918), he attempted to devise a logic around 1910, a "non-Aristotelian" one, that would have been compatible with his views, but he did not succeed in doing so until the summer of 1917 when he worked out the details of his three-valued logic.[19] It is worth mentioning, however, that already as part of his studies of the problem of induction and probability (Łukasiewicz 1913), he considered a many-valued logic which we examine on p. 145. His views regarding determinism and causality appeared in Łukasiewicz 1906 and later in the inauguration address he delivered as rector of the University of Warsaw at the beginning of the academic year 1922/1923 (Łukasiewicz 1961) and in the detailed analysis Łukasiewicz 1930.[20] These philosophical problems motivated his three-valued logic and were a driving force for his investigations,

---

[16] The principle of contradiction says that two contradictory statements can not both be true at the same time. The principle is sometimes formalized in propositional logic by demanding that the formula $\neg(\varphi \wedge \neg\varphi)$, called the law of contradiction but sometimes the law of noncontradiction, be a tautology.

[17] Translator's note in *The Review of Metaphysics* 24 (1971): 485.

[18] T. Kotarbiński, "Zagadnienie istnienia przyszłości," *Przegląd Filozoficzny* 16 (1913) 74–92, trans. R. Rand, "The problem of the existence of the future," *Polish Review* 13, no. 3 (1968): 7–22; see also section 2.1 in Malinowski 1993. Tadeusz Kotarbiński (1886–1981), a Polish philosopher and logician, was a pupil of Twardowski and an important figure of the Lvov-Warsaw school.

[19] See the quotation on p. 12. Later on, Łukasiewicz preferred to use "non-Chrysippean" instead of "non-Aristotelian" because "Chrysippus appears to have been the first logician to consciously set up and stubbornly defend . . . that every proposition is either true or false" (Łukasiewicz 1930, 175).

[20] Łukasiewicz was rector of the University of Warsaw twice, in 1922/1923 and in 1931/1932. In addition, he was dean of the Faculty of Arts in 1916 and vice-rector of the university in 1917; see Łukasiewicz 1994.

which is clear from his publications. He considered these problems fundamentally important and deeply connected to logic, and considered a precise analysis by means of mathematical logic the right tool to cope with them. These circumstances gave his three-valued logic the status of a rather important and seriously motivated alternative to classical logic, which was important for its further development.

To better understand his views, we now examine his 1922 address (Łukasiewicz 1961).[21] Łukasiewicz first introduces the idea of determinism by examples and proceeds to make precise what he calls the *thesis of determinism*, namely the belief that *if A is b at instant t, then it is true at any instant earlier than t that A is b at t*:

> Nobody who adopts this belief can treat the future differently from the past. If everything that is to occur and become true at some future time is true already today, and has been true from all eternity, the future is as much determined as the past. . . . We are only the puppets in the universal drama. There remains for us nothing else to do but watch the spectacle and patiently await its end. . . .
>
> This is a strange view and by no means obvious. However, there are two arguments of considerable persuasive power which have been known for a long time and which seem to support determinism. One of them, originating with Aristotle, is based on the logical principle of the excluded middle, and the other, which was known to the Stoics, on the physical principle of causality.

He then presents these arguments, that is to say, shows how the thesis of determinism may be derived from either the *principle of the excluded middle* or *the principle of causality*. Afterwards, he presents a careful analysis of these arguments and their various modifications and shows that they are erroneous. He also points out that Aristotle's analyses actually undermine the principle of bivalence and remarks that this principle cannot be proved, only believed in, and that it is not self-evident. Then he argues for a third truth value which is to be assigned to a certain category of propositions including the propositions about future facts. He calls such propositions indeterminate and mentions that he presented an outline of a logic with a third truth-value in Łukasiewicz 1920. The thesis of determinism cannot be obtained in his three-valued logic because:

> [W]e can assign such values to variables "A", "b", and "t" that its antecedent changes into a true sentence and its consequent into an indeterminate one, that is into a sentence having the third truth-value. This always happens when the cause of the fact that A is b at a future instant t does not yet exist today. A conditional whose antecedent is true and consequent indeterminate cannot be accepted as true; for truth can imply only truth. The logical argument which seems to support determinism falls decisively.

Łukasiewicz then arrives at a conclusion:

> In my view, the age-old arguments in support of determinism do not withstand the test of critical examination. This does not at all imply that determinism is a false view; the falsehood of the arguments does not demonstrate the falsehood of the thesis. Taking advantage of my preceding critical examination, I should like to state only one thing, namely that determinism is not a view better justified than indeterminism.
>
> . . . I may declare myself for indeterminism. I may assume that not the whole future is determined in advance. . . . It is well that it should be so. . . . We may believe that we are not merely

---

[21]The quotes below in this section are from the English translation in Borkowski 1970 (pp. II3–I4, I27).

> passive spectators of the drama but also its active participants. . . . We can ourselves somehow shape the future of the world in accordance with our designs. I do not know how this is possible, but I believe that it is.

Łukasiewicz three-, n-, and infinitely-valued logics Ł$_3$, Ł$_n$, and Ł$_\infty$

Leaving the University of Warsaw in 1918 to serve as head of the Department of Higher Schools in the Polish Ministry of Education and from January until December of 1919 as Minister of Religious Denominations and Public Education in the cabinet of Prime Minister Paderewski, Łukasiewicz announced his three-valued logic in his Farewell Lecture (Łukasiewicz 1918).[22] He presented this logic in some detail at the 207th scientific meeting of the Polish Philosophical Society in Lvov on June 19, 1920. An outline of this presentation also appeared in a short paper (Łukasiewicz 1920) which we now examine.

Łukasiewicz introduced the logic, which we denote Ł$_3$, by defining the truth functions of connectives on a three-element set $\{0, 1/2, 1\}$ of truth values.[23] He did so in a way "which deviates least from 'two-valued' logic" by preserving the values of these functions on classical truth values and defined the truth functions of the connectives of implication, →, and equivalence, ↔, according to

→	0	1/2	1
0	1	1	1
1/2	1/2	1	1
1	0	1/2	1

and

↔	0	1/2	1
0	1	1/2	0
1/2	1/2	1	1/2
1	0	1/2	1

In fact, he used "<" and "=" to denote the truth functions of → and ↔, called = identity, and defined it by extending the classical "principle of identity" using "$(0 = 1/2) = (1/2 = 0) = (1 = 1/2) = (1/2 = 1) = 1/2, (1/2 = 1/2) = 1$." This, somewhat ambiguous, definition may be explained by pointing out that the term "identity" and the symbol "=" were being used by Łukasiewicz for the connective of equivalence at that time, hence the above expression actually means $(0 \leftrightarrow 1/2) = (1/2 \leftrightarrow 0) = (1 \leftrightarrow 1/2) = (1/2 \leftrightarrow 1) = 1/2, (1/2 \leftrightarrow 1/2) = 1$.[24] Furthermore, Łukasiewicz defined the truth functions ¬, ∨, and ∧ of negation, ¬, disjunction, ∨, and conjunction, ∧, in terms of → and 0 by[25]

$$\neg a = a \rightarrow 0, \quad a \vee b = (a \rightarrow b) \rightarrow b, \quad \text{and} \quad a \wedge b = \neg(\neg a \vee \neg b),$$

which means that these truth functions are given by the following tables:

φ	¬φ
0	1
1/2	1/2
1	0

∨	0	1/2	1
0	0	1/2	1
1/2	1/2	1/2	1
1	1	1	1

∧	0	1/2	1
0	0	0	0
1/2	0	1/2	1/2
1	0	1/2	1

---

[22] See p. 12 in chapter 2 for quotations from this lecture.

[23] He actually used "2" instead of "1/2."

[24] See e.g. p. 98 in J. Łukasiewicz, "Two-valued logic," in Borkowski 1970, 98–109.

[25] He denoted these truth functions $'$, $\cdot$, and $+$.

As a matter of fact, $\leftrightarrow\cdot$ is also definable by $\rightarrow\cdot$ and 0 since $a \leftrightarrow\cdot b = (a \rightarrow\cdot b) \wedge\cdot (b \rightarrow\cdot a)$. Note also that at about the same time, the same logical system, different only in the choice of its basic connectives, was invented by Bernays whose work and motivations—completely different from Łukasiewicz's—are discussed on p. 134 below.

Łukasiewicz observed that in $Ł_3$, some classical tautologies may take the truth value of $1/2$, such as the law of excluded middle, $\varphi \vee \neg\varphi$, or the law of contradiction, $\neg(\varphi \wedge \neg\varphi)$. He also observed that some may even have the truth value 0 in L_3, such as $\neg(\varphi \leftrightarrow\cdot \neg\varphi)$, which amounts to having $a \leftrightarrow\cdot \neg a = 1$ for some truth value a, and added that "This accounts for the fact that in the three-valued logic there are no antinomies." These observations are easily seen because $1/2 \vee \neg\cdot 1/2 = 1/2$, $1/2 \wedge \neg\cdot 1/2 = 1/2$, and $1/2 \leftrightarrow\cdot \neg\cdot 1/2 = 1$.[26] The concluding paragraph sums up Łukasiewicz's views on the prospects of $Ł_3$:[27]

> [T]he three-valued logic has above all theoretical importance as an endeavour to construct a system of non-Aristotelian logic. Whether that new system of logic has any practical importance will be seen only when logical phenomena, especially those in the deductive sciences, are thoroughly examined, and when the consequences of the indeterministic philosophy, which is the metaphysical substratum of the new logic, can be compared with empirical data.

Łukasiewicz logic was subsequently investigated intensively by members of the Warsaw school but the results were not published during the 1920s. An exception is Łukasiewicz 1923, a short report of Łukasiewicz's lecture presented at the 232nd meeting of the Polish Philosophical Society in Lvov on October 14, 1922, in which he examines 192 tautologies of classical logic in chapters *1–*5 of volume I of Whitehead and Russell's *Principia Mathematica* (1910–13) through the lens of many-valued logic. Importantly, he introduced for this purpose the truth function $\rightarrow\cdot$ of implication defined on the real unit interval $[0, 1]$ by

$$a \rightarrow\cdot b = \min(1, 1 - a + b), \tag{4.4}$$

extending thus the definition of implication in $Ł_3$ to a logical system $Ł_{\aleph_1}$ which involves $L = [0, 1]$ as the set of truth values. The other, derived connectives are defined as in $Ł_3$, i.e.:

$\neg\varphi$	as	$\varphi \rightarrow 0$,	hence	$\neg a = 1 - a$,	(4.5)
$\varphi \vee \psi$	as	$(\varphi \rightarrow \psi) \rightarrow \psi$,	hence	$a \vee b = \max(a, b)$,	(4.6)
$\varphi \wedge \psi$	as	$\neg(\neg\varphi \vee \neg\psi)$,	hence	$a \wedge b = \min(a, b)$,	(4.7)
$\varphi \leftrightarrow \psi$	as	$(\varphi \rightarrow \psi) \wedge (\psi \rightarrow \varphi)$,	hence	$a \leftrightarrow b = 1 - |a - b|$.	(4.8)

---

[26] In his important book, Rescher (1969, 27) incorrectly attributes the discovery of a classical tautology which can take the truth value 0 in $Ł_3$ to Turquette, the tautology being $\neg(\varphi \rightarrow \neg\varphi) \vee \neg(\neg\varphi \rightarrow \varphi)$, i.e. a formula essentially the same as $\neg(\varphi \leftrightarrow \neg\varphi)$ from Łukasiewicz's 1920 paper, and adds "This refutes the contention—encountered in various places in the literature—that no two-valued tautology can take on the truth-value F [i.e. 0] in $Ł_3$. This erroneous claim was made by A. N. Prior in the *Encyclopedia of Philosophy* . . . and echoed by the present writer in *Topics in Philosophical Logic* (Dordrecht: Kluwer, 1968)."

[27] English translation is from p. 88 in Borkowski 1970.

For every formula φ, Łukasiewicz defines what might be called the degree to which φ is a tautology by

$$\|\varphi\| = \min\{\|\varphi\|_e \mid e \text{ is a truth evaluation}\},$$

where $\|\varphi\|_e$ is the truth value of φ in a truth evaluation e mapping propositional symbols to $[0, 1]$. Łukasiewicz observed that of the 192 classical tautologies φ, sixty have the value $\|\varphi\| = 1/2$, three have the value 0, and the rest have the value 1. Interestingly, he observed that since the deduction rules of *Principia Mathematica* preserve the property of having the value 1, many-valued logic can be used to show independence of some axioms from other axioms. This idea had already been discovered independently by Bernays in his *Habilitationsschrift* in 1918, the same year that Łukasiewicz announced his three-valued logic, as acknowledged in n. 12 on p. 136 in Łukasiewicz and Tarski 1930.[28] According to this note, Łukasiewicz applied the idea in 1924, following a suggestion by Tarski, and proved some independence results for the axiom systems of propositional logic of Whitehead and Russell as well as Hilbert. He published these results without proofs in 1925.[29] Using many-valued logics for proving independence of axioms of propositional logic became a standard subject in textbooks on mathematical logic, see e.g. § 19 in Church 1956. The idea is the following. To show that a formula φ (an axiom) cannot be proved from formulas $\psi, \ldots$ (other axioms), it is sufficient to find a set L of truth values, its subset $D \subseteq L$ of designated truth values, and truth functions of connectives in such a way that each of the formulas among $\psi, \ldots$ is a tautology, i.e. takes on a truth value in D for every truth evaluation, φ is not a tautology, and all of the deduction rules preserve tautologicity in that whenever the premises of a deduction rule are tautologies, its consequence is a tautology as well. Then, all provable formulas are tautologies and hence φ is not provable. As an example, we can easily show using $Ł_3$, for which $D = \{1\}$, that the axiom (B.3) of classical propositional logic in appendix B cannot be proven from the remaining two, (B.2) and (B.4). Namely, as one easily verifies, while (B.2) and (B.4) are tautologies in $Ł_3$, (B.3) is not because it takes on the value $1/2$ when $\|\varphi\|_e = 1/2$, $\|\psi\|_e = 1/2$, and $\|\chi\|_e = 0$. Observing that *modus ponens* preserves tautologicity finishes the argument.

Some of the results on Łukasiewicz logic obtained by the Warsaw school are described in the paper Łukasiewicz and Tarski 1930. Before presenting them, we examine Łukasiewicz 1930 which appeared in the same issue as Łukasiewicz and Tarski 1930. In this paper, Łukasiewicz argues cautiously in support of a non-bivalent logic and provides additional motivations for his many-valued logics. He considers modal propositions such as "it is possible that P," "it is not possible that P," "it is necessary that P," and treats the modality "possible" as a unary connective denoted M.[30] He

---

[28] Bernays's remarkable work, which contains several many-valued logics, is examined in more detail on pp. 134ff.

[29] "Démonstration de la compatibilité des axiomes de la théorie de la déduction" [Demonstration of compatibility of the axioms of the theory of deduction], *Annales de la Société Polonaise de Mathématique* 3 (1925): 149.

[30] As we shall see below, he treats M as a truth-functional connective. This might seem poor from a modern point of view, but recall that the modern Kripke semantics of modal logic appeared about 30 years later.

starts by presenting three groups of theorems he identified in the history of logic. The groups are exemplified by three theorems, which read:

Thm. I   If it is not possible that P, then not-P.

Thm. II   If it is supposed that not-P, then it is
              (on this assumption) not possible that P.

Thm. III   For some P: it is possible that P and it is possible that not-P.

Łukasiewicz explains how to understand these theorems to comply with their understanding by the ancient logicians to whom he refers. He then formalizes Thm. I and Thm. II in classical propositional logic and Thm. III in the so-called extended propositional calculus worked out by Polish logicians.[31] Most importantly, he shows that Thm. II and Thm. III are incompatible in that the resulting system is inconsistent and allows one to prove every formula. None of the two apparent solutions is acceptable: if one keeps Thm. II and rejects Thm. III, then φ is equivalent to $M\varphi$ for any φ, i.e. introducing modal propositions to logic does not make sense; if one keeps Thm. III and rejects Thm. II, one may prove $M\varphi$ for any φ, meaning that every proposition is possible which is not a good option either. Łukasiewicz thus argues for a third truth value, recalls the basics of his three-valued logic, and says that "the desired equations [i.e. definitions of the truth functions] I obtained on the basis of detailed considerations, which were more or less plausible to me." He shows that in order for Thm. I, Thm. II, and Thm. III to be compatible in the three-valued logic, the unary connective M, representing "possible," is uniquely determined and definable in terms of $\neg$ and $\rightarrow$ as

$$M\varphi \quad \text{meaning} \quad \neg\varphi \rightarrow \varphi,$$

and credits Tarski for this definition. That is, the truth function of M is given by:

φ	$M\varphi$
0	0
1/2	1
1	1

In particular, Thm. I, Thm. II, and Thm. III remain valid in the resulting system. Łukasiewicz actually says that Thm. II is "in a certain sense valid," although not as a simple implication. He means that even though $\neg\varphi \rightarrow \neg M\varphi$ is not a tautol-

---

[31]From the point of view of modal logic, his remarks regarding the formalization of Thm. II are of particular interest. They amount to the question of whether Thm. II is properly formalized by the formula $\neg\varphi \rightarrow \neg M\varphi$. Łukasiewicz says that to represent Thm. II in the language of propositional logic a different way seems impossible to him. In fact, Thm. II is more naturally read as a rule "from $\neg\varphi$ infer $\neg M\varphi$" and the equivalence of these two representations amounts to the question of whether the deduction theorem "$\neg\varphi \rightarrow \neg M\varphi$ is provable iff $\neg M\varphi$ is provable from $\neg\varphi$." The question regarding whether the deduction theorem should be valid in modal logic has later been repeatedly discussed, see e.g. R. Hakli and S. Negri, "Does the deduction theorem fail for modal logic?" *Synthese* 187 (2012): 849–67, hence Łukasiewicz may be considered as touching on this problem at an early stage. Concerning the extended propositional calculus, see e.g. Woleński 2004, and the references therein.

ogy (this is easily checked with $\|\varphi\|_e = 1/2$), the inference "from $\neg\varphi$ infer $\neg M\varphi$" is obtained in $Ł_3$ because it suffices to apply *modus ponens* twice to $\neg\varphi$ and the formula $\neg\varphi \rightarrow (\neg\varphi \rightarrow \neg M\varphi)$ which is a tautology of $Ł_3$. Although not explicitly mentioned, a consequence of this argument—nowadays a well-known property (Gottwald 2001, Hájek 1998b), cf. n. 31 on p. 118—is that the classical deduction theorem does not hold in $Ł_3$ because otherwise, the obtained rule "from $\neg\varphi$ infer $\neg M\varphi$" would imply that $\neg\varphi \rightarrow \neg M\varphi$ is a tautology. Łukasiewicz summarizes the situation as follows:[32]

> All the traditional theorems for modal propositions have been established free of contradiction in the three-valued propositional calculus, . . .

He then claims that one can proceed in a similar manner in any of his finitely- or infinitely-valued calculi, and expresses his preference for the latter:

> It was clear to me from the outset that among all the many-valued systems only two can claim any philosophical significance: the three-valued and the infinite-valued ones. For if values other than "0" and "1" are interpreted as "the possible," only two cases can reasonably be distinguished; either one assumes that there are no variations in degree of the possible and consequently arrives at the three-valued system; or one assumes the opposite, in which case it would be most natural to suppose (as in the theory of probabilities) that there are infinitely many degrees of possibility, which leads to the infinitely-valued propositional calculus. I believe that the latter system is preferable to all others. Unfortunately this system has not yet been investigated sufficiently.[33]

Another remark concerns the results on the extended many-valued propositional calculus, whose theorems neither contain nor are contained in the theorems of classical propositional logic and which thus presents an

> example of a consistent logical system which is as different from the ordinary two-valued systems as any non-Euclidean geometry is from the Euclidean . . .[34]

Łukasiewicz concludes by saying:

> [I]t seems to me that the philosophical significance of the systems of logic treated here might be at least as great as the significance of non-Euclidean systems of geometry.

Łukasiewicz and Tarski's report

The report by Łukasiewicz and Tarski (1930) represents an important survey of results obtained by members of the Warsaw seminar during the 1920s. The results, most of which had not been published before, appear without proofs and are attributed to the persons who obtained them.[35] In addition to Łukasiewicz logic, the report also describes results on classical logic.

---

[32] For this and the next three quotations, see Borkowski 1970 (pp. 172, 173, 175, 176).

[33] Łukasiewicz (1953) developed a new, four-valued system of modal logic in which negation may be regarded the same as in $Ł_4$ but implication is different. He expressed a preference for this four-valued logic over the three- and infinitely-valued systems and admitted: "This opinion [in Łukasiewicz 1930], as I see it today, was wrong." (Borkowski 1970, 371). The four-valued logic is analyzed and discussed in J. M. Font and P. Hájek, "On Łukasiewicz's four-valued modal logic," *Studia Logica* 70 (2002): 157–82.

[34] Section 5 of the report Łukasiewicz and Tarski 1930 describes investigations of this calculus.

[35] Unfortunately, these results have often been attributed mistakenly to both Łukasiewicz and Tarski. It clearly follows from the remarks in Łukasiewicz and Tarski 1930 and other writings, e.g. Łukasiewicz 1930, that the invention of the finitely- and infinitely-valued logics, extending the three-valued logic, is due to Łukasiewicz alone.

The Warsaw school followed the principle that a well-designed logical system should be simple and elegant. It is in the spirit of this principle that Łukasiewicz (1929a,b) introduced his now-famous notation for writing logical formulas.[36] In this notation, one does not need auxiliary symbols such as parentheses. As symbols of negation, conjunction, disjunction, implication, and equivalence, Łukasiewicz used N, K, A, C, and E, respectively. Connective symbols are put first, followed by their arguments. Hence, we have Cpq for $p \to q$ or $CCNpNqCqp$ for $(\neg p \to \neg q) \to (q \to p)$. This notation was used in most papers on Łukasiewicz logic until the 1960s, but has subsequently been abandoned in favor of the infix notation which we use in this book. The desire for simplicity is exemplified by demands of properties such as a small number of primitive concepts and axioms or independence of axioms. Many important results of the Warsaw school were actually directly inspired by these demands, see e.g. Woleński 2004.

The report contains Łukasiewicz's definitions of finitely- and infinitely-valued propositional logics.[37] He considered the following sets of truth values:

$$\text{Ł}_n = \{0, 1/n-1, \ldots, n-2/n-1, 1\}, \quad \text{for } n = 1, 2, 3, \ldots,$$
$$\text{Ł}_{\aleph_1} = [0, 1], \quad \text{i.e. the real unit interval,}$$
$$\text{Ł}_{\aleph_0} = [0, 1] \cap \mathbb{Q}, \quad \text{i.e. the rationals in } [0, 1].$$

In particular,

$$\text{Ł}_2 = \{0, 1\}, \ \text{Ł}_3 = \{0, 1/2, 1\}, \ \text{Ł}_4 = \{0, 1/3, 2/3, 1\}, \ \text{Ł}_5 = \{0, 1/4, 1/2, 3/4, 1\}, \text{ etc.}$$

The set of designated truth values is $D = \{1\}$ and $\neg$ and $\to$ with their truth functions defined by (4.5) and (4.4), respectively, are taken as basic connectives. Other connectives are defined as in (4.6)–(4.8). Clearly, Ł_3 and $\text{Ł}_{\aleph_1}$ are the calculi from Łukasiewicz 1920 and Łukasiewicz 1923.

Of primary interest then were the sets of tautologies. One easily gets $\text{Taut}(\text{Ł}_{\aleph_0}) \supseteq \text{Taut}(\text{Ł}_{\aleph_1})$ and due to the continuity of $\neg$ and $\to$, also $\text{Taut}(\text{Ł}_{\aleph_0}) \subseteq \text{Taut}(\text{Ł}_{\aleph_1})$. These facts were known to Lindenbaum who proved the following generalization:

Theorem 4.1. Let Ł_∞ be any infinite subset of $[0, 1]$ containing 1 and closed w.r.t. $\neg$ and $\to$ defined by (4.5) and (4.4).[38] Then the tautologies of Ł_∞ coincide with those of $\text{Ł}_{\aleph_0}$. In particular, $\text{Taut}(\text{Ł}_{\aleph_0}) = \text{Taut}(\text{Ł}_{\aleph_1})$.

As a consequence, we write Ł_∞ or even Ł instead of $\text{Ł}_{\aleph_0}$ and $\text{Ł}_{\aleph_1}$ and call Ł_∞ the infinitely-valued Łukasiewicz logic provided the set of truth values does not matter. The relationship between the tautologies was solved as follows:

---

[36] The notation later became known as the Polish notation. It is useful particularly for processing on computers.

[37] The definitions are given in terms of logical matrices (see p. 108 above). We present them in the way we have adopted in this book.

[38] An example of such Ł_∞, different from both $\text{Ł}_{\aleph_0}$ and $\text{Ł}_{\aleph_1}$, is the set of all reals in $[0, 1]$ which are of the form $a + bq$, where q is a fixed irrational number and a and b are arbitrary rationals.

Theorem 4.2. (a) $\text{Taut}(Ł_2)$ *is the set of all tautologies of classical propositional logic;* (b) $\text{Taut}(Ł_m) \subseteq \text{Taut}(Ł_n)$ *if and only if $n-1$ is a divisor of $m-1$;* (c) $\text{Taut}(Ł_\infty) = \bigcap_{n=1}^{\infty} \text{Taut}(Ł_n)$.

All claims are Łukasiewicz's except for the "⇒"-part of (b) which is Linden-baum's. Another issue examined in the report was *completeness with respect to a consequence operator* (p. 110 above). Following Tarski's investigations on deductive systems (Tarski 1930a), the operator Cn of consequence was considered, i.e.

$$Cn(T) = \bigcap \{S \mid S \supseteq T \text{ and } S \text{ is closed w.r.t. } \textit{modus ponens} \text{ and substitution}\}$$

for any set T of formulas.[39] Lindenbaum proved the following result:

Theorem 4.3. If $\varphi \notin \text{Taut}(Ł_3)$, *then* $\text{Taut}(Ł_3) \cup \{\varphi\}$ *is either equivalent to* $\text{Taut}(Ł_2)$ *or inconsistent.*

Theorem 4.3 reveals a situation which is not known in classical logic—adding to its axioms any formula that is not a classical tautology results in an inconsistent system. In this sense, classical logic is complete with respect to consequence. Now, $Ł_3$, which itself is a consistent logical system, is not complete. This problem has been studied further. Tarski has shown that theorem 4.3 holds true for any $Ł_n$ provided $n-1$ is prime. The problem for other n was open until Alan Rose's papers, to which we now turn. Recall first that according to Tarski 1930a,b, the *ordinal degree of completeness* (w.r.t. Cn) of T, in symbols $\gamma(T)$, is the least ordinal number α for which there does not exist a strictly increasing sequence of type α of consistent sets S of formulas that contain T and are closed, i.e. $Cn(S) = S$; the *cardinal degree of completeness* (w.r.t. Cn) of T, in symbols $g(T)$, is the cardinality of all closed sets S of formulas containing T. In terms of this notion, a direct consequence of Tarski's generalization of theorem 4.3 is that if $n-1$ is prime, the cardinal as well as the ordinal degree of completeness of $Ł_n$ is 3. The following theorem is due to Rose. The case of $Ł_n$ is proved in Rose 1952a with a correction in Rose 1969, the case of $Ł_\infty$ in Rose 1953.

Theorem 4.4.

(a) *For $n \geq 2$, the ordinal degree of completeness of $Ł_n$ is $d(n-1)+1$, i.e.* $\gamma(\text{Taut}(Ł_n)) = d(n-1)+1$, *where $d(n-1)$ is the number of divisors of $n-1$ (including 1 and $n-1$).*
(b) *The cardinal degree of completeness of $Ł_\infty$ is $\aleph_0$, i.e.* $g(\text{Taut}(Ł_\infty)) = \aleph_0$.[40]
(c) *The ordinal degree of completeness of $Ł_\infty$ is ω, i.e.* $\gamma(\text{Taut}(Ł_\infty)) = \omega$.

---

[39] *Modus ponens* and substitution are the same rules as in classical propositional logic (appendix B). Alternatively, $Cn(T)$ is the set of formulas provable from T using *modus ponens* and substitution.

[40] $\aleph_0$ is the least infinite cardinal, i.e. the cardinal number of the set of all nonnegative integers. In (c), ω is the least infinite ordinal, i.e. the order type of the set of all nonnegative integers.

The problem of the cardinal degree of completeness of Ł_n was later solved by Tokarz (1974) as follows. Let $c(n) = \{a_1, \ldots, a_k\}$ where $n - 1 = a_1 > a_2 > \cdots > a_k = 2$ are all the divisors of $n - 1$. Let $N_{c(n)}(a_i)$ denote the number of all subsequences s of $a_1, \ldots, a_k$ satisfying that $b \geq a_i$ for every b in s and that for every $a_i \neq a_j$ in s, a_i is not a divisor of a_j. Tokarz proved that

$$g(\text{Taut}(\text{Ł}_n)) = 1 + \sum_{a_i \in c(n)} N_{c(n)}(a_i).$$

An important problem is that of a complete axiomatization. For Ł_3 it was obtained by Wajsberg in the 1920s and later published in Wajsberg 1931. Its axioms are:

$$\varphi \to (\psi \to \varphi) \tag{4.9}$$

$$(\varphi \to \psi) \to ((\psi \to \chi) \to (\varphi \to \chi)) \tag{4.10}$$

$$(\neg\psi \to \neg\varphi) \to (\varphi \to \psi) \tag{4.11}$$

$$((\varphi \to \neg\varphi) \to \varphi) \to \varphi \tag{4.12}$$

With the rules of *modus ponens* and substitution and the usual notion of provability Wajsberg proved the following theorem:[41]

Theorem 4.5. A formula φ is provable from (4.9)–(4.12) iff φ is a tautology of Ł_3.

Łukasiewicz and Tarski 1930 contains Wajsberg's axiomatization along with several other interesting results, some of which never appeared with proofs. In particular, we mention a theorem due to Lindenbaum according to which Ł_n is axiomatizable for every $n = 1, 2, 3, \ldots$ Łukasiewicz and Tarski mention that the theorem has an effective proof which enables one to obtain the particular axiomatizations. Explicit axiomatizations of the logics Ł_n later indeed appeared (see p. 156 and p. 183 below). Another interesting theorem, which documents the meticulous nature of these investigations, says that for every $n \geq 2$, there exists an axiomatization of Ł_n containing exactly m independent axioms, for any $m \geq 1$, and in every such axiomatization, there appear at least three distinct symbols of propositions.

Importantly, the report also contains a conjecture that the infinitely-valued Łukasiewicz logic Ł_∞ is axiomatized by the following axioms:[42]

$$\varphi \to (\psi \to \varphi) \tag{4.13}$$

$$(\varphi \to \psi) \to ((\psi \to \chi) \to (\varphi \to \chi)) \tag{4.14}$$

$$(\neg\psi \to \neg\varphi) \to (\varphi \to \psi) \tag{4.15}$$

$$((\varphi \to \psi) \to \psi) \to ((\psi \to \varphi) \to \varphi) \tag{4.16}$$

$$((\varphi \to \psi) \to (\psi \to \varphi)) \to (\psi \to \varphi) \tag{4.17}$$

This conjecture was confirmed by Wajsberg (1935), who mentioned on p. 240 that the proof was presented in the Faculty of Mathematics at the University of Warsaw

---

[41] As in the classical case (appendix B), one may clearly consider (4.9)–(4.12) as axiom schemes and omit the rule of substitution.

[42] Notice that (4.13)–(4.15) are identical to (4.9)–(4.11), respectively.

and "wird nächstens wahrscheinlich als Publikation der Warschauer Gesellschaft für Wissenschaft erscheinen" ["will soon appear probably as a publication of the Warsaw Scientific Society"]. Wajsberg's proof, however, never appeared in print.

Completeness (axiomatizability) of $Ł_\infty$, Chang's proof, and MV-algebras

For the first time, a proof of axiomatizability of the infinitely-valued Łukasiewicz logic $Ł_\infty$ by (4.13)–(4.17) appeared in Rose and Rosser 1958, definitely confirming Łukasiewicz's conjecture. Interestingly, Carew A. Meredith, a student of Łukasiewicz in Dublin, and Chen Chung Chang proved independently that (4.17) is provable from (4.13)–(4.16), and may hence be dropped. Their papers, Meredith 1958 and Chang 1958b, appeared immediately after Rose and Rosser's paper in the same issue. According to the editor's note in Meredith 1958, Meredith obtained the result in 1955, while Chang did so in 1956. The latter fact is confirmed by Chang (1998). It is worth noting that (4.17), which may be rewritten on account of (4.6) as $(\varphi \to \psi) \vee (\psi \to \varphi)$, is the axiom of prelinearity which plays an important role in fuzzy logic (p. 194 below). Taking the dispensability of (4.17) into account, Rose and Rosser's proof essentially established the following theorem:

Theorem 4.6. A formula φ is provable from (4.13)–(4.16) iff φ is a tautology of $Ł_\infty$.

The independence of the remaining four axioms was later established by Turquette (1963). Rose and Rosser's proof utilizes linear inequalities and an important theorem due to McNaughton which we examine below.

Shortly after Rose and Rosser, Chang worked out another proof of theorem 4.6. Chang obtained his PhD in mathematics from UC Berkeley in 1955 under the supervision of Tarski, who came to the United States in August 1939, shortly after the invasion of Poland. Even though Tarski was one of the leading figures in the early developments of Łukasiewicz logic, it was Rosser who stimulated Chang's interest.[43] This happened in the spring of 1956 when Rosser gave a seminar based on the results later published in Rose and Rosser 1958. According to Chang 1998, he attended this seminar but had "no idea where the arguments were leading up to." When he was about to give up in trying to follow them, the idea occurred to him that another approach to prove theorem 4.6 might be possible, namely one analogous to the algebraic method for proving completeness of classical logic via the Lindenbaum algebra. In classical logic, such an algebra is a Boolean algebra and the method is based on the theory of Boolean algebras.[44] For Łukasiewicz logic, however, a new kind of algebra would be needed. Eventually, Chang succeeded in his effort and developed so-called *MV-algebras*, with "MV" from "many-valued," and the algebraic proof of

---

[43] The historical information we provide in this paragraph is based on Chang 1998.

[44] The method was developed for predicate logic by Rasiowa and Sikorski (1950), who also mention in the last section that a similar method may be used for propositional logic. For propositional logic, the method explicitly appeared, though in a slightly different fashion and without a reference to Rasiowa and Sikorski, in J. Łoś, "An algebraic proof of completeness for the two-valued propositional calculus," *Colloquium Mathematicae* 2 (1951): 236–40.

completeness for Ł_∞ in Chang 1958a, 1959. Interestingly, Chang (1998) says: "[I]f I had been able to follow Rose and Rosser's arguments, then there would be no MV-algebras (at least not by me)." Even though Chang's two papers may seem to present a work smoothly carried out according to a detailed plan, the reality was different. The proof for classical logic is based on the results, essentially due to Stone (1936), regarding maximal ideals in Boolean algebras, in particular the fact that a factor algebra by a maximal ideal is just the two-element Boolean algebra and that for every element $a \neq 1$ there exists a maximal ideal containing a. Although Chang obtained several deep results on MV-algebras and their ideals, including those on maximal ideals, the needed notion eluded him. At the beginning of 1957, he sent off the manuscript of Chang 1958a only because his way of proving the dependence result in Chang 1958b, which itself was a big surprise, gave some hope for the usefulness of MV-algebras. Nevertheless, Chang's failure to prove completeness was "a disappointment to me at that time" (Chang 1998). This all changed during the Conference on Logic held in the summer of 1957 at Cornell when Dana Scott, after listening to Chang's talk based on Chang 1958a, suggested the notion of a prime ideal for MV-algebras. While in Boolean algebras this notion coincides with that of a maximal ideal, for MV-algebras these notions are different. In particular, it follows that factor algebras by prime ideals are just the linearly ordered MV-algebras, which fact along with further suggestions made by Scott provided the decisive impetus for Chang's eventual success in proving the completeness theorem by the end of 1957 (Chang 1959). Since Chang's study proved to be of great importance for subsequent developments in fuzzy logic, we examine his method in some detail.

A (Chang) *MV-algebra*[45] is an algebra $\mathbf{L} = \langle L, +, \cdot, ^-, 0, 1 \rangle$, where $+$ and $\cdot$ are binary operations, $^-$ is unary, and 0 and 1 are elements of L, satisfying

$$a + b = b + a, \qquad\qquad\qquad a \cdot b = b \cdot a,$$
$$a + (b + c) = (a + b) + c, \qquad\qquad a \cdot (b \cdot c) = (a \cdot b) \cdot c,$$
$$a + a^- = 1, \qquad\qquad\qquad a \cdot a^- = 0,$$
$$a + 1 = 1, \qquad\qquad\qquad a \cdot 0 = 0,$$
$$a + 0 = a, \qquad\qquad\qquad a \cdot 1 = a,$$
$$(a + b)^- = a^- \cdot b^-, \qquad\qquad (a \cdot b)^- = a^- + b^-,$$
$$a = (a^-)^-,$$
$$0^- = 1,$$

and, when $\sqcap$ and $\sqcup$ are defined by

$$a \sqcup b = (a \cdot b^-) + b \quad \text{and} \quad a \sqcap b = (a + b^-) \cdot b,$$

---

[45] We use "Chang MV-algebra" to distinguish it from other, equivalent definitions of MV-algebras, which appeared later. The other definitions are discussed in section 4.4.

also satisfying

$$a \sqcup b = b \sqcup a, \qquad\qquad a \sqcap b = b \sqcap a,$$
$$a \sqcup (b \sqcup c) = (a \sqcup b) \sqcup c, \qquad\qquad a \sqcap (b \sqcap c) = (a \sqcap b) \sqcap c,$$
$$a + (b \sqcap c) = (a + b) \sqcap (a + c), \qquad\qquad a \cdot (b \sqcup c) = (a \cdot b) \sqcup (a \cdot c).$$

Chang showed that the relation $\leq$ defined on any MV-algebra by $a \leq b$ iff $a \sqcup b = b$ is a partial order with respect to which the operations $+$, $\cdot$, $\sqcup$, and $\sqcap$ are isotone. Moreover, $\langle L, \sqcap, \sqcup, 0, 1 \rangle$ is a bounded lattice in which $\sqcap$ and $\sqcup$ are the infimum and supremum w.r.t. $\leq$. The first example of an MV-algebra provided by Chang is the algebra of classes of provably equivalent formulas of $Ł_\infty$. That is,

$$L = \{ [\varphi]_\equiv \mid \varphi \text{ is a formula of } Ł_\infty \} \quad \text{with} \quad \varphi \equiv \psi \text{ iff } \vdash \varphi \leftrightarrow \psi,$$

where $\vdash$ is the provability from (4.13)–(4.16). Now, $\equiv$ is an equivalence which is a congruence in that if $\varphi \equiv \psi$ then $\neg \varphi \equiv \neg \psi$, and if $\varphi_1 \equiv \psi_1$ and $\varphi_2 \equiv \psi_2$, then $\varphi_1 \to \varphi_2 \equiv \psi_1 \to \psi_2$. As a consequence, one may introduce the operations $+$, $\cdot$, $-$, 0, and 1 on L by $[\varphi]_\equiv + [\psi]_\equiv = [\neg \varphi \to \psi]_\equiv$, $[\varphi]_\equiv \cdot [\psi]_\equiv = [\neg(\varphi \to \neg\psi)]_\equiv$, $[\varphi]_\equiv^- = [\neg\varphi]_\equiv$, $0 = [\varphi]_\equiv$, where φ is such that $\vdash \neg\varphi$, and $1 = [\varphi]_\equiv$, where φ is such that $\vdash \varphi$. Chang proved that $\mathbf{L} = \langle L, +, \cdot, -, 0, 1 \rangle$—the *Lindenbaum algebra of $Ł_\infty$* (cf. p. 158)—is an MV-algebra. Another example of an MV-algebra pointed out by Chang, in fact a family of them, consists of the sets $Ł_n$, $Ł_{\aleph_0}$, or $Ł_{\aleph_1}$ (p. 120) equipped with the operations

$$a + b = \min(1, a + b), \tag{4.18}$$
$$a \cdot b = \max(0, a + b - 1), \tag{4.19}$$
$$a^- = 1 - a. \tag{4.20}$$

The algebra on $Ł_{\aleph_1}$ is called the *standard MV-algebra*, denoted $[0,1]_Ł$. The operations $+$ and $\cdot$ defined by (4.18) and (4.19) are nowadays called the Łukasiewicz disjunction and conjunction, respectively, although Łukasiewicz never used them. The earliest use of these terms is probably due to Frink (1938), who calls them "Łukasiewicz-Tarski arithmetical sum and product."[46] Still another example is what is now called the *MV-algebra of infinitesimals* (Cignoli, D'Ottaviano, and Mundici 2000). The second part of Chang 1958a provides a deeper algebraic study of MV-algebras and considerations regarding the completeness of $Ł_\infty$.

In the 1959 paper, Chang utilized MV-algebras to work out a new proof of theorem 4.6. Note first that the basic operations of Chang's MV-algebras are different from those corresponding to the basic connectives $\neg$ and $\to$ of $Ł_\infty$. However, one easily checks that on $[0,1]$, Łukasiewicz's truth functions and Chang's MV-algebra operations (4.18)–(4.20) are mutually definable via

$$a + b = \neg a \to b, \quad a \cdot b = \neg(a \to \neg b), \quad a^- = \neg a, \tag{4.21}$$

---

[46] The terms "(logical) sum" and "(logical) product" were commonly used at that time for what we now call the truth function of disjunction and conjunction, respectively.

and, conversely,

$$\neg a = a^- \quad \text{and} \quad a \to b = a^- + b.$$

Correspondingly, one may alternatively consider as basic the connectives $\oplus, \otimes, \neg, \overline{0}$, and $\overline{1}$, which are mutually definable with Łukasiewicz's $\neg$ and $\to$ when considering

$$\varphi \oplus \psi, \quad \varphi \otimes \psi, \quad \overline{0}, \quad \text{and} \quad \overline{1} \quad \text{as shorthands for} \qquad (4.22)$$

$$\neg \varphi \to \psi, \quad \neg(\varphi \to \neg\psi), \quad \neg(p \to p), \quad \text{and} \quad p \to p,$$

with p being an arbitrary propositional symbol, and

$$\varphi \to \psi \text{ a shorthand for } \neg\varphi \oplus \psi. \qquad (4.23)$$

The new connectives, $\oplus, \otimes, \neg, \overline{0}$, and $\overline{1}$, are then directly interpretable in MV-algebras by $+, \cdot, {}^-, 0$, and 1.[47] It is now immediate that a formula φ over $\neg$ and $\to$ is a tautology of $Ł_\infty$ iff its translation according to (4.23) is a $[0,1]_Ł$-tautology iff for the term t_φ in the language of MV-algebras corresponding to the translation of φ it holds that the equation $t_\varphi = 1$ is valid in $[0,1]_Ł$. For instance, for φ being $p \to (q \to p)$ we get that $p \to (q \to p)$ is a tautology of $Ł_\infty$ iff the translation $\neg p \oplus (\neg q \oplus p)$ is a $[0,1]_Ł$-tautology iff the equation $x^- + (y^- + x) = 1$ is valid in $[0,1]_Ł$. Chang's proof of theorem 4.6 is based on two theorems revealing the role of MV-chains, i.e. linearly ordered MV-algebras which by definition satisfy $a \le b$ or $b \le a$ for every a and b. The first is:

Theorem 4.7. *Every MV-algebra is a subdirect product of MV-chains.*

This theorem is proved by means of the prime ideals mentioned above. Since a subdirect product is a particular subalgebra of the direct product, any equation $t_1 = t_2$ valid in every MV-chain is valid in every MV-algebra. The second is:

Theorem 4.8. *An equation $t_1 = t_2$ in the language of MV-algebras is valid in the standard MV-algebra $[0,1]_Ł$ if and only if it is valid in every MV-chain.*

This theorem is based on deep results from the theory of ordered Abelian groups (cf. n. 161 on p. 189 below) and on Chang's construction that makes it possible to identify MV-chains with certain ordered Abelian groups. For example, the standard MV-algebra $[0,1]_Ł$ may thus be identified with the group $\langle \mathbb{R}, + \rangle$ of real numbers. On the one hand, $[0,1]_Ł$ is constructed as an interval in $\langle \mathbb{R}, + \rangle$ with operations obtained from those of $\langle \mathbb{R}, + \rangle$ by (4.18)–(4.20). On the other hand, $\langle \mathbb{R}, + \rangle$ may be constructed, though in a more involved way, from $[0,1]_Ł$ so that the two constructions are mutually inverse. Now, the soundness part of theorem 4.6 is easy to prove. Namely, since the axioms (4.13)–(4.16) are tautologies and since *modus ponens* preserves tautologicity, every provable formula is a tautology of $Ł_\infty$. Chang proved the opposite direction as follows. Let φ be a tautology of $Ł_\infty$. According to the above observation, $t_\varphi = 1$ is valid in $[0,1]_Ł$. Due to theorem 4.8, $t_\varphi = 1$ is

---

[47]Note in passing that $\oplus$ and $\otimes$, considered as additional connectives defined by (4.22), appear already in Rose and Rosser 1958, where they are denoted as B and L.

valid in every linearly ordered MV-algebra. Now, due to theorem 4.7 and its consequence mentioned above, $t_\varphi = 1$ is valid in every MV-algebra. In particular, $t_\varphi = 1$
is valid in the above Lindenbaum algebra L of $Ł_\infty$. The latter fact implies that for
the evaluation e sending every propositional symbol p to the class $[p]_\equiv$, one has
$\|\varphi\|_e^L = 1^L$. But since $\equiv$ is a congruence, one obtains $\|\varphi\|_e^L = [\varphi]_\equiv$. Taking into
account that 1^L is $[p \to p]_\equiv$, we have $[\varphi]_\equiv = [p \to p]_\equiv$. The definition of $\equiv$ now
yields $\vdash \varphi \leftrightarrow (p \to p)$, whence also $\vdash (p \to p) \to \varphi$. Finally, applying *modus ponens*
to $\vdash p \to p$ and the last formula yields that $\vdash \varphi$, i.e. φ is provable.

    Several other proofs of theorem 4.6 appeared subsequently and we mention
them below in this chapter. Note also that Chang's method has been adopted for
the finitely-valued Łukasiewicz logics $Ł_n$ by Grigolia (p. 183 below).

Functional incompleteness of Łukasiewicz logics, McNaughton's theorem

Contrary to classical logic, none of the Łukasiewicz logics is functionally complete.
This is because Łukasiewicz's truth functions of $\neg$ and $\to$ extend classical ones and
hence assign 0 or 1 whenever the arguments are 0 or 1. Hence, no combination of
them may express any function assigning a value different from 0 and 1 to arguments
that are 0 or 1, e.g. $f(0,1) = 1/2$. Two natural questions thus arise. First, can one add
to $\neg$ and $\to$ additional connectives so that the resulting system becomes functionally complete? Second, what functions may be obtained as the truth functions of
formulas in Łukasiewicz logics?

    The first question was studied by Słupecki (1936), who showed that expand-
ing $Ł_3$ by a unary connective T whose truth function is constantly equal to $1/2$, i.e.
$T(0) = T(1/2) = T(1) = 1/2$, and adding to the axioms for $Ł_3$ the axioms

$$T\varphi \to \neg T\varphi \quad \text{and} \quad \neg T\varphi \to T\varphi, \qquad (4.24)$$

one obtains the following theorem:[48]

*Theorem 4.9. The system $\{\neg, \to, T\}$ is functionally complete; a formula φ involving
$\neg$, $\to$, and T is provable from (4.9)–(4.12) and (4.24) iff φ is a tautology.*

    Słupecki obtained further results on representations of truth functions to which
we return on p. 157. His result on $Ł_3$ has been generalized for $Ł_n$ in Rosser and Turquette 1952. The problem of when adding a constant function makes $Ł_n$ functionally complete has been solved completely by Evans and Schwartz (1958):

*Theorem 4.10. $Ł_n$ expanded by a constant function $T_{i/n-1}$ taking on the value $i/n-1$ is
functionally complete iff $n-1$ and $\min(n-i, i+1) - 1$ are relatively prime.*

    As one easily checks, adding $T_{1/n-1}$ or $T_{n-2/n-1}$ always results in a functionally com-
plete system, both yielding Słupecki's result for $Ł_3$.

---

[48]According to Słupecki, "T" stands for "tertium."

The second question was solved by Robert McNaughton (1924–2014). Call a function $f : [0,1]^n \to [0,1]$ *representable in* $Ł_\infty$ if there exists a formula φ of $Ł_\infty$ for which $f = \varphi$, see (4.1). Observe first that the truth function of $\neg$ is a polynomial, namely $f(x_1) = 1 - x_1$. That of $\to$ is described by two polynomials, $f_1(x_1, x_2) = 1 - x_1 + x_2$ and the constant polynomial $f_2(x_1, x_2) = 1$, in that it equals $f_1(x_1, x_2)$ for $x_2 \leq x_1$ and $f_2(x_1, x_2)$ for $x_2 > x_1$. The above polynomials have all their coefficients equal to 1. This need not be the case because the truth function f of the formula $p \otimes p \otimes \cdots \otimes p$ with n occurrences of p is the polynomial $f(x_1) = n \cdot x_1 - (n-1)$ with integer coefficients. McNaughton (1951) proved that this kind of characterization yields all functions representable in $Ł_\infty$:

Theorem 4.11. *A function* $f : [0,1]^n \to [0,1]$ *is representable in* $Ł_\infty$ *if and only if it is continuous and there exist polynomials*

$$f_i(x_1, \ldots, x_n) = \sum_{j=1}^{n} a_{ij} x_j + b_i, \quad i = 1, \ldots, m,$$

with integer coefficients a_{ij} *and* b_i *such that for every* $a_1, \ldots, a_n \in [0,1]$ *there exists* i *such that* $f(a_1, \ldots, a_n) = f_i(a_1, \ldots, a_n)$.

Using theorem 4.11, he also characterized the finitely-valued case:

Theorem 4.12. *A function* $f : Ł_m^n \to Ł_m$ *is representable in* $Ł_m$ *if and only if for all truth values* $k_1/m-1, \ldots, k_n/m-1 \in Ł_m$, *every common divisor of* $k_1, \ldots, k_n, m-1$ *is a divisor of* k, *where* $k/m-1 = f(k_1/m-1, \ldots, k_n/m-1)$.

Notice that for $m = 2$, theorem 4.12 implies that every function $f : \{0, 1\}^n \to \{0, 1\}$ is representable in $Ł_2$, i.e. in classical propositional logic. In addition, the condition in theorem 4.12 may naturally be reformulated to prove that f is representable in $Ł_m$ if and only if for every $Ł_q \subseteq Ł_m$, f maps $Ł_q^n$ into $Ł_q$.

Łukasiewicz-Moisil algebras

In the late 1930s, the Romanian mathematician Grigore C. Moisil (1906–1973) initiated a study of certain algebraic structures related to the finitely-valued Łukasiewicz logics which are nowadays known as *Łukasiewicz-Moisil algebras*.[49] Along with Chang's MV-algebras, these algebras are the main algebraic structures connected to Łukasiewicz logic. Moisil (1940, 1941) introduced his three-, four-, and generally n-valued algebras as having the lattice operations, an operation of negation, and certain unary operations called Chryssipean endomorphisms, and investigated logics corresponding to these algebras in the 1960s. In 1968, he also introduced more general, possibly infinitely-valued algebras. According to Georgescu, Iorgulescu, and Rudeanu 2006 (p. 83), Moisil was thinking of the infinitely-valued algebras

---

[49] See Georgescu, Iorgulescu, and Rudeanu 2006.

a long time before but it was only after the appearance of fuzzy set theory that he obtained solid motivations to publish his ideas. Because it was shown that for $n \geq 5$, the Łukasiewicz implication cannot be expressed in such n-valued algebra,[50] Cignoli (1982) has shown that true counterparts of Łukasiewicz logics are obtained when extending the n-valued Łukasiewicz-Moisil algebras with certain binary operations. The theory of Łukasiewicz-Moisil algebras has been developed to a great extent by Moisil and his pupils, including Georgescu and Rudeanu, as well as other mathematicians and logicians, most notably of the Bahía-Blanca school of António Monteiro (1907–1980) in Argentina. For further information, see Boicescu et al. 1991.

This ends our visit to the early developments of the propositional Łukasiewicz logics before the 1960s. Those of the predicate case are described in section 4.2.4. Further contributions are covered subsequently, particularly in section 4.4.

4.2.2 Gödel finitely- and infinitely-valued logics

In a short note, Gödel (1932) introduced a family of many-valued connectives which later became of great importance for fuzzy logic. Kurt Gödel (1906–1978), arguably the greatest logician of the 20th century,[51] used this family to show that there is no finite structure of truth values adequate for Heyting's axiomatization of intuitionistic logic.

Intuitionism is an important direction in the foundations of mathematics. Its roots may be found in the works of several prominent mathematicians, including Kronecker, Lebesgue, and Poincaré. It is, however, the Dutch mathematician Luitzen Egbertus Jan Brouwer (1881–1966) who is clearly considered its founder.[52] Intuitionism considers mathematics a result of constructive mental activity, rather than a formal system in which truth may be captured by a formal notion of proof based on the principles of classical logic. It rejects mechanical applications of classical logical rules in deriving mathematical theorems. A case in point is Brouwer's criticism and rejection of the law of excluded middle, $\varphi \vee \neg\varphi$. In classical mathematics, this law allows one to prove a statement such as "the number x is rational" by deriving a contradiction from its negation. To prove "the number x is rational," intuitionists require instead that two integers, p and q, be found for which $x = p/q$, i.e. a construction of x must be given. Intuitionists thus restrict classical means of proving theorems and thus restrict the set of provable mathematical claims. Even though the very conception of intuitionism precludes considering intuitionistic mathematics as a system with formally codified axioms and deduction rules, Brouwer's exam-

---

[50] Observation by A. Rose in 1957; see the introduction in R. Cignoli, "Algebras de Moisil de orden n" (PhD diss., Bahía Blanca: Universidad Nacional del Sur, 1969).

[51] See e.g. J. W. Dawson Jr., *Logical Dilemmas: The Life and Work of Kurt Gödel* (Wellesley, MA: Peters, 1997) and J. L. Casti and W. DePauli, *Gödel: A Life of Logic* (Cambridge, MA: Basic Books, 2001).

[52] A seminal work is Brouwer's 1907 thesis whose English translation, "On the foundations of mathematics," is available in A. Heyting, ed., *L. E. J. Brouwer: Collected Works. 1. Philosophy and Foundations of Mathematics* (Amsterdam: North-Holland, 1975), 11–101. A good introduction to intuitionism is Heyting 1956, probably the most popular book on intuitionism.

ination of the rules of correct logical practice led to a debate over what had been called a Brouwerian logic and to the development of intuitionistic logic.[53] Soon the problem of describing intuitionistic logic as a formal logical calculus appeared. Such a calculus emerged from the work of Valerii Glivenko (1897–1940) and Arend Heyting (1898–1980), a student of Brouwer who later took a leading role in the intuitionist movement. Heyting proposed his calculus in a paper which was awarded a prize by the Dutch Mathematical Association in 1928 and whose reworked first part later appeared as Heyting 1930.[54] This calculus, now called the *intuitionistic propositional logic*, is based on the following axioms:

$$\varphi \rightarrow (\varphi \wedge \varphi) \tag{4.25}$$

$$(\varphi \wedge \psi) \rightarrow (\psi \wedge \varphi) \tag{4.26}$$

$$(\varphi \wedge \psi) \rightarrow ((\varphi \wedge \chi) \rightarrow (\psi \wedge \chi)) \tag{4.27}$$

$$((\varphi \rightarrow \psi) \rightarrow (\psi \rightarrow \chi)) \rightarrow (\varphi \rightarrow \chi) \tag{4.28}$$

$$\varphi \rightarrow (\psi \rightarrow \varphi) \tag{4.29}$$

$$(\varphi \wedge (\varphi \rightarrow \psi)) \rightarrow \psi \tag{4.30}$$

$$\varphi \rightarrow (\varphi \vee \psi) \tag{4.31}$$

$$(\varphi \vee \psi) \rightarrow (\psi \vee \varphi) \tag{4.32}$$

$$((\varphi \rightarrow \chi) \wedge (\psi \rightarrow \chi)) \rightarrow ((\varphi \vee \psi) \rightarrow \chi) \tag{4.33}$$

$$\neg\varphi \rightarrow (\varphi \rightarrow \psi) \tag{4.34}$$

$$((\varphi \rightarrow \psi) \wedge (\varphi \rightarrow \neg\psi)) \rightarrow \neg\varphi \tag{4.35}$$

Importantly, Heyting was not attempting to build a logical system representing intuitionistic mathematics as a whole. He described his goal thus (Heyting 1930):[55]

> It is in principle impossible to set up a system of formulas that would be equivalent to intuitionistic mathematics, for the possibilities of thought cannot be reduced to a finite number of rules set up in advance. . . . the attempt to reproduce the most important parts of mathematics in formal language is justified exclusively by the greater conciseness . . .

Heyting proved independence of the axioms and proved that the law of excluded middle cannot be derived from them. He showed the latter by providing semantics

---

[53] The term "Logique Brouwerienne," French for "Brouwerian logic," was coined by R. Wavre. An examination of the emergence of intuitionistic logic is provided in part IV of Mancosu 1998.

[54] There is an apparent mutual influence between Glivenko and Heyting. In note 3 to his 1929 paper, Glivenko acknowledges that Heyting made him aware of the appropriateness of certain axioms which Glivenko employed; see Glivenko, "Sur quelques points de la logique de M. Brouwer" [On some points of the logic of Mr. Brouwer], *Académie Royale de Belgique, Bulletins de la Classe des Sciences* 5 (1929): 183–88, trans. P. Mancosu and W. P. van Stigt, in Mancosu 1998, 301–5. On the other hand, speaking of his logical calculus, Heyting says ". . . one obtains the logic devised by Mr. Glivenko, and which I have developed in more detail in a recent paper"; see Heyting, "Sur la logique intuitionniste" [On intuitionistic logic], *Académie Royale de Belgique, Bulletins de la Classe des Sciences* 16 (1930): 957–63, trans. P. Mancosu and W. P. van Stigt, in Mancosu 1998, 306–10.

[55] English translation is from p. 311 in Mancosu 1998.

based on the following three-valued structure:[56]

φ	$\neg\varphi$	$\rightarrow$	0	1/2	1	$\wedge$	0	1/2	1	$\vee$	0	1/2	1
0	1	0	1	1	1	0	0	0	0	0	0	1/2	1
1/2	0	1/2	0	1	1	1/2	0	1/2	1/2	1/2	1/2	1/2	1
1	0	1	0	1/2	1	1	0	1/2	1	1	1	1	1

The appropriateness of the axioms of intuitionistic logic was, however, being assessed with respect to an informally described semantics, later known as the BHK (Brouwer, Heyting, Kolmogorov) interpretation, whose basic idea is that propositions correspond to problems or expectations. Nevertheless, the question of whether there exists a many-valued semantics adequate for intuitionistic logic gained attention. In his important contribution, Gödel (1932) proved the following theorem:

Theorem 4.13. There is no finite structure of truth values such that the tautologies are just the provable formulas of intuitionistic propositional logic.

　　Furthermore, there is an infinite chain of systems between the intuitionistic and classical propositional logics, ordered by inclusion of the sets of their tautologies.

Gödel's argument goes as follows. By contradiction, assume a system with $n \geq 2$ truth values in which tautologies are just the formulas provable from the axioms (4.25)–(4.35). Consider the formulas γ_m of the form

$$\bigvee_{1 \leq i < j \leq m} p_i \leftrightarrow p_j,$$

where $p_i \leftrightarrow p_j$ is a shorthand for $(p_i \rightarrow p_j) \wedge (p_j \rightarrow p_i)$. Then γ_{n+1} is a tautology, i.e. assumes a designated truth value for every evaluation e. Indeed, since e uses at most n truth values, there exist $i \neq j$ for which $e(p_i) = e(p_j)$. Now, γ_{n+1}, rewritten as $(p_i \leftrightarrow p_j) \vee \varphi$, has the same truth value as $(p \leftrightarrow p) \vee \varphi$. As $(p \leftrightarrow p) \vee \varphi$ is provable from intuitionistic axioms, it is a tautology and hence takes on 1 in e. Hence, γ_{n+1} also takes on 1 in e. Since e is arbitrary, γ_{n+1} is a tautology and hence provable due to the assumption. To establish the first part of theorem 4.13, it suffices to check that none of $\gamma_2, \gamma_3, \gamma_4, \ldots,$ is provable from the axioms, a contradiction to the provability of γ_{n+1}. To verify this fact, Gödel introduced for every $n = 2, 3, 4\ldots,$ an n-valued system G_n with the set $L_n = \{0, 1/n{-}1, \ldots, n{-}2/n{-}1, 1\}$ of truth values, designated truth value 1, and truth functions defined by:[57]

$$\neg a = \begin{cases} 1 & \text{if } a = 0, \\ 0 & \text{if } a \neq 0, \end{cases} \tag{4.36}$$

$$a \rightarrow b = \begin{cases} 1 & \text{if } a \leq b, \\ b & \text{if } a > b, \end{cases} \tag{4.37}$$

---

[56] One observes that with 1 as a designated element, (4.25)–(4.35) are all tautologies and *modus ponens* preserves tautologicity. Hence, all intuitionistically provable formulas are tautologies. On the other hand, $p \vee \neg p$ representing the law of excluded middle is not a tautology in the three-valued semantics, because for an evaluation $e(p) = 1/2$, its truth value is $1/2$. Hence, $p \vee \neg p$ is not provable from Heyting's axioms.

[57] In fact, Gödel used $\{1, 2, \ldots, n\}$ as the set of truth values with 1 and n representing truth and falsity. Hence, in the original formulas, Gödel's has max in the definition of $\wedge$, min in the definition of $\vee$, and so on.

$$a \wedge b = \min(a, b), \tag{4.38}$$

$$a \vee b = \max(a, b). \tag{4.39}$$

Notice that G_2 and G_3 coincide with classical propositional logic and with the above three-valued system of Heyting, respectively. Since all axioms (4.25)–(4.35) are tautologies in any G_n and since *modus ponens* preserves tautologicity, all formulas provable from the axioms are tautologies as well. On the other hand, γ_n is not a tautology of G_n since it assumes the truth value $n-2/n-1$ when different truth values are assigned to different p_is in γ_n. Therefore, γ_n is not provable from the axioms, hence the first part of theorem 4.13. The second part now follows from the fact that $\mathrm{Taut}(G_{i+1}) \subseteq \mathrm{Taut}(G_i)$ and the above argument, due to which γ_{i+1} is a tautology of G_i but not of G_{i+1} for each i, hence the systems $G_2, G_3, G_4, \ldots$, have mutually distinct sets of tautologies.

This argument lets us see the reason for Gödel's choice of truth functions of connectives but also that he could not have used the then-available Łukasiewicz's connectives because they do not satisfy the intuitionistic axioms. Note also that Jaśkowski (1936) found a single infinite structure providing adequate semantics for Heyting's axioms.[58]

Inspired by Gödel's results, Dummett (1959) proposed an important extension to infinite sets of truth values.[59] He defined the truth functions by Gödel's formulas (4.36)–(4.39) on the countably infinite set $L_{\aleph_0} = [0, 1] \cap \mathbb{Q}$ and showed that a complete axiomatization of the resulting logic $G_{\aleph_0}$ is obtained by adding the axiom

$$(\varphi \rightarrow \psi) \vee (\psi \rightarrow \varphi) \tag{4.40}$$

to the original Heyting's axioms:[60]

Theorem 4.14. φ *is provable from* (4.25)–(4.35) *and* (4.40) *iff* φ *is a tautology of* $G_{\aleph_0}$.

Dummett showed that with any infinite bounded chain L instead of $L_{\aleph_0}$, the set of tautologies does not change. As a result, the tautologies of the system $G_{\aleph_1}$ in which $L = [0, 1]$ coincide with those of $G_{\aleph_0}$. He also showed that these are just the tautologies common to $G_2, G_3, G_4, \ldots$ He thus proved the following theorem:

Theorem 4.15. *For any infinite chain L bounded by 0 and 1 with the truth functions* (4.36)–(4.39), *we have* $\mathrm{Taut}(G_L) = \mathrm{Taut}(G_{\aleph_0}) = \mathrm{Taut}(G_{\aleph_1}) = \bigcap_{n=1}^{\infty} \mathrm{Taut}(G_n)$.

Furthermore, he applied Gödel's argument to show that there is no finite structure of truth values adequate for (4.25)–(4.35) and (4.40). He also proved that adding to the intuitionistic axioms the axiom $\neg\varphi \vee \neg\neg\varphi$, (4.40) remains unprovable.

---

[58] He expounded only a rough idea of his proof. A detailed proof appears in S. J. Surma, "Jaśkowski's matrix criterion for the intuitionistic propositional calculus," in *Studies in the History of Mathematical Logic*, ed. S. J. Surma (Wroclaw: Polish Academy of Sciences, 1973), 87–121.

[59] The infinitely-valued logic, which we call *Gödel logic*, is sometimes called *Gödel-Dummett logic*.

[60] In fact, Dummett used $L = \{0, 1, 2, \ldots, \omega\}$, with 0 and ω representing truth and falsity, and used the original Gödel's formulas in which max appears in the definition of $\wedge$ and likewise for the other connectives, cf. n. 57 above. As we mention later, the sets of tautologies corresponding to these two structures coincide. Axiom (4.40), called now the *axiom of prelinearity* (p. 194 below), plays an important role in contemporary fuzzy logic; see section 4.4.

The systems $G_3, G_4, \ldots$, and $G_{\aleph_0}$, are examples of *intermediate logics*, i.e. logics whose sets of tautologies are between those of classical and intuitionistic propositional logics. They represent one stream of research inspired by Gödel's note. On the other hand, removing (4.34) results in a so-called minimal logic.[61]

Several other types of semantics have also been proposed for intuitionistic logic, including the topological semantics introduced by Tarski (1938).[62] This is a particular case of semantics based on Heyting algebras, an important generalization of Boolean algebras. A *Heyting algebra* is a bounded lattice $\mathbf{L} = \langle L, \wedge^L, \vee^L, 0^L, 1^L \rangle$ such that for every $a, b \in L$, the set $\{c \in L \mid a \wedge^L c \leq b\}$ has a greatest element denoted $a \to^L b$ and called the pseudo-complement of a relative to b.[63] Letting furthermore $\neg^L a = a \to^L 0^L$, we may use L as a set of truth values and the operations $\wedge^L, \ldots$ as the truth functions of logical connectives $\wedge, \ldots$ We thus obtain the notion of a formula true in $\mathbf{L}$ as a formula φ always assuming 1^L in every evaluation e of propositional symbols in L. Open sets of a topology on X always form a Heyting algebra, $\mathbf{X}$, with $A \wedge^X B = A \cap B, A \vee^X B = A \cup B, A \to^X B = i((X - A) \cup B), \neg^X A = i(X - A), 0^X = \emptyset$, and $1^X = X$, where i is the topology's interior operator. For intuitionistic logic, Heyting algebras play the same role as Boolean algebras do in classical logic. Their role as well as that of the topological semantics is the following:[64]

Theorem 4.16. φ is provable from Heyting's axioms (4.25)–(4.35) iff φ is true in every Heyting algebra iff φ is true in every finite Heyting algebra iff φ is true in every Heyting algebra of open sets of a topological space iff φ is true in the Heyting algebra of open sets of the usual topology on $\mathbb{R}$.

Dummett's infinitely-valued variant of Gödel logic was later investigated by Alfred Horn (1918–2001). Horn is probably best known for what is nowadays called Horn sentences, which became fundamentally important for logic programming.[65] His papers, however, spanned various topics and include e.g. his famous conjecture regarding eigenvalues of matrices.[66] In Horn 1969, he made advances in the infinitely-valued Gödel logic and presented results for the predicate case (p. 153 below). For the propositional case, he provided a simple, algebraic proof of Dummett's completeness result (theorem 4.14). He relied on the then already available

[61] I. Johansson, "Der Minimalkalkül, ein reduzierter intuitionistischer Formalismus" [Minimal calculus, a reduced intuitionistic formalism], *Compositio Mathematica* 4 (1936): 119–36.

[62] Important among these semantics is the relational semantics due to S. Kripke, "Semantic analysis of intuitionistic logic I," in *Formal Systems and Recursive Functions*, ed. J. N. Crossley and M. A. E. Dummett (Amsterdam: North-Holland, 1965): 92–130.

[63] Also called *pseudo-Boolean algebra* (Rasiowa and Sikorski 1963), *relatively pseudo-complemented lattice*, or *Brouwer lattice* (though the last term also denotes a dual notion).

[64] See A. Tarski, "Der Aussagenkalkül und die Topologie" [Sentential calculus and topology], *Fundamenta Mathematicae* 31 (1938): 103–34; J. C. C. McKinsey and A. Tarski, "On closed elements in closure algebras," *Ann Math* 47 (1946): 122–62; J. C. C. McKinsey and A. Tarski, "Some theorems about the sentential calculi of Lewis and Heyting," *J Symb Logic* 13 (1948): 1–15; and (Rasiowa and Sikorski 1963, chapter IX).

[65] "On sentences which are true of direct unions of algebras," *J Symb Logic* 16 (1951): 14–21.

[66] "Eigenvalues of sums of Hermitian matrices," *Pacific J Math* 12 (1962): 225–41. The conjecture was proved in 1998 by A. A. Klyachko and, independently, in 1999 by A. Knutson and T. Tao.

Rasiowa and Sikorski 1963 with its extensive algebraic analysis of intuitionistic logic in terms of Heyting algebras and introduced the notion of an L-*algebra* as a Heyting algebra satisfying the prelinearity axiom (4.40), i.e. $(a \rightarrow^L b) \vee^L (b \rightarrow^L a) = 1^L$.[67] Dummett's infinite algebra for $G_{\aleph_0}$ is an example of an L-algebra. Horn proved the following subdirect representation theorem:

Theorem 4.17. A Heyting algebra is an L-algebra iff it is a subdirect product of chains, i.e. linearly ordered Heyting algebras.

Using this theorem he obtained a simple proof of theorem 4.14. He proved:

Theorem 4.18. The following conditions are equivalent for any formula φ:

    (a) φ *is provable from the axioms* (4.25)–(4.35) *and* (4.40).
    (b) φ *is true in every* L-*algebra.*
    (c) φ *is true in every Heyting chain.*
    (d) φ *is true in some infinite Heyting chain.*
    (e) φ *is true in every finite Heyting chain.*

4.2.3 Other propositional logics

Bernays's Habilitationsschrift

As was mentioned in chapter 2 (see also p. 117), one of the first many-valued logics was due to Paul Bernays, who described and utilized them in order to prove the independence of axioms of classical propositional logic. The idea of using many-valued logics for this purpose was later discovered independently by Łukasiewicz and is described on p. 117. Unlike Łukasiewicz and Post, Bernays seems not to have been interested in many-valued logics per se. Rather, he conceived of them solely as a tool for proving independence and considered them only in this context.

Bernays worked as an assistant of the famous mathematician David Hilbert in Göttingen, where he came at Hilbert's invitation in the fall of 1917.[68] His duties, unlike those of other assistants at that time, included discussions with Hilbert on foundational topics and the preparation of lecture notes for Hilbert's classes for the winter term 1917/1918. Since Bernays did not have a permanent position and depended on stipends, Hilbert encouraged him to obtain the *venia legendi*—a permission to teach classes. Bernays thus prepared his *Habilitationsschrift* (Bernays 1918) and submitted it along with the application for habilitation on July 9, 1918. The *Habilitationsschrift* overlaps significantly with the lecture notes, which he had prepared, parts of which were later published in Bernays 1926.[69] These parts include the independence proofs, which are the most interesting for us and are discussed in

---

[67] The same notion was later introduced by Hájek (1998b) under the name G-algebra; see section 4.4.

[68] The historical information regarding Bernays is based mainly on Zach 1999.

[69] D. Hilbert, "Prinzipien der Mathematik" [Principles of mathematics] (lecture notes by P. Bernays, Winter-Semester 1917/1918). Unpublished typescript, Bibliothek, Mathematisches Institut, Universität Göttingen.

more detail below. The parts that overlap with the lecture notes contain some fundamental results in classical propositional logic, which have usually been attributed to Post (1921), who obtained them independently (cf. p. 136 below).[70] These include, most importantly, completeness w.r.t. deductive consequence (p. 109 above) of an axiom system of propositional logic equivalent to and only slightly different from the system of *Principia Mathematica* (Whitehead and Russell 1910–13),[71] its syntactico-semantical completeness, and a procedure deciding whether a given formula of propositional logic is a tautology. These results are essentially due to Hilbert, as noted by Bernays (1918) himself. However, as Zach (1999) convincingly argues, there are substantial additions by Bernays, both in the lecture notes and the *Habilitationsschrift*, in particular an explicit formulation of propositional logic based on a clear distinction of its syntax and semantics, and the formulation of syntactico-semantical completeness.

Bernays's independence results, which are not contained in the lecture notes, are presented in § 5 of his *Habilitationsschrift*. Bernays first shows that one of the five axioms of the system under scrutiny is derivable and proceeds by showing that the four remaining ones are independent. He explains and employs the method we present on p. 117, crediting Ernst Schröder with the idea behind the method. He mentions Schröder's independence result whose proof is based on this idea,[72] and presents a simple proof of this result in an appendix to the *Habilitationsschrift*. It is worth noting that as early as 1905, Hilbert had already used arithmetical interpretations of logical formulas to prove independence.[73] In his proof, Bernays introduces six systems of many-valued logic, called systems I–VI. Other systems, VII–XI, are introduced in his discussion of deduction rules in § 6, and System XII is used to prove Schröder's independence result. All the systems are finitely valued, contain up to five truth values, with one or two designated values. Bernays was the first to use the concept of designated values—in Bernays 1926 he called them "ausgezeichnete Werte," from which the term "designated value" actually comes. Due to the essence of the method, Bernays was actually looking for a structure of truth values and a set of designated truth values with respect to which all axioms are true except the one to be proved independent. While most of the systems do not seem very significant per se, system I is important. Denoting disjunction and negation by $\oplus$ and $\neg$, and truth values by 0, $1/2$, and 1 (instead of Bernays's $\cdot$ and $\overline{a}$, and E, A, and O, respectively), the system has 1 as a designated value and truth functions of connectives as follows:

---

[70] A proper acknowledgment of Hilbert and Bernays's work in this respect has only recently been given in Zach 1999, which presents evidence that these and related results were obtained by Hilbert and Bernays before Post.

[71] The system appears in chapters *1–*5 of volume I of Whitehead and Russell 1910–13.

[72] E. Schröder, *Algebra der Logik*, vol. 1 (Leipzig: Teubner, § 12, p. 282 and Anhang 4–5), 616–43.

[73] D. Hilbert, "Logische Prinzipien des mathematischen Denkens" [Logical principles of mathematical reasoning] (lecture notes by E. Hellinger, Vorlesung, Sommersemester 1905). Unpublished manuscript, Bibliothek, Mathematisches Institut, Universität Göttingen.

$\oplus$	0	1/2	1		φ	$\neg\varphi$
0	0	1/2	1		0	1
1/2	1/2	1	1		1/2	1/2
1	1	1	1		1	0

Clearly, these are the truth functions of Łukasiewicz's three-valued disjunction (4.18) and negation. Since Bernays defines $\varphi \to \psi$ as a shorthand for $\overline{\varphi} \oplus \psi$, the truth function of $\to$ in Bernays's logic is just the three-valued Łukasiewicz implication to which Łukasiewicz referred in his Farewell Lecture of 1918 and which appeared in print in Łukasiewicz 1920. Łukasiewicz (1918) mentions that he worked his logic out "last summer," i.e. in 1917. Therefore, Bernays probably found this system later than Łukasiewicz because he did not arrive in Göttingen until the fall of 1917. Nevertheless, the first explicit printed description of this three-valued system is Bernays's because he submitted his *Habilitationsschrift* on July 9, 1918.

One might wonder why Bernays (and Hilbert) published their results only several years later. One possible answer is offered by the interview from 1977 with Bernays, mentioned in n. 5 in Zach 1999, according to which investigations of foundations connected to mathematical logic were not taken very seriously at that time. Be that as it may, the fact remains that some of the first many-valued logics including what is now called the three-valued Łukasiewicz logic were defined in a rather modern way by Bernays in 1918.

Post n-valued logic

In 1921, the American mathematician Emil Post presented a finitely-valued logic, along with other important results on propositional logic (Post 1921). Post obtained his results as part of his doctoral thesis at Columbia University and presented them at a meeting of the American Mathematical Society held on April 24, 1920. The abstract of his talk appeared in Post 1920b. Post's logic has in general n truth values and is different from Łukasiewicz's $Ł_n$ as well as from Bernays's systems. His motivations were mainly mathematical.

Post's general aim was, like Bernays's, to investigate further the system of propositional logic that appeared in *Principia Mathematica* (see n. 71 above), and to examine other possible systems of propositional logic that could be developed in the same spirit. Post's paper has four main parts. In the first part, he obtained important results regarding propositional logic, including functional completeness, syntactico-semantical completeness w.r.t. the axioms of Whitehead and Russell, and completeness w.r.t. deductive consequence, which are now considered fundamental results in logic (cf. p. 134 above on the independently obtained contributions of Hilbert and Bernays). In the second part, the possibility of choosing another system of basic connectives, different from negation and disjunction, is considered. The third presents a conception of a general propositional logic based on arbitrary connectives, deduction rules, and axioms. The considerations on pp. 176–79, regarded by Post as "merely an introduction to the general theory" in fact represent an attempt to develop a syntax of a general, possibly many-valued propositional logic. The semantic facet of a class of such many-valued logics is the content of the fourth part of Post's paper. Before presenting it in detail, note that he envisaged broad ramifications of the general logics (Post 1921, 164):[74]

> One class of such systems, and we study these in detail, seems to have the same relation to ordinary logic that geometry in a space of an arbitrary number of dimensions has to the geometry of Euclid. Whether these "non-Aristotelian" logics and the general development which includes them will have a direct application we do not know; but we believe that inasmuch as the theory of elementary propositions is at the base of the complete system of 'Principia,' this broadened outlook upon the theory will serve to prepare us for a similar analysis of that complete system, and so ultimately of mathematics.

Post considered n truth values and denoted them $t_1, \ldots, t_n$. For convenience, we denote them $1, \ldots, n$. In this notation, 1 represents truth and n represents falsity. The truth functions of negation and disjunction, $\neg$ and $\vee$, are given by

$$\neg a = \begin{cases} a+1 & \text{for } a < n, \\ 1 & \text{for } a = n, \end{cases} \quad \text{and} \quad a \vee b = \min(a, b).$$

The negation thus represents a cyclic shift of truth values, and may concisely be described by means of division modulo n as $\neg a = a(\bmod n) + 1$. It is clearly seen that for $n = 2$, in which case 1 and 2 represent truth and falsity, $\neg$ and $\vee$ essentially become the truth functions of classical negation and disjunction. For higher n, however, the situation is rather different. For instance, with $n = 4$ we obtain the following truth functions:

φ	$\neg\varphi$		$\vee$	1	2	3	4		φ	$\neg\varphi$		$\vee$	0	$1/3$	$2/3$	1
1	2		1	1	1	1	1		0	1		0	0	$1/3$	$2/3$	1
2	3		2	1	2	2	2	or	$1/3$	0		$1/3$	$1/3$	$1/3$	$2/3$	1
3	4		3	1	2	3	3		$2/3$	$1/3$		$2/3$	$2/3$	$2/3$	$2/3$	1
4	1		4	1	2	3	4		1	$2/3$		1	1	1	1	1

if $1, \ldots, n$ are represented by the normalized values $1, \ldots, 1/n{-}1, 0$, respectively, so that truth and falsity correspond to 1 and 0. Interestingly, Post speaks of "higher truth-values" and clearly alludes to the idea that truth values are ordered according to truthfulness. In addition, he provides an interpretation of his n-valued system in terms of classical logic. In this interpretation, the truth value i is represented by an $(n-1)$-tuple $\langle a_1, \ldots, a_{n-1} \rangle$ of 0s and 1s, starting with $i - 1$ 0s followed by $n - i$ 1s. For instance, in the case $n = 4$, the truth values 1 and 2 are represented by $\langle 1, 1, 1 \rangle$ and $\langle 0, 1, 1 \rangle$, respectively. Therefore, the n truth values are represented by sequences of propositions of classical logic, as Post puts it, with the provision that the higher

---

[74] Several writings mistakenly describe Post's results as having a purely formal motivation. For instance, Bolc and Borowik (1992, xi) say that "Post's way of treating the problem was pure formalism." The following quotation presents several remarks in Post's paper regarding analogies with n-dimensional geometric space, which, as well as the notes presented below, show that such claims are erroneous.

the truth value, the larger the number of true propositions in the sequence. The n-valued truth functions $\neg^{\cdot}$ and $\vee^{\cdot}$ can then be described using the classical truth functions $\neg^2$ and $\vee^2$ as

$$\neg^{\cdot}\langle a_1,\ldots,a_{n-1}\rangle = \langle \neg^2 a_{n-1},\ldots,\neg^2 a_1\rangle,$$
$$\langle a_1,\ldots,a_{n-1}\rangle \vee^{\cdot} \langle b_1,\ldots,b_{n-1}\rangle = \langle a_1 \vee^2 b_1,\ldots,a_{n-1} \vee^2 b_{n-1}\rangle.$$

Post provides an argument proving the following claim which is only mentioned in the paper without explicitly being stated as a theorem:

Theorem 4.19. For every n, the system consisting of the truth functions $\neg^{\cdot}$ and $\vee^{\cdot}$ is functionally complete.[75]

Next, Post provides a brief description of an axiomatization of the n-valued logic, generalizing the syntactico-semantical completeness of the classical propositional logic which he obtained in the first part of his paper. Interestingly, he considers tautologies with respect to sets $\{1,\ldots,k\}$, $k < n$, of designated truth values — a concept introduced independently a little earlier by Bernays (p. 135 above). That is, his result says that a formula of the n-valued logic is a tautology, i.e. takes on truth value $\leq k$ for every truth evaluation, if and only if it is provable from the axioms he provides. Post was the first to systematically consider designated truth values and notions, such as tautology, relative to them. His axioms are also shown to be complete in the sense that adding any nonprovable formula makes them inconsistent.

Further developments, Post algebras

Although Post did not return to it in his future publications, his n-valued logic considerably influenced further developments in many-valued logics, in particular those regarding functional completeness and algebras of many-valued logics. The general notion of Boolean algebra, of which the two-element Boolean algebra implicitly present in the formulation of classical logic is a particular example, appears as an algebraic counterpart of classical logic. In a sense, the n-element algebra on $\{1,\ldots,n\}$ equipped with $\neg^{\cdot}$ and $\vee^{\cdot}$, as defined by Post, replaces the two-element Boolean algebra. A natural question thus arises of what are the appropriate algebras, generalizing the n-element algebra of Post, that for Post logic play the same role as Boolean algebras play for classical logic. Rosenbloom (1942) answered this question by his notion of Post algebra. A disadvantage of this approach is a large number of complicated conditions postulated for his two basic operations. Epstein (1960) presented a simpler definition which is equivalent to Rosenbloom's. Later on, various definitions have been proposed of this kind including the following by Dwinger (1977). A *Post*

---

[75]In "Definition of Post's generalized negative and maximum in terms of one binary operation," *American J Math* 58 (1936): 193–94, D. L. Webb shows that Post's n-valued logic may be formulated using a single binary connective, thus generalizing the well-known case $n = 2$ of classical logic. This led to further investigations of Sheffer-like functions, see e.g. I. G. Rosenberg, "On generating large classes of Sheffer functions," *Aequationes Mathematicae* 17, no. 1 (1978): 164–81, and the references therein.

algebra of order n is an algebra

$$\langle L, \sqcap, \sqcup, \neg, D_1, \ldots, D_{n-1}, c_0, \ldots, c_{n-1} \rangle \quad \text{where}$$

(a) $\langle L, \sqcap, \sqcup, \neg, c_0, c_{n-1} \rangle$ forms a *De Morgan algebra*, i.e. $\langle L, \sqcap, \sqcup, c_0, c_{n-1} \rangle$ forms a bounded distributive lattice satisfying $\neg\neg a = a$, $\neg 0 = 1$, $\neg 1 = 0$, $\neg(a \sqcap b) = \neg a \sqcup \neg b$, and $\neg(a \sqcup b) = \neg a \sqcap \neg b$;

(b) $c_0 \leq c_1 \leq \cdots \leq c_{n-1}$;

(c) $D_i(a \sqcup b) = D_i(a) \sqcup D_i(b)$ and $D_i(a \sqcap b) = D_i(a) \sqcap D_i(b)$;

(d) $D_i(a) \sqcup \neg D_i(a) = c_{n-1}$ and $D_i(a) \sqcap \neg D_i(a) = c_0$;

(e) if $i \leq j$ then $D_j(a) \leq D_i(a)$;

(f) $D_i(\neg a) = \neg D_{n-1-i}(a)$;

(g) if $i \leq j$ then $D_i(c_j) = c_{n-1}$, if $i > j$ then $D_i(c_j) = c_0$;

(h) $a = (c_1 \sqcap D_1(a)) \sqcup \cdots \sqcup (c_{n-1} \sqcap D_{n-1}(a))$.

An interesting example of Post algebras of order n, presented in Epstein 1960, is obtained by taking for L the set $Ł_n^U$ of all $Ł_n$-sets in a given universe U, i.e. mappings $A : U \rightarrow \{0, 1/n-1, \ldots, 1\}$, and by letting $\sqcap$ and $\sqcup$ be the min-intersection and max-union, $\neg$ be the standard complement, i.e. $(\neg A)(u) = 1 - A(u)$, c_i be the constant function representing the truth value $i/n-1$, i.e. $c_i(u) = i/n-1$ for each $u \in U$, and $D_i(A)$ be the characteristic function of the set

$$\{u \in U \mid A(u) \geq i/n-1\}.$$

In terms of fuzzy sets, $D_i(A)$ is the $i/n-1$-cut of A and condition (h) is the formula for representation of fuzzy sets by their cuts invented later by Zadeh (section 3.2 and p. 271). This example is important for the theory of Post algebras because, as proved in Wade 1945, every Post algebra of order n is isomorphic to some Post algebra of $Ł_n$-sets (Wade 1945, Epstein 1960). In particular, Epstein (1960) proved that every Post algebra is isomorphic to the lattice of all continuous functions of some Boolean space into an n-element chain.[76]

Kleene's three-valued logic

Stephen C. Kleene (1909–1994) introduced a three-valued logic as a part of his studies in computability theory (Kleene 1938). Computability theory emerged in the 1930s in the works of Alonzo Church, Kurt Gödel, and Alan M. Turing. They attempted to make precise the informal notion of computability and proposed three different formalizations—that of recursivity inspired by Herbrand and formulated by Gödel, λ-definability of Church, and computability by formal machines, nowadays called Turing machines, after Turing.[77] Kleene, who was a student of Church,

---

[76] A Boolean space is a compact Hausdorff topological space in which the sets that are open and closed form a base. Epstein's result is a generalization of the famous Stone theorem for Boolean algebras; see M. H. Stone, "The theory of representations for a Boolean algebra," *Trans AMS* 40 (1936): 37–111.

[77] Gödel, "Über formal unentscheidbare Sätze der Principia Mathematica und verwandter Systeme I." [On formally undecidable theorems of Principia Mathematica and related systems] *Monatshefte für Mathematik und*

soon made important contributions to computability theory, e.g. the now-classic theorem establishing the equivalence of the three above-mentioned formal notions of computability.[78] Later on, Kleene introduced the term "Church thesis" for the claim that the formal notion of computable function coincides with the informal notion of function computable by an algorithm. While the early concepts of computability referred to total functions and relations, Kleene introduced and studied the fundamentally important notions of partial recursive functions and relations.[79] Unlike total recursive functions and relations, the partial ones need not be defined for every argument.

Kleene's motivation was the following. According to a claim established by Gödel, for any two (total) recursive relations, the relations obtained from them by the operations based on classical logical connectives, e.g. the complement, intersection, and union, are (total) recursive as well. The corresponding claim for partial recursive relations was not immediate. In particular, it was not obvious what the operations of complement, union, intersection, etc., should be for partial relations. For this purpose, Kleene introduced a third truth value, $1/2$, and interpreted it as "undefined."[80] Clearly, any partial relation may then be represented by its three-valued characteristic function R such that $R(u) = 0$ means that R is defined and false for u, $R(u) = 1$ means that R is defined and true for u, and $R(u) = 1/2$ means that R is undefined for u. Kleene (1938) defined the truth functions as:

φ	$\neg\varphi$		$\vee$	0	$1/2$	1		$\wedge$	0	$1/2$	1
0	1		0	0	$1/2$	1		0	0	0	0
$1/2$	$1/2$		$1/2$	$1/2$	$1/2$	1		$1/2$	0	$1/2$	$1/2$
1	0		1	1	1	1		1	0	$1/2$	1

$\rightarrow$	0	$1/2$	1		$\leftrightarrow$	0	$1/2$	1
0	1	1	1		0	1	$1/2$	0
$1/2$	$1/2$	$1/2$	1		$1/2$	$1/2$	$1/2$	$1/2$
1	0	$1/2$	1		1	0	$1/2$	1

$$(4.41)$$

His reason was the desired property that the partial relations obtained from partial recursive relations by means of these three-valued connectives be partial recursive. For instance, if R and S are partial recursive relations, the partial relation $R \vee S$ de-

Physik 38 (1931): 173–98; Gödel, "On Undecidable Propositions of Formal Mathematical Systems" (notes by S. C. Kleene and J. B. Rosser on lectures at the Institute for Advanced Study, Princeton, NJ, mimeographed, 30 pp., 1934); Church, "An unsolvable problem of elementary number theory," *American J Math* 58 (1936): 345–63; Church, "A note on the Entscheidungsproblem," *J Symb Logic* 1 (1936): 40–41. Turing, "On computable numbers, with an application to the Entscheidungsproblem," *Proc London Math Soc* 42 (1937): 230–65. Corrections, ibid., 43 (1938): 544–46.

[78] Important contributions to many-valued logics have been made by another student of Church and a fellow student of Kleene, J. B. Rosser.

[79] See the historical comments in Kleene 1952 and the foreword by M. Beeson to the 2009 reprint of Kleene 1952 by Ishi Press International, Bronx, NY.

[80] Kleene used "f," "t," and "u" in place of "0," "1," and "1/2."

fined by $(R \vee S)(u) = R(u) \vee S(u)$ is partial recursive as well. This claim is presented in a small note on p. 153 in Kleene 1938.

Kleene employs his three-valued logic and elaborates on it further in his famous book Kleene 1952. He presents conditions which should be obeyed by all reasonable truth functions of three-valued connectives suitable for the purpose described in the preceding paragraph and calls such truth functions *regular*. The truth functions in (4.41) are regular and Kleene calls them *strong*, as opposed to the *weak* truth functions which he defines as:

φ	$\neg\varphi$		$\vee$	0	$1/2$	1		$\wedge$	0	$1/2$	1	
0	1		0	0	$1/2$	1		0	0	$1/2$	0	
$1/2$	$1/2$		$1/2$	$1/2$	$1/2$	$1/2$		$1/2$	$1/2$	$1/2$	$1/2$	(4.42)
1	0		1	1	$1/2$	1		1	0	$1/2$	1	

$\rightarrow$	0	$1/2$	1		$\leftrightarrow$	0	$1/2$	1
0	1	$1/2$	1		0	1	$1/2$	0
$1/2$	$1/2$	$1/2$	$1/2$		$1/2$	$1/2$	$1/2$	$1/2$
1	0	$1/2$	1		1	0	$1/2$	1

The regularity conditions demand that any column of a table contain 1 in the row $1/2$, only if the column contains 1 in all rows, and analogously for rows containing 1. Kleene justifies the term "strong" by observing that the strong truth functions are the strongest regular extensions of the classical ones in that they have a 0 or a 1 in every position where every three-valued regular extension of the classical truth function can have a 0 or a 1. In addition to the primary interpretation of $1/2$ as "undefined," Kleene suggests and briefly discusses other interpretations, such as "unknown" and "undecidable whether true or false."

Dienes (1949) examined a new implication in Łukasiewicz logic, $\rightarrow_D$, so that $\varphi \rightarrow_D \psi$ stands for $(\neg\varphi \rightarrow \psi) \rightarrow \psi$. The truth function of this implication satisfies:

$$a \rightarrow_D b = \neg a \vee b, \text{ i.e. } a \rightarrow_D b = \max(1-a, b). \qquad (4.43)$$

These definitions present a generalization of Kleene's strong truth functions, with which they coincide on $Ł_3$. In fact, while mentioning some well-known classically equivalent formulas which are also equivalent for his strong connectives, Kleene actually observed that $\varphi \rightarrow \psi$ is equivalent to $\neg\varphi \vee \psi$ (Kleene 1952, 336). An implication whose truth function is defined by (4.43) is known as the Kleene-Dienes implication in the literature on fuzzy logic.[81]

Kleene's three-valued logic became a classic in the literature on many-valued logic and an inspiration for further research, particularly in partial logics. A natural deduction system for Kleene's (predicate) logic and its syntactico-semantical completeness theorem was obtained by Kearns.[82]

---

[81] This term was probably coined in Bandler and Kohout 1978.

[82] J. T. Kearns, "Vagueness and failing sentences," *Logique et Analyse* 17, no. 67–68 (1974): 301–15; and "The strong completeness of a system for Kleene's three-valued logic," *Z Math Logik Grundlagen Math* 25 (1979): 61–68. A complete axiomatization of Kleene's logic is also provided in section 2.4 in Urquhart 1986.

Bochvar's three-valued logic

The Russian logician Dmitri A. Bochvar introduced a three-valued logic to obtain a calculus suitable for analysis of paradoxes in classical logic such as the Russell paradox. In Bochvar 1938, he attempted to provide a system in which it is possible to prove formally that certain statements are meaningless.[83] The third truth value, $1/2$, employed in addition to the classical truth values 0 and 1, is interpreted as "meaningless" or "nonsense."[84] The paper has three parts. In the first, Bochvar proposed a three-valued propositional logic which we examine in detail in what follows. In the remaining two parts, certain restricted and extended predicate logics corresponding to the three-valued propositional logic are introduced and applied to the analysis of Russell and Grelling-Nelson paradoxes.[85]

Bochvar motivates his propositional logic by semantic considerations. In particular, he distinguishes between the *internal* and *external* forms of assertion. For instance, corresponding to the internal forms "φ" and "φ or ψ" are the external forms "φ is true" and "φ is true or ψ is true." He emphasizes a crucial semantic difference between the two forms, namely that the two forms have in general different truth values. For instance, if φ represents a meaningless statement, then the internal form "not φ" is still meaningless while the corresponding external form "φ is false" is not meaningless because it is false. The distinction between the internal and external forms reflects itself in the fact that Bochvar's propositional logic contains for every so-called internal connective the corresponding external connective. In particular, negation $\neg_I$ and conjunction $\wedge_I$ are the basic internal connectives by which those of disjunction, implication, and equivalence are defined as shorthands:

$$\varphi \vee_I \psi \quad \text{for} \quad \neg_I(\neg_I\varphi \wedge_I \neg_I\psi),$$
$$\varphi \rightarrow_I \psi \quad \text{for} \quad \neg_I(\varphi \wedge_I \neg_I\psi),$$
$$\varphi \leftrightarrow_I \psi \quad \text{for} \quad (\varphi \rightarrow_I \psi) \wedge_I (\psi \rightarrow_I \varphi).$$

The truth functions of the internal connectives are described by the following tables:

φ	$\neg_I\varphi$	$\wedge_I$	0	$1/2$	1	$\vee_I$	0	$1/2$	1
0	1	0	0	$1/2$	0	0	0	$1/2$	1
$1/2$	$1/2$	$1/2$	$1/2$	$1/2$	$1/2$	$1/2$	$1/2$	$1/2$	$1/2$
1	0	1	0	$1/2$	1	1	1	$1/2$	1

---

[83] This paper is now available in English translation by Merrie Bergmann; see the entry for Bochvar 1938 in the list of references. Unfortunately, it is incorrectly stated in the translator's introduction that the original Russian version was published in 1937. It is also incorrectly stated in Rescher 1969 and Urquhart 1986 that the original version was published in 1939.

[84] Bochvar uses the Russian word "бессмысленность" and denotes 0, 1, and $1/2$ as F, T, and N, respectively.

[85] Bochvar speaks of Weyl's paradox; the Grelling-Nelson paradox is sometimes mistakenly attributed to Hermann Weyl.

$\to_I$	0	1/2	1
0	1	1/2	1
1/2	1/2	1/2	1/2
1	0	1/2	1

$\leftrightarrow_I$	0	1/2	1
0	1	1/2	0
1/2	1/2	1/2	1/2
1	0	1/2	1

Note that these truth functions are the same as Kleene's weak truth functions, cf. (4.42). Using the unary connectives A and D of external assertion and external denial, whose truth functions are given as

φ	$A\varphi$
0	0
1/2	0
1	1

and

φ	$D\varphi$
0	1
1/2	0
1	0

,

the external connectives of conjunction $\wedge_E$, disjunction $\vee_E$, implication $\to_E$, being of the same strength $\leftrightarrow_E$, equivalence $\equiv_E$, being meaningless $\downarrow$, and being not true $\neg_E$, are defined as shorthands:

$\varphi \wedge_E \psi$ for $A\varphi \wedge_I A\psi$, $\varphi \vee_E \psi$ for $A\varphi \vee_I A\psi$,

$\varphi \to_E \psi$ for $A\varphi \to_I A\psi$, $\varphi \leftrightarrow_E \psi$ for $A\varphi \leftrightarrow_I A\psi$,

$\varphi \equiv_E \psi$ for $(\varphi \leftrightarrow_E \psi) \wedge_I (\neg_I \varphi \leftrightarrow_E \neg_I \psi)$, $\neg_E \varphi$ for $\neg_I A\varphi$,

$\downarrow \varphi$ for $\neg_I(A\varphi \vee_I D\varphi)$.

One easily checks that the truth functions of the external connectives are described by the following tables:

$\wedge_E$	0	1/2	1
0	0	0	0
1/2	0	0	0
1	0	0	1

$\vee_E$	0	1/2	1
0	0	0	1
1/2	0	0	1
1	1	1	1

$\to_E$	0	1/2	1
0	1	1	1
1/2	1	1	1
1	0	0	1

$\leftrightarrow_E$	0	1/2	1
0	1	1	0
1/2	1	1	0
1	0	0	1

$\equiv_E$	0	1/2	1
0	1	0	0
1/2	0	1	0
1	0	0	1

φ	$\downarrow \varphi$
0	0
1/2	1
1	0

φ	$\neg_E \varphi$
0	1
1/2	1
1	0

The formulas are defined as usual in propositional logic with $\neg_I, A, D$, and $\wedge_I$ being the primitive connectives. Tautologies are defined as formulas which always have the value 1, contradictions as those which never assume 1. Bochvar provides several theorems regarding his logic. Among them are the observations that no classical formula, which is a formula based on $\neg_I$ and $\wedge_I$ and hence contains only the internal connectives, is a tautology. Furthermore, $\downarrow \varphi$ is not equal in strength, and thus also not equivalent, to any classical formula. He also observes that the fragment consisting of formulas involving the external connectives $\neg_E$, $\wedge_E$, $\vee_E$, $\rightarrow_E$, and $\leftrightarrow_E$ is equivalent to classical propositional logic. This is because the external connectives do not distinguish 0 and 1/2, and because upon identification of 0 and 1/2, the truth functions of external connectives coincide with those of classical logic.

The restricted and extended propositional logics are developed to enable analysis of paradoxes. In particular, the extended logic is built in the style of Hilbert and Ackermann's extended predicate logic (Hilbert and Ackermann 1928, chapter IV). Recall that Russell's paradox involves a formula R such that $R \leftrightarrow \neg R$ is established as a valid formula, which is a contradiction in classical logic.[86] The formula $R \leftrightarrow \neg R$ may be derived in the classical extended predicate logic of Hilbert and Ackermann by defining R to be Pd($\neg$Pd) where Pd is a unary predicate defined by letting Pd(ϕ) be equivalent to $\phi(\phi)$ for any second-order predicate ϕ. Bochvar shows that in his three-valued extended predicate logic, the above-mentioned derivation of $R \leftrightarrow \neg R$ is not available with $\neg_I$ and $\leftrightarrow_I$ replacing $\neg$ and $\leftrightarrow$ in the classical derivation. This is because one of the formulas employed in the derivation, namely $\varphi \leftrightarrow_I \varphi$, is not a tautology of the three-valued logic. Instead, however, one may proceed with the tautology $\varphi \equiv_E \varphi$ and thus obtain $R \equiv_E \neg_I R$. Since $(\varphi \equiv_E \neg_I\varphi) \equiv_E \downarrow \varphi$ is a tautology in Bochvar's logic, the properties of $\equiv_E$ yield $\downarrow R$ tautologous, proving that R, the formula on which Russell's paradox is based, is meaningless in Bochvar's sense. We shall return to the problem of paradoxes in section 5.4.2.

In section III, Bochvar (1938) mentions that whether his extended predicate logic is consistent is an open question, but adds that his attempts to derive a contradiction in it were in vain. Interestingly, Church points out in his review of Bochvar 1938 that Bochvar's three-valued logic is inconsistent because another form of Russell's paradox still appears in it, with the external negation $\neg_E$ in place of $\neg_I$.[87] However, in a correction to this review (see n. 87), Church admits that his claim regarding inconsistency was erroneous and adds: "On the contrary, the suggested alternative to the theory of types is far from devoid of interest."[88] Later on, Bochvar (1943) returned to the problem of consistency of his three-valued extended logic and related systems, presented a constructive consistency proof of a fragment of his logic, and outlined a method for a nonconstructive consistency proof.[89] The choice of the fragment is important since the fragment corresponds to the formalism needed to construct Russell's paradox.

Bochvar's idea of a third truth value representing "meaningless" was further developed by several other logicians (see p. 147 below). In particular, axiomatizations of Bochvar's and related logics were studied by Finn.[90]

---

[86] See section 5.4.2 for more information.

[87] Church, *J Symb Logic* 4 (1939): 98–99, with correction in *J Symb Logic* 5 (1940): 119.

[88] Unfortunately, Church's original claim regarding the inconsistency of Bochvar's logic has uncritically been accepted in some works on many-valued logics, e.g. in the handbook chapter Urquhart 1986.

[89] "К вопросу о парадоксах математической логики и теории множества" [To the question of paradoxes of mathematical logic and the theory of sets], *Matematicheskii Sbornik* 15, no. 57 (1944): 369–84.

[90] V. K. Finn, "Об аксиоматизации некоторых трехзначных логик" [On axiomatization of some three-valued

Probability logics

In a sense, classical logic is concerned with inferences that lead from certainly true statements to certainly true statements. It comes as no surprise that the idea of extending classical logic to account for "probably true" instead of "certainly true" has occupied the minds of logicians and mathematicians for centuries. The idea is part of a long-standing quest for a relationship of logic and probability which appears in the work of Jakob Bernoulli, Lambert, Bolzano, Boole, De Morgan, and Peirce.[91] The works of these scholars predate rather extensive research on this topic in the 20th century, particularly in its second half.

From the point of view of many-valued logic, it seems natural to assign to propositions their probabilities and interpret them as truth values of these propositions. An early approach along this line of reasoning was developed by Hugh MacColl (1877–98, 1906) and we examined it in section 2.1.[92]

Assignments of probabilities to logical formulas appear in Łukasiewicz's logical theory of probability. Łukasiewicz (1913) assigns to a propositional function with one variable over a finite set of individuals, which he calls an "indefinite proposition," the ratio between the number of individuals for which the proposition is true and the number of all individuals, and calls this ratio the truth value of the proposition. Thus, if $1, 2, \ldots, 6$ are all the individuals considered, then "x is greater than 4" is assigned $1/3$. The truth value of φ is thus naturally interpreted as the probability of φ and we denote it $\|\varphi\|_P$. Łukasiewicz then postulates the following three axioms:

$\|\varphi\|_P = 0$ if and only if φ is a classical contradiction,

$\|\varphi\|_P = 1$ if and only if φ is a classical tautology,

$\|\psi\|_P = \|\varphi\|_P + \|\neg\varphi \wedge \psi\|_P$ whenever $\varphi \to \psi$ is a classical tautology,

and derives from them a number of theorems, such as:

if φ and ψ are logically equivalent then $\|\varphi\|_P = \|\psi\|_P$,

$\|\varphi\|_P + \|\neg\varphi\|_P = 1$,

$\|\varphi \vee \psi\|_P = \|\varphi\|_P + \|\psi\|_P - \|\varphi \wedge \psi\|_P$,

$\|\varphi \vee \psi\|_P = \|\varphi\|_P + \|\psi\|_P$ if and only if $\varphi \wedge \psi$ is a classical contradiction.

Łukasiewicz's axioms and theorems, or their variants, later reappeared in several writings on probabilistic semantics of propositional logic. Most of them presuppose, as Łukasiewicz does, classical propositional logic from which the language

logics], *Nauchno-tekhnicheckaia Informacia* 2, no. 11 (1971): 16–20; V. K. Finn, "Аксиоматизация некоторых трехзначных исчислений высказываний и их алгебр" [Axiomatizations of some three-valued propositional calculi and their algebras], in *Filosofia v Sovremennom Mire: Filosofia i Logika* (Moscow: Nauka, 1974), 398–438.

[91] J. Bernoulli, *Ars Conjectandi* [The art of conjecturing] (Basel: Thurneysen Brothers, 1713); J. H. Lambert, *Neues Organon* [New Organon] (Leipzig, 1764); B. Bolzano, *Wissenschaftslehre* [Theory of science] (Sulzbach: Seidel, 1837); Boole 1847, 1854; A. De Morgan, *Formal Logic* (London: Taylor and Walton, 1847); C. S. Peirce, "A theory of probable inference," in *Studies in Logic*, ed. C. S. Peirce (Boston, MA: Little, Brown, 1883).

[92] About MacColl, see M. Astroh and S. Read, eds., *Hugh MacColl and the Tradition of Logic*. Special issue, *Nordic J Philosophical Logic* 3 (1998): i–vi, 1–234;

and the notions of contradiction, tautology, and the like are borrowed. An important connection to the underlying classical logic consists in an easy-to-see fact that a propositional formula φ is a classical tautology if and only if it is a probabilistic tautology in that $\|\varphi\|_P = 1$ for every function $\|\cdot\|_P$ satisfying the axioms.

In the 1930s, the relationship between probability and many-valued logic was investigated by the Polish logician and philosopher Zygmunt Zawirski (1934a,b, 1935) and the German mathematician Hans Reichenbach (1932, 1935). Both of them had to defend their view of a probability logic as a useful and viable kind of many-valued logic against criticism coming from important figures of the Polish school. One point of this criticism was based on the observation by Mazurkiewicz, mentioned in Tarski's critique (1935) and briefly presented on pp. 3–4 in Mazurkiewicz 1932, that *probability logic is not truth functional*, which was considered a serious defect because truth functionality was required of proper many-valued logics (Tarski 1935). This important fact may be explained using the example of disjunction. That the probability $\|\varphi \vee \psi\|_P$ cannot be determined from the probabilities $\|\varphi\|_P$ and $\|\psi\|_P$ alone is not difficult to see on intuitive grounds, for in determining $\|\varphi \vee \psi\|_P$, the relationship between φ and ψ is substantial. Consider for instance the following two cases. First, let $\|\varphi\|_P = 0.5$ and let ψ be $\neg\varphi$. Then $\|\varphi\vee\psi\|_P = \|\varphi\vee\neg\varphi\|_P = 1$, because $\varphi\vee\neg\varphi$ is a tautology. Second, let $\|\varphi\|_P = 0.5$ and let ψ be φ. Then $\|\varphi \vee \psi\|_P = \|\varphi \vee \varphi\|_P = \|\varphi\|_P = 0.5$, because $\varphi \vee \varphi$ is logically equivalent to φ. In both cases, $\|\varphi\|_P = 0.5$ and $\|\psi\|_P = 0.5$ (in the first case this follows from $\|\neg\varphi\|_P = 1 - \|\varphi\|_P$) but still, the probabilities $\|\varphi \vee \psi\|_P$ are different. Another, perhaps more fundamental, point of criticism came from the view, propounded by Ajdukiewicz and Mazurkiewicz, that the argument of a probability function should not be a proposition itself but rather a proposition and a set of propositions, thus probability should be relativized to a set of propositions.[93] The criticism was rejected by Reichenbach (1935–36). Note that the non-truth-functionality of connectives in probability logic was well known to Reichenbach, who designed his logic in such a way that the truth functions of connectives had an additional parameter, u, representing the probability of ψ given that φ is true. For instance, he defines $\|\varphi \vee \psi\|_P = \|\varphi\|_P + \|\psi\|_P - \|\varphi\|_P \cdot u$.

To conclude, note that the debate over a broadly conceived interplay between logic and probability in the first half of the 20th century was quite rich.[94] Rather than attempt to cover it in full, we restrict our focus to aspects relevant to fuzzy logic, most importantly to considering degrees of probability as truth values and the resulting non-truth-functionality of connectives, to which we return in section 4.5.

---

[93] K. Ajdukiewicz, *Główne zasady metodologii i logiki formalnej* [Fundamental principles of the methodology of science and of formal logic] (Warsaw: 1928), mimeographed; (Mazurkiewicz 1932, 1934). This view is close to Keynes's; see n. 94.

[94] Important works include F. P. Ramsey, "Truth and probability," in *The Foundations of Mathematics and Other Logical Essays*, ed. R. B. Braithwaite (London: Kegan, Paul, Trench, Trubner, 1931), 156–98, written in 1926; J. M. Keynes, *Treatise on Probability* (London: Macmillan, 1921); R. Carnap, *Logical Foundations of Probability* (Chicago, IL: Univ. Chicago Press, 1950); H. Jeffreys, *Theory of Probability* (London: Oxford Univ. Press, 1961).

Other logics

Several other logics with a third truth value interpreted as "meaningless," "non-sense," or "significant," appeared after Bochvar's paper. These include Halldén's logic of nonsense and further logics inspired by it such as those in the works of Åqvist, Segerberg, Piróg-Rzepecka, and Goddard and Routley.[95] Note also that Bochvar's three-valued logic has been generalized to the case of n truth values by Bochvar and Finn.[96]

Another direction concerns applications to quantum mechanics which started with Zawirski 1931b. In their pathbreaking paper, Birkhoff and von Neumann (1936) founded what is now called a logic of quantum mechanics or a quantum logic. The basic idea is to identify experimental propositions by subsets of a phase-space. While in classical mechanics, the experimental propositions form a Boolean algebra, the algebra corresponding to experimental propositions in quantum mechanics does not conform to the Boolean laws and is more general. Related contributions include Reichenbach's (1944) logic with its third truth value corresponding to statements that are called meaningless in the Bohr-Heisenberg interpretation. Research in this area grew considerably in the following years.[97]

4.2.4 Predicate logics

Important contributions

Examining the research in many-valued predicate logics until the 1960s, one may identify several significant contributions. One of the first many-valued predicate logics that appeared in the literature was Bochvar's three-valued logic in 1938 (p. 142). Another early work is Rosser's paper, which nevertheless never appeared in print.[98] Another important contribution is Rosser and Turquette's method of axiomatization for a broad class of finitely-valued propositional and predicate logics which is the main content of their influential book Rosser and Turquette 1952. We present the propositional part on pp. 155ff. They consider generalized quantifiers whose application yields formulas of the form $(Qx_1,\ldots,x_k)(\varphi_1,\ldots,\varphi_l)$, binding at once

---

[95] S. Halldén, *The Logic of Nonsense* (Uppsala: Uppsala Univ. Press, 1949); L. Åqvist, "Reflections on the logic of nonsense," *Theoria* 28 (1962): 138–57; K. Segerberg, "A contribution to nonsense-logics," *Theoria* 31 (1965): 199–217; K. Piróg-Rzepecka, *Systemy Nonsense-Logics* [Systems of nonsense logics] (Warsaw: Państwowe Wydawnictwo Naukowe, 1977); L. Goddard and R. Routley, *The Logic of Significance and Context* (Edinburgh: Scottish Academic Press, 1973).

[96] E.g. Bochvar and Finn, "О многозначных логиках, допускающих формализацию анализа антиномий 1." [On many-valued logics admitting formalization of the analysis of antinomies 1.], in *Issledovania po matematicheskoi lingvistike, matematicheskoi logike i informacionnym jazykam* (Moscow: Nauka, 1972), 238–95.

[97] For further information, see K. Engesser, D. M. Gabbay, and D. Lehmann, eds., *Handbook of Quantum Logic and Quantum Structures* (Amsterdam: Elsevier, 2007).

[98] "The introduction of quantification into a three-valued logic," preprinted from the *J Unified Science* (*Erkenntnis*) for the members of the Fifth Int. Congress for the Unity of Science, Cambridge, MA., 1939, 6 pp.; the paper was scheduled for vol. 9 but never appeared in print; according to A. Church's review in the *J Symb Logic* 4 (1939): 170, "A system of logic [in Rosser's paper] is considered, based on an n-valued propositional calculus . . . [which covers] also quantification, descriptions, classes."

a finite number of variables x_i and formulas φ_j. This includes the universal and existential quantifiers, $\forall$ and $\exists$, as particular cases with $k = l = 1$. For their proof of completeness, they adapted to their general setting the method developed for classical predicate logic by Henkin (1949). Another direction is inspired by considerations of the axiom of comprehension in many-valued set theory. This topic is discussed in detail in chapter 5. Of great importance are the results obtained in infinitely-valued predicate Łukasiewicz and Gödel logics, to which we now turn.

Infinitely-valued predicate Łukasiewicz logic

The predicate Łukasiewicz logic with truth degrees in $[0, 1]$ has been thoroughly investigated since the late 1950s. This logic, $\text{Ł}\forall$, is obtained from the propositional Łukasiewicz logic basically in the manner described on p. 110. It contains $\neg$ and $\rightarrow$ as basic propositional connectives, symbols of relations, and the symbol $\exists$ of existential quantifier. Formulas with other connectives, $\varphi \vee \psi$, $\varphi \wedge \psi$, $\varphi \oplus \psi$, and $\varphi \otimes \psi$, are considered as shorthands for (4.6), (4.7), $\neg\varphi \rightarrow \psi$, and $\neg(\varphi \rightarrow \neg\psi)$, respectively, and $(\forall x)\varphi$ as a shorthand for $\neg(\exists x)\neg\varphi$.[99] The basic problem was to find an axiomatization of all $[0, 1]_{\text{Ł}}$-tautologies, i.e. tautologies w.r.t. the standard Łukasiewicz algebra $[0, 1]_{\text{Ł}}$. These are formulas taking on the truth degree 1 in every $[0, 1]_{\text{Ł}}$-structure $\mathbf{M}$, provided $\exists$ is interpreted according to (4.3). In this case, $\forall$ obeys (4.2). In general, one may consider tautologies w.r.t. a structure $\mathbf{L}$ different from $[0, 1]_{\text{Ł}}$, and even w.r.t. a class $\mathscr{L}$ of structures, i.e. formulas whose truth degree is 1 in every $\mathbf{L}$-structure $\mathbf{M}$ with $\mathbf{L} \in \mathscr{L}$. Such sets are denoted $\text{Taut}^{\mathbf{L}}(\text{Ł}\forall)$ and $\text{Taut}^{\mathscr{L}}(\text{Ł}\forall)$, respectively (cf. p. 109). In this notation, the above set of tautologies is $\text{Taut}^{[0,1]_{\text{Ł}}}(\text{Ł}\forall)$.

Thoralf Skolem (1887–1963) seems to have been the first who explicitly considered $\text{Ł}\forall$. Skolem (1957) was motivated by questions regarding the axiom of comprehension which we examine in section 5.2.3. Some positive results toward the problem of axiomatization were obtained in a doctoral thesis by Joseph D. Rutledge (1959), a student of Rosser at Cornell. Some of these results appeared in Rutledge 1960. For one, Rutledge succeeded in obtaining a syntactico-semantically complete axiomatization of the monadic part of $\text{Ł}\forall$, i.e. a part in which only unary relation symbols are allowed. For this purpose, he used and significantly generalized Chang's method (1959). Furthermore, Rutledge proved the following theorem, generalizing theorem 4.2 (d) of the propositional case:[100]

Theorem 4.20. Letting $\mathbf{L}_n$ *denote the standard Łukasiewicz n-valued algebra, we have* $\text{Taut}^{[0,1]_{\text{Ł}}}(\text{Ł}\forall) = \bigcap_{n=1}^{\infty} \text{Taut}^{\mathbf{L}_n}(\text{Ł}\forall)$.

Since it was known, due to Rosser and Turquette, that every finitely-valued predicate Łukasiewicz logic is axiomatizable by finitely many axioms, Rutledge's results provided hope for the axiomatizability of infinitely-valued predicate logic.

---

[99] One may, however, pick other connectives as basic, e.g. $\neg$ and $\oplus$, as in Rutledge 1960.

[100] Rutledge (1960), however, does not mention the propositional results of Łukasiewicz and Tarski 1930.

Further results on Ł∀ were obtained by Louise S. Hay (1935–1989), another student of Rosser at Cornell, in her master's thesis (Hay 1959). Hay later described the circumstances of her work as follows:[101]

> Like a good fifties wife, I followed him [Hay's husband], first doing a Master's thesis to have something to show for my two years. (I must admit I was glad to leave—I lacked confidence that I could finish a Ph. D. at that point.) My thesis advisor, J. Barkley Rosser, was extremely helpful, both in getting a visiting job for me at Oberlin (here the 'old-boy network' worked in my favor), and in leaving me with detailed instructions for proving a theorem in infinite-valued logic which would constitute a thesis, while he took off for a summer vacation trip around the world with his family. As it happened, I found a counterexample to the main lemma, which made the thesis publishable under my own name.

Selected results of Hay's thesis appeared in Hay 1963. Hay considers an axiomatic system with the four axiom schemes (4.13)–(4.16) of Łukasiewicz infinitely-valued propositional logic, and offers the following five schemes regarding quantification:

$$((\exists x)\varphi) \otimes ((\exists x)\varphi) \to (\exists x)(\varphi \otimes \varphi), \tag{4.44}$$

$$\varphi \to (\exists x)\psi \tag{4.45}$$

  if φ results by replacing all free occurrences of x in ψ by y

  and no such occurrence is in a subformula of ψ of the form $(\exists y)\chi$,

$$(\exists x)\varphi \to (\exists y)\psi \tag{4.46}$$

  if ψ results by replacing all free occurrences of x in φ by y

  and φ results by replacing all free occurrences of y in ψ by x,

$$(\forall x)(\varphi \to \psi) \to ((\exists x)\varphi \to \psi) \tag{4.47}$$

  if x is not free in ψ,

$$(\varphi \to (\exists x)\psi) \to (\exists x)(\varphi \to \psi) \tag{4.48}$$

  if x is not free in φ,

and the classical rules of *modus ponens* and generalization (appendix B). In what follows, let $\vdash$ denote provability, and let $m\varphi$ and φ^m denote $\varphi \oplus \cdots \oplus \varphi$ and $\varphi \otimes \cdots \otimes \varphi$, both with m occurrences of φ, respectively. Hay's axioms are easily seen to be sound, i.e. every provable formula is a $[0,1]_Ł$-tautology. Hay was unable to prove completeness but proved the following theorem:

Theorem 4.21. If $\|\varphi\|_{M,v} > 0$ *for any* $[0,1]_Ł$-*structure* **M** *and valuation* v *then* $\vdash m\varphi$ *for some* $m \geq 1$. *In particular, if* φ *is a tautology, i.e.* $\varphi \in \text{Taut}^{[0,1]_Ł}(\forall Ł)$, *then* $\vdash m\varphi$ *for some* $m \geq 1$.

At about the same time, important results were also obtained by Chang and his student at UCLA, Lawrence Peter Belluce.[102] Belluce was a graduate student of

---

[101] L. Hay, "How I became a mathematician (or how it was in the bad old days)," *Newsletter of the Association for Women in Mathematics* 19, no. 5 (1989): 8–10.

[102] Part of the historical information in this paragraph is based on personal communication with L. P. Belluce.

mathematics in 1959 whom Chang hired as a research assistant and considered a potential PhD student. Chang presented him with the problem of completeness of
the predicate Łukasiewicz logic. Chang suggested he follow the elegant algebraic
method of Rasiowa and Sikorski (1950) for classical predicate logic. The method is
based on the following theorem on the existence of prime ideals in Boolean algebras
known as the R-S lemma (cf. n. 44 on p. 123 above).

*Theorem 4.22. Let $a, a_n, a_{n,i}$ be elements of a Boolean algebra L, where i ranges over
an (arbitrary) set I_n, $n = 1, 2, \ldots$, such that $a_n = \sup_{i \in I_n} a_{n,i}$ in L and $a \neq 1$ where 1
is the largest element of L. Then there exists a prime ideal K of L preserving all the
suprema $\sup_{i \in I_n} a_{n,i}$ such that $a \in K$.*

Preserving the suprema in theorem 4.22 means that $[a_n]_K = \sup_{i \in I_n} [a_{n,i}]_K$ in
the factor algebra L/K. Chang gave Belluce the problem of obtaining an analogous theorem for MV-algebras. One difficulty was that the proof of the R-S lemma
in Rasiowa and Sikorski 1950 was topological and Chang and Belluce had no corresponding topological results for MV-algebras. Fortunately, Tarski gave an algebraic proof of the R-S lemma. Tarski's proof seems never to have been published
but Feferman provided a hint to it in his 1952 review.[103] Belluce eventually proved
an analogous theorem for prime ideals in MV-algebras, but only for the Lindenbaum MV-algebra rather than for all MV-algebras. This was in fact what was needed
because the R-S lemma in Rasiowa and Sikorski's proof is effectively used only for
the Lindenbaum algebra. From the analogy of the R-S lemma, however, Belluce
and Chang were only able to obtain what they called *weak completeness*, which we
present next. They used a set of axioms and deduction rules basically the same as
Hay (1959), about whose work they had some information, which probably came
from Rosser, Hay's advisor, who according to Belluce's recollections met with both
Belluce and Chang in Chang's house. Reciprocally, Rosser probably informed Hay
about Belluce and Chang's results. Rosser was indeed very well informed about the
subject and in 1960 published his valuable paper Rosser 1960, explaining lucidly the
concepts, problems, and proof methods in propositional and predicate Łukasiewicz
logics. At about the same time Belluce proved the analogy of the R-S lemma, Belluce
and Chang obtained important information from E. P. Specker about Scarpellini's
negative axiomatizability result which we present below. Eventually, their results
were announced in Belluce 1960 and Belluce and Chang 1960, and were later published in the authors' joint paper Belluce and Chang 1963.

The weak completeness refers to L-structures **M** where **L** may not only be the
standard algebra $[0, 1]_\text{Ł}$ but a general MV-chain, i.e. a linearly ordered MV-algebra.
This raises a slight problem because in evaluating $(\exists x)\varphi$, the required supremum

---

[103]S. Feferman, "A proof of the completeness theorem of Gödel by H. Rasiowa, R. Sikorski," *J Symb Logic* 17
(1952): 72. The R-S lemma (actually its dual version for filters) with Tarski's proof is found as *Tarski's lemma* in J.
L. Bell and A. B. Slomson, *Models and Ultraproducts. An Introduction* (Amsterdam: North-Holland, 1969), 21.

may not exist in $\mathbf{L}$ and the truth degree $\|(\exists x)\varphi\|_{\mathbf{M},v}$ may thus be undefined. That is why Belluce and Chang introduce the following notions.[104] They call a formula φ

- *valid in $\mathbf{M}$ and v* if $\|\varphi\|_{\mathbf{M},v} = 1$ whenever $\|\varphi\|_{\mathbf{M},v}$ is defined;
- *strongly valid* if for every MV-chain $\mathbf{L}$, $\mathbf{L}$-structure $\mathbf{M}$ and M-valuation v, φ is valid in $\mathbf{M}$ and v;
- *valid* if $\|\varphi\|_{\mathbf{M},v} = 1$ for each $[0,1]_{\text{Ł}}$-structure $\mathbf{M}$ and M-valuation v, i.e. if $\varphi \in \text{Taut}^{[0,1]_{\text{Ł}}}(\text{Ł}\forall)$.

Belluce and Chang's weak completeness theorem is the following claim in which a sentence is a formula with no free variables and consistency of T means that for no φ do we have $T \vdash \varphi$ and $T \vdash \neg\varphi$.

Theorem 4.23. Let T be a consistent set of sentences. Then $T \vdash \varphi$ if and only if φ is valid in each $\mathbf{L}$-structure $\mathbf{M}$ and valuation v such that $\mathbf{L}$ is an MV-chain and every $\psi \in T$ is valid in $\mathbf{M}$ and v. In particular, $\vdash \varphi$ if and only if φ is strongly valid.

As shown in Hájek, Paris, and Shepherdson 2000a, the soundness part of this theorem is not correct.[105] To obtain a correct version, one has to resort to what later became known as safe structures (p. 198) in the definition of strong validity.

Chang and Belluce also proved that every consistent set T of sentences is satisfiable in that for each $\varphi \in T$ one has $\|\varphi\|_{\mathbf{M},v} = 1$ for some $[0,1]_{\text{Ł}}$-structure $\mathbf{M}$ and valuation v, and obtained an improvement of Hay's result (theorem 4.21), according to which $\vdash \varphi \oplus \varphi$ whenever $\|\varphi\|_{\mathbf{M},v} > 1/2$ for any $[0,1]_{\text{Ł}}$-structure $\mathbf{M}$ and valuation v. This result implies that for a tautology φ we have $\vdash \varphi \oplus \varphi$, while Hay's result only ensures $\vdash m\varphi$ for some $m \geq 1$. In her journal paper, Hay (1963) presents yet another generalization: $\|\varphi\|_{\mathbf{M},v} > 1/m$ implies $\vdash m\varphi$. Another interesting result was announced in Belluce 1960 and presented with a proof in Belluce and Chang 1963 and also in Hay 1963:

Theorem 4.24. If we add to Hay's system the infinitary deduction rule

$$\frac{\varphi \oplus \varphi^n, n = 1, 2, 3, \ldots}{\varphi},$$

we obtain a complete axiomatization of the tautologies in $\text{Taut}^{[0,1]_{\text{Ł}}}(\text{Ł}\forall)$.

A solution to the unsuccessful attempts to obtain finitary completeness for $\text{Ł}\forall$ appeared in about 1960 in work of Bruno Scarpellini, a student of Ernst P. Specker at ETH Zürich. Scarpellini's fundamental result was announced by Rosser (1960, 152), who "received a private communication from Specker" about it, and was published with proof in Scarpellini 1962:

---

[104] We have rephrased slightly the original formulation.

[105] In n. 58 on p. 279 in Hájek 1998b, this observation is announced and attributed to Shepherdson. The observation appeared in Hájek, Paris, and Shepherdson 2000a where the authors also observed that the notion of truth degree of $(\exists x)\varphi$ in Belluce and Chang 1963 is different from the usual one. See p. 198 below.

Theorem 4.25. The set $\mathrm{Taut}^{[0,1]_{\text{Ł}}}\mathrm{Ł}(\forall)$ *of all tautologies of the infinitely-valued predicate Łukasiewicz logic is not recursively enumerable (hence not axiomatizable).*

This result implies that there is no algorithm that would print the formulas in $\mathrm{Taut}^{[0,1]_{\text{Ł}}}(\text{Ł}\forall)$ one after another in such a way that every tautology eventually appears in the printed list. It immediately follows that neither Hay's system nor any other system with finitely many axiom schemes, or more generally a recursive set of axioms, and finitary deduction rules provides an adequate axiomatization of Ł∀. Otherwise, one could utilize the system to generate mechanically all proofs and all provable formulas, hence all $[0,1]_{\text{Ł}}$-tautologies, and thus obtain an algorithm listing all $[0,1]_{\text{Ł}}$-tautologies. In this sense, the *infinitely-valued predicate Łukasiewicz logic is not axiomatizable*. Scarpellini's result reveals a fundamental difference of Ł∀ from the classical predicate logic because, as is well known, the latter is recursively axiomatizable (Gödel 1930).

Scarpellini's proof is based on a certain recursive mapping of formulas $\varphi_{\text{Ł}}$ of Łukasiewicz logic to formulas φ of classical predicate logic that has the property that φ is satisfiable in a finite structure if and only if $\varphi_{\text{Ł}}$ may attain a truth degree > 0. Invoking the famous Trakhtenbrot's theorem, saying that the set of classical predicate formulas satisfiable in finite structures is not decidable, Scarpellini observes that the set of formulas that are not satisfiable in finite structures is not recursively enumerable.[106] It then follows that the set of formulas of Łukasiewicz logic that are identically false is not recursively enumerable, from which one immediately obtains theorem 4.25 because tautologies are just negations of the latter formulas.

At about the same time, axiomatizability of some many-valued logics was investigated by Andrzej Mostowski, a former student of Kuratowski and Tarski. His approach is rather general. Mostowski (1961a) assumes that the truth degrees form a partially ordered set which is bicompact in its order topology, and considers finite as well as linearly ordered infinite sets. Furthermore, he considers generalized quantifiers in case of finite sets of truth degrees and the existential and universal ones based on suprema and infima for infinite sets.[107] Interestingly, he obtains general results of axiomatizability without explicitly listing the axioms. These include parts of the results on axiomatizability of finitely-valued logics by Rosser and Turquette (1952), which Mostowski presents for illustration. The results are only existential and are based on Tychonoff's theorem of topology. It follows from theorem 5.2 in Mostowski 1961a that for every rational number $0 \leq r < 1$, the set $\mathrm{Taut}^{[0,1]_{\text{Ł}}}_{(r,1]}(\text{Ł}\forall)$, i.e. formulas always assuming a truth degree in $(r,1]$, is recursively enumerable. Notice that the case of $r = 0$, i.e. $\mathrm{Taut}^{[0,1]_{\text{Ł}}}_{(0,1]}(\forall\text{Ł})$, follows from Hay's theorem 4.21. Closely related to Scarpellini's result is Mostowski 1961b, in which it is shown that addi-

---

[106] В. А. Trakhtenbrot, "Невозможность алгорифма для проблемы разрешимости на конечных классах" [The impossibility of an algorithm for the decidability problem on finite classes], *Doklady Akademii Nauk SSSR* 70, no. 4 (1950): 569–72.

[107] We return to Mostowski 1957, now a classic paper on generalized quantifiers, in section 4.5.

tion of other quantifiers than those based on suprema and infima to the systems of Mostowski 1961a may lead to a loss of recursive enumerability of tautologies, and hence to loss of axiomatizability.

Results in axiomatizability of tautologies $\mathrm{Taut}_D^{[0,1]_\text{Ł}}(Ł\forall)$ of $Ł\forall$ with respect to sets D of designated truth degrees, including the above-discussed $D = \{1\}$ and $D = (r, 1]$, were obtained by Belluce (1964). He proved that $\mathrm{Taut}_D^{[0,1]_\text{Ł}}(Ł\forall)$ is recursively enumerable for $D = (0, 1]$, exhibiting a proof different from Hay's and then obtained the above-mentioned Mostowski's generalization with $D = (r, 1]$ for rational r. Unlike Mostowski, Belluce was able to provide an explicit axiomatization of $\mathrm{Taut}_{(r,1]}^{[0,1]_\text{Ł}}(Ł\forall)$. Defining as *weak recursive real* (wrr) a real $r \in [0, 1]$ for which there exist recursive functions p and q such that $p(n) \leq q(n) > 0$ for all $n = 1, 2, \ldots$, and $r = \inf\{p(n)/q(n) \mid n = 1, 2, \ldots\}$, he obtained the following results which document the intricacy of predicate Łukasiewicz logic:

Theorem 4.26.

(a) $\mathrm{Taut}_{(r,1]}^{[0,1]_\text{Ł}}(Ł\forall)$ *is recursively enumerable iff r is a wrr.*

(b) *If r is a wrr then* $\mathrm{Taut}_{[r,1]}^{[0,1]_\text{Ł}}(Ł\forall)$ *is recursively enumerable iff* $\mathrm{Taut}_{[r,1]}^{[0,1]_\text{Ł}}(Ł\forall) = \mathrm{Taut}_{(r,1]}^{[0,1]_\text{Ł}}(Ł\forall)$.

(c) *If r is not a wrr then* $\mathrm{Taut}_{[r,1]}^{[0,1]_\text{Ł}}(Ł\forall)$ *is not recursively enumerable.*

Łukasiewicz logic was subject to intensive investigations after the 1960s, and we return to this in section 4.4. Note that Thiele (1958) obtained completeness for the finitely-valued predicate Łukasiewicz logics.

Infinitely-valued predicate Gödel logic

Contrary to Łukasiewicz logic, the infinitely-valued predicate Gödel logic, $G\forall$, is axiomatizable as proved by Horn (1969). As in the propositional case, Horn uses L-algebras (p. 134) as the structures of truth values and strongly relies on the methods developed by Rasiowa and Sikorski (1963) for intuitionistic logic. Horn's system is basically obtained from the propositional Gödel logic by adding the symbols $\forall$ and $\exists$ of quantifiers. Propositional connectives are interpreted as in the propositional case. The symbols of relations are interpreted by L-relations where L is an arbitrary L-algebra, not necessarily complete as a lattice. Quantifiers are interpreted according to (4.2) and (4.3) with the provision that the possible nonexistence of infima and suprema is handled as follows.[108] Horn calls an L-structure **M** *suitable* if for every M-valuation v of individual variables, all the required infima and suprema needed according to (4.2) and (4.3) exist in L.[109] A formula φ is called an L-tautology if for every suitable L-structure **M** and every M-valuation v we have $\|\varphi\|_{\mathbf{M},v} = 1$. Horn

---

[108] See the similar treatment in the weak completeness theorem of Belluce and Chang (1963).

[109] This is exactly what Hájek (1998b) calls a *safe* L-structure in the context of his logic BL$\forall$, cf. pp. 198ff.

also considers what we call the standard Gödel algebra on the unit interval $[0,1]$ of all reals and $[0,1] \cap \mathbb{Q}$ of all rationals, in which the truth functions are given by (4.36)–(4.39). Horn's axioms are those of the propositional case—for which he uses axioms (T_1)–(T_{11}) of intuitionistic propositional logic from Rasiowa and Sikorski 1963 (p. 379)—plus prelinearity (4.40) and a new axiom scheme,

$$(\forall x)(\varphi \vee \psi) \rightarrow (\varphi \vee (\forall x)\psi), \qquad (4.49)$$

provided x does not appear in φ. The system involves the six deduction rules (r_1)–(r_6) from chapter V in Rasiowa and Sikorski 1963. The first two are the rules of *modus ponens* and substitution, i.e. "from φ infer $\varphi(x/y)$," where $\varphi(x/y)$ denotes the result of substituting the variable y for x in φ. The other rules are

$$\frac{\varphi \rightarrow \psi}{(\exists y)\varphi(x/y) \rightarrow \psi} \quad \text{if } x \text{ does not occur in } \psi \text{ and } y \text{ in } \varphi,$$

$$\frac{\psi \rightarrow \varphi}{\psi \rightarrow (\forall y)\varphi(x/y)} \quad \text{with the same condition as above,}$$

$$\frac{(\exists y)\varphi(x/y) \rightarrow \psi}{\varphi \rightarrow \psi} \quad \text{if } y \text{ does not occur in } \varphi,$$

$$\frac{\psi \rightarrow (\forall y)\varphi(x/y)}{\psi \rightarrow \varphi} \quad \text{with the same condition as above.}$$

Horn then obtains, using an approach inspired by the algebraic method of Rasiowa and Sikorski 1950, the following completeness theorem:

Theorem 4.27. The following conditions are equivalent for any formula φ:

(a) φ *is provable in* G$\forall$.
(b) φ *is an* L*-tautology for every Heyting chain* L.[110]
(c) φ *is an* L*-tautology for every countable Heyting chain* L.
(d) φ *is an* L*-tautology where* L *is the standard Gödel algebra on* $[0,1]$.
(e) φ *is an* L*-tautology where* L *is the standard Gödel algebra on* $[0,1] \cap \mathbb{Q}$.

Subsequent investigations in the Gödel logic are examined in section 4.4.

Continuous model theory of Chang and Keisler

In the early 1960s, Chang initiated a study of model theory for the infinitely-valued Łukasiewicz predicate logic. Model theory was a subject of thorough investigations at that time.[111] As Chang used some recent results of H. Jerome Keisler in his first works (Chang 1961), it was only logical that the two started to work jointly on the topic, first in the spring of 1961 at Berkeley, then in the summer of 1962 at Princeton,

---

[110] Heyting chains, i.e. linearly ordered Heyting algebras, are just linearly ordered L-algebras.
[111] A standard reference in model theory for classical logic is Chang and Keisler 1990.

and a year later at Madison, Wisconsin. They announced their first results in Chang and Keisler 1962 and later published a book of 165 pages (Chang and Keisler 1966).[112]

The main feature of *continuous model theory*, as the authors call it, is the assumption that truth values form a compact Hausdorff topological space $\langle L, \mathscr{L} \rangle$ equipped with a certain quasiorder and containing distinct elements 0 and 1 representing falsity and truth. These components along with truth functions of connectives and quantifiers, which are functions $c : L^k \to L$ and $q : 2^L \to L$ and which are required to be continuous with respect to the topology, constitute a structure **L** of truth values, called a *continuous logic* by the authors. This means that, in a sense, the structure of truth values is represented by their neighborhoods which have the ability to separate truth values, and that logical connectives, quantifiers and the (quasi)ordering respect this structure—certainly an interesting conception. Formulas and their truth values in **L**-structures **M**, which involve L-valued relations, are defined as usual using a truth-functional semantics. Any structure **M** induces a mapping in L^Σ from the set Σ of sentences to L assigning to a sentence φ its truth value in **M**. This mapping from the class $\mathscr{M}$ of all **L**-structures to L^Σ and the product topology on L^Σ induce in a standard way a topology on $\mathscr{M}$, called the *elementary topology*, which is of fundamental importance. In particular, since **L** is compact and Hausdorff, one may prove the following theorem, which generalizes the compactness theorem of classical predicate logic and plays a major role in continuous model theory.

Theorem 4.28. The elementary topology of any continuous logic is compact.

Chang and Keisler obtained various advanced results generalizing those of classical model theory. Their methods are very different from the classical ones and provide new insight into the classical notions and proofs. Important parts such as proofs and axiomatizability are, however, not developed in their book.

Importantly, both classical predicate logic and the infinitely-valued Łukasiewicz predicate logic appear as particular examples of continuous logic. Further interesting examples may also be obtained and the reader is referred to Chang and Keisler 1966. To conclude, Chang and Keisler's results and, perhaps more importantly, their conceptions and the methods developed, are highly relevant to fuzzy logic in the narrow sense. However, they do not seem to have been widely recognized, which is perhaps due to the fact that up until now the main focus in fuzzy logic has been on proof theory rather than model theory.

4.2.5 Further developments

Axiomatization

As was mentioned in section 4.2.3, Post outlined a method for axiomatization of finitely-valued functionally complete propositional logics with an arbitrary num-

---

[112] According to a personal communication from Belluce, after completing Belluce and Chang 1963, Belluce asked Chang if they should develop model theory for Łukasiewicz logic but Chang's answer was negative.

ber of designated truth values. An explicit axiomatization of a large class of functionally complete logics was provided by Słupecki (1939a). Nevertheless, the problem for general finitely-valued logics which may be functionally incomplete, such as Łukasiewicz's, was open. An axiomatization of a large class of such logics was found by Rosser and Turquette (1945).[113] It was simplified in Rosser and Turquette 1950, extended to predicate logic in Rosser and Turquette 1948, 1951, and included in the highly influential book Rosser and Turquette 1952.

We now present a simplified version of their propositional system which we denote RT. This is a general n-valued logic with the set $L = \{0, 1/n-1, \ldots, n-2/n-1, 1\}$ of truth values such that $s/n-1, \ldots, 1$ are designated for some $0 < s \leq n-1$.[114] Its language contains a binary connective $\Rightarrow$ and unary ones, $J_0, J_{1/n-1}, \ldots, J_1$, whose truth functions satisfy the conditions discussed below, or such connectives are definable in RT. $\Rightarrow$ plays the role of implication, $J_0, \ldots, J_{s-1/n-1}$ play the role of negation, and $J_{s/n-1}, \ldots, J_1$ play the role of identity in classical logic. Let us now define recursively $\Gamma_0 \varphi_i \psi$ as ψ and $\Gamma_{k+1} \varphi_i \psi$ as $\varphi_{k+1} \Rightarrow \Gamma_k \varphi_i \psi$. That is, $\Gamma_3 \varphi_i \psi$ denotes $\varphi_3 \Rightarrow (\varphi_2 \Rightarrow (\varphi_1 \Rightarrow \psi))$ and the like. RT has the following axioms:

$$\varphi \Rightarrow (\psi \Rightarrow \varphi) \tag{4.50}$$

$$(\varphi \Rightarrow (\psi \Rightarrow \chi)) \Rightarrow (\psi \Rightarrow (\varphi \Rightarrow \chi)) \tag{4.51}$$

$$(\varphi \Rightarrow \psi) \Rightarrow ((\psi \Rightarrow \chi) \Rightarrow (\varphi \Rightarrow \chi)) \tag{4.52}$$

$$(J_a(\varphi) \Rightarrow (J_a(\varphi) \Rightarrow \psi)) \Rightarrow (J_a(\varphi) \Rightarrow \psi) \tag{4.53}$$

$$\Gamma_n(J_{n-i/n-1}(\varphi) \Rightarrow \psi)\psi \tag{4.54}$$

$$J_a(\varphi) \Rightarrow \varphi \text{ for } a = s/n-1, \ldots, 1 \tag{4.55}$$

$$\Gamma_k J_{a_i}(\varphi_i) J_a(c(\varphi_1, \ldots, \varphi_k)) \text{ for each } k\text{-ary connective } c, \tag{4.56}$$

$$\text{every } a_1, \ldots, a_k \in L, \text{ and } a = c'(a_1, \ldots, a_k)$$

and has *modus ponens* for $\Rightarrow$ as a deduction rule. Observe that (4.50)–(4.52) are familiar classical tautologies and that (4.53)–(4.56) describe the properties of truth values and logical connectives. For instance, (4.56) may be seen as saying that if φ_i assumes the truth value a_i for $i = 1, \ldots, k$, then $c(\varphi_1, \ldots, \varphi_k)$ assumes $c'(a_1, \ldots, a_k)$.

The truth functions of $\Rightarrow$ and J_as are *plausible* if every axiom (4.50)–(4.56) assumes a designated truth value, and if $a \Rightarrow' b$ is undesignated whenever a is designated and b is undesignated, for every $a, b \in L$. The latter condition assures that *modus ponens* is sound. It is easy to check that $\Rightarrow$ and J_as are plausible if they satisfy *standard conditions*: (a) $a \Rightarrow' b$ is undesignated if and only if a is designated and b is undesignated, (b) $J_a'(b)$ is designated if and only if $a = b$. Rosser and Turquette proved the following result, which complements Wajsberg's general axiomatizability results announced in Łukasiewicz and Tarski 1930:

---

[113] Słupecki 1939a does not refer to Post 1921, and Rosser and Turquette 1945 does not refer to Słupecki 1939a.

[114] Actually, Rosser and Turquette use $L = \{1, \ldots, M\}$ with 1 corresponding to truth and M to falsity.

Theorem 4.29. Let $\Rightarrow$ and J_as satisfy the standard conditions or, more generally, be plausible. Then a formula φ is provable in RT if and only if φ always assumes a designated truth value.

Note that this theorem applies in particular to classical logic. Namely, if we let $n = 2$ and $s = 1$, let $\Rightarrow$ be the classical implication, $J_0(\varphi)$ be $\neg\varphi$ and $J_1(\varphi)$ be $\neg\neg\varphi$, conditions (a) and (b) apply. Likewise, theorem 4.29 applies whenever the propositional logic is functionally complete because in this case, one can always define $\Rightarrow$ and J_as that satisfy the standard conditions. The situation is not so straightforward for Łukasiewicz n-valued logic $Ł_n$. Nevertheless, the authors show how one may define J_as in $Ł_n$ such that $J'_a(b) = 1$ if $a = b$ and $J'_a(b) = 0$ if $a \neq b$ for every $a, b \in L$.[115] Letting furthermore $\overline{\varphi}$ denote $J_0(\varphi) \vee \cdots \vee J_{s-1/n-1}(\varphi)$ where $\vee$ is defined by (4.6), and letting $\varphi \Rightarrow \psi$ denote $\overline{\varphi} \vee \psi$, it is easy to check that $\Rightarrow$ and J_as satisfy the standard conditions. If 1 is the only designated degree, i.e. $s = n - 1$, one may even use the Łukasiewicz implication $\rightarrow$ for $\Rightarrow$ because $\rightarrow$ and J_as are plausible in this case, which is not so for $s < n - 1$. This way, one obtains axiomatizations of logics $Ł_n$, though not as elegant as those found specifically for $Ł_n$s (section 4.2.1). Note also that theorem 4.29 cannot be used this way to obtain an axiomatization of Gödel logics G_n because $J_{s-1/n-1}$ with standard properties is not definable in G_n.[116] The method of Rosser and Turquette has often been used later, e.g. in the interesting papers of Rose (1951b,a, 1952) who examined logics whose truth degrees form lattices. A different approach was developed by Schröter (1955). Contrary to RT, it is based on sequents and does not assume restrictions on connectives.[117]

Functional completeness and definability of functions

Questions regarding definability of functions by logical connectives have also received considerable attention. Rescher 1969 contains on pp. 313–15 a comprehensive bibliography on this topic up to the 1960s. In addition to the results on Łukasiewicz and Post logics mentioned above, many others were obtained. Since functional completeness was considered a desirable property, several finitely-valued logics have been proposed which obey this property, adding thus to the early papers (Post 1921, Słupecki 1936) examined above. Among them are the well-known three-valued logic of Słupecki (1946) and the n-valued logics of Słupecki (1939a) and Sobociński (1936). In addition, definability of functions in finitely-valued logics was studied. Post (1921) observed that a set of logical connectives is functionally complete whenever all unary and binary functions are definable in this set. An important criterion was obtained by Słupecki (1939b), who proved that if all unary functions are definable

---

[115] That such J_as are definable in $Ł_n$ follows from theorem 4.12.

[116] Assume the contrary. Then $J'_{s-1/n-1}(s-1/n-1) = 1$ because $J'_{s-1/n-1}(s-1/n-1)$ needs to be designated and because for every unary f definable in G_n, $f(a) \in \{0, a, 1\}$. It then easily follows that $J'_{s-1/n-1}(1) = 1$, a contradiction because 1 is designated.

[117] Chapter 7 in Gottwald 2001 provides a good exposition of Schröter's system.

then the logic is functionally complete if and only if a binary function is definable whose truth table fulfills the following conditions: (a) at least one row does not have all its entries identical and the same for columns, i.e. the function essentially depends on two arguments; (b) every truth value appears in the table. A condition for definability of all unary functions was obtained by Piccard, according to which it is sufficient that the following unary functions be definable:[118]

$$f(i/n-1) = \begin{cases} i-1/n-1 & \text{for } i > 0, \\ 1 & \text{for } i = 0, \end{cases} \qquad \begin{aligned} g(1) &= n-2/n-1, \\ g(n-2/n-1) &= 1, \\ g(a) &= a \text{ otherwise,} \end{aligned} \qquad \begin{aligned} h(0) &= 1, \\ h(a) &= a \text{ otherwise.} \end{aligned}$$

A general criterion for functional completeness was obtained for a three-valued logic by Jablonski (1958) and for general n-valued logics eventually by Rosenberg (1970).

Logical matrices and further metalogical contributions

As mentioned in section 4.2.1, Polish logicians produced major contributions to several areas between the world wars. Of fundamental importance are their metalogical contributions. A major figure in this respect was Tarski, who defined the abstract notion of a *logical matrix* (Łukasiewicz and Tarski 1930, n. 5) and developed it further in Tarski 1938, put forward the view of a deductive system as a *consequence operator* (Tarski 1930a,b), and proposed his *conception of truth* (Tarski 1936).[119] All these contributions played significant roles in the development of fuzzy logic in the narrow sense.

Related to logical matrices is the important concept of *Lindenbaum algebra*, also called *Lindenbaum-Tarski algebra*, which is an algebra constructed as follows. Let L be the set of all formulas of a propositional logic and define for every n-ary logical connective c an n-ary operation c^L on L by $c^L(\varphi_1, \ldots, \varphi_n) = c(\varphi_1, \ldots, \varphi_n)$. Let furthermore D be the set of all provable formulas. Consider the algebra $\mathbf{L} = \langle L, c^L, \ldots \rangle$ and define a relation $\equiv$ of *provable equivalence* on L defined by

$$\varphi \equiv \psi \text{ if and only if both } \varphi \to \psi \text{ and } \psi \to \varphi \text{ are provable.}$$

If $\equiv$ is a congruence on $\mathbf{L}$, we may consider the factor algebra $\mathbf{L}/\equiv$ and the matrix $\langle \mathbf{L}/\equiv, D/\equiv \rangle$. The matrix $\langle \mathbf{L}, D \rangle$, or $\langle \mathbf{L}/\equiv, D/\equiv \rangle$, is called the Lindenbaum matrix and the algebra $\mathbf{L}$, or $\mathbf{L}/\equiv$, the Lindenbaum algebra. Under certain natural conditions one can prove the important property that

$$\varphi \text{ is provable iff } \varphi \text{ is a tautology w.r.t. the Lindenbaum matrix } \langle \mathbf{L}, D \rangle,$$

where being a tautology means that $\|\varphi\|_e \in D$ for every evaluation of propositional symbols in L, and analogously for $\langle \mathbf{L}/\equiv, D/\equiv \rangle$.

---

[118] S. Picard, "Sur les fonctions définies dans les ensembles finis quelconques" [On functions defined in arbitrary finite sets], *Fund Math* 24 (1935): 298–301.

[119] For the notion of logical matrix, see n. 8 on p. 108 above. A comprehensive treatment of logic from the point of view of consequence operators is provided by R. Wójcicki, *Theory of Logical Calculi: Basic Theory of Consequence Operators* (Dordrecht: Kluwer, 1988). Tarski's contributions are presented in Woodger 1956.

The very idea and first results are due to Lindenbaum, who worked them out during 1926–27.[120] There is, however, only indirect evidence of his results because Lindenbaum never published them. Łukasiewicz and Tarski (1930) attribute theorem 3 in their paper to Lindenbaum, which asserts a variant of the above property, but the construction of $\langle L, D\rangle$ is not described in this paper. Tarski (1935–36) constructs a Boolean algebra from formulas of propositional logic and introduces the provable equivalence, as well as the factor algebra by this equivalence but with no reference to Lindenbaum. It was indeed due to Tarski that the method and Lindenbaum's name became widely known. McKinsey writes of "an unpublished method of Lindenbaum" and adds "the method was explained to me by Professor Tarski" when referring to the first part of his construction of a matrix, whose elements are formulas of which the provable ones are designated.[121] The second part consists in observing that the provable equivalence ≡ is a congruence on this matrix and in factorizing this matrix. Thus, McKinsey attributes to Lindenbaum only the first part of his construction, while the factorization by provable equivalence seems to be McKinsey's own invention.[122] This corresponds to Henkin and Tarski's remark according to which the factor algebras were introduced in Tarski 1935–36 and "the historical justification . . . for calling these algebras Lindenbaum algebras seems to be incorrect."[123] Lindenbaum's idea may in a broader view be considered as the idea of constructing semantic structures from syntactic concepts. This idea was later applied to predicate logic and became a standard method in logic.[124] Logical matrices and Lindenbaum algebras represent a precursor of the modern use of algebraic methods in logic (Rasiowa and Sikorski 1963, Rasiowa 1974, Blok and Pigozzi 1989, Font and Jansana 1996).

4.2.6　The mid-1960s and Zadeh's idea of fuzzy sets

Around the mid-1960s, many-valued logic was a branch of logic with a history of contributions going back to the pioneering publications of the 1920s. The first monograph devoted to many-valued logic by Rosser and Turquette (1952) was later augmented by the philosophically oriented book by Zinov'ev (1963), the lucid, brief exposition by Ackermann (1967), and finally the comprehensive treatise by Rescher (1969) which became—along with its chronological list of references amounting to

---

[120] S. J. Surma, "The concept of Lindenbaum algebra," in *Studies in the History of Mathematical Logic*, ed. S. J. Surma (Wroclaw: Ossolineum, 1973), 239–53.

[121] J. C. C. McKinsey, "A solution to the decision problem for the Lewis systems S2 and S4, with an application to topology," *J Symb Logic* 6 (1941): 117–34.

[122] In "The concept of Lindenbaum algebra" (n. 120 above), Surma does not distinguish what in particular McKinsey refers to in his comment regarding Lindenbaum.

[123] L. Henkin and A. Tarski, "Cylindric algebras," in *Lattice Theory*, ed. R. P. Dilworth (Providence, RI: American Mathematical Society, 1961), 83–113.

[124] J. Łoś, "O matrycach logicznych" [On logical matrices], *Prace Wroclawskiego Towarzystwa Naukowego* B, no. 19 (1949), 42 pp.; L. Henkin, "The completeness of first-order functional calculus," *J Symb Logic* 14 (1949): 159–66; Rasiowa and Sikorski 1950; L. Rieger, "On countable generalized σ-algebras, with a new proof of Gödel's completeness theorem," *Czech Mathematical J* 1, no. 76 (1951): 29–40.

nearly fifty pages—a standard reference for many years to come. While—as we have shown above in this section—some of the many-valued logics had proper motivations and their additional truth values had meanings attached to them, some were introduced as mere technical tools to serve a certain purpose. In addition, as the fields advanced, new contributions started to be increasingly disconnected from the original motivations and various logics were simply developed as abstract logical calculi. The lack of clear interpretations of additional truth values in most of the existing contributions resulted in doubts regarding the meaningfulness of many-valued logics. Even though abstract investigations may result in valuable findings, criticism of the pursuit of pure theory with no apparent interpretation or applications is surely well grounded. On the other hand, a clear interpretation provides theory with a much needed connection to real-world phenomena which may suggest interesting problems for the theory itself. The situation described above and the issues involved call for a detailed examination, provided in section 5.4.1. For our present purpose, it is important to state that the situation changed significantly in the mid-1960s when Zadeh published his idea of fuzzy sets and fuzzy logic. Even though, as explained in chapter 2, others had published similar ideas previously, it was Zadeh who was able to make a clear case for the idea, soon pursued by many, that vague predicates are naturally modeled by fuzzy sets. He thus paved the way to fuzzy logic, i.e. a logic whose propositions may involve vague predicates such as "tall" or "similar." The intuitively appealing interpretation of numbers in $[0, 1]$ as degrees of truth of such propositions was in sharp contrast with most of the interpretations suggested thus far, not to mention studies with no interpretation of truth values at all. Even though truth values were ascribed, implicitly or explicitly, a comparative character in several previous studies (section 5.4.1), Zadeh's interpretation of *truth values* in $[0, 1]$ *as truth degrees* of fuzzy propositions, as well as a strong appeal of Zadeh's ideas for numerous applications, represent a significant landmark in the development of many-valued logic. His ideas eventually led to the development of fuzzy logic in the narrow sense as a particular branch of many-valued logic whose agenda is driven by the aim to model reasoning in natural language (cf. p. 322). As we shall see, fuzzy logic naturally encompasses both Łukasiewicz and Gödel logic and, in fact, the emergence of the idea of fuzzy logic provided an impetus for further studies of these logics. The development of fuzzy logic in the narrow sense since the appearance of Zadeh's ideas is the subject of the following sections of this chapter.

4.3   Fuzzy logics with graded consequence

Zadeh's ideas served as inspiration for new logical investigations which gradually led to the establishment of fuzzy logic in the narrow sense. Some of these investigations appeared very soon. Among them, an original conception based on the idea of graded consequence and reasoning from partly true premises has a distinguished place. This led to significant early developments examined in the present section.

4.3.1 Goguen's logic of inexact concepts

Goguen's papers of 1967 and 1968–69

Shortly after the publication of Zadeh's seminal paper (1965a), Joseph Goguen, Zadeh's student at Berkeley, published two long papers—the "L-fuzzy sets" (Goguen 1967) and "The logic of inexact concepts" (Goguen 1967)—which had a major impact on the subsequent development of fuzzy logic.[125] The latter is the first paper in which the term "fuzzy logic" is used in its present meaning.[126] These papers have broad ramifications and a truly visionary character (Goguen 1967, 145):

> This paper explores the foundations of, generalizes, and continues the work of Zadeh . . . The significance of this work may lie more in its point of view than in any particular results. The theory is still young, and no doubt many concepts have yet to be formulated, while others have yet to take their final form. However, it should now be possible to visualize the outlines of the theory.

While Goguen's emphasis is on conceptional and foundational issues, he worked out his ideas to a considerable level of mathematical detail. What clearly separates Goguen's papers from most of the preceding investigations in many-valued logic is the attention to natural, nonformalistic motivations and applications and their careful and detailed exposition. The motivations and applications come mainly from the realm of human reasoning and, in particular, vagueness. What separates them from the preceding work on vagueness is a solid mathematical and logical treatment. Considered from today's perspective, several of Goguen's ideas were groundbreaking and led to important investigations in logic, mathematics, and philosophy.

Goguen anticipated possible misconceptions and emphasized that the calculus of "inexact predicates" itself is by no means inexact or fuzzy in any way (Goguen 1967, 147). Furthermore, he was cautious enough to clearly differentiate this calculus from the calculus of probability (Goguen 1968–69, 327, 333):

> A number of theories seem to resemble our logic of inexact concepts. Perhaps probability is closest in spirit . . . But the manipulations allowed in probability theory are different from those our examples suggest for fuzzy sets. In fact, the theory is a calculus of vagueness . . . rather than likelihood. It is a sort of multi-valued logic but its operations differ from those traditional for multi-valued logics. . . .
>
>   We are not concerned with the likelihood that a man is short, after many trials; we are concerned with the shortness of one observation.

Here Goguen refers both to formal differences from probability theory and to the simple but fundamentally important fact that fuzzy logic and probability theory

---

[125] See also pp. 27ff. above and section 5.1.4 for other aspects of Goguen's work.

[126] The terms "fuzzy sets and logic" and "fuzzy logic" appear on pp. 337 and 359 of Goguen 1968–69, respectively. "Fuzzy logic" is the title of Marinos 1966, which is concerned with the analysis and synthesis of many-valued logic circuits. Marinos 1966 is the first paper ever in which the term "fuzzy logic" appears, Goguen 1968–69 is the second.

are constructed to model completely different phenomena, namely vagueness and randomness. The fact that these misconceptions persisted only confirms Goguen's bright and visionary intellect.

Apparently, Goguen was not aware of the existing work on Łukasiewicz logic. He refers only to intuitionistic logic when introducing the algebraic structures of his logic and omits completely the vast number of results and the existing books on many-valued logics. As a result, he reinvents some already established definitions for predicate many-valued logics, such as (4.2) and (4.3). Nevertheless, his conception of deduction in the presence of truth degrees and the algebraic structures of truth degrees that he proposed are greatly original and crucially important for fuzzy logic.

Deduction from partially true assumptions

Goguen derived the conception of deduction from his analysis—inspired by Black (1963)—and resolution of the well-known sorites paradox.[127] The following observations are crucial (Goguen 1968–69, 336, 371):

> Just as propositions . . . are no longer either 'true' or 'false', but can be intermediate, deductions are no longer 'valid' or 'invalid'. . . .
> [T]he present methods provide a framework for an 'inexact mathematics' in which we can apply approximately valid deductive procedures to approximately true hypotheses. The 'logic of inexact concepts' then helps assess the validity of the final conclusions.

This is best illustrated by the rule of *modus ponens* which he examines (Goguen 1968–69, 356):

> First, how is deduction used? We have a truth value $[P]$ and a truth value $[P \Rightarrow Q]$, and we want to estimate the truth value $[Q]$;[128]

Thus, in contrast to the ordinary *modus ponens* saying "from φ and $\varphi \to \psi$ infer ψ," Goguen proposes a more general rule,

from φ with its truth degree a and $\varphi \to \psi$ with its truth degree c $\qquad$ (4.57)

infer ψ with a lower bound $\underline{b}$ of its truth degree.

Here $\underline{b}$ is computed from a and c. In particular, he proposes using a truth function $\otimes$ of a many-valued conjunction for this purpose and thus his *modus ponens* may be displayed as[129]

$$\frac{\varphi \text{ with degree } a, \; \varphi \to \psi \text{ with degree } c}{\psi \text{ with degree } a \otimes c}. \qquad (4.58)$$

In the classical case, in which $L = \{0, 1\}$, Goguen's rule may be identified with the ordinary *modus ponens*, since then $a = 1$ and $c = 1$ is the only combination of $a, c \in \{0, 1\}$ resulting in a nonzero estimate $a \otimes c$ and, therefore, (4.58) may be represented in the simple form $\frac{\varphi, \varphi \to \psi}{\psi}$.

---

[127] Further information about this and other paradoxes is provided in section 5.4.2.

[128] Goguen denotes formulas by P and Q, implication by $\Rightarrow$, and truth value of P by $[P]$.

[129] Though Goguen himself did not display the rule in such symbolic form.

Modus ponens of the form (4.58) has the intuitively appealing and desired property that the result of "a long chain of only slightly unreliable deductions can be very unreliable" (Goguen 1968–69, 327). For suppose we have formulas $\varphi_0, \varphi_1, \ldots, \varphi_n$ for which the degree of validity of φ_0 is a and that of $\varphi_{i-1} \to \varphi_i$ is c_i for every $i = 1, \ldots, n$.[130] Then we may infer φ_1 to degree $a \otimes c_1$ by a single application of (4.58) to φ_0 with a and $\varphi_0 \to \varphi_1$ with c_1, and then also φ_2 to degree $a \otimes c_1 \otimes c_2$ from φ_1 with $a \otimes c_1$ and $\varphi_1 \to \varphi_2$ with c_2, and so on. After n steps, we obtain φ_n to degree $a \otimes c_1 \otimes \cdots \otimes c_n$. Since the $\otimes$-product gets generally smaller with a larger number of factors, the degree $a \otimes c_1 \otimes \cdots \otimes c_n$ of the conclusion φ_n may be much smaller than any of the degrees $a, c_1, \ldots, c_n$ of the assumptions. This is concretely seen if L is the unit interval $[0, 1]$ and $\otimes$ is the usual product $\cdot$ of numbers—an original truth function of conjunction proposed by Goguen to resolve the sorites paradox. For instance, even if $a = 1$ and $c_i = 0.95$ for all i, i.e. φ_0 is fully true and every $\varphi_i \to \varphi_{i+1}$ is almost fully true, the degree to which we infer φ_n may be quite small with just 15 inference steps, because

$$\text{if } a = 1 \text{ and } c_i = 0.95, \text{ we infer } \varphi_{15} \text{ with } a \cdot c_1 \cdots c_{15} = 0.95^{15} \approx 0.46, \qquad (4.59)$$

rendering φ_n very unreliable. Goguen's conception initiated an important direction in fuzzy logic which we examine in section 4.3.2.

Residuated lattices as natural structures of truth degrees

The second important contribution of Goguen (1968–69) was his proposal of residuated lattices as appropriate structures of truth degrees for fuzzy logic, which he nicely justified from a logical viewpoint. Goguen first derives the usual product on the unit interval $[0, 1]$ as an appropriate truth function of conjunction that resolves the sorites paradox. Then he argues that $[0, 1]$ is not the only reasonable set because one naturally also needs nonlinearly ordered ones, e.g. in multicriteria decisions, with degrees such as $\langle 0, 0.9, 0.5 \rangle \in [0, 1]^3$. The set L of truth degrees should thus be partially ordered, bounded by 0 and 1 representing falsity and truth, respectively, and, moreover, form a complete lattice $\langle L, \wedge, \vee, 0, 1 \rangle$.[131] The existence of arbitrary infima $\wedge$ and suprema $\vee$ is needed to define the semantics of quantifiers by (4.2) and (4.3) as well as intersections and unions of fuzzy sets. As a generalization of the product conjunction on $[0, 1]$, Goguen proposes an associative operation $\otimes$ on L satisfying some reasonable properties derived basically from the considerations regarding the graded *modus ponens* (4.58). He thereby comes to the requirement that the structure $\langle L, \otimes, \wedge, \vee, 0, 1 \rangle$ should form a *completely lattice ordered semigroup* for which 1 is neutral w.r.t. $\otimes$. He thus requires that

---

[130] One may think of the degree of validity as a label attached to a formula—a degree to which the formula has been assumed or established by deduction. Goguen does not clearly distinguish this concept, a syntactic one, from the semantic concept of a truth degree of a formula, partly due to his semi-formal considerations regarding *modus ponens*. A proper distinction and formulation of rules in Goguen's conception is due to Pavelka; see section 4.3.2.

[131] Using general partially ordered sets as sets of truth degrees was also mentioned in Zadeh 1965a.

(a) $\langle L, \wedge, \vee, 0, 1 \rangle$ be a complete lattice,

(b) $\langle L, \otimes, 1 \rangle$ be a semigroup with neutral element 1, thus $\otimes$ satisfies associativity, i.e. $a \otimes (b \otimes c) = (a \otimes b) \otimes c$, and the neutrality of 1, i.e. $a \otimes 1 = 1 \otimes a = a$,

(c) $\otimes$ be distributive over arbitrary suprema, i.e.

$$a \otimes \bigvee_{j \in J} b_j = \bigvee_{j \in J} (a \otimes b_j). \tag{4.60}$$

Furthermore, he requires that implication $\to$ be the *residuum* of $\otimes$, i.e.

$$a \to b = \bigvee \{ c \mid a \otimes c \leq b \}, \tag{4.61}$$

and observes that $\otimes$ and $\to$ are connected via a so-called *adjointness* property, i.e.

$$a \otimes b \leq c \text{ if and only if } b \leq a \to c. \tag{4.62}$$

Partially ordered semigroups, the rule of residuation, adjointness, and the role of (4.60) were well known in the late 1960s. A study of such structures began with the study of ideals in ring theory by Krull[132] and, independently, Ward and Dilworth (1939), and Dilworth (1939). Several results on these structures were available in Birkhoff 1948, to which Goguen himself refers, and were not new in this regard. It was also known that Goguen's conditions (a)–(c) may equivalently be rephrased in terms of Ward and Dilworth's (1939) residuated lattices, a notion that was later used by many authors; see sections 4.3.2 and 4.4. A *complete residuated lattice* is a structure

$$\mathbf{L} = \langle L, \wedge, \vee, \otimes, \to, 0, 1 \rangle$$

satisfying the above conditions (a) and (b) of Goguen, commutativity for $\otimes$, i.e. $a \otimes b = b \otimes a$, and the adjointness property (4.62). Now, it can be proved that in a structure $\langle L, \otimes, \wedge, \vee, 0, 1 \rangle$ satisfying (a) and (b) with commutative $\otimes$, condition (c) is equivalent to the existence of $\to$ that satisfies (4.62).[133]

Truly original was Goguen's presentation and justification of residuated lattices as structures of truth degrees.[134] Here associativity of $\otimes$ is a natural requirement since we want every formula $(\varphi \otimes \psi) \otimes \chi$ to have the same truth value as $\varphi \otimes (\psi \otimes \chi)$. For similar reasons, one might require commutativity of $\otimes$, which Goguen considers as a further possible restriction. Neutrality of 1 is required since we want a conjunction of any formula φ and a fully true formula to have the same value as φ. Condition (4.60) guarantees that Goguen's *modus ponens*, (4.58), is

(a) *sound* in that whenever $a \leq \|\varphi\|$ and $c \leq \|\varphi \to \psi\|$ then $a \otimes c \leq \|\psi\|$; in words, whenever φ and $\varphi \to \psi$ are true at least to degrees a and c, respectively, then $a \otimes c$ indeed provides a lower estimation of the truth degree of ψ; and

(b) *strong* in that the lower estimation $a \otimes c$ of $\|\psi\|$ is large.

---

[132] W. Krull, "Axiomatische Begründung der allgemeinen Idealtheorie" [Axiomatic foundation of the general theory of ideals], *Sitzungsberichte der Physikalisch-medizinischen Sozietät zu Erlangen* 56 (1924): 47–63.

[133] Most fuzzy logics assume commutativity as a natural property. For the noncommutative case see p. 208 below.

[134] Our justification slightly differs from Goguen's in that it is somewhat more direct and complete. The main points are, however, Goguen's.

Indeed, (4.60) ensures that with $\to$ defined by (4.61), we have $\|\varphi\| \otimes (\|\varphi\| \to \|\psi\|) \leq \|\psi\|$. Now, isotony of $\otimes$ and $\|\varphi \to \psi\| = \|\varphi\| \to \|\psi\|$ yield $a \otimes c \leq \|\varphi\| \otimes (\|\varphi\| \to \|\psi\|) \leq \|\psi\|$, verifying soundness. Furthermore, the definition of $\to$ by (4.61) implies that the rule is strong. For suppose in particular $a = \|\varphi\|$, $b = \|\psi\|$, and thus $c = \|\varphi \to \psi\| = a \to b$. The lower estimation of b obtained by *modus ponens* is then $a \otimes (a \to b)$ and it cannot be larger given the restriction imposed by soundness. Because to make the lower estimation of b larger, one would need to define the value of $a \to b$ differently from (4.61), say equal to x, and we would need $a \otimes x > a \otimes (a \to b)$. But if x is a candidate for the definition of $a \to b$ which is admissible in that $a \otimes x$ would still be a lower estimation of b, i.e. $a \otimes x \leq b$, then $x \leq a \to b$ due to (4.61). Isotony of $\otimes$ then yields $a \otimes x \leq a \otimes (a \to b)$, contradicting the assumption $a \otimes x > a \otimes (a \to b)$. Hence (4.61) guarantees that *modus ponens* is strong.[135]

Further contributions

Let us at least briefly mention some further contributions of Goguen 1968–69. Relatively standard is his definition of negation as a pseudo-complement, i.e. $\neg a = a \to 0$. However, he also considers residuated lattices with an additional, involutive negation which may be seen as a precursor to Esteva et al. 2000. In the last section, Goguen considers what he calls "inexact quantifiers" such as "some," "most," or "few," and mentions some interesting "nearly valid" deduction rules involving these quantifiers, such as "from 'most A are B and most B are C' infer 'most A are C'." Worth mentioning are also Goguen's ideas, examined in detail in section 5.1.4 and other parts of chapter 5, which significantly contributed to the development of mathematics based on fuzzy logic.

4.3.2　Pavelka-style fuzzy logic

Goguen's results served as a profound inspiration to a number of people, including Jan Pavelka (1948–2007). In a series of three papers (Pavelka 1979), entitled "On fuzzy logic I, II, III," he formalized and substantially developed Goguen's ideas and worked out a general logical framework—a kind of abstract fuzzy logic which allows for truth-functional but also non-truth-functional semantics. Inspired by Tarski's view (p. 158 above), he emphasized the notion of consequence which plays a fundamental role in his approach. He also developed a propositional fuzzy logic within this framework and obtained important results for it. A logic developed along the lines of Goguen and Pavelka's ideas is nowadays called a *Pavelka-style*

---

[135] A justification of residuated lattices which involves, instead of (4.60), the equivalent property of adjointness (4.62) is presented in Hájek 1998b, Novák et al. 1999, and Belohlavek 2002c.

logic,[136] even though the fundamental step made by Goguen would well justify the term "Goguen-Pavelka-style logic."

Pavelka obtained his results as part of his doctoral studies at the Charles University in Prague.[137] He learned about fuzzy logic in the interdisciplinary Katětov seminar held there. According to Aleš Pultr, Pavelka's advisor, Pavelka was a very gifted and hard-working student who aimed for perfection in every aspect of his work. After he defended his CSc. thesis in 1978,[138] he intended to continue working at the university, which he did for about a year. After this period there was no position suitable for him any more. He therefore left the university, disappointed that working in academia was no longer possible, to work for the company ČKD Polovodiče in Prague as a programmer, something he eventually came to enjoy.[139] After leaving the university, he never returned to fuzzy logic but continued to maintain contacts with academia and followed to a certain extent further developments in the field. He stayed with ČKD Polovodiče until after the Velvet Revolution in 1989, when he founded and directed a private company DCIT Ltd., specializing in information technology. He was fully devoted to growing the company, which he did with considerable success. At the same time, he had been teaching practically oriented computer science courses at Charles University, until his premature death in 2007, at the age of 59. We now examine Pavelka's contributions in detail.

Pavelka's abstract fuzzy logic

Pavelka assumes an abstract set $\mathscr{F}$ of *formulas* which may need not have an inner structure—they may be built by means of connectives but in general they are just abstract entities.[140] The truth degrees form a complete lattice $\mathbf{L} = \langle L, \leq \rangle$ possibly equipped with further operations. They are primarily interpreted as degrees of truth in the sense of fuzzy logic, but any other meaningful interpretation including probabilistic is possible. A common approach in logic to define semantics is to specify a collection of semantic structures and a rule assigning to every formula φ and every structure $\mathbf{S}$ a truth degree $\|\varphi\|_{\mathbf{S}}$ in L. Every $\mathbf{S}$ thus induces a mapping $E_{\mathbf{S}} : \mathscr{F} \to L$ by $E_{\mathbf{S}}(\varphi) = \|\varphi\|_{\mathbf{S}}$. Each such $E_{\mathbf{S}}$ is thus an L-set (fuzzy set with truth degrees in L) of formulas. If we abstract from this and consider as primitive the mappings in $E_{\mathbf{S}} \in L^{\mathscr{F}}$ instead of the structures $\mathbf{S}$, we arrive at Pavelka's conceptual viewpoint:

---

[136] Sometimes also a *fuzzy logic with evaluated syntax* (Novák et al. 1999) or *graded-style fuzzy logic*. See also p. 200 below which addresses a related confusion regarding Pavelka-style logic and so-called rational Pavelka logic.

[137] The historical information in this paragraph is based on recollections of Aleš Pultr, Pavelka's wife Emilia Pavelková, and Pavelka's colleagues from DCIT Ltd.

[138] "Reziduované svazy a logika s neurčitostí" [Residuated lattices and logic with uncertainty] (PhD diss., 1977, defended February 12, 1978). At that time, the Czechoslovak degree equivalent to the PhD was the "candidate of sciences" degree, abbreviated by "CSc."

[139] According to recollections of Vilém Novák, in the 1980s Pavelka concluded that theoretical mathematics was of little use and he preferred to do practical things.

[140] We use a slightly different notation and terminology; moreover, we make explicit some notions that are only implicitly present in Pavelka 1979.

an **L**-*semantics* for $\mathscr{F}$ is an arbitrary set $\mathscr{S}$ of *L*-sets of formulas, i.e. $\mathscr{S} \subseteq L^{\mathscr{F}}$. Elements $E \in \mathscr{S}$ may be thought of as truth *evaluations*. We indeed denote $E(\varphi)$ also $\|\varphi\|_E$ and call it the truth degree of φ in E to retain the intended meaning.

For illustration, consider two very different examples of L-semantics. First, let $\mathscr{F}$ be the set of all formulas of the infinitely-valued Łukasiewicz logic, **L** be $[0,1]$ with its natural order, and define the L-semantics as

$$\mathscr{S} = \{E \in [0,1]^{\mathscr{F}} \mid \text{for some evaluation } e : E(\varphi) = \|\varphi\|_e \text{ for each } \varphi \in \mathscr{F}\}.$$

In this case, $E \in \mathscr{S}$ are just the evaluations induced by evaluations of propositional symbols and $E(\varphi)$ is the truth degree of φ. As a second example, let $\Omega \neq \emptyset$ be a finite set with $\omega \in \Omega$ called elementary events, **L** be $[0,1]$ again, and $\mathscr{F} = 2^{\Omega}$. Then

$$\mathscr{S} = \{E \in [0,1]^{\mathscr{F}} \mid \text{for some probability distribution } p : E(\varphi) = \sum_{\omega \in \varphi} p(\omega)\}$$

is an L-semantics. In this case, $E \in \mathscr{S}$ are just the probability measures on Ω and $E(\varphi)$ is the probability of φ.

In classical logic, a theory is understood as a set of formulas. It may represent assumptions about some domain. In Pavelka-style logic, a *theory* T over $\mathscr{F}$ is any *L*-set of formulas, i.e. $T \in L^{\mathscr{F}}$. This corresponds to Goguen's idea that an assumption may be true only to a certain degree. We say that $E \in \mathscr{S}$ is a *model* of a theory T if $T \subseteq E$, i.e. if $T(\varphi) \leq E(\varphi)$ for every φ. This means that every formula φ is true in E at least to the degree prescribed by T. For instance, the *L*-set

$$T = \{^{0.8}/young(John), ^{0.9}/baby(Jim), \ldots, ^1/baby(x) \rightarrow young(x)\}$$

is a theory which asserts that *John* is young to degree (at least) 0.8 and so on. Returning to our example with probabilities, let $\Omega = \{1, \ldots, 6\}$ represent tossing a die. With a fair die, we have $p(1) = \cdots = p(6) = 1/6$ for the corresponding probability distribution p. However, for a biased die we might have $p_b(1) = \cdots = p_b(4) = 1/6$, $p_b(5) = 1/12$, $p_b(6) = 3/12$, representing a situation in which 6 is more likely at the expense of 5. The theory $T = \{\langle 1, 1/12\rangle, \ldots, \langle 6, 1/12\rangle\}$ requires the probability of each result i be at least $1/12$, i.e. the die cannot be biased too much. Clearly, the evaluation E corresponding to p_b is a model of T, while E corresponding to a fake die containing only even numbers is not as $E(i) = 0 \not\geq 1/12 = T(i)$ for $i = 1, 3, 5$.

For an L-semantics $\mathscr{S}$ for $\mathscr{F}$, a theory T, and a formula $\varphi \in \mathscr{F}$, the *degree* $\|\varphi\|_T^{\mathscr{S}}$ *to which* φ *semantically follows* from T is defined by

$$\|\varphi\|_T^{\mathscr{S}} = \bigwedge\{\|\varphi\|_E \mid E \in \mathscr{S} \text{ is a model of } T\},$$

where $\bigwedge$ is the infimum in **L**. The corresponding *semantic consequence operator* $C_{\mathscr{S}} : L^{\mathscr{F}} \rightarrow L^{\mathscr{F}}$ induced by $\mathscr{S}$ is defined by

$$[C_{\mathscr{S}}(T)](\varphi) = \|\varphi\|_T^{\mathscr{S}}, \text{ i.e. } C_{\mathscr{S}}(T) = \bigcap\{E \in \mathscr{S} \mid T \subseteq E\}.$$

$C_{\mathscr{S}}$ is indeed a closure operator in $\langle L^{\mathscr{F}}, \subseteq \rangle$, i.e. satisfies for any $T, T_1, T_2 \in L^{\mathscr{F}}$:

$$T \subseteq C_{\mathscr{S}}(T),$$

$$T_1 \subseteq T_2 \text{ implies } C_{\mathscr{S}}(T_1) \subseteq C_{\mathscr{S}}(T_2), \tag{4.63}$$

$$C_{\mathscr{S}}(T) = C_{\mathscr{S}}(C_{\mathscr{S}}(T)).$$

While $C_{\mathscr{S}}$ is basically defined according to a common pattern, the corresponding syntactic consequence and the underlying notion of proof is based on that of the deduction rule inspired by Goguen's *modus ponens* (4.57). An *n*-ary *deduction rule* for $\mathscr{F}$ and L is a pair $R = \langle R_{\text{syn}}, R_{\text{sem}} \rangle$ consisting of a partial function $R_{\text{syn}} \colon \mathscr{F}^n \to \mathscr{F}$ (syntactic part) and a function $R_{\text{sem}} \colon L^n \to L$ (semantic part). Such a rule may be visualized as[141]

$$\frac{\langle \varphi_1, a_1 \rangle, \ldots, \langle \varphi_n, a_n \rangle}{\langle R_{\text{syn}}(\varphi_1, \ldots, \varphi_n), R_{\text{sem}}(a_1, \ldots, a_n) \rangle}.$$

Such a rule R enables us to infer that formula $R_{\text{syn}}(\varphi_1, \ldots, \varphi_n)$ is valid to degree (at least) $R_{\text{sem}}(a_1, \ldots, a_n)$ if we know that each φ_i is valid to degree (at least) a_i. Pavelka assumes that the rules preserve nonempty suprema in that $R_{\text{sem}}(\ldots, \bigvee_i a_i, \ldots) = \bigvee_i R_{\text{sem}}(\ldots, a_i, \ldots)$. Examples are

$$\frac{\langle \varphi, a \rangle, \langle \varphi \to \psi, b \rangle}{\langle \psi, a \otimes b \rangle}, \qquad \frac{\langle \varphi, a \rangle}{\langle (\forall x)\varphi, a \rangle}, \qquad \text{and} \qquad \frac{\langle \varphi, a \rangle, \langle \varphi, b \rangle}{\langle \varphi, a \vee b \rangle}. \tag{4.64}$$

The first one is Goguen's *modus ponens* (4.57), the second one a generalization of a well-known classical rule, and the third one is degenerate in classical logic.

A theory $T \in L^{\mathscr{F}}$ is *closed w.r.t. a set $\mathscr{R}$ of rules* if

$$R_{\text{sem}}(T(\varphi_1), \ldots, T(\varphi_n)) \leq T(R_{\text{syn}}(\varphi_1, \ldots, \varphi_n))$$

for every *n*-ary rule R in $\mathscr{R}$ whenever R_{syn} is defined. This condition says "in degrees" that whatever may be inferred by rules in $\mathscr{R}$ from T is in T already. Pavelka shows that the set of all theories closed w.r.t. any $\mathscr{R}$ is closed with respect to intersections. We may then consider a fixed L-set $A \in L^{\mathscr{F}}$ of *logical axioms* and consider for a given theory $T \in L^{\mathscr{F}}$ the intersection $C_{A,\mathscr{R}}(T)$ of all theories that are closed w.r.t. $\mathscr{R}$ and contain both A and T, i.e.

$$C_{A,\mathscr{R}}(T) = \bigcap \{ T' \in L^{\mathscr{F}} \mid A \subseteq T', T \subseteq T', \ T' \text{ closed w.r.t. } \mathscr{R} \}. \tag{4.65}$$

Note that as with theories, we deal with degrees $A(\varphi)$ to which formulas φ are considered logical axioms. It then turns out:

Theorem 4.30. The mapping $T \mapsto C_{A,\mathscr{R}}(T)$ in (4.65) is a closure operator in $L^{\mathscr{F}}$. $C_{A,\mathscr{R}}(T)$ is the least theory closed w.r.t. $\mathscr{R}$ and containing A and T.

While $C_{A,\mathscr{R}}$ is syntactically based, it still is not based on a suitable notion of proof. Pavelka, inspired by Goguen 1968–69, proposed a notion of proof that involves pairs $\langle \varphi, a \rangle$ consisting of a formula $\varphi \in \mathscr{F}$ and a truth degree $a \in L$, which we call *(L-)weighted formulas*. An *(L-)weighted proof* of $\langle \varphi, a \rangle$ from a theory T using logical axioms in A and deduction rules in $\mathscr{R}$ is a sequence $\langle \varphi_1, a_1 \rangle, \ldots, \langle \varphi_n, a_n \rangle$ such that φ_n is φ, $a_n = a$, and for each $i = 1, \ldots, n$,

---

[141]Pedantically considered and following a common practice in logic, one should represent the truth degrees $a \in L$ by symbols $\bar{a}$ in the language, and thus work with $\langle \varphi, \bar{a} \rangle$ instead of $\langle \varphi, a \rangle$.

-   $a_i = T(\varphi_i)$, or
-   $a_i = A(\varphi_i)$, or
-   $\langle \varphi_i, a_i \rangle$ is obtained from some $\langle \varphi_j, a_j \rangle$'s, $j < i$, by some rule $R \in \mathscr{R}$, in that $\varphi_i = R_{\mathrm{syn}}(\varphi_{i_1}, \ldots, \varphi_{i_k})$ and $a_i = R_{\mathrm{sem}}(a_{i_1}, \ldots, a_{i_k})$ for some $i_1, \ldots, i_k \leq i$.

In such a case, we write $T \vdash^{A,\mathscr{R}} \langle \varphi, a \rangle$ and call $\langle \varphi, a \rangle$ *provable* from T using A and $\mathscr{R}$. φ is then called *provable to degree* (at least) a from T using A and $\mathscr{R}$. Finally, the *degree of provability* of φ from T using A and $\mathscr{R}$, denoted by $|\varphi|_T^{A,\mathscr{R}}$, is defined as

$$ |\varphi|_T^{A,\mathscr{R}} = \bigvee \{ a \mid T \vdash^{A,\mathscr{R}} \langle \varphi, a \rangle \}, \tag{4.66} $$

i.e. as the supremum of all $a \in L$ for which there exists an L-weighted proof of $\langle \varphi, a \rangle$. This notion presents a significant difference from the ordinary case in which all proofs are equally important. In Pavelka-style logic, we may obtain proofs $\ldots, \langle \varphi, a_i \rangle$, $i = 1, 2, \ldots$, with different degrees a_i. It is then natural to take the "best proof"—the one with the largest value. However, such a proof may not exist. For instance, it may happen that while there exist proofs $\ldots, \langle \varphi, 1 - \varepsilon \rangle$ for arbitrarily small $\varepsilon > 0$, no proof $\ldots, \langle \varphi, 1 \rangle$ exists. It is thus natural to define the degree of provability as the supremum (4.66), i.e. the best upper bound. Having found a particular proof $\ldots, \langle \varphi, a \rangle$, we only have a lower estimation $a \leq |\varphi|_T^{A,\mathscr{R}}$. Notice that from this view, good proofs are short and use strong inference rules—a property that certainly has an intuitive appeal. Pavelka proved that his notion of provability corresponds to $C_{A,\mathscr{R}}$, which is therefore called the *syntactic consequence operator*:

Theorem 4.31. $[C_{A,\mathscr{R}}(T)](\varphi) = |\varphi|_T^{A,\mathscr{R}}$ *for every theory T and any formula φ.*

A tuple $\mathsf{L} = \langle \mathscr{F}, \mathbf{L}, \mathscr{S}, A, \mathscr{R} \rangle$ with components as discussed above may then be called an *abstract fuzzy logic*.[142] The concepts of soundness and completeness then naturally assume the following forms. L is

-   *sound* (in Pavelka's sense) if $|\varphi|_T^{A,\mathscr{R}} \leq \|\varphi\|_T^{\mathscr{S}}$ for every $T \in L^{\mathscr{F}}$ and $\varphi \in \mathscr{F}$;
-   *complete* (in Pavelka's sense) if for every theory $T \in L^{\mathscr{F}}$ and formula $\varphi \in \mathscr{F}$,

$$ |\varphi|_T^{A,\mathscr{R}} = \|\varphi\|_T^{\mathscr{S}}. \tag{4.67} $$

Pavelka observes that L is sound whenever every rule $R \in \mathscr{R}$ is sound w.r.t. $\mathscr{S}$, meaning that every $E \in \mathscr{S}$ is closed w.r.t. R, and $A \subseteq E$ for every $E \in \mathscr{S}$. For example, if $\mathbf{L}$ is a residuated lattice, Goguen's *modus ponens*, i.e. the first rule in (4.64), is sound w.r.t. the corresponding truth-functional semantics because soundness then amounts to $\|\varphi\|_e \otimes \|\varphi \to \psi\|_e \leq \|\psi\|_e$, which is easily verified: $\|\varphi \to \psi\|_e = \|\varphi\|_e \to \|\psi\|_e$ and $\|\varphi\|_e \otimes (\|\varphi\|_e \to \|\psi\|_e) \leq \|\psi\|_e$ is a direct consequence of adjointness.

Pavelka's "graded" notions generalize their ordinary counterparts such as that of deduction rule, soundness, and completeness. The novelty in Goguen and Pavelka's

---

[142] This term is consistent with Hájek 1998b and Gerla 2001, although Gerla does not include A and $\mathscr{R}$ and considers instead as primitive a general operator of syntactic consequence. Moreover, Gerla assumes that syntactic and semantic consequences coincide for an abstract fuzzy logic, while for Pavelka, this property needs to be postulated (see completeness in Pavelka's sense below). Pavelka himself did not have any name for L.

approach consists in taking consistently the graded nature of not only the validity of formulas, but also other metalogical notions. While Pavelka himself probably considered these notions as preparatory for parts II and III of his paper, we shall see in section 4.5.3 that they provide a useful framework for various calculi for reasoning about different kinds of uncertainty.[143]

Pavelka-style propositional logics

In parts II and III, Pavelka (1979) investigates in the framework of his abstract logic certain propositional fuzzy logics. Each such logic involves a fixed complete residuated lattice $\mathbf{L} = \langle L, \wedge', \vee', \otimes', \rightarrow', 0, 1 \rangle$ as the structure of truth degrees (p. 164 above). The logic has four basic connectives, $\wedge, \vee, \rightarrow$, and $\otimes$, interpreted by $\wedge', \vee', \rightarrow'$, and $\otimes'$, respectively. In addition, for every truth degree $a \in L$ it contains a nullary connective $\bar{a}$, i.e. a symbol of the truth degree. Moreover, Pavelka considers equivalence and negation, defined as shorthands: $\varphi \leftrightarrow \psi$ for $(\varphi \rightarrow \psi) \wedge (\psi \rightarrow \varphi)$ and $\neg \varphi$ for $\varphi \rightarrow \bar{0}$. He calls $\leftrightarrow$ biimplication and its truth function, $\leftrightarrow'$, *biresiduum*. Furthermore, k-ary connectives c are allowed, intended for use in applications, whose truth functions c' are *fitting*, i.e. satisfying

$$(a_1 \leftrightarrow' b_1)^{p_1} \otimes' \cdots \otimes' (a_k \leftrightarrow' b_k)^{p_k} \leq c'(a_1, \ldots, a_k) \leftrightarrow' c'(b_1, \ldots, b_k)$$

for all $a_i, b_i \in L$, where p_i are exponents fixed for c and the powers $(a_i \leftrightarrow' b_i)^{p_i}$ are with respect to $\otimes'$. This condition ensures that any congruence θ on $\mathbf{L}$ is also a congruence w.r.t. any fitting operation on $\mathbf{L}$. Pavelka calls a residuated lattice $\mathbf{L}$ along with a set of fitting operations an *enriched residuated lattice* based on $\mathbf{L}$.

Pavelka then comes to the problem of the axiomatizability of propositional logics within an abstract fuzzy logic in which evaluations $E \in \mathscr{S}$ are induced in a truth-functional manner by evaluations of propositional symbols and satisfy $E(\bar{a}) = a$ for all $a \in L$. He considers two important cases. First, logics with L being a finite chain of truth degrees, for which he obtains a positive result which we mention below. Second, logics with L being the real unit interval $[0, 1]$ for which he obtains:

Theorem 4.32. If $\rightarrow'$ is not continuous then the logic is not axiomatizable.

Continuity in theorem 4.32 means continuity of $\rightarrow'$ as a real function but, in fact, Pavelka proved the theorem for a general complete residuated chain and continuity with respect to the order topology on L with open basis given by intervals $(a, b) = \{c \in L \mid a < c < b\}$. According to Menu and Pavelka 1976, continuity of $\rightarrow'$ implies continuity of $\otimes'$. As Pavelka recalls, the important result on classification of t-norms by Mostert and Shields (1957) implies that there exist $2^{\aleph_0}$ nonisomorphic continuous t-norms $\otimes'$. Now, in view of theorem 4.32, Menu and Pavelka (1976) were able to show:

---

[143] The subtitle of part I reads "Many-valued rules of inference"; there Pavelka presents several notions without giving them names.

Theorem 4.33. *If a residuated lattice on* $[0,1]$ *has a continuous residuum* $\rightarrow$*, then its adjoint* $\otimes$ *is the Łukasiewicz t-norm or its isomorphic copy.*

Hence, of all the $2^{\aleph_0}$ candidates, only the standard Łukasiewicz algebra $[0,1]_{\text{Ł}}$ with its isomorphic copies remains as the structure for which a complete axiomatization in Pavelka style may exist. Pavelka indeed found such an axiomatization. In summary, he proved the following theorem on Pavelka-style propositional logics:

Theorem 4.34. *Consider an enriched residuated lattice based on a complete residuated lattice* **L**.

(a) *If L is a finite chain, then the logic is axiomatizable, i.e. there exist A and* $\mathscr{R}$ *with* (4.67).
(b) *If L is the real unit interval then the logic is axiomatizable iff* **L** *is isomorphic to the standard Łukasiewicz chain* $[0,1]_{\text{Ł}}$.

Partly because Pavelka was not aware of the existing work on Łukasiewicz logics, the axiomatizations he provides to justify theorem 4.34 are rather complex, with over thirty axioms, and we thus omit details. Pavelka's proof of completeness relies on the algebraic methods developed by Rasiowa and Sikorski (1963) and utilizes the Lindenbaum algebra of classes of provably equivalent formulas with provable equivalence of φ and ψ meaning that the degree of provability of $\varphi \leftrightarrow \psi$ from a theory T (an L-set of formulas) equals 1. Pavelka-style Łukasiewicz logic was later significantly simplified by Hájek (1995a), who, however, does not work in the framework of abstract logic and "simulates" Pavelka-style propositional logics in the ordinary logical framework (p. 200 below), and by Novák (see the next section).

Pavelka's result on the nonaxiomatizability of propositional fuzzy logics other than Łukasiewicz's certainly presents a serious limitation to Pavelka-style logics. Nevertheless, these limitations should not be overemphasized. There exist other practically motivated calculi that have been developed within the framework of Pavelka's abstract fuzzy logic and some of them, along with related developments, are the subject of the next section.

4.3.3 Further developments

Pavelka-style predicate Łukasiewicz logic

At the close of part III, Pavelka (1979) explains: "Added in proof: In 1977 the results of this paper were extended by the author to first order predicate calculi." Very briefly, he indicates the basic proof ideas, again inspired by Rasiowa and Sikorski 1963. Pavelka's extension, however, has never been published.

Predicate Łukasiewicz logic in Pavelka style, i.e. an extension of Pavelka's propositional logic, was studied by Vilém Novák, who presented it in Novák 1987, 1990. In view of Pavelka's theorem 4.34, Novák assumes the real unit interval $L = [0,1]$

equipped with the standard Łukasiewicz operations or a finite Łukasiewicz chain. Note that Pavelka provided an axiomatization for propositional logics for every finite residuated chain, hence Novák imposes a further restriction. The propositional part of Novák's logic is derived from Pavelka's work. It contains the symbols $\bar{a}$ of each truth degree $a \in L$ and a symbol $\rightarrow$ of implication, leaving $\wedge$, $\vee$, and $\otimes$ derived ones as on p. 148 above. The language further contains symbols of variables, relations, functions, and the symbol $\forall$ of the general quantifier, with $(\exists x)\varphi$ being a shorthand for $\neg(\forall x)\neg\varphi$. The L-structures are defined as usual—the symbols of relations and functions are interpreted by L-relations and ordinary functions, respectively. Terms, formulas, values of terms, and truth degrees of formulas are defined as usual as well (p. 111), given that $\bar{a}$ is interpreted by a. As in the case of Pavelka's propositional logics, these notions define an L-semantics for the set $\mathscr{F}$ of formulas in which the evaluations $E \in \mathscr{S}$ correspond to evaluations of formulas in L-structures. An additional specification of an L-set A of axioms and a set $\mathscr{R}$ defines an abstract fuzzy logic $\mathsf{L} = \langle \mathscr{F}, \mathsf{L}, \mathscr{S}, A, \mathscr{R} \rangle$ in Pavelka's sense, which automatically yields the notions of theory, model, semantic consequence, the notion of proof, completeness, etc. While Pavelka points out that the notions of his propositional logics are particular instances of those of his abstract logic, Novák does not present the predicate logic from this perspective. Instead, he presents it as an extension of Pavelka's propositional logic and its notions. For instance, when presenting an analogy of theorem 4.31 for the predicate logic, he says that the proof is a verbatim repetition of the proof of theorem 16 from part I of Pavelka 1979, while it is in fact a particular instance of theorem 16 and hence needs no proof.[144]

In particular, Novák proved a Pavelka-style completeness for the first-order fuzzy logic. His proof is inspired by Pavelka's adaptation of the method of Rasiowa and Sikorski (1963), uses a smaller system of axioms compared to Pavelka's for propositional logics, but is still complex and we do not present it here. A simplification of the system along with a new proof of completeness based on Henkin's method of models built from constants was presented in Novák 1995a. This was inspired by Hájek's (1995a) simplification of Pavelka-style propositional logic (see p. 200), which itself was later extended by Hájek (1997) to a simplification of Novák's original system. Further results, including a deduction theorem, the notion of consistency and its variants, completeness of theories, Herbrand's theorem, and some basic model-theoretical results, are available in Novák et al. 1999, which provides a comprehensive treatment of the Pavelka-style predicate Łukasiewicz logic.

As is clear from many of Novák's publications, including Novák 1995b, 1996, 2012, Novák et al. 1999, he pushed the idea that fuzzy logic in the narrow sense should be taken as a formal basis for fuzzy logic in the broad sense and for appli-

---

[144] This conceptual difference might partly have contributed to a widespread misconception that Pavelka-style logic means just the propositional and predicate logics of Pavelka and Novák, while in fact it represents a general logical framework encompassing various, even non-truth-functional logical systems involving degrees of truth; see below in this section and section 4.5.3.

cations. He was a vigorous proponent of this idea and had solid practical experience with real-world applications, particularly of fuzzy control. This interesting feature, probably also due to his engineering background, distinguishes him from most other contributors to fuzzy logic in the narrow sense. Important to note is also the fact that Novák was the pioneer of fuzzy logic in the former Czechoslovakia and due to his influence as well as his early book on fuzzy sets, several people in the Czech Republic and Slovakia decided to pursue research in fuzzy logic.[145]

Pavelka-style fuzzy logic is also the subject of a series of papers by Mingsheng Ying. In Ying 1991, he considers a general form of deduction theorem in abstract fuzzy logic containing an implication-like connective. Ying 1992a examines ultraproducts in Pavelka-style predicate fuzzy logic with enriched residuated lattices as structures of truth degrees and with certain generalized quantifiers. Compactness and the Löwenheim-Skolem property are studied in Ying 1992b. Pavelka-style propositional Łukasiewicz logic is also the subject of several papers by Turunen, e.g. Turunen 1995 and Turunen, Öztürk, and Tsoukiás 2010.

Graded consequence relations and operators

The foundational work of Tarski (1930a, 1935–36) on the notion of logical consequence as well as Pavelka's (1979) explicit usage of consequence operators motivated studies of graded consequence relations. These are fuzzy relations assigning to a collection A of formulas and a formula φ an element in a complete lattice L of truth degrees. Such an element, denoted $[C(A)](\varphi)$ or, more frequently, $A \vdash \varphi$, is interpreted as the degree to which φ follows from A. The first paper on this topic is Chakraborty 1988; other early contributions include Castro and Trillas 1991; Trillas and Alsina 1993; Castro 1994; Castro, Trillas, and Cubillo 1994; Gerla 1994a; Chakraborty 1995; and Gerla 1996. Formulas in these works are members of an abstract set $\mathscr{F}$ and the operators in question are only required to satisfy the three basic conditions (4.63) plus possibly a variant of the compactness property. We therefore discuss these works in detail within the context of general fuzzy closure structures in chapter 5 (p. 288).

Gerla's work on abstract fuzzy logic

Several important contributions are due to Giangiacomo Gerla, who worked out Pavelka's abstract approach to fuzzy logic in a number of papers and in his important book Gerla 2001. Accordingly, Gerla studies general closure operators that may play the role of consequence operators of various fuzzy logics. This way, he obtains results for general classes of fuzzy logics as well as particular fuzzy logics includ-

---

[145] *Fuzzy množiny a jejich aplikace* [Fuzzy sets and their applications] (Prague: Státní nakladatelství technické literatury, 1990); the book appeared in English as Novák 1989.

ing truth-functional and also probabilistic and other non-truth-functional logics. In this section, we present a general outline of Gerla's framework and results.[146]

Gerla's basic notions are slightly more general than Pavelka's. He starts with the notion of *abstract logic*—a triplet $\langle V, C, \mathscr{S} \rangle$ where V is a complete lattice, C a closure operator in V called *deduction operator*, and $\mathscr{S}$ a subset of V called *abstract semantics*. It is assumed that $\mathscr{S}$ does not contain V's largest element and that syntactico-semantical completeness holds in the sense that C coincides with the *semantic consequence operator* which is the closure operator $C_{\mathscr{S}}$ induced by $\mathscr{S}$, i.e. $C_{\mathscr{S}}(T) = \bigwedge \{ E \in \mathscr{S} \mid T \leq E \}$ for $T \in V$. Most important is the case where for a given complete lattice L of truth degrees and an abstract set $\mathscr{F}$ of formulas, V is the set $L^{\mathscr{F}}$ of all L-sets of formulas equipped with inclusion $\subseteq$. In particular, if $L = [0, 1]$ then C being a closure operator in V means that it is a fuzzy closure operator as considered by Pavelka, i.e. it satisfies (4.63). In this case, one writes just $\langle \mathscr{F}, C, \mathscr{S} \rangle$. In particular, for $L = \{0, 1\}$ we obtain *abstract crisp logic* and related notions such as abstract crisp semantics, while for $L = [0, 1]$ we obtain what Gerla calls an *abstract fuzzy logic*, *abstract fuzzy semantics*, etc. Note that Gerla restricts his arguments to $L = [0, 1]$. Accordingly, we assume this same restriction. Other than that, Gerla's notions are basically congruent with those of Pavelka.

Since *fuzzy closure operators*, i.e. closure operators $C : [0, 1]^{\mathscr{F}} \to [0, 1]^{\mathscr{F}}$, and *fuzzy closure systems*, i.e. closure systems in $\langle [0, 1]^{\mathscr{F}}, \subseteq \rangle$, have a fundamental role, Gerla studies them to a great extent.[147] Particularly important are *continuous* closure operators $C : L^{\mathscr{F}} \to L^{\mathscr{F}}$ which satisfy $C(\sup \mathscr{T}) = \sup_{T \in \mathscr{T}} C(T)$ for any directed subset $\mathscr{T} \subseteq L^{\mathscr{F}}$.[148] For $L = \{0, 1\}$, continuity is equivalent to the ordinary notion of compactness, sometimes called algebraicity, i.e. the property that $\varphi \in C(T)$ iff $\varphi \in C(T')$ for some finite $T' \subseteq T$. For fuzzy sets, such a characterization is due to Murali (1991) (cf. p. 285): a fuzzy closure operator C is continuous iff it is *algebraic* in that $C(T) = \bigcup \{ C(T') \mid T' \text{ is finite and } T' \ll T \}$, where $T' \ll T$ means that $T'(\varphi) < T(\varphi)$ whenever $T'(\varphi) > 0$. Furthermore, C is called *logically compact* if the set of all *consistent* Ts, i.e. such that $C(T)$ is not the largest element in $L^{\mathscr{F}}$, is inductive in that it is closed w.r.t. suprema of directed subsets. In such a case, every consistent T is contained in some maximal C-closed theory. Alternatively, C is logically compact iff for any $T \in [0, 1]^{\mathscr{F}}$, T is consistent iff every finite $T' \ll T$ is consistent. Continuity and logical compactness are in general different properties.

A number of Gerla's results relate to the question of an extension of a given crisp logic to a fuzzy logic, which amounts to the problem of extending ordinary closure operators and systems to fuzzy ones (Gerla 1994a, Biacino and Gerla 1996). For this purpose, Gerla defines the *canonical extension* of an operator $C : \{0, 1\}^{\mathscr{F}} \to \{0, 1\}^{\mathscr{F}}$

---

[146]In addition, we present his results on fuzzy closure operators in section 5.3.2, on non-truth-functional fuzzy logics in section 4.5.3, and on computational complexity in section 4.6.2.

[147]Fuzzy closure operators and systems are further examined in chapter 5 on pp. 288ff.

[148]Meaning that for $T_1, T_2 \in \mathscr{T}$, there exists $T \in \mathscr{T}$ with $T_1, T_2 \subseteq T$.

as the operator $C^* : [0,1]^{\mathscr{F}} \to [0,1]^{\mathscr{F}}$ defined for any $T \in [0,1]^{\mathscr{F}}$ by

$$C^*(T) = \bigcup_{a \in [0,1]} a \wedge C(^a T), \text{ i.e. } [C^*(T)](\varphi) = \bigvee \{a \in [0,1] \mid \varphi \in C(^a T)\}.$$

That is, the a-cut of $C^*(T)$ is obtained by applying C to the a-cut of T. Analogously, one extends a system $\mathscr{S} \subseteq \{0,1\}^{\mathscr{F}}$ of sets to a system $\mathscr{S}^* \subseteq [0,1]^{\mathscr{F}}$ of fuzzy sets by

$$\mathscr{S}^* = \{E \in [0,1]^{\mathscr{F}} \mid {}^a E \in \mathscr{S} \text{ for every } a \in [0,1]\}.$$

Canonical extensions have convenient properties. For example, C^* extends C in that both yield the same results for crisp subsets of $\mathscr{F}$. Moreover, C is a closure operator iff C^* is a fuzzy closure operator. The extensions naturally commute in that if C and $\mathscr{S}$ are a closure operator and a closure system, and if $D_{\mathscr{S}}$ and $\mathscr{F}^D$ denote the (fuzzy) closure operator and system induced by $\mathscr{F}$ and D, respectively, then $(C_{\mathscr{S}})^* = C_{\mathscr{S}^*}$ and $(\mathscr{S}_C)^* = \mathscr{S}_{C^*}$, and hence also $C^* = C_{\mathscr{S}_{C^*}}$ and $\mathscr{S}^* = \mathscr{S}_{C_{\mathscr{S}^*}}$. One may prove that C is ordinarily compact iff C^* is continuous. Gerla proved a number of properties regarding extensions of crisp logics, including:

Theorem 4.35. Let $\langle \mathscr{F}, C, \mathscr{S} \rangle$ be an abstract crisp logic. Then $\langle \mathscr{F}, C^*, \mathscr{S}^* \rangle$ is an abstract fuzzy logic, called the canonical extension of $\langle \mathscr{F}, C, \mathscr{S} \rangle$. Moreover, the fuzzy semantics $\mathscr{S}^*$ is equivalent to $Q(\mathscr{S}) = \{a \vee E \mid E \in \mathscr{S}, a \in [0,1)\}$.

An abstract fuzzy logic (with $L = [0,1]$) in the sense of Pavelka represents a particular case of Gerla's abstract fuzzy logics, namely logics $\langle \mathscr{F}, C, \mathscr{S} \rangle$, for which there exists a *fuzzy Hilbert system* $\langle A, \mathscr{R} \rangle$, i.e. a fuzzy set $A \in [0,1]^{\mathscr{F}}$ of axioms and a set $\mathscr{R}$ of (Pavelka-style) deduction rules, such that C coincides with the syntactic consequence operator $C_{A,\mathscr{R}}$ (4.65). These logics are the best-known Pavelka-style logics and include those described in the preceding sections and in section 4.5. Gerla also proved that these are just fuzzy logics with continuous deduction operators:

Theorem 4.36. A fuzzy closure operator C is continuous iff there exists a fuzzy Hilbert system $\langle A, \mathscr{R} \rangle$ such that $C = C_{A,\mathscr{R}}$.

Interesting examples of such logics result as follows. Let $\langle A, \mathscr{R} \rangle$ be a crisp Hilbert system, i.e. $A \in \{0,1\}^{\mathscr{F}}$ and $\mathscr{R}$ is a set of ordinary deduction rules. For every n-ary $R \in \mathscr{R}$, consider the Pavelka-style deduction rule $R^* = \langle R_{\text{syn}}, R_{\text{sem}} \rangle$ where $R_{\text{syn}} = R$ and $R_{\text{sem}}(a_1, \ldots, a_n) = a_1 \wedge \cdots \wedge a_n$. The fuzzy Hilbert systems $\langle A, \mathscr{R}^* \rangle$, called *canonical extensions*, obey the following properties.

Theorem 4.37. (a) The fuzzy consequence operator $C_{A,\mathscr{R}^*}$ of the canonical extension $\langle A, \mathscr{R}^* \rangle$ of a crisp Hilbert system $\langle A, \mathscr{R} \rangle$ is the canonical extension of the ordinary consequence operator $C_{A,\mathscr{R}}$; (b) the set of $C_{A,\mathscr{R}^*}$-closed theories is the canonical extension of the set of $C_{A,\mathscr{R}}$-closed theories.

Moreover, consequence operators of the canonical extensions admit a simple characterization in terms of strong cuts ${}^{a+}A = \{\varphi \mid A(\varphi) > a\}$ of theories:

Theorem 4.38. Let C be a fuzzy closure operator which maps crisp sets to crisp sets. C is the consequence operator of a canonical extension of a crisp Hilbert system iff (a) *C is continuous and* (b) *$C(^{a+}T) = {}^{a+}C(T)$ for every $a \in [0,1]$ and $T \in [0,1]^{\mathscr{F}}$.*

Of the several examples of extensions of crisp Hilbert systems, we present in section 4.5.3 the logic of necessities which results as a canonical extension of classical propositional logic.

Gerla examines the following way of obtaining fuzzy closure operators from ordinary ones. For a system $\mathscr{C} = (C_a)_{a \in [0,1]}$ of operators $C_a : \{0,1\}^{\mathscr{F}} \to \{0,1\}^{\mathscr{F}}$, the operator $C_{\mathscr{C}} : [0,1]^{\mathscr{F}} \to [0,1]^{\mathscr{F}}$ defined by

$$[C_{\mathscr{C}}(T)](\varphi) = \bigvee \{a \in [0,1] \mid \varphi \in C_a(^aT)\}$$

is called the *operator associated with* $(C_a)_{a \in [0,1]}$. This construction generalizes the canonical extension since if all C_as are equal to C, $C_{\mathscr{C}}$ is just C^*. A natural condition is that $(C_a)_{a \in [0,1]}$ be a chain, i.e. C_0 is constantly equal to $\mathscr{F}$ and $a \leq b$ implies $C_a(T) \supseteq C_b(T)$ for any T, or even a continuous chain in which case the latter condition is replaced by $C_b(T) = \bigcap_{a < b} C_a(T)$ for any T. Some basic properties are:

Theorem 4.39. Let $\mathscr{C} = (C_a)_{a \in [0,1]}$ be a chain of closure operators. (a) *$C_{\mathscr{C}}$ is a fuzzy closure operator.* (b) *The system $\mathscr{C}' = (C'_a)_{a \in [0,1]}$ defined by $C'_b(T) = \bigcap_{a < b} C_a(T)$ is a continuous chain and $C_{\mathscr{C}} = C_{\mathscr{C}'}$.*

If $\mathscr{C}$ is not a chain, $C_{\mathscr{C}}$ need not be idempotent but we may consider the fuzzy closure operator $\overline{C_{\mathscr{C}}}$ generated by $C_{\mathscr{C}}$, i.e. the least fuzzy closure operator greater than or equal to $C_{\mathscr{C}}$. If a fuzzy closure operator C is of such a form, it is called *stratified* and if, moreover, $\mathscr{C}$ is a chain, C is called *well-stratified*. These notions have their counterparts in terms of closure systems. The *system of fuzzy sets associated with* a system $\mathscr{K} = (\mathscr{S}_a)_{a \in [0,1]}$ of $S_a \subseteq \{0,1\}^{\mathscr{F}}$ is defined by $\mathscr{S}_{\mathscr{K}} = \{T \in [0,1]^{\mathscr{F}} \mid {}^aT \in \mathscr{S}_a$ for each $a \in (0,1]\}$. One may prove that if S_as are closure systems, $\mathscr{S}_{\mathscr{K}}$ is a fuzzy closure system which is called *stratified*. $\mathscr{S}_{\mathscr{K}}$ is well stratified if $\mathscr{K}$ is, moreover, a chain in that $\mathscr{S}_0 = \mathscr{F}$ and $a \leq b$ implies $\mathscr{S}_a \subseteq \mathscr{S}_b$. One then obtains:

Theorem 4.40. A fuzzy closure operator is (well) stratified iff the corresponding fuzzy closure system is (well) stratified.

An example of fuzzy logic with a stratified (even well stratified) fuzzy consequence operator is the similarity logic developed in Biacino, Gerla, and Ying 2000 and Ying 1994. On the other hand, many fuzzy logics, including the truth-functional ones to which we now turn, are not stratified.

Gerla considers a propositional logic with connectives including $\neg$, $\wedge$, and $\vee$, whose truth functions extend the classical ones, and considers the semantics $\mathscr{S}$ consisting of all *truth-functional* evaluations $E \in [0,1]^{\mathscr{F}}$, i.e. satisfying for any n-ary connective c and formulas $\varphi_i \in \mathscr{F}$: $E(c(\varphi_1, \ldots, \varphi_n)) = c(E(\varphi_1), \ldots, E(\varphi_n))$. He shows that a fuzzy logic with a truth-functional semantics is never stratified, hence truth-functional fuzzy logics are of a very different kind compared to the various extensions of crisp logics. Gerla examines the problem of *axiomatizability* of a truth-functional semantics $\mathscr{S}$, in particular the question of whether there exists a fuzzy Hilbert system $\langle A, \mathscr{R} \rangle$ such that the operators $C_{\mathscr{S}}$ and $C_{A,\mathscr{R}}$ of semantic and syntactic consequence coincide (note that no finitary or recursive properties are required at this general level). He calls a truth-functional semantics *continuous* if all the connectives are continuous as real functions and proves an important theorem, which is in accordance with Pavelka's theorem 4.34.

Theorem 4.41. Let $\mathscr{S}$ be a truth-functional semantics. (a) If $\mathscr{S}$ is continuous then it is axiomatizable. (b) If $\mathscr{S}$ is logically compact and axiomatizable then it is continuous.

Completely different from the truth-functional fuzzy logics are Gerla's probabilistic logics. These are examined in section 4.5.3.

Further contributions

That Pavelka-style fuzzy logic represents a useful framework for developing various kinds of logical calculi may additionally be demonstrated by further contributions. One is represented by a series of papers by Belohlavek and Vychodil on various kinds of dependencies in data and the logical calculi for reasoning about such dependencies, see e.g. Belohlavek and Vychodil 2015. The second one is Vojtáš 2001 in which a Pavelka-style complete fuzzy logic programming system is presented. Both these contributions are parts of broader activities in the respective areas of data analysis and logic programming (section 6.9). Worth noting is also the equational fragment of fuzzy predicate logic initiated in Belohlavek 2002b, 2003a. These investigations are examined in section 5.3.2.

4.4 Fuzzy logics based on t-norms and their residua

Both the infinitely-valued Łukasiewicz and Gödel logics examined in sections 4.2.1 and 4.2.2 may be looked at as logics in which conjunction and implication are interpreted by a t-norm and its residuum. In the first case, it is the Łukasiewicz t-norm, $a \otimes b = \max(0, a + b - 1)$, in the second the minimum t-norm, $a \otimes b = \min(a, b)$. Even though the original motivations for the two logics are completely different from those for fuzzy logic, they both may be considered as particular fuzzy logics of a broader class—the class of *fuzzy logics whose conjunction and implication are interpreted by a t-norm and its residuum*. A systematic development of these logics, which started around the mid-1990s, is the topic of this section.

4.4.1 Fuzzy logics based on t-norms until the mid-1990s

Connectives for fuzzy logic

Beginning in the late 1970s, a rich variety of prospective truth functions of connectives for fuzzy logic became the subject of numerous investigations. Many results were thus obtained, in particular regarding relationships between various properties of truth functions and representation theorems for particular classes of truth functions. As a rule, however, these studies did not pay attention to deductive aspects, axiomatizability and many other questions traditionally studied in mathematical logic. In fact, many of the proposed truth functions of fuzzy logic connectives were rather ad hoc because they were looked at too narrowly—mainly with regard to the question of how they generalized classical truth functions—and not in the broader context of logic. Exceptions to this were the developments in Pavelka-style logic (section 4.3.2), the continuing contributions to the Łukasiewicz, Gödel, and other many-valued logics, which were not concerned however with the idea of fuzzy logic, as well as a few other contributions which are surveyed below.

Already in his 1965 paper, Zadeh proposed min, max, and $1-x$ as the truth functions of conjunction, disjunction, and negation.[149] He also mentioned the number-theoretic product, $a \cdot b$, and its dual, $1-((1-a)\cdot(1-b)) = a+b-a\cdot b$, as possible alternative functions. Investigations of these and other functions became a subject of many studies. Among the early ones were experimental studies of the appropriateness of particular fuzzy logic connectives in certain cognitive tasks, which are surveyed in section 5.4.1 (p. 332).

In their stimulating paper, Bellman and Giertz (1973) offered proofs of uniqueness of min and max under certain assumptions as the truth functions for the intersection and union of fuzzy sets. They also discussed important properties of negations. In particular, they considered continuous and strictly decreasing functions $\neg$ in $[0,1]$ satisfying $\neg(0) = 1$, $\neg(1) = 0$, which later became known as *strict negations*, as well as *strong negations* which are moreover *involutive*, i.e. also satisfy $\neg\neg(a) = a$. The Łukasiewicz connectives as viable connectives for fuzzy logic were propounded by Giles (1976), who called them *bold connectives*. Hamacher (1978) observed redundancy in Bellman and Giertz's assumptions and proposed a family of connectives alternative to min and max, later called the Hamacher family. The view that reasonable truth functions of fuzzy logic conjunction and disjunction are represented by t-norms and t-conorms appeared in Alsina, Trillas, and Valverde 1980, 1983, and Dubois and Prade 1980a,b, but t-norms as conjunctions were already used in Anthony and Sherwood 1979 in the context of fuzzy groups (p. 284). There is also some evidence that discussions on the role of t-norms as conjunctions took place in the 1979 Linz Seminar. The connection to the theory of t-norms provided links to such important concepts and results as additive generators of Archimedean t-norms

---

[149] In fact, he proposed them as functions underlying the intersection, union, and complement of fuzzy sets.

(Ling 1965) and representation of continuous t-norms (Mostert and Shields 1957) which we discuss in sections 4.4.2 and 5.3.1. Broad classes of truth functions were thus codified. Note, however, that Goguen (1968–69) had already proposed resid- uated lattices as structures of truth degrees and that these cover all left-continuous t-norms as special cases (section 4.3.1 and theorem 4.49). But Goguen's work was not generally recognized in the early 1980s. Trillas (1979) obtained a representation theorem for strong negations in the spirit of Ling 1965, according to which strong negations are just functions of the form $\neg(a) = f^{-1}(1 - f(a))$ where f is an auto- morphism of $[0, 1]$; Esteva and Domingo 1980 is another early paper on this topic. In addition to bringing attention to t-norms, Alsina, Trillas, and Valverde (1980, 1983) considered *De Morgan triplets*, whose logics we consider in the next section.[150] These are triplets $\langle \otimes, \oplus, \neg \rangle$ where $\neg$ is a strong negation, $\otimes$ is a t-norm, and $\oplus$ its dual t-conorm, i.e. $a \oplus b = \neg(\neg a \otimes \neg b)$. The above papers, along with other early contributions to connectives based on t-norms and t-conorms, such as Yager 1980a, Dombi 1982, Klement 1982c, and Weber 1983, were followed by many further con- tributions; see Fodor and Yager 2000, Gottwald 2001 for further information.

While the properties of t-norms are intuitively acceptable as basic conditions for truth functions of conjunction, the situation is different with truth functions $\rightarrow$ of implication. That $\rightarrow$ extends classical implication and obeys monotonicity condi- tions is usually considered a basic requirement (Fodor and Yager 2000, Gottwald 2001, Kitainik 1993):

$$0 \rightarrow 0 = 0 \rightarrow 1 = 1 \rightarrow 1 = 1, \quad 1 \rightarrow 0 = 0, \tag{4.68}$$

$$a \leq b \text{ implies } a \rightarrow c \geq b \rightarrow c, \tag{4.69}$$

$$a \leq b \text{ implies } c \rightarrow a \leq c \rightarrow b. \tag{4.70}$$

It clearly follows that such $\rightarrow$ also satisfies $0 \rightarrow a = 1$ and $a \rightarrow 1 = 1$. Several other conditions obeyed by classical implication have been considered such as:

$$1 \rightarrow a = a, \tag{4.71}$$

$$a \rightarrow a = 1, \tag{4.72}$$

$$a \leq b \text{ iff } a \rightarrow b = 1, \tag{4.73}$$

$$a \rightarrow (b \rightarrow c) = b \rightarrow (a \rightarrow c), \tag{4.74}$$

$$a \rightarrow b \geq b, \tag{4.75}$$

$$a \rightarrow b = \neg b \rightarrow \neg a \text{ for a given negation } \neg, \text{ and} \tag{4.76}$$

$$\text{continuity of } \rightarrow. \tag{4.77}$$

These conditions are not independent. For instance, (4.70), (4.73), and (4.74) imply any of (4.68)–(4.76); see Fodor and Roubens 1994. The above and other properties of implications were examined in a number of pioneering contributions including Bandler and Kohout 1980a, Baldwin and Pilsworth 1980, Trillas and Valverde 1981,

---

[150] The authors did not use the term "De Morgan triplet." This term appeared in Klement 1981.

1985b, Smets and Magrez 1987, Fodor 1991, Kitainik 1993, and Fodor and Roubens 1994. For more information see Fodor and Yager 2000, Gottwald 2001.

As is well known, classical implication may be equivalently defined by negation, disjunction, and conjunction in a number of ways. For instance, $a \rightarrow b$ equals $\neg a \oplus b$ but also $\neg(a \otimes \neg b)$. In fuzzy logic, these definitions need not be equivalent—the result depends on the properties of the truth functions involved. Of the many possible definitions of implication in fuzzy logic, R- and S-implications play a prominent role.[151] The *residuated implication*, or *R-implication*, associated with a t-norm $\otimes$ is defined by $a \rightarrow b = \bigvee\{c \mid a \otimes c \leq b\}$. This is the well-known definition (4.61) of residuum used in a more general context by Goguen (1968–69); see section 4.3. The *S-implication* associated with a t-conorm $\oplus$ and a strong negation $\neg$ is defined by $a \rightarrow b = \neg a \oplus b$. These two kinds of implications behave rather differently. For example, for the S-implication defined by max as $\oplus$ and $1-x$ as $\neg$, we obtain $a \rightarrow b = 1$ if and only if $a = 0$ or $b = 1$, i.e. $\rightarrow$ does not satisfy the useful property (4.73) valid for each R-implication. Early papers examining R- and S-implications are Trillas and Valverde 1981, 1985a, Dubois and Prade 1984, and Weber 1983. R-implications also appeared in solving fuzzy relational equations (Sanchez 1976, Pedrycz 1982). Worth noting is that this context has a practical rationale—the R-implications naturally emerge because their adjointness property w.r.t. a given t-norm is needed in solving these equations. The R- and S-implications were characterized as follows:

Theorem 4.42. A function $\rightarrow$ is an R-implication associated with some left-continuous t-norm iff it satisfies (4.69), (4.73), (4.74), and the function mapping x to $a \rightarrow x$ is right-continuous for each a (Miyakoshi and Shimbo 1985, Fodor and Roubens 1994).

A function $\rightarrow$ is an S-implication associated with some t-conorm and strong negation iff it satisfies (4.71), (4.74), and (4.76) (Trillas and Valverde 1981, 1985a).

One easily checks that the Łukasiewicz implication, $a \rightarrow b = \min(1, 1-a+b)$, is both an R- and S-implication. Furthermore, the following theorem (Smets and Magrez 1987, Fodor and Roubens 1994) shows its uniqueness up to an isomorphism (cf. theorem 4.33).

Theorem 4.43. A function $\rightarrow$ satisfies (4.70), (4.73), and (4.74) and is continuous iff for some automorphism f of $[0,1]$ we have $a \rightarrow b = f^{-1}(\min(1, 1-f(a)+f(b)))$.

Let us finally mention that the above research on connectives for fuzzy logic falls into a broader context encompassing more general functions in $[0,1]$, such as various aggregation operators examined in chapter 3.

Fuzzy logics based on De Morgan triplets

Zadeh's (1965a) suggestion to use min, max, and $1-x$ leads to the natural idea of examining a logic with truth degrees in $[0,1]$ in which these operations are truth

---

[151] We omit the so-called QL-implications which are sometimes also considered (Gottwald 2001).

functions of conjunction, $\otimes$, disjunction, $\oplus$, and negation, $\neg$, i.e.

$$\|\varphi \otimes \psi\|_e = \min(\|\varphi\|_e, \|\psi\|_e),$$
$$\|\varphi \oplus \psi\|_e = \max(\|\varphi\|_e, \|\psi\|_e),$$
$$\|\neg\varphi\|_e = 1 - \|\varphi\|_e.$$

Such a logic was explored by Lee and Chang (1971)—one of the early logical papers inspired by Zadeh's writings. The authors examine the notions of tautology, inconsistent formula, normal form, entailments, implicants, and related subjects, and provide several interesting observations. They call a formula φ fuzzily valid if $\|\varphi\|_e \geq 0.5$ and fuzzily inconsistent if $\|\varphi\|_e \leq 0.5$ for each evaluation e, and observe that φ is fuzzily valid if and only if it is a classical tautology and observe the same for fuzzy inconsistency and classical contradiction. They also provide an algorithm for computing prime implicants of a formula with entailment of ψ by φ, meaning that $\|\psi\|_e \geq a$ whenever $\|\varphi\|_e \geq a$ for every $a \in [0,1]$. The same logic is also a subject of Lee 1972, in which a predicate version is considered with quantifiers interpreted by infima and suprema, the primary focus of which is on the resolution principle. Worth noting in this context is Preparata and Yeh 1972, another early contribution in a logical vein.

Clearly, $\langle\min, \max, 1 - x\rangle$ is but a particular De Morgan triplet and one may study the general case. The corresponding propositional logics, with basic connectives $\neg$ and $\otimes$ in which $\varphi \oplus \psi$ and $\varphi \to \psi$ are defined as $\neg(\neg\varphi \otimes \neg\psi)$ and $\neg\varphi \oplus \psi$, respectively, were studied by Butnariu, Klement, and Zafrany (1995).[152] $\to$ is then the S-implication associated with $\oplus$ and $\neg$. They proved the following form of *compactness*. A set T of formulas is K-*satisfiable* for $K \subseteq [0,1]$ if there exists an evaluation e such that $\|\varphi\|_e \in K$ for each $\varphi \in T$. If every finite subset of T is K-satisfiable, T is called finitely K-satisfiable. Now, the authors proved that if K is a closed subset of $[0,1]$, T is satisfiable iff it is finitely satisfiable. Moreover, if T is $\{a\}$-satisfiable for some $a \in [0,1]$ then there exists a largest a with this property.

As far as axiomatizability is concerned, the logics based on De Morgan triplets have a somewhat strange behavior. For instance, for many triplets including $\langle\min, \max, 1 - x\rangle$ there is no formula always assuming the truth degree 1 and hence no tautology in this sense. Let $V_\varphi = \{\|\varphi\|_e \mid e \text{ an evaluation}\}$ denote the set of all truth degrees to which φ may evaluate. Butnariu, Klement, and Zafrany (1995) proved:

Theorem 4.44. *For a logic based on a De Morgan triplet $\langle\otimes, \oplus, 1 - x\rangle$ such that $\otimes$ has no zero divisors, i.e. $a \otimes b = 0$ implies $a = 0$ or $b = 0$, the following holds true for any formula φ:*

(a) φ *is a classical tautology iff for some $a > 0$, $V_\varphi = [a, 1]$;*
(b) φ *is a classical contradiction iff for some $b < 1$, $V_\varphi = [0, b]$;*
(c) φ *is neither a classical tautology nor a contradiction iff $V_\varphi = [0, 1]$.*

---

[152] In fact, $\oplus$ instead of $\otimes$ was considered as a basic connective in this paper.

Theorem 4.44 does not apply to the infinitely-valued Łukasiewicz logic because the Łukasiewicz t-norm has zero divisors (e.g. $0.4 \otimes 0.3 = \max(0, 0.4 + 0.3 - 1) = 0$). In general, the nature of the values a from theorem 4.44 (a) is complex and these values cannot be made more specific. Namely, Hekrdla, Klement, and Navara (2003) considered the Frank t-norms $\otimes_\lambda$ which are defined as $a \otimes_\lambda b = \log_\lambda \left(1 + \frac{(\lambda^a - 1)(\lambda^b - 1)}{\lambda - 1} \right)$ for $\lambda \in [0, \infty]$. These t-norms cover the basic continuous t-norms because as is well known (Schweizer and Sklar 1983), the limit cases for $\lambda = 0$, $\lambda = 1$, and $\lambda = \infty$, yield the minimum, the product, and the Łukasiewicz t-norm. The authors proved that for $0 < \lambda < \infty$, the set of all a such that $V_\varphi = [a, 1]$ for some classical tautology φ is a dense subset of $[0, 1]$. Nevertheless, for $\langle \min, \max, 1 - x \rangle$, which is a popular choice of connectives, Butnariu, Klement, and Zafrany (1995) proved that theorem 4.44 may be strengthened as follows:

(a) φ is a classical tautology iff $V_\varphi = [0.5, 1]$;
(b) φ is a classical contradiction iff $V_\varphi = [0, 0.5]$.

This result in a sense justifies the above-mentioned notion of fuzzily valid formula by Lee and Chang (1971). The unpleasant consequence for the logics of De Morgan triplets is that the notion of tautology is troublesome in these logics. One might define a tautology as any formula with $\|\varphi\|_e > 0$ for each e, thus taking $(0, 1]$ as the set of designated truth values. Such tautologies are easily axiomatizable because in view of theorem 4.44, they are just the tautologies of classical propositional logic. However, such axiomatization—virtually identifying the fuzzy logic based on De Morgan triplets with classical logic—is certainly not very appealing. Taking as designated the truth degrees in $(a, 1]$ or $[a, 1]$ for some fixed $a > 0$ does not help in general either, because the above results imply that for $\langle \min, \max, 1 - x \rangle$ such tautologies do not exist if $a > 0.5$ and coincide with classical tautologies again if $a \leq 0.5$.

Łukasiewicz logics

Łukasiewicz logics and in particular MV-algebras have generated considerable interest ever since Chang's stimulating papers (1958a, 1959).[153] Worth noting is Scott's (1974) alternative proof of completeness of the infinitely-valued Łukasiewicz logic. Scott not only gave a new proof but also provided a new semantics for Ł. He proposed using multiple classical evaluations indexed by elements of an ordered Abelian group instead of multiple truth values and argued that the new proof "replaces the idea in Chang 1959 of extending an MV-algebra to a group and then using arguments about the first-order theory of ordered Abelian groups. It would seem to be a more direct approach." A similar approach to semantics of many-valued logics appeared in Urquhart 1973; see also Urquhart 1986 (pp. 101ff.).

Several equivalent definitions of MV-algebras appeared and MV-algebras and related structures were carefully investigated. Mangani (1973) found a simpler def-

---

[153] For simplicity we denote the infinitely-valued Łukasiewicz logic denoted previously $Ł_{\aleph_1}$ by Ł.

inition of Chang MV-algebras which is currently in use. A (Mangani) MV-algebra may be viewed as an algebra $\langle L, \oplus, \neg, 0 \rangle$ such that $\oplus$ is an associative and commutative operation with a neutral element 0 satisfying:[154]

$$\neg\neg a = a, \quad a \oplus \neg 0 = \neg 0, \quad \text{and} \quad \neg(\neg a \oplus b) \oplus b = \neg(\neg b \oplus a) \oplus a.$$

If $\langle L, +, \cdot, \neg, 0, 1 \rangle$ is a Chang MV-algebra then the reduct $\langle L, +, \neg, 0 \rangle$ is a Mangani MV-algebra; conversely we obtain from a Mangani MV-algebra $\langle L, \oplus, \neg, 0 \rangle$ a Chang MV-algebra $\langle L, \oplus, \otimes, \neg, 0, 1 \rangle$ by putting $a \otimes b = \neg(\neg a \oplus \neg b)$ and $1 = \neg 0$ and these constructions are one-to-one. Font, Rodríguez, and Torrens (1984) introduced Wajsberg algebras—another equivalent notion which is convenient because of its direct correspondence to the original axioms (4.13)–(4.16) for Ł. A *Wajsberg algebra* is an algebra $\langle L, \rightarrow, \neg, 1 \rangle$ satisfying

$$1 \rightarrow a = a, \qquad\qquad (a \rightarrow b) \rightarrow ((b \rightarrow c) \rightarrow (a \rightarrow c)) = 1,$$

$$(a \rightarrow b) \rightarrow b = (b \rightarrow a) \rightarrow a, \qquad (\neg a \rightarrow \neg b) \rightarrow (b \rightarrow a) = 1.$$

Given a Chang MV-algebra $\langle L, +, \cdot, \neg, 0, 1 \rangle$, we obtain a Wajsberg algebra $\langle L, \rightarrow, \neg, 1 \rangle$ by putting $a \rightarrow b = a^- + b$. Conversely, we get a Chang MV algebra from a Wajsberg algebra by (4.21) when taking $\neg 1$ and 1 for 0 and 1. Further structures equivalent to MV-algebras include bounded commutative BCK-algebras (Iséki and Tanaka 1978), bricks (Bosbach 1981), S-algebras (Buff 1985), CN-algebras (Komori 1981), and Ł-algebras (Lacava 1979).

Grigolia (1973, 1977) adopted Chang's method for finitely-valued Łukasiewicz logics $Ł_n$ and obtained an explicit list of axioms and a completeness theorem for each $Ł_n$. Naturally, the algebras corresponding to $Ł_n$, called MV_n-algebras by Grigolia, are MV-algebras satisfying additional axioms. The resulting axiomatization of $Ł_n$ consists of the four Łukasiewicz axioms, (4.13)–(4.16), plus $n\varphi \rightarrow (n-1)\varphi$, and if $n > 3$ then for any $1 < j < n - 1$ such that j does not divide $n - 1$, it also includes the axioms $(n-1)((\neg\varphi)^j \oplus (\varphi \otimes ((j-1)\varphi)))$. In those axioms, $k\varphi$ and φ^k stand for $\varphi \oplus \cdots \oplus \varphi$ and $\varphi \otimes \cdots \otimes \varphi$ (both k times). Independently, using McNaughton's (1951) results, a different axiomatization of $Ł_n$ was obtained by Tokarz (1974). Schwartz (1977, 1980) studied polyadic MV-algebras, generalizing Halmos's polyadic Boolean algebras, which correspond to predicate Łukasiewicz logics. Wójcicki (1973) studied entailment in Łukasiewicz finitely- and infinitely-valued logics. He also observed that for the infinitely-valued Ł, it is not the case that a set T of formulas entails a formula φ iff some finite $T' \subseteq T$ entails φ. Buff (1985) proved that the word problem for MV-algebras is solvable but their elementary theory is undecidable.

Even though various results on MV-algebras and Łukasiewicz logic appeared, systematic study did not begin until about the mid-1980s, i.e. almost thirty years after Chang's papers, when connections to other mathematical theories started to be revealed. In these developments, Daniele Mundici was a leading figure. His several

---

[154] That commutativity follows from the other conditions of Mangani's definition and that when dropping commutativity, one obtains a nonredundant system of conditions, was shown by Kolařík (2013).

important contributions from the late 1980s stimulated the interest of several other researchers. In Mundici 1986a, he provided an interpretation of approximately finite-dimensional (AF) C^*-algebras by formulas of Łukasiewicz logic.[155] In this interpretation, AF C^*-algebras are assigned lattice-ordered Abelian groups (Abelian l-groups) with strong unit which are then assigned MV-algebras. The MV-algebras thus arising are isomorphic to Lindenbaum algebras corresponding to theories in Łukasiewicz logic. Mundici obtained a nice result on the classification of AF C^*-algebras in terms of properties of the above-mentioned theories regarding their decidability and partial decidability. For the second assignment, Mundici defined the following mapping, Γ, examined also in Mundici 1986b. For an Abelian l-group $\langle G, +, 0, \leq \rangle$ with strong unit u let $\Gamma(G, u)$ be the algebra on the set $[0, u] = \{a \in G \mid 0 \leq a \leq u\}$ with operations defined by $a \oplus b = (a+b) \wedge u$ and $\neg a = u - x$.[156] Mundici proved the following nontrivial generalization of Chang's result for linearly ordered MV-algebras:

Theorem 4.45. Γ is a natural equivalence between the category of Abelian l-groups with strong unit and the category of MV-algebras.

In particular, this means that $\Gamma(G, u)$ is always an MV-algebra and that every MV-algebra is an image of a unique (up to an isomorphism) group with strong unit. In Mundici 1987, he proved that in the infinitely- as well as in every finitely-valued Łukasiewicz logic, the satisfiability problem is NP-compete, i.e. not efficiently computable unless P = NP. The satisfiability problem consists of deciding whether for a given propositional formula φ there exists an evaluation e for which $\|\varphi\|_e > 0$. In another paper, Mundici (1994) provides an elementary, constructive proof of McNaughton's theorem (p. 128 above). A number of Mundici's papers were written with coauthors. With Roberto Cignoli, an important member of Monteiro's Bahía Blanca school (p. 129), he worked out an elementary proof of Chang's completeness theorem (Cignoli and Mundici 1997). At that time, several other proofs of completeness for the Łukasiewicz logic were available. We have mentioned already the first proof by Rose and Rosser (1958), the celebrated proof by Chang (1959), and the proof by Scott (1974). Other proofs were Cignoli's (1993) based on a representation of free Abelian l-groups and Panti's (1995) geometric proof using a theorem on elimination of points of indeterminacy in toric varieties. Cignoli himself made other significant contributions to Łukasiewicz logic in the early 1990s and—as we shall see—later also to more general fuzzy logics. Among the first papers on further structural studies of MV-algebras in the late 1980s were those by Belluce (1986, 1992)—Chang's

---

[155] C^*-algebras are algebras of continuous linear operators on a complex Hilbert space and play an important rôle in functional analysis; W. Arveson, *An Invitation to C^*-algebra* (New York and Heidelberg: Springer, 1976) provides an introduction. An AF C^*-algebra is an inductive limit of a sequence of finite-dimensional C^*-algebras; see O. Bratteli, "Inductive limits of finite dimensional C^*-algebras," *Trans AMS* 171 (1972): 195–234.

[156] The strong unit u is such that for every $a \in G$ there exists a positive integer n with $a \leq nu$. In fact, Mundici (1986a) used Chang's original definition and dealt thus with more basic operations.

student who made important discoveries in the early 1960s (section 4.2.4)—who studied so-called semisimple MV-algebras, which are such that 0 is the only common element of all maximal ideals. Belluce explicitly mentioned a connection between Łukasiewicz logic and what he called the bold fuzzy set theory, as an analogy to the connection between Boolean algebras and ordinary set theory. Another important figure in that period was Antonio Di Nola, who made a number of contributions to the theory of MV-algebras in the early 1990s and has continued to make significant contributions ever since. Among them is an important representation theorem for MV-algebras (Di Nola 1993), according to which every MV-algebra is (isomorphic to) the algebra of $[0,1]^*$-valued functions over some set for some ultrapower $[0,1]^*$ of $[0,1]$. MV-algebras may hence be thought of as algebras of fuzzy sets with truth degrees in $[0,1]^*$.

MV-algebras became a subject of thorough investigations beginning in the mid-1980s. See Cignoli, D'Ottaviano, and Mundici 2000 for significant results obtained through the late 1990s. Later developments are examined in section 4.6.3.

Gödel logics

Kubin (1979) obtained an axiomatization of the finitely-valued propositional Gödel logics by adding further axioms, including Dummett's axiom (4.40), to those of intuitionistic logic. The predicate Gödel logic, which we examined on p. 153, was investigated by Takeuti and Titani (1984). This was almost fifteen years after Horn (1969). Interestingly, Takeuti and Titani did not refer to Horn 1969 nor to Dummett 1959, and were thus probably not aware of these previous contributions. They developed a Gentzen-style calculus for the infinitely-valued predicate Gödel logic and proved its completeness. An alternative proof of completeness was published by Takano (1987), who also proved that one of the deduction rules used by Takeuti and Titani was redundant. Takano was already aware of Horn's system and proved its equivalence with that of Takeuti and Titani. Another completeness theorem for a variant of the predicate Gödel logic is due to Corsi (1992). Since the early 1990s, Gödel logic has been subject to intensive research by Mathias Baaz and his group. Baaz (1996) introduced the $\triangle$ connective, which has turned out to play an important role in later developments (p. 204). There are also the findings regarding the role of sets of truth degrees in infinitely-valued Gödel logics. While, as proved by Dummett (1959), the sets of tautologies of the propositional Gödel logic are the same for every infinite bounded chain used as a set of truth degrees (theorem 4.15), this is not so for entailment. In fact, there are infinitely many infinitely-valued propositional Gödel logics with mutually different entailment relations and only some such logics satisfy the compactness property (Baaz and Zach 1998). In addition, while we have a complete axiomatization of the infinitely-valued predicate Gödel logic for the set $L = [0,1]$ of truth degrees (theorem 4.27), there exist infinite subsets $L \subseteq [0,1]$ for which the logic has a different set of tautologies and is even nonaxiomatizable (Baaz, Leitsch, and Zach 1996).

An important source of information on Łukasiewicz and Gödel logics was Siegfried Gottwald's book (1989) written in German, which was later substantially revised, extended, and published in English (Gottwald 2001).[157]

Monoidal logic

Several important contributions to fuzzy logic are due to Ulrich Höhle. These include Höhle 1995a, in which he thoroughly examined the *monoidal logic*, ML, as well as Höhle 1994 to which we return several times in the subsequent sections. In this logic, the truth functions of connectives are interpreted by operations in residuated lattices. Höhle's paper (1995a) actually provides a thorough algebraic study of residuated lattices in the context of fuzzy logic. Following Birkhoff 1948, he calls them *integral, commutative, residuated l-monoids*. Höhle defines residuated lattices essentially as structures $\mathbf{L} = \langle L, \wedge, \vee, \otimes, \to, 0, 1 \rangle$ such that $\langle L, \wedge, \vee, 0, 1 \rangle$ forms a bounded lattice, $\langle L, \otimes, 1 \rangle$ forms a commutative monoid, and $\otimes$ and $\to$ are bound by the adjointness condition (4.62). He regards residuated lattices as a common framework for a number of logics, including Łukasiewicz logic whose MV-algebras are just residuated lattices satisfying $a \vee b = (a \to b) \to b$, intuitionistic logic whose Heyting algebras are residuated lattices satisfying $a = a \otimes a$, and even Girard's linear logic (Girard 1987), whose algebras are just residuated lattices satisfying the law of double negation, $a = (a \to 0) \to 0$. Höhle defines his propositional monoidal logic ML as a logic with $\neg, \wedge, \vee, \otimes$, and $\to$ as its basic connectives, which has the axioms:

$$(\varphi \to \psi) \to ((\psi \to \chi) \to (\varphi \to \chi)) \qquad \varphi \to (\varphi \vee \psi)$$
$$\psi \to (\varphi \vee \psi) \qquad (\varphi \to \chi) \to ((\psi \to \chi) \to ((\varphi \vee \psi) \to \chi))$$
$$(\varphi \wedge \psi) \to \varphi \qquad (\varphi \otimes \psi) \to \varphi$$
$$(\varphi \wedge \psi) \to \psi \qquad (\varphi \otimes \psi) \to (\psi \otimes \varphi)$$
$$(\chi \to \varphi) \to ((\chi \to \psi) \to (\chi \to (\varphi \wedge \psi))) \qquad \varphi \otimes (\psi \otimes \chi) \to (\varphi \otimes \psi) \otimes \chi$$
$$(\varphi \to (\psi \to \chi)) \to ((\varphi \otimes \psi) \to \chi) \qquad ((\varphi \otimes \psi) \to \chi) \to (\varphi \to (\psi \to \chi))$$
$$(\varphi \otimes \neg\varphi) \to \psi \qquad (\varphi \to (\varphi \otimes \neg\varphi)) \to \neg\varphi$$

and *modus ponens* as the only deduction rule. Given a residuated lattice $\mathbf{L}$, the notion of truth degree $\|\varphi\|_e$ of formula φ in an evaluation e is defined as usual. Höhle established the following completeness theorem:

Theorem 4.46. A formula φ is provable in ML *iff $\|\varphi\|_e = 1$ for each residuated lattice $\mathbf{L}$ and each $\mathbf{L}$-evaluation e.*

---

[157] Siegfried Gottwald (1943–2015) made important early contributions in particular to the theory of fuzzy sets and relations and fuzzy relational equations. Later he also contributed to axiomatic systems of fuzzy logic.

Höhle (1994) also developed a predicate monoidal logic, ML$\forall$, with semantics defined over *complete* residuated lattices as the structures of truth degrees. ML$\forall$ is the usual extension of ML by symbols of functions, relations, and quantifiers (p. 110). It has two additional axioms,

$$(\forall x)\varphi \rightarrow \varphi(x/t) \text{ and } \varphi(x/t) \rightarrow (\exists x)\varphi,$$

in which x is assumed free in φ and $\varphi(x/t)$ results by a substitution of the term t for x in φ. Moreover, there are two new deduction rules,

$$\frac{\varphi \rightarrow \psi}{\varphi \rightarrow (\forall x)\psi} \text{ and } \frac{\varphi \rightarrow \psi}{(\exists x)\varphi \rightarrow \psi},$$

in which x is assumed not to be free in ψ and φ, respectively. It is worth noting that using the structures of truth degrees that are complete lattices for a predicate logic is different from what was later adopted in predicate fuzzy logics but also from earlier developments by Rasiowa; cf. pp. 197ff. below and particularly the negative result by Montagna and Sacchetti (2003, 2004). The feasibility of Höhle's approach is due to the fact that the Dedekind-MacNeille completion of the Lindenbaum algebra of ML$\forall$ is again a complete residuated lattice.

Interestingly, Höhle (1994, 1995a) did not refer to Goguen or to Pavelka, both of whom considered complete residuated lattices. Nevertheless, his ML$\forall$ is actually a kind of predicate logic for Goguen's semiformally presented logic of inexact concepts (section 4.3.1). Höhle, however, did not consider degrees of entailment or deduction from partially true premises.

4.4.2 Product logic

An important result of Mostert and Shields (1957), proved actually in a general setting of so-called I-semigroups, says that each t-norm that is continuous as a real function may be obtained from three basic ones—the Łukasiewicz, the Gödel, and the product t-norm—by a construction called *ordinal sum*. This means that for every continuous t-norm ∘, there exists a finite or countably infinite set of nonoverlapping subintervals (a_i, b_i) of $[0, 1]$ such that the restriction of ∘ to any (a_i, b_i) is either the Łukasiewicz or the product t-norm, and if $a, b \in [0, 1]$ are not contained in any (a_i, b_i) then $a \circ b = \min(a, b)$.[158] In a sense, one drops on the diagonal of the square $[0, 1]^2$—the domain of ∘—the little nonoverlapping squares $[a_i, b_i]^2$, labels each of them as Łukasiewicz or product, and declares that outside these squares the result is minimum. This construction yields all continuous t-norms.

While the logics with conjunction interpreted by the Łukasiewicz and Gödel t-norms were examined long before the inception of fuzzy logic (sections 4.2.1 and 4.2.2), the situation with the product t-norm was different. This t-norm is presently also called the Goguen t-norm because it was used for its convenient properties by Goguen (1968–69). It was, nevertheless, not until the paper by Hájek, Godo, and

---

[158] A comprehensive treatment of t-norms is found in Klement, Mesiar, and Pap 2000.

Esteva (1996) that a deeper logical investigation of a logic with product conjunction appeared. The primary motivation of the authors was to find a logic for the third basic t-norm, but they did not refer to Goguen. The proposed logic, called the *product logic* and denoted Π, is a propositional logic with two binary connectives, the conjunction $\otimes$ and implication $\rightarrow$, and two truth constants, $\overline{0}$ and $\overline{1}$ denoting (full) falsity and truth.[159] Π has a truth-functional semantics in which $\otimes$ and $\rightarrow$ are interpreted by the product t-norm $\otimes^{\cdot}$ and its residuum $\rightarrow^{\cdot}$, i.e.

$$a \otimes^{\cdot} b = a \cdot b \quad \text{and} \quad a \rightarrow^{\cdot} b = \begin{cases} 1 & \text{for } a \leq b, \\ b/a & \text{for } a > b. \end{cases}$$

Further connectives are defined as abbreviations:

$$\neg\varphi \text{ for } \varphi \rightarrow \overline{0}, \qquad \varphi \leftrightarrow \psi \text{ for } (\varphi \rightarrow \psi) \otimes (\psi \rightarrow \varphi), \qquad (4.78)$$

$$\varphi \wedge \psi \text{ for } \varphi \otimes (\varphi \rightarrow \psi), \qquad \varphi \vee \psi \text{ for } ((\varphi \rightarrow \psi) \rightarrow \psi) \wedge ((\psi \rightarrow \varphi) \rightarrow \varphi), \qquad (4.79)$$

and their truth functions are

$$\neg^{\cdot} a = \begin{cases} 1 & \text{for } a = 0, \\ 0 & \text{for } a > 0, \end{cases} \qquad a \leftrightarrow^{\cdot} b = \min(a \rightarrow^{\cdot} b, b \rightarrow^{\cdot} a),$$

$$a \wedge^{\cdot} b = \min(a, b), \qquad a \vee^{\cdot} b = \max(a, b).$$

The axioms of Π are

$$\varphi \rightarrow (\psi \rightarrow \varphi) \qquad (4.80)$$

$$(\varphi \rightarrow \psi) \rightarrow ((\psi \rightarrow \chi) \rightarrow (\varphi \rightarrow \chi)) \qquad (4.81)$$

$$\varphi \rightarrow \overline{1}, \quad \overline{0} \rightarrow \varphi \qquad (4.82)$$

$$(\varphi \otimes \psi) \rightarrow (\psi \otimes \varphi) \qquad (4.83)$$

$$((\varphi \otimes \psi) \otimes \chi) \rightarrow (\varphi \otimes (\psi \otimes \chi)) \qquad (4.84)$$

$$(\varphi \otimes (\psi \otimes \chi)) \rightarrow ((\varphi \otimes \psi) \otimes \chi) \qquad (4.85)$$

$$(\varphi \rightarrow (\psi \rightarrow \chi)) \rightarrow ((\varphi \otimes \psi) \rightarrow \chi) \qquad (4.86)$$

$$((\varphi \otimes \psi) \rightarrow \chi) \rightarrow (\varphi \rightarrow (\psi \rightarrow \chi)) \qquad (4.87)$$

$$(\varphi \rightarrow \psi) \rightarrow (\varphi \otimes \chi \rightarrow \psi \otimes \chi) \qquad (4.88)$$

$$\neg\neg\varphi \rightarrow ((\varphi \otimes \chi \rightarrow \psi \otimes \chi) \rightarrow (\varphi \rightarrow \psi)) \qquad (4.89)$$

$$(\chi \rightarrow \varphi) \rightarrow ((\chi \rightarrow \psi) \rightarrow (\chi \rightarrow (\varphi \wedge \psi))) \qquad (4.90)$$

$$(\varphi \rightarrow \chi) \rightarrow ((\psi \rightarrow \chi) \rightarrow ((\varphi \vee \psi) \rightarrow \chi)) \qquad (4.91)$$

$$(\varphi \rightarrow \psi) \vee (\psi \rightarrow \varphi) \qquad (4.92)$$

$$(\varphi \wedge \neg\varphi) \rightarrow \overline{0} \qquad (4.93)$$

and again *modus ponens* is the only deduction rule. Axioms (4.89) and (4.93), called the axioms of *cancellation* and *contradiction* (or *pseudo-complementation*), are par-

---

[159] Originally this was denoted by ΠL. The symbol Π was introduced in Hájek 1998b.

ticularly worth mentioning because, as we shall see later, they separate the product logic from other logics based on continuous t-norms.[160]

Hájek, Godo, and Esteva (1996) proved the following completeness for Π, in which $\|\varphi\| = 1$ means that φ assumes the truth degree 1 in every truth evaluation:

Theorem 4.47. A formula φ of Π is provable iff $\|\varphi\| = 1$.

This theorem has an algebraic proof inspired by Chang's proof (p. 123 above). Instead of MV-algebras, the authors introduced Π-*algebras* which are residuated lattices satisfying the counterparts of axioms (4.89), (4.92), and (4.93), i.e. $\neg\neg a \otimes (a \otimes c \to^{\cdot} b \otimes c) \leq (a \to^{\cdot} b), (a \to^{\cdot} b) \vee (b \to^{\cdot} a) = 1$, and $(a \wedge \neg a) = 0$. Π-algebras obey essential properties needed for the proof. Namely, the unit interval $[0, 1]$ with the product t-norm and its residuum form a Π-algebra, the *standard Π-algebra* $[0, 1]_{\Pi}$, and so does the Lindenbaum algebra of Π. Moreover, every Π-algebra is a subdirect product of linearly ordered Π-algebras and these may be associated with linearly ordered Abelian groups in a similar manner as with MV-algebras. A result on ordered Abelian groups[161] then enables one to prove the crucial analogy to theorem 4.8, which is the last ingredient needed for the completeness proof:

Theorem 4.48. An equation $t_1 = t_2$ is valid in the standard Π-algebra $[0, 1]_{\Pi}$ iff it is valid in every linearly ordered Π-algebra.

4.4.3   Hájek's logic BL and *Metamathematics of Fuzzy Logic*

Petr Hájek and his program

The development of fuzzy logic was significantly influenced by Petr Hájek—the most important and influential contributor to fuzzy logic in the narrow sense. His works on fuzzy logic since around the mid-1990s culminated in the remarkable book *Metamathematics of Fuzzy Logic* (Hájek 1998b). This book not only introduced the very important fuzzy logic BL, but also systematized earlier results and examined many interesting logical questions in the setting of fuzzy logic, some of which had never been examined before. As Hájek puts it in the preface:

> This book presents a systematic treatment of deductive aspects and structures of fuzzy logic understood as many valued logic *sui generis*. Some important systems of real-valued propositional and predicate calculus are defined and investigated. The aim is to show that fuzzy logic as a logic of imprecise (vague) propositions does have well developed formal foundations and that most things usually named "fuzzy inference" can be naturally understood as logical deduction.

---

[160] The two axioms may be replaced by a single one, see p. 197 below.

[161] A universal sentence of the first-order theory of linearly ordered Abelian groups is true in the additive group of reals iff it is true in every linearly ordered Abelian group; see Y. S. Gurevich and A. I. Kokorin, "Универсальная эквивалентность упорядоченных абелевых групп" [Universal equivalence of ordered Abelian groups], *Algebra i Logika* 2 (1963): 37–39. In fact, Chang (1959) used a similar theorem which he derived from Tarski's result that the theory of divisible ordered Abelian groups is complete; A. Tarski, "Sur les ensembles définissables de nombres réels" [On definable sets of real numbers], *Fund Math* 17 (1931): 210–39.

The book contains many deep results, some of them new, some known but presented in a new perspective. Above all, Hájek presents many open problems and promising research directions almost always accompanied by preliminary results. The book, comprising 299 pages, is cleanly written and published in a Kluwer's well-recognized series, Trends in Logic. Two more features, which we examine in more detail below, are worth emphasizing. First, Hájek attempts as much as possible to approach the agenda of fuzzy logic via standard logical analysis. Second, he carefully pays attention to developments in fuzzy logic in the broad sense. He considers it an important source of inspiration for fuzzy logic in the narrow sense and attempts to produce results that are relevant to it. All this makes the book a comprehensive and visionary piece of work. The book offers a tempting program in logic which has since been vigorously pursued.

Hájek's success in writing his book in a remarkably short period of time was no coincidence. In the early 1990s, when he started his research in fuzzy logic, he was already a first-class logician with deep knowledge not only of logic but also of various other relevant areas such as uncertainty, data analysis, and artificial intelligence. Hájek was born on February 6, 1940, in Prague.[162] After completing his secondary education, he entered the Faculty of Mathematics and Physics at the Charles University in Prague. He graduated in 1962 with a thesis in algebra under the supervision of the Czech algebraist Vladimír Kořínek (1899–1981). As a deeply religious person in communist Czechoslovakia, he was not allowed to obtain a position at the university. Being an excellent student, he was offered a position at the Institute of Mathematics of the Czechoslovak Academy of Sciences.[163] He has remained at the academy to the present, switching later to the academy's Institute of Computer Science of which he was the director from 1992 until 2000. He obtained from the academy his CSc. degree (candidate of science), the Czechoslovak equivalent of PhD, in 1965 under the guidance of the renowned Czech logician Ladislav Rieger (1916–1963) and, later, Karel Čulík (1926–2002), who was appointed Hájek's supervisor after Rieger's death in 1963. In 1990 he also obtained his DrSc. degree (doctor of science), the highest scientific degree in Czechoslovakia. Since childhood, Hájek was interested in music. He eventually obtained a masters' degree from the Academy of Performing Arts in Prague, specializing in organ. For many years, he was an organist in the St. Clement Anglican Episcopal Church in Prague.

In the course of his career, Hájek worked in several areas of mathematics, logic, and computer science, and left significant marks in each of them. The first was set theory, which was also the topic of his CSc. thesis, *Models of Set Theory with Individuals*. He was a participant in Petr Vopěnka's well-known Prague set the-

---

[162] The bibliographical information provided here is largely based on Z. Haniková, "Petr Hájek: A scientific biography," in Montagna 2015, 21–38. Additional information about Hájek's life and work may be found in Montagna 2015 and P. Cintula, Z. Haniková, and V. Švejdar, eds., *Witnessed Years: Essays in Honour of Petr Hájek* (London: College Publ., 2009).

[163] Renamed after the splitting of Czechoslovakia in 1993 to the Academy of Sciences of the Czech Republic.

ory seminar, whose members made important contributions to classical set theory and, in later years, also nonclassical ones—in particular the theory of semisets and Vopěnka's remarkable alternative set theory. Hájek actually coauthored a book on semisets (Vopěnka and Hájek 1972). Beginning in the 1970s, he pursued arithmetic—the part of logic connected with Gödel's famous incompleteness results. A vigorous group met at regular seminars on this topic. Eventually, Hájek and Pudlák, his former student and an important figure in that group, published a book on arithmetic in 1993, bearing the same name as Hájek's DrSc. thesis submitted in 1988.[164] The results and the book brought its authors and the Prague group worldwide recognition, partly because the topic was in the mainstream of mathematical logic. Shortly after his CSc. studies, Hájek and his colleagues, Metoděj Chytil who initiated the effort along with Ivan Havel, began to develop an original method for analysis of data describing objects and their yes-or-no attributes whose aim was to produce all interesting statements such as "most patients with symptoms x and y have also symptom z." The method was called GUHA, as an acronym for "General Unary Hypotheses Automaton," Hájek was the group's main logician. Hájek later coauthored two books on GUHA, one theoretical, the other mainly for GUHA users.[165] In the 1990s, the method was rediscovered, with much less developed theoretical foundations, by the data mining community and has been the subject of considerable research ever since.[166] Partly because of the needs to guide GUHA users in applications, Hájek became interested in expert systems and uncertainty management in the early 1980s. His collaborative research resulted again in a book.[167] The interest in uncertainty as well as his acquaintance with some previous research efforts in fuzzy logic naturally stimulated yet another shift of Hájek's interest. From the early 1990s, he became increasingly interested in fuzzy logic, to which he shortly became fully devoted. In a sense, he made it his mission to show that, as he nicely puts it (Hájek 1998b, 5):

- *Fuzzy logic is not a poor man's logic nor poor man's probability.* Fuzzy logic (in the narrow sense) is a reasonably deep theory.
- *Fuzzy logic is a logic.* It has its syntax and semantics and notion of consequence. . . .
- *There are various systems of fuzzy logic, not just one.* We have one basic logic (BL) and three of its most important extensions: Łukasiewicz logic, Gödel logic and the product logic.
- *Fuzzy logic in the narrow sense is a beautiful logic, but is also important for applications:* it offers foundations.

Hájek's deep knowledge of pure and applied logic as well as his ability to work very hard were certainly decisive factors in fulfilling his mission. Very important, how-

---

[164] P. Hájek and P. Pudlák, *Metamathematics of First-Order Arithmetics* (Berlin: Springer, 1993).

[165] Hájek and Havránek 1978 and P. Hájek, T. Havránek, and M. Chytil, *Metoda GUHA: Automatická tvorba hypotéz* [The GUHA method: Automated formation of hypotheses] (Prague: Academia, 1983).

[166] A good overview is provided by P. Hájek, M. Holeňa, and J. Rauch, "The GUHA method and its meaning for data mining," *J Computer and System Sciences* 76 (2010): 34–48.

[167] P. Hájek, T. Havránek, and R. Jiroušek, *Uncertain Information Processing in Expert Systems* (Boca Raton: CRC Press, 1992).

ever, was his stature as an established logician. Having served as president of the Kurt Gödel Society (1996–2003, 2009–15) and the first vice-president of the International Union for History and Philosophy of Science, Division of Logic (1995–99), Hájek, indeed, is an internationally important figure in mathematical logic. He was soon able to convince several colleagues with whom he had previously collaborated in arithmetic and other areas, along with many other fellow logicians as well as young researchers, to consider fuzzy logic in the narrow sense as a serious and exciting topic for research. What also needs to be mentioned is Hájek's extremely kind personality, his patience and readiness to explain and help. For his long-term excellent contributions, Hájek received several awards including the Medal of Merit from the president of the Czech Republic in 2006.

Hájek's basic logic

Among the first of Hájek's publications on fuzzy logic are Hájek 1994, 1995a, Hájek, Godo, and Esteva 1995, as well as several other papers surveyed in section 4.5.4, in which he clarified the position of fuzzy logic among various theories of uncertainty and some previous works in fuzzy logic through the lenses of mathematical logic. Having axiomatized the product logic, as described in section 4.4.2, Hájek embarked on the project of developing his *basic logic*. This logic, denoted BL, may be regarded as the *logic of continuous t-norms*. In particular, BL generalizes all the three previously developed logics based on the three basic continuous t-norms—the Łukasiewicz logic Ł, Gödel logic G, and product logic Π—which are obtained by adding further axioms to BL. In the same sense, BL generalizes classical logic.

The language of (propositional) BL contains propositional symbols, symbols $\otimes$ and $\rightarrow$ of connectives of conjunction (sometimes called strong conjunction) and implication, and the symbol $\overline{0}$ of falsity. Formulas are defined the usual way and further connectives, $\neg$, $\wedge$, $\vee$, and $\leftrightarrow$, are defined as in (4.78) and (4.79).[168]

A natural way to arrive at BL is the following. Each continuous t-norm $*$ determines a many-valued propositional logic L_* with the same language as BL. Namely, the t-norm and its residuum, $\Rightarrow_*$ (pp. 163ff.), which is defined by

$$a \Rightarrow_* b = \bigvee \{c \mid a * c \leq b\},$$

determine a structure of truth degrees in $[0, 1]$ in which $\otimes$, $\rightarrow$, and $\overline{0}$ are interpreted by $*$, $\Rightarrow_*$, and the number 0. This structure may be used to interpret formulas the truth-functional way. Thus, a truth evaluation e of propositional symbols in $[0, 1]$ extends to formulas according to the rules $\|\overline{0}\|_e = 0$, $\|\varphi \otimes \psi\|_e = \|\varphi\|_e * \|\psi\|_e$, and $\|\varphi \rightarrow \psi\|_e = \|\varphi\|_e \Rightarrow_* \|\psi\|_e$. For different t-norms, we generally obtain different sets of tautologies, i.e. formulas which assume 1 in every evaluation. For instance, while $\varphi \leftrightarrow \varphi \otimes \varphi$ is a tautology for the Gödel t-norm, this fails to be so for both the

---

[168] Note that Hájek's definition of $\varphi \leftrightarrow \psi$ as $(\varphi \rightarrow \psi) \otimes (\psi \rightarrow \varphi)$ differs from some of the previous definitions, including Łukasiewicz 1923, Pavelka 1979, and Höhle 1995a, which use $(\varphi \rightarrow \psi) \wedge (\psi \rightarrow \varphi)$. These are in general different but in BL they are equivalent due to prelinearity (Belohlavek 2002c).

Łukasiewicz and product t-norm. One may now attempt to axiomatize and further examine the logics L_*. Clearly, for the Łukasiewicz, Gödel, and product t-norm, L_* is—up to the choice of basic connectives—just the Łukasiewicz logic Ł, the Gödel logic G, and the product logic Π. One may also attempt to capture all continuous t-norms, thus to develop a logic whose provable formulas are the tautologies common to all L_*s. As we shall see, Hájek's BL is just such a logic.

Hájek's axioms of BL are:

$$(\varphi \rightarrow \psi) \rightarrow ((\psi \rightarrow \chi) \rightarrow (\varphi \rightarrow \chi)) \tag{4.94}$$

$$(\varphi \otimes \psi) \rightarrow \varphi \tag{4.95}$$

$$(\varphi \otimes \psi) \rightarrow (\psi \otimes \varphi) \tag{4.96}$$

$$(\varphi \otimes (\varphi \rightarrow \psi)) \rightarrow (\psi \otimes (\psi \rightarrow \varphi)) \tag{4.97}$$

$$(\varphi \rightarrow (\psi \rightarrow \chi)) \rightarrow ((\varphi \otimes \psi) \rightarrow \chi) \tag{4.98}$$

$$((\varphi \otimes \psi) \rightarrow \chi) \rightarrow (\varphi \rightarrow (\psi \rightarrow \chi)) \tag{4.99}$$

$$((\varphi \rightarrow \psi) \rightarrow \chi) \rightarrow (((\psi \rightarrow \varphi) \rightarrow \chi) \rightarrow \chi) \tag{4.100}$$

$$\overline{0} \rightarrow \varphi \tag{4.101}$$

These axioms along with *modus ponens* as the only deduction rule yield the usual notion of provability (p. 109). Axioms (4.94)–(4.96) represent the transitivity of →, weakening for ⊗, and commutativity of ⊗. Lehmke (2004) showed by a computer prover that (4.95) is redundant and may be omitted. Cintula (2005b) proved the same for (4.96). Finally, Chvalovský (2012), again using a computer prover, showed that both (4.95) and (4.96) may be removed simultaneously and that the remaining axioms are mutually independent. (4.97) asserts the commutativity of ∧ and relates to condition (4.105) of the definition of BL-algebras and hence to the continuity of the truth function of ⊗ discussed in the next paragraph. (4.98) and (4.99) express the adjointness of ⊗ and →. (4.100) may be seen as an analogy of the proof by cases in that if χ follows from $\varphi \rightarrow \psi$ then if χ follows from $\psi \rightarrow \varphi$ then one may infer χ. It turns out that (4.100) is equivalent to axiom (4.40), i.e. $(\varphi \rightarrow \psi) \vee (\psi \rightarrow \varphi)$— an axiom in Łukasiewicz's conjectured axiomatization of his infinitely-valued logic (p. 123) as well as in Dummett's axiomatization of the Gödel logic (p. 132). This axiom is called the *prelinearity* axiom because it directly corresponds to condition (4.106) of the definition of BL-algebras, $(a \rightarrow b) \vee (b \rightarrow a) = 1$, which Hájek termed prelinearity. (4.101) says that falsity implies any formula.

To have an algebraic counterpart of BL, which plays the same role for BL that Boolean algebras do for classical logic, Hájek introduced BL-algebras. A *BL-algebra* is an algebra $L = \langle L, \wedge, \vee, \otimes, \rightarrow, 0, 1 \rangle$ such that

$\langle L, \wedge, \vee, 0, 1 \rangle$ is a bounded lattice (with partial order denoted $\leq$), (4.102)

$\langle L, \otimes, 1 \rangle$ is a commutative semigroup with a neutral element 1, (4.103)

$a \otimes b \leq c$ iff $b \leq a \rightarrow c$, (4.104)

$$a \wedge b = a \otimes (a \rightarrow b), \tag{4.105}$$

$$(a \rightarrow b) \vee (b \rightarrow a) = 1 \tag{4.106}$$

hold for every $a, b, c \in L$. One can see that (4.102)–(4.104) say that L is a residuated lattice (p. 164). The additional conditions, (4.105) and (4.106), are the conditions of *divisibility* and *prelinearity* mentioned in the previous paragraph. Divisibility expresses the continuity of the truth function of $\otimes$ (cf. theorem 4.49) and this name was considered by Höhle (1995a) in the context of his monoidal logic—an important predecessor of BL. Prelinearity is called the algebraic strong De Morgan law by Höhle (1995a). For each t-norm $*$ one may consider the algebra

$$[0, 1]_* = \langle [0, 1], \min, \max, *, \Rightarrow_*, 0, 1 \rangle.$$

If $*$ is continuous, $[0, 1]_*$ is called a *t-algebra* (Hájek 1998a). The continuity properties of $*$ may be characterized by the following theorem. Note that the *left-continuity* of $*$ means that $(\bigvee_i a_i) * (\bigvee_j b_j) = \bigvee_{i,j} (a_i * b_j)$ for every increasing sequences $a_1, a_2, \ldots \in [0, 1]$ and $b_1, b_2, \ldots \in [0, 1]$.

Theorem 4.49. For a binary operation $*$ it holds that $[0, 1]_*$ is a residuated lattice iff $*$ is a left-continuous t-norm (Menu and Pavelka 1976). Moreover, a t-norm $*$ is continuous iff $[0, 1]_*$ is a BL-algebra (Höhle 1995a).

The t-algebras are particular examples of *BL-chains*, i.e. BL-algebras in which $\leq$ is a linear order.[169] Höhle (1995a), inspired by Chang (1958a, 1959), worked out a theory of filters for residuated lattices that Hájek was able to adopt for BL-algebras. A *filter* in L is a nonempty subset $F \subseteq L$ for which $a \otimes b \in F$ whenever $a, b \in F$, and $b \in F$ whenever $a \leq b$ for some $a \in F$. A filter F is *prime* if $a \rightarrow b \in F$ or $b \rightarrow a \in F$ for any $a, b \in L$. This is directly inspired by Chang's notion of prime ideal. For BL-algebras, Hájek's notion of primality is equivalent to Höhle's, according to which F is prime if $a \vee b \in F$ implies $a \in F$ or $b \in F$ (Belohlavek 2002c). Höhle (1995a) proved that a residuated lattice L satisfies prelinearity if and only if it is a subdirect product of linearly ordered residuated lattices.[170] Moreover, the factors may be chosen such that they satisfy all the identities satisfied by L.

Let us now denote the classes of all BL-algebras, all BL-chains, and all t-norm algebras by $\mathscr{BL}$, $\mathscr{BL}^l$, and $\mathscr{BL}^t$. Each BL-algebra L may be used as a structure of truth degrees for BL as usual (p. 108)—we have L-evaluations e mapping propositional symbols to L and they extend truth-functionally to all formulas φ using the operations in L, yielding truth degrees $\|\varphi\|_e^L$ in L. One also has the notion of tautology and entailment: φ is an L-*tautology* if $\|\varphi\|_e^L = 1$ for every L-evaluation e; it is a $\mathscr{BL}$-*tautology* if it is an L-tautology for each $L \in \mathscr{BL}$ and likewise for $\mathscr{BL}^l$ and $\mathscr{BL}^t$. Moreover, φ semantically follows from a *theory* T (set of formulas) with respect to $\mathscr{BL}$, in symbols $T \models^{\mathscr{BL}} \varphi$, if for every $L \in \mathscr{BL}$ we have $\|\varphi\|_e^L = 1$

---

[169] Linearity means that $a \leq b$ or $b \leq a$ for every a, b.

[170] For Heyting algebras, the same theorem was proved by Horn (1969); see theorem 4.17.

whenever e is a model of T, i.e. $\|\psi\|_e^L = 1$ for each $\psi \in T$, and likewise for $\mathcal{BL}^l$ and $\mathcal{BL}^t$. Hájek proved that the Lindenbaum algebra $\mathbf{L}_{BL}$ of classes of provably equivalent formulas of BL is indeed a BL-algebra and, using Höhle's subdirect representation theorem mentioned above, obtained in a standard manner the following completeness for BL:

Theorem 4.50. For a formula φ of BL, the following conditions are equivalent.

(a) φ *is provable in* BL.
(b) φ *is an* L-*tautology for each BL-algebra* L.
(c) φ *is an* L-*tautology for each BL-chain* L.

In symbols, $\vdash_{BL} \varphi$ iff $\models^{\mathcal{BL}} \varphi$ iff $\models^{\mathcal{BL}^l} \varphi$. Hájek also proved a *strong completeness* saying that for any theory T,

$$T \vdash_{BL} \varphi \quad \text{iff} \quad T \models^{\mathcal{BL}} \varphi \quad \text{iff} \quad T \models^{\mathcal{BL}^l} \varphi. \qquad (4.107)$$

Nevertheless, these results did not answer in full the problem of whether BL is indeed the logic of continuous t-norms. Hájek announced a step toward a solution in his *Metamathematics of Fuzzy Logic*. Details have been published in Hájek 1998a, where he presented two additional axioms (α^2 stands for $\alpha \otimes \alpha$),

$$(\varphi \to (\varphi \otimes \psi)) \vee (\chi \to \varphi) \vee (\psi \to \chi) \vee ((\varphi \otimes \psi) \to (\varphi \otimes \psi)^2) \qquad (4.108)$$
$$\vee \, [(\varphi \to (\varphi \otimes \psi)) \to (\chi \to (\chi \otimes \psi))], \text{ and}$$

$$((\varphi \wedge \psi) \to \varphi^2) \vee ((\varphi \wedge \psi) \to (\varphi \otimes \psi)) \vee [((\psi \to \varphi^2) \to \varphi^2) \to \psi], \qquad (4.109)$$

and proved via the analysis of the structure of BL-chains that if one adds to BL axioms (4.108) and (4.109), the resulting logic, $\text{BL}^\#$, is a logic of continuous t-norms in that $\vdash_{BL^\#} \varphi$ iff $\models^{\mathcal{BL}^t} \varphi$, i.e. the provable formulas are just the tautologies common to all t-algebras. The problem of whether (4.108) and (4.109) are provable in BL had been left open in Hájek 1998a. It was answered in the positive by Cignoli et al. (2000), who proved the following theorem:

Theorem 4.51. Any of the conditions in theorem 4.50 holds iff φ is a $[0, 1]_*$-tautology for each continuous t-norm $*$. Hence, BL is the logic of continuous t-norms.

This theorem is called the *standard completeness* for BL because it refers to semantics based on the *standard* set $[0, 1]$ of truth degrees. We thus have the same situation as for MV-algebras except that while there exists a unique up-to-isomorphism MV-algebra on $[0, 1]$—the standard Łukasiewicz algebra $[0, 1]_Ł$—there are infinitely many nonisomorphic BL-algebras $[0, 1]_*$. Interestingly, the completeness theorems for Łukasiewicz logic, Ł, Gödel logic, G, and the product logic, Π, obtain as corollaries of theorem 4.51. The *strong standard completeness*, i.e. that

$$T \vdash_{BL} \varphi \quad \text{iff} \quad T \models^{\mathcal{BL}^t} \varphi, \qquad (4.110)$$

fails for BL, contrary to the strong completeness (4.107). Nevertheless, the *finite strong standard completeness*, which asserts (4.110) for finite T, holds.

Hájek (1998b) also introduced the following useful notion. A *schematic exten-sion* of BL is a logic C that results from BL by adding a finite or infinite set of axiom schemata as additional axioms.[171] A BL-algebra **L** is a C-*algebra* if all of the axioms of C are **L**-tautologies. The strong completeness for BL generalizes to schematic ex-tensions. Thus, denoting by $\mathscr{C}$ and by $\mathscr{C}^l$ the classes of all C-algebras and linearly ordered C-algebras, and by $\vdash_C$ the provability of C, we obtain

$$T \vdash_C \varphi \quad \text{iff} \quad T \models^{\mathscr{C}} \varphi \quad \text{iff} \quad T \models^{\mathscr{C}^l} \varphi. \tag{4.111}$$

Łukasiewicz, Gödel, and product logic as schematic extensions of basic logic

All of the logics Ł, G, and Π, which correspond to the three basic continuous t-norms (p. 187), obtain as schematic extensions of BL. We start with Ł—the infinitely-valued Łukasiewicz logic (section 4.2.1). Ł may be conceived of as a schematic extension of BL by a single axiom,

$$\neg\neg\varphi \to \varphi. \tag{4.112}$$

Hájek originally used the fourth Łukasiewicz axiom (4.16). Axiom (4.112) was sug-gested by Olivetti and proved equivalent by Ciabattoni and Cicalese. Hájek (1998b) first shows that the original system of the four Łukasiewicz axioms, (4.13)–(4.16), proves the same formulas as the extension of BL by (4.112).[172] Then he develops Chang's proof of completeness in the framework of BL and BL-algebras. He de-fines MV-algebras as BL-algebras satisfying $\neg\neg a = a$ and shows that they are poly-nomially equivalent to Wajsberg algebras (p. 183) and thus also to Chang's MV-algebras (pp. 124, 183): if $\langle L, \wedge, \vee, \otimes, \to, 0, 1\rangle$ is (Hájek's) MV-algebra, then putting $\neg a = a \to 0$ and $1 = \neg 0$, we obtain a Wajsberg algebra $\langle L \to, \neg, 1\rangle$. Conversely, from $\langle L \to, \neg, 1\rangle$ we obtain (Hájek's) MV-algebra by putting $a \otimes b = \neg(a \to \neg b)$, $a \wedge b = a \otimes (a \to b)$, and $a \vee b = (a \to b) \to b$. Eventually, a slight extension of Chang's theorem 4.8 in the setting of Hájek's MV-algebras and the completeness theorem (4.111) yield the following finite strong standard completeness for Ł, first established by Hay (1963):

$$T \vdash_Ł \varphi \quad \text{iff} \quad T \models^{\mathscr{MV}} \varphi \quad \text{iff} \quad T \models^{\mathscr{MV}^l} \varphi \quad \text{iff} \quad T \models^{[0,1]_Ł} \varphi \tag{4.113}$$

holds for any finite theory T. In (4.113), which extends the standard completeness (theorem 4.6), $\mathscr{MV}$ and $\mathscr{MV}^l$ denote the classes of all (Hájek's) MV-algebras and MV-chains, and $[0, 1]_Ł$ is the standard Łukasiewicz algebra. Hájek also shows, orig-inally in Hájek 1995a, that strong standard completeness fails because there exist in-finite theories T such that $T \models^{[0,1]_Ł} \varphi$ but $T' \not\models^{[0,1]_Ł} \varphi$ for any finite $T' \subseteq T$. As mentioned on p. 183, the failure of strong standard completeness of Ł was already observed by Wójcicki (1973).

---

[171]Each axiom schema is the set of all formulas that may be obtained from it by substitution.

[172]The basic connectives of the two systems differ, hence one has to define the undefined connectives. Hájek considers $\to$ and $\overline{0}$ the basic connectives for the system based on (4.13)–(4.16) and defines $\varphi \otimes \psi$ as $\neg(\varphi \to \neg\psi)$. Note that in the original treatments (Łukasiewicz and Tarski 1930, Rose and Rosser 1958, Chang 1958a), the basic ones are $\neg$ and $\to$. Clearly, the choice is only a matter of preference.

Similarly, Gödel logic, G, obtains as a schematic extension of BL by the axiom

$$\varphi \to (\varphi \otimes \varphi). \tag{4.114}$$

G-algebras are thus BL-algebras with an idempotent conjunction, i.e. $a \otimes a = a$. These are exactly the Heyting algebras satisfying prelinearity which Horn (1969) introduced as L-algebras (p. 134).[173] Denoting by $\mathscr{G}$ and $\mathscr{G}^l$ the classes of all G-algebras and G-chains, Hájek obtains completeness, in this case for arbitrary T:

$$T \vdash_G \varphi \quad \text{iff} \quad T \models^{\mathscr{G}} \varphi \quad \text{iff} \quad T \models^{\mathscr{G}^l} \varphi \quad \text{iff} \quad T \models^{[0,1]_G} \varphi.$$

He also shows that the classical propositional logic is a schematic extension of BL (and also of G) by the *law of excluded middle*,

$$\varphi \vee \neg\varphi.$$

The product logic, Π, is obtained as an extension of BL by two axioms, (4.89) and (4.93). Cintula (2001a) proved that they may be replaced by a single axiom with only two propositional variables, $\neg\neg\varphi \to ((\varphi \to \varphi \otimes \psi) \to (\psi \otimes \neg\neg\psi))$, and that no single axiom using only one variable is sufficient. Another such single axiom, due to Cintula, Hájek, and Noguera 2011, is $\neg\varphi \vee ((\varphi \to \varphi \otimes \psi) \to \psi)$. The corresponding algebras comprising the class $\mathscr{P}$ are equivalent to Π-algebras from Hájek, Godo, and Esteva 1996, cf. section 4.4.2, and as in the case of Ł, they obey finite strong standard completeness, i.e. for any finite T,

$$T \vdash_\Pi \varphi \quad \text{iff} \quad T \models^{\mathscr{P}} \varphi \quad \text{iff} \quad T \models^{\mathscr{P}^l} \varphi \quad \text{iff} \quad T \models^{[0,1]_\Pi} \varphi.$$

The class $\mathscr{BL}$ of all BL-algebras forms a variety (Hájek 1998b), i.e. a class of all algebras satisfying a certain set of equations.[174] Esteva, Godo, and Montagna (2004) proved that for any continuous t-norm $*$, the subvariety generated by the t-norm algebra $[0,1]_*$ is finitely axiomatizable, i.e. may be described by a finite number of equations. This means that the corresponding logical calculus L_*, which proves exactly the $[0,1]_*$-tautologies, may be obtained as a schematic extension of BL by a finite number of axioms. They also provide an algorithm generating such equations (axioms) for a given $*$ presented in a canonical, finitary manner. Haniková (2014) proved that every subvariety of $\mathscr{BL}$ is generated by some finite class of continuous t-norms. It follows that for any class C of continuous t-norms, the obviously defined logic L_C, i.e. the logic which shares common properties of the t-norms $*$ in C, is finitely axiomatizable.

Predicate basic logic and predicate Ł, G, and Π

Chapter 5 of Hájek's book deals with predicate fuzzy logics. BL∀ is obtained from BL as usual (section 4.1): Its language contains symbols of individual variables, $x, \ldots$, relations, $r, \ldots$ (at least one), propositional connectives and truth constants of BL,

---

[173] Hájek seems not to have been aware of this fact.
[174] $x \otimes y = y \otimes x$ is an equation which is satisfied by all BL-algebras; the equation $x \to y = y \to x$ is not.

quantifiers $\forall$ and $\exists$, and possibly constants, $c, \ldots$, but other symbols of functions are not considered.[175] *Terms* and *formulas* are defined as usual, and so are the concepts of **L**-*structure* $\mathbf{M} = \langle M, (r^{\mathbf{M}})_r, (c^{\mathbf{M}})_c \rangle$—the **L**s now being BL-algebras, of *value* $\|t\|_{\mathbf{M},v}$ *of term* t, and *value* $\|\varphi\|_{\mathbf{M},v}$ (or *truth degree*) *of formula* φ in an M-valuation v of variables (section 4.1). The quantifiers $\forall$ and $\exists$ are interpreted by infima and suprema in **L**, i.e. $\|(\forall x)\varphi\|_{\mathbf{M},v} = \bigwedge \{\|\varphi\|_{\mathbf{M},v'} \mid v' =_x v\}$ and, dually, $\|(\exists x)\varphi\|_{\mathbf{M},v} = \bigvee \{\|\varphi\|_{\mathbf{M},v'} \mid v' =_x v\}$. In a sense, BL$\forall$ relates to BL the same way as classical predicate logic relates to the classical propositional one.

Since BL-algebras are not required to be complete lattices, it may happen that $\|(\forall x)\varphi\|_{\mathbf{M},v}$ or $\|(\exists x)\varphi\|_{\mathbf{M},v}$ are not defined. Clearly, this cannot happen if **L** is a t-norm algebra, because every t-norm algebra $[0,1]_*$ is complete as a lattice. Hájek thus calls a general **L**-structure **M** *safe* if $\|\varphi\|_{\mathbf{M},v}$ is defined for every φ and v. If **M** is safe, we put $\|\varphi\|_{\mathbf{M}} = \bigwedge \{\|\varphi\|_{\mathbf{M},v} \mid v \text{ an } M\text{-valuation}\}$. φ is an **L**-*tautology* if $\|\varphi\|_{\mathbf{M}} = 1$ for each safe **L**-structure **M**. An **L**-*model* of a theory T is a safe **L**-structure **M** such that $\|\varphi\|_{\mathbf{M}} = 1$ for each formula $\varphi \in T$.

The axiomatic system of BL$\forall$, or more generally C$\forall$ for a schematic extension C of BL (p. 196), has the axioms of C for propositional connectives, the axioms

$(\forall x)\varphi \rightarrow \varphi(t)$	provided t is substitutable for x in φ,	(4.115)
$\varphi(t) \rightarrow (\exists x)\varphi(x)$	provided t is substitutable for x in φ,	(4.116)
$(\forall x)(\psi \rightarrow \varphi) \rightarrow (\psi \rightarrow (\forall x)\varphi)$	provided x is not free in ψ,	(4.117)
$(\forall x)(\varphi \rightarrow \psi) \rightarrow ((\exists x)\varphi \rightarrow \psi)$	provided x is not free in ψ,	(4.118)
$(\forall x)(\varphi \vee \psi) \rightarrow ((\forall x)\varphi \vee \psi)$	provided x is not free in ψ	(4.119)

for quantifiers, and the deduction rules of *modus ponens* and generalization, $\frac{\varphi}{(\forall x)\varphi}$. Notice that (4.118) is the same as Hay's (1963) axiom (4.47) and that (4.119) is the same as Horn's (1969) axiom (4.49). Hájek worked out completeness by adaptation of Henkin's method. Worth mentioning is Hájek's concept of complete theory as a set T of formulas such that $T \vdash (\varphi \rightarrow \psi)$ or $T \vdash (\psi \rightarrow \varphi)$ for every closed φ and ψ, generalizing thus the ordinary notion which requires $T \vdash \varphi$ or $T \vdash \neg\varphi$ for any closed φ.[176] He thus obtained:

*Theorem 4.52. For every theory T and formula φ, C$\forall$ proves φ iff for each linearly ordered C-algebra **L** and each **L**-model **M** of T we have $\|\varphi\|_{\mathbf{M}} = 1$.*

Notice that theorem 4.52 is restricted to safe structures (models are safe by definition) and linearly ordered BL-algebras. Both restrictions deserve closer examination. One predecessor of safeness is Chang and Belluce's (1963) *strong validity*. As noted in n. 105 on p. 151 above, this notion is wrong in that Chang and Belluce's theorem 4.23 needs validity with respect to safe structures to become correct. Namely,

---

[175]Constants are actually symbols of nullary functions. Function symbols are considered in Hájek 2000.

[176]Complete theories were later called *linear theories* (Cintula 2006, Hájek and Cintula 2006) to avoid confusion with the notion of a complete theory as a maximal consistent theory.

Hájek, Paris, and Shepherdson (2000a) showed that there is a provable closed formula φ in $Ł\forall$, an MV-chain L, and an L-structure M such that $\|\varphi\|_M$ is defined but $\|\varphi\|_M < 1$, i.e. φ is provable but not strongly valid. They also proved the same for $\Pi\forall$ and observed that for $G\forall$, strong validity coincides with validity with respect to safe structures. Actually, Rasiowa and Sikorski's (1963, chapter VI) *realization* of a formalized language of first-order in a set and in a generalized algebra is just Hájek's safe structure, rephrased in a different framework. As noted in n. 109 on p. 153, Hájek's safe structures are exactly Horn's (1969) *suitable* realizations with which Horn worked in predicate Gödel logic. Horn does not attribute this notion to Rasiowa and Sikorski (1963), but since he uses their formalism of predicate logics and explicitly acknowledges this fact, he was no doubt inspired by Rasiowa and Sikorski's notion. Hájek was not aware of these predecessors. In fact, the above concept of realization also appears in an important book by Rasiowa (1974, supplement, p. 368) and only in Hájek 2006a does he state this fact and attributes its observation to Cintula.

In principle, safeness could be avoided if one used only complete C-chains in theorem 4.52, in which arbitrary infima and suprema exist. However, as shown by Montagna and Sacchetti (2003, 2004), this option would not be a good choice: the set of all formulas φ for which $\|\varphi\|_M = 1$ for each complete BL-chain L and L-structure M is not recursively axiomatizable (since it is not arithmetical, cf. section 4.6.2).

As for the restriction to linearly ordered BL-algebras in theorem 4.52, Esteva et al. (2003) proved that it may not be dropped: there exists a nonlinearly ordered BL-algebra L (a G-algebra in fact) and a safe L-structure M for which axiom (4.119) has truth degree < 1. They also showed that if one removes axiom (4.119), the resulting logic, $BL\forall^-$, is complete with respect to semantics over *all* BL-algebras:[177]

Theorem 4.53. For every theory T and formula φ, $BL\forall^-$ proves φ iff for each BL-algebra L and each L-model M of T we have $\|\varphi\|_M = 1$.

As was discovered later, see e.g. Hájek 2006a, Cintula and Hájek 2010, theorem 4.53 is a particular case of general completeness theorems for implicative logics of Rasiowa 1974. Rasiowa's remarkable work is therefore highly relevant for fuzzy logic in the narrow sense.

In his book, Hájek also presents three extensions of $BL\forall$, the $Ł\forall$, $G\forall$, and $\Pi\forall$. As for $Ł\forall$, he recalls its nonaxiomatizability for the standard semantics due to Scarpellini's theorem 4.25, develops the predicate version of his rational Pavelka logic which we examine below, and presents some of the results of Hay, Chang, and Belluce examined in section 4.2.4. As for $\Pi\forall$, Baaz et al. (1998) proved:

---

[177]In fact, the authors formulate the theorem for some prominent logics C, such as MTL, BL, Ł, G, and Π.

Theorem 4.54. *There exists a faithful embedding of* Ł∀ *in* Π∀. *Hence,* Π∀ *is not axiomatizable for the standard semantics due to theorem 4.25.*[178]

Hájek (2004a) actually found an unprovable tautology for the standard semantics in Π∀. For G∀, Hájek obtains a strong standard completeness with respect to $[0,1]_G$. That is, for C = G, the two equivalent conditions in theorem 4.52 may be completed by a third one saying that $\|\varphi\|_M = 1$ for each $[0,1]_G$-model of T, i.e.

$$T \vdash \varphi \text{ in G}\forall \quad \text{iff} \quad T \models^{[0,1]_G} \varphi.$$

This strengthens the equivalence of (a) and (d) in Horn's theorem 4.27 and accompanies the strong standard completeness in Gentzen style obtained by Takeuti and Titani (p. 185).

Predicate fuzzy logics were later investigated further in a number of papers. Some developments are mentioned in the forthcoming sections. Hájek himself coauthored a number of contributions (Hájek and Cintula 2006, Hájek 2007a,b, 2010). For more information, see the surveys Cintula and Hájek 2010 and Cintula, Hájek, and Noguera 2011. Predicate fuzzy logics are obviously relevant for applications. Hájek takes this issue seriously and we discuss it in section 4.6.1.

Rational Pavelka logic

Hájek reviewed both Pavelka's and Novák's theses and was therefore familiar with what is now called Pavelka-style logic. In one of his earliest papers on fuzzy logic, Hájek (1995a) demonstrates that Pavelka's (1979) axiomatization, leading to the completeness of Pavelka-style Łukasiewicz propositional logic (p. 171 below), may be simplified considerably. The result is reproduced in Hájek 1998b in which the system is called the *rational Pavelka logic*, RPL, and which we use for our discussion here.

There is widespread confusion about this logic. The key point in Pavelka-style completeness is equality of degree of semantic entailment and degree of provability. While Pavelka proves this in the framework of his abstract fuzzy logic (p. 166 above), Hájek does so, as we are about to see, in the framework of the Łukasiewicz logic Ł as a schematic extension of his basic logic BL. Importantly, degree of provability is a genuine concept in Pavelka's abstract logic. On the other hand, Hájek's notion of degree of provability is a derived concept in the logic BL which itself has the ordinary, bivalent concept of provability. In a sense, Hájek simulates Pavelka's degree of provability in his logic BL for one particular instance of abstract fuzzy logic—the Pavelka-style Łukasiewicz logic. Hájek's simulation is remarkably elegant, fruitful and has stimulated many subsequents developments. Nevertheless, there exist many useful abstract fuzzy logics for which such simulation is not possible.[179]

---

[178] A faithful embedding (or faithful interpretation) of logic L_1 in L_2 is a mapping (translation) τ of formulas of L_1 to formulas of L_2 such that φ is provable in L_1 iff its translation $\tau(\varphi)$ is provable in L_2.

[179] Hájek is well aware of this situation—he devotes a section of his book to Pavelka's abstract fuzzy logic and frequently comments on its capability to accommodate even non-truth-functional semantics in his works.

A crucial observation in Hájek's simulation is this. Add to the language of Ł truth constants $\bar{a}$ for reals $a \in [0,1]$ and postulate $||\bar{a}||_e = a$ for any truth $[0,1]$-evaluation e. Then, since $b \to c = 1$ iff $b \le c$, any formula φ satisfies

$$a \le ||\varphi||_e \text{ iff } ||\bar{a} \to \varphi||_e = 1, \text{ i.e. } \varphi \text{ is} \ge a\text{-true iff } \bar{a} \to \varphi \text{ is 1-true.}$$

Formulas $\bar{a} \to \varphi$ may thus be regarded as counterparts to Pavelka's weighted formulas $\langle \varphi, a \rangle$. This is the trick Hájek used to capture partially true formulas.

The language of RPL is that of Ł extended by the truth constants $\bar{a}$ for *rationals* $a \in [0,1]$. Such a language is then countable, contrary to Pavelka's treatment of the Łukasiewicz logic in which there is a truth constant for every real in $[0,1]$. Truth evaluations e assign *reals* in $[0,1]$ to propositional symbols and extend to formulas by the usual rules extended by $||\bar{a}||_e = a$. The axioms of RPL are those of Ł plus two bookkeeping axioms (Hájek's term),

$$(\bar{a} \to \bar{b}) \leftrightarrow \overline{a \to b} \quad \text{and} \quad \neg \bar{a} \leftrightarrow \overline{\neg a}. \tag{4.120}$$

Thus we have $(\overline{0.5} \to \overline{0.3}) \leftrightarrow \overline{0.8}$ or $\neg \overline{0.1} \leftrightarrow \overline{0.9}$ as axioms. RPL has *modus ponens* and ordinary provability $\vdash$. Hájek defines the truth degree of φ over T, $||\varphi||_T = \inf\{||\varphi||_e \mid e \text{ a model of } T\}$, where being a model means $||\psi||_e = 1$ for each $\psi \in T$, and defines the *provability degree* of φ over T by

$$|\varphi|_T = \sup\{a \mid T \vdash \bar{a} \to \varphi\}.$$

Notice that $||\varphi||_T$ is just Pavelka's degree of entailment from a crisp T. He then obtains completeness à la Pavelka for RPL:

Theorem 4.55. For each theory T and formula φ of RPL, $|\varphi|_T = ||\varphi||_T$.

Worth noting is that any theory T in Pavelka's sense, i.e. a fuzzy set of formulas, for which $T(\varphi)$ are rationals may be represented in RPL by an ordinary theory $T' = \{\overline{T(\varphi)} \to \varphi \mid \varphi \text{ a formula}\}$. Hájek makes several further observations including the following theorem, saying among other things that provability degrees from finite theories are actually attained:

Theorem 4.56. If T is a finite theory of RPL, then $||\varphi||_T$ is rational. Moreover, $T \vdash (\overline{||\varphi||_T} \to \varphi)$, hence $||\varphi||_T$ is the largest a for which $T \vdash \bar{a} \to \varphi$.

Hájek also presents RPL∀—the *rational Pavelka predicate logic* and a completeness theorem for it fully analogous to theorem 4.55. Notice that since Ł∀ is not recursively axiomatizable (theorem 4.25), no system, including Ł∀, satisfies that T entails φ iff T proves φ. However, the completeness for RPL∀ yields that T entails φ iff T proves $\bar{a} \to \varphi$ for any $a < 1$ in RPL∀. Hájek, Paris, and Shepherdson (2000b) have shown that RPL∀ is in fact a conservative extension of Ł∀, i.e. a formula φ of Ł∀ is provable in Ł∀ iff it is provable in RPL∀, and that the same is true for the *real Pavelka logic* which has truth constants for all reals in $[0,1]$. Another result says that the notion of provability degree of RPL∀ may already be expressed in Ł∀. This may seem unexpected because no truth constants other than $\overline{0}$ are present in Ł∀. Due to Pavelka's theorem 4.34, a full analogy of RPL for logics other than Ł is impossible. Hájek nevertheless discusses certain possibilities for G and Π.

Further topics concerning the Metamathematics of Fuzzy Logic

We now briefly comment on further important topics addressed in Hájek's book. Among them is computational complexity which we discuss in section 4.6.2, generalized quantifiers and modalities which we present in section 4.5.4, as well as logical analysis of problems arising in applications of fuzzy logic that are addressed by fuzzy logic in the broad sense. Hájek takes these foundational issues for fuzzy logic seriously and addresses them several times in his book and other writings. We discuss these matters in section 4.6.1.

In addition to the numerous open problems mentioned in *Metamathematics*, Hájek wrote several overview papers encouraging others to pursue fuzzy logic as a research program (Hájek 1999, 2006b, Hájek and Paris 1997).

4.4.4 Logics related to BL

Monoidal t-norm based logic—the logic of left-continuous t-norms

The continuity of t-norms seems a natural property as it means that tiny changes in the truth degrees of two propositions result in a tiny change of the truth degree of their conjunction. This justifies Hájek's logic BL. On the other hand, Goguen's justification (section 4.3.2) leads to the requirement that $[0,1]_* = \langle [0,1], \min, \max, *, \Rightarrow_*, 0, 1 \rangle$ be a complete residuated lattice. According to theorem 4.49, this happens if and only if $*$ is a left-continuous t-norm. A question thus arises of what is the logic of left-continuous t-norms. Such a logic was proposed by Esteva and Godo (2001) who call it the *monoidal t-norm based logic*—MTL. It results from BL by adding $\wedge$ to its language as a basic connective and replacing axiom (4.97) by

$$\varphi \wedge \psi \rightarrow \varphi, \quad \varphi \wedge \psi \rightarrow \psi \wedge \varphi, \quad \text{and} \quad \varphi \otimes (\varphi \rightarrow \psi) \rightarrow \varphi \wedge \psi.$$

The addition of $\wedge$ is a consequence of the fact that $\wedge$ is not definable as in BL, and in fact, not definable in MTL by $\otimes$, $\rightarrow$, and $\overline{0}$ at all (Cintula, Hájek, and Horčík 2007). MTL is weaker than BL, which is its axiomatic extension by the axiom of divisibility (4.97). On the other hand, MTL is a schematic extension of Höhle's monoidal logic (p. 186) by axiom (4.100) or, equivalently by the axiom of prelinearity (4.40). In fact, Esteva and Godo (2001) mention that they changed their original name, "quasi-basic logic," after a discussion in October 1999 with Hájek and Gottwald and favored the new terminology inheriting "monoidal" from Höhle's logic, as well as "t-norm based" from Gottwald 1999. The corresponding *MTL-algebras* are just residuated lattices satisfying prelinearity. Esteva and Godo were able to prove for MTL the analogs of Hájek's results for BL, in particular the completeness theorem with respect to MTL-algebras and MTL-chains, i.e. theorem 4.50 with "BL" replaced by "MTL."[180] Their conjecture of standard completeness was eventually proved by Jenei and Montagna (2002):

Theorem 4.57. A formula φ is provable in MTL iff it is a $[0,1]_*$-tautology for each left-continuous t-norm $*$. Hence, MTL is the logic of left-continuous t-norms.[181]

Esteva and Godo (2001) also obtained strong completeness theorems for the predicate version MTL∀ and its extensions, essentially by applying the methods from Hájek 1998b.

Hierarchy of logics based on (left-)continuous t-norms

MTL became an important logic in the hierarchy of logics related to t-norms to which we now turn. Esteva and Godo (2001) themselves studied some schematic extensions of MTL. Among them is IMTL—the *involutive* MTL—which extends MTL by the axiom $\neg\neg\varphi \to \varphi$, forcing *involutive* negation, i.e. $\neg\neg a = a$. The results on schematic extensions of MTL mentioned at the end of the last section yield the general completeness w.r.t. IMTL-algebras. Esteva et al. (2002) obtained, by adapting ideas from Jenei and Montagna 2002, the standard completeness for ITML. An alternative axiomatization of IMTL was obtained by Gottwald and Jenei (2001).

Another logic examined in Esteva and Godo 2001 is NM—the *logic of nilpotent minimum*—which extends MTL by the axioms

$$((\varphi \otimes \psi) \to \bar{0}) \vee ((\varphi \wedge \psi) \to (\varphi \otimes \psi)) \quad \text{and} \quad \neg\neg\varphi \to \varphi. \qquad (4.121)$$

The nilpotent minimum, introduced by Fodor (1995), is an example of a left-continuous t-norm that is not continuous. If n is a strong negation (p. 178), then the *nilpotent minimum* corresponding to n is defined by

$$a *_n b = \begin{cases} 0 & \text{for } a \leq n(b), \\ \min(a, b) & \text{otherwise.} \end{cases}$$

All nilpotent minimums are isomorphic to $*_{1-x}$—the one corresponding to the standard negation $n(a) = 1 - a$—for which $[0,1]_{*_{1-x}}$ is called the standard NM-algebra. Esteva and Godo proved standard completeness for NM—a formula is provable in NM iff it is a $[0,1]_{*_{1-x}}$-tautology—as well as for NM∀. A logic equivalent to NM was independently developed by Wang (1999).

Esteva and Godo (2001) also examined the logic of *weak nilpotent minimums*, WNM, which is an extension of MTL by the first axiom of (4.121). While all the MV-algebras on $[0,1]$ are mutually isomorphic and the same holds for the Π-algebras, G-algebras, as well as the NM-algebras, there does not exist a unique up-to-an-isomorphism WNM algebra on $[0,1]$. Instead, there is a whole family of such

---

[180] In fact, they proved this for $T = \emptyset$.

[181] Hájek (2002) observed that the proof actually works for strong standard completeness, i.e. for provability from possibly countably infinite theories.

nonisomorphic WNM-algebras on $[0, 1]$ and the authors proved standard completeness with respect to this family.

Interesting logics may be obtained as extensions of BL and MTL by axiom (4.93) and are denoted SBL and SMTL, respectively. It is easily seen that (4.93) means that the negation $\neg$ is the Gödel negation, or equivalently, that $a \otimes b = 0$ iff $a = 0$ or $b = 0$. Hence, SBL and SMTL may be considered as the logics of continuous and left-continuous t-norms with *Gödel negation*. The Gödel negation is sometimes called the strict negation, whence the letter "S." SBL was introduced in Esteva et al. 2000. SMTL, which is weaker than SBL (Esteva et al. 2002), is discussed in Hájek 2002 and in Esteva et al. 2002. As schematic extensions of BL and MTL, these logics satisfy completeness theorems w.r.t. SBL- and SMTL-algebras, respectively. SBL obeys finite strong standard completeness w.r.t. the SBL-algebras on $[0, 1]$ as proved in Esteva et al. 2002—an interesting paper relating finite strong completeness to certain embedding properties of MTL-algebras. Strong standard completeness fails for SBL. Using a modification of the ideas from Jenei and Montagna 2002, SMTL was proved to be strongly standard complete in Esteva et al. 2002.[182]

Further results on MTL were obtained by Hájek (2002), who also examined what happens when MTL is extended by axioms (4.112), (4.114), and (4.89) plus (4.93), which in the case of BL lead to Łukasiewicz, Gödel, and the product logics Ł, G, and Π. Clearly, for (4.112) one obtains IMTL discussed above. Hájek observed that it is weaker than Ł but that adding the fourth axiom of Łukasiewicz, (4.16), yields Ł. An extension of MTL by (4.114) already yields G. Finally, MTL extended by (4.89) plus (4.93), denoted ΠMTL, is weaker than Π. The problem of standard completeness of ΠMTL was considered by Esteva et al. (2002), who obtained what they called *rational completeness*, i.e. completeness w.r.t. ΠMTL-algebras on the rationals in $[0, 1]$. The problem was finally solved by Horčík (2005, 2007), who showed that ΠMTL obeys finite strong standard completeness but not strong standard completeness.

Further extensions have also been considered, e.g. by axioms of *n-contraction* (Ciabattoni 2002), $\varphi^{n-1} \to \varphi^n$, where φ^k is $\varphi \otimes \cdots \otimes \varphi$ (k times), generalizing axiom (4.114) of idempotence, or *n-nilpotence* (Gispert and Torrens 2002), $\neg\varphi^n \vee \varphi$, generalizing the law of excluded middle, $\varphi \vee \neg\varphi$. Extension by $\bigvee_{i=1}^{n}(\varphi_{i-1} \to \varphi_i)$ ensures that every chain has at most n elements (Cintula et al. 2009).

The hierarchy of some of the logics discussed so far is depicted in figure 4.1. The higher in the hierarchy, the stronger the logic. For instance, BL is stronger than MTL, which means that it results by adding additional axioms to those of MTL. Classical logic is the strongest logic and ML is the weakest of the depicted logics.

Fuzzy logics with additional connectives

Interesting fuzzy logics have been obtained by adding further connectives. Hájek (1998b) examined the addition of "Baaz's delta"—a unary connective Δ previously

---

[182] The proof may be adapted for strong standard completeness, cf. n. 181 on p. 203 above.

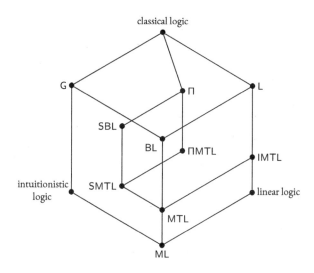

Figure 4.1: Hierarchy of selected t-norm based logics and other prominent logics.

studied for the Gödel logic by Baaz (1996). For a schematic extension C of BL, the logic C_Δ adds to C the following axioms of Baaz (1996):

$$\Delta\varphi \to \varphi \qquad \Delta\varphi \to \Delta\Delta\varphi \qquad\qquad \Delta(\varphi \to \psi) \to (\Delta\varphi \to \Delta\psi) \qquad (4.122)$$

$$\Delta\varphi \vee \neg\Delta\varphi \qquad \Delta(\varphi \vee \psi) \to (\Delta\varphi \vee \Delta\psi) \qquad\qquad\qquad (4.123)$$

and the rule of necessitation, $\frac{\varphi}{\Delta\varphi}$, as a new deduction rule. A C_Δ-algebra is a C-algebra with an additional unary operation Δ satisfying

$$\Delta a \leq a, \qquad\qquad \Delta a \leq \Delta\Delta a, \qquad\qquad \Delta a \otimes \Delta(a \leq b) \leq \Delta b,$$

$$\Delta a \vee \neg\Delta a = 1, \qquad \Delta(a \vee b) \leq (\Delta a \vee \Delta b), \qquad \Delta 1 = 1.$$

One can easily observe that C_Δ-chains are just the C-algebras in which

$$\Delta a = \begin{cases} 1 & \text{for } a = 1, \\ 0 & \text{for } a < 1. \end{cases} \qquad\qquad (4.124)$$

Hájek established strong completeness for C_Δ, i.e. the obvious analogy of (4.111).

As part of his attempt to address issues of fuzzy logic in the broad sense, Hájek (2001b) considers the linguistic hedge "very" (cf. chapter 3). He conceives of it as a unary connective vt and $vt\varphi$ is read as "φ is very true." The expansion of BL thus obtained, denoted BL_{vt}, has three new axioms,

$$vt\varphi \to \varphi$$

$$vt(\varphi \to \psi) \to (vt\varphi \to vt\psi)$$

$$vt(\varphi \vee \psi) \to (vt\varphi \vee vt\psi)$$

and a new deduction rule, $\frac{\varphi}{vt\varphi}$. The connective vt is interpreted by *truth-stressing hedges*, which are functions vt on BL-algebras satisfying $vt(1) = 1$, $vt(a) \leq a$,

$vt'(a \to b) \le vt'(a) \to vt'(b)$, and $vt'(a \lor b) \le (vt'(a) \lor vt'(b))$. An example of such a hedge on any L is $vt'(a) = a \otimes' a$ or more generally $a \otimes' \cdots \otimes' a$ (k times). For the product logic, this yields $a \cdot a$ (square of a)—a hedge discussed by Zadeh. Another example is Baaz's delta described in (4.124). Hájek obtained strong completeness for BL_{vt} as well as completeness for the obvious extensions of $BL\forall$ and $RPL\forall$. Vychodil (2006) studied a dual case—the *truth-depressing hedges* as unary connectives interpreted as "slightly true." He defined them relative to truth-stressing hedges and thus expanded BL_{vt} by a unary connective *st*. With $\varphi \to st\varphi$, $st\varphi \to \neg vt \neg \varphi$, and $vt(\varphi \to \psi) \to (st\varphi \to st\psi)$ as new axioms and no new deduction rule, he obtained strong completeness. Linguistic hedges as unary logic connectives became a subject of many subsequent investigations, see e.g. Esteva, Godo, and Noguera 2013. Hedges have also proved useful in parameterizing algebraico-logical approaches to reasoning about data (Belohlavek and Vychodil 2012).

Esteva et al. (2000) axiomatized an expansion of the logic SBL (p. 204 above) by additional negation, $\sim$, whose truth function is involutive. This logic, $SBL_\sim$, is a schematic extension of $BL_\triangle$, with $\triangle \varphi$ actually defined as $\neg \sim \varphi$, by the axioms

$$\sim\sim \varphi \leftrightarrow \varphi, \quad \neg\varphi \to \sim \varphi, \quad \text{and} \quad \triangle(\varphi \to \psi) \to (\sim \psi \to \sim \varphi).$$

The deduction rules are those of $BL_\triangle$. The $SBL_\sim$ algebras are structures of truth degrees where an involutive negation is available, although it is not defined by the residuum as usual.

An important logic, ŁΠ, results by "putting the Łukasiewicz and product logics together" as its authors say in Esteva and Godo 1999. The language has two conjunctions—the Łukasiewicz $\otimes$ and the product $\odot$—as well as the corresponding implications, $\to_\otimes$ and $\to_\odot$, and contains the truth constant $\overline{0}$. Closely related is ŁΠ1/2 (Montagna 2000), which contains in addition the truth constant $\overline{1/2}$. Both logics have been intensively investigated (Esteva, Godo, and Montagna 2001; Cintula 2001b,c, 2003, 2005a). A simple axiom system for ŁΠ (Cintula 2003) consists of the axioms of Ł, the axioms of Π, and three additional axioms,

$$\triangle(\varphi \to_\otimes \psi) \to_\otimes (\varphi \to_\odot \psi)$$
$$\triangle(\varphi \to_\odot \psi) \to_\otimes (\varphi \to_\otimes \psi)$$
$$\varphi \odot \neg_\otimes \psi \leftrightarrow_\otimes \varphi \otimes_\otimes (\varphi \odot \psi)$$

where $\neg_\otimes$, $\leftrightarrow_\otimes$, and $\neg_\odot$ are defined as in Ł and Π, and $\triangle\chi$ is $\neg_\odot \neg_\otimes \chi$. The deduction rules are *modus ponens* for $\to_\otimes$ and $\frac{\varphi}{\triangle\varphi}$. ŁΠ1/2 needs one more axiom, $\overline{1/2} \leftrightarrow_\otimes \neg_\otimes \overline{1/2}$. Both ŁΠ and ŁΠ1/2 enjoy general completeness w.r.t. the obviously defined ŁΠ- and ŁΠ1/2-algebras (Esteva and Godo 1999), finite strong standard completeness, and both conservatively extend Ł, Π, as well as G (Esteva, Godo, and Montagna 2001). ŁΠ and ŁΠ1/2 are very expressive. They contain the three logics of basic continuous t-norms, Ł, G, and Π, but also several other logics. For example, every rational truth constant is definable in ŁΠ1/2 and the truth functions definable in ŁΠ1/2 include not only those definable by the three basic continuous t-norms, but also those definable by any of their finite ordinal sums (Cintula 2003).

Horčík and Cintula (2004) axiomatized the *product-Łukasiewicz logic*, PŁ, which results by adding to the Łukasiewicz logic with basic connectives $\otimes$, $\rightarrow$, and $\bar{0}$ a product conjunction, $\odot$, but not the product implication as in the case of ŁΠ. With other connectives defined as usual in Ł, plus $\chi \ominus \theta$ defined as $\chi \otimes \neg \theta$, the authors propose as axioms of PŁ the axioms of Ł plus

$$(\varphi \odot \psi) \ominus (\varphi \odot \chi) \leftrightarrow \varphi \odot (\psi \ominus \chi), \quad \varphi \odot (\psi \odot \chi) \leftrightarrow (\varphi \odot \psi) \odot \chi,$$
$$\varphi \rightarrow \varphi \odot \bar{1}, \quad \varphi \odot \psi \rightarrow \varphi, \quad \varphi \odot \psi \rightarrow \psi \odot \varphi,$$

and *modus ponens* as the only deduction rule. PŁ then enjoys strong completeness w.r.t. the semantics given by naturally defined *PŁ*-algebras as well as *PŁ*-chains. However, standard completeness w.r.t. to $[0,1]_S$, which is the standard MV-algebra $[0,1]_Ł$ expanded by the ordinary product, fails for PŁ. The authors therefore define PŁ′ which results from PŁ by adding a new deduction rule, $\frac{\neg(\varphi \odot \varphi)}{\neg \varphi}$. PŁ′ satisfies the general completeness w.r.t. *PŁ′*-algebras and chains, but also the finite strong standard completeness w.r.t. $[0,1]_S$. While *PŁ*-algebras form a variety of algebras, the *PŁ′*-algebras, defined as *PŁ*-algebras for which $a \odot a = 0$ implies $a = 0$, only form a quasivariety, i.e. they cannot be defined by identities. The authors also examine the expansion of PŁ by Baaz's Δ operator, a kind of rational Pavelka logic RPPŁ, and the corresponding predicate logics. Also worth noting in this context is the logic developed by Takeuti and Titani (1992), which was shown to coincide with one of the predicate logics considered in Horčík and Cintula 2004.

Fuzzy logics with truth constants

Pavelka's pioneering work on abstract fuzzy logics led to a number of contributions which are surveyed in section 4.3.3. Here we are concerned with logics initiated by Hájek's rational Pavelka logic (p. 200). For Pavelka-style completeness of such logics, theorem 4.34 presents a fundamental limitation.[183] The limitation may be overcome by using infinitary deduction rules as in Esteva et al. 2000, where the authors prove Pavelka-style completeness of the *rational product logic*. This logic, RΠL, results from Π by adding truth constants for rationals in $[0,1]$, bookkeeping axioms (p. 201), and has *modus ponens* and a new, *infinitary* rule $\frac{\varphi \rightarrow \bar{a}, a \in (0,1] \cap \mathbb{Q}}{\varphi \rightarrow \bar{0}}$ which takes care of the problematic discontinuity of $\rightarrow$ at $\langle 0,0 \rangle$. Moreover, the infinitary rule also guarantees that if $|\varphi|_T = 1$ then $\bar{1} \rightarrow \varphi$ and thus φ is also actually provable.[184]

In view of the limitation due to Pavelka's theorem 4.34, one may also be interested in the ordinary rather than Pavelka-style completeness of logics with truth

---

[183] This limitation, however, need not apply if one considers just fragments such as those in Belohlavek 2002b and Belohlavek and Vychodil 2005; see p. 177 above and p. 286 below.

[184] Without the infinitary rule, $|\varphi|_T = 1$ means that $\bar{a} \rightarrow \varphi$ is provable for each $a < 1$ only.

constants. In addition to the above-mentioned ŁΠ^1/2, logics with rational truth constants have been considered for the Gödel and weak nilpotent minimum logics (Esteva, Godo, and Noguera 2006) and the product logic (Savický et al. 2006) for which various general as well as standard completeness results were proved. General logics have been examined by Esteva et al. (2007) and Esteva, Godo, and Noguera (2010), who consider expansions of the logic L$_*$ of an arbitrary continuous t-norm $*$ by constants from a given countable subalgebra of the t-algebra $[0, 1]_*$.

Weaker logics and generalizations

Several logics that are weaker than those mentioned above have been advanced. One way to obtain such logics is to relax some properties that are usually required. For example, logics with *noncommutative conjunction* and the corresponding algebraic structures have been considered e.g. in Flondor, Georgescu, and Iorgulescu 2001; Hájek 2003a,b; Jenei and Montagna 2003; and Kühr 2003. The lack of commutativity implies that there are two residua, $\rightarrow$ and $\leftarrow$, which are bound with $\otimes$ by the following form of adjointness: $a \otimes b \leq c$ iff $b \leq a \rightarrow c$ iff $a \leq c \leftarrow b$. Correspondingly, the axiomatic systems of noncommutative logics are considerably more involved. Examples of noncommutative conjunctions are *left-continuous pseudo-t-norms*, i.e. binary operations $*$ in $[0, 1]$ which are left-continuous in both arguments, associative, isotone, and have 1 as a neutral element. An example of a noncommutative left-continuous t-norm is obtained by putting for $0 < p < q < 1$,

$$a * b = \begin{cases} 0 & \text{if } a \leq p \text{ and } b \leq q, \\ \min(a, b) & \text{otherwise.} \end{cases}$$

Interestingly, Mostert and Shields 1957 implies that any continuous pseudo-t-norm is already commutative, hence a t-norm.

Another generalization is represented by the logic UL of *left-continuous conjunctive uninorms* (Metcalfe and Montagna 2007). Uninorms, introduced by Yager and Rybalov (1996) as a common generalization of t-norms and t-conorms, are binary operations on $[0, 1]$ which are commutative, associative, isotone, and have a neutral element e. Clearly, $e = 1$ and $e = 0$ lead to t-norms and t-conorms, respectively. Each uninorm is either conjunctive in that $0 * 1 = 0$ or disjunctive in that $0 * 1 = 1$ (Yager and Rybalov 1996). An example of a uninorm, which is neither a t-norm nor a t-conorm, is given by

$$a * b = \begin{cases} \min(a, b) & \text{if } a + b \leq 1, \\ \max(a, b) & \text{otherwise.} \end{cases}$$

Weaker logics may also be obtained by considering fragments of the basic logics, i.e. logics with formulas not containing certain connectives. A comprehensive study of various fragments of the main fuzzy logics mentioned above is provided in Cintula, Hájek, and Horčík 2007.

Systemization and theory of fuzzy logics

As we have seen, there exists a rich variety of fuzzy logics. Many of them share important properties. The situation therefore calls for systemization and a general theory of fuzzy logics. Some first steps toward delineation of a broad class of fuzzy logics were made in Běhounek and Cintula 2006a, Cintula 2006. Hájek and Cintula (2006) introduced the notion of *core fuzzy logic* which is a logic L satisfying the following three properties:

- L expands MTL: the language of L contains that of MTL and the provability of L includes that of MTL;
- L satisfies the *local deduction theorem*: for every theory T and formulas φ and ψ, we have $T, \varphi \vdash \psi$ iff there exists a positive integer n such that $T \vdash \varphi^n \rightarrow \psi$;[185]
- L satisfies the *congruence condition*: for all φ, ψ, and χ, we have $\varphi \leftrightarrow \psi \vdash \chi(\varphi) \leftrightarrow \chi(\psi)$, ensuring desirable properties of the Lindenbaum algebra.[186]

Core fuzzy logics include most of the logics presented above, e.g. MTL, BL, Ł, G, Π, the rational Pavelka logic, and also all of their schematic extensions and expansions by truth constants. On the other hand, intuitionistic logic is not a core fuzzy logic because prelinearity, $(\varphi \rightarrow \psi) \vee (\psi \rightarrow \varphi)$, is not provable in it. Classical logic is a core fuzzy logic, the strongest one in that adding any formula not already provable yields an inconsistent logic. Many properties can now be proved for the general notion of a core fuzzy logic, including various completeness theorems.

Inspired by Rasiowa (1974), a still broader class, that of *weakly implicative logics*, was introduced by Cintula (2006) and further developed by Cintula and Noguera (2010). These are logics whose language contains a binary connective $\rightarrow$ satisfying some structurally important properties, namely $\vdash \varphi \rightarrow \varphi, \varphi, \varphi \rightarrow \psi \vdash \psi, \varphi \rightarrow \psi, \psi \rightarrow \chi \vdash \varphi \rightarrow \chi$, and $\varphi \rightarrow \psi, \psi \rightarrow \varphi \vdash \chi(\varphi) \rightarrow \chi(\psi)$. Rasiowa's implicative logics are just the weakly implicative logics satisfying $\varphi \vdash \psi \rightarrow \varphi$. Most of the above discussed logics including their fragments are then special instances of what Cintula and Noguera call semilinear algebraically implicative logics.

A number of general results on *completeness of fuzzy logics* were obtained in Cintula et al. 2009. For a core fuzzy logic L and a class $\mathcal{K}$ of algebras of L, they considered the three kinds of completeness dealt with above: L is

- *strongly complete w.r.t.* $\mathcal{K}$ provided $T \vdash \varphi$ iff $T \models^{\mathcal{K}} \varphi$ for any T and φ;
- *finitely strongly complete w.r.t.* $\mathcal{K}$ provided $T \vdash \varphi$ iff $T \models^{\mathcal{K}} \varphi$ for any finite T and any φ;
- *complete w.r.t.* $\mathcal{K}$ provided $\vdash \varphi$ iff $\models^{\mathcal{K}} \varphi$ for any φ.

---

[185]The term "local" refers to the fact that n depends on the formulas involved. Fuzzy logics in general do not satisfy the classical deduction theorem, i.e. $T, \varphi \vdash \psi$ iff $T \vdash \varphi \rightarrow \psi$. In fact, the only schematic extension of MTL which satisfies it is the Gödel logic. Interestingly, the logic NM of nilpotent minimum (p. 203) satisfies $T, \varphi \vdash \psi$ iff $T \vdash \varphi \otimes \varphi \rightarrow \psi$ (Dubois et al. 2007).

[186]$\chi(\varphi)$ and $\chi(\psi)$ result by substitution of φ and ψ for the same propositional symbol in χ.

Thus, for instance, BL is strongly complete w.r.t. $\mathcal{K} = \mathcal{BL}^l$ (BL-chains), cf. (4.107); BL is finitely strongly complete but not strongly complete w.r.t. $\mathcal{K} = \mathcal{BL}^t$ (t-algebras), cf. (4.110). Notice also that the concept of standard completeness used above corresponds to the case where $\mathcal{K}$ consists of a single or several L-algebras on $[0, 1]$. The authors obtained various algebraic characterizations of these completeness properties. For instance, L is complete w.r.t. $\mathcal{K}$ iff the class $\mathcal{L}$ of all L-algebras is just the variety generated by $\mathcal{K}$. Furthermore, L is finitely strongly complete iff $\mathcal{L}$ is the quasivariety generated by $\mathcal{K}$ iff every L-chain can be embedded into an ultraproduct of some members of $\mathcal{K}$. The authors also consider semantics given by various kinds of chains. In addition to the standard set $[0, 1]$, they also consider the chain $[0, 1] \cap \mathbb{Q}$ of all rationals in $[0, 1]$, finite chains, and the chains consisting of hyperreals, i.e. ultrapowers of $[0, 1]$, which were considered for the first time by Di Nola (1993) in his representation theorem for MV-algebras (p. 185 above).

4.5 Fuzzy logic and uncertainty

4.5.1 Degrees of truth vs. belief and truth functionality

Degrees of truth and degrees of belief

Degrees of truth must be distinguished from degrees of belief which subsume degrees of various kinds of uncertainty such as those measured by probability theory, possibility theory (Dubois and Prade 1988), or Dempster-Shafer theory (Shafer 1976). Although the calculi dealing with degrees of truth have some similarities to those dealing with degrees of belief, such as using the real unit interval $[0, 1]$ for the scale of degrees, there exist important differences, most notably *truth functionality*. While degrees of truth are truth functional in most logical calculi including classical logic, degrees of belief are not. An example of the non-truth-functionality of probabilities, known already to Mazurkiewicz (1932), is shown on p. 146 above.[187]

More important than the formal similarities and differences, however, is the fact that these two types of calculi model two different phenomena—vagueness and uncertainty. Degrees of truth are used to model vague notions while degrees of belief express uncertainty resulting from lack of evidence. "Peter is tall" is vague but not uncertain. "Peter comes tomorrow" is uncertain but not vague. We may assign 0.7 or other degrees to both sentences but in the first case, we do so to account for the vagueness of the concept *tall* while in the second to account for our uncer-

---

[187] One needs to be careful when speaking of truth functionality of a logical calculus. A given logical calculus may be given two equivalent semantics, i.e. having the same notion of consequence and hence of tautology. Still one semantics may be truth functional, the other not. See J. Marcos, "What is a non-truth-functional logic?" *Studia Logica* 92 (2009): 215–40; one of the first papers related to this issue is N. Rescher, "Quasi-truth-functional systems of propositional logic," *J Symb Logic* 27 (1962): 1–10; see also A. Avron and I. Lev, "Non-deterministic multi-valued structures," *J Logic Comput* 15 (2005): 241–61. One should thus speak of truth functionality of semantics, or of truth functionality of evaluations of formulas, rather than truth functionality of a given logic.

tainty about Peter's actual appearance tomorrow. In spite of this fundamentally grounded distinction, a lot of confusion has resulted from a failure to recognize it. This includes the assumption that truth degrees in fuzzy logic must be probabilities, as well as claims that probability theory is superior to fuzzy logic, which implicitly make the wrong assumption that the aim of fuzzy logic is to handle probabilistic uncertainty. Such claims resemble comparing apples with oranges. Logical calculi for reasoning about beliefs and uncertainty nevertheless represent a research direction with a long tradition as explained in section 4.2.3.

We now clarify some basic issues. Propositions are subject to beliefs. The truth of propositions may admit two or more degrees, as do beliefs in propositions. In general, however, we deal with a *degree of belief that a given proposition φ has a particular degree of truth*. When φ is crisp, i.e. admits only two truth degrees, we commonly speak of degrees of belief in φ, meaning actually degrees of belief that the truth degree of φ is 1. In general, φ may be a formula of a certain fuzzy logic and we may thus consider degrees of belief that the truth degree of φ is, say, 0.8.

A degree of belief in (a crisp) φ may be interpreted as the truth degree of the statement "φ is believed." This allows the possibility of applying many-valued logical connectives to degrees of beliefs and hence of reasoning about complex propositions involving beliefs such as "if φ is believed and $\varphi \to \psi$ is believed then ψ is believed," using a suitable fuzzy logic. Alternatively, one may consider only qualitative, two-valued statements involving beliefs. One option is to consider formulas $P\varphi$, read as "φ is (sufficiently) probable," or more generally "φ is believed," meaning that the probability of φ exceeds a given threshold value. "Being probable" and "being believed" behave like a modality, similarly as "being necessary" in classical modal logics. A pioneering contribution along this line is Hamblin 1959. Another option, proposed by Segerberg (1971) and Gärdenfors (1975), is to consider formulas representing comparative statements such as "φ is believed at least as much as ψ is."

It is apparent that the various types of uncertainty and the various types of propositions that are subject to belief yield numerous prospective logics. Due to the above remarks, such logics are closely related to modal logic as well as to generalized quantifiers.[188] We thus examine some significant fuzzy logics of belief and related investigations in modal logics and generalized quantifiers.[189]

---

[188] Modal logic deals with modal connectives, classic ones are "necessary" and "possible." Also, many other language constructs related to reasoning about time-dependent propositions or about knowledge, may be treated as modalities. Modal logic goes back to Aristotle and other Greek philosophers and was further developed by the Scholastics. Its modern treatment was developed by Clarence I. Lewis (1883–1964), its contemporary semantics, based on the notion of a possible world, is due to Saul Kripke. A classic book is Hughes and Cresswell 1996.

Generalized quantifiers are quantifiers such as "for many" in the statement "for many x: φ." Pioneering work is due to Mostowski (1957) and P. Lindström, "First order predicate logic with generalized quantifiers," *Theoria* 32 (1966): 186–95. Good accounts may be found in M. Krynicki, M. Mostowski, and L. W. Szczerba, eds., *Quantifiers: Logics, Models, and Computing* (Dordrecht: Kluwer, 1995) and Peters and Westerståhl 2006.

[189] One reason to present the topics together is that modalities may be seen as hidden quantifiers (Hájek 1998b).

Probability logics

Our presentation would not be complete without mentioning some directions in probability logic which are not our primary interest but are closely related. As in the early approaches (p. 145), probability logics usually consider formulas of classical logic, propositional or predicate. Most research has focused on how uncertainties flow from premises to conclusions in deductive inferences. For instance, one is interested in determining tight bounds on $P(\varphi)$ from the bounds on $P(\varphi_1), \ldots, P(\varphi_n)$, given that $\varphi_1, \ldots, \varphi_n \models \varphi$ in the sense of classical logic. Adams (1998) and Hailperin (1996) provide detailed information on this topic. Paris (1994) considers Dempster-Shafer belief and possibilistic belief, which we examine in section 4.5.2. Alternatively, one may study logics in which the classical logic language is enhanced with various kinds of probabilistic operators. The formulas then include $P(\varphi) \geq a$, expressing that the probability of φ is at least a. Halpern (2003) provides an extensive treatment of this topic.

4.5.2 Possibilistic logic

Possibilistic logic has been developed by Didier Dubois, Henri Prade, and their collaborators in a number of papers, including the surveys Dubois, Lang, and Prade 1994; Lang 2000; Dubois and Prade 2004; and Dubois et al. 2007. It is a particular logic of belief in which formulas are assigned degrees of possibility and necessity. It is thus closely related to possibility theory. We now present an important conceptual framework (Dubois, Lang, and Prade 1991; Dubois et al. 2007) which helps to understand further the relationship of degrees of truth and belief.

Suppose φ represents a proposition about reality and D is our description of reality. D may be complete, in which case there is exactly one state s of affairs compatible with D, or incomplete, in which case there are at least two such states. The proposition represented by φ may be two- or many-valued and we assume that its truth degree $\|\varphi\|$ is given in each particular state s. We may now consider the following situations. First, assume D is complete. Then the unique state s compatible with D determines the truth degree a of φ. In this case we are certain that $\|\varphi\| = a$, i.e. our degree of belief in $\|\varphi\| = a$ is 1 and our degree of belief in $\|\varphi\| = b$ for $b \neq a$ is 0. Second, assume D is not complete and φ is two-valued. Then there may be two distinct states, s_1 and s_2, compatible with D such that $\|\varphi\| = 1$ in s_1 but $\|\varphi\| = 0$ in s_2. Our degree of belief in $\|\varphi\| = 1$ is equal to the degree of belief that given D, the actual state s belongs to the set S of states in which $\|\varphi\| = 1$. Such degree of belief is equal to the measure $\mu(S)$ of S's certainty which underlies our belief. Third, assume D is not complete and φ is many-valued. Then, analogously to the previous case, for any truth degree a, our belief in $\|\varphi\| = a$ is equal to the measure $\mu(S)$ of the set of states in which $\|\varphi\| = a$. In this case, every degree of truth $a \in L$ is assigned a degree of belief in the scale K of degrees of belief. We thus obtain a fuzzy set in L with truth degrees in K, which in case $L = K = [0, 1]$ may be interpreted as a fuzzy truth value.

As Dubois and Prade repeatedly emphasized, this is how Zadeh's notion of a fuzzy truth value, as propounded e.g. in Zadeh 1979a, should properly be understood.

Possibilistic logic fits the second situation. Formulas of its basic version to which we restrict our discussion here are pairs $\langle \varphi, a \rangle$ where φ is a formula of classical propositional logic and a a real in $[0, 1]$.[190] The intended meaning of $\langle \varphi, a \rangle$ is that the necessity of φ is at least a. Such formulas are thus $[0, 1]$-weighted formulas in the sense of Pavelka's abstract logic and we shall see that this is no coincidence. The semantics is as follows. Let W be the set of all classical truth evaluations w of the given propositional language and recall that $w \models \varphi$ means that $\|\varphi\|_w = 1$. Let π be a normalized possibility distribution on W, i.e. a mapping $\pi : W \to [0, 1]$ such that there exists $w \in W$ for which $\pi(w) = 1$. The *possibility of a propositional formula* φ given by π, $\Pi_\pi(\varphi)$, and the *necessity* $N_\pi(\varphi)$ of φ given by π are defined by

$$\Pi_\pi(\varphi) = \bigvee \{ \pi(w) \mid w \models \varphi \} \quad \text{and} \quad N_\pi(\varphi) = 1 - \Pi_\pi(\neg \varphi).$$

$\Pi_\pi(\varphi)$ is thus just the value of the possibility measure induced by π of the set of all evaluations in W which make φ true and similarly for $N_\pi(\varphi)$. As a result, Π_π and N_π satisfy the counterparts of properties of possibility and necessity measures: denoting a contradiction and a tautology by $\bar{0}$ and $\bar{1}$, respectively, we have

$$\Pi_\pi(\bar{0}) = 0, \ \Pi_\pi(\bar{1}) = 1, \qquad\qquad N_\pi(\bar{0}) = 0, \ N_\pi(\bar{1}) = 1,$$
$$\Pi_\pi(\varphi \wedge \psi) \leq \min(\Pi_\pi(\varphi), \Pi_\pi(\psi)), \qquad N_\pi(\varphi \wedge \psi) = \min(N_\pi(\varphi), N_\pi(\psi)),$$
$$\Pi_\pi(\varphi \vee \psi) = \max(\Pi_\pi(\varphi), \Pi_\pi(\psi)), \qquad N_\pi(\varphi \vee \psi) \geq \max(N_\pi(\varphi), N_\pi(\psi)).$$

The inequalities, which may be proper, indicate that possibilistic logic is not truth functional. If $\varphi \models \psi$ in classical logic, then $\Pi_\pi(\varphi) \leq \Pi_\pi(\psi)$ and $N_\pi(\varphi) \leq N_\pi(\psi)$. As a consequence, if φ and ψ are classically equivalent, then $\Pi_\pi(\varphi) = \Pi_\pi(\psi)$ and $N_\pi(\varphi) = N_\pi(\psi)$. Furthermore, we say that π *satisfies* $\langle \varphi, a \rangle$, in symbols $\pi \models \langle \varphi, a \rangle$, if $N_\pi(\varphi) \geq a$. A possibilistic formula $\langle \varphi, a \rangle$ *semantically follows* from T, in symbols $T \models \langle \varphi, a \rangle$, if every normalized possibility distribution that satisfies every formula in T also satisfies $\langle \varphi, a \rangle$. One is naturally interested in the supremum $\bigvee \{ a \in [0, 1] \mid T \models \langle \varphi, a \rangle \}$, which we denote $\|\varphi\|_T$ because it may be viewed as the degree to which φ semantically follows from T. It has been proved (Dubois and Prade 2004) that if $T = \{ \langle \varphi_1, a_1 \rangle, \ldots, \langle \varphi_n, a_n \rangle \}$, then the possibility distribution π_T defined by

$$\pi_T(w) = \begin{cases} 1 & \text{if } w \models \varphi_1 \wedge \cdots \wedge \varphi_n, \\ \min\{ 1 - a_i \mid w \models \neg\varphi_i, i = 1, \ldots, n \} & \text{otherwise,} \end{cases}$$

is the least specific among all possibility distributions that satisfy T and $\|\varphi\|_T$ is the necessity of φ corresponding to π_T.

An axiomatization of (predicate) possibilistic logic is due to Lang (1991). In the propositional case, one may consider axioms of classical propositional logic, each

---

[190] The version of possibilistic logic we present has been extended in a number of ways, e.g. to account for partial inconsistency and nonmonotonic inference (Dubois et al. 2007).

with weight 1. Deduction rules are a graded form of *modus ponens* and weakening:

$$\frac{\langle\varphi,a\rangle,\langle\varphi\to\psi,a\rangle}{\langle\psi,a\wedge b\rangle}\quad\text{and}\quad\frac{\langle\varphi,a\rangle}{\langle\varphi,b\rangle}\text{ with }b\leq a.$$

The usual notion of provability, $\vdash$, yields completeness:

Theorem 4.58. $T\vdash\langle\varphi,a\rangle$ *iff* $T\models\langle\varphi,a\rangle$.

Let us now examine relationships to Pavelka-style logic.[191] Whether a theory is a set T of pairs $\langle\varphi,a\rangle$ or a fuzzy set T of formulas such that $T(\varphi)=a$ is clearly a *façon de parler*, hence theories in possibilistic logic may be identified with theories in Pavelka's sense. The degree $\|\varphi\|_T$ may be understood as the degree of entailment in Pavelka's sense. Namely, one easily verifies that for the $\langle[0,1],\leq\rangle$-semantics

$$\mathscr{S}_{\text{pos}}=\{N\mid N=N_\pi\text{ for some normalized }\pi:W\to[0,1]\}$$

it holds that $\|\varphi\|_T=\bigwedge\{N(\varphi)\mid N\in\text{Mod}(T)\}$. Clearly, both deduction rules and proofs in possibilistic logic are deduction rules and $[0,1]$-weighted proofs in Pavelka's sense. One may thus examine the corresponding degree of provability $|\varphi|_T=\bigvee\{a\mid T\vdash\langle\varphi,a\rangle\}$ and notice that this degree does not change if we omit the second deduction rule. Therefore, denoting by $\mathscr{F}$, $\mathscr{R}_{\text{Pos}}$, and A_{Pos} the set of classical propositional formulas, the deduction rules, and the axioms (with the same provision as for theories) of possibilistic logic, possibilistic logic may be identified with Pavelka's abstract fuzzy logic $\langle\mathscr{F},\langle[0,1],\leq\rangle,\mathscr{S}_{\text{Pos}},\mathscr{R}_{\text{Pos}},A_{\text{Pos}}\rangle$. It is then easily seen that due to theorem 4.58, this logic is Pavelka-style complete, i.e. $|\varphi|_T=\|\varphi\|_T$.

Gerla (1994a, 2001) arrived at possibilistic logic from his studies in abstract fuzzy logic. He considers the canonical extension of a Hilbert system of classical logic (p. 175). The resulting logic, *generalized necessity logic* L_{gn}, is just the axiomatic system of possibilistic logic except that its second rule is not considered because it is superfluous when it comes to degrees of entailment. Gerla shows that closed theories of this canonical extension are just fuzzy sets N of formulas for which $N(\bar{1})=1$ for $\bar{1}$ being a tautology, $N(\varphi\wedge\psi)=\min(N(\varphi),N(\psi))$, and $N(\varphi)=N(\psi)$ whenever φ and ψ are classically equivalent. The adjective "generalized" refers to the fact that $N(\bar{0})=0$ for a contradiction $\bar{0}$ may not be satisfied. However, $N(\bar{0})=0$ becomes satisfied if we add the collapsing rule defined by $R_{\text{syn}}(\varphi)=\varphi$ for any contradiction φ, and $R_{\text{sem}}(a)=1$ if $a>0$ and $=0$ if $a=0$. This logic, L_{n}, is called the *necessity logic* and its evaluations are exactly the same as those of formulas in possibilistic logic.

---

[191]Comments on these relationships are found in the writings of Dubois and Prade, e.g. Dubois and Prade 2004, but they are not accurate and are not based on a proper understanding of Pavelka-style logics. Gerla (2001) provides results on a logic with possibilistic semantics, which we examine below, but does not cover connections to Dubois and Prade's possibilistic logic and its notions. These connections are examined in R. Belohlavek, "Pavelka-style fuzzy logic in retrospect and prospect," *Fuzzy Sets Syst* 281 (2015): 61–72.

4.5.3 Gerla's probabilistic fuzzy logics

The necessity logic, L_n, described in the previous section, is but one of the logics concerned with degrees of belief obtained by Gerla as particular examples of Pavelka's abstract fuzzy logics. We now examine three more logics of this kind. All of them obey Pavelka-style completeness; see Gerla 2001 for further information on their properties. In these logics, degrees in $[0,1]$ are assigned to formulas φ of classical propositional logic and are interpreted as truth degrees of "φ is believed." Naturally then, all these logics are non-truth-functional. They share the property that tautologies are assigned 1 and that classically equivalent formulas are assigned the same degrees. Gerla thus makes the assumption that formulas of the respective abstract fuzzy logics are the elements of the Lindenbaum algebra of classical propositional logic and, in fact, still more generally a Boolean algebra $\mathbf{B} = \langle B, \wedge, \vee, \neg, \bar{0}, \bar{1}\rangle$.

The *logic of superadditive measures* (Gerla 1997), L_{sa}, has as its semantics the set $\mathscr{S}_{sa}$ of all constant-sum superadditive measures on $\mathbf{B}$. A superadditive measure is a mapping $p : \mathbf{B} \to [0,1]$ satisfying $p(\bar{1}) = 1$ and $p(\varphi \vee \psi) \geq p(\varphi) + p(\psi)$ whenever $\varphi \wedge \psi = \bar{0}$. If, moreover, $p(\varphi) + p(\neg\varphi) = 1$, p is called a constant-sum superadditive measure. Such measures are more general than probability measures and are known as characteristic functions of n-person games.[192] Gerla considers $\bar{1}$ with degree 1 as the only axiom and two deduction rules,

$$\frac{\langle \varphi, a\rangle, \langle \psi, b\rangle}{\langle \varphi \vee \psi, a \oplus b\rangle} \quad \text{and} \quad \frac{\langle \varphi, a\rangle, \langle \psi, b\rangle}{\langle \bar{0}, c(a,b)\rangle},$$

which are both defined for $\varphi \wedge \psi = \bar{0}$ and in which $\oplus$ is the Łukasiewicz disjunction, $a \oplus b = \min(1, a + b)$, and $c(a, b) = 1$ for $a + b > 1$ and $= 0$ for $a + b \leq 1$.

The semantics of the *logic* L_{ul} *of upper-lower probabilities* is the set $\mathscr{S}_{ul}$ of all upper-lower probabilities on $\mathbf{B}$, i.e. superadditive measures whose dual function $p^\perp(\alpha) = 1 - p(\neg\alpha)$ satisfies subadditivity, i.e. $p^\perp(\varphi \vee \psi) \leq p^\perp(\varphi) \vee p^\perp(\psi)$.[193] The axiomatic system of L_{ul} results by extending L_{sa} by the so-called rule of conjunction,

$$\frac{\langle \varphi, a\rangle, \langle \psi, b\rangle}{\langle \varphi \wedge \psi, a \otimes b\rangle},$$

where $\otimes$ is the Łukasiewicz conjunction.

The third logic (Gerla 1994c, 1997, 2001), L_p, called by its author the *logic of envelopes* or *probability logic*, has as its semantics the set $\mathscr{S}_p$ of all finitely additive probabilities on $\mathbf{B}$. This choice, instead of σ-additivity, is natural because there are no σ-additive measures on the Lindenbaum algebra of the classical propositional logic.[194] The name of the logic comes from the fact that the semantic closure $C_{\mathscr{S}_p}(T)$ of any theory T is exactly the *lower envelope* of T, i.e. the intersection of all prob-

---

[192] G. Owen, *Game Theory* (New York: Academic Press, 1982).

[193] The term "upper-lower probability" usually denotes the pair $\langle p, p^\perp \rangle$; see A. P. Dempster, "Upper and lower probabilities induced by a multivalued mapping," *Annals Mathematical Statistics* 38 (1967): 325–39.

[194] A. Horn and A. Tarski, "Measures in Boolean algebras," *Trans AMS* 64 (1948): 467–97.

abilities p for which $T \leq p$ by pointwise ordering.[195] The axiomatic system of L_p consists of $\bar{1}$ with weight 1 as the only axiom and the following deduction rules. Let an h-k-connective be the h-ary function C^k in B defined as follows: C^0 is constantly equal to $\bar{1}$; for $k > 0$, and $\varphi_1, \ldots, \varphi_h \in B$, C^k is defined by

$$C^k(\varphi_1, \ldots, \varphi_h) = \bigvee \{\varphi_{i_1} \wedge \cdots \wedge \varphi_{i_k} \mid i_j \in \{1, \ldots, h\} \text{ are distinct}\}.[196]$$

The h-m-k-rule is the deduction rule

$$\frac{\langle \varphi_1, a_1 \rangle, \ldots, \langle \varphi_h, a_h \rangle}{\left\langle C^k(\varphi_1, \ldots, \varphi_h), \mathrm{n}\left(\frac{a_1 + \cdots + a_h - k + 1}{m - k + 1}\right)\right\rangle},$$

defined for those φ_is for which m is the largest number k such that $C^k(\varphi_1, \ldots, \varphi_h) \neq \bar{0}$, and such that $\mathrm{n}(a) = 0$ for $a < 0$, $= 1$ for $a > 1$, and $= a$ for $a \in [0, 1]$. For $h \geq m$, the h-m-collapsing rule is the deduction rule

$$\frac{\langle \varphi_1, a_1 \rangle, \ldots, \langle \varphi_h, a_h \rangle}{\langle \bar{0}, c(a_1, \ldots, a_h) \rangle}$$

defined for the same φ_is as the h-m-k-rule, and such that $c(a_1, \ldots, a_h) = 1$ for $a_1 + \cdots + a_h > 1$ and $= 0$ otherwise. Gerla puts forward an interesting hypothesis that since logical deductions should admit reasoning under incomplete information, lower envelopes may in fact be more adequate than probability as a fundamental concept in probabilistic logics and supports this by examples.

Denoting by $\mathcal{T}_*$ the set of all consistent closed theories of L_*, we obtain a natural hierarchy of these logics:

$$\mathcal{T}_n \subseteq \mathcal{T}_p \subseteq \mathcal{T}_{ul} \subseteq \mathcal{T}_{sa}.$$

Interestingly, Gerla showed that no abstract fuzzy logic exists whose closed theories coincide with Dempster-Shafer belief measures.

4.5.4 Belief, modality, and quantifiers in fuzzy logic

Various calculi dealing with degrees of belief have been developed in papers coauthored by Hájek. His *Metamathematics of Fuzzy Logic* has a long chapter on this topic and outlines several research directions. We now examine these developments.

Two-valued modal logics of qualitative (comparative) belief

Logics for comparative reasoning about belief may be built as two-valued modal logics as follows. The formulas are like classical propositional formulas with ⊲ as an additional connective. Formulas $\varphi \triangleleft \psi$ are interpreted as saying that the belief in ψ is at least as large as that in φ. The semantics is defined in terms of *Kripke structures with a measure of belief* which are triplets $\mathbf{K} = \langle W, e, \mu \rangle$, where W is a nonempty

---

[195] P. Walley and T. L. Fine, "Towards a frequentist theory of upper and lower probability," *Annals of Statistics* 10 (1982): 741–61.

[196] These functions were originally considered by Horn and Tarski (see n. 194 above).

set of *possible worlds*, e assigns to every $w \in W$ and propositional symbol p a truth value $e(w, p) \in \{0, 1\}$, and μ is a measure of belief on W assigning numbers in $[0, 1]$ to certain subsets of W. The truth value $\|\varphi\|_{\mathbf{K},w}$ of a formula in $\mathbf{K}$ and $w \in W$ is defined as follows: for a propositional symbol p, $\|p\|_{\mathbf{K},w} = e(w, p)$; the truth values of $\neg\varphi$ and $\varphi \wedge \psi$ are defined classically; and for $\varphi \triangleleft \psi$ we proceed as follows. Let $Sat_{\mathbf{K}}(\varphi) = \{w \in W \mid \|\varphi\|_{\mathbf{K},w} = 1\}$ and put

$$\|\varphi \triangleleft \psi\|_{\mathbf{K},w} = 1 \text{ iff } \mu(Sat_{\mathbf{K}}(\varphi)) \leq \mu(Sat_{\mathbf{K}}(\psi))$$

where both $\mu(Sat_{\mathbf{K}}(\varphi))$ and $\mu(Sat_{\mathbf{K}}(\psi))$ are defined. For a class $\mathcal{K}$ of Kripke structures $\mathbf{K}$, one may now examine its tautologies.

Harmanec and Hájek (1994) developed a logic, called the *comparative belief function modal propositional logic* (CBMPL) in Hájek 1998b, over the class $\mathcal{K}_{\mathrm{DS}}$ of Kripke structures with Dempster-Shafer belief measures. Its axioms consist of classical propositional axioms and the following ones, where $\square\varphi$ stands for $(\neg\varphi \triangleleft \bar{0})$:

$$(\varphi \triangleleft \psi) \to ((\psi \triangleleft \chi) \to (\varphi \triangleleft \chi)) \qquad\qquad (\varphi \triangleleft \psi) \vee (\psi \triangleleft \varphi)$$

$$\neg(\bar{1} \triangleleft \bar{0}) \qquad\qquad\qquad\qquad\qquad\qquad \square(\varphi \to \psi) \to (\varphi \triangleleft \psi)$$

$$[\square(\varphi \to \psi) \wedge \square\neg(\psi \wedge \chi)] \to [((\psi \vee \chi) \triangleleft (\varphi \vee \chi)) \to (\psi \triangleleft \varphi)]$$

$$(\varphi \triangleleft \psi) \to \square(\varphi \triangleleft \psi) \qquad\qquad\qquad \neg(\varphi \triangleleft \psi) \to \square\neg(\varphi \triangleleft \psi)$$

With *modus ponens* and $\frac{\varphi \to \psi}{\varphi \triangleleft \psi}$ as deduction rules, the authors proved completeness:

Theorem 4.59. φ *is provable iff* $\|\varphi\|_{\mathbf{K},w} = 1$ *in every world w of any $\mathbf{K} \in \mathcal{K}_{\mathrm{DS}}$.*

They also proved that CBMPL is a conservative extension of the well-known modal logic KD45 of belief. Hájek (1998b) also developed a logic over the class $\mathcal{K}_{\mathrm{poss}}$ of Kripke structures with measures of possibility—the *comparative possibilistic modal propositional logic* (CPMPL)—which differs from CBMPL in that the fourth and fifth axioms are replaced by $(\varphi \to \psi) \to ((\varphi \vee \chi) \triangleleft (\psi \triangleleft \chi))$—and proved the same kind of completeness for it.

A different kind of logics was studied by Esteva et al. (1997). They employ *Kripke structures with similarity*, $\mathbf{K} = \langle W, e, s \rangle$, in which W and e are as above and $s : W \times W \to G \subseteq [0, 1]$, called similarity, is a fuzzy equality w.r.t. a given t-norm-like operation $\otimes$ on G (p. 277 below). The language is that of classical propositional logic extended with unary modal connectives $\Diamond_a^c$ and $\Diamond_a^o$ for every $a \in G$ and formulas are defined as usual. Dual modalities, $\square_a^c$ and $\square_a^o$, are defined via standard abbreviations: $\square_a^*$ stands for $\neg\Diamond_a^*\neg$. Validity of formulas is two-valued Kripke-style except that the truth of $\Diamond_a^*\varphi$ is defined as $\|\Diamond_a^c\varphi\|_{\mathbf{K},w} = 1$ iff $\bigvee_{\|\varphi\|_{\mathbf{K},w'}=1} s(w, w') \geq a$, meaning that the truth degree of the proposition "there exists a world w' similar to w in which φ is true" is at least a, and analogously for $\Diamond_a^o\varphi$ with "$\geq$" replaced by "$>$." While the axiomatization of the general case is left as an open problem in Esteva et al. 1997, the authors present completeness results for two cases, one with finite G and the other with G dense in $[0, 1]$ and $\otimes$ being min. Note also that Hájek (1998b, chapter 8) considers a different kind of modality, $\square_a$, for which $||\square_a \varphi||_{K,w} = 1$ means that $||\varphi||_{K,v} = 1$ for each v with $s(w,v) \geq a$.

Fuzzy modal logics and quantitative belief

The above modal logics were two-valued. For one, the underlying logic was classical. Second, the modalities were two-valued. We now present fuzzy modal logics in which these features are abandoned.[197]

The first logic (Hájek and Harmancová 1996, Hájek 1998b) is a generalization of the well-known system S5 (Hughes and Cresswell 1996) which is also known as the *logic of knowledge*. S5 is due to Clarence I. Lewis (Lewis and Langford 1932, appendix II). In general, one may consider for the logic BL or its schematic extension C (section 4.4), the fuzzy modal logic S5(C) whose formulas are built from the connectives of BL and two modalities, $\square$ and $\Diamond$. A Kripke structure for S5(C) is a triplet $K = \langle W, e, L \rangle$ where W is a nonempty set of possible worlds, e a truth evaluation assigning to every $w \in W$ and every propositional symbol p a degree $e(w,p) \in L$ and L a C-algebra. Truth degrees $||\varphi||_{K,w}$ are defined by the usual rules such as $||p||_{K,w} = e(w,p)$ or $||\varphi \otimes \psi||_{K,w} = ||\varphi||_{K,w} \otimes ||\psi||_{K,w}$, and two new rules,

$$||\square\varphi||_{K,w} = \bigwedge_{v \in W} ||\varphi||_{K,v} \quad \text{and} \quad ||\Diamond\varphi||_{K,w} = \bigvee_{v \in W} ||\varphi||_{K,v},$$

provided the infima and suprema exist (otherwise undefined). Many observations on S5(C) are provided in Hájek 1998b. Hájek and Harmancová (1996) also consider S5($[0,1]_\text{Ł}$). This logic, called RPL$\square$, results by adding $\square$ to the rational Pavelka logic, defining $\Diamond$ as a shorthand for $\neg\square\neg$, and the above rules for semantics extended by $||\bar{r}||_{K,w} = r$ for rationals $r \in [0,1]$. The axioms consist of those of RPL, S5, two axioms analogous to two rules in Fitting 1991,

$$(\bar{r} \to \square\varphi) \leftrightarrow \square(\bar{r} \to \varphi) \quad \text{and} \quad (\bar{r} \to \Diamond\varphi) \leftrightarrow \Diamond(\bar{r} \to \varphi),$$

and the axiom $(\Diamond\varphi \otimes \Diamond\psi) \leftrightarrow \Diamond(\varphi \otimes \psi)$. The authors obtain a Pavelka-style completeness, in which $|\varphi| = \bigvee\{r \in [0,1] \mid \vdash \bar{r} \to \varphi\}$ is the degree of provability and $||\varphi|| = \bigwedge\{||\varphi||_{K,w} \mid K = \langle W, e, [0,1]_\text{Ł}\rangle$ is a Kripke structure and $w \in W\}$ the degree of tautologicity of φ:

Theorem 4.60. For every formula φ of RPL$\square$ *we have* $|\varphi| = ||\varphi||$.

Next we examine logics in which the *many-valued modalities* are related to measures of uncertainty. Hájek, Harmancová, and Verbrugge (1995) proposed two kinds of *comparative possibilistic logics* whose underlying logic is the n-valued propositional Łukasiewicz logic and which has a many-valued binary modal connective $\vartriangleleft$ such that $\varphi \vartriangleleft \psi$ means that ψ is (almost) at least as possible as φ. Their complete axiomatization and further theorems complement other results revealing close links between logics with possibilistic semantics, as well as tense and other modal logics.[198]

---

[197] Although prior contributions exist, two pioneering works in many-valued modal logics are Fitting 1991, 1992.
[198] See for example Bendová and Hájek 1993, Hájek 1995b, and Hájek et al. 1994.

We turn to logics with a unary modal connective P representing probability or other kinds of belief in which formulas involving $P\varphi$ are many-valued, i.e. are no longer two-valued as a result of thresholding discussed in section 4.5.1. In the present case, a formula of the form $P\varphi$ is assigned a degree of belief interpreted as the truth degree of the proposition "φ is probable," or more generally "φ is believed." φ itself may be two-valued or many-valued. The former option appears when considering probabilities of ordinary events, the latter when the events are fuzzy. Both options make good sense. The degrees of belief may thus be manipulated by many-valued connectives. This scenario leads to interesting fuzzy logical calculi that we now examine.

Hájek, Godo, and Esteva (1995) proposed a *fuzzy probability modal logic* which may in general be formulated over a schematic extension C of BL.[199] This propositional logic, FP(C), has connectives $\otimes$, $\rightarrow$, $\wedge$, $\vee$, and $\neg$, the truth constants $\overline{0}$ and $\overline{1}$, a unary modal connective P, and two kinds of formulas. The *Boolean* ones result from propositional symbols and truth constants using the connectives $\wedge$, $\vee$, and $\neg$; the *modal* ones result from the formulas of the form $P\varphi$, with φ being a Boolean formula, by using the connectives and truth constants of C. In particular, they consider FP(RPL), i.e. C is the rational Pavelka logic, in which case the atomic modal formulas also contain $\overline{r}$ for each rational $r \in (0,1)$. Truth of formulas is considered with respect to the class $\mathcal{K}_{\text{prob}}$ of Kripke structures $\langle W, e, \mu \rangle$ with measures (p. 217) in which μ is a finitely additive probability on a field of subsets of W such that μ is defined on the set $\{w \in W \mid e(p,w) = 1\}$ for each p. The truth value $\|\varphi\|_K$ is independent of a particular world and is defined as follows. For Boolean formula φ, the definition is classical and yields 0 or 1; $\|P\varphi\|_K = \mu(\{w \in W \mid \|\varphi\|_{K,w} = 1\})$ and $\|\overline{r}\|_K = r$; for other compound modal formulas, one employs rules such as $\|\varphi \otimes \psi\|_K = \|\varphi\|_K \otimes \|\psi\|_K$ and the Łukasiewicz connectives. The axioms are those of classical propositional logic for Boolean formulas, of RPL for modal formulas, and the following axioms for P:

$$P\varphi \rightarrow (P(\varphi \rightarrow \psi) \rightarrow P\psi)$$
$$P(\neg\varphi) \longleftrightarrow \neg P\varphi$$
$$P(\varphi \vee \psi) \longleftrightarrow [(P\varphi \rightarrow P(\varphi \wedge \psi)) \rightarrow P\psi]$$

for any Boolean φ and ψ. A *modal theory* is a set T of modal formulas. As usual, K is a model of T if $\|\varphi\|_K = 1$ for each $\varphi \in T$. With the degree of semantic consequence $\|\varphi\|_T = \bigwedge\{\|\varphi\|_K \mid K \in \mathcal{K}_{\text{prob}} \text{ a model of } T\}$ and the degree of provability $|\varphi|_T$, defined as usual in RPL, the authors establish completeness:

Theorem 4.61. For every modal formula φ and modal theory T, $|\varphi|_T = \|\varphi\|_T$.

The present setting nicely accommodates the modality "many" (Hájek 1998b). The resulting logic, FMany(RPL), has the same formulas as FP(RPL) and semantics

---

[199] Our presentation is based on Hájek 1998b, where this name is used.

defined over finite structures $\mathbf{K} = \langle W, e \rangle$ with e as above and a different meaning of P for which one uses relative frequency:

$$\|P\varphi\|_{\mathbf{K}} = \frac{\mathrm{card}(\{w \in W \mid \|\varphi\|_{\mathbf{K},w} = 1\})}{\mathrm{card}(W)}.$$

$P\varphi$ is thus naturally interpreted as "for many worlds, φ" or "usually, φ." With the same axioms as for FP(RPL), Hájek obtains the same kind of completeness. Hájek, Godo, and Esteva 1995 and Hájek 1998b also provide a logic for conditional probabilities, for which purpose they work with an extension of RPL by the product conjunction $\odot$. The connection consists in that the probability of ψ conditioned by φ is $\geq a$ iff the truth degree of $\bar{a} \odot P(\varphi) \rightarrow P(\varphi \wedge \psi)$ is 1. Hájek, Godo, and Esteva (1995) also show that the same kind of completeness is established for the *fuzzy necessity modal logic*, FN(RPL), whose semantics is defined by the class $\mathscr{K}_{\mathrm{nec}}$ of structures with necessity measures. Denoting the modality by N, the axioms result by replacing the above axioms for P by

$$N(\varphi \rightarrow \psi) \rightarrow (N\varphi \rightarrow N\psi), \quad \neg N(\bar{0}), \quad \text{and} \quad N(\varphi \wedge \psi) \longleftrightarrow (N\varphi \wedge N\psi).$$

The modal logics presented in this section were all propositional. We now present one predicate extension developed in Hájek 1998b. Consider an extension of the predicate rational Pavelka logic RPL∀ by a new quantifier, $\int$ (read "probably"), and extend the definition of formulas so that if φ is a formula and x a variable, then $\int \varphi dx$ is a formula in which x is bound. The semantics is defined over structures $\mathbf{M}$ of RPL∀ with a finite and countable domain M to which one adds a probability measure $\mu : M \rightarrow [0,1]$ satisfying $\sum_{m \in M} \mu(m) = 1$. Such $\mathbf{M}$s are called *probabilistic structures*. For a fuzzy set $f : M \rightarrow [0,1]$, define the *integral* $\int f \, d\mu$ of f by

$$\int f \, d\mu = \sum_{m \in M} f(m) \cdot \mu(m),$$

where $\cdot$ is the usual product. Notice that if f is crisp, it represents an event and $\int f \, d\mu$ is the probability of f. A general fuzzy set f may be looked at as representing a vaguely described event, which applies to m to degree $f(m)$, and $\int f \, d\mu$ is then naturally interpreted as the probability of the fuzzy event f, a concept introduced by Zadeh (1968a) which we examine in section 5.3.5. The truth degree of $\int \varphi dx$ in a given probabilistic structure $\mathbf{M}$ and valuation v of variables is then defined by

$$\| \int \varphi dx \|_{\mathbf{M},v} = \int Sat_{\mathbf{M},v}(\varphi, x) \, d\mu,$$

where $Sat_{\mathbf{M},v}(\varphi, x)$ is a fuzzy set in M defined for any $m \in M$ by $[Sat_{\mathbf{M},v}(\varphi, x)](m) = \|\varphi\|_{\mathbf{M},v(x/m)}$—the truth degree of φ in $\mathbf{M}$ when x is assigned m. This logic, RPL∀$\int$, has the axioms of RPL∀ with the new notion of formula and the following new axioms in which χ does not contain x as a free variable:

$$\int \chi dx \longleftrightarrow \chi$$
$$\int (\neg \varphi) dx \longleftrightarrow \neg \int \varphi dx$$
$$\int (\varphi \rightarrow \psi) dx \rightarrow (\int \varphi dx \rightarrow \int \psi dx)$$

$$\int(\varphi \oplus \psi)dx \longleftrightarrow ((\int \varphi dx \rightarrow \int(\varphi \otimes \psi)dx) \rightarrow \int \psi dx)$$
$$(\forall x)\varphi \rightarrow \int \varphi dx$$
$$\int(\int \varphi dx)dy \longleftrightarrow \int(\int \varphi dy)dx$$

The deduction rules are those of RPL$\forall$ and the derived rules include

$$\frac{\varphi}{\int \varphi dx} \quad \text{and} \quad \frac{\varphi \rightarrow \psi}{\int \varphi dx \rightarrow \int \psi dx}.$$

Hájek (1998b) leaves open the problem of completeness with respect to probabilistic structures but provides valuable observations on RPL$\forall\int$. Nevertheless, he provides a completeness result for semantics over *weak probabilistic structures*. These are like the probabilistic ones except that μ is replaced by the so-called weak integral. Let $\mathscr{F} \subseteq [0,1]^M$ be a system of fuzzy sets containing all constant functions with a rational value and closed with respect to pointwise extensions of truth functions of logical connectives, i.e. if $f, g \in \mathscr{F}$ then $f \otimes g \in \mathscr{F}$, and similarly for the other connectives of RPL$\forall\int$. A *weak integral* on $\mathscr{F}$ is a mapping I assigning each f a number I$f\,dx$ such that the following analogs of the above axioms for $\int$ hold:

$$\text{I}f_c dx = c, \qquad \text{I}(1-f)dx = 1 = \text{I}f\,dx, \qquad \text{I}(f \rightarrow g) \le (\text{I}f\,dx \rightarrow \text{I}g\,dx),$$
$$\text{I}(f \oplus g) = \text{I}f\,dx + \text{I}g\,dx - \text{I}(f \otimes g)dx, \qquad \text{I}(\text{I}h\,dx)dy = \text{I}(\text{I}h\,dy)dx,$$

for each constant function f_c with value c and each function $h : M \times M \rightarrow [0,1]$ whose projections $h(a,y)$ and $h(x,a)$ are both in $\mathscr{F}$ for every $a \in [0,1]$ and where I$h\,dx$ is the unary function assigning to each a the integral I$h(x,a)dx$ of the projection $h(x,a)$ and similarly for I$h\,dy$. With the usual RPL-notion of degree of provability, $|\varphi|_T$, Hájek obtained the following completeness theorem:

Theorem 4.62. For every formula φ and theory T of RPL$\forall\int$, $|\varphi|_T = \|\varphi\|_T$, where $\|\varphi\|_T = \bigwedge\{\|\varphi\|_M \mid M$ a weak probabilistic structure which is a model of $T\}$.

This completes our overview of selected approaches to reasoning about belief using modal logics and logics with generalized quantifiers in a fuzzy setting. The literature contains many other approaches to generalized quantifiers, usually called fuzzy quantifiers, about which one may consult Glöckner 2006 and the numerous references therein. These are mainly semantically based and often defined in an ad hoc way. As Hájek mentions, further logical analysis of generalized quantifiers seems to be a promising task. This includes not only unary quantifiers but also binary and more general ones, in particular the multitudinal quantifiers involved in statements of the form "many φs are ψs," which are important in data analysis. Hájek's comments, however brief, are also very insightful due to his fundamental role in the development of the logico-statistical data-analytical method GUHA (Hájek and Havránek 1978) in which generalized quantifiers play a crucial role.

4.6 Miscellaneous issues

In this section, we examine some directions which were only briefly mentioned previously. Further topics, including various theories over fuzzy logic, are discussed in chapter 5. For topics only briefly touched upon in this book, additional information may be found in Galatos et al. 2007, Gottwald 2001, Gerla 2001, Hájek 1998b, Novák et al. 1999, and the recent handbook Cintula, Hájek, and Noguera 2011.

4.6.1 Relationship to applications

We now address the question of the relationship of fuzzy logic in the narrow sense to fuzzy logic in the broad sense and to applications of fuzzy logic generally. This question has been addressed repeatedly by Zadeh. It is also thoroughly discussed in Hájek's *Metamathematics of Fuzzy Logic*, parts of which are actually inspired by it. Hájek's view is apparent from the following remark addressed to researchers involved in fuzzy logic applications and soft computing (Hájek 1998b, vii):

> As a matter of fact, most of these are not professional logicians so that it can easily happen that an application, clever and successful as it may be, is presented in a way which is logically not entirely correct or may appear simple-minded. (Standard presentations of the logical aspects of fuzzy controllers are the most typical example.) This fact would not be very important if only the *bon ton* of logicians were harmed; but it is the opinion of the author (who is a mathematical logician) that a better understanding of the strictly logical basis of fuzzy logic (in the usual broad sense) is very useful for fuzzy logic appliers since if they know better what they are doing, they may hope to do it better. Still more than that: a better mutual understanding between (classical) logicians and researchers in fuzzy logic promises to lead to deeper cooperation and new results.

Fuzzy control mentioned here by Hájek is indeed frequently discussed in debates on the role of fuzzy logic in the narrow sense in applications, see e.g. Dubois and Prade 1991, Gerla 2001, Gottwald 1993, Hájek 1998b, Höhle 1996, Klawonn and Novák 1996, and Novák et al. 1999. One reason is that fuzzy control is one of the most visible applications of fuzzy logic. Another is that its terminology involves some typically logical terms—"fuzzy if-then rules" (p. 358 below), "(compositional) rule of inference" (p. 76 above), and, in particular, "generalized *modus ponens*."[200] What follows is a brief outline of a logical analysis of these notions worked out in chapter 7 in Hájek 1998b and in Godo and Hájek 1999.

---

[200] The compositional rule of inference is introduced in Zadeh 1975a,b and in part III of Zadeh 1975c. Generalized *modus ponens* is introduced and thoroughly discussed in part III of Zadeh 1975c, but is also mentioned in Zadeh 1975a under the name "compositional *modus ponens*." In essence, both the compositional rule of inference and the generalized *modus ponens* may be viewed as a semantic rule that allows us to infer from a fuzzy set $A : U \rightarrow [0,1]$ and a fuzzy relation $R : U \times V \rightarrow [0,1]$ a new fuzzy set $B = A \circ R$ in V. Here, $\circ$ is the composition of A and R defined by $(A \circ R)(v) = \bigvee_{u \in U} A(u) \otimes R(u,v)$ and $\otimes$ is an appropriate truth function of conjunction. R is thought of as a given fuzzy relation or as a fuzzy relation constructed from other fuzzy sets, e.g. by $R(u,v) = \neg C(u) \oplus D(v)$ where $C : U \rightarrow [0,1]$ and $D : V \rightarrow [0,1]$ are fuzzy sets which appear in a statement "if x is C then y is D." Zadeh depicts such a rule symbolically as $\dfrac{u \text{ is } A, u \text{ and } v \text{ are } R}{v \text{ is } A \circ R}$ or $\dfrac{u \text{ is } A, \text{ if } u \text{ is } C \text{ then } v \text{ is } D}{v \text{ is } A \circ R}$, and claims that this rule generalizes classical *modus ponens*, whence the term "generalized *modus ponens*." This is, nevertheless, somewhat misleading because Zadeh's rule in fact generalizes only a *particular instance* of classical *modus ponens*, namely $\dfrac{u \text{ is } A, \text{ if } u \text{ is } A \text{ then } v \text{ is } B}{v \text{ is } B}$ (in fact, it generalizes a semantic counterpart of this instance).

Central in fuzzy control are expressions of the form "X is A" where X is a name of a variate, such as "temperature," and A is a name of a fuzzy set in a given domain M associated with X.[201] Hájek proposes two ways of approaching such expressions in a many-sorted predicate fuzzy logic in which the Ms are domains. In the first approach, which we present, the name X is considered a symbol of an object constant interpreted as the actual value of the variate. "X is A" then becomes the closed formula $A(X)$, where A is a symbol of a unary relation. An if-then rule "if X is A then Y is B" may be interpreted by the formula $A(X) \rightarrow B(Y)$. Hence, one uses a many-sorted L-structure $\mathbf{M}$ with domains $M_X, M_Y, \ldots$, unary L-relations $A, B, \ldots$, and elements $X \in M_X, Y \in M_Y$, etc., interpreting A, B, X, Y, etc.

In the present context, Zadeh's compositional rule of inference says that if A_* is a fuzzy set representing the value of the input variate X and R is a fuzzy relation representing the if-then rules, one infers output B_* representing the value of Y by

$$B_*(v) = \bigvee_{u \in M_X} A_*(u) \otimes R(u,v).$$

This semantically defined inference may properly be understood as deduction. Namely, the last equality means that the formula $(\forall y)(B_*(y) \leftrightarrow (\exists x)(A_*(x) \otimes R(x,y)))$ of the predicate logic BL$\forall$, denoted $Comp$, assumes the truth degree 1 in $\mathbf{M}$. Hájek then shows that the formula

$$Comp \otimes A_*(X) \otimes R(X,Y) \rightarrow B_*(Y)$$

is provable in BL$\forall$. Hence, if one assumes $Comp$, then $B_*(Y)$ is provable from $A_*(X)$ and $R(X,Y)$, which yields the deduction rule

$$\frac{Comp, \ A_*(X), \ R(X,Y)}{B_*(Y)}.$$

Moreover, one may prove that if $Comp$ is true in $\mathbf{M}$ then B_* is the smallest of the fuzzy sets B_o for which the inference is sound in that $A_*(u) \otimes R(u,v) \leq B_o$. This is how the compositional rule of inference may be understood as a semantic counterpart of syntactic deduction. Zadeh's *generalized modus ponens* may in a similar manner be understood as deduction, namely as a semantic counterpart of the deduction rule

$$\frac{Comp_{MP}, \ A_*(X), \ A(X) \rightarrow B(Y)}{B_*(Y)},$$

where $Comp_{MP}$ is the formula $(\forall y)(B_*(y) \leftrightarrow (\exists x)(A_*(x) \otimes (A(x) \rightarrow B(y))))$. This is a valuable approach since without it, whether and how the suggestively termed Zadeh's generalized *modus ponens* is indeed related to deduction in the standard logical sense is not clear. Hájek's analyses make an important point—several notions of fuzzy logic in the broad sense may be clarified in the framework of fuzzy logic in the narrow sense whereby their logical nature thus becomes apparent. Still, the role of fuzzy logic in the narrow sense is wider than the possibility just discussed to recast and clarify notions and we examine this further in chapter 5.

---

[201] Hájek suggests using "variate" to avoid confusion with the term "variable" in the sense of predicate logic.

4.6.2 Computability and complexity

Questions of the effectiveness of formal systems, studied nowadays within the field of computability and computational complexity, go back at least to Leibniz's *characteristica universalis* and *calculus ratiocinator*—a universal language and a framework for logical computations.[202] Computability—the basic question of which is what can in principle be computed—has its roots in the fundamental discoveries of Gödel and Turing on the undecidability of first-order logic (p. 139 and n. 77 on p. 139). Complexity concerns the resources—time and space (i.e. computer memory)—needed for solving a particular problem. Early contributions to complexity include the groundbreaking work by Hartmanis and Stearns, who provided definitions of time and space complexity and proved some fundamental theorems.[203] The field started to flourish rapidly after the publication of the famous result on NP-completeness proved independently by Cook and Levin, and that of Karp establishing NP-completeness of 21 other important combinatorial problems.[204] Logic offers many important computational problems which have been thoroughly examined.

Results on complexity problems for fuzzy logic form a rich body, which started with Scarpelini's theorem followed by Belluce's results (p. 152, p. 153 above). The result of Scarpellini has later been strengthened by Ragaz (1981, 1983), who proved that the set of tautologies of the predicate Łukasiewicz logic is Π_2-complete.[205] For propositional fuzzy logics, we already mentioned the NP-completeness of satisfiability of the Łukasiewicz logics due to Mundici (1987). Complexity results form a substantial part of the early paper on fuzzy logic by Hájek (1995a) and have been vigorously pursued ever since.

We now present selected results, beginning with propositional fuzzy logics. As in the classical case, satisfiability of logical formulas is the basic problem here. The situation, however, is different from classical logic in several respects. One is that in classical logic, φ is not satisfiable iff $\neg\varphi$ is a tautology. An easy consequence is that deciding satisfiability is just as difficult as deciding tautologicity. In fuzzy logic, this is different. For this reason, one distinguishes the following basic problems for a propositional logic L and the notion of a standard algebra:[206] SAT (satisfiability)

---

[202] We assume familiarity with the notions used below, covered for example in M. Sipser, *Introduction to the Theory of Computation*, 2nd ed. (Boston: Thomson Course Technology, 2006); D. Kozen, *Theory of Computation* (London: Springer, 2006); H. Rogers Jr., *Theory of Recursive Functions and Effective Computability* (Maidenhead: McGraw-Hill, 1967). For Leibniz, see L. E. Loemker, ed., *Leibniz: Philosophical Papers and Letters* (Dordrecht: Reidel, 1969).

[203] J. Hartmanis and R. E. Stearns, "On the computational complexity of algorithms," *Trans AMS* 117 (1965): 285–306.

[204] S. A. Cook, "The complexity of theorem proving procedures," in *Proc ACM Symp Theory of Computing* (New York: ACM, 1971), 151–58. L. A. Levin, "Универсальные задачи перебора" [Universal search problems], *Problemy Peredachi Informatsii* 9, no. 3 (1973): 115–16. R. M. Karp, "Reducibility among combinatorial problems," in *Complexity of Computer Computations*, ed. R. E. Miller and J. W. Thacher (New York: Plenum Press, 1972), 85–103.

[205] Alternative proofs were obtained later; see M. Goldstern, "The complexity of fuzzy logic," http://arxiv.org/abs/math/9707205 (1997), accessed August 1, 2015, and Hájek 1998b.

[206] Standard algebras for particular fuzzy logics are discussed earlier in this chapter.

logic	SAT	SAT_{pos}	TAUT	TAUT_{pos}
Ł∀	Π_1-c.	Σ_2-c.	Π_2-c.	Σ_1-c.
G∀	Π_1-c.	Π_1-c.	Σ_1-c.	Σ_1-c.
Π∀	n.-ar.	n.-ar.	n.-ar.	n.-ar.
BL∀	n.-ar.	n.-ar.	n.-ar.	n.-ar.
SMTL∀	Π_1-c.	Π_1-c.	Σ_1-c.	Σ_1-c.
ΠMTL∀	Π_1-h.	Π_1-h.	Σ_1-h.	Σ_1-h.
IMTL∀	Π_1-c.	Π_1-c.	Σ_1-c.	Σ_1-c.
MTL∀	Π_1-c.	Π_1-c.	Σ_1-c.	Σ_1-c.
classical∀	Π_1-c.	Π_1-c.	Σ_1-c.	Σ_1-c.

Table 4.1: Complexity results for predicate fuzzy logics with standard semantics ("c." and "n.-ar." stand for "complete" and "nonarithmetical").

involves deciding for a given formula φ whether there exists a standard algebra $\mathbf{L}$ and an L-evaluation e for which $\|\varphi\|_e = 1$; SAT_{pos} is a modification in which one replaces $\|\varphi\|_e = 1$ by $\|\varphi\|_e > 0$; TAUT (tautologicity) consists of deciding for φ whether for each standard algebra $\mathbf{L}$ and each L-evaluation e one has $\|\varphi\|_e = 1$; TAUT_{pos} is again a modification with $\|\varphi\|_e > 0$. As usual, we identify these problems with sets of formulas for which the answer is positive, i.e. SAT is the set of all satisfiable formulas, etc. The complexity results for the SAT-problems are the following. SAT and SAT_{pos} are NP-complete for Ł, G, Π, BL, SMTL, ΠMTL, and NP-hard for IMTL and MTL. For Ł, the result is due to Mundici (1987); for G and Π see Baaz et al. 1998; for BL see Baaz et al. 2001; for SMTL see Horčík and Terui 2011; for ΠMTL see Horčík 2006; for IMTL and MTL see Blok and van Alten 2002, and Horčík and Terui 2011. For a predicate fuzzy logic L∀ with a notion of standard algebra, the corresponding problems are defined as in the propositional case except L-evaluations e are replaced by safe $\mathbf{L}$-structures. Thus, SAT consists of deciding for a sentence φ whether there exists a standard algebra $\mathbf{L}$ and a safe $\mathbf{L}$-structure $\mathbf{M}$ for which $\|\varphi\|_{\mathbf{M}} = 1$, etc. While for propositional logics, the problems are computationally expensive but decidable, for predicate logics they are undecidable, as in the case of classical logic. One is therefore interested in degrees of undecidability in terms of arithmetical hierarchy and in terms of the well-known classes Σ_i and Π_i of sets of natural numbers defined by formulas of arithmetic. The results obtained in arithmetical complexity of predicate fuzzy logics are shown in table 4.1. Notice that while G∀ is as undecidable as the classical predicate logic, Ł∀ is more undecidable and Π∀ is even nonarithmetical. In fact, the result for SAT of Π∀ was the first nonarithmeticity result in fuzzy logic (Hájek 2001a). Notice also the surprising difference of BL∀ from MTL∀. Of the many papers on complexity of predicate fuzzy logics, we emphasize Hájek's monograph (1998b) as well as Hájek 2001a, 2004b, Montagna 2001, 2005, and Montagna and Ono 2002, which may be regarded as pioneering works, where most of the results in table 4.1 are proved. The result for TAUT_{pos} of Ł$\forall$ is due to Mostowski (1961a). In addition to the above problems, their variations and other problems have also been considered, for which see Haniková 2011 and Hájek, Montagna, and Noguera 2011.

A different direction is taken in the recent paper by Hájek (2013c) on Gödel's incompleteness theorems for a certain weak arithmetic over BL$\forall$. The problems thus set represent fundamental questions regarding fuzzy logic as a possible foundation for mathematics based on it and their future examination is highly important.

Another fundamental line of research in computability aspects of fuzzy logic has been pursued by Gerla and Biacino. Their results have appeared in several papers including Biacino and Gerla 1987, 1989, 2002, Gerla 1982, 1987, 1989, 2006, as well as in the book Gerla 2001. Related results, often based on different versions of the fuzzified notions involved, are contained in several works including Li 2009, Syropoulos 2014, and Wiedermann 2002. Generally speaking, the aim is to examine fundamental computability notions from a fuzzy logic point of view, such as recursive and recursively enumerable set or arithmetical hierarchy. The first contributions of this sort are due to Santos (1970, 1976) who, inspired by Zadeh 1968b, introduced the notions of Turing fuzzy algorithm, Markov normal fuzzy algorithms and fuzzy programs, and proved that they are equivalently strong. These generalizations yield a corresponding notion of recursively enumerable fuzzy set. Different definitions are due to Harkleroad (1984). Biacino and Gerla (1987) provide generalizations of Santos's and Harkleroad's notions, which seem more appropriate. To illustrate the character of these investigations, let a set U admit a coding, i.e. a computable one-one function of U to $\mathbb{N}$. A fuzzy set $A : U \rightarrow [0,1]$ is *recursively enumerable* if there exists a recursive function $h : U \times \mathbb{N} \rightarrow [0,1]_{\mathbb{Q}}$, where $[0,1]_{\mathbb{Q}} = [0,1] \cap \mathbb{Q}$, such that for each $u \in U$, $h(u,n)$ is increasing in n and $\lim_{n\to\infty} h(u,n) = A(u)$. Such a definition enjoys some interesting properties. For instance, A is recursively enumerable iff the set $\{\langle u,a \rangle \in U \times [0,1]_{\mathbb{Q}} \mid A(u) > a\}$ is classically recursively enumerable. Moreover, recursively enumerable fuzzy sets in U are closed under finite min-intersections and max-unions. Interestingly, an ordinary subset $V \subseteq U$ is an a-cut of some recursively enumerable fuzzy subset iff it belongs to Π_2 in the arithmetical hierarchy. On the other hand, recursive enumerability of a fuzzy set is not equivalent to (ordinary) recursive enumerability of its a-cuts: there exists a recursively enumerable fuzzy set whose a-cuts are not recursively enumerable, as well as a fuzzy set that is not recursively enumerable but has recursively enumerable a-cuts. Interesting considerations regard decidability of fuzzy sets as well.

Biacino and Gerla introduce a kind of arithmetical hierarchy for fuzzy sets and thus obtain classes of fuzzy sets, Σ_i, Π_i, $i = 0, 1, \ldots$ They follow the alternative classical definition of the hierarchy, define Σ_0 and Π_0 to be the class of fuzzy relations with truth degrees in $[0,1]_{\mathbb{Q}}$ which are computable as functions and define Σ_i and Π_i by projections. For instance, if $A \in U^{p+1} \rightarrow [0,1]$ is in Σ_i, then its inf-projection

$B \in U^p \to [0, 1]$ is in Π_{i+1} where $B(u_1, \ldots, u_n) = \inf\{A(u_0, u_1, \ldots, u_n) \mid u_0 \in U\}$, and similarly for sup-projections which lead from Π_i to Σ_{i+1}. A number of ordinary properties carry over to this hierarchy, but conceptually new phenomena, hidden in the classical case, emerge due to the involvement of infima, suprema, and limits.

Questions of effectiveness of abstract fuzzy logics are examined in Biacino and Gerla 2002, and Gerla 2006. They call $C : [0,1]^{\mathscr{F}} \to [0,1]^{\mathscr{F}}$ an *enumeration operator* if it is determined by some recursively enumerable set of fuzzy sequents, i.e. pairs $\langle A, \varphi \rangle$ where A is a finite fuzzy set of formulas and φ a formula. They prove that these operators are continuous (p. 174 above), preserve recursive enumerability, and provide further results. They also call an abstract fuzzy logic *effective* if its closure operator is an enumeration operator, opening the way to important problems. For instance, they consider *effective fuzzy Hilbert systems* (p. 175 above) as those $\langle A, \mathscr{R} \rangle$ for which the fuzzy set A of axioms is recursively enumerable, the syntactic parts R_{syn} of rules in $\mathscr{R}$ are effectively coded recursive functions, and the semantic parts R_{sem} are certain extensions of effectively coded partial recursive functions d in $[0,1]_{\mathbb{Q}}$. In particular, the extension R_{sem} of an n-ary d is defined for any reals $a_i \in [0,1]$ by $R_{\mathrm{sem}}(a_1, \ldots, a_n) = \sup\{d(q_1, \ldots, q_n) \mid q_i < a_i \text{ and } d \text{ is defined for } \langle q_1, \ldots, q_n \rangle\}$. In fact, any continuous t-norm taken as R_{sem} may be obtained this way. Biacino and Gerla obtained the following counterpart of theorem 4.36.

Theorem 4.63. A fuzzy closure operator C is effective iff there exists an effective fuzzy Hilbert system $\langle A, \mathscr{R} \rangle$ such that $C = C_{A,\mathscr{R}}$.

Scarpellini's theorem 4.25 yields the seemingly paradoxical fact that there exists an effective fuzzy Hilbert system with a decidable fuzzy set of axioms whose fuzzy set of provable formulas, i.e. $C_{A,\mathscr{R}}(\emptyset)$, is a recursively enumerable fuzzy set the 1-cut of which is, however, not a recursively enumerable set. This fact, however, is due to the nature of the notion of recursive enumerability of fuzzy sets.

4.6.3 Further developments

Similarity

Similarity is one of the most important phenomena present in human reasoning. Interesting models of similarity are obtained by fuzzifying classical axioms of identity. The notions thus resulting along with their logical ramifications are examined in chapter 5. Here we mention two logics in which similarity plays a fundamental role. Both are inspired by the question of how to handle formally intuitively appealing inference patterns involving similarity such as: from "φ is approximately true" and "φ approximately implies ψ" infer "approximately ψ." While this type of reasoning is nicely captured by Goguen's conception and Pavelka-style logics, a different account, initiated by Ruspini (1991), has been explored in Esteva, Garcia-Calvés, and Godo 1994 and in Dubois et al. 1997, and cast in a multimodal logic setting in Esteva et al. 1997, which is examined in section 4.5.4. Note also that similarity-based semantics for nonmonotonic inference is examined in Godo and Rodríguez 2002. The second approach was proposed by Ying (1994) as a logic wherein one may apply inference rules even if their premises are only approximately matched, which requires that a similarity be defined on formulas. To assess the approximate match, one assumes that there is a similarity defined on formulas of classical logic. For instance, if φ' and φ are similar to degree 0.8, one may infer ψ to degree 0.8 from φ' and $\varphi \rightarrow \psi$. Such logic was later developed as a particular Pavelka-style logic in Biacino, Gerla, and Ying 2000.

Proof theory and automated proving

Considerable effort in logic has been devoted to automated proving.[207] For this purpose, Hilbert-style logics, which we have already examined, are not convenient. Instead, various other kinds of calculi have been proposed—sequent calculi, tableaux methods, and resolution-based methods.[208] These problems have a clear application appeal and, in fact, logic programming—a resolution-based system of automated theorem proving—has became very popular in the area of expert systems and has been used in many real-world applications. Some approaches advocated for logic programming in the setting of fuzzy logic are discussed in section 6.8. An early investigation of sequent calculi in many-valued logic is due to Schröter (1955); see p. 157 above. Sequent calculi for fuzzy logics were studied in a number of papers starting with Avron 1991, which examined Gödel logic. Further results were obtained in Metcalfe, Olivetti, and Gabay 2004 for the product logic, in Metcalfe, Olivetti, and Gabay 2005 for Łukasiewicz logic, in Baaz, Ciabattoni, and Montagna 2004 for MTL, and in Bova and Montagna 2008 for BL. See also Metcalfe and Montagna 2007, the book Metcalfe, Olivetti, and Gabay 2008 devoted to proof theory for fuzzy logic, Metcalfe 2011, as well as Gottwald 2001, and Dubois et al. 2007. It should be mentioned that in this framework it is necessary to work with hypersequents, which are finite multisets of sequents introduced by Avron.[209] Early works on tableaux methods for many-valued logics include Carnielli 1987. These methods have been further developed by Hähnle (1994) and embrace Łukasiewicz logic.

Algebraic structures of truth degrees

It is clear from many examples in this chapter that fuzzy logic in the narrow sense has an algebraic character. Various algebras, such as MV-algebras, BL-algebras, or

---

[207] See e.g. M. Fitting, *First-Order Logic and Automated Theorem Proving*, 2nd ed. (New York: Kluwer, 1996).

[208] Sequent calculi were invented by G. Gentzen, "Untersuchungen über das logische Schließen. I, II" [Investigations on logical consequence. I, II], *Mathematische Zeitschrift* 39 (1935): 176–210, 39 (1935): 405–31. Tableaux methods were introduced by E. W. Beth, "Semantic entailment and formal derivability," *Mededelingen der Koninklijke Nederlandse Akademie van Wetenschappen, Afdeling Letterkunde* 18, no. 3 (1955): 309–42. Resolution-based methods are due to M. Davis and H. Putnam, "A computing procedure for quantification theory," *J ACM* 7, no. 3 (1960): 201–15; and J. A. Robinson, "A machine-oriented logic based on the resolution principle," *J ACM* 12, no. 1 (1965): 23–41.

[209] A. Avron, "A constructive analysis of RM," *J Symb Logic* 52, no. 4 (1987): 939–51.

MTL-algebras, which in fuzzy logic replace the Boolean algebras of classical logic, are fundamental in answering questions pertaining to fuzzy logics. Apart from playing this role, these algebras appear to be interesting mathematical objects per se, as are Boolean algebras—a view propounded by Mundici since his early works on MV-algebras (p. 183ff.). It is therefore natural that algebras of fuzzy logic have been the subject of numerous, continuing investigations. Good resources include Cignoli, D'Ottaviano, and Mundici 2000; Di Nola and Leuştean 2011; and Mundici 2011 for MV-algebras; Busaniche and Montagna 2011 for BL-algebras; and Galatos et al. 2007 for residuated lattices. There has also been considerable research on t-norms and other aggregation operations; see, for instance, Klement, Mesiar, and Pap 2000 and Grabisch et al. 2009.

Representation of functions by fuzzy logic connectives

We have already mentioned several results on functional completeness and the representation of functions by logical connectives obtained for Łukasiewicz logic and general many-valued logics (section 4.2.1 and pp. 157ff.; see also p. 184). The problems at stake have later been considered for several other fuzzy logics, including the Gödel logic (Aguzzoli and Gerla 2002, Gerla 2000) and the product logic (Aguzzoli and Gerla 2002, Cintula and Gerla 2004). For a recent exposition see Aguzzoli, Bova, and Gerla 1998.

Fuzzy logics in the context of other logics

From a broader perspective, fuzzy logics may be seen as particular members of a broad class of logics called *substructural logics*. These logics include the fuzzy logics discussed above—Ł, G, Π, BL, MTL and their extensions—as well as ML (p. 186) and its extensions such as intuitionistic logic, linear logic, and other logics including classical full Lambek calculus, relevance logic, and, clearly, classical logic. For more information about these, see Galatos et al. 2007 and the references therein.

Higher-order and other fuzzy logics

In addition to the predicate logics examined above, two kinds of higher-order fuzzy logics have been worked out and we present them briefly in section 5.2.2. Fuzzifications of various bivalent logics, such the intuitionistic fuzzy logic of Takeuti and Titani (1984) mentioned on p. 185 above or the fuzzy modal logics mentioned in section 4.5 have also been subject of a number of contributions. Worth noting is the book by Gabbay (1999) whose chapter 12, entitled "How to make your logic fuzzy," applies the method of fibring for this purpose.

Chapter 5

Mathematics Based on Fuzzy Logic

5.1 Principal issues and outline of development

5.1.1 What is mathematics based on fuzzy logic?

Throughout history, various approaches to mathematics have been made according to its conception and subject matter. Among them, classical mathematics, also called standard mathematics, undoubtedly represents the mainstream approach.[1] Classical mathematics is based on sets and classical logic. This means that classical mathematics studies notions that may be defined in terms of the notion of set, and that mathematical reasoning is governed by the laws of classical logic. Classical mathematics thus, ipso facto, admits the principle of bivalence—membership in sets as well as validity of mathematical statements about sets, and thus about all mathematical objects, are two-valued.

With the acceptance of intermediate truth degrees the situation changes substantially—statements about mathematical objects now admit degrees of truth. As these include statements about membership, the objects include fuzzy sets. Moreover, since classical logic is not appropriate for reasoning with statements that admit degrees of truth, it needs to be replaced by a new one and fuzzy logic is an obvious candidate in this respect. The admission of truth degrees thus has broad ramifications, brings about nontrivial considerations, and—as is shown in this chapter—leads to an important and interesting mathematics which we call *mathematics based*

---

[1] The others include mainly intuitionistic mathematics and various versions of constructivism. See, for example, S. Shapiro, ed., *The Oxford Handbook of Philosophy of Mathematics and Logic* (New York: Oxford Univ. Press, 2005) and p. 129 above.

on fuzzy logic.[2] Our aim in this chapter is to provide a comprehensive examination of this subject as well as an overview of its development.

Before we proceed, let us make a few important remarks to prevent possible misunderstanding.[3] First, by admitting degrees of truth and hence admitting fuzzy sets, one does not exclude from consideration ordinary sets nor the other objects of classical mathematics. Consequently, notions and statements of mathematics based on fuzzy logic may concern not only fuzzy sets but also ordinary sets and classical structures. Second, the statements about fuzzy sets and fuzzy relations may be both many-valued, i.e. admit classical as well as intermediate truth degrees, and bivalent, i.e. admit classical truth values 0 and 1 only. An example of the former is the statement "u is in A," in which u is an element of a given universe and A is a fuzzy set in this universe. In this case, the truth degree of the statement simply equals $A(u)$. Another example is the statement "A is included in B," provided we interpret inclusion as a many-valued relation assigning to fuzzy sets A and B in the universe U, e.g., the truth degree $\bigwedge_{u \in U} A(u) \to B(u)$, where $\to$ is a (many-valued) truth function of implication. A simple example of a bivalent statement about fuzzy sets is the statement "the membership degree of u in A exceeds 0.3." Another one is the above statement "A is included in B," provided we interpret inclusion as a bivalent relation that applies to A and B if $A(u)$ is less than or equal to $B(u)$ for each u in U and does not apply otherwise. Third, even statements about ordinary sets may be many-valued as well as bivalent. The latter is clear—examples are statements in classical mathematics such as "A contains at least one element." As an example of the former, consider a subset A of a given universe U. We may then consider the statement "A is large" as a many-valued statement having the truth degree $\frac{|A|}{|U|}$.

Thus conceived, mathematics based on fuzzy logic naturally extends classical mathematics.[4] Clearly, a part of mathematics based on fuzzy logic has to deal with fuzzy sets themselves. This part—which may aptly be called a theory of fuzzy sets— is of fundamental importance because fuzzy sets appear in one way or another in all the other parts. This opens the problem of foundations of mathematics based on fuzzy logic. Not surprisingly, this problem leads to delicate considerations involving a number of topics, such as the various facets of the relationship of mathematics based on fuzzy logic to classical mathematics, intuitive and axiomatic fuzzy set theories, a possibility of other foundational approaches such as category theoretic ones, as well as the role of fuzzy logic in the narrow sense for mathematics based on fuzzy logic.

---

[2] Alternatively, one could use the term "fuzzy mathematics," which is sometimes used in the literature. Even though this term is shorter, it has inappropriate connotations—the mathematics in question has precise definitions, theorems and proofs, and is not fuzzy in this respect.

[3] We do so informally since the formal treatments of the issues involved are irrelevant at this stage and since we return to these issues in greater detail below in this chapter.

[4] This claim is to be understood in the intuitive sense explained in the preceding paragraph. Foundational aspects are discussed below in this chapter; see also section 7.4.

Note that intuitive fuzzy set theory is closely related to fuzzy logic in the broad sense.[5] Zadeh (1996a) even characterizes the latter as "almost synonymous with the theory of fuzzy sets." As mentioned in chapter 3, fuzzy logic in the broad sense actually surpasses such characterization and is concerned with methods based on fuzzy sets that emulate human reasoning. Nevertheless, intuitive fuzzy set theory has indeed been developed as part of it to a considerable extent.

In our treatment, we proceed as follows. In the rest of this section, we provide a structured conceptual view of the major issues pertaining to mathematics based on fuzzy logic—we discuss the role of foundations and applications and provide an outline of its development. In section 5.2, we examine existing contributions to foundations and in section 5.3 to selected areas of mathematics based on fuzzy logic. Section 5.4 concludes this chapter by examining various semantic questions pertaining to fuzzy logic and the mathematics based on it.

5.1.2   The problem and role of foundations

In section 2.3 we mentioned that in the 1980s, fuzzy logic and fuzzy set theory were criticized on several grounds including the claim that it lacked solid foundations. As we show in section 5.1.4, this claim was basically correct at that time because, except for a few papers, contributions to the foundations as a result of focused, systematic work only began to emerge in the 1990s. Part of our purpose in this chapter is to establish that due to these contributions, the claim that fuzzy logic lacks proper foundations is no longer justified.

This said, we need to add that this does not mean that the problem of foundations for fuzzy logic and the mathematics based on it has been completely resolved. Look at classical mathematics. As is well known, the problem of its foundations has a long history.[6] The several approaches and views which have evolved include mathematical arguments as well as, implicit or explicit, philosophical stands. These are sometimes complementary but often mutually incompatible. Even though several fundamental results have been obtained and several conceptions have been developed, the problem of foundations of classical mathematics is too subtle to assert that it has been satisfactorily settled. Given the groundbreaking character of fuzzy logic's rejection of the principle of bivalence and its admission of degrees of truth, it is apparent that the same is true of foundations of mathematics based on fuzzy logic. To put it bluntly, this matter cannot be satisfactorily resolved for everyone, on principle. On the other hand, it is legitimate to examine the extent to which the foundations of mathematics based on fuzzy logic have been developed, which is our goal here.

---

[5] See section 5.1.2.

[6] See e.g. E. W. Beth, *The Foundations of Mathematics*, rev. ed. (Amsterdam: North-Holland, 1964); Mancosu 1998; and S. Shapiro, ed., *The Oxford Handbook of Philosophy of Mathematics and Logic*, op. cit.

We start by returning to the above-mentioned claims from the 1980s regarding lack of foundations. To be more specific, the proponents of such claims argued that while there existed a growing number of papers exploring various mathematical theories involving the notion of fuzzy set, and while these papers frequently used the term "fuzzy logic," it was not clear what fuzzy logic actually was from a logical perspective. This was in sharp contrast to the situation with classical mathematics which is rooted in classical logic. This point was valid in the 1980s, though it must be added that important contributions, such as those of Goguen, Pavelka, and even earlier contributions to many-valued logic (sections 4.3 and 4.2), were available, albeit perhaps not well known at that time. The point, nevertheless, has lost its validity because, as shown in chapter 4, fuzzy logic in the narrow sense has undergone rapid development since the 1990s and now represents a coherent body of logical calculi. Some of them provide nontrivial generalizations of established bivalent logics, including classical logic. Some—such as the non-truth-functional ones—represent an even more radical departure from classical logic.

To proceed, we need to recall the key issues pertaining to foundations of classical mathematics. Modern foundational studies go back to the development of calculus. Beginning with the 17th-century explorations by Newton and Leibniz, calculus at first relied on intuitive geometric ideas. With investigation of more and more advanced topics, this became untenable. Calculus had to be put on a solid basis. Around 1830, Cauchy made a significant step forward by developing a new way of proving results in calculus rigorously using a small number of properties of real numbers. Nevertheless, the very concept of real number was still approached via geometric intuition about the real line. The first formal definitions of the concept of real number, which both appeared in 1872, are due to Richard Dedekind (1831–1916) and Georg Cantor (1845–1918). According to Dedekind, real numbers are certain sets, defined by so-called cuts of rational numbers, while Cantor defined them as equivalence classes of certain sequences of rationals. Both definitions thus relied on the notion of set. Due to Dedekind's and Cantor's work, reducing mathematical concepts to sets eventually became a standard practice in mathematics. The first systematic considerations regarding sets are actually due to Bernard Bolzano (1781–1848).[7] His work, interesting particularly due to his ideas about infinity, did not become widely known. The birth of modern set theory is marked by Cantor's 1874 paper in which he establishes the denumerability of the set of all algebraic numbers, but the nondenumerability of the set of all real numbers, as well as a series of other papers he published over the following decade, culminating with his early monograph *Grundlagen einer allgemeinen Mannigfaltigkeitslehre*.[8] Although Can-

---

[7] *Wissenschaftslehre* [A theory of science] (Sulzbach: Seidelsche Buchhandlung, 1837); *Paradoxien des Unendlichen* [Paradoxes of the infinite] (Leipzig: C. H. Reclam, 1851).

[8] "Ueber eine Eigenschaft des Inbegriffes aller reellen algebraischen Zahlen" [On a property of the collection of all real algebraic numbers], *J für die reine und angewandte Mathematik* 77 (1874): 258–62. J. W. Dauben, *Georg Cantor. His Mathematics and Philsophy of the Infinite* (Cambridge, MA: Harvard University Press, 1979).

tor proved many results about sets, the concept of set itself was conceived in an intuitive manner and was usually described in philosophical terms such as an aggregate or totality at that time. Cantor himself defined a set as "any collection M of definite well-distinguished objects m of our perception or our thought (which are called the 'elements' of M) into a whole."[9] Theories of sets based upon such an intuitive understanding of the notion of set constitute *intuitive* (or *naive*) set theories but these soon led to serious difficulties. At the end of the 1890s, Cantor and Burali-Forti (among others) discovered certain contradictions, later called paradoxes, pointing to the problems when considering for example the set of all sets.[10]

In 1901, Bertrand Russell (1872–1970) discovered another such contradiction while working on his *Principles of Mathematics* (Russell 1903).[11] The contradiction was more direct than the previous ones. Since Russell included an extensive discussion of the contradiction along with an appendix providing an outline of a possible solution of the contradiction in the *Principles of Mathematics*, and since the contradiction pointed out a fundamental inconsistency in Gottlob Frege's (1848–1925) monumental *Grundgesetze der Arithmetik*,[12] the paradoxes became an important impetus for further examination of foundations of mathematics. *Russell's paradox* may be explained as follows. Consider the set n of all sets that are not members of themselves, i.e. the set $n = \{x \mid x \notin x\}$, and call the sets in n normal. Is n normal? That is, do we have $n \in n$? According to classical logic, we either have $n \in n$ or $n \notin n$. If $n \in n$, then n being an element of n must have the property of the elements in n, i.e. $n \notin n$—a contradiction. If $n \notin n$, then n satisfies the property of the elements in n, whence we have $n \in n$—a contradiction again.[13] In response to such paradoxes and the ensuing foundational crisis, three basic approaches developed. One is represented by intuitionism (p. 129) which is not relevant for our purposes here. The other two are the theory of types and axiomatic set theory.

The *theory of types* was proposed by Russell in appendix B to the *Principles of Mathematics*, which was entitled "The doctrine of types." Russell added further developments in 1908.[14] The basic idea is that mathematical objects are assigned types: the primary objects are of type 0, their properties are of type 1, the properties of these properties are of type 2, etc., and no properties that do not belong to the thus-established hierarchy are allowed. Under this view, Russell's paradox disap-

---

[9] "Beiträge zur Begründung der transfiniten Mengenlehre," *Mathematische Annalen* 46 (1895): 481–512. trans. P. E. B. Jourdain, *Contributions to the Founding of the Theory of Transfinite Numbers* (Chicago: Open Court, 1915).

[10] They proved that the set of all ordinal numbers is well ordered. Therefore, its order type should be an ordinal number larger than all ordinal numbers in this set, establishing the contradiction. Cantor also obtained the related contradiction regarding the set of all cardinal numbers.

[11] According to B. Russell, "My Mental Development," in *The Philosophy of Bertrand Russell*, 3rd ed., ed. P. A. Schilpp (New York: Tudor, 1951), 3–20.

[12] [The basic laws of arithmetic] (Jena: Verlag Hermann Pohle, Band I 1893, Band II 1903).

[13] In fuzzy logic, such reasoning does not yield a contradiction. The question of whether Cantor-style intuitive fuzzy set theory is consistent is an interesting open problem (p. 254 below). See also section 5.4.2 on paradoxes.

[14] "Mathematical logic as based on the theory of types," *American J Math* 30 (1908): 222–62.

pears since the property on which it is based is not allowed. The theory of types later underwent considerable further development by Carnap, Ramsey, Church, Quine, Tarski, and others.[15] This led to a variety of formal systems, including various higher-order logics which are considered as offering possible foundations for classical mathematics. Developments of such systems from the point of view of fuzzy logic are examined in section 5.2.2.

The second approach to eliminating the paradoxes of set theory is based on the idea that the theory of sets needs to be based on suitable axioms. This eventually led to *axiomatic set theory*—an extensively studied field connected to many fundamental discoveries in logic and mathematics.[16] The first set of such axioms is due to Ernst Zermelo (1871–1953).[17] Zermelo, who also independently discovered Russell's paradox, proposed several axioms including what is now called the axiom of *restricted comprehension* (or *separation*), which basically says that for every set y and every property, there exists a set containing exactly the objects of y having this property. The argument of Russell's paradox then only yields that the set of all sets does not exist. Namely, if this set, say v, existed, then taking "$x \notin x$" as the property of x, the axiom of separation yields the existence of the set $\{x \mid x \in v \text{ and } x \notin x\}$. Denoting this set n, we obtain, as in Russell's paradox, that $n \in n$ if and only if $n \notin n$. Zermelo's system of axioms was further extended independently by Adolf Abraham Fraenkel (1891–1965) and Thoralf Skolem (1887–1963) who proposed to add the so-called axiom of *replacement* (saying that the image of a set under a definable function is a set). They also suggested treating Zermelo's axiomatic set theory as a theory in predicate logic with identity and $\in$ as its only nonlogical symbols, and suggested that the notion of property be conceived as a formula with one free variable—this is basically how the theory is still considered today. Zermelo's axioms further included the axioms of *extensionality* (two sets are equal iff they contain the same elements), *elementary sets* (there exists a set containing no elements—the empty set; for every set x there exists a set containing just x; for all sets x and y there exists a set containing exactly x and y), *union* (for every set x there exists a set whose elements are just the elements of the elements of x), *powerset* (for every set x there exists a set whose elements are just the subsets of x), *infinity* (there exists a set x containing the empty set such that if $y \in x$ then $y \cup \{y\} \in x$, i.e. there exists an infinite set), and also the problematic axiom of *choice* (for every set s there exists a function f such that for each nonempty $x \in s$, $f(x)$ belongs to x, i.e. f selects from x one of its elements). Replacing this axiom by the so-called axiom of *regularity* (or *foundation*: for every nonempty set x there exists its element which is disjoint with x) proposed in 1917 by Dmitri Mir-

---

[15] J. Collins, *A History of the Theory of Types: Developments after the Second Edition of 'Principia Mathematica'* (Saarbrücken: Lambert Academic Publ., 2012).

[16] On the development of set theory, see J. Ferreirós, *Labyrinth of Thought: A History of Set Theory and Its Role in Modern Mathematics* (Basel: Birkhäuser, 2007). For a comprehensive account of axiomatic set theory, see Jech 2003.

[17] "Untersuchungen über die Grundlagen der Mengenlehre. I" [Investigations on the foundations of set theory. I], *Mathematische Annalen* 65 (1908): 261–81.

imanov (1861–1945) and further examined by John von Neumann (1903–1957), one obtains the system now called Zermelo's system and denoted Z. Adding to it the axiom of *replacement*, one obtains what is now called the *Zermelo-Fraenkel system* and denoted ZF. By ZFC one denotes ZF with the axiom of choice. The Zermelo-Fraenkel system is widely accepted, often with the axiom of choice, as a foundation for classical mathematics. Another system was worked out by Bernays who built on the previous work of von Neumann. Bernays's axiom system was further developed by Gödel, i.e. the *Gödel-Bernays* (also von Neumann-Gödel-Bernays) *system* denoted GB. The Gödel-Bernays system is formalized as essentially containing two kinds of variables, one for sets and one for classes (sets are classes that are elements of other sets). ZF and GB may be considered as equally strong in that GB is a conservative extension of ZF—a statement in the language of ZF is provable in ZF if and only if it is provable in GB. We examine axiomatizations of fuzzy sets inspired by ZF and GB in section 5.2.3. In addition to ZF and GB, several other axiomatic set theories exist.[18]

Much later, another approach to the foundations of mathematics emerged— *category theory*, pioneered by Samuel Eilenberg (1913–1998) and Saunders Mac Lane (1909–2005). Category theory, which since then has been considerably developed, represents the viewpoint that mathematics is the study of mathematical structures and structure-preserving mappings. Category-theoretic foundations of fuzzy logic are examined in section 5.2.4.

Partly in response to the problems with intuitive set theory and the intuitive way of doing mathematics, David Hilbert (1862–1943) proposed in the early 1920s what is now called *Hilbert's program*—a proposal to build firm foundations of mathematics in the form of axiomatic systems for various parts of mathematics such that any true mathematical statement be provable and such that the consistency of the system be provable by finitary methods.[19] He hoped to solve thereby "the foundational questions in mathematics as such once and for all" (Mancosu 1998, 228). That this goal cannot be attained became clear when Gödel published his famous *incompleteness theorems*.[20] Gödel's first incompleteness theorem implies that no consistent extension of Zermelo-Fraenkel set theory, or any other theory strong enough to formalize arithmetic, decides any formula, i.e. is such that for any formula φ, either φ or $\neg\varphi$ is provable. His second incompleteness theorem implies that the consistency of Zermelo-Fraenkel set theory, as well as any axiomatic theory strong enough to formalize arithmetic, cannot be proved within the theory itself. This unexpected result came as a shock and revealed the fundamental limits of Hilbert's formalistic approach. Since Zermelo-Fraenkel set theory encompasses the majority of mathematical practice, there is little hope for proving its consistency. On the other hand,

---

[18] In section 7.5 we mention Vopěnka's *alternative set theory*, which is in a sense radically different from classical set theories in that its purpose is to account for unsharply delineated collections, being thus relevant to fuzzy sets.

[19] van Heijenoort 1967 and Mancosu 1998.

[20] "Über formal unentscheidbare Sätze der Principia Mathematica und verwandter Systeme I"; see n. 77 on p. 139.

even though Hilbert's program is not feasible in its full scope, many further developments in logic and foundations of mathematics may certainly be regarded as its reasonable continuation despite these limitations.[21]

For an ordinary mathematician, who is not a specialist in set theory or logic, foundational aspects need not be and usually are not of fundamental concern. Part of the reason is that in most of classical mathematics as it is practiced, foundational concerns—the paradoxes, incompleteness, etc.—never play a role. Most ordinary mathematicians do not work in terms of axiomatic set theory. Instead, they regard sets as intuitive, naive concepts and tacitly assume that if needed, all definitions and results can be phrased and proved in the framework of ZF or another axiomatic version of set theory. The fact that such theories are widely believed to be consistent and are broadly accepted as providing a reasonable foundation justifies the view that mathematics proceeds on a secure basis.

As for the problem of foundations of mathematics based on fuzzy logic, the fundamental question is whether or not reasonable foundations may be developed. Here we are not talking about the interpretation of this question as discussed on p. 234, namely whether there is any solid logic to which the term "fuzzy logic" refers. This question has been answered affirmatively above. Rather, we consider foundations in the sense just outlined. Classical mathematics naturally offers directions for possible answers and these are explored in detail in section 5.2. As we shall see, a number of interesting views and approaches have been developed which justify the claim that reasonable foundations for mathematics based on fuzzy logic exist or, rather, may be developed. Some of these contributions are quite recent and the whole effort is still in progress. Given the groundbreaking character and diverse ramifications of the admission of degrees of truth by fuzzy logic, no doubt some aspects of the problem of foundations, perhaps even fundamental ones, have yet to be thoroughly explored. We return to this point in chapter 7.

There is also the additional question of what significance such possible foundations may have for researchers working in mathematics based on fuzzy logic or even for researchers who just use it. The basic answer is the same as with classical mathematics—they provide a secure basis. In addition, however, there is an important practical aspect of the role of fuzzy logic in the narrow sense, apparent to a much lesser extent in classical mathematics, which we examine in section 5.2.1.

5.1.3 The problem and role of applications

No less important than foundations are applications. Here we are not concerned with particular real-world applications of fuzzy logic and mathematics based on it— these are surveyed in chapter 6. Our purpose here is to examine the role of applications for the development of mathematics based on fuzzy logic.

---

[21]R. Zach, "Hilbert's program then and now," in *Philosophy of Logic. Handbook of the Philosophy of Science*, vol. 5, ed. D. Jacquette (Amsterdam: Elsevier, 2006), 411–47.

As we saw in chapter 2, Zadeh's primary motivation for introducing fuzzy sets was to have a notion which could be used more appropriately than classical sets to represent concepts such as "high temperature," "low inflation," or "being similar," which are crucial for human reasoning. Most existing applications of fuzzy logic are indeed based on modeling such concepts by fuzzy sets. In fact, the development of fuzzy logic in the broad sense has been driven by the goal to obtain mathematical models that emulate human reasoning. Mathematics based on fuzzy logic may, however, involve any kind of notion based on fuzzy sets. Any such notion is a possible object of investigation—one may form theories involving such notions and prove theorems about them. Mathematical practice is, however, different. It is not a formalistic enterprise. For a mathematical notion to be of interest, there needs to be a solid motivation for introducing it. The motivation may come from applications or from mathematics itself. While the situation is in principle the same as in classical mathematics, there is an important aspect that is new.

This aspect involves the possibility of developing mathematics based on fuzzy logic with the following reasoning. Classical mathematics developed to the current stage with its organization into areas such as algebra, geometry, topology, or calculus for a good reason. Since fuzzy sets generalize classical sets, it is desirable—the reasoning goes—to systematically "fuzzify" classical mathematics, hence to develop "fuzzy algebra," "fuzzy geometry," "fuzzy topology," and so on. Mathematics based on fuzzy logic obtained this way will have a greater scope of applicability. Perhaps even more: since classical mathematics offers tools for a broad variety of applications, it is likely that the mathematics thus obtained will offer an appropriate tool whenever a problem involving concepts with unsharp, i.e. fuzzy boundaries is encountered. As such—the reasoning concludes—the systematic generalization of classical mathematics from a fuzzy logic point of view seems a reasonable research program. Let us call this program the *fuzzifier's temptation* program. It is a matter of fact that too many existing papers may be viewed as following this program.

There are several fundamental problems with the *fuzzifier's temptation*. For one, there is a priori no guarantee that by a fuzzification of a useful mathematical notion, even though mathematically correct, one will always obtain a notion that is useful again. Good examples include the various existing fuzzifications of algebraic notions.[22] Second, what does it mean to fuzzify a certain notion, theory, or theorem? There are two aspects to this question. The first, important but only technical, is how to correctly obtain proper generalizations of classical notions, theories, and theorems. We shall show, particularly in section 5.2.1, that fuzzy logic in the narrow sense is very useful in this respect. Much more important is the second aspect, namely the problem of what is to be made fuzzy in the new notion or theory. Consider, as an example, the mathematical notion of a group. Does a fuzzification of this notion mean that we replace the support set of a group by a fuzzy set? Or shall

---

[22] We return to this and other examples later in this chapter.

we instead replace the group operation by a kind of fuzzy operation? Or just fuzzify the property of associativity or some other property of groups? As a rule, as in this case, several options exist, and knowledge of the classical notion itself does not help in answering these questions.

The point we are getting at is that both problems of the *fuzzifier's temptation* have a common cause—detachment from applications and the resulting lack of proper motivations. Proper motivations are typically derived from a perceived specific need in modeling the external world, such as formation of unsharp clusters or decision making in the presence of linguistically described information. They may also be derived from pure mathematical considerations, such as the need for a generalization of two or more existing particular notions. In any case, solid motivations are a reasonable guarantee of usefulness of the resulting theories.

Related to this issue is the possibility that when developed from applications, mathematics based on fuzzy logic will naturally have a structure different from that of classical mathematics—in some areas, fuzzification may simply make less sense than in others. Motivations derived from application needs also provide guidance for resolving the second problem of the fuzzifier's temptation. The answer to the question of how to obtain a generalized, fuzzified version of a classical notion naturally derives from the application needs at hand as well as from a careful examination of properties of the fuzzified notions and their interactions with other notions. In addition, it is often the case that in the presence of intermediate degrees, new important phenomena appear. While applications help us clarify such phenomena along with their significance and the problems to which they naturally lead, formalistic fuzzification detached from applications does not—as a rule the new phenomena are either trivial and thus hidden or do not make sense in the classical case.

Clearly, the very question of whether a given motivation is solid or not is delicate and, admittedly, a matter of degree. Our point, however, is that complete detachment from applications and a purely mechanical fuzzification of classical mathematical results is not a meaningful or appropriate way of doing mathematics based on fuzzy logic. Unfortunately, many existing contributions are of just this sort. This has certainly contributed to the wrong impression that mathematics based on fuzzy logic is simply a formalistic enterprise devoid of proper motivations and meaning. This situation has been reinforced by the emergence of too many journals and publication venues of varying quality in the last decade or two—a social phenomenon that is not specific to fuzzy logic but concerns science generally.

From the present viewpoint, the development of mathematics based on fuzzy logic is best viewed as a continuing process. This process needs to have solid foundations—a secure formal basis with its principles and results regarding both the possibilities and limitations of mathematics based on fuzzy logic. The development of particular mathematical theories is a continuing activity which should primarily be directed by application needs and justified by proper motivations rather than by a desire simply to fuzzify the whole of classical mathematics.

5.1.4 Outline of development

The emergence of mathematics based on fuzzy logic may be identified with that of fuzzy logic in the mid-1960s, which is described along with some earlier closely related approaches in section 2.2. In this section we are solely concerned with the development of mathematics based on fuzzy logic.

General aspects

We start by emphasizing a substantial difference between the development of mathematics based on fuzzy logic and other directions in mathematics, which regard foundational questions such as intuitionistic mathematics.[23] While the primary motivation for intuitionistic mathematics was the issue of foundational concerns and while its development has thus been carried out by pure mathematicians, the situation with fuzzy logic and the mathematics based on it was different. For Zadeh, the main concern was the inadequacy of classical sets to model human reasoning. Even though he repeatedly addressed various conceptual issues, he did so from the viewpoint of human reasoning and was not primarily interested in mathematical foundations. To illustrate his points as well as to make tentative first steps in developing the mathematical models he envisioned, Zadeh always began by proposing simple concepts and he strove for intuitive clarity rather than mathematical rigor and generality. This approach corresponded well to his initial purpose as well as to his background in electrical engineering and applied mathematics. Such an attitude is actually characteristic for the first period of development of mathematics based on fuzzy logic, which has to a large extent been carried out by applied mathematicians, electrical and system engineers, and computer scientists. Even though exceptions existed, foundational mathematical studies regarding fuzzy logic and the mathematics based on it only began to be systematically developed in the late 1980s and early 1990s. Those studies also revisited early contributions which—even though frequently opening conceptually new possibilities—were often ad hoc and not properly developed. The early contributions were thus polished and treated in the emerging theoretical frameworks and, importantly, developed to a fuller extent in ways that often significantly clarified the topics in question. In addition, connections to related developments were discovered and began to be examined, such as those to the many contributions to many-valued logic predating Zadeh's seminal papers. To establish such connections was important not only for historical reasons but also for a proper placement and understanding of the new contributions in the context of mathematics as a whole.

---

[23]Our comparison is not to imply any deeper parallel between mathematics based on fuzzy logic and intuitionistic mathematics. These have very different motivations and are based on different conceptions.

Early developments of mathematics based on fuzzy logic have often been subject to criticism.[24] In some respects, such criticism has reflected the specific nature of early developments as described in the previous paragraph. A frequent criticism, addressed in section 5.1.2, consisted in claiming that fuzzy logic lacked foundations. This was basically correct in the early phase and was due to the fact that foundational mathematical issues were of little concern for most of the early contributors, but eventually this point lost its validity for the reasons described in section 5.1.2.

A different point of criticism consisted in claiming that the existing contributions often ignored relevant work in mathematics.[25] In several cases, the contributors were not aware of existing related approaches, which were often developed in different areas with completely different motivations. Such unawareness is certainly natural. In later developments the previously unrecognized approaches became standard references. In addition, however, there was more serious criticism, namely that even though some contributions may have been useful for engineering applications, their quality from a mathematical perspective was questionable. In particular, it was pointed out that while the proponents present fuzzy logic and fuzzy set theory as a foundational move—as an alternative to classical logic and classical set theory—the early contributions lacked the depth and elegance of their classical counterparts as well as those of other foundational approaches. Moreover, as the criticisms went, no comparisons were made by the proponents of fuzzy logic to other existing foundational approaches, such as topos theory, which naturally offer room for examination of alternative set theories, and no references were made to the existing approaches in already published books and other works on fuzzy logic.[26] In fact, some of the early contributions to fuzzy logic did contain foundational considerations—for example, the two important papers by Goguen (1967, 1968–69) examined previously in section 4.3.1 and below in this chapter. However, such contributions were exceptions and most publications did not pay attention to foundational issues. This again may be seen as a consequence of the nature of the early days when books were written by applied mathematicians and engineers rather than experts in logic, set theory, and foundations of mathematics. Moreover, while such terms as "theory," "foundations," or "new approach" may well seem reasonable in a given context to an application-oriented researcher, they may seem wholly inappropriate and exaggerated to a researcher specializing in foundations of mathematics. Such criticisms nevertheless eventually lost their substance in large part due to the emergence of various kinds of foundational studies which are examined in chapter 4 and section 5.2.

---

[24] Criticisms of fuzzy logic are examined in chapter 2. In this section we are concerned with specific aspects pertaining to mathematics based on fuzzy logic.

[25] This kind of argument is propounded e.g. in M. A. Arbib's review of Zadeh et al. 1975, Kaufmann 1975, and Negoiţă and Ralescu 1975a in *Bull AMS* 83 (1977): 946–51; E. G. Manes's review of Kandel and Lee 1979 in *SIAM Review* 23 (1981): 271–73; and E. G. Manes's review of Dubois and Prade 1980a in *Bull AMS* 7 (1982): 603–12.

[26] The reviews in the preceding n. 25 contain examples of such criticisms.

There is, however, yet another significant phenomenon accompanying the development of mathematics based on fuzzy logic, namely the poor quality of many papers. This phenomenon, which can be observed since the early years of its development, has been critically commented upon by the opponents of fuzzy logic as well as by its contributors. In particular, too many papers only present simple fuzzifications of classical mathematical notions and results without presenting proper motivations for them. Authors often restrict themselves to superficial justifications, claiming that their generalizations enable one to cope with uncertainty, leaving virtually untouched any serious consideration of why such fuzzifications are actually needed and what their advantage may be over the existing approaches. Quite often they do not present any justifications at all, but only refer to existing, often poorly justified works, and either further develop these works or generalize them, for example from the framework of standard fuzzy sets to other types of fuzzy sets, such as the intuitionistic ones and their various generalizations. The situation is amplified by the fact mentioned above that, as a rule, fuzzification of a given notion or theorem may be carried out in several ways. The basic reason for this situation, we believe, is a combination of three factors. First is the *fuzzifier's temptation* (p. 239) and the virtually unconstrained opportunity to fuzzify. Second is the pressure in academia to publish, the "publish or perish" rule, and the large and still growing number of publication venues, as well as the varying quality of the review process. Last but not least is the simplicity of the basic idea of fuzzy logic and the resulting low cost of entry to the area for potential contributors. All these factors resulted in a high number of superficial publications. This phenomenon pertains not only to the development of mathematics based on fuzzy logic but also to applications of fuzzy logic.

Early contributions and key directions

With his notion of fuzzy set, a clear and appealing motivation for it, and a few examples of fuzzified classical mathematical notions, Zadeh (1965a) set the stage for development of mathematics based on fuzzy logic. A distinguished place in the early developments of mathematics based on fuzzy logic is occupied by the work of Goguen, whose contributions to fuzzy logic in the narrow sense are discussed in section 4.3. His visionary papers, Goguen 1967, 1968–69, are outstanding for their lucid style, as well as for a number of cleanly treated conceptual but also technical issues. They became highly influential in further developments of mathematics based on fuzzy logic and we thus discuss here some of their further aspects.

Goguen was well aware of various possible misconceptions—later indeed frequently occurring in the literature—and devoted considerable attention to preventing them. Thus, in addition to clearly distinguishing the theory of fuzzy sets from the theory of probability on conceptual grounds (p. 161 above), he also made sure that his theory, which he termed a "logic of inexact quantities," was by no means inexact or fuzzy in any way (Goguen 1967, 147):

> We have produced a "logic of inexact quantities" within the framework of modern pure mathematics; the results concerning fuzzy sets are proved as rigorous mathematical theorems. Yet we hold that these results are for "inexact quantities."

Goguen (1968–69, 327) makes it clear that his metalevel is based on classical logic: "Our models are typical purely exact constructions, and we use ordinary exact logic and set theory ... in their development." He adds: "It is hard to see how we can study our subject at all rigorously without such assumptions." These foundational questions were only seriously considered much later, and we discuss them in section 5.2. Interestingly, Goguen (1968–69, 326) mentions Boolean-valued models of set theory as other examples of nonclassical sets and claims that applications of fuzzy sets provide an answer to Cohen and Hersch's question of whether non-Cantorian set theories have nonmathematical applications.[27] On the other hand, Goguen (1968–69, 337) shows great insight by issuing the following warning:[28]

> Ordinary set theory and logic have been of greatest importance in providing a convenient language for mathematical thought. They have not made the exercise of creative intelligence unnecessary either in mathematics or its applications. Similarly we should not expect more of fuzzy sets and logic than that they facilitate the development and study of models in the inexact sciences, and that they be an interesting area for pure mathematical investigation.

Goguen's general view of fuzzy sets in which the set of truth values forms a partially ordered set, denoted L, is also conceptually important.[29] He provided a nice justification of its further desirable properties and gave practical reasons for allowing L to be nonlinearly ordered. In particular, he argued for a structure of a complete residuated lattice (p. 163 above), which became a basic structure in later developments. The symbol "L" used by Goguen for the lattice of truth degrees as well as his term "L-fuzzy set" or just "L-set" became adopted in later developments. He was well aware that his conception would allow for other than the originally intended interpretations of fuzzy sets (Goguen 1967, 146): "Because of the generality of the mathematical definition, some important applications of fuzzy sets do not involve the intuitive concept of fuzziness at all," of which there later appeared many examples.[30] It is also appropriate to mention at this point some further issues of conceptual significance discussed by Goguen (1967, 1968–69), such as his view of the particular structure L of truth degrees as a parameter of the theory of fuzzy sets, whose borderline case $L = \{0, 1\}$ may be identified with the theory of classical sets; his usage of the adjective "crisp" as denoting fuzzy sets whose membership degree is

---

[27] P. J. Cohen and R. Hersch, "Non-Cantorian set theory," *Scientific American* 217, no. 6 (1967): 104–16. Nonclassical sets are discussed below in section 5.2.3.

[28] Of the many examples when researchers mistakenly expected more than what Goguen suggests we refer to Belohlavek et al. 2009; see also the related discussion of fuzzy logic in psychology in chapter 6.

[29] This option was already mentioned in a footnote in Zadeh 1965a.

[30] E.g. the various probabilistic logics which are particular instances of Pavelka's abstract fuzzy logic.

0 or 1;[31] his recognition of the fundamental importance of a nearness relation on the universe with respect to which the fuzzy sets considered are compatible; his emphasis on the ordinal nature of the set of truth values and its topology; his awareness of measurement theory—an area of mathematical psychology—as a proper conceptual tool for considerations regarding the truth degrees and their assignment; his consideration of context as determining e.g. the right choice of the truth functions of logical connectives; his general considerations regarding the problem of fuzzification; his recognition of the importance of fuzzy properties of fuzzy relations, including in particular the degree of inclusion of fuzzy sets; his recognition of category theory and categories of fuzzy sets as an important conceptual means of treating foundational problems; his detailed outline of a general theory of fuzzy relations; as well as his considerations regarding higher types of fuzzy sets and interval-valued fuzzy sets. As we shall see, all these issues proved to be of vital importance in further developments.

Goguen's contributions represent the first attempts to develop foundations of fuzzy logic and the mathematics based on it. Even though these investigations started to take place on a larger scale around the late 1980s, several important contributions appeared earlier. These include contributions to fuzzy logic in the narrow sense which are examined in sections 4.3 and 4.4, the contributions to axiomatic theories of fuzzy sets which have roots in the various contributions to many-valued sets that predate Zadeh's papers, as well as contributions to categories of fuzzy sets which go back to Goguen 1967. Apart from being important in themselves, these and later foundational developments which are examined in section 5.2 helped cultivate the otherwise often ad hoc contributions to mathematics based on fuzzy logic. This influence is particularly apparent since the early 1990s.

The important but rare contributions to foundations in the late 1960s and early 1970s mentioned above accompanied a gradually growing number of contributions to various particular areas of mathematics based on fuzzy logic. These areas, most of which emerged in the 1970s, include the intuitive theory of fuzzy sets and relations, which clearly has a distinguished role, closure structures, algebra, topology, quantities and mathematical analysis, probability, statistics, geometry as well as several areas of applied mathematics such as optimization. These areas were being subsequently developed, often to an advanced stage and with major reconsiderations of earlier contributions. We examine the respective developments in section 5.3.

Interestingly, a number of contributions to mathematics based on fuzzy logic were motivated by semantic questions, such as those regarding the meaning of the notion of truth degree, its possible interpretations, its psychological relevance, the adequacy of fuzzy logic as a tool for modeling vagueness, as well as the ability of fuzzy logic to resolve certain paradoxes. These questions are the subject of section 5.4.

---

[31] A fuzzy set A in a universal set U is *crisp* if $A(u) = 0$ or $A(u) = 1$ for every element u of U. In the literature on fuzzy logic, "crisp set" often means an ordinary set, i.e. a classical set. More generally, "crisp" indicates that no truth degrees other than 0 and 1 are admitted.

5.2 Foundations of mathematics based on fuzzy logic

5.2.1 The role of fuzzy logic in the narrow sense

The role of fuzzy logic in the narrow sense for mathematics based on fuzzy logic is in principle similar to the role of classical logic for classical mathematics and as such has several facets. Most fundamental is the foundational aspect. Thus, one may take a given fuzzy logic, such as one of the predicate or higher-order fuzzy logics examined in chapter 4 and section 5.2.2, and develop theories within these logics, for example a theory of fuzzy equivalence relations, fuzzy partial orders, fuzzy topologies, and so on. Even more basically, one may develop within a given predicate fuzzy logic an axiomatic fuzzy set theory and develop mathematics within this axiomatic theory in the same sense as classical mathematics may be developed in classical axiomatic set theory. In addition, there are two aspects of more practical nature. The first—discussed in section 4.6.1 and on p. 234—relates to the possibility offered by fuzzy logic in the narrow sense of clarifying from a logical viewpoint the various contributions in which the term "fuzzy logic" is used in a rather informal way and in particular of answering the question "where is there actually any logic in these contributions?" The second aspect, to which we now turn, relates to the very practice of researchers working in mathematics based on fuzzy logic.

In classical mathematics, the intuitive way of working with sets is accompanied by mostly informal usage of classical logic. Thus, mathematicians express their definitions, theorems, and proofs in ordinary language. For instance, one says "if u is related to v and v is related to w then u is related to w" when expressing the transitivity of a given relation R, or more technically:

$$\text{"for every } u, v, \text{ and } w \text{ in } U, \text{ if } \langle u, v \rangle \text{ belongs to } R \tag{5.1}$$
$$\text{and } \langle v, w \rangle \text{ belongs to } R \text{ then } \langle u, w \rangle \text{ belongs to } R."$$

Due to the intuitive clarity of such a claim one need not resort to the formal apparatus of classical logic—its informal use is sufficient. When intermediate truth degrees come into play and R is a fuzzy relation with truth degrees in L, things get more complex. The propositions "$\langle u, v \rangle$ belongs to R," "$\langle v, w \rangle$ belongs to R," and "$\langle u, w \rangle$ belongs to R" are then each a matter of degree, the above informal approach cannot be used, and one needs to resort to the apparatus of fuzzy logic more explicitly. What is thus the meaning of (5.1)? Fuzzy logic in the narrow sense offers the following general and conceptually clean approach.[32] First, one fixes a language of an appropriate fuzzy logic and expresses the verbally described condition (5.1) by

---

[32] This approach appears explicitly particularly in the works of Gottwald from the mid-1970s on, e.g. Gottwald 1979, 1993, 2001, of Höhle from the mid-1980s on, e.g. Höhle 1987, and later of Hájek (1998b), Belohlavek (2002c), and Běhounek and Cintula (2005). To some extent it is also present in Klaua's works, e.g. Klaua 1965, 1966.

a formula of this language.[33] In our case, we may take the predicate logic BL$\forall$ or some of its variants and its language containing a single binary relation symbol r.[34] (5.1) is then expressed by the formula which we denote φ_{Tra}:

$$\forall x \forall y \forall z (r(x,y) \otimes r(y,z) \rightarrow r(x,z)). \tag{5.2}$$

Second, one takes a structure $\mathbf{L}$ of truth degrees and an $\mathbf{L}$-structure $\mathbf{M} = \langle M, r^{\mathbf{M}} \rangle$ for the language in which $M = U$ and in which the symbol r is interpreted by the fuzzy relation R, i.e. $r^{\mathbf{M}} = R$. For instance, $\mathbf{L} = \langle L, \wedge, \vee, \otimes, \rightarrow, 0, 1 \rangle$ may be a complete BL-algebra.[35] The fact that (5.1) is satisfied by the fuzzy relation R is then naturally understood as saying that φ_{Tra} is true in $\mathbf{M}$, i.e. that $\|\varphi_{Tra}\|_{\mathbf{M}} = 1$. The principles of predicate fuzzy logic yield that this occurs exactly when for every $u, v, w \in U$ we have $R(u,v) \otimes R(v,w) \rightarrow R(u,w) = 1$, which is equivalent to:

for every u, v, w in U we have $R(u,v) \otimes R(v,w) \leq R(u,w)$.   (5.3)

In this sense, (5.3) is the meaning of (5.1) for fuzzy relations. We might thus say that a fuzzy relation R, or $\mathbf{L}$-relation to make the structure of truth degrees explicit, is transitive if it satisfies (5.3) and thereby obtain the definition of transitivity for fuzzy relations. Clearly, for particular cases of $\mathbf{L}$, (5.3) obtains particular forms. Thus, for the standard Gödel and Łukasiewicz algebra, $R(u,v) \otimes R(v,w) \leq R(u,w)$ reads $\min(R(u,v), R(v,w)) \leq R(u,w)$ and $\max(0, R(u,v) + R(v,w) - 1) \leq R(u,w)$, respectively, which express Zadeh's (1971b) and Ruspini's (1982) notions of transitivity.

Importantly, this approach offers a way to *fuzzify classical concepts*: write down the definition of a classical concept by a logical formula and interpret it according to the principles of fuzzy logic. In the above example: the definition of transitivity, (5.1), is written down as (5.2); its interpretation and requirement that (5.2) be true yields the definition (5.3) of transitivity of a fuzzy relation.

Thus we obtain from a definition Def$_2$ of a classical concept regarding classical relations R, such as the definition of transitivity, a definition Def$_{\mathbf{L}}$ regarding fuzzy relations with truth degrees in $\mathbf{L}$. Notice that Def$_{\mathbf{L}}$ is then, by the very nature of this fuzzification process, indeed a *generalization of the classical definition* Def$_2$ in that applying Def$_{\mathbf{L}}$ in the classical setting yields the classical definition. Namely, "classical setting" means $\mathbf{L}$ is the two-element Boolean algebra $\mathbf{2}$, so in the classical setting Def$_{\mathbf{L}}$ is Def$_2$ which is the classical definition we started with.

Compare this now methodologically to how definitions of fuzzified concepts were being found in the early stage of fuzzy logic. Let Def$_2$ again be a definition

---

[33] This may be a predicate fuzzy logic or some higher-order logic—whichever is sufficiently powerful to formalize the given situation but simple enough to work with. Clearly, one may prefer to select a logic powerful enough to formalize most if not all of the conceivable mathematical considerations, in which case a choice of some higher-order fuzzy logic (section 5.2.1) would be appropriate. From a certain perspective, the particular choice is not essentially important.

[34] See section 4.4.3 for the notions needed below.

[35] We thus only assume that $\wedge$, $\vee$, $\otimes$, $\rightarrow$, 0, and 1 satisfy the axioms of BL-algebras. In looking for the meaning of (5.1), we thus proceed on a general level. In a sense, it is not our business to choose the truth functions at this point. We may leave the choice open thus getting the benefit that what we do will be valid for any particular choice.

of a concept regarding classical relations. The then implicitly used rule may explicitly be described as follows: find a definition Def_L regarding fuzzy relations such that any classical relation R satisfies Def_2 if and only if its characteristic function χ_R, which is a crisp fuzzy set, satisfies Def_L. This rule, however—basically because it lacks a firmer rooting in logic and is in this sense too liberal—amplifies the arbitrariness in the process of fuzzification. As an example, consider the following definition: a fuzzy relation R is *transitive in a new sense* if $R(u,v) > 0$ and $R(v,w) > 0$ implies $R(u,w) > 0$ for every u, v, w. This definition is substantially different from the usual (5.3) in that there exist fuzzy relations which obey one of these definitions but not the other. Both are, however, generalizations of the classical concept of transitivity. One might object that the above definition of transitivity in a new sense is obviously not right because it is "too crisp" and that one would hardly propose such a definition as reasonable. In the face of such an objection just recall that Zadeh's (1971b) definition of antisymmetry requiring that $R(u,v) > 0$ and $R(v,u) > 0$ imply $u = v$ is exactly of the same sort and became part of standard textbooks on fuzzy sets before problems with it became recognized (cf. p. 278 below).

To be sure, we are not saying that fuzzy logic in the narrow sense offers "the right fuzzification" of a given classical concept nor that "the right fuzzification" exists. We only say that it offers a reasonable and methodologically clean way to define new concepts which involve fuzzy sets and in particular a way to fuzzify classical concepts. A fuzzification of classical concepts is not unique on principle. As an example, let it be noted that the above concept of transitivity in a new sense may also be obtained via the approach offered by fuzzy logic in the narrow sense. Indeed, the concept is described, using Baaz's Δ (p. 204), by the formula

$$\forall x \forall y \forall z ((\neg\Delta\neg r(x,y) \otimes \neg\Delta\neg r(y,z)) \rightarrow \neg\Delta\neg r(x,z)), \tag{5.4}$$

which in the classical case, $\mathbf{L} = \mathbf{2}$, is equivalent to (5.2) and hence also defines classical transitivity.[36] The issue at stake is the well-known fact that what may be equivalent in classical logic need not be equivalent in fuzzy logic—formulas (5.2) and (5.4) being a case in point. Nevertheless, it should now be clear that the way via logic formulas makes the process and the resulting concept more transparent. From this perspective, one might regard the concept of transitivity in the new sense as somewhat strange compared to the usual transitivity of a fuzzy relation because formula (5.4) is more complex and in a sense strange compared to (5.2). In particular, the appearance of Baaz's Δ makes it "too crisp." Note, however, that some crispness inside a fuzzified notion may be desirable in certain situations.[37] In general, which of the two or more possible but nonequivalent fuzzifications one should take and work with is another question. In fact, both may be useful, differently for different pur-

---

[36] In the classical case, Δ is the identity function: $\Delta(1) = 1$ and $\Delta(0) = 0$.

[37] For example, a fuzzy topology, encountered below in section 5.3.3, is defined as a *crisp* set τ of fuzzy sets—the open ones—satisfying certain properties. One might, however, also consider a *fuzzy* set τ of fuzzy sets with $\tau(A)$ interpreted as the degree to which A is open.

poses and while one of them may obey a generalization of some classical theorem, the other need not. The question is a matter of a careful examination of the motivations for the given concept, its applicability, as well as explorations of its properties and relationships to other concepts.

Reliance on fuzzy logic in the narrow sense makes it natural to consider fuzzy properties of fuzzy relations and, more generally, to consider theorems involving such properties.[38] For instance, instead of just declaring a fuzzy relation R transitive if it makes the transitivity formula φ_{Tra} in (5.2) true to degree 1, one may in general consider the truth degree of formula (5.2) for R, i.e. the degree $\|\varphi_{Tra}\|_M$ in the L-structure $M = \langle M = U, r^M = R \rangle$ defined above, and call it *degree of transitivity* of R. The basic rules of fuzzy logic yield that for this degree, $Tra(R)$, we have

$$Tra(R) = \bigwedge\nolimits_{u,v,w \in U}(R(u,v) \otimes R(v,w) \to R(u,w)).$$

The transitivity of R, i.e. (5.3), is then equivalent to $Tra(R) = 1$. One may then obtain various properties such as $\bigwedge_i Tra(R_i) \leq Tra(\bigwedge_i R)$, which is naturally interpreted as *saying in degrees* that the intersection $\bigwedge_i R_i$ of transitive fuzzy relations R_i is a transitive relation. Notice that as a corollary we obtain that if $Tra(R_i) = 1$ for each i, then $Tra(\bigwedge_i R) = 1$, which may be interpreted as *saying crisply* the same and which is in fact the commonly presented generalization of the well-known classical property that the intersection of transitive relations is transitive. The permeation of truth degrees to the metalevel, as illustrated above, is an interesting phenomenon. For instance, as illustrated in Belohlavek and Funioková 2004, while in the classical setting transitivity represents a sharp gap between nontransitive tolerances and equivalences, in a fuzzy setting the concept of degree of transitivity makes it possible to see nontransitive tolerances and fuzzy equivalences as boundary cases of a continuous spectrum of fuzzy relations which are naturally transitive to a greater or lesser extent.[39]

We conclude this section by illustrating that the logical viewpoint offers a conceptually clean perspective on various particular notions commonly defined for fuzzy sets and offers solutions to some natural problems. As an example, let us mention that both the mappings defined by Zadeh's extension principle and the mappings assigning to input fuzzy sets the output fuzzy sets according to the compositional rule of inference, even though usually presented as completely separate and with no connections between them, are simply mappings defined by first-order logical formulas, and hence both obey the general properties of such mappings.[40] In particular, a mapping $\overline{F} : L^U \to L^U$ obtained by the extension principle from an ordinary mapping $F : U \to U$ is just the mapping defined by the formula $(\exists x)(r(x) \otimes (f(x) \approx y))$ in the following sense. Denote this formula $\varphi(y)$ and fix a structure M with uni-

---

[38] Graded properties of fuzzy relations have systematically been considered by Gottwald (see the expositions in Gottwald 1993, 2001) and later in Belohlavek 2002c, Běhounek and Cintula 2005, and in some subsequent works, e.g. Belohlavek and Funioková 2004, and Běhounek and Daňková 2006a.

[39] A tolerance is a reflexive and symmetric relation.

[40] See chapter 3 and also section 4.6.1 and p. 273 for both notions.

verse U, in which f is interpreted by F, r by an arbitrary fuzzy set $A \in L^U$, and $\approx$ by crisp equality. Then $\overline{F}(A)$ is just the fuzzy set represented by $\varphi(y)$ in that for each $u \in U$, $[\overline{F}(A)](u)$ equals the truth degree $\|\varphi(y)\|_{M,v}$ of $\varphi(y)$ in $\mathbf{M}$ and a valuation v of variables which maps y to u, i.e.

$$[\overline{F}(A)](u) = \|\varphi(y)\|_{M,v}.$$

General considerations like those above are naturally made using fuzzy logic in the narrow sense and natural questions regarding fuzzy sets and relations are then answered by model-theoretical results (Di Nola and Gerla 1986, Gottwald 2001, Belohlavek 2002c). For illustration, Belohlavek 2007 contains what may be seen as a partial answer to the often posed critical remark that while fuzzy sets are intended to model inexact concepts, they do so by exactly specified membership degrees and thus impose artificial precision (cf. p. 344 below). The general result in Belohlavek 2007 implies that in most of the commonly used fuzzy models, this exact specification of membership degrees does not matter and thus fuzzy logic models are robust in this respect, confirming mathematically the practitioners' common experience.

5.2.2 Higher-order logic approaches

The motivation for classical higher-order logics is to be more expressive than with first-order ones.[41] For instance, while a first-order logic allows quantification over individuals only, second-order logics allow quantification over sets of individuals and over relations, whereas third-order logics allow quantification over sets of sets of individuals, etc. General higher-order logics enable quantification over arbitrarily nested sets. The price paid for this greater expressiveness is that some fundamental properties of first-order logic are lost.

The first developed higher-order fuzzy logic is Novák's (2005) *fuzzy type theory*, FTT, which is essentially a fuzzified version of Church's simple type theory. Even though Novák's paper is restricted to the logical calculus itself, the author mentions at the end that "Whether FTT can be used as a tool for a deeper development of fuzzy mathematics is a question of future research." For one such further examination see Novák and Dvořák 2011. Fuzzy type theory has subsequently been developed, see e.g. Novák 2012 for initial steps in its model theory.

In connection with the preceding section we now examine the *fuzzy class theory*, FCT, of Běhounek and Cintula (2005), which provides a framework for their methodological manifesto (Běhounek and Cintula 2006b). Rather than developing a fully general formal theory, such as an axiomatic fuzzy set theory of the kind we examine in section 5.2.3, they aim at developing a sufficiently expressive yet simple formal framework (Běhounek and Cintula 2005, 35):

> We observed that real-world applications of fuzzy sets need only a small portion of set-theoretical concepts. The central notion in fuzzy sets is the membership of elements (rather than fuzzy sets)

---

[41] A good introduction is provided by P. B. Andrews, *An Introduction to Mathematical Logic and Type Theory: To Truth through Proof*, 2nd ed. (Dordrecht: Kluwer, 2002).

into a fuzzy set. In the classical setting, the theory of the membership of atomic objects into sets is called ... *class theory*. It is a theory with two sorts of individuals—objects and classes—and one binary predicate—the membership of objects into classes. In this paper we develop a *fuzzy class theory*. The classes in our theory correspond exactly to Zadeh's fuzzy sets.

FCT is a theory over a two-sorted version of a predicate fuzzy logic with identity $=$, such as $Ł\Pi\forall$ or $Ł\Pi^{1}/_{2}\forall$ (see pp. 206ff. above). These logics were chosen because of their expressive power, e.g. the availability of all three basic—and in fact all—continuous t-norms and their residua and of truth constants for all rationals. The two sorts of FCT are for *objects* and *classes*. Object and class variables are denoted $x, y, \ldots$ and $X, Y, \ldots$, respectively. The only relation symbol in FCT, except for $=$, is $\in$, denoting object-in-class membership. Thus $x \in X$ is allowed, while $x \in x$, $X \in Y$, and $Y \in x$ are not. In addition to the logical axioms of $Ł\Pi\forall$ or $Ł\Pi^{1}/_{2}\forall$, the main axiom of FCT is the following version of the axiom of comprehension:

$$(\exists X)\Delta(\forall x)(x \in X \leftrightarrow \varphi(x)), \tag{5.5}$$

where φ is an arbitrary formula not containing X. Note that Baaz's delta, Δ, which is available in $Ł\Pi\forall$ and $Ł\Pi^{1}/_{2}\forall$, plays a highly important role in FCT. Namely, recall that in the standard semantics, i.e. $L = [0,1]$, its truth function is given by $\Delta(1) = 1$ and $\Delta(a) = 0$ for $a < 1$. Δ thus allows one to *demand full truth*: $\Delta\varphi$ has two possible truth degrees—1 when φ is fully true and 0 when the truth degree of φ is less than 1. Δ may also be used to *demand crispness*: the formula $\Delta(\varphi \vee \neg\varphi)$ assumes truth degree 1 (and otherwise 0) if and only if φ is crisp in that it assumes the truth degree 1 or 0. Now, the appearance of Δ in (5.5) enforces the actual existence of a class such that the degree of membership of each object, referred to by x, in this class *equals* the truth degree of $\varphi(x)$. By omitting Δ, the axiom could be true even if such a class does not exist.[42] In addition, the presence of Δ in (5.5) allows one to work appropriately with comprehension terms, i.e. terms $\{x \mid \varphi(x)\}$, which are introduced by the axioms $y \in \{x \mid \varphi(x)\} \leftrightarrow \varphi(y)$. Another axiom considered is the axiom of extensionality,

$$(\forall x)\Delta(x \in X \leftrightarrow x \in Y) \rightarrow X = Y.$$

The *standard Zadeh model* **M** of FCT over a universe M, which also proves the consistency of FCT, is obtained as follows: object variables range over M, class variables range over the set $[0,1]^M$ of all fuzzy sets in M, the truth degree $\|x \in X\|_{\mathbf{M},v}$ of formula $x \in X$ in **M** and valuation v is $[v(X)](v(x))$, i.e. the value of the fuzzy set $v(X) \in [0,1]^M$ at the element $v(x) \in M$, and this extends to all formulas as usual.

Operations with fuzzy sets are approached in FCT by comprehension terms $\{x \mid \varphi(x \in X_1, \ldots, x \in X_n)\}$ where $\varphi(p_1, \ldots, p_n)$ is an arbitrary *propositional formula* inducing the term. Now, the operation of standard union, $\cup$, is induced by $p_1 \vee p_2$ and is represented by the term $\{x \mid x \in X_1 \vee x \in X_2\}$ in that in the standard Zadeh model **M**, if v assigns to X_1 and X_2 the fuzzy sets A_1 and A_2, then $A_1 \cup A_2$ is just the fuzzy set

---

[42] Because there could be an infinite collection of classes for which the respective membership degrees of the object would approximate, but not be equal to, the truth degree of $\varphi(x)$.

to which $\{x \mid x \in X_1 \lor x \in X_2\}$ evaluates, i.e. $A_1 \cup A_2 = \|\{x \mid x \in X_1 \lor x \in X_2\}\|_{\mathbf{M},v}$. Interestingly, one can obtain in this way the operations of a-cut, kernel, or support as well. For instance, the operation of a-cut is induced by $\Delta(\overline{a} \to p)$. In a similar fashion, one may represent syntactically various relations and fuzzy relations defined on fuzzy sets which commonly appear in the literature. For example, the graded inclusion S, which we examine in section 5.3.1, is induced by $p_1 \to p_2$ and is represented by the formula $(\forall x)(x \in X_1 \to x \in X_2)$, denoted $X_1 \subseteq_{\otimes} X_2$, in that with the same conditions as for $\cup$ one has $S(A_1, A_2) = \|(\forall x)(x \in X_1 \to x \in X_2)\|_{\mathbf{M},v}$. The properties of such operations, relations, and fuzzy relations defined for fuzzy sets, are obtained by certain easily derived metatheorems using the corresponding properties of the underlying propositional logic, such as ŁΠ or ŁΠ½, which is a nice feature of FCT. Thus, the propositional provability of $(p \to q) \to ((p \otimes r) \to (q \otimes r))$ yields the provability in FCT of $(X \subseteq_{\otimes} Y) \to ((X \cap_{\otimes} Z) \subseteq_{\otimes} (Y \cap_{\otimes} Z))$, from which one obtains by correctness of FCT that $S(A,B) \leq S(A \otimes C, B \otimes C)$ for all fuzzy sets A, B, and C—a well-known property of graded inclusion.

Běhounek and Cintula (2005) then outline how FCT may be extended to account for fuzzy relations and for higher-level classes, i.e. classes of classes, classes of classes of classes, etc., as well as for an additional structure on the universe, such as that of a field.[43] In several subsequent papers they attempt to develop within FCT certain areas, most notably fuzzy relations (Běhounek, Bodenhofer, and Cintula 2008; Běhounek and Daňková 2006a).

Essential in motivating FCT is the idea of developing mathematics based on fuzzy logic within a framework of some formal system of fuzzy logic. Thus, operations, relations, and properties of fuzzy sets and relations, as well as new mathematical concepts, such as fuzzy equivalence or fuzzy topology, are described by logical formulas. Even though this idea is not new and appeared implicitly as well as explicitly earlier, e.g. in the works of Höhle, Gottwald, and Bělohlávek (section 5.2.1), Cintula and Běhounek pushed this idea further. They took advantage of then-available formal fuzzy logics and offered a way to develop mathematics based on fuzzy logic on a purely syntactic level. This is surely an elegant and sound way. The authors also regard FCT as providing a general method of fuzzification of classical mathematical concepts and theories—one basically takes a logical formula φ defining a concept, and works with it within FCT. If crispness in the definition is desirable at some point, one uses Δ to handle it. Běhounek and Cintula (2005, 54) claim that "In contemporary fuzzy mathematics the methodology of fuzzification of concepts is somewhat sketchy and non-consistent . . ." and that their approach "promises that no crispness will be unintentionally 'left behind'. . . . [S]ome features . . . may *intentionally* be left crisp. . . . The advantage of the proposed approach is that we always know *which* features are left crisp." While we basically agree with this point of view,

---

[43] Even then, the corresponding variant of FCT remains first-order. In a sense, FCT—like the Henkin-type higher-order logics—simulates the naturally higher-order considerations within a first-order framework.

the significance of FCT as a methodology of fuzzification should not be exaggerated. As discussed in detail in sections 5.1.3 and 5.2.1, the real challenge in finding useful concepts, which fuzzify the classical ones, and their interesting properties consists of a careful examination of their significance in relation to applications and their relationships to other concepts. The given framework within which this happens is for the main part of limited importance.

5.2.3 Set-theoretic approaches

The first examinations of axiomatic theories of many-valued sets appeared in the mid-1950s and were connected to paradoxes in classical set theory (pp. 235ff.). Skolem (1957), inspired by Rosser's remark in his *Logic for Mathematicians*[44] on Bochvar's logic for analysis of paradoxes (pp. 142ff. above), observed that already in the three-valued Łukasiewicz logic, $Ł_3\forall$, Russell's paradox disappears because the formula $x \in x \leftrightarrow \neg(x \in x)$ is no longer contradictory.[45] It might thus seem that in the three-valued Łukasiewicz logic, one may add the *axiom of (full) comprehension:*[46]

$$\forall z_1 \cdots z_n \exists y \forall x (x \in y \leftrightarrow \varphi(x, z_1, \ldots, z_n)), \tag{5.6}$$

without losing consistency. Namely, the above observation implies that the instance of (5.6) with φ being $\neg(x \in x)$, i.e. $\exists y \forall x (x \in y \leftrightarrow \neg(x \in x))$, is consistent if consistency of a set T of formulas means the existence of a model $\mathbf{M}$ and valuation v such that $\|\psi\|_{\mathbf{M},v} = 1$ for each $\psi \in T$. The axiom of full comprehension may be seen as a principle of set formation saying in particular that every property represented by a formula $\varphi(x)$ determines a set—the set of all individuals making $\varphi(x)$ true. In classical logic, Russell's paradox shows that this axiom leads to inconsistency. Classical axiomatic set theories, such as ZF, avoid Russell's paradox by not allowing this too liberal a principle and allowing instead only the *axiom of restricted comprehension.*[47] Skolem's observation therefore suggests that in addition to restricting comprehension, another way to avoid inconsistency a la Russell's paradox is to make the logic many-valued. However, Skolem also showed that while Russell's paradox no longer represents inconsistency in $Ł_3\forall$, there are new inconsistencies such as the one arising from comprehension with φ being $(x \in x) \to ((x \in x) \to \neg(x \in x))$. Furthermore, while the latter is no longer an inconsistency in $Ł_4\forall$, comprehension leads to inconsistency even in $Ł_4\forall$ and in fact in any finitely-valued Łukasiewicz logic $Ł_n\forall$.[48]

---

[44](New York: McGraw-Hill, 1953); the remark appears on p. 202 of Rosser's book.

[45]Because generally, if φ (such as $x \in x$) assumes the truth value $1/2$, then $\varphi \leftrightarrow \neg\varphi$ assumes 1. We mentioned on p. 116 that the latter fact was pointed out already by Łukasiewicz himself, who noted that "in the three-valued logic there are no antinomies." This is also observed in Moh 1954 which we discuss in section 5.4.2.

[46]Sometimes called unrestricted comprehension. This is actually an axiom *scheme* in which φ is an arbitrary formula of a language containing $\in$, all of whose free variables are among $x, z_1, \ldots, z_n$ and are distinct from y.

[47]This axiom scheme, also called separation, whose general form is $\forall z_1 \cdots z_n \forall z \exists y \forall x (x \in y \leftrightarrow x \in z \land \varphi(x, z_1, \ldots, z_n))$, only says that for each set, represented by z, the individuals *in* z making φ true form a set for every n-tuple of parameters represented by $z_1, \ldots, z_n$.

[48]Skolem only conjectured this property for $n > 3$; the property is claimed without proof in Chang 1963.

Skolem thus turned to the infinitely-valued predicate Łukasiewicz logic and proved the following theorem about his axiomatic set theory:[49]

Theorem 5.1. The set of formulas (5.6) with φ quantifier free is consistent.

Skolem also mentioned the problem of generalizing this theorem to general formulas φ involving quantifiers. For certain cases but not the general one, a generalization was obtained by Chang (1963) and Fenstad (1964), but the general problem of consistency of Skolem's set theory with full comprehension remained open for a long time. Eventualy, White (1979) presented a proof of consistency. Recently, however, Kazushige Terui found a flaw in White's proof, hence the problem at present still remains open.[50] As Hájek (2013b, 183) subsequently remarked:

> Now we have just to hope that [this fuzzy set theory] is consistent, (Similarly as we believe that the Zermelo-Fraenkel axiomatic set theory ZF is consistent; . . .)[51]

Interestingly, Skolem (1957, 7) asked whether it is possible to develop mathematics in a set theory of this kind and included several considerations in this regard, particularly on arithmetic.[52] He published two more papers on set theory in many-valued logic but it was not until the mid-2000s that it was more thoroughly examined by Hájek and later by Yatabe. In the meantime, however, several related contributions appeared and we refer to Hájek 2005 for references and a short overview. Since full comprehension is used in Cantor's intuitive set theory, Hájek calls Skolem's theory the *Cantor-Łukasiewicz set theory*. In particular, Hájek conceives of it as a predicate Łukasiewicz logic (section 4.4.3) with one binary relation symbol, $\in$, in which terms are object variables x and expressions $\{x \mid \varphi(x,\dots)\}$ with $\varphi(x,\dots)$ being a formula, atomic formulas are $t \in s$ with terms t and s, and which has full comprehension in that for each formula $\varphi(x)$ in which u is not free, the formula $u \in \{x \mid \varphi(x,\dots)\} \leftrightarrow \varphi(u,\dots)$ is an axiom.[53] The main result of Hájek 2005 is negative: if one introduces natural numbers, ω, and postulates a simple schema of induction, one gets inconsistency. Inconsistency results, particularly for standard Łukasiewicz logic, were later obtained by Yatabe (2007). Worth noting is that if one defines the Leibniz equality, $=$, by $x = y \leftrightarrow \forall z(x \in z \leftrightarrow y \in z)$ and the extensional equality, $=_e$, by $x =_e y \leftrightarrow \forall z(z \in x \leftrightarrow z \in y)$, one may prove that $=$ is crisp, that $x = y \rightarrow x =_e y$ is provable, but adding $x = y \leftrightarrow x =_e y$ leads to inconsistency, i.e. provability of all formulas, which is essentially due to results of Grishin and Cantini

---

[49] At that time, no studies of predicate Łukasiewicz logic were available, hence Skolem's study is the first, cf. p. 148. He used the rationals of $[0, 1]$ as truth values.

[50] A flaw in R. B. White's article "The consistency of the axiom of comprehension in the infinite-valued predicate logic of Lukasiewicz" (unpublished note, July 27, 2014). On March 27, 2015, Terui informed us that he had found another gap in White's proof.

[51] As Hájek (2005) remarked, in Gödel or product logic, such a theory is inconsistent.

[52] "Wird es möglich sein, die übliche Mathematik auf einer Mengenlehre dieser Art aufzubauen?"

[53] In fact, terms $\{x \mid \varphi(x,\dots)\}$ are terms in an extended language of the basic one which includes only individual variables as terms, similarly as class terms are considered in axiomatic set theories.

on full comprehension in certain weaker logics.[54] Later on, Hájek (2013a) further examined these two equalities, showed that there exist infinitely many sets extensionally equal to ω, and studied an arithmetic weaker than the inconsistent one.

Rasiowa (1964) presented a second-order logic representing a generalization of a formal theory of fields of sets. The structures of truth degrees are called $\mathscr{S}$-algebras, which are general algebras subsuming e.g. Heyting algebras, which Rasiowa explicitly mentions, but also residuated lattices. In particular, she notes that if U is a set and L an $\mathscr{S}$-algebra, then each element A of L^U may be interpreted as an L-subset of U (in her symbols, "$\mathscr{A}$-subset of I") and $A(u)$ may be interpreted as the truth value of the proposition "u belongs to A." Rasiowa's concept thus covers Zadeh's concept of fuzzy set as well as the basic algebra of fuzzy sets. She also considers, even though she does not define them in particular terms, graded inclusion and equality of her L-valued sets. For a semantics given by a certain class of $\mathscr{S}$-algebras, she obtains a completeness theorem for her logic.

A series of studies in the mid and late 1960s by Klaua represents another interesting stream of research. Klaua, whose first paper appeared the same year as Zadeh's seminal paper, was not aware of Zadeh's work and was inspired by discussions after a colloquium of Menger in East Berlin in the first half of the 1960s.[55] Klaua developed within classical set theory a cumulative hierarchy of fuzzy sets.[56] For simplicity, he restricted himself to Łukasiewicz connectives mostly on finite sets L of truth values.[57] In Klaua 1965, he starts with a set U of urelements of an appropriate classical set theory and a fuzzy equality id, which is reflexive, symmetric, transitive w.r.t. the Łukasiewicz $\otimes$, and separating (p. 274). With the standard powerset operation P, he defines his hierarchy with ω (the least infinite ordinal) levels by

$$V_0 = U \times \{0\}, \quad V_{n+1} = V_n \cup P(V_n) \times \{1\}, \quad V = \bigcup_{n < \omega} V_n.$$

V is the set of elements of Klaua's set theory over U, V_0 contains the individuals, and $V - V_0$ are Klaua's many-valued sets over U. On V, he defines a graded membership, $\in_m$, graded inclusion, $\subseteq_m$, and graded identity, $=_m$, by induction on rank as follows: $x \in_m y = \sup_{z \in \pi_1(y)} x =_m z$ with π_1 denoting the first projection; $x \subseteq_m y = \inf_{z \in \pi_1(x)} z \in_m y$; and for $=_m$, $x =_m y$ equals $id(\pi_1(x), \pi_1(y))$ if x, y are in V_0, 0 if x and y are of different ranks, and $\min(x \subseteq_m y, y \subseteq_m x)$ if $x, y \in V - V_0$

[54] V. N. Grishin, Предикатные и теоретико-множественные исчисления, основанные на логике без сокращений [Predicate and set-theoretic calculi based on logic without contractions], *Izvestiya Akademii Nauk SSSR Seriya Matematicheskaya* 45 (1981): 47–68; A. Cantini, "The undecidability of Grišin's set theory," *Studia Logica* 74 (2003): 345–68.

[55] Gottwald 2006 (p. 216). In his first paper, Klaua (1965) thanks his thesis advisor, who was Schröter (p. 157 above), for initiating his interest in the topic. A reference to Zadeh is in Klaua 1970.

[56] Classical cumulative hierarchies are used as models of classical axiomatic set theories. An example is the von Neumann universe $V = \bigcup_{\alpha \text{ ordinal}} V_\alpha$, where $V_0 = \emptyset$, $V_{\alpha+1} = P(V_\alpha)$—the power set of V_α—for a successor ordinal $\alpha + 1$, and $V_\beta = \bigcup_{\alpha < \beta} V_\alpha$ for a limit ordinal β. Each set has its *rank* in the hierarchy, e.g. $\emptyset$ has rank 0.

[57] The infinite-valued case was later studied by K.-U. Jahn, "Über einen Ansatz zur mehrwertigen Mengenlehre unter Zulassung kontinuumvieler Wahrheitswerte" [On an approach to many-valued set theory admitting a continuum of truth values], (master's thesis, Leipzig: Univ. Leipzig, 1969). Schwartz (1972) worked out a generalization for general algebras of truth degrees.

are of equal rank. Thus, for instance, $x \in_m y$ is intuitively understood as the degree to which there exists z in y which is equal to x. Klaua shows several results, among them that certain versions of axioms of extensionality and comprehension hold, and that a natural algebra may be developed in which one has several natural laws such as the transitivity of graded inclusion, $(x \subseteq_m y) \otimes (y \subseteq_m z) \leq x \subseteq_m z$, or both forms of extensionality of $\in_m$,

$$(x \in_m y) \otimes (y =_m z) \leq (x \in_m z) \quad \text{and} \quad (x =_m y) \otimes (y \in_m z) \leq (x \in_m z), \quad (5.7)$$

which later appeared in the fuzzy set literature but with a different meaning. Klaua's many-valued sets are conceptually different from fuzzy sets. Interestingly, they are very close to Boolean-valued models of set theory, which started in the late 1960s, and are also close to the notion of L-sets considered by Höhle in category-theoretic considerations, which has its roots in Boolean-valued models (section 5.2.4). A year later, Klaua (1966) took a different approach, already suggested in his 1965 paper, in which the many-valued sets in the cumulative hierarchy are essentially fuzzy sets in Zadeh's sense. Starting again from a set U of urelements, he defines by transfinite induction (cf. n. 56):

$$V_0 = U, \quad V_{\alpha+1} = V_\alpha \cup S_\alpha, \quad V_\beta = \bigcup_{\alpha < \beta} V_\alpha, \quad \text{and} \quad V = \bigcup_\alpha V_\alpha,$$

where S_α is the set of all functions $A : V_\alpha \to L$, i.e. fuzzy sets, such that for each $\beta < \alpha$ there exists γ with $\beta < \gamma \leq \alpha$ and $x \in V_\gamma - \bigcup_{\delta < \gamma} V_\delta$ such that $A(x) \neq 0$. This last condition prevents doubling—it forbids adding to $V_{\alpha+1}$ fuzzy sets which have the same support but larger domain than some fuzzy set in V_α. The elements of $V - V_0$ are new many-valued sets. Klaua then again defines graded relations, $\in_m$, $\subseteq_m$, and $=_m$, this time as follows: $x \in_m y$ equals $y(x)$ if y is not in U and x is in the domain of y, and 0 otherwise; $x \subseteq_m y$ equals $\bigwedge_z (x(z) \to y(z))$ for $x, y \in V - V_0$, and 0 otherwise; $x =_m y = (x \subseteq_m y) \wedge (y \subseteq_m x)$. Note that $\subseteq_m$ was independently introduced in Goguen 1968–69. Klaua proved many properties of his theory of fuzzy sets, some of which only appeared considerably later in fuzzy set literature, such as e.g. $(x \subseteq_m y) \otimes (u \subseteq_m v) \leq (x \cap u \subseteq_m y \cap v)$. Compared to his first approach, however, some forms of extensionality such as the second law in (5.7) were missing. This was because graded identity of fuzzy sets was not taken into account in Klaua's cumulative hierarchy. This problem was examined by Gottwald (1976–77, 1979), Klaua's student and later a leading figure in the theory of fuzzy sets. This approach has subsequently been worked out in further detail; see Gottwald 2006.

    Investigations in axiomatic set theory related to Paul Cohen's method of forcing, which he used to prove the independence of the famous Continuum Hypothesis, led to the development of *Boolean-valued models* (Bell 2005) invented in the mid-1960s independently by Robert Solovay and Petr Vopěnka.[58] While ordinary

---

[58] According to Scott's foreword to Bell 2005, Solovay observed in 1965 that forcing can be seen as assigning Boolean truth values to formulas. Scott, who joined Solovay in pursuing Boolean models in 1965, authored the widely circulated "Boolean-valued models for set theory," mimeographed notes for the 1967 American Mathematical Society Symposium on axiomatic set theory, which were planned to be extended to a joint paper with

models of classical axiomatic set theories consist of sets, Boolean-valued models consist, for a given complete Boolean algebra, of so-called B-valued sets which are functions assigning elements of B to elements of a given set. B-valued sets thus generalize classical sets because characteristic functions of the latter are just the B-valued sets for $B = \{0, 1\}$. The concept of a B-valued set may clearly be seen as a particular instance of Goguen's (1967) general concept of a fuzzy set with degrees in a complete lattice. Very different, however, is the motivation behind B-valued sets, hence also the interpretation of truth degrees. Within ZFC, Boolean-valued models, $V^{(B)}$, are obtained as cumulative hierarchies of B-valued sets where B is a complete Boolean algebra conceived of as a set in ZFC. They are defined inductively as follows: $V_0^{(B)} = \{\emptyset\}$, for a successor ordinal $\alpha + 1$, $V_{\alpha+1}^{(B)}$ is the set of all functions from $V_\alpha^{(B)}$ to B, for a limit ordinal β, $V_\beta^{(B)} = \bigcup_{\alpha < \beta} V_\alpha^{(B)}$, and finally, $V^{(B)} = \bigcup_\alpha V_\alpha^{(B)}$. Notice the striking similarity to Klaua's cumulative hierarchy of fuzzy sets in the preceding paragraph which seems not to have been recognized in the literature. On $V^{(B)}$, one defines the B-valued relations of membership and equality, denoted $|| \cdot \in \cdot ||$ and $|| \cdot = \cdot ||$, by recursion on a certain well-founded relation as

$$||u \in v|| = \bigvee\nolimits_{y \in \mathrm{dom}(v)} (v(y) \wedge ||u = y||) \quad \text{and}$$

$$||u = v|| = \bigwedge\nolimits_{y \in \mathrm{dom}(v)} (v(y) \rightarrow ||y \in u||) \wedge \bigwedge\nolimits_{x \in \mathrm{dom}(u)} (u(x) \rightarrow ||x \in v||).$$

This definition of truth values in B is then extended truth functionally, interpreting $\forall$ and $\exists$ by the infimum, $\bigwedge$, and supremum, $\bigvee$, in B, to obtain the truth value $||\varphi||$ of the arbitrary formula φ of axiomatic set theory in $V^{(B)}$. One then obtains that every axiom of ZFC assumes the truth value 1 in $V^{(B)}$. In other words:

Theorem 5.2. $V^{(B)}$ is a (Boolean-valued) model of ZFC.

To show that the Continuum Hypothesis is not provable in ZFC it then suffices to construct B in such a way that the Continuum Hypothesis is false in $V^{(B)}$.

From the axiomatic point of view, Boolean-valued models as well as Klaua's cumulative hierarchies of non-Boolean fuzzy sets are developed in classical axiomatic set theories. This is not true of the set theory initiated by Skolem but this theory differs significantly from the classical ZF by having unrestricted comprehension. Axiomatic set theories with the usual comprehension, which are thus more ZF-like, within logics other than classical ones represent a different challenge with great conceptual importance and interesting, conceptually new problems. Pioneering in this sense are axiomatic set theories in intuitionistic logic, in particular ZF, examined for the first time at the 1971 Cambridge Summer School in Mathematical Logic by Friedman (1973) and Myhill (1973)[59] and later by other authors as well, e.g.

Solovay. The paper was never written. Regarding the notes, Scott says "in writing them my main role was that of an expositor." In the literature, Boolean-valued models are commonly attributed to Solovay, Scott, and Vopěnka. According to Jech (2003, 224), Vopěnka first used—in 1964—open sets in a topological space and arrived, jointly with P. Hájek, at values in a Boolean algebra in 1967.

[59] According to Myhill, the first part of the paper is a joint work with Friedman.

Powell (1975) and Grayson (1979). Various fundamental questions have been studied, such as relative consistency of classical and various versions of intuitionistic ZF: is the consistency of classical ZF implied by the consistency of intuitionistic ZF and vice versa? Closely related is Myhill's constructive set theory which results from the intuitionistic set theory by restrictions on set-theoretic axioms. Interestingly, parts of mathematics, such as analysis or measure theory, based on axiomatic intuitionistic or constructive set theory, have been thoroughly examined.[60] This endeavor is thought of as a foundation for Bishop's constructive mathematics.[61] Especially important from the point of view of fuzzy logic are the papers by Takeuti and Titani (1984, 1992) entitled "Intuitionistic fuzzy logic and intuitionistic fuzzy set theory" and "Fuzzy logic and fuzzy set theory." They build on the above-mentioned work of Powell (1975) and Grayson (1979). In particular, the second paper develops axiomatic set theory in Gödel logic expanded by Łukasiewicz connectives. In addition, the paper is an early advanced contribution to fuzzy logic in the narrow sense which employs for good reasons novel logical connectives. In particular, it employs the globalization connective, $\square$, in that $\square\varphi$ means at the metalevel that "the truth value of φ is 1," a property that has been used in many later papers on fuzzy logic. Note also a related earlier but not very developed approach by Zhang (1983). With Hájek's predicate fuzzy logic BL$\forall$ at their disposal and inspired by the above mentioned works and, importantly, also by Shirahata's paper, Hájek and Haniková (2001, 2003) developed an axiomatic fuzzy set theory.[62] This theory, FST, is a theory in BL$\forall\Delta$ with equality = and with a binary relation symbol $\in$.[63] That is, it has the axioms of BL$\forall$, the axioms (4.122)–(4.123) for Δ, and the axioms of equality and congruence for $\in$. FST has the following additional axioms:

$$\forall x\forall y(x = y \leftrightarrow (\Delta(x \subseteq y) \otimes \Delta(y \subseteq x))) \qquad \text{(extensionality)}$$

$$\exists x\Delta\forall y\neg(y \in x) \qquad \text{(empty set)}$$

$$\forall x\forall y\exists z\Delta\forall u(u \in z \leftrightarrow u = x \vee u = y) \qquad \text{(pair)}$$

$$\forall x\exists z\Delta\forall u(u \in z \leftrightarrow \exists y(u \in y \otimes y \in x)) \qquad \text{(union)}$$

$$\forall x\exists z\Delta\forall u(u \in z \leftrightarrow \Delta(u \subseteq x)) \qquad \text{(weak power)}$$

$$\exists z\Delta(\emptyset \in z \otimes \forall x \in z(x \cup \{x\} \in z)) \qquad \text{(infinity)}$$

$$\forall x\exists z\Delta\forall u(u \in z \leftrightarrow (u \in x \otimes \varphi(u, x))) \qquad \text{(separation)}$$

$$\forall x\exists z\Delta(\forall u \in x\exists v\varphi(u, v) \rightarrow \forall u \in x\exists v \in z\varphi(u, v)) \qquad \text{(collection)}$$

$$\Delta\forall x(\forall y \in x\varphi(y) \rightarrow \varphi(x)) \rightarrow \Delta\forall x\varphi(x) \qquad \text{($\in$-induction)}$$

$$\forall x\exists z(\text{crisp}(z) \otimes \Delta(x \subseteq z)) \qquad \text{(support)}$$

[60] E.g. J. Myhill, "Constructive set theory," *J Symb Logic* 40 (1975): 347–82; H. Friedman, "Set-theoretical foundations for constructive analysis," *Ann Math* 105 (1977): 1–28.

[61] E. Bishop, *Foundations of Constructive Analysis* (New York: McGraw-Hill, 1967). The book opened a new era of intuitionistic mathematics; see M. Beeson, *Foundations of Constructive Mathematics* (Berlin: Springer, 1985).

[62] M. Shirahata, "Phase-valued models of linear set theory" (preprint, 1999).

[63] In fact, in an arbitrary schematic extension of BL$\forall\Delta$ (p. 198), which we omit for simplicity.

where φ is any formula which in the axioms of separation and collection must not contain z free, $\text{crisp}(z)$ is defined by

$$\text{crisp}(z) \longleftrightarrow \forall y(\Delta(x \in y) \vee \Delta\neg(x \in y)),$$

and $\emptyset$, $\cup$, $\{\,\}$ (definable function symbols) and $\subseteq$ are as usual. Hájek and Haniková first construct in ZFC, i.e. in the classical Zermelo-Fraenkel axiomatic set theory with the axiom of choice, for a given BL_Δ-algebra L a cumulative hiearchy $V^{(L)}$ of fuzzy sets, i.e. L-valued functions, analogously as described above for the Boolean-valued $V^{(B)}$. Then they define, again analogously, the truth evaluation $\|\cdot\|$ of formulas of ZFC with values in L. Worth noticing is a difference from the Boolean case: for the atomic formulas they define

$$\|u = v\| = \begin{cases} 1 & \text{if } u \text{ equals } v, \\ 0 & \text{otherwise.} \end{cases}$$

This crispness is forced by the fact that in certain schematic extensions of FST, such as those based on Łukasiewicz or the product logic, the equality is provably crisp. Correspondingly, they define

$$\|u \in v\| = \begin{cases} v(u) & \text{for } u \in \text{dom}(v), \\ 0 & \text{otherwise.} \end{cases}$$

They then prove that $\|\varphi\| = 1$ for every axiom φ of FST. In this sense, they proved:

Theorem 5.3. $V^{(L)}$ is a model of FST.

    Thus in particular, FST is consistent if ZFC is. Then they construct within FST a class H, i.e. a formula in FST with one free variable, of hereditary finite crisp sets and construct for every formula φ of ZF a formula φ^H of FST, the relativization of φ to H, in such a way that if φ is provable in ZF then φ^H is provable in FST. H is defined in terms of the class HCT of hereditary crisp transitive sets by:

$$H(x) \longleftrightarrow \text{crisp}(x) \otimes \exists y \in \text{HCT}(x \subseteq y),$$

where

$$\text{HCT}(x) \longleftrightarrow \text{crisp}(x) \otimes \forall u \in x(\text{crisp}(u) \otimes u \subseteq x).$$

In this sense, they proved:

Theorem 5.4. H is an inner model of ZF in FST. For any axiom φ of ZF, its translation φ^H is provable in FST.

    Hence ZF is interpretable in FST, which also means that ZF is consistent if FST is. Later on, Hájek and Haniková (2013) returned to FST and provided an interpretation of Titani's (1999) lattice-valued set theory in a modification of FST.

    Lake (1976) outlined the possibility of utilizing von Neumann's axiomatization of set theory, taking as primitive the notion of function for a common axiomatization of fuzzy sets and multisets.[64] This idea was thoroughly developed in an inter-

---

[64]English translation of J. von Neumann, "An axiomatization of set theory" in J. van Heijenoort, ed., *From*

esting but not well-known paper by Blizard (1989) who considers generalized multisets, called msets, as mappings assigning to elements nonnegative reals. His system is based on a two-sorted classical predicate logic, one sort for msets and one for
the reals, and includes among other symbols a ternary membership ϵ with $\epsilon(x, y, z)$
meaning "x is in y with multiplicity z" and constants for nonnegative reals and for
the number -1.[65] The axioms include axioms for msets and also axioms for the field
of reals. This theory with multiplicities restricted to $(0, 1]$ then yields an axiomatic
theory of fuzzy sets. Both theories are shown to contain a model of ZFC and to be
consistent relative to the consistency of ZFC.

There also appeared other classical-logic-based axiomatic systems for fuzzy sets
but mostly without deeper results. An indication of the first of such works is a short
abstract by Barone Netto.[66] Chapin (1974, 1975) proposed an axiomatization within
classical predicate logic with a ternary membership symbol ϵ with $\epsilon(x, y, z)$ interpreted as "x is in y to degree at least z." He includes 14 axioms, some of them of
varying strength, and establishes some usual theorems of the algebra of fuzzy sets
on these grounds. The axioms include axioms which have their counterparts in the
classical axioms of ZF but also axioms which do not. Of the latter kind is the axiom

$$(\forall x y z u)[(\epsilon(x, y, z) \wedge u \subseteq z) \to \epsilon(x, y, u)].$$

A planned continuation dealing with ordinal and cardinal numbers never appeared.
Another axiomatization within classical predicate logic is due to Weidner (1974, 1981)
who criticizes Chapin on conceptual grounds and includes a ternary membership
symbol ϵ, with $\epsilon(x, y, z)$ now meaning "x is in y to degree z" and binary symbol
$\leq$ intended for comparison of fuzzy sets which are truth degrees. It is shown that
the developed axiomatic theory with nine axioms proves various commonly used
properties of fuzzy sets, that it is consistent relative to consistency of ZF, and that
Boolean-valued universes of set theory are models of it. Neither graded inclusion
nor graded equality is discussed. For other axiomatic approaches, e.g. of Novák
(1980) inspired by GB and having a ternary membership relation, Prati (1988) based
on a quaternary membership relation and aimed at axiomatization of generally conceived "fuzzy objects," Toth (1993) with a two-sorted language with a sort for truth
degrees, and Demirci and Çoker (1983) with two binary membership symbols of
belonging to a fuzzy set and belonging to a crisp class based on Bernays's axiomatization of sets and classes, see Gottwald 2006. All these axiomatizations are based on
classical logic and none discusses graded inclusion or equality.

Frege to Gödel: A Source Book in Mathematical Logic, 1879–1931 (Cambridge, MA: Harvard Univ. Press, 1967),
393–413.

An element may appear in a multiset several times.

[65] Notice that the axiomatic systems mentioned above are based on many-valued logics.

[66] A. Barone Netto, "Fuzzy classes," *Notices AMS* 15 (1968): 945. The author claims to have developed a theory
of fuzzy classes as a first-order theory which contains classical set theory. No specifics are given.

5.2.4 Category-theoretic approaches

Category theory and its foundational role

The concept of category is due to Samuel Eilenberg and Saunders Mac Lane.[67] Central to category theory is the view, going back to the 1930s, that mathematics is a study of structures and structure-preserving mappings and that it is these notions which should be taken as primitive in mathematics. A *category* consists of a collection of *objects* and *arrows*, representing structures and their morphisms. To every arrow f two objects are assigned, its domain X and codomain Y, written as $f : X \to Y$. It is required that for every two arrows $f : X \to Y$ and $g : Y \to Z$, there exists an arrow $f \circ g : X \to Z$, their *composition*, that for every object X there exists an arrow $id_X : X \to X$, the *identity* of X, and that for $f : X \to Y$, $g : Y \to Z$, and $h : Z \to W$, one always has $id_X \circ f = f = f \circ id_X$ and $f \circ (g \circ h) = (f \circ g) \circ h$. Sets and mappings, groups and group homomorphisms, topological spaces and continuous functions are all examples of categories. In a sense, category theory looks at how mathematical objects are interconnected via morphisms rather than at how they are structured in terms of being composed of other objects. Until the early 1960s, category theory was mainly used as a convenient language and relied in virtually every aspect on set theory. The situation changed particularly due to F. William Lawvere, who introduced in his PhD thesis the idea of category of categories as a new foundation for mathematics, providing an alternative to set theory.[68] Later, he provided a first-order axiomatization of the category of sets—by adding eight further axioms to the conditions of Eilenberg and Mac Lane, which in fact were first-order axioms—and the category of categories.[69] Being first-order definable, such categories are called *elementary*. Lawvere's stand is well illustrated by the following statement from p. 1 of his 1966 paper:

> In the mathematical development ... one sees clearly the rise of the conviction that the relevant properties of mathematical objects are those which can be stated in terms of their abstract structure rather than in terms of the elements which the objects were thought to be made of. The question thus naturally arises whether one can give a foundation for mathematics which expresses wholeheartedly this conviction concerning what mathematics is about and in particular in which classes and membership in classes do not play any role ... A foundation of the sort we have in mind would seemingly be much more natural and readily usable than the classical one.

Later on, Osius proved that there exists an extension of the elementary theory of categories in which the ZFC axiomatic set theory (p. 237 above) is interpretable in

---

[67]"General theory of natural equivalences," *Trans AMS* 58 (1945): 231–94. The authors regarded the concept of category as secondary, their main emphasis was on functors and natural transformations. The considerations arose from their work on homology groups. A classic textbook is S. Mac Lane, *Categories for the Working Mathematician*, 2nd ed. (New York: Springer, 1998). For topos theory, see R. Goldblatt, *Topoi: The Categorical Analysis of Logic* (Amsterdam: Noth-Holland, 1984).

[68]*Functorial Semantics of Algebraic Theories and Some Algebraic Problems in the Context of Functorial Semantics of Algebraic Theories* (PhD diss., Columbia Univ., 1963).

[69]"An elementary theory of the category of sets," *Proc Natl Acad Sci USA* 52 (1964): 1506–11 (the result says that any *complete* category satisfying the axioms is equivalent to the category of sets); "The category of categories as a foundation for mathematics," *Proc Conf Categorical Algebra* (New York: Springer, 1966), 1–20.

such a way that the interpretation of any theorem of ZFC is provable in this extension.[70] Since the converse interpretation is also possible, these two theories are equivalent. The question of whether category theory may be considered as a foundation for mathematics nevertheless has several facets and many objections to its affirmative answer have been advanced, among which is Feferman's argument that some variant of set theory must be prior to any categorical foundation.[71] Osius's extension was in fact an extension of the axiomatic theory of topoi. Topoi play a crucial role in connections of category theory and logic. The concept of *topos*,[72] emerging from Grothendieck's work on algebraic geometry and in particular on categories of sheaves of sets on a topological space, was defined in the early 1960s by Grothendieck and Verdier.[73] Topos is a category which behaves like the category of sets and has a suitable notion of localization. Grothendieck's concept is what is now called *Grothendieck's topos*. This is a stronger notion than that of *elementary topos* which refers to categories satisfying certain first-order axioms proposed by Lawvere and Tierney in search of axioms for categories of sets.[74] An elementary topos is a category which has the following (in parentheses, we provide intuitive meanings in terms of sets): an initial object (empty set), a terminal object (one-element set), binary products (Cartesian products may be formed), binary coproducts (disjoint unions may be formed), equalizers (set of elements x for which $f(x) = g(x)$ for $f, g : X \to Y$), coequalizers (quotient sets may be formed), exponentials (the set of all functions from X to Y may be formed for every X and Y), and a subobject classifier (the set $\{0, 1\}$ which acts as a set of truth values). Having a subobject classifier, Ω, is crucial. Morphisms $X \to \Omega$ act like subobjects of X and Ω may be thought of as a generalized set of truth values. In a sense, working with topoi means working with certain but not all means available in classical set theory and hence in classical mathematics.

A stream of thought independent of Grothendieck's and Lawvere's emerged from Dana Scott's work on Boolean-valued models of set theory (p. 256) and the idea of replacing Boolean by Heyting algebras.[75] Scott's considerations lead to topoi

---

[70] G. Osius, "Categorical set theory: A characterization of the category of sets," *J Pure Appl Algebra* 4 (1974): 79–119.

[71] S. Feferman, "Categorical foundations and foundations of category theory," in *Logic, Foundations of Mathematics, and Computability Theory*, ed. R. E. Butts and J. Hintikka (Dordrecht: Reidel, 1977), 149–69. Also J. L. Bell, "Category theory and the foundations of mathematics," *British J for the Philosophy of Science* 32 (1981): 349–58; G. Hellman, "Does category theory provide a framework for mathematical structuralism?" *Philosophia Mathematica* 11 (2003), 129–57.

[72] From the Greek word "τόπος" which means place.

[73] M. Artin, A. Grothendieck, and J.-L. Verdier, eds., "Séminaire de Géométrie Algébrique du Bois Marie 1963–64: Théorie des topos et cohomologie étale des schémas," *Lect Notes Math* 305 (1972): 269–70.

[74] F. W. Lawvere, "Applications of elementary topos to set theory" (lecture, 1970); M. Tierney, "Elementary topos" (lecture, 1970); report about these lectures is provided by J. W. Gray, "The meeting of the midwest category seminar in Zurich: August 24–30, 1970," *Lect Notes Math* 195 (1971): 248–55.

[75] D. S. Scott, "Extending the topological interpretation to intuitionistic analysis, Part I," *Compositio Mathematica* 20 (1968): 194–210, "Part II," in *Intuitionism and Proof Theory*, ed. J. Myhill, A. Kino, and R. E. Vesley (Amsterdam: North-Holland, 1970), 235–55.

structures and the concept of Heyting-valued sets (Fourman and Scott 1979, Scott 1979) which play an important role in the developments we examine next. Fourman and Scott's approach generalizes an earlier influential approach by Higgs who described a category suitable for a categorical treatment of Boolean-valued sets.[76]

An interesting and highly nontrivial feature is that in any topos one may define arrows which play the role of truth functions of logical connectives on Ω. In this sense, each topos "internalizes" a logic. It is a striking fact that, in general, Ω becomes a Heyting algebra and thus the internal logic of topoi is intuitionistic logic. For certain topoi, such as the category of sets, Ω may even be a Boolean algebra and hence the internal logic is classical. Not only this fact but the whole enterprise of category theory poses interesting problems for fuzzy logic which we examine in what follows.

Category theory and fuzzy logic

The first category-theoretic considerations regarding fuzzy sets appeared in the early contributions by Goguen. In Goguen 1967, he used the language of categories to display some results about fuzzy relations, which practice has later been followed.[77] He also introduced in his considerations regarding fuzzification a construction to obtain from any category $\mathbf{C}$ another category, $L^{\mathbf{C}}$, whose objects are those of $\mathbf{C}$ and whose morphisms are fuzzy sets of morphisms of $\mathbf{C}$.[78] He also introduced the first category of fuzzy sets, $\mathbf{S}(L)$. For a complete residuated lattice L, $\mathbf{S}(L)$ has as objects the pairs $\langle U, A \rangle$ where U is a set and A an L-set in U, i.e. $A : U \to L$, and as morphisms $R : \langle U, A \rangle \to \langle V, B \rangle$ the L-relations $R : U \times V \to L$ satisfying $\bigvee_{u \in U}(A(u) \otimes R(u,v)) \leq B(v)$, along with the $\circ$-composition.[79] Stout (1995) later established that this category is a *quasitopos* (contrary to topos, a quasitopos may only have a *weak* subobject classifier). In his PhD thesis (Goguen 1968), parts of which were later published as Goguen 1969, 1974, he introduced another category, $\mathbf{Set}(L)$. For a partially ordered set L, objects of $\mathbf{Set}(L)$ are pairs $\langle U, A \rangle$ where U is a set and A an L-set in U and morphisms $f : \langle U, A \rangle \to \langle V, B \rangle$ are functions f of U to V satisfying $A(u) \leq B(f(u))$.[80] He proved that if L is a complete lattice, $\mathbf{Set}(L)$ satisfies Lawvere's axioms of category of sets and gave a categorical characterization of the categories $\mathbf{Set}(L)$ for L being a completely distributive lattice. Essentially the same category was considered by Ralescu (1978) in his attempts to systematize fuzzi-

---

[76] Described in an unpublished but frequently cited manuscript, D. Higgs, "A category approach to Boolean valued set theory" (preprint, Univ. of Waterloo, 1973); later extended to Heyting-valued sets in his "Injectivity in the topos of complete Heyting algebra valued sets," *Canadian J Math* 36 (1984): 550–68.

[77] For instance, the fact that taking sets as objects and taking fuzzy relations between sets as morphisms forms a category means that composition of fuzzy relations is associative and has a neutral element. Note for example the large part of Novák et al. 1999 devoted to categorical treatment or several papers by Močkoř, e.g. Močkoř 1999, which utilize category-theoretic language.

[78] Goguen was probably not aware that his definition of composition of morphisms is essentially a definition of the extension principle for functions with two arguments. He established some properties for it. The principle was then available for unary mappings from Zadeh 1965a with no further results. This fact seems unknown.

[79] $(R \circ S)(u,w) = \bigvee_{v \in V} R(u,v) \otimes S(v,w)$.

[80] If L is a complete residuated lattice, $\mathbf{Set}(L)$ is essentially a subcategory of $\mathbf{S}(L)$.

fication of classical concepts in the spirit of level-cut representation (Negoiță and Ralescu 1975b). Interestingly, Pultr (1976) showed how to turn **Set**(L) with L being a residuated lattice into a monoidal closed category[81]—generally an important kind of category which in the setting of fuzzy sets reflects the presence of a possibly nonidempotent conjunction.

Note also that Eytan (1981) described a category of Heyting-algebra-valued sets, similar to Goguen's **S**(L), and claimed that it forms a topos, which was disproved by Pitts.[82] Eytan nevertheless initiated the study of relationships between fuzzy sets and topoi. Stout (1991) and Wyler (1995) surveyed the then available categories of complete Heyting-algebra-valued sets and put them in correspondence with categories of fuzzy sets. The categories they considered contained as objects either pairs $\langle U, A \rangle$ where $A : U \to L$ or pairs $\langle U, E \rangle$ where $E : U \times U \to L$ is a binary L-relation interpreted as a kind of fuzzy equality and satisfying

$$E(u,v) = E(v,u) \qquad \text{(symmetry)}, \qquad (5.8)$$

$$E(u,v) \wedge E(v,w) \leq E(u,w) \qquad \text{(transitivity)}. \qquad (5.9)$$

$E(u,u) = 1$ need not hold and, according to Scott (1979), $\epsilon(u) = E(u,u)$ is interpreted as the extent of u or degree of existence of u. The objects $\langle U, A \rangle$ were considered by Goguen, while the $\langle U, E \rangle$ were studied by Higgs and, independently, Fourman with Scott. Naturally, each $\langle U, E \rangle$ determines a corresponding $\langle U, A_E \rangle$ by putting $A_E(u) = E(u,u)$ and each $\langle U, A \rangle$ induces its corresponding $\langle U, E_A \rangle$ where $E_A(u,v) = A(u)$ if $u = v$ and $E_A(u,v) = 0$ if $u \neq v$. Morphisms were regarded either as ordinary mappings or, more generally, fuzzy relations both satisfying possibly additional restrictions. The latter approach with objects $\langle U, E \rangle$ of the second type is due to the above-mentioned works by Higgs and due to Fourman and Scott (1979), and morphisms $f : \langle U, E \rangle \to \langle V, F \rangle$ in this approach are L-relations $f : U \times V \to L$ satisfying

$$E(u',u) \wedge f(u,v) \wedge F(v,v') \leq f(u',v') \qquad \text{(extensionality)},$$

$$f(u,v) \wedge f(u,v') \leq F(v,v') \qquad \text{(single-valuedness)},$$

$$E(u,u) = \bigvee_{v \in V} f(u,v) \qquad \text{(totality)}.$$

Fourman and Scott (1979) require strictness, meaning $f(u,v) \leq \epsilon(u) \wedge \epsilon(v)$, but this follows from single-valuedness and totality. The morphisms are composed by $(f \circ g)(u,w) = \bigvee_v f(u,v) \wedge g(v,w)$. Due to Higgs, and independently Fourman and Scott (1979), we have the following important theorem about this category, Ω-**Set**, which is hence termed the *Higgs topos*.

Theorem 5.5. Ω-**Set** *forms a topos.*

---

[81]S. Eilenberg and G. M. Kelly, "Closed categories," in *Proc Conf on Categorical Algebra* (Berlin: Springer-Verlag, 1966), 421–562.

[82]A. M. Pitts, "Fuzzy sets do not form a topos," *Fuzzy Sets Syst* 8 (1982): 101–4. Later it became clear that Eytan's category is the category of subobjects of constant objects in the Higgs topos (Stout 1991).

Higgs and Fourman with Scott also proved that Ω-**Set** is categorically equivalent to the important category of sheaves on L. The former approach with the same objects has as morphisms $f : \langle U, E \rangle \to \langle V, F \rangle$ the functions $f : U \to V$ satisfying

$$E(u, u') \leq F(f(u), f(u')). \tag{5.10}$$

Alternatively, Monro considered as morphisms the classes of these mappings modulo the equivalence $\sim$ defined by $f \sim g$ if $E(u, u) \leq F(f(u), g(u))$.[83] He denoted this category $\mathbf{Mod}^0(L)$ and proved:

Theorem 5.6. $\mathbf{Mod}^0(L)$ *forms a quasitopos.*

With objects $\langle U, A \rangle$ both functions and fuzzy relations were considered in the above-mentioned works by Goguen. The restriction in the above-mentioned approaches to Heyting algebras—the structures of intuitionistic logic—and the suggestions in some literature that Heyting algebras should preferably be used for fuzzy sets were not acceptable from a fuzzy logic point of view because they do not account for nonidempotent conjunctions.[84]

Surpassing the intuitionistic framework became the central topic of investigations by Höhle (1991, 1992, 1995b, 2000, 2005, 2007). Equipped with his thorough experience with many facets of the mathematics of fuzzy logic, Höhle authored several papers on categories of fuzzy sets which showed how the results regarding Heyting-algebra-valued sets due to Higgs and Fourman with Scott generalize to the setting of structures of truth degrees in which conjunction need not be idempotent. Höhle chose for L the so-called *GL-monoids*, which are complete residuated lattices satisfying divisibility, i.e. $a \leq b$ implies $a = b \otimes c$ for some c, or *GL-algebras*, which also satisfy $a \otimes \bigwedge_i b_i = \bigwedge_i (a \otimes b_i)$, and started to pursue the problem of finding a suitable category $\mathbf{C}_L$ of L-sets or some other reasonable category to whose objects the L-sets are naturally related. In particular, he required that the category be finitely complete; have a subobject classifier, Ω, and arrow true, t, such that strict and extensional L-valued maps admit internalization as morphisms with codomain Ω; have even uniquely classifiable subobjects that are classifiable by Ω and t; and be categorically equivalent with the Higgs topos if L is a complete Heyting algebra.[85] The seemingly suitable category having as objects the pairs $\langle U, E \rangle$ where $E : U \times U \to L$ is a separated, *global* L-valued equality in that it satisfies (5.8), (5.9), but also $E(u, u) = 1$ and separation, i.e. $u = v$ whenever $E(u, v) = 1$, and as morphisms mappings satisfying (5.10) with the usual composition of mappings, is not suitable because it does not allow for a unique classification of extremal subobjects (Höhle 2000). Höhle indeed found the needed category, the *category of sheaves over a GL-algebra*. Its construction is based on an important category, L-**Set**, which is

[83]G. P. Monro, "Quasitopoi, logic and Heyting-valued models," *J Pure Appl Algebra* 42 (1986): 141–64.

[84]See e.g. Wyler 1995. Exceptions were the above-mentioned work by Pultr and a predecessor of Höhle's later work, U. Cerruti and U. Höhle, "An approach to uncertainty using algebras over a monoidal closed category," *Rendiconti del Circolo Matematico di Palermo* 12, no. 2 (1986): 47–63.

[85]Strictness and extensionality of $A : \langle U, E \rangle \to L$ mean (5.13) and (5.11).

inspired by Ω-**Set**, and a notion of singleton which is inspired by the corresponding notion from Fourman and Scott 1979. *L*-**Set** has for objects so-called (*local*) *L-valued sets* which are pairs $\langle U, E \rangle$ with U a set and E a local *L*-valued equality which is an *L*-relation $E : U \times U \to L$ satisfying

$$E(u,v) \leq E(u,u) \wedge E(v,v) \qquad \text{(strictness)},$$

$$E(u,v) = E(v,u) \qquad \text{(symmetry)},$$

$$E(u,v) \otimes (E(v,v) \to E(v,w)) \leq E(u,w) \qquad \text{(transitivity)}.$$

In this context, separation requires that $E(u,u) \vee E(v,v) \leq E(u,v)$ implies $u = v$. It is easily seen that every (separated) global *L*-valued set is a (separated) local *L*-valued set and that if L is a complete Heyting algebra, then the local *L*-valued sets are just the *L*-valued sets in the sense of Higgs and Fourman and Scott, cf. (5.8) and (5.9). Morphisms $f : \langle U, E \rangle \to \langle V, F \rangle$ are mappings $f : U \to V$ satisfying extensionality, i.e. (5.10), and the following condition of strictness:

$$E(u,u) = F(f(u), f(u)).$$

A *singleton* of $\langle U, E \rangle$ in the present setting is a fuzzy set $A : U \to L$ satisfying

$$A(u) \otimes (E(u,u) \to E(u,v)) \leq A(v) \qquad \text{(extensionality)}, \qquad (5.11)$$

$$A(u) \otimes (\bigvee_{w \in U} A(w) \to A(v)) \leq E(u,v) \qquad \text{(singleton property)}, \qquad (5.12)$$

which also implies that A is strict in that

$$A(u) \leq E(u,u). \qquad (5.13)$$

Note that in particular every $u \in U$ and $a \in L$ determine the singleton $s_{u,a}$ by $s_{u,a}(v) = (E(u,u) \to a) \otimes E(u,v)$ and its particular form (for $a = 1$) s_u by $s_u(v) = E(u,v)$, which in case of global *L*-valued equality coincide with the usually defined singletons w.r.t. fuzzy equality relations examined below, cf. (5.26). The construction of Höhle's category $\mathbf{C}_L$ of sheaves over L proceeds by constructing an *L*-valued set of singletons, a particular monad $\mathbf{T}$ of this *L*-valued set, based on which the sheaves over L are then defined as particular $\mathbf{T}$-algebras. In addition to the above-mentioned properties it is important to note that the truth functions of L, such as the Łukasiewicz negation $1 - a$ in case L is the standard MV-algebra, are obtained as truth arrows in $\mathbf{C}_L$. For further details see Höhle 2000.

The conditions of local *L*-valued equality, separation, and that of singleton are important by themselves because they generalize the usual conditions of transitivity, separation, and singleton to a setting which is very reasonable from a category-theoretic point of view. In this connection it is interesting to note that Höhle 2007 provides numerous examples from various areas and demonstrates how sheaves over complete Heyting algebras provide foundations for and help clarify various issues in mathematics based on fuzzy logic with idempotent conjunction.

5.3 Selected areas of mathematics based on fuzzy logic

In this section we cover significant directions in mathematics based on fuzzy logic. Due to the large number of existing contributions we need to be selective, both in terms of including references and covering existing directions. For each of the presented directions, we describe its motivations, historical background including initial contributions, its development, as well as the basic technical notions involved and—primarily for illustration—some of the results obtained. The directions are grouped in sections which correspond to fields of classical mathematics. The lengths of these sections should not, however, be regarded as indicating the extent or maturity of the directions covered. Nor should the groupings be interpreted as representing a view that mathematics based on fuzzy logic is the result of a systematic fuzzification of the fields of classical mathematics. The significance of particular areas and thus a natural structure of mathematics based on fuzzy logic may be different from those of classical mathematics. For instance, while homogeneous second-order partial differential equations of elliptic type form a significant part of classical mathematics, their fuzzified counterparts may well remain insignificant because there may be no solid motivations for their development or their actual applications. From this perspective, our grouping by sections is primarily a matter of arbitrary convenience.

5.3.1 Sets and relations

The notion of fuzzy set is the central notion of mathematics based on fuzzy logic and its significance is analogous to the significance of the notion of set for classical mathematics. By introducing this notion, Zadeh (1965a) set the stage for developing a theory of fuzzy sets and fuzzy relations. While related foundational aspects regarding fuzzy sets are covered in section 5.2, here we deal with the intuitive theory of fuzzy sets understood as mappings A of an ordinary set U to a set L of truth degrees, such as $[0, 1]$. Having the basic concepts of fuzzy sets covered in section 3.2, we now examine further aspects.

Basic issues in fuzzy sets

The seminal work by Goguen (1967, 1968–69), whose conceptual contributions to mathematics based on fuzzy logic are discussed on pp. 243ff., is of great importance for the present topic. Goguen assumes a clean algebraic view and assumes that the set L of truth degrees should form a structure **L** more general than the two-element Boolean algebra of classical logic, namely that $\mathbf{L} = \langle L, \wedge, \vee, \otimes, \rightarrow, 0, 1 \rangle$ be a complete residuated lattice (section 4.3.1). Conceptually important is his view (Goguen 1967, 149) that **L** may be regarded as a parameter. One may thus develop a general theory of **L**-sets and **L**-relations whose results are valid for every **L**, or every **L** from a given class $\mathscr{L}$, such as the class of all BL-algebras or Heyting algebras.[86] A given condition may then have different meanings for different Ls. Thus the meaning of transitivity of a fuzzy relation R, expressed by $R(u,v) \otimes R(v,w) \leq R(u,w)$, becomes $\min(R(u,v), R(v,w)) \leq R(u,w)$ if L is the standard Gödel algebra, but becomes $\max(0, R(u,v) + R(v,w) - 1) \leq R(u,w)$ for the standard Łukasiewicz algebra. It took time before this view became common—in many of the early contributions, results were proved separately for different structures of truth degrees which were mainly represented by different t-norms. In this view, (characteristic functions of) classical sets and relations are just *all* L-sets and L-relations for the particular case of L being the two-element Boolean algebra, i.e. $L = \{0,1\}$. This is in contrast with a different view of classical sets, namely as *particular* L-sets for any L: the *crisp* L-sets, i.e. the mappings $A : U \rightarrow L$ for which $A(u) \in \{0,1\}$ for every u. The latter view is the original view of Zadeh (1965a), who considered it for $L = [0,1]$.

Related is also Goguen's natural motivation for considering simultaneously different Ls and in particular his emphasis on morphisms $h : \mathbf{L}_1 \rightarrow \mathbf{L}_2$. This view was later adopted by Di Nola and Gerla (1986) and Belohlavek (2002c), where the authors consider general fuzzy relational systems with possibly different Ls.[87] For instance, a morphism then consists of two mappings, $f : U_1 \rightarrow U_2$ of universes and $g : L_1 \rightarrow L_2$ of sets of truth degrees, and similar ideas are applied to congruences, factor structures, products, and substructures; see also Cintula and Hájek 2010. This view is also reflected in the distinction of fixed-basis and variable-basis fuzzy topology (p. 297 below), in which "basis" refers to L.

Goguen also observed that the set L^U of all L-sets in U is naturally equipped with operations and relations inherited from those on L in a componentwise manner. Thus, for instance, the truth function $\otimes$ of conjunction of L induces the corresponding operation, denoted by slight abuse of notation also by $\otimes$, which is defined for $A, B \in L^U$ by $(A \otimes B)(u) = A(u) \otimes B(u)$.[88] In this case, $A \otimes B$ is also called the $\otimes$-intersection of A and B and is often denoted $A \cap_\otimes B$. Likewise, the partial order $\leq$ on L extends to L^U and one has $A \leq B$ iff $A(u) \leq B(u)$ for each $u \in U$, in which case one commonly uses $A \subseteq B$ and calls $\subseteq$ the *inclusion* of fuzzy sets. Goguen observed that the thus obtained structure $\mathbf{L}^U = \langle L^U, \wedge, \vee, \otimes, \rightarrow, 0, 1 \rangle$ of all fuzzy sets $A : U \rightarrow L$ is in fact the direct product of Ls indexed by U and somewhat vaguely says that its operations "obey any law valid in L." More precisely, the following theorem follows from results of classical model theory:[89]

---

[86] Goguen presents his view somewhat less explicitly than we do here. Note also that he uses the terms "*L-fuzzy set*" and "*L-set*"; we use "L-set" to make explicit the structure on L.

[87] Di Nola and Gerla were inspired by Rasiowa and Sikorski (1963), who consider varying sets of truth values as well.

[88] Note that the operation $\otimes^\mathbf{L}$ on L induces the operation $\otimes^{\mathbf{L}^U}$ on L^U by $(A \otimes^{\mathbf{L}^U} B)(u) = A(u) \otimes^\mathbf{L} B(u)$. We adopt the convenient practice of omitting the superscripts $^\mathbf{L}$ and $^{\mathbf{L}^U}$ if no confusion arises.

[89] See proposition 6.2.2 in Chang and Keisler 1990. This result is observed in Schwartz 1972. The concerned formulas are first-order formulas with function symbols for $\otimes$ and other operations and relation symbol for $\leq$.

Theorem 5.7. $\mathbf{L}^U$ *satisfies all Horn formulas satisfied by* $\mathbf{L}$.

Thus, $\subseteq$ is a partial order on $\mathbf{L}^U$ since $\leq$ is a partial order on $\mathbf{L}$, $\otimes$ on $\mathbf{L}^U$ is commutative whenever $\otimes$ on $\mathbf{L}$ is, simply because the properties in question may be described by Horn formulas. More generally, if $\mathbf{L}$ is a residuated lattice then so is $\mathbf{L}^U$ and the same holds for being a BL-algebra, MV-algebra, or other structure expressible by Horn formulas. This observation makes it possible to obtain automatically properties for fuzzy sets from the respective properties of structures of truth degrees.[90]

In addition, further operations, relations, and fuzzy relations were defined for fuzzy sets. Also, various characteristics were defined for fuzzy sets, such as the height or measure of fuzziness mentioned in section 3.1. It is important to note that some of these notions make sense for classical sets while some do not. Examples of the former are the above-mentioned inclusion of fuzzy sets or the height of fuzzy set defined by $h(A) = \bigvee_{u \in U} A(u)$. Examples of the latter are the averages of fuzzy sets mentioned in section 3.3, e.g. the simple average $[h(A, B)](u) = \frac{A(u)+B(u)}{2}$. The conceptual difference is that the language in which the former notions are defined is a language of a class $\mathscr{L} = \{\mathbf{L} \mid \cdots\}$ of structures which contains the two-element Boolean algebra $\mathbf{2}$, and hence the notions make sense for classical sets. Since they make sense for classical sets, they are in fact *generalizations* of classical notions to fuzzy sets, namely to $\mathbf{L}$-sets with $\mathbf{L} \in \mathscr{L}$, even though they may not be commonly used for classical sets. This is because while useful for general fuzzy sets, these notions may be degenerate for classical sets. For example, the a-cut of a (characteristic function of a) classical set A is the whole universe for $a = 0$ and A itself for $a = 1$ which are the only options because there is no other truth degree in the classical case. Of a similar nature is the concept of height. Hence, while making sense for fuzzy sets, these notions are not recognized for classical sets. It is in this sense that these notions may be regarded as new notions appearing in the theory of fuzzy sets. The discovery of such notions is a part of the process of developing mathematics based on fuzzy logic and the concepts of a-cut and height are just particular examples. On the other hand, notions which do not make sense for classical sets, such as averages, are defined in a language of a class $\mathscr{L}$ of structures which does not contain the two-element Boolean algebra. This language typically expands the language of Boolean algebras by symbols which do not have a reasonable interpretation in Boolean algebras, such as those for addition or division. In this sense, they are new notions appearing in the theory of fuzzy sets but they are *not generalizations* of classical ones.

Particularly important are the *degrees of inclusion* and *equality* of fuzzy sets,

$$S(A, B) = \bigwedge_{u \in U} (A(u) \to B(u)) \text{ and} \tag{5.14}$$

$$A \approx B = \bigwedge_{u \in U} (A(u) \leftrightarrow B(u)), \tag{5.15}$$

---

[90] One thus need not prove them from scratch as e.g. in J. G. Brown, "A note on fuzzy sets," *Inf Control* 18 (1971): 32–39, or A. De Luca and S. Termini, "Algebraic properties of fuzzy sets," *J Math Anal Appl* 40 (1972): 373–86.

which generalize inclusion and equality of fuzzy sets. $S(A,B)$ is naturally interpreted as the truth degree of the proposition "for each $u \in U$, if u is in A then u is in B" and similarly for $A \approx B$. If $\rightarrow$ is the residuum and thus satisfies that $a \rightarrow b = 1$ iff $a \leq b$, then $S(A,B) = 1$ and $A \approx B = 1$ are equivalent to $A \subseteq B$, as defined above, and $A = B$, respectively. Hence $\subseteq$ is just the 1-cut of the fuzzy relation S, and the same for $=$ and $\approx$. For crisp fuzzy sets and in particular for $L = \{0,1\}$, S may be identified with $\subseteq$ and thus gives nothing new. Interestingly, for residuated structures of truth degrees, S and $\approx$ satisfy natural generalizations of properties known for ordinary sets such as

$$S(A,B) \otimes S(B,C) \leq S(A,C) \quad \text{and} \quad (A \approx B) \otimes (B \approx C) \leq A \approx C \qquad (5.16)$$

or $(A \approx B) \otimes S(B,C) \otimes (C \approx D) \leq S(A,D)$, $S(A, \bigwedge_i B_i) = \bigwedge_i S(A,B_i)$, and many other generalizations of classical properties of inclusion and equality of sets (Gottwald 2001, Belohlavek 2002c). Note that (5.16) may be interpreted as claiming that if A is approximately included in B and B is approximately included in C then A is approximately included in C. In fact, (5.16) says that the fist-order formula representing such a claim assumes 1 as its truth degree. Another way of looking at (5.16), which is virtually trivial in the classical case, is that the application of $\otimes$ to $S(A,B)$ and $S(B,C)$ provides a lower bound of $S(A,C)$. Clearly, the same considerations pertain to the other inequality in (5.16).

The degree $S(A,B)$ was introduced by Goguen (1968–69) but it appeared earlier along with several results in the early studies of fuzzy sets by Klaua (1966) and the related later work by Gottwald (1976).[91] In fact, a precursor of $S(A,B)$ was already discussed by Black (1937), who considered but did not definitely define a degree $i(B,A)$ of deviation from full inclusion $B \supseteq A$.[92] Clearly, i and S are conceptually related by $i(B,A) = 1 - S(A,B)$. Black even considers a form of transitivity for i: he asserts $i(C,A) \leq i(C,B) + i(B,A)$ which for $i = 1 - S$ is equivalent to $S(A,B) + S(B,C) - 1 \leq S(A,C)$, which is equivalent to (5.16) for $\otimes$ being the Łukasiewicz t-norm. The degree of inclusion (5.14) and its slight variations later reappeared in works by Giles (1976) who proves some results for it, Bandler and Kohout (1980a,c) who examine inclusions defined by (5.14) for various functions $\rightarrow$ including nonresiduated implications, and Willmott (1986) who considers the transitivity of S in the sense of (5.16) for Łukasiewicz $\otimes$ and various implications including those from Bandler and Kohout 1980a. In the late 1980s, graded inclusions along with implications (pp. 179ff.) were examined by Kitainik, mostly in papers written in Russian. Kitainik also considered axiomatic conditions for graded inclusions and representation theorems (Kitainik 1993). Axiomatic conditions were independently considered by Sinha and Dougherty (1993). Conceptually different is the approach by Wygralak (1983), who construes the degree of equality of fuzzy sets as a pair of truth degrees and applies this view as well to operations with fuzzy sets.

---

[91] Klaua's and Gottwald's works are examined in section 5.2.3.

[92] Black suggested that $i(B,A)$ be based on the ratio $|\{u \mid B(u) < A(u)\}|/|\{u \mid B(u) > A(u)\}|$.

Goguen (1968–69) also observed that for (5.14), a single element u for which $A(u) \to B(u) = 0$ causes $S(A,B) = 0$, which might be seen as desirable but also as disadvantageous in some cases. He thus suggested that S be alternatively based on other than the infimum aggregation of the truth degrees $A(u) \to B(u)$. Such an approach was indeed, without reference to Goguen, later developed by Kosko (1986b), who defined for standard fuzzy sets in a finite universe

$$S(A,B) = [\textstyle\sum_{u \in U} \max(0, A(u) - B(u))]/|A|,$$

where $|A| = \sum_{u \in U} A(u)$, and observed that then for the min-intersection $\cap$, one obtains $S(A,B) = \frac{|A \cap B|}{|A|}$. Notice, however, that Goguen's hypothetical objection applies to classical set inclusion as well. Therefore, Goguen's, Kosko's and other proposals of alternative graded inclusions are in fact also proposals for alternatives to classical inclusion and are motivated by a phenomenon orthogonal to many-valuedness. Similar reasoning resulted in many proposals of degrees of equality of fuzzy sets, which are discussed below in connection with fuzzy equivalences.

Representations of fuzzy sets

Interesting questions arise when one asks for relationships between (intuitively understood) fuzzy sets and classical sets and, more generally, between various notions of mathematics based on fuzzy logic and classical mathematics. This leads us to the problem of representation of fuzzy sets.

Already in his seminal paper of 1965, Zadeh defined for a fuzzy set A in U and truth degree a the ordinary set $^aA = \{u \in U \mid a \le A(u)\}$, which we called the a-cut in section 3.2.[93] Zadeh (1971b) observed that the system $\{^aA \mid a \in [0,1]\}$ is nested in that $a \le b$ implies $^aA \supseteq {}^bA$ and that A is reconstructed by

$$A = \bigcup\nolimits_{a \in (0,1]} a \cdot {}^aA, \text{ where } (a \cdot {}^aA)(u) = \begin{cases} a & \text{for } u \in {}^aA, \\ 0 & \text{for } u \notin {}^aA. \end{cases}$$

Negoiţă and Ralescu (1975b) developed this representation further. They assumed fuzzy sets with truth degrees in a complete lattice satisfying for every $a \in L$ and $K \subseteq L$ that $a < \bigvee K$ implies $a \le b$ for some $b \in K$—a condition verified by any chain, thus also $[0,1]$. They characterized systems of a-cuts as systems $\mathscr{A} = \{A_a \subseteq U \mid a \in L\}$ satisfying

$$A_0 = U \quad \text{and} \quad A_{\bigvee K} = \bigcap\nolimits_{a \in K} A_a \text{ for every } K \subseteq L. \tag{5.17}$$

Such systems are naturally ordered by putting $\mathscr{A} \sqsubseteq \mathscr{B}$ iff $A_a \subseteq B_a$ for each $a \in L$ and form a complete lattice $\langle \mathscr{C}(U), \sqsubseteq \rangle$. Negoiţă and Ralescu (1975b) proved the following representation theorem:[94]

Theorem 5.8. The mapping assigning to every fuzzy set $A \in L^U$ the system of all its a-cuts is an isomorphism between the complete lattices $\langle L^U, \subseteq \rangle$ of all fuzzy sets in U

---

[93]Zadeh (1971b) calls ^aA the a-level-set of A. In optimization, this notion is known as an a-superlevel set.

[94]In fact, they established a dual isomorphism between $\langle L^U, \subseteq \rangle$ and the set of so-called L-flou subsets of U, a notion generalizing flou sets by Gentilhomme (1968). What we present is easily extracted from their proof.

and $\langle \mathscr{C}(U), \subseteq \rangle$ of all systems satisfying (5.17). Its inverse mapping assigns to a system $\mathscr{A} \in \mathscr{C}(U)$ the fuzzy set A defined by $A(u) = \bigvee \{a \in L \mid u \in A_a\}$.

Note that (5.17) implies that $\mathscr{A}$ is nested. Later on, other characterizations of systems of a-cuts have been considered. For instance, Gerla (2001) characterizes them—in case $L = [0,1]$—as nested systems satisfying $A_0 = U$ and $A_a = \bigcap_{b<a} A_b$; Belohlavek (2002c)—in case L is an arbitrary complete lattice—as systems which are nested and such that for each $u \in U$ the set $\{a \in L \mid u \in A_a\}$ has a greatest element. On their common instances these characterizations are equivalent.

The representation by a-cuts became important in connection with fuzzy quantities and the extension principle, a topic examined in section 5.3.4, but also as a conceptual link between classical notions and their fuzzified versions. An example of such a link was presented by Negoiță and Ralescu (1975b), who established an isomorphism between the lattice of all fuzzy subgroups of a given group G (section 5.3.2) and the lattice of $[0, 1]$-indexed systems of classical subgroups of G. In view of theorem 5.8, the essence of this relationship is the fact that A is a fuzzy subgroup of G iff each ^aA is a classical subgroup of G. Similar observations were soon made for other situations. Bandler and Kohout (1993) thus called a property, such as symmetry, which applies to both fuzzy and ordinary relations, *cutworthy* if any fuzzy relation R has this property if and only if each a-cut ^aR has this property as a classical relation. For example, symmetry, asserting that $R(u,v) = R(v,u)$ for every u, v, is cutworthy because it applies iff each ^aR is symmetric as a classical relation. The same holds for transitivity if defined by $R(u,v) \wedge R(v,w) \leq R(u,w)$. However, transitivity ceases to be cutworthy if defined w.r.t. a general t-norm $\otimes$, different from $\wedge$. Many observations on cuts of fuzzy relations are contained in De Baets and Kerre 1994. Moreover, Klir and Yuan (1995) speak of cutworthy operations. For instance, min-intersection of fuzzy sets is cutworthy in that $^a(A \cap B) = {}^aA \cap {}^aB$. The complement of a fuzzy set, $\overline{A}$, however, is not cutworthy. The tempting option to define properties of fuzzy relations by properties of their cuts or to define operations with fuzzy sets in a cut-by-cut manner has therefore serious limitations. Nevertheless, even though not cutworthy, the standard negation obeys $^a\overline{A} = \overline{\bigcup_{1-a<b} {}^bA}$, and thus the a-cut of the complement of the fuzzy set A is expressed in terms of classical complement and certain b-cuts of A, but in a more complicated manner than plain cutworthiness. Belohlavek (2003b) proposed a general approach. Within the framework of predicate fuzzy logic, he showed that cutworthiness of both properties and operations discussed in the literature may be seen as a general property that he called cutworthiness of logical formulas. In particular, he presented a theorem enabling one to describe constructively the a-cut of an output fuzzy set in terms of cuts of the arguments for any operation on fuzzy sets definable by logical formulas as well as to describe properties of fuzzy relations definable by logical formulas in terms of classical properties of their cuts. This approach is applicable to the examples above as well as to others including the important extension principle.

Cartesian representation is another representation of fuzzy sets by classical sets. It is much less known and appeared independently in the following works. Lowen (1978, 453) used it implicitly to define a functor from the category of fuzzy topological spaces to the category of classical topological spaces (p. 297 below); Head (1995), again implicitly, in his metatheorem regarding fuzzy subalgebras; Gerla (1994b) in representing fuzzy closure operators by classical ones; and Belohlavek (2002c) in several representation theorems regarding closure structures and Galois connections. The idea is to represent a fuzzy set $A \in L^U$ by the subset $\lfloor A \rfloor$ of the Cartesian product $U \times L$—the area below A—defined by[95]

$$\lfloor A \rfloor = \{\langle u, a \rangle \in U \times L \mid a \leq A(u)\}.$$

From $\lfloor A \rfloor$, A is retrieved by the mapping assigning to any subset $B \subseteq U \times L$ the fuzzy set $\lceil B \rceil(u) = \bigvee\{b \mid \langle u, b \rangle \in B\}$. For further information see Belohlavek 2002c and Gerla 2001.

In several papers (Fortet and Kambouzia 1976, Orlov 1978, Höhle 1981, Goodman 1980, Wang and Sanchez 1982), the idea of considering *fuzzy sets as random sets* has appeared.[96] The concept of random set differs from the notion of random variable in that it assigns to experimental outcomes sets of values rather than individual values. The primary view of such a concept is that it accounts for situations in which the exact value of the original, underlying random variable is not known exactly. More precisely, a random set in a universe U is essentially a measurable set-valued function $S : \Omega \to 2^U$ for some probability spaces $\langle \Omega, \mathscr{B}, P \rangle$ and $\langle 2^U, \mathscr{C}, S^{-1} \circ P \rangle$ where the σ-algebra $\mathscr{C}$ contains at least the classes of all sets containing u, for each $u \in U$.[97] A given random set S naturally induces a fuzzy set A_S by putting

$$A_S(u) = P\{\omega \in \Omega \mid u \in S(\omega)\},$$

and this is actually the only fuzzy set one-point-equivalent to it. This relationship between fuzzy sets and random sets, which is not one-to-one, has been studied along with questions regarding interpretation and other problems in a number of papers. In particular, see the overview in Goodman 1998.

Extension principles

In addition to Zadeh's extension principle, which we present in section 3.2 and further examine in sections 5.2.1 and 5.3.4, a number of further extension principles which allow one to extend functions from U to V, or from subsets of U to subsets of V, to functions assigning to fuzzy sets in U fuzzy sets in V, or which can act in similar scenarios, have been proposed. Some of these are mentioned in section 5.3.4; see also Dubois and Prade 2000 (pp. 50ff., 500). Note also that extensions of closure operators by Gerla are examined in section 4.3.2, pp. 173ff.

---

[95] In theory of functions, $\lfloor A \rfloor$ is known as the *hypograph* of A.

[96] Also in an unpublished 1976 paper by Goodman, see Goodman 1998.

[97] For a general concept see G. Matheron, *Random Sets and Integral Geometry* (New York: Wiley, 1975).

Fuzzy relations

Fuzzy relations represent a concept of fundamental importance for mathematics based on fuzzy logic as well as for applications of fuzzy logic. They appear in several parts of this book, particularly in section 3.5, in which we examine some basic issues. It is worth noting that research on fuzzy relations has steadily revealed the importance of residuated structures of truth degrees and in general the role of fuzzy logic in the narrow sense for mathematics based on fuzzy logic. We now continue our exploration of fuzzy relations by examining certain aspects of fuzzy equivalences and orderings, perhaps the most important types of fuzzy relations.

Fuzzy equivalences

As mentioned in section 3.5, fuzzy equivalences were introduced by Zadeh (1971b) under the name *similarity relations* as reflexive, symmetric, and transitive fuzzy relations. In the general context of residuated lattices **L**, these conditions require

$$E(u, u) = 1, \tag{5.18}$$

$$E(u, v) = E(v, u), \tag{5.19}$$

$$E(u, v) \otimes E(v, w) \leq E(u, w), \tag{5.20}$$

for any $u, v, w \in U$. In this case E is also called an $\otimes$-*fuzzy equivalence*. We made several remarks on fuzzy equivalences in section 5.2.1 where we also mentioned that from a logical viewpoint, (5.18)–(5.20) mean exactly that the classical axioms of reflexivity, symmetry, and transitivity are true to degree 1. Zadeh's similarity relations are a particular case of fuzzy equivalences for $L = [0, 1]$ and $\otimes = \min$.[98] The probabilistic indistinguishability relations of Menger (1951b), see p. 26, are just fuzzy equivalences for $L = [0, 1]$ and $\otimes$ being the product. For $L = [0, 1]$ and the Łukasiewicz t-norm for $\otimes$ we obtain Ruspini's (1982) likeness relations, but note that this case as well as the minimum and product are already contained in Bezdek and Harris 1978. For a general t-norm $\otimes$, we obtain the indistinguishability relations of Trillas (1982). For general complete residuated lattices we obtain the L-nearness from the early paper by Pultr (1982). It later became common to call *fuzzy equality* a fuzzy equivalence satisfying the condition

$$E(u, v) = 1 \text{ implies } u = v, \tag{5.21}$$

called *separation*. Interestingly, for a complete Boolean algebra **L**, fuzzy equalities are just the Boolean-valued equalities used by Scott (1967). An important example is the fuzzy equality $\approx$ on the set L^U of all fuzzy sets defined by (5.15).

Even though classical equivalences are sometimes considered as models of similarity, they are inadequate for this purpose because they do not account for the basic fact that similarity comes in grades. Another problem is transitivity condition of classical equivalences which is connected to the Poincaré paradox examined in sec-

---

[98]Zadeh (1971b) already considered fuzzy equivalences for $\otimes$ other than min and noted explicitly the product.

tion 5.4.2. This is why similarities were mostly modeled by various distance functions with the understanding that smaller distance means greater similarity. The move from classical to fuzzy logic makes the notion of equivalence relation a much more realistic model of similarity. While the role of transitivity in this context is examined in section 5.4.2, we now consider the apparent similarities of the conditions (5.18)–(5.20) and (5.21) for fuzzy equivalences to the notions of metric and ultrametric.[99] The fact that via the notion of fuzzy equivalence, fuzzy logic has a natural means of modeling the similarity phenomenon without the necessity of resorting to the extralogical notion of metric, is of great conceptual importance.

Zadeh (1971b) observed that there is a one-to-one relationship between min-equivalences E and $[0,1]$-valued ultrametrics δ on a set U given by $E = 1 - \delta$, and this relationship has further been generalized by Valverde (1985). Valverde (1985) also proved a general theorem relating pseudometrics and fuzzy equivalences for Archimedean t-norms $\otimes$.[100] This theorem has been generalized by De Baets and Mesiar (2002) who proved:

Theorem 5.9. If E is a $\otimes$-fuzzy equality for a t-norm $\otimes$ with an additive generator f, then $\delta = f \circ E$ is a metric on U. If δ is a metric on U and $\otimes$ a continuous Archimedean t-norm with an additive generator f, then $E = f^{(-1)} \circ \delta$ is a $\otimes$-fuzzy equality. The same holds for fuzzy equivalences and pseudometrics.

Thus, as $f(x) = -\log(x)$ is a generator of the product t-norm, $\cdot$, and $f^{(-1)}(y) = e^{-y}$, any metric δ induces a $\cdot$-equality $E(u,v) = e^{-\delta(u,v)}$, a fact observed for $\delta(u,v) = |u - v|$ by Zadeh (1971b). Since $f(x) = 1 - x$ is a generator of the Łukasiewicz t-norm $\otimes$ and $f^{(-1)}(y) = \max(0, 1-y)$, any metric δ induces a $\otimes$-equality $E(u,v) = \max(0, 1 - \delta(u,v))$.

Valverde (1985) also obtained an important representation theorem for fuzzy equivalences. This theorem, present also in a previous paper (Trillas and Valverde 1984), is connected to the famous *Leibniz rule* according to which two objects are

---

[99] A pseudometric on a set U is a mapping $\delta : U \times U \to [0, \infty)$ which satisfies $\delta(u, u) = 0$, $\delta(u, v) = \delta(v, u)$, and $\delta(u, v) + \delta(v, w) \geq \delta(u, w)$ (triangle inequality); δ is called an ultrapseudometric if it satisfies $\max(\delta(u, v), \delta(v, w)) \geq \delta(u, w)$ (stronger than triangle inequality); a metric (ultrametric) is a pseudometric (ultrapseudometric) if $\delta(u, v) = 0$ implies $u = v$.

[100] Recall important facts regarding t-norms (Klement, Mesiar, and Pap 2000). A strictly decreasing function $f : [0, 1] \to [0, \infty]$ with $f(1) = 0$ for which $x, y \in f([0, 1])$ implies $x + y \in f([0, 1])$ or $x + y > f(0)$ is called an *additive generator*. A *pseudoinverse* of a nonincreasing function $f : [0, 1] \to [0, \infty]$ is the function $f^{(-1)} : [0, \infty] \to [0, 1]$ defined by $f^{(-1)}(y) = \sup\{x \in [0, 1] \mid f(x) > y\}$. If f is an additive generator, then the function $\otimes : [0, 1]^2 \to [0, 1]$ defined by

$$a \otimes b = f^{(-1)}(f(a) + f(b)) \tag{5.22}$$

is a t-norm and f is its additive generator. If f is an additive generator of $\otimes$ then the continuity of f is equivalent to the continuity of $\otimes$. Each t-norm with an additive generator is *Archimedean*, i.e. for each $a, b \in (0, 1)$ there exists n such that $a^n < b$ where a^n stands for $a \otimes \cdots \otimes a$ with n times a. For continuous t-norms, this Archimedean property is equivalent to $a \otimes a < a$ for each $a \in (0, 1)$. Not all t-norms have an additive generator and min is an important example. However, it holds that a function $\otimes : [0, 1]^2 \to [0, 1]$ is a continuous Archimedean t-norm iff there exists a continuous additive generator f for which (5.22) holds.

identical if and only if they posses the same properties.[101] If properties are construed by fuzzy sets, then fuzzy equivalences are just the relations generated by the Leibniz rule as shown by the next theorem in which **L** is any complete residuated lattice and $\leftrightarrow$ is its biresiduum:[102]

Theorem 5.10. $E : U \times U \to L$ is a fuzzy equivalence if and only if there exists a set $\mathscr{S}$ of fuzzy sets $A : U \to L$ such that $E(u,v) = \bigwedge_{A \in \mathscr{S}}(A(u) \leftrightarrow A(v))$.

    Another important relationship concerns fuzzy equivalences and fuzzy partitions, which are also discussed in section 3.5. Several notions of a fuzzy partition have appeared in the literature, including one implicitly present in Ruspini 1969; see p. 68 above. Such approaches were mostly inspired by pattern recognition needs, were formulated in terms of arithmetic rather than logical operations, and did not yield a relationship between partitions and equivalences known from the classical case. Such a relationship is obtained with the following notion, which is clearly a generalization of the classical notion of partition. A *fuzzy partition* on a set U is a set $\mathscr{P}$ of fuzzy sets in U satisfying the following conditions: (P1) for each $A \in \mathscr{P}$ there exists $u \in U$ such that $A(u) = 1$; (P2) for each $u \in U$ there exists $A \in \mathscr{P}$ such that $A(u) = 1$; (P3) for every $A, B \in \mathscr{P}$ and $u \in U, A(u) \otimes B(u) \leq A \approx B$, see (5.14). This simple notion of fuzzy partition is from Belohlavek 2002c. A condition equivalent to (P3)—saying that the degree of $\otimes$-intersection of A and B is not greater than the degree of equality—appears in Gottwald 1993, but the notion of fuzzy partition there has certain disadvantages, e.g. the lack of a one-to-one correspondence between fuzzy partitions and equivalences. Gottwald's condition also appears in a similar context on p. 197 in Kruse, Gebhardt, and Klawonn 1994 and in De Baets and Mesiar 1998. In the latter paper, a definition equivalent to (P1)–(P3) is presented whose conditions, unlike (P1)–(P3), are not independent. Note also that Höhle (1998) has a more general definition of a fuzzy partition in the context of local L-valued equalities which naturally accommodates the notion of fuzzy partition and fuzzy equivalence on a fuzzy set. Now, for a fuzzy partition $\mathscr{P}$, define a fuzzy relation $E_{\mathscr{P}}$ by $E_{\mathscr{P}}(u,v) = A(v)$ where $A \in \mathscr{P}$ is such that $A(u) = 1$. According to Zadeh (1971b), for a fuzzy equivalence E on U and $u \in U$ let $[u]_E$ denote the fuzzy set in U, called the *class of u*, defined by $[u]_E(v) = E(u,v)$. The set $U/E = \{[u]_E \mid u \in U\}$ of all classes of U is called the *factor set* of U modulo E. One then obtains:

Theorem 5.11. The mappings $E \mapsto U/E$ and $\mathscr{P} \mapsto E_{\mathscr{P}}$ are mutually inverse bijections between the sets of all fuzzy equivalences and all fuzzy partitions on U.

---

[101]"Discourse on metaphysics," in *G. W. Leibniz: Philosophical Papers and Letters*, 2nd ed. trans. and ed. L. Loemker, (Dordrecht: Reidel, 1969), 303–30. In its original formulation the rule says that no two distinct things exactly resemble each other.

[102]Valverde proved this for $\otimes$ being a continuous t-norm. The generalized version is found in Belohlavek 2002c, along with a variant regarding weights expressing importance of attributes.

Many further problems connected with fuzzy equivalences, including their isometries, dimensions, bases, or aggregation of fuzzy equivalences, which were studied by the above-mentioned researchers as well as by Alsina, Castro, Boixader, Jacas, Ovchinnikov, Recasens, and Yager, may be found in the book Recasens 2010.

Fuzzy equalities

Similarity represents one major way of interpreting fuzzy equivalences. The other is to look at them, in particular at fuzzy equalities, as many-valued identities. From this point of view, a fuzzy equality E is understood as a fundamental indiscernibility inherently present on the underlying universe set U.[103] All the fuzzy relations and functions considered on U should then naturally be compatible with E. Such compatibility is naturally expressed by the classical axioms of equality, which are the following first-order formulas of a language with the symbol $\approx$ of equality, the last two of which are required for arbitrary symbols f of function and r of relation:

$$x \approx x, \quad x \approx x \rightarrow y \approx y, \quad (x \approx y) \otimes (y \approx z) \rightarrow (x \approx z),$$
$$(x_1 \approx y_1) \otimes \cdots \otimes (x_n \approx y_n) \rightarrow f(x_1, \ldots, x_n) \approx f(y_1, \ldots, y_n),$$
$$(x_1 \approx y_1) \otimes \cdots \otimes (x_n \approx y_n) \otimes r(x_1, \ldots, x_n) \rightarrow r(y_1, \ldots, y_n).$$

Alternatively, the last two axioms are called the axioms of *identity, compatibility, congruence,* or *extensionality*.[104] It is easy to see that if $\approx$, f, and r are interpreted by an L-relation $\approx'$, ordinary function f', and L-relation r' in U, then in a residuated lattice **L**, the axioms assume the truth degree 1 if and only if $\approx'$ is a fuzzy equivalence which satisfies for every $u_i, v_i \in U$,[105]

$$(u_1 \approx v_1) \otimes \cdots \otimes (u_n \approx v_n) \leq f'(u_1, \ldots, u_n) \approx f'(v_1, \ldots, v_n), \qquad (5.23)$$
$$(u_1 \approx v_1) \otimes \cdots \otimes (u_n \approx v_n) \otimes r'(u_1, \ldots, u_n) \leq r'(v_1, \ldots, v_n). \qquad (5.24)$$

Many-valued equalities with variants of the equality axioms seem to have appeared for the first time in the work of Rasiowa[106] and, in the context of fuzzy logic, in Thiele 1958. They also appear in the early work on Boolean models of set theory by Scott (1967) and explicitly in the above form in his pioneering work on identity and existence in intuitionistic logic (Fourman and Scott 1979, Scott 1979), which is mentioned in section 5.2.4. The equality axioms appeared in fuzzy logic in the late 1990s in works by Höhle (1996), Novák (see e.g. Novák et al. 1999), and Hájek (1998b). In particular, given a set with fuzzy equality or equivalence, $\langle U, \approx \rangle$, *exten-*

---

[103] If identity is considered as an extreme case of similarity, the second interpretation may be regarded as an extreme case of the first.

[104] In the context of many-valued logic, extensionality was first mentioned by Skolem (1957) within Łukasiewicz logic. The above axioms are first considered in Rasiowa and Sikorski 1963. Later they appeared in the works of Scott, Higgs, and Fourman (section 5.2.4).

[105] Recall that we distinguish a symbol s, such as $\otimes$, $\approx$, or f, from its interpretation s', i.e. $\otimes'$, $\approx'$, or f', and, following common practice, drop the superscript if there is no danger of confusion.

[106] H. Rasiowa, "Algebraic models of axiomatic theories," *Fund Math* 41 (1954): 291–310.

sional fuzzy sets in U, i.e. fuzzy sets $A : U \to L$ satisfying

$$A(u) \otimes (u \approx v) \leq A(v) \tag{5.25}$$

became of interest (Kruse, Gebhardt, and Klawonn 1994; Klawonn and Castro 1995), because (5.25) represents the natural condition that may verbally be described as "if u is in A and u is indistinguishable from v then v belongs to A." In particular, it has been observed that the operator

$$A \mapsto C_{\approx}(A), \text{ where } [C_{\approx}(A)](v) = \bigvee\nolimits_{u \in U} A(u) \otimes (u \approx v),$$

which assigns to A the least extensional fuzzy set containing A, is a closure operator which was studied by Klawonn and Castro (1995) and later by several other researchers. Systems of extensional fuzzy sets and the closure operators $C_{\approx}$ were axiomatized in Klawonn and Castro 1995 and Belohlavek 2002c. Of particular importance, both for theory and applications, is the notion of singleton w.r.t. fuzzy equality $\approx$, a particular case of which are fuzzy sets $s_u : U \to L$ defined for $v \in U$ by

$$s_u(v) = u \approx v \quad \text{or more generally} \quad s_{u,a}(v) = a \otimes (u \approx v) \tag{5.26}$$

for fixed $u \in U$ and $a \in L$. Singletons and their foundational role go back to the above-mentioned Fourman and Scott 1979 and Scott 1979. These ideas were accommodated to the setting of fuzzy logic by Höhle (pp. 266ff. above). Note that if $\approx$ is crisp, and hence represents classical identity, then s_u and $s_{u,a}$ are the commonly used singletons $\{^1/u\}$ and $\{^a/u\}$.

For further information about fuzzy logics with equality, extensional fuzzy sets and their role in mathematics based on fuzzy logic, see particularly Belohlavek 2002c, Gottwald 2001, and Recasens 2010.

Fuzzy orderings

Fuzzy orderings were considered for the first time by Menger (1951b), who denoted them as P and interpreted $P(u, v) \in [0, 1]$ as the probability that u precedes v. He introduced transitivity as for his probabilistic indistinguishabilities, i.e. as transitivity w.r.t. product t-norm (p. 274 and p. 26), and asymmetry in the form $P(u, v) + P(v, u) = 1$. Independently and along with fuzzy equivalences, Zadeh (1971b) defined *fuzzy partial orderings* as binary fuzzy relations $\preceq : U \times U \to [0, 1]$ on a set U which are reflexive, transitive, and antisymmetric. While he naturally defined reflexivity and transitivity as for fuzzy equivalences, i.e. by

$$u \preceq u = 1 \quad \text{and} \quad (u \preceq v) \otimes (v \preceq w) \leq (u \preceq w),$$

he defined antisymmetry, which is a considerably more delicate property, by requiring that

$$u \preceq v > 0 \text{ and } v \preceq u > 0 \text{ imply } u = v. \tag{5.27}$$

Clearly, crisp fuzzy partial orderings are precisely the characteristic functions of classical partial orderings and from this viewpoint, Zadeh's concept generalizes the classical one.[107] In fact, Zadeh only considered $\otimes = \min$, due to which he was able to observe that fuzzy partial orders are just fuzzy relations $\preceq$ whose a-cuts $^a\!\preceq$ are classical partial orders. Zadeh also defined various further notions, such as that of linear fuzzy ordering, by additionally requiring completeness meaning that $u \preceq v > 0$ or $v \preceq u > 0$ for every $u, v \in U$, for which he was able to prove a generalization of Szpilrajn's theorem, according to which every fuzzy partial order $\preceq$ may be extended to a linear one, $\preceq_1$, in that $u \preceq v > 0$ implies $u \preceq v = u \preceq_1 v$.[108]

Zadeh's notion of fuzzy partial ordering and related notions became the starting point of many studies in *fuzzy preference* modeling, which aims at modeling preferences that come in degrees. This area started with Orlovsky 1978, has been investigated particularly in the early 1990s in the works of Fodor, Ovchinnikov, and Roubens, and has continued to be investigated ever since. The book by Fodor and Roubens (1994) provides a comprehensive account of the most important investigations until the mid-1990s.

In the works on fuzzy preferences, Zadeh's notions regarding orderings were reexamined and some redefined, such as that of completeness of which the new proposed version was $(u \preceq v) \oplus (v \preceq u) = 1$ for a suitable t-conorm. What remained unrealized for quite a long time was that Zadeh's notion of antisymmetry, and hence his notion of fuzzy partial order—which in the meantime started to appear in textbooks on fuzzy sets—is not a good generalization of the classical notion of partial order. The reason is that in Zadeh's formulation of antisymmetry, order is fuzzy but equality remains crisp. As a consequence, several fundamental properties which hold in classical mathematics are lost when the notion of partial order is understood in Zadeh's sense, which is a serious shortcoming given the importance of the notion of partial order in mathematics. The inclusion relation $\subseteq$, which is a textbook case of a classical partial order, serves as a simple example. Namely, its fuzzified version—Goguen's graded inclusion S defined by (5.14)—is not a fuzzy partial order in Zadeh's sense even though it otherwise has nicely analogous properties to those of classical inclusion. In particular, S violates Zadeh's antisymmetry.[109] Thus while in the classical case $\langle 2^U, \subseteq \rangle$ is an important partially ordered set, this no longer holds for $\langle L^U, S \rangle$ in the generalized setting of fuzzy logic with Zadeh's notion of fuzzy partial order. An answer to the legitimate question of why this happened and why Zadeh's notion kept reappearing even in influential textbooks is that the notion was not obtained from natural examples. Rather, it was obtained by a formalistic fuzzification for which the examples were constructed ex post facto.[110] Furthermore, the concept was not put on a test—while classical partial order is employed in many fundamen-

---

[107] Classical orderings are binary relations $\leq$ satisfying for every u, v, w the conditions $u \leq u$ (reflexivity); if $u \leq v$ and $v \leq w$ then $u \leq w$ (transitivity); and if $u \leq v$ and $v \leq u$ then $u = v$ (antisymmetry).

[108] E. Szpilrajn, "Sur l'extension de l'ordre partiel" [On extension of a partial order], *Fund Math* 16 (1930): 386–89.

[109] Consider $U = \{u\}$, $A = \{^1/u\}$, and $B = \{^{0.5}/u\}$. Then $S(A, B) = 0.5 > 0$, $S(B, A) = 1 > 0$ but $A \neq B$.

[110] This is rather unusual for Zadeh. As a rule, he derives his concepts from natural examples.

tal mathematical constructions, almost no such constructions were exercised with fuzzy partial orders.

The above facts were recognized independently by Höhle and Blanchard (1985), Höhle (1987), whose work remained virtually unnoticed for a long time,[III] Bodenhofer (1998, 2000), and Belohlavek (2001b, 2002c, 2004).[112] These authors proposed new notions of a fuzzy partial order for which they had different motivations. In fact, the notion proposed by Höhle and Blanchard is the same as Bodenhofer's except that Bodenhofer uses $L = [0, 1]$ with left-continuous t-norms, while Höhle and Blanchard use certain complete residuated lattices as structures of truth degrees. A different notion was proposed by Höhle (1987) and by Belohlavek.[113] In fact, Höhle used $L = [0, 1]$ with left-continuous t-norms and a general notion of fuzzy equality possibly not satisfying the usual reflexivity, $(u \approx u) = 1$, while Belohlavek used complete residuated lattices and assumes reflexivity of $\approx$. This is the definition we use below. Thus let $\approx$ be a fuzzy equality on U as defined on p. 274 and assume that **L** is a complete residuated lattice. According to Höhle and Blanchard (1985) and Bodenhofer, a fuzzy ordering on $\langle U, \approx \rangle$ is a fuzzy relation $\preceq$ satisfying $u \approx v \leq u \preceq v$, $\otimes$-transitivity, and the following form of antisymmetry:

$$(u \preceq v) \otimes (v \preceq u) \leq (u \approx v). \tag{5.28}$$

According to Höhle (1987) and Belohlavek, a fuzzy ordering on $\langle U, \approx \rangle$ is a fuzzy relation $\preceq$ which is compatible with $\approx$ in the sense of (5.24), i.e. $(u' \approx u) \otimes (u \preceq v) \otimes (v \approx v') \leq (u' \preceq v')$, and satisfies the usual reflexivity, $\otimes$-transitivity, and the following form of antisymmetry:

$$(u \preceq v) \wedge (v \preceq u) \leq (u \approx v). \tag{5.29}$$

It is easily shown that antisymmetry is the only condition in which these two notions differ. Let us also note that Zadeh's antisymmetry, (5.27), and hence his notion of fuzzy order, is a particular case of both (5.28) and (5.29), namely in the particular and restrictive case of $\approx$ being the crisp equality. Unlike Zadeh's case, the graded inclusion S now becomes a fuzzy ordering on the set L^U of all fuzzy sets equipped with $\approx$ defined by (5.15), for both of the new approaches. While the condition (5.29) is stronger than (5.28) and hence less general, fuzzy orderings with (5.29) have been subject to many investigations revealing that generalizations of several fundamental constructions and theorems known for classical orders, to which we turn next, can naturally be obtained using this notion.[114]

[III]Perhaps because the authors used a rather abstract setting, unusual in writings on fuzzy sets at that time.

[112]Both Bodenhofer and Belohlavek presented their preliminary results in February 1998 at the 4th Fuzzy Sets Theory and its Applications conference in Liptovský Ján. They were both influenced by Höhle's work on fuzzy equalities.

[113]Interestingly, Höhle does not refer to his previous, different approach from Höhle and Blanchard 1985.

[114]For an attempt using fuzzy orderings with (5.28), see P. Martinek, "On generalization of fuzzy concept lattices based on change of underlying fuzzy order," in *Proc 6th Conf on Concept Lattices and Their Applications*, ed. R. Belohlavek and S. O. Kuznetsov (Olomouc: Palacký Univ. Press), 207–15.

Belohlavek's original motivation was to generalize in a fuzzy setting the basic results and methods of formal concept analysis, which is a method of analysis of data
in the form of a relation between objects and attributes.[115] The theory of formal
concept analysis is largely based on the theory of ordered sets. In fact, the equation
"formal concept analysis = general theory of complete lattices, Galois connections,
and closure structures + interpretation of partial order as hierarchy of concepts"
accurately describes the situation as well as the motivation of its founder, Rudolf
Wille, to reinterpret and develop lattice theory in the spirit of its original motivations.[116] For a binary fuzzy relation $I : X \times Y \to L$, with X and Y being usually
interpreted as sets of objects and attributes, respectively, and $I(x,y)$ as the degree to
which the object x has the attribute y, Belohlavek (1999) defined the pair of mappings $\uparrow_I : L^X \to L^Y$ and $\downarrow_I : L^Y \to L^X$ by

$$A^{\uparrow_I}(y) = \bigwedge_{x \in X}(A(x) \to I(x,y)) \quad \text{and} \quad B^{\downarrow_I}(x) = \bigwedge_{y \in Y}(B(y) \to I(x,y)), \quad (5.30)$$

and put $\mathscr{B}(X,Y,I) = \{\langle A,B \rangle \in L^X \times L^Y \mid A^{\uparrow_I} = B, B^{\downarrow_I} = A\}$. The set $\mathscr{B}(X,Y,I)$ of
fixpoints of $\uparrow_I$ and $\downarrow_I$ is called the *fuzzy concept lattice* of I because its elements $\langle A,B \rangle$
are naturally interpreted as fuzzy concepts with extent A and intent B. One may fur-
thermore equip $\mathscr{B}(X,Y,I)$ with a partial order $\leq$ by putting $\langle A_1,B_1 \rangle \leq \langle A_2,B_2 \rangle$ iff
$A_1 \subseteq A_2$ or, equivalently, $B_1 \supseteq B_2$, mimicking a hierarchy of concepts. For $L = \{0,1\}$,
this yields the long-known classical construction which is of fundamental impor-
tance: the classical $\mathscr{B}(X,Y,I)$ with $\leq$ is always a complete lattice and every complete lattice is isomorphic to some $\mathscr{B}(X,Y,I)$. Furthermore, if $\sqsubseteq$ is a partial order
on U, then $\mathscr{B}(U,U,\sqsubseteq)$ is just its Dedekind-MacNeille completion, i.e. the least
complete lattice into which $\langle U,\sqsubseteq \rangle$ embeds.[117] In particular, the construction of
$\mathscr{B}(\mathbb{Q},\mathbb{Q},\leq_{\mathbb{Q}})$ from the rationals, $\mathbb{Q}$, and their natural order, $\leq_{\mathbb{Q}}$, is just the construction of real numbers due to Dedekind.[118] To obtain generalizations of these
results, Belohlavek introduced the above notion of fuzzy order and observed that
upon putting $\langle A_1,B_1 \rangle \preceq_{\mathscr{B}} \langle A_2,B_2 \rangle = S(A_1,A_2)$ and $\langle A_1,B_1 \rangle \approx_{\mathscr{B}} \langle A_2,B_2 \rangle = A_1 \approx A_2$,
$\preceq_{\mathscr{B}}$ is a fuzzy partial order on the set $\mathscr{B}(X,Y,I)$ with fuzzy equality $\approx_{\mathscr{B}}$. More-
over, he introduced the following generalization of the notion of complete lattice.
For a fuzzy partially ordered set $\langle U, \approx, \preceq \rangle$ and a fuzzy set $A \in L^U$, the lower cone
of A is the fuzzy set $\mathscr{L}(A) \in U$ defined by $[\mathscr{L}(A)](u) = \bigwedge_{v \in U}(A(v) \to (u \preceq v))$
and similarly for the upper cone $\mathscr{U}(A)$. The infimum and supremum of a fuzzy set
$A \in L^U$ is the fuzzy set $\inf(A) \in L^U$ and $\sup(A) \in L^U$ defined by

$$\inf(A) = \mathscr{L}(A) \wedge \mathscr{U}\mathscr{L}(A) \quad \text{and} \quad \sup(A) = \mathscr{U}(A) \wedge \mathscr{L}\mathscr{U}(A).$$

---

[115] B. Ganter and R. Wille. *Formal Concept Analysis: Mathematical Foundations* (Berlin: Springer, 1999). A different but in a sense rather incomplete generalization to a fuzzy setting is due to Burusco and Fuentes-González
(1994); independent of Belohlavek's, but earlier and closely related, is the approach by Pollandt (1997), who, however, did not develop the fundamental order-theoretic structures in a fuzzy setting.

[116] R. Wille, "Restructuring lattice theory: An approach based on hierarchies of concepts," in *Ordered Sets*, ed.
I. Rival (Dordrecht and Boston: Reidel, 1982), 445–70.

[117] For the notions involved, see e.g. Birkhoff 1948 or the above mentioned book by Ganter and Wille.

[118] In fact, an analogous construction of fuzzy reals was the main aim for Höhle (1987); see p. 305 below.

Now, $\langle U, \approx, \preceq \rangle$ is called a *completely lattice fuzzy ordered set* if for every fuzzy set $A \in L^U$, both inf(A) and sup(A) are singletons with respect to $\approx$. With this notion, Belohlavek was able to generalize the classical results and obtained a theorem fully characterizing the structure of the fuzzy partially ordered sets $\mathcal{B}(X, Y, I)$. Furthermore, he generalized the classical theorem on Dedekind-MacNeille completions. With suitable notions of isomorphism, embedding, preservation, and "being least," parts of these results relevant to our discussion are the following generalizations of the fundamental results by Birkhoff and MacNeille:[119]

Theorem 5.12. (a) $\langle U, \approx, \preceq \rangle$ *is a completely lattice fuzzy ordered set iff it is isomorphic to some* $\langle \mathcal{B}(X, Y, I), \approx_{\mathcal{B}}, \preceq_{\mathcal{B}} \rangle$. (b) *For each fuzzy partially ordered set* $\langle U, \approx, \preceq \rangle$, $\langle \mathcal{B}(U, U, \preceq), \approx_{\mathcal{B}}, \preceq_{\mathcal{B}} \rangle$ *is the least completely lattice fuzzy ordered set to which* $\langle U, \approx, \preceq \rangle$ *embeds in such a way that infima and suprema are preserved.*

Given the role of complete lattices in mathematics, it is only natural that these investigations have led to many subsequent contributions, which improved and developed them further. Interestingly, it turns out that similar structures have been examined in the theory of enriched categories.[120] For further information see e.g. Georgescu and Popescu 2004, Krupka 2010, Martinek 2011, Yao and Lu 2009, Zhang 2010, and Zhao and Zhang 2008.

5.3.2 Algebra

Fuzzy subalgebras and substructures

The research on so-called fuzzy subalgebras and other substructures began with the paper "Fuzzy groups" by Rosenfeld (1971) and represents a significant direction in applying fuzzy sets to algebraic concepts. As we shall see below, the notion of fuzzy subalgebra derives from the concept of a fuzzy set being closed with respect to an operation and there exist natural examples of this notion.

In his paper, Rosenfeld (1971) introduced the notion of a fuzzy subgroupoid and a fuzzy subgroup and established a number of results for these notions. All these results are straightforward and easy to prove but the paper is conceptually significant. In the abstract, Rosenfeld writes: "The concept of a fuzzy set, introduced in Zadeh 1965a, was applied in Chang 1968 to generalize some of the basic concepts of general topology. The present note constitutes a similar application to the elementary theory of groupoids and groups." Except for this rather general claim, the paper does not contain any specific motivation or example of the new notions. Rosenfeld calls a *fuzzy subgroupoid* of a classical groupoid $\langle U, \cdot \rangle$[121] any fuzzy set $A : U \to [0, 1]$ which

---

[119] H. M. MacNeille, "Partially ordered sets," *Trans AMS* 42 (1937): 416–60; G. Birkhoff, *Lattice Theory*, vol. 25 (Providence, RI: American Mathematical Society, 1940). Belohlavek's results overlap with those of Höhle (1987).

[120] E.g. K. R. Wagner, "Liminf convergence in Ω-categories," *Theor Comput Sci* 184 (1997): 61–104; (Zhang 2010).

[121] I.e. U is a set and $\cdot$ is a binary operation on U. A groupoid is called a semigroup if $\cdot$ is associative. A monoid is a semigroup with a neutral element e, i.e. $e \cdot u = u \cdot e = u$ holds for each u in U. A group is a monoid in which

satisfies for every $u, v \in U$ the inequality

$$\min(A(u), A(v)) \leq A(u \cdot v). \tag{5.31}$$

If U is a group, then A is its *fuzzy subgroup* if it is its fuzzy subgroupoid satisfying

$$A(u) \leq A(u^{-1}) \tag{5.32}$$

for every $u \in U$. In spite of the paper's title, Rosenfeld's idea is actually to generalize the concept of subgroupoid and subgroup rather than the concept of groupoid and group, which is also apparent from the way he treats these notions in the paper. Even though Rosenfeld does not mention it, let us note that from the point of view of fuzzy logic in the narrow sense, the notion of fuzzy subgroupoid obtains as follows (cf. section 5.2.1). Consider a first-order language with a binary function symbol $\circ$ and a unary relation symbol r. The basic rules of semantics of classical logic yield that a subset A of M is a classical subgroupoid of $\langle M, \cdot \rangle$ if and only if the formula

$$\forall x \forall y (r(x) \otimes r(y) \to r(x \circ y)) \tag{5.33}$$

is true in the structure $\mathbf{M} = \langle M, r^{\mathbf{M}}, \circ^{\mathbf{M}} \rangle$ in which the interpretations $r^{\mathbf{M}}$ and $\circ^{\mathbf{M}}$ of symbols r and $\circ$ equal A and $\cdot$, respectively. Now, the notion of fuzzy subgroupoid obtains by taking the same formula and interpreting it in a first-order fuzzy logic with the set $L = [0, 1]$ of truth degrees and min as the truth function of $\otimes$. Namely, a fuzzy set $A : M \to [0, 1]$ is a fuzzy subgroupoid of $\langle M, \cdot \rangle$ iff (5.33) is true to degree 1 in the L-structure $\mathbf{M} = \langle M, r^{\mathbf{M}}, \circ^{\mathbf{M}} \rangle$ in which $r^{\mathbf{M}}$ and $\circ^{\mathbf{M}}$ equal A and $\cdot$, respectively. In a sense, the notion of fuzzy subgroupoid carries the same idea as the classical notion, namely the meaning of the formula (5.33) which reads "for every x and y, if x and y are in r then $x \circ y$ is in r." In this view, this as well as several other notions regarding various kinds of substructures are automatically obtained by moving from classical logic to an appropriate fuzzy logic. Rosenfeld expresses his view of generalization differently by a theorem saying that if A is a crisp fuzzy set which is a characteristic function of a set S_A, then A is a fuzzy subgroupoid of U iff S_A is a classical subgroupoid of U. Rosenfeld's results include the following claims.

Theorem 5.13. Each a-cut ^aA of a fuzzy subgroupoid is a classical subgroupoid. The intersection of any system of fuzzy subgroupoids is a fuzzy subgroupoid. A homomorphic preimage of a fuzzy subgroupoid A is a fuzzy subgroupoid; the same is true for images if for each $V \subseteq U$ there exits $u \in V$ such that $A(u) = \bigvee_{v \in V} A(v)$. For a fuzzy subgroup, $A(u) = A(u^{-1})$ and $A(u) \leq A(e)$ for every $u \in U$.

The first paper further developing Rosenfeld's concept was Katsaras and Liu 1977, which examines fuzzy subspaces of vector spaces. It defines a fuzzy subspace of a vector space U over a field K as a fuzzy set $A : U \to [0, 1]$ satisfying $A + A \subseteq A$ and $k \cdot A \subseteq A$ for each $k \in K$, where the operations $+$ and $\cdot$ are based on Zadeh's extension principle. The authors show that this definition is equivalent to the condition

---

for every u in U there exists v in U, denoted u^{-1} and called the inverse of u, for which $u \cdot v = v \cdot u = e$. We follow the common usage of writing just U instead of $\langle U, \cdot \rangle$.

in Rosenfeld's style, namely that $\min(A(u), A(v)) \leq A(u+v)$ and $A(u) \leq A(k \cdot u)$ for each u, v, and k. They examine variations of properties examined by Rosenfeld but also new ones, such as those regarding convexity, and define factor spaces. They also utilize the then-recent notion of fuzzy topology (section 5.3.3), define fuzzy topological vector spaces as vector spaces U with a fuzzy topology τ on U such that the vector addition and scalar multiplication are continuous w.r.t. the product topologies based on τ and a topology on the field of scalars, and examine fuzzy subspaces of fuzzy topological vector spaces. Independently, fuzzy topologies on algebras as well as fuzzy substructures of such fuzzy topological algebras appeared in a comprehensive treatment by Foster (1979). Important examinations appeared in the works by Anthony and Sherwood (1979, 1982). These are among the first in the literature on fuzzy sets that proposed using t-norms instead of just min to combine truth degrees in a conjunctive manner.[122] Namely, the authors generalized Rosenfeld's notion of fuzzy subgroup by replacing (5.31) by

$$A(u) \otimes A(v) \leq A(u \circ v),$$

a more general condition in which $\otimes$ is a t-norm, strengthened (5.32) by $A(u) = A(u^{-1})$, and, in Anthony and Sherwood 1982, added $A(e) = 1$. Importantly, the authors provided the first examples of fuzzy subgroups since the invention of this concept by Rosenfeld in 1971. The examples are as follows. Let U be a group, Ω be a set of its classical subgroups, and for each $u \in U$ let $\Omega_u = \{V \in \Omega \mid u \in V\}$. Let $\mathscr{B}$ be a σ-algebra of subsets of Ω containing all the sets Ω_u for $u \in U$ and let P be a probability measure on $\mathscr{B}$. One may then consider the fuzzy set $A_P : U \to [0,1]$ defined by $A_P(u) = P(\Omega_u)$ for each $u \in U$. The truth degree $A_P(u)$ is then interpreted as the probability that a subgroup selected randomly from Ω contains the element u. In general, A_P is a fuzzy subgroup w.r.t. the Łukasiewicz t-norm $\otimes$ but not w.r.t. min. If Ω is linearly ordered by set inclusion, then A_P is a fuzzy subgroup w.r.t. min, i.e. in Rosenfeld's sense. This example is further examined in Anthony and Sherwood 1982.

Of the other early contributions, let us mention Das 1981, which adds observations regarding a-cuts, e.g. a fuzzy set A is a fuzzy subgroup iff ^aA is an ordinary subgroup for each $a \leq A(e)$. Particularly important are contributions by Gerla, who studied general as well as particular types of fuzzy subalgebras in his explorations of Pavelka-style logic and fuzzy closure operators which we examine elsewhere (see particularly pp. 173ff. and 288ff.). Di Nola and Gerla (1987) provide a thorough examination of general fuzzy subalgebras studied also in Biacino and Gerla 1984, which they call L-algebras and define for a complete lattice L as L-sets $A : U \to L$ in a universe U of a general algebra which satisfy $A(u_1) \wedge \cdots \wedge A(u_n) \leq A(f(u_1, \ldots, u_n))$ for every n-ary operation of the algebra. They examine the category whose objects are the triplets $\langle U, A, L \rangle$ and study the notions of morphism, congruence, factor algebra

---

[122] This fact is virtually unknown. The others are mentioned at the beginning of section 4.4.1.

in the spirit of Di Nola and Gerla 1986 (see p. 250 above) and thus provide a general approach to the notion of fuzzy subalgebra. Gerla (1985) observed a particular case of a general relationship (Gerla 2001, section 6.7) according to which fuzzy subalgebras are just closed theories in certain Pavelka-style logics. There is also the important paper by Murali (1991) who, independently of Di Nola and Gerla, studied fuzzy subalgebras for the particular case $L = [0, 1]$. Murali introduced the notion of an *algebraic fuzzy closure system* as a fuzzy closure system whose associated fuzzy closure operator is algebraic in the sense explained on p. 174 above and proved:[123]

Theorem 5.14. A fuzzy closure system $\mathscr{S} \subseteq [0, 1]^U$ is algebraic iff it is inductive in that the supremum in $\langle [0, 1]^U, \subseteq \rangle$ of every nonempty chain $\mathscr{C} \subseteq \mathscr{S}$ belongs to $\mathscr{S}$. All fuzzy subalgebras of a given algebra form an algebraic fuzzy closure system.

As we noted above, the first natural example of a fuzzy substructure was that of a fuzzy subgroup in Anthony and Sherwood 1979. Other examples appeared later. They include fuzzy substructures of lattices, such as *fuzzy filters* (Biacino and Gerla 1984) which are defined as fuzzy subsets $A : U \to [0, 1]$ of a lattice $\langle U, \sqcap, \sqcup \rangle$ that are nondecreasing and satisfy $A(u) \wedge A(v) \leq A(u \sqcap v)$ and whose natural examples appear in fuzzy topology (section 5.3.3 below). For instance, the axioms (5.47)–(5.49) of Šostak imply that his fuzzy topology is a particular fuzzy filter in the lattice $U = \langle [0, 1], \subseteq \rangle$. It is also easy to see that the generalized necessities on a Boolean algebra B with the largest element $\overline{1}$ are just the fuzzy filters N on U satisfying $N(\overline{1}) = 1$.[124] Formato, Gerla, and Scarpati (1999) found an interesting relationship between fuzzy subgroups and fuzzy equivalence relations, both with respect to a t-norm $\otimes$. Consider for a nonempty set X the symmetric group Σ_X, i.e. the group of all bijective mappings on X with the operation $\circ$ of composition of mappings. For a binary fuzzy relation R on X, let A_R be the fuzzy set in Σ_X defined by $A_R(f) = \bigwedge_{x \in X} R(x, f(x))$ for each $f \in \Sigma_X$. Thus, $A_R(f)$ is interpreted as the truth degree of "for each $x \in X$, x is R-related with $f(x)$." Conversely, if A is a fuzzy set in Σ_X, let R_A be the fuzzy relation in X defined by $R_A(x, y) = \bigvee \{A(f) \mid f \in \Sigma_X, f(x) = y\}$. Hence, $R_A(x, y)$ is interpreted as the truth degree of "there exists f in A which maps x to y." The relationship is:

Theorem 5.15. If R is a fuzzy equivalence then A_R is a fuzzy subgroup. If A is a fuzzy subgroup then R_A is a fuzzy equivalence. Moreover, $R = R_{A_R}$ and $A \subseteq A_{R_A}$.

Since its publication, Rosenfeld's paper has been cited by hundreds of other papers on various kinds of fuzzy subalgebras and substructures, many of which are expounded in Mordeson and Malik 1998.[125] Most of these contributions, however,

---

[123] In view of the preceding remarks, it follows that Gerla's theorem 4.36 which completely characterizes algebraic fuzzy closure operators, is a generalization of the first part of Murali's theorem 5.14.

[124] The generalized necessities are defined on p. 214 above. In the present context, they are conceived of as functions on the Boolean algebra B rather than on a set of formulas; cf. the view in section 4.5.3.

[125] As of May 2015, the Scopus database shows over 850 citations of Rosenfeld 1971.

present examinations of various kinds of fuzzy substructures of many types of algebras with no natural motivations and examples, follow the patterns and proof methods established by earlier contributors, and are of little value. This is a typical result of the fuzzifier's temptation mentioned on p. 239. This situation led to a metatheorem obtained by Head (1995) showing that several of the results on fuzzy subalgebras, in fact, follow in a relatively simple manner from the corresponding results for ordinary algebras.

Algebras with fuzzy equalities

A different approach was initiated by Belohlavek (2002b,c, 2003a) and further developed e.g. in Belohlavek and Vychodil 2005, 2006a,b, and Vychodil 2007. The idea is to look at the concept of algebra from the point of view of fuzzy logic in the narrow sense and to develop universal algebra, thus a part of model theory for fuzzy logic, in this setting.[126] For a type $\langle F, \sigma \rangle$ and complete residuated lattice $\mathbf{L}$, an *algebra with fuzzy equality*, or an $\mathbf{L}$-algebra for short, is a structure $\mathbf{M} = \langle M, \approx^{\mathbf{M}}, F^{\mathbf{M}} \rangle$ such that $\langle M, F^{\mathbf{M}} \rangle$ is an ordinary algebra of type $\langle F, \sigma \rangle$,[127] $\approx^{\mathbf{M}}$ is a fuzzy equality on M (p. 274 above), and each operation $f^{\mathbf{M}} \in F^{\mathbf{M}}$ is *compatible* with $\approx^{\mathbf{M}}$ in that

$$(m_1 \approx^{\mathbf{M}} n_1) \otimes \cdots \otimes (m_k \approx^{\mathbf{M}} n_k) \leq f^{\mathbf{M}}(m_1, \ldots, m_k) \approx^{\mathbf{M}} f^{\mathbf{M}}(n_1, \ldots, n_k)$$

for every k-ary operation $f^{\mathbf{M}}$ and every $m_i, n_i \in M$. The degrees $m \approx^{\mathbf{M}} n$ are interpreted as degrees of similarity of m and n. From the logical point of view, algebras with fuzzy equalities are just structures of a predicate fuzzy logic with a language containing $\approx$ as a single relation symbol, $f \in F$ as function symbols, and in which the structure $\mathbf{L}$ of truth degrees may be any complete residuated lattice. In a sense, they thus represent an extreme fragment of predicate fuzzy logic with fuzzy equality in which there are, except for $\approx$, no symbols of relations. Furthermore, the compatibility condition means that the formula verbally described as "if x_i and y_i are pairwise similar for $i = 1, \ldots, k$ then $f(x_1, \ldots, x_k)$ and $f(y_1, \ldots, y_k)$ are similar" is true. Hence, algebras with fuzzy equalities may be regarded as algebras whose support is equipped with information regarding similarity of its elements and whose operations preserve similarity in that similar inputs are mapped to similar outputs. Clearly, for $L = \{0, 1\}$ algebras with fuzzy equalities may be identified with classical algebras. There are many examples of such algebras (Belohlavek and Vychodil 2005), including the fuzzy concept lattices examined on p. 281 above.

The idea of algebras with fuzzy equalities is closely related to that of metric algebras.[128] A *metric algebra* is a triplet $\mathbf{M} = \langle M, \rho^{\mathbf{M}}, F^{\mathbf{M}} \rangle$ consisting of an ordinary alge-

---

[126] Universal algebra studies the abstract notion of algebra. A classic book on universal algebra is by S. Burris and H. P. Sankappanavar, *A Course in Universal Algebra* (New York: Springer, 1981).

[127] I.e. F is a set of function symbols, $\sigma(f)$ is the arity of f for each $f \in F$, and $F^{\mathbf{M}}$ consists of functions on M interpreting function symbols, i.e. for each $f \in F$, $f^{\mathbf{M}}$ is an $\sigma(f)$-ary function on M.

[128] N. Weaver, "Generalized varieties," *Algebra Universalis* 30 (1993): 27–52; "Quasi-varieties of metric algebras," *Algebra Universalis* 33 (1995): 1–9.

bra $\langle M, F^M \rangle$ and a metric ρ^M on M. Of interest are metric algebras whose operations f^M are equicontinuous, meaning in a certain sense that close values are mapped to close ones.[129] Although compatibility in algebras with fuzzy equalities and equicontinuity in metric algebras represent different types of constraints, some relationships have been explored (Belohlavek and Vychodil 2005, section 2.9). For example, in the particular case of $\otimes$ being a continuous Archimedean t-norm (which excludes min), every algebra with fuzzy equality may naturally be transformed to a metric algebra.

Many results on algebras with fuzzy equalities in the spirit of classical results in universal algebra are expounded in the book Belohlavek and Vychodil 2005. Among them are those regarding basic structural notions such as subalgebras, morphisms, congruences, factorization, direct products, as well as advanced topics including subdirect product representation, direct unions and limits, reduced products, free algebras, varieties and other classes such as surreflective classes, semivarieties, and quasivarieties, Mal'cev conditions, equational and Horn fuzzy logics with their completeness theorems, and further topics. For illustration, we now present a generalization of two of Birkhoff's fundamental results for universal algebra.[130]

A *fuzzy equational logic* is a particular kind of Pavelka's abstract fuzzy logic (section 4.3.2), which assumes an arbitrary complete residuated lattice $\mathbf{L}$ as the structure of truth degrees, and has as its formulas the identities $t \approx s$ with t and s being ordinary terms. Algebras $\mathbf{M}$ with fuzzy equalities serve as semantic structures. The concept $\|t \approx s\|_\mathbf{M}$ of truth degree of $t \approx s$ in $\mathbf{M}$ is defined as the infimum over all valuations v of the truth degrees $\|t \approx s\|_{\mathbf{M},v}$ which are defined as usual in predicate fuzzy logic. This gives the notions of model and the *degree* $\|t \approx s\|_T$ *of semantic entailment of $s \approx t$ from a theory T*, i.e. from a fuzzy set of identities. The logic has the following Pavelka-style deduction rules:

$$\frac{}{\langle t \approx t, 1 \rangle}, \quad \frac{\langle t \approx t', a \rangle}{\langle t' \approx t, a \rangle}, \quad \frac{\langle t \approx t', a \rangle, \langle t' \approx t'', b \rangle}{\langle t \approx t'', a \otimes b \rangle},$$

$$\frac{\langle t \approx t', a \rangle}{\langle s \approx s', a \rangle}, \quad \frac{\langle t \approx t', a \rangle}{\langle t(x/r) \approx t'(x/r), a \rangle},$$

where $t, t', \ldots$ are terms, $a, b \in L$, t appears as a subterm in s and s' results from s by substitution of one occurrence of t by t'. This yields the notion of *degree* $|t \approx s|_T$ *of provability of $t \approx s$ from a theory T*. The following Pavelka-style completeness theorem (Belohlavek 2002b) generalizes the classic Birkhoff result (n. 130 on p. 287):

Theorem 5.16. For any theory T and any identity $t \approx s$: $\|t \approx s\|_T = |t \approx s|_T$.

---

[129] f^M is equicontinuous if for each $\varepsilon > 0$ there exists $\delta > 0$ such that for each valuation v of variables x_i, y_i, if the formulas $\rho(x_1, y_1) \preceq 0, \ldots, \rho(x_k, y_k) \preceq 0$ are δ-true then $\rho(f(x_1, \ldots, x_k), f(y_1, \ldots, y_k)) \preceq 0$ is ε-true; $\rho(t, s) \preceq a$ is γ-true for v if $\rho^M(\|t\|_{\mathbf{M},v}, \|s\|_{\mathbf{M},v}) \leq a + \gamma$.

[130] G. Birkhoff, "On the structure of abstract algebras," *Proc Cambridge Philosophical Society* 31 (1935): 433–54. Birkhoff developed a logic for reasoning with identities, known as *equational logic*, and proved its completeness; he also proved his *variety theorem*: a class of algebras is described by a set of identities iff it forms a variety, i.e. is closed under the formation of subalgebras, homomorphic images, and direct products.

With the notions of subalgebra, homomorphism, and direct product naturally modified for L-algebras, i.e. algebras with fuzzy equalities, one may call a class $\mathcal{K}$ of L-algebras a *variety* if it is closed under the formation of subalgebras, homomorphic images, and direct products. Moreover, $\mathcal{K}$ is called *equational* if it is just the class of all models of some fuzzy set T of identities. The following theorem then generalizes Birkhoff's variety theorem:

Theorem 5.17. A class of L-algebras forms a variety iff it is equational.

A different generalization of Birkhoff's variety theorem was later obtained by Cintula and Hájek (2010) within predicate core fuzzy logics (section 4.4.3). Further results are found in Belohlavek and Vychodil 2005. This book also contains open problems, among them the question of whether a reasonable approach can be taken which combines algebras with fuzzy equalities and fuzzy subalgebras. For the particular case of ⊗ being ∧, one such approach is examined in Budimirović et al. 2014.

Other developments

In addition to fuzzy subalgebras and algebras with fuzzy equalities, several other approaches have appeared in the literature. In particular, Demirci's *vague algebras* have been developed since the late 1990s; see, for instance, Demirci 1999, 2003. Vague algebras are based on Demirci's fuzzy functions which are conceived of as certain fuzzy relations on sets equipped with fuzzy equivalences with which they are required to be compatible. Ordinary functions can be seen as their particular instances. This makes Demirci's vague algebras more general than algebras with fuzzy equalities, even though Demirci develops them only with binary operations. Most of the structural notions and results available for algebras with fuzzy equalities have, however, not been developed for vague algebras and it is far from obvious how such developments could be carried out. Nevertheless, the idea of generalizing the very concept of operations on algebras is interesting and worth mentioning.

Fuzzy closure operators and related structures

Closure operators and related structures, such as closure systems, interior operators and systems, or Galois connections, are fundamental for many areas of mathematics including set theory, algebra, topology, and geometry.[131] They are also of fundamental importance in logic. The notion of *fuzzy closure operator* on a set U, as commonly used, denotes a mapping $C : L^U \to L^U$, with L being $[0,1]$ or a general complete lattice, for which

---

[131] The notion of closure operator is due to E. H. Moore, *Introduction to a Form of General Analysis* (New Haven: Yale Univ. Press, 1910), 53–88.

$$A \subseteq C(A), \tag{5.34}$$

$$A \subseteq B \text{ implies } C(A) \subseteq C(B), \text{ and} \tag{5.35}$$

$$C(A) = C(C(A)) \tag{5.36}$$

hold for any fuzzy sets $A, B \in L^U$. However, important variants and generalizations have been advanced and are examined below.

In the literature on fuzzy sets, fuzzy closure operators appeared for the first time in Lowen 1976. In fact, Lowen imposed additional conditions, namely $C(A \cup B) = C(A) \cup C(B)$ and $C(a_U) = a_U$, where a_U is a constant fuzzy set with $a_U(u) = a$ for each $u \in U$, and observed that the sets of fixpoints of such operators are just the closed sets of fuzzy topologies he examined. Lowen's conditions generalize the well-known Kuratowski axioms for topological closure.[132] Mashhour and Ghanim (1985) examined even more general operators and required only (5.34), $C(\emptyset) = \emptyset$, and $C(A \cup B) = C(A) \cup C(B)$. For ordinary sets, these are exactly the so-called Čech closure operators which play an important role in investigations of proximity in general topology.[133] Fuzzy topology (section 5.3.3) represents one area in which particular fuzzy closure and interior operators were studied. Independently, fuzzy closure operators in the above sense appeared in Pavelka 1979. As we show in section 4.3.2 (pp. 167ff.), they play a fundamental role in Pavelka-style fuzzy logic. The third area where they appeared early was fuzzy subalgebras (p. 282), in particular in Biacino and Gerla 1984. Later, fuzzy closure operators and related structures appeared in a number of studies which we now examine.

Note first that fuzzy closure operators in the sense of (5.34)–(5.36) are the backbone of Gerla's studies of Pavelka-style fuzzy logic examined in section 4.3.2. Since in general no restrictions beyond (5.34)–(5.36) are imposed, the majority of these results may be regarded as a theory of fuzzy closure operators motivated by their interpretation as deduction operators of abstract logics. Some results naturally concern particular types of fuzzy closure operators, such as the important algebraic ones in Murali's sense (pp. 174 and 285).

Fuzzy closure operators in the sense of (5.34)–(5.36) are exactly the closure operators in the lattice $\langle L^U, \subseteq \rangle$.[134] Similarly, the related notion of a *fuzzy closure system* as a system $\mathscr{S} \subseteq L^U$ that is closed w.r.t. arbitrary intersections is just a particular notion of a closure system in a lattice. Some relationships, which have appeared in the literature for fuzzy closure operators, are thus particular cases of established results in a lattice-theoretic framework (Birkhoff 1948, chapter 5). For instance, given a fuzzy closure operator C and system $\mathscr{S}$, the set $\mathscr{S}_C \subseteq L^U$ and operator

---

[132] K. Kuratowski, "Sur l'operation $\overline{A}$ de l'Analysis Situs" [On the operation $\overline{A}$ of Analysis Situs], *Fund Math* 3 (1922): 182–99.

[133] E. Čech, *Topological Spaces*, rev. ed. (Prague: Academia, 1966).

[134] A closure operator in a lattice $\langle V, \leq \rangle$ is a mapping $c : V \to V$ for which $v \leq c(v)$, $v \leq w$ implies $c(v) \leq c(w)$, and $c(v) = c(c(v))$ (Birkhoff 1948).

$C_{\mathscr{S}} : L^U \to L^U$ defined by

$$\mathscr{S}_C = \{A \mid A = C(A)\} \quad \text{and} \quad C_{\mathscr{S}}(A) = \bigcap\{B \mid B \in \mathscr{S}, A \subseteq B\} \tag{5.37}$$

are a fuzzy closure system and a fuzzy closure operator, respectively, and the mappings $C \mapsto \mathscr{S}_C$ and $\mathscr{S} \mapsto C_{\mathscr{S}}$ are mutually inverse bijections.

A different notion of fuzzy closure operator emerged from Belohlavek's examinations of lattices, Galois connections, and other structures in the setting of fuzzy logic over an arbitrary complete residuated lattice L. These structures were motivated by formal concept analysis but they in fact appear in a number of other areas (pp. 281ff.). Properties of the fundamental operators $\uparrow_I$ and $\downarrow_I$ defined by (5.30) are axiomatized by the following notion. An L-*Galois connection* between X and Y is a pair $\uparrow : L^X \to L^Y$ and $\uparrow : L^Y \to L^X$ satisfying

$$A \subseteq A^{\uparrow\downarrow}, \qquad\qquad B \subseteq B^{\downarrow\uparrow}, \tag{5.38}$$

$$S(A_1, A_2) \leq S(A_2^{\uparrow}, A_1^{\uparrow}), \qquad\qquad S(B_1, B_2) \leq S(B_2^{\downarrow}, B_1^{\downarrow}), \tag{5.39}$$

where $S(\cdot, \cdot)$ is Goguen's degree of inclusion (5.14). Notice the monotony conditions using S which are stronger than the frequently occurring variant using "if $A_1 \subseteq A_2$ then $A_2^{\uparrow} \subseteq A_1^{\uparrow}$." Both are equivalent for $L = \{0, 1\}$, in which case they yield the classical notion of Galois connection, but only the stronger version leads to the following generalization of Ore's classical result (Belohlavek 1999):[135]

Theorem 5.18. The mapping sending I to $\langle \uparrow_I, \downarrow_I \rangle$ is a bijection between the set of all L-relations and L-Galois connections between X and Y.

These considerations lead to the notion of L-*closure operator* (Belohlavek 2001a) as a mapping $C : L^U \to L^U$ satisfying (5.34) and (5.36), but instead of (5.35) ensuring that C preserves inclusion, a stronger condition, $S(A, B) \leq S(C(A), C(B))$, ensures that C preserves graded inclusion. For each L-Galois connection $\langle \uparrow, \downarrow \rangle$, the composite mapping $\uparrow\downarrow$ is an L-closure operator and each L-closure operator is of this form, thus preserving the classical results. As a common generalization of these two approaches, based on the bivalent and graded inclusions, $\subseteq$ and S, Belohlavek (2001a) proposed the following notions. Let $\emptyset \neq K \subseteq L$ be an $\leq$-filter, i.e. $b \in K$ whenever $a \in K$ and $a \leq b$. An L_K-*closure operator* in U is a mapping $C : L^U \to L^U$ satisfying (5.34), (5.36), and

$$S(A, B) \leq S(C(A), C(B)) \text{ whenever } S(A, B) \in K.$$

The last condition may be read as "if the degree of inclusion of A in B is high then that of $C(A)$ in $C(B)$ is no smaller" and generalizes the two above conditions which are its particular cases for $K = \{1\}$ and $K = L$. Independently, a concept equivalent to the notion of an L_K-closure operator for the particular case $K = L$ was discovered by Rodríguez et al. (2003), who called it an implicative closure operator and examined it in the context of consequence operators to which we turn next. With the

---

[135] O. Ore, "Galois connexions," *Trans AMS* 55 (1944): 493–513.

concept of shift defined for $a \in L$ and $A \in L^U$ by $(a \to A)(u) = a \to A(u)$, one may define the notion of $\mathbf{L}_K$-*closure system* as a set $\mathscr{S} \subseteq L^U$ closed under intersections and shifts in that

$$A_i \in \mathscr{S} \text{ implies } \bigcap_i A_i \in \mathscr{S} \quad \text{and} \quad a \in K, A \in \mathscr{S} \text{ imply } a \to A \in \mathscr{S}.$$

One then obtains the following relationship (Belohlavek 2001a):

Theorem 5.19. The mappings $C \mapsto \mathscr{S}_C$ and $\mathscr{S} \mapsto C_{\mathscr{S}}$ defined by (5.37) are mutually inverse bijections between $\mathbf{L}_K$-closure operators and systems.

Note also that Yao later took the natural next step and generalized these considerations to the setting of fuzzy ordered sets; see e.g. Yao and Lu 2009.

We now examine studies of graded consequence relations, briefly mentioned in section 4.3.2 on p. 173, which represent another area where fuzzy closure operators appear. Initial work here is due to Chakraborty (1988, 1995), who first studied the notions below for $L = [0, 1]$ and $\otimes = \min$ and later for a complete residuated lattice $\mathbf{L} = \langle L, \wedge, \vee, \otimes, \to, 0, 1 \rangle$. Chakraborty attempted to generalize the notion of consequence relation as a binary relation $\vdash$ between sets of formulas and formulas in a given abstract set $\mathscr{F}$ satisfying for any $A, B \subseteq \mathscr{F}$ and $\varphi \in \mathscr{F}$ that $\varphi \in A$ implies $A \vdash \varphi$; $A \vdash \varphi$ implies $A \cup B \vdash \varphi$; and the important *cut property*, i.e. $A \vdash \psi$ for each $\psi \in B$ and $A \cup B \vdash \varphi$ implies $A \vdash \varphi$.[136] Chakraborty, who was not aware of Pavelka's work,[137] introduced *graded consequence relations* as fuzzy relations $\vdash$ between $2^{\mathscr{F}}$ and $\mathscr{F}$, i.e. $\vdash \in L^{2^{\mathscr{F}} \times \mathscr{F}}$, satisfying the following conditions:

$$\text{if } \varphi \in A \text{ then } (A \vdash \varphi) = 1,$$
$$\text{if } A \subseteq B \text{ then } (A \vdash \varphi) \leq (B \vdash \varphi) \text{ for any } \varphi \in \mathscr{F}, \text{ and}$$
$$\left(\bigwedge_{\psi \in B} (A \vdash \psi) \right) \otimes (A \cup B \vdash \varphi) \leq (A \vdash \varphi).$$

Each $A \vdash \varphi$ is interpreted as the degree to which formula φ is a consequence of A. Gerla (1996) showed the following relationship to closure operators.

Theorem 5.20. For $L = [0, 1]$ and $\otimes = \min$, a fuzzy relation $\vdash$ between $2^{\mathscr{F}}$ and $\mathscr{F}$ is a graded consequence relation iff there exists a well-stratified fuzzy closure operator C (p. 176) such that for every $A \subseteq \mathscr{F}$ and $\varphi \in \mathscr{F}$,

$$(A \vdash \varphi) = [C(A)](\varphi). \tag{5.40}$$

He also noted that in view of Pavelka's conception, it is natural to consider $A \vdash \varphi$ even for fuzzy sets A. Such fuzzy relations have indeed been studied by Castro, Trillas, and Cubillo (1994), who call them *fuzzy consequence relations* if they satisfy for $A, B \in L^{\mathscr{F}}$ and $\varphi \in \mathscr{F}$ the following conditions:[138]

$$A(\varphi) \leq (A \vdash \varphi), \tag{5.41}$$

---

[136] D. J. Shoesmith and T. J. Smiley, *Multiple Conclusion Logic* (New York: Cambridge Univ. Press, 1978).

[137] According to M. K. Chakraborty and S. Basu, "Introducing grade to some metalogical notions," in *Fuzzy Sets, Logics and Reasoning About Knowledge*, ed. D. Dubois, H. Prade, and E. P. Klement (Dordrecht: Kluwer), 85–99.

[138] In the original treatment, the third condition was different. The present condition is due to Rodríguez et al. (2003). One may prove that the present conditions are equivalent to the original ones.

$$\text{if } A \subseteq B \text{ then } (A \vdash \varphi) \leq (B \vdash \varphi) \text{ for any } \varphi \in \mathscr{F}, \text{ and} \qquad (5.42)$$

$$\text{if } B(\psi) \leq (A \vdash \psi) \text{ for each } \psi \in \mathscr{F} \text{ then } (A \cup B \vdash \varphi) \leq (A \vdash \varphi). \qquad (5.43)$$

They showed that such relations also embrace existing examples not covered by Chakraborty's notion and proved that (5.40) for $A \in L^{\mathscr{F}}$ and $\varphi \in \mathscr{F}$ provides a one-to-one relationship between fuzzy consequence relations and fuzzy closure operators in the sense of (5.34)–(5.36). However, as observed in Rodríguez et al. 2003, the notion of fuzzy consequence relation is not a generalization of Chakraborty's because when restricted to ordinary sets, Chakraborty's cut property is stronger than Castro et al.'s (5.43). Theorem 5.20 and Gerla's observation that consequences corresponding to general fuzzy closure operators are not graded consequences in Chakraborty's sense opened the question of what kind of consequence operators correspond to general fuzzy closure operators. Inspired by this question, Belohlavek (2002a) proposed a general notion of $\mathbf{L}_K$-*consequence relation*, with K a $\leq$-filter in L as above, as a fuzzy relation $\vdash$ satisfying (5.41), (5.42), and the following cut property:

$$S(B, A \vdash \cdot) \otimes (A \cup B \vdash \varphi) \leq (A \vdash \varphi) \text{ whenever } S(B, A \vdash \cdot) \in K$$

in which $A \vdash \cdot$ denotes the fuzzy set for which $(A \vdash \cdot)(\varphi) = A \vdash \varphi$. This notion may be seen as the right counterpart (Belohlavek 2002a):

Theorem 5.21. (5.40) *provides a one-to-one correspondence between* $\mathbf{L}_K$-*consequence relations and* $\mathbf{L}_K$-*closure operators.*

Furthermore, $\mathbf{L}_K$-consequence relations become classical consequence relations for $L = \{0, 1\}$, Castro et al.'s fuzzy consequence relations for $K = \{1\}$, and Chakraborty's graded consequences when restricted to classical sets. Independently and under the name implicative consequence relations, $\mathbf{L}_L$-consequence relations were found and studied in Rodríguez et al. 2003.[139] Consequence relations appear in a number of additional studies which relate them to further concepts; see e.g. Castro and Trillas 1991, Elorza and Burillo 2003.

5.3.3 Topology

Examinations of concepts and results of classical topology from a fuzzy logic point of view started with Chang 1968. The so-called fuzzy topology thus represents one of the earliest areas of mathematics based on fuzzy logic. It also represents one of the most advanced areas in which the existing approaches and results are far beyond obvious generalizations of classical concepts.

Recall that a (classical) topology on a set X is a collection τ of its subsets which contains the empty set, contains X, and is closed w.r.t. arbitrary unions and w.r.t. finite intersections.[140] The sets in τ are called open and their complements closed.

---

[139]Note that it is wrongly claimed in Rodríguez et al. 2003 that the implicative relations generalize both Chakraborty's and Castro et al.'s consequence relations. While the first is true, the opposite of the second claim is valid: implicative relations are a particular case of Castro et al.'s consequences.

[140]J. L. Kelley, *General Topology* (Princeton, NJ: Van Nostrand, 1955) is a classic book.

Accordingly, Chang (1968) defined a *fuzzy topology* on a set X as a collection τ of standard fuzzy sets in X, i.e. $\tau \subseteq [0,1]^X$, satisfying

$$\emptyset \in \tau, \ X \in \tau, \tag{5.44}$$

$$A_i \in \tau, i \in I, \text{ implies } \bigvee_{i \in I} A_i \in \tau, \tag{5.45}$$

$$A, B \in \tau \text{ implies } A \wedge B \in \tau, \tag{5.46}$$

where with a slight abuse of notation we denote by $\emptyset$ and X the empty and full fuzzy sets, respectively, i.e. $\emptyset(u) = 0$ and $X(u) = 1$ for each $u \in X$, and $\bigvee_{i \in I} A_i$ and $A \wedge B$ denote pointwise union and intersection based on the supremum $\bigvee$ and infimum $\bigwedge$ in $[0,1]$. The pair $\langle X, \tau \rangle$ is then called a *fuzzy topological space*. The fuzzy sets $A \in \tau$ are called *open* and Chang calls *closed* the standard complements $\neg A = 1 - A$ of open fuzzy sets. Chang introduced some straightforward generalizations of certain classical notions, such as neighborhood, continuity of mappings and compactness, and proved simple results for them.

Early contributions to fuzzy topology include Goguen 1973, Wong 1973, 1974a,b, Michálek 1975, Hutton 1975, Weiss 1975, Lowen 1976, and Lowen 1977, as well as other papers by these authors. These early works mainly attempted to generalize for fuzzy topology some basic results of classical topology. Several of the subsequent contributions, e.g. Hutton 1975, assumed that the standard set $[0,1]$ of truth degrees is replaced by a complete lattice L and that the standard negation is replaced by an order-reversing involutive function. An important conception is due to Goguen (1973) who approached fuzzy topology in the spirit of his pioneering paper (Goguen 1967) and suggested using a complete residuated lattice $\mathbf{L}$ with a possibly noncommutative truth function $\otimes$ of conjunction. Goguen thus defines the notion of an $\mathbf{L}$-topological space as a pair $\langle X, \tau \rangle$ where $\tau \subseteq L^X$ satisfies the above conditions except that the last one is replaced by

$$A, B \in \tau \text{ implies } A \otimes B \in \tau.$$

Chang's definition is a particular case for $L = [0,1]$ and $\otimes$ being $\wedge$ and, importantly, classical notion of topology is a particular case for $L = \{0,1\}$. Goguen observed that properties of $\mathbf{L}$-topologies substantially depend on those of $\mathbf{L}$.

Various fuzzifications of classical topological concepts evolved as further aspects and connections have been taken into account. For example, consider the important notion of *compactness*.[141] A notion of a fuzzy topological space $\langle X, \mu \rangle$ being *compact* was presented in Chang 1968 meaning that each cover $\mu \subseteq \tau$ of X contains a finite subset which is again a cover of X, where μ being a *cover* means that $\bigvee_{A \in \mu} A(u) = 1$ for each $u \in X$. This obvious generalization of classical compactness is subject to a number of problems. For instance, even a one-point space may not be compact according to this notion, e.g. $\langle \{u\}, \{\{^a/u\} \mid a \in [0,1]\} \rangle$. Furthermore, it follows from Goguen 1973 that Chang's fuzzy topological spaces do not satisfy Tychonoff's

---

[141] An ordinary topological space $\langle X, \tau \rangle$ is compact if every cover of X contains a finite subset which is still a cover of X. A cover of X is a collection μ of open sets, i.e. $\mu \subseteq \tau$, whose union is X, i.e. $\bigcup_{A \in \mu} A = X$.

theorem, which Goguen examined with his general notion of L-topology.[142] Interestingly, Goguen (1973) obtained a necessary and sufficient condition on L regarding the validity of Tychonoff's theorem. Goguen uses the following notions, some of which he adopted from the above paper by Chang (1968): cover and compactness are as by Chang, but for L instead of $[0,1]$; a *base* of $\langle X, \tau \rangle$ is a subset $\beta \subseteq \tau$ such that for every $A \in \tau$ there is $\mu \subseteq \beta$ with $A = \bigvee_{B \in \mu} B$; a *subbase* is a subset $\sigma \subseteq \tau$ such that $\{A_1 \otimes \cdots \otimes A_n \mid A_i \in \sigma, n \in \mathbb{N} \cup \{0\}\}$ is a base; one proves that each $\sigma \subseteq L^X$ is a subbase of a unique topology which is said to be *generated* by σ; a *product* of L-topological spaces $\langle X_i, \tau_i \rangle$, $i \in I$, is the space $\langle X, \tau \rangle$ where $X = \prod_{i \in I} X_i$ and τ is the L-topology generated by the subbase $\sigma = \{\pi_i^{-1}(A_i) \mid A_i \in \tau_i, i \in I\}$, where $\pi_i : X \to X_i$ are projections and $[\pi_i^{-1}(A_i)](u) = A_i(\pi_i(u))$. Furthermore, Goguen calls the top element 1 of L x-*isolated* where x is a cardinal, if $\bigvee_{i \in I} a_i < 1$ whenever $|I| \leq x$ and $1 > a_i \in L$ for each $i \in I$. Goguen obtained the following generalization of Tychonoff's theorem:

Theorem 5.22. Every product of x compact L-topological spaces is compact iff 1 is x-isolated.

Since 1 is not $\aleph_0$-compact in $[0,1]$, there exists an infinite collection of Chang's fuzzy topological spaces whose product is not compact. Furthermore, since 1 is x-isolated in $\{0 \leq 1\}$, the classical Tychonoff theorem is a consequence of theorem 5.22.

That fuzzy topological spaces need not always satisfy Tychonoff's theorem was seen as unacceptable by some. Thus Lowen (1976) suggested a different notion of a fuzzy topological space in which Chang's condition $\emptyset \in \tau$ and $X \in \tau$ was replaced by a stronger requirement $a_X \in \tau$ for each $a \in [0,1]$ where a_X is the constant fuzzy set defined by $a_X(u) = a$. He also observed how such topologies are defined in terms of fuzzy closure and interior operators (p. 289 above). Lowen calls $\langle X, \tau \rangle$ *fuzzy compact* if each $a_X \in \tau$ is fuzzy compact, where the latter condition means that whenever $\bigvee_{A \in \mu} \geq a_X$ for some $\mu \subseteq \tau$ then for each $\varepsilon > 0$ there exists a finite $\beta \subseteq \mu$ for which $\bigvee_{A \in \beta} \geq a_X - \varepsilon$. Lowen (1977) thus proved:

Theorem 5.23. The product of any family of fuzzy compact fuzzy topological spaces (in Lowen's sense) is fuzzy compact.

Gantner, Steinlage, and Warren (1978) shared Lowen's view that Tychonoff's theorem should be preserved but pointed out that contrary to Chang's notion, the characteristic functions of open sets of a classical topology do not form a fuzzy topology in Lowen's sense, which they regarded as deficient. Instead, they proposed the following notion of compactness in a general setting in which L need not be $[0,1]$ but forms instead a completely distributive lattice. For an L-topological space $\langle X, \tau \rangle$

---

[142] Tychonoff's theorem has been described as "probably the most important single theorem of general topology"; see J. L. Kelley, *General topology*, op. cit., 143. The theorem says that a product of compact topological spaces is compact.

and $a \in L$, they call a collection $\mu \subseteq \tau$ an a-shading if for each $u \in X$ there exists $A \in \mu$ with $A(u) > a$; a subset of an a-shading that is also an a-shading is called an a-subshading. Now, $\langle X, \tau \rangle$ is called a-compact if every a-shading contains a finite a-subshading. With this notion, every finite fuzzy topological space is a-compact for any $a \in L$. Denoting L^* the set of all $a \in L$ comparable to every $b \in L$ and having the property that $a < b, c$ implies $a < b \wedge c$, the following theorem is a generalization of the classical Tychonoff theorem (Gantner, Steinlage, and Warren 1978):

Theorem 5.24. Let $a \in L^*$. The product of any family of a-compact fuzzy topological spaces is a-compact.

A further examination of the various notions of compactness is found in Lowen 1978; more recent ones in Šostak 1996 and Höhle and Šostak 1999, which include further approaches by Šostak, Wang Guo-jun, Ying-ming Liu, and others.

The early research also introduced an interesting notion of fuzzy unit interval due to Hutton (1975). Hutton formulated his notion when examining Urysohn's lemma for fuzzy topological spaces. He needed a suitable fuzzy topology generalizing the classical topology on $[0, 1]$. Hutton's notion was later extended to the notion of a fuzzy real line in Gantner, Steinlage, and Warren 1978, which generalizes the usual topology of reals. Given a completely distributive lattice $\langle L, \leq \rangle$ with order-reversing involution $'$, consider the set of all monotone decreasing functions $\lambda : \mathbb{R} \to L$ satisfying $\lambda(t) = 1$ for $t < 0$ and $\lambda(t) = 0$ for $t > 1$ and the equivalence relation $\sim$ defined on them by

$$\lambda \sim \varkappa \quad \text{iff} \quad \text{for each } t : \bigwedge_{s<t} \lambda(s) = \bigwedge_{s<t} \varkappa(s) \text{ and } \bigvee_{s>t} \lambda(s) = \bigvee_{s>t} \varkappa(s).$$

A *fuzzy unit interval*, $[0, 1](L)$, consists of all equivalence classes $[\lambda]$ of such functions. A *fuzzy real line*, $\mathbb{R}(L)$, extends $[0, 1](L)$ in that it is the set of equivalence classes $[\lambda]$ defined as for $[0, 1](L)$ but on all monotone decreasing functions $\lambda : \mathbb{R} \to L$ satisfying $\bigwedge_{t \in \mathbb{R}} \lambda(t) = 0$ and $\bigvee_{t \in \mathbb{R}} \lambda(t) = 1$. A fuzzy topology on $[0, 1](L)$, and similarly on $\mathbb{R}(L)$, is defined by taking as its subbasis the set of all fuzzy sets L_t and R_t for $t \in \mathbb{R}$ given by

$$L_t([\lambda]) = (\bigwedge_{s<t} \lambda(s))' \quad \text{and} \quad R_t([\lambda]) = (\bigvee_{s>t} \lambda(s)).$$

The set $\mathbb{R}$ of reals may be embedded in $\mathbb{R}(L)$ by assigning to any $r \in \mathbb{R}$ the class $[\lambda_r]$ of λ_r such that $\lambda_r(t) = 1$ for $t < r$ and $\lambda_r(t) = 0$ for $t > r$. It is easily seen that in the bivalent case, i.e. $L = \{0, 1\}$, the subbasis may be identified with the classical subbasis consisting of intervals (r, ∞) and $(-\infty, r)$ where $r \in \mathbb{R}$, and hence the fuzzy topology of the fuzzy real line becomes the usual topology of reals. The notion of fuzzy real line provides an alternative model of the intuitive notion of fuzzy number and we return to this idea in section 5.3.4.

Since the early 1980s, the number of contributions to fuzzy topology has grown rapidly, totalling more than 600 publications according to the comprehensive survey Šostak 1989, and to more than 1000 by another survey, Šostak 1996, where it is noted that these publications are of "quite different value and originality." The number of publications on fuzzy topology has continued to grow ever since and includes the monographs Liu and Luo 1997, Höhle 2001, the authoritative volume Höhle and Rodabaugh 1999 and its follow-up companion Rodabaugh and Klement 2003. Important contributions have been made by many researchers of which we also mention in addition to the above names from the 1970s—in alphabetical order— Burton, Eklund, Höhle, Katsaras, Klein, Kotzé, Kubiak, Ying-ming Liu, Martin, Rodabaugh, Šostak, Wang Guo-jun, and Warren. Many of the basic notions of classical topology have been investigated as well as the very notion of fuzzy topology. Particularly noteworthy is Šostak's (1985) concept of fuzzy topology on $\check{X}$ as a fuzzy set, rather than a set, of fuzzy sets in X satisfying

$$\tau(\emptyset) = 1, \tau(X) = 1, \tag{5.47}$$

$$\bigwedge_{i \in I} \tau(A_i) \leq \tau(\bigvee_{i \in I} A_i), \tag{5.48}$$

$$\tau(A) \wedge \tau(B) \leq \tau(A \wedge B). \tag{5.49}$$

$\tau(A)$ is interpreted as the degree to which the fuzzy set A is open, and one can see that Chang's original notion may be regarded as a particular case in which $\tau(A)$ is 0 or 1 for every A. These two notions play a fundamental role in modern treatments (Höhle and Šostak 1999). In particular, these treatments use structures based on so-called *complete quasi-monoidal lattices*, which are complete lattices L equipped with an additional binary operation $\otimes$ interpreted as a truth function of conjunction and which are more general than complete residuated lattices. By an L-topology one means a fuzzy topology in the sense of Goguen's generalization of Chang's notion, i.e. a subset $\tau \subseteq L^X$ satisfying (5.44)–(5.46) with $\wedge$ replaced by $\otimes$, while by an L-fuzzy topology one means Šostak's notion, i.e. a fuzzy set $\tau : L^X \to L$ satisfying (5.47)–(5.49) with $\otimes$ in place of $\wedge$. The spaces satisfying Lowen's condition that all the constant fuzzy sets $a_X, a \in L$, belong to τ are called weakly stratified in this context. Note that the view of fuzzy topology as a fuzzy set of fuzzy sets or a fuzzy set of sets appeared later in Ying 1991–93, 1993, where the author put forward the approach of explicitly developing topology within fuzzy logic in the narrow sense.

Further examinations of the notion of fuzzy topology include the associated interior and closure operators, as well as compactness and normality mentioned above. The notion of point in fuzzy topology and the related concept of separation have proven to be particularly peculiar and have been approached in a number of ways. Related are examinations of the local structure of fuzzy topological spaces and in particular the concept of neighborhood. Interesting examinations concern the notion of convergence to which two approaches have been developed, one based on the concept of net and the other on that of filter. Filter theory led to further examinations of neighborhoods and spaces whose structure may be recovered from the neighborhood systems. The notions of metric and metrizability, to which several approaches emerged, have proved to be of special interest. Various connections to probabilistic metric spaces (Schweizer and Sklar 1983) and probabilistic topologies (Höhle 1978b) are also worth mentioning.

Category theory soon became a working language in fuzzy topological spaces because, on the one hand, it was necessary to compare the various notions of fuzzy topology, and thus to compare the respective categories of fuzzy topological spaces. On the other hand, it was necessary to compare these categories to the category TOP of classical topological spaces. Pioneering in this respect was the work of Lowen (1976), who found a functor ω mapping TOP isomorphically to the category of his fuzzy topological spaces in such a way that a given classical topological space $\langle X, \tau \rangle$ is assigned $\langle X, \omega(\tau) \rangle$, where the fuzzy topology is the set of all lower semicontinuous mappings from $\langle X, \tau \rangle$ to $[0, 1]$. It was again Lowen (1978) who found a functor G from the category of fuzzy topological spaces to TOP, assigning to a fuzzy topological space $\langle X, \tau \rangle$ the classical space $G(X, \tau) = \langle X \times (0, 1], G(\tau) \rangle$, where $G(\tau)$ is the topology generated by the subbase

$$\{\lceil A \rceil_< \mid A \in \tau\} \text{ where } \lceil A \rceil_< = \{\langle u, a \rangle \in X \times (0, 1] \mid 0 \leq a < A(u)\}. \qquad (5.50)$$

This important functor was independently described by Santos in an unpublished manuscript.[143] These functors along with many other functors between the various categories of fuzzy topological spaces were thoroughly investigated in the literature; see Höhle and Šostak 1999 and Rodabaugh 1999 for overviews.

Two related views which have been developed are known as *fixed-basis fuzzy topology* and *variable-basis fuzzy topology*. Here the term "basis" refers to the lattice L of truth degrees involved. In the first view, expounded in the comprehensive study by Höhle and Šostak (1999), fuzzy topological spaces are considered with a fixed L while in the second, L may vary within a theory. Thus, in the first view a given category contains spaces with a common L, while in the second a category contains spaces with various lattices L and thus, for example, the notion of morphism involves morphisms between lattices. The latter view was pioneered by Hutton (1975) and is examined in the comprehensive study by Rodabaugh (1999). The two views are particular manifestations of the corresponding general feature of the theory of fuzzy sets which is mentioned on p. 268.

Also worth noting is the problem, repeatedly observed in fuzzy topology, of whether fuzzy topology is capable of contributing to solutions of mathematical problems for which classical topology is not. One of the first examples was provided by Lowen (1985a), who showed that fuzzy topologies on classical metric spaces provide new insights on these spaces relevant for approximation theory. Further examples are found e.g. in Höhle 2001, 2003.

---

[143] E. S. Santos, "Topology versus fuzzy topology," (preprint, Youngstown, OH: Youngstown State Univ., 1977).

5.3.4 Quantities and mathematical analysis

A large part of classical mathematics deals with exact quantities. Such quantities are represented by the concept of number. The very idea of fuzzy logic, however, leads to considerations involving fuzzy quantities represented by expressions like "approximately 30."[144] Fuzzy sets in the real line seem natural models of such quantities. This is the basic reason why these fuzzy sets have played a special role in the development of fuzzy logic and the mathematics based on it since the appearance of Zadeh's seminal paper. In this section we provide an overview of developments regarding fuzzy quantities including parts of mathematical analysis with fuzzy quantities replacing numbers. Note, however, that fuzzy quantities are utilized in several other areas and therefore also appear in other sections of this chapter.

Fuzzy quantities

The idea of considering whole sets of numbers instead of particular numbers and using them in computation is not new in mathematics and appears in a relatively modern form with studies of Young (1931) who used sets of limits of a given function in case the limit was not unique.[145] Young considered intervals and other sets of numbers and defined generalizations of arithmetic operations and relations on them. For instance, she defined the sum $[a_1, a_2] + [b_1, b_2]$ as the set of all numbers $a + b$ where $a \in [a_1, a_2]$ and $b \in [b_1, b_2]$. These rules later reappeared in the context of *interval analysis*, which emerged in the 1950s in connection with analysis of errors in computations on digital computers independently in the works of Dwyer, Warmus, Sunaga, and Moore, who referred to Dwyer's book.[146] The basic arithmetic operations for intervals introduced by Young are reinvented in these works. Later on, interval analysis became widely known due to the book of Moore (1966). Common to fuzzy quantities and the above works is the idea of approximate representation. But they differ with respect to the gradedness of membership in fuzzy quantities which derives from the motivation to model inexact quantities employed by humans.

---

[144] The term "fuzzy quantity" was probably first coined in Mareš 1977.

[145] Rosalind Cecilia Hildegard Tanner (née Young, 1900–1992), an English mathematician. Young attributes the idea of a set of limits to her father's paper, W. H. Young, "Sulle due funzioni a più valori costituite dai limiti d'una funzione di variabile reale a destra ed a sinistra di ciascun punto" [On the two many-valued functions obtained from the right and left limits of a function of one variable at every point], *Atti della Reale Accademia dei Lincei. Rendiconti, classe di scienze fisiche, matematiche e naturali* 17, no. 5 (1908): 582–87.

[146] P. S. Dwyer, *Linear Computations* (New York: Wiley, 1951). Dwyer uses the term "approximate number" and "range number," the latter meaning a real interval, and refers to H. M. Walker and V. Sanford, "The accuracy of computation with approximate numbers," *Ann Mathematical Statistics* 5 (1934): 1–12, which also considers intervals. M. Warmus, "Calculus of approximations," *Bulletin de l'Académie Polonaise des Sciences* 4, no. 5 (1956): 253–57. T. Sunaga, "Theory of interval algebra and its application to numerical analysis," *RAAG Memoirs* 2 (1958): 29–46; reprinted in *Japan J Indust Appl Math* 26 (2009): 125–43. R. E. Moore, "Automatic error analysis in digital computation" (Technical Report Space Div. Report LMSD 84821, Lockheed Missiles and Space Co., 1959).

A recent book is J. L. Gustafson, *The End of Error: Unum Computing* (Boca Raton, FL: CRC Press, 2015).

A fuzzy quantity is usually understood as a fuzzy set A in the universe $\mathbb{R}$, or generally $\mathbb{R}^n$, having some further, intuitively appealing or technically convenient properties. These properties emerged from the early studies examined in this section, in particular in connection with the extension principle (p. 273, and also chapter 3) upon which a computation with fuzzy quantities is usually based. With further studies, particularly of metric and measurability questions in the context of fuzzy random variables (section 5.3.5), the significance of the particular properties of fuzzy quantities has been recognized. Thus in general, a *fuzzy quantity* $A : \mathbb{R}^n \to [0,1]$ is usually required to

- be *normal* in that $A(u) = 1$ for some u;
- be *upper semicontinuous* at each point;
- have often as well a *bounded support set*, i.e. the set $\{u \in U \mid A(u) > 0\}$.

Note that upper semicontinuity generalizes the property of continuity and may equivalently be phrased in terms of cuts: A is upper semicontinuous iff for each $a \in [0,1]$, the a-cut ${}^aA$ is a closed set.[147] Many contributions are restricted to continuous fuzzy quantities with the disadvantage that such restriction excludes characteristic functions of compact subsets of $\mathbb{R}^n$, and thus excludes ordinary intervals.

Considerable interest has been shown in *convex fuzzy quantities*, i.e. those satisfying $A(\lambda u + (1-\lambda)v) \geq \min(A(u), A(v))$ for each $u, v \in \mathbb{R}^n$ and $\lambda \in [0,1]$. Convexity of A, introduced by Zadeh (1965a), means that each a-cut ${}^aA$ is a convex subset of $\mathbb{R}^n$. Hence, convexity of fuzzy quantities in $\mathbb{R}$ means that the a-cuts are intervals, open or closed. Convexity of a fuzzy quantity corresponds to the idea of a quantity having a membership decreasing with a growing distance from the quantity's core values. *Fuzzy intervals*, which are convex fuzzy quantities in $\mathbb{R}$, and *fuzzy numbers*, which are fuzzy intervals for which there exists a unique u with $A(u) = 1$, have become important subjects.[148] In this view, fuzzy intervals are simply normal fuzzy quantities with bounded supports whose a-cuts are closed intervals, or equivalently, fuzzy quantities A with a bounded support whose core ${}^1A$ forms a closed interval $[a, b]$ such that A is right-continuous on $(-\infty, a]$ and left-continuous on $[b, \infty)$.

Crucial for these considerations are Zadeh's papers, particularly Zadeh 1965a which contains the basic formulation of the extension principle, calls attention to fuzzy sets in $\mathbb{R}^n$, and defines for them the notion of convexity, as well as Zadeh 1975c, which uses the terms "extension principle" and "fuzzy number," develops the extension principle more comprehensively, and demonstrates how to apply it for the addition of fuzzy numbers. Rather soon, further important papers appeared.

---

[147] Upper semicontinuity is a well-known property of real functions; A is upper semicontinuous at u_0 if for each $\varepsilon > 0$ there exists a neighborhood N of u_0 w.r.t. the natural metric such that $A(u) \leq A(u_0) + \varepsilon$ for each $u \in N$.

The result stated here is basically due to R. Baire, *Leçons sur les fonctions discontinues, professées au Collège de France* [Lessons on discontinuous functions, professed at the College of France] (Paris: Gauthier-Villars, 1905).

[148] The condition of bounded support is sometimes dropped for fuzzy intervals. The various existing notions of fuzzy number and interval are, by and large, special instances of the notions presented here.

Mizumoto and Tanaka (1976a) studied for fuzzy sets in $\mathbb{R}$ the basic arithmetic operations, $+, -, \cdot$, and $/$ obtained by the extension principle.[149] Thus, for instance for fuzzy numbers $A, B \in [0,1]^{\mathbb{R}}$, their sum $A + B \in [0,1]^{\mathbb{R}}$ is defined by

$$(A+B)(u) = \bigvee_{u=v+w} A(v) \wedge B(w),$$

where v and w range over $\mathbb{R}$. They observed several properties, such as the preservation of normality by all operations and convexity by $+$, $-$, and $\cdot$, but not $/$. They also observed that the operations $+$ and $\cdot$ are associative, commutative, have the singletons $\{^1/0\}$ and $\{^1/1\}$, denoted commonly by 0 and 1, respectively, as their neutral elements, but also that there are no inverse elements w.r.t. $+$ and $\cdot$ and that the distributive law $A \cdot (B + C) = A \cdot B + A \cdot C$ holds for *positive* fuzzy numbers, i.e. those with $A(u) = 0$ for $u < 0$, but not in general.[150] They also turned to the important problem of ordering of fuzzy numbers and presented results obtained in their other paper (Mizumoto and Tanaka 1976b). They defined the operations $\sqcap$ and $\sqcup$ on fuzzy sets in $\mathbb{R}$ derived from the operations of minimum and maximum on $\mathbb{R}$ by the extension principle, i.e.[151]

$$(A \sqcap B)(u) = \bigvee_{u=v \sqcap w} A(v) \wedge B(w) \quad \text{and} \quad (A \sqcup B)(u) = \bigvee_{u=v \sqcup w} A(v) \wedge B(w), \quad (5.51)$$

and proved the following theorem.[152]

Theorem 5.25. Consider the operations $\sqcap$ and $\sqcup$ defined by (5.51) and the binary relations $\sqsubseteq_{\sqcap}$ and $\sqsubseteq_{\sqcup}$ defined by $A \sqsubseteq_{\sqcap} B$ iff $A \sqcap B = A$ and $A \sqsubseteq_{\sqcup} B$ iff $A \sqcup B = B$.

(a) For arbitrary fuzzy sets in $\mathbb{R}$, $\sqcap$ and $\sqcup$ are idempotent, commutative, and associative; hence the relations $\sqsubseteq_{\sqcap}$ and $\sqsubseteq_{\sqcup}$ are partial orders.

(b) Convex fuzzy sets are closed w.r.t. $\sqcap$ and $\sqcup$ and satisfy the distributive laws.

(c) Normal convex fuzzy sets are closed w.r.t. $\sqcap$ and $\sqcup$ and satisfy also the absorption laws. Hence, they form a distributive lattice in which $\sqsubseteq_{\sqcap}$ and $\sqsubseteq_{\sqcup}$ coincide.

Nguyen (1978) considered several problems related to the extension principle, in particular when applied to extending the arithmetic and other operations with fuzzy sets in $\mathbb{R}$. He obtained the following general result:[153]

---

[149] They defined fuzzy numbers as fuzzy sets in $\mathbb{R}$ and focused on convex and normal ones. In fact, the term "fuzzy number" is used with a variety of meanings and very often for what is nowadays called a fuzzy interval, such as in Dubois and Prade 1978, another early paper.

[150] That is, there exist fuzzy numbers A such that no B satisfies $A + B = \{^0/1\}$; similarly for $\cdot$.

[151] To emphasize the logic of this definition, we denote in the definientia by $\sqcap$ and $\sqcup$ the min and max in $\mathbb{R}$ (universe) and by $\wedge$ and $\bigvee$ the infimum and supremum in $[0,1]$ (scale of truth degrees).

[152] Recall the two possible views on the notion of lattice (Birkhoff 1948), which provide a perspective on this theorem. First, a lattice may be viewed as an algebra L with two binary operations, $\sqcap$ and $\sqcup$, which are commutative, associative, idempotent (which condition may be dropped due to its redundancy) and satisfy the absorption laws $x \sqcap (x \sqcup y) = x$ and $x \sqcup (x \sqcap y) = x$. Second, a lattice may be viewed as a set L with a partial order relation $\sqsubseteq$ for which the sup and inf exist for two-element subsets. From the first notion one obtains the second by putting $x \sqsubseteq y$ iff $x \sqcap y = x$ (equivalently, iff $x \sqcup y = y$); conversely then by putting $x \sqcap y = \inf\{x,y\}$ and $x \sqcup y = \sup\{x,y\}$. In general, lattices are not distributive, i.e. need not satisfy $x \sqcap (y \sqcup z) = (x \sqcap y) \sqcup (x \sqcap z)$ nor its dual law.

[153] This is mentioned, although somewhat unclearly and without proof, already in Zadeh 1975c (part I, p. 237).

Theorem 5.26. For $f : U \times V \to W$ and $A \in [0,1]^U, B \in [0,1]^V$, we have

$$[f(A,B)](w) = \bigvee\{a \in [0,1] \mid w \in f(^aA, ^aB)\},$$

i.e. the system $\{f(^aA, ^aB) \mid a \in [0,1]\}$ provides a representation of $f(A,B)$.

Nguyen observed that $^a[f(A,B)] \supseteq f(^aA, ^aB)$ is always the case and that the equality, $^a[f(A,B)] = f(^aA, ^aB)$, need not hold and is actually obtained if and only if for each $w \in W$, the supremum $\bigvee_{w=f(u,v)} A(u) \wedge B(v)$ is attained. Because continuous functions, and thus in particular $+$ or $\cdot$, attain suprema of their values on compact sets, he was able to obtain the following important result, which holds in particular for fuzzy quantities as defined above.

Theorem 5.27. If $f : \mathbb{R} \times \mathbb{R} \to \mathbb{R}$ is continuous and $A, B \in [0,1]^{\mathbb{R}}$ are upper semicontinuous and have compact closures of their supports, then for each $a \in [0,1]$ one has

$$^a[f(A,B)] = f(^aA, ^aB).$$

Later on, Fullér and Keresztfalvi (1991) generalized Nguyen's result for arbitrary upper semicontinuous (i.e. right continuous) t-norms $\otimes$ in place of min and showed that for f thus defined, i.e. $[f(A,B)](w) = \bigvee_{w=f(u,v)} A(u) \otimes B(v)$, one has

$$^a[f(A,B)] = \bigcup_{a \leq b \otimes c} f(^bA, ^cB).$$

Various modifications of Nguyen's result appeared later and provided bases for representation of and computing with fuzzy quantities by computers; see e.g. Dubois et al. 2000; Kruse, Gebhardt, and Klawonn 1994. In another early paper, Dubois and Prade (1978) defined fuzzy numbers as fuzzy sets whose membership function is 0 on $[-\infty, a]$, increasing on $[a, b]$, equals 1 on $[b, c]$, then decreases on $[c, d]$ and remains 0 on $[d, \infty]$, showed closedness of such fuzzy sets for $+$ and $\cdot$, along with some further properties.[154] They also introduced the so-called *L-R-fuzzy numbers*, which are such that their increasing and decreasing parts have a uniform shape and thus may be represented by three numbers as parameters only. Such particular fuzzy sets became useful in fuzzy logic in the broad sense and in applications because of the possibility of representing them as well as computing with them easily (section 3.4).

Other early contributions to arithmetic operations with fuzzy quantities—both different from Zadeh's extension principle and not referring to it—are due to Jain (1976a), who proposed what can be seen as a modified extension principle in which sup is replaced by the t-conorm dual to the product t-norm, and Mareš (1977), who considers additions on fuzzy sets in a rather general setting, namely fuzzy sets in the universe U which forms a group w.r.t. the group operation $+$ and a given topology on U. Furthermore, he assumes that the fuzzy sets are measurable and integrable functions in a measure space $\langle U, \mathcal{U}, \mu \rangle$ such that the σ-algebra $\mathcal{U}$ contains all closed sets of the topology. Mareš essentially defined addition $A + B$ as the truncation by

---

[154] Note that they did not refer to Zadeh for the extension principle and spoke instead of Gaines's formula, due to Gaines (1976, 637), who provided the formula for the principle without reference to Zadeh.

1 of the convolution of A and B, i.e. convolution $(A + B)(u) = \min(1, \int A(u - v) \cdot B(v)\,d\mu(v))$, which is very different from addition based on the extension principle. Later on, Mareš (1989) also examined the addition, $+$, of rational fuzzy quantities, i.e. $U = \mathbb{Q}$, based on the extension principle from the point of view of the fact that $+$ does not have inverse elements, hence even though $\mathbb{Q}$ with $+$ forms a group, the rational fuzzy quantities do not. He defined an equivalence relation $\sim$ on fuzzy quantities in such a way that $A \sim B$ if there exist symmetric fuzzy quantities C_1 and C_2 such that $\overline{A} + \overline{C_1} = \overline{B} + \overline{C_2}$, where being symmetric means that $C_i(r) = -C_i(r)$ for each r and where $\overline{D}$ is the normalized version of D defined by $\overline{D}(r) = \frac{D(r)}{\sup_{s \in \mathbb{Q}} D(s)}$. He also proved that $\sim$ is a group congruence. Hence, upon $\sim$ fuzzy quantities form a group and, in particular, inverse elements exist. Many more results are available in his book Mareš 1994.

The above two approaches by Jain and Mareš represent rather radical departures from the standard extension principle, i.e. $[f(A,B)](w) = \bigvee_{w=f(u,v)} A(u) \wedge B(v)$. One less radical departure consists of replacing $\wedge$ by a more general t-norm $\otimes$, i.e. to consider $[f(A,B)](w) = \bigvee_{w=f(u,v)} A(u) \otimes B(v)$ as in the above-mentioned result by Fullér and Keresztfalvi (1991). This generalized form was proposed by Dubois and Prade (1981), but the particular case of addition of probability distribution functions based on a left-continuous t-norm is studied in Schweizer 1975. The arithmetic of fuzzy quantities thus generalized differs from that based on the classical extension principle in several respects and has been studied in a number of papers, including Dubois and Prade 1981 which contains formulas for computing additions of L-R-numbers based on the three basic continuous t-norms, and several further papers, e.g. by Fullér and Keresztfalvi, Gebhardt, Dug-hun Hong, Mesiar, and Marková-Stupňanová, which concern the case of more general t-norms, preservation of shapes of fuzzy intervals, and further properties, for which see Dubois et al. 2000. Another generalization concerns constraints in the extension principle. Zadeh (1975c) considers constraints expressed by a binary fuzzy relation R and suggests a generalized version of the extension principle that is based on the formula $[f(A,B)](w) = \bigvee_{w=f(u,v)} (A(u) \wedge B(v) \wedge R(u,v))$. Such constraints may restore some desired properties which fail with the basic version of the principle. For example, while the binary function of subtraction of fuzzy intervals does not satisfy $A - A = 0$, this identity is restored by adding the constraint via R defined by $R(u,v) = 1$ iff $u = -v$. These constraints are directly connected to constrained fuzzy arithmetic (section 3.4).

A more substantial departure is inspired by problems with solving equations such as $A + X = C$, for which the difference $C - A$ based on the extension principle need not be a solution. Such problems were observed already by Mizumoto and Tanaka (1976a) and reappeared in a number of papers. Sanchez (1984) observed that the problem of solving equations $A + X = B$ is a particular case of solving fuzzy relational equations, on which topic he wrote the influential paper Sanchez 1976. He considered a general scenario in which $* : U \times V \to W$ is a binary function and the problem is to determine a solution of $A * X = C$, where $A \in [0,1]^U$ and $C \in [0,1]^W$ are given fuzzy sets and $A * X$ results from the extension principle. The above problem $A + X = C$ for fuzzy quantities is a special case. Among other results, he proved the following theorem in which $\rightarrow$ denotes Gödel implication:

Theorem 5.28. $A * X = C$ *has a solution iff the fuzzy set* $A \backslash_* C$ *defined by*

$$(A \backslash_* C)(v) = \bigwedge \{A(u) \rightarrow C(w) \mid u \in U, w = u * v\}$$

is its solution. If $A * X = C$ *is solvable,* $A \backslash_* C$ *is its greatest solution, and the set of solutions forms a sup-semilattice w.r.t. inclusion of fuzzy sets.*

The connection to fuzzy relational equations is the following. For $A \in [0,1]^U$, consider the fuzzy relation $A^* \in [0,1]^{V \times W}$ defined by $A^*(v, w) = \bigvee \{A(u) \mid u \in U, u * v = w\}$. Then one may check that $A * X = X \circ A^*$ with $\circ$ being the sup-min composition of fuzzy relations, hence the solutions of $A * X = C$ coincide with those of the fuzzy relational equation $X \circ A^* = C$. The result (Sanchez 1976) says that the greatest solution of $X \circ A^* = C$, if it exists, is $A^* \triangleleft C$, i.e. $(A^* \triangleleft C)(v) = \bigwedge \{A^*(v, w) \rightarrow C(w) \mid w \in W\}$, and it may be checked that $A^* \triangleleft C$ is just $A \backslash_* C$. For $*$ being $+$, the operation $\backslash_+$ plays a role of alternative subtraction and has been examined in further papers such as Biacino and Lettieri 1989.

Numerous later studies have been devoted to various practically motivated questions, including numerical characteristics of fuzzy intervals, such as the mean value, measures of specificity and fuzziness or defuzzifications. The above-mentioned observation by Mizumoto and Tanaka (1976a) regarding the lack of certain properties in the arithmetic of fuzzy intervals, which were known in classical interval analysis, led to many investigations, particularly with respect to the problem that while, for instance, the expressions $x \cdot (y + z)$ and $x \cdot y + x \cdot z$ represent the same functions of reals, they represent two different functions of fuzzy intervals due to the lack of distributivity for fuzzy intervals. These investigations went along with those in interval computations and we refer to Kreinovich et al. 1997 as well as the more recent Kreinovich 2013; Kreinovich, Chiangpradit, and Panichkitkosolkul 2012; and Xiang and Kreinovich 2013. Practical motivations regarding feasibility of representation and computation led to a number of investigations. These include further studies of the above-mentioned L-R-fuzzy numbers and intervals and further parameterized representations, computing with fuzzy intervals using the parameterized representations, approximation of general fuzzy intervals or quantities by those from a given parameterized class of fuzzy intervals or quantities, as well as approximations using the representation by a-cuts. For further information and references see Dubois et al. 2000.

Another practically motivated problem which has received considerable attention is ranking of fuzzy intervals. Various approaches have been developed, including those based on comparing ordinary numbers obtained as characteristics of fuzzy intervals and those based on defining fuzzy ordering-like relations, generalizing classical interval orders, on the bases of which a ranking is subsequently performed.[155] An overview with an extensive list of references on this topic can be found again in Dubois et al. 2000.

Mathematically motivated notions of fuzzy number

The above notions of fuzzy number, interval, and quantity were motivated by the view that a fuzzy number basically represents a vaguely delineated collection of numbers linguistically described as "approximately 5," or the like. Such notions, however, do not obey certain mathematical properties one might wish to demand in certain situations. We now survey approaches to fuzzy numbers that were primarily motivated by certain mathematical considerations.

The first two are closely related and were developed independently in Hutton 1975; Gantner, Steinlage, and Warren 1978; Rodabaugh 1982; Lowen 1983; and Höhle 1978a, 1981. In the context of his work on probabilistic metric spaces, Höhle (1978a, 1981) introduced *fuzzy nonnegative real numbers* as functions A from $\mathbb{R}^+ = [0, \infty)$ to $[0, 1]$ satisfying $A(0) = 0$, $\bigvee_{t \in \mathbb{R}^+} A(t) = 1$, and $A(t) = \bigvee_{s \in [0,t)} A(s)$.[156] We denote the set of such As by H and note that it is just the set of all probability distribution functions on $\mathbb{R}^+$. With a partial ordering defined by $A \sqsubseteq B$ iff $A(t) \geq B(t)$ for each $t \in \mathbb{R}^+$, H becomes a complete lattice. $\mathbb{R}^+$ embeds to H by sending each $t \in \mathbb{R}^+$ to δ_t defined by $\delta_t(s) = 0$ for $s \leq t$ and $= 1$ for $s > t$, i.e. δ_t is the characteristic function of the open interval (t, ∞) which is to be seen as the representation of the real t in this approach. The truth degree $A(t)$ is naturally interpreted as the truth degree of "(fuzzy number) A is no larger than (ordinary real number) t." The structure is nicely related to the set $H^{(-1)}$ of all pseudoinverses of the members in H. A *pseudoinverse* of $A \in H$ (cf. n. 100 on p. 275) is the function $A^{(-1)} : [0, 1] \to \mathbb{R}^+$ defined by $A^{(-1)}(s) = \bigvee_{A(t) < s} t$. The set $H^{(-1)}$ consists of all left-continuous nondecreasing functions from $[0, 1]$ to $\mathbb{R}^+$ mapping 0 to 0. When equipped with a pointwise order $\leq$ and a pointwise addition $+$, i.e. $(A^{(-1)} + B^{(-1)})(t) = A^{(-1)}(t) + B^{(-1)}(t)$, it becomes an isomorphic copy of H:

Theorem 5.29. The mapping $^{(-1)}$ is an involutive order-preserving isomorphism of $\langle H, \sqsubseteq, \overline{+} \rangle$ onto $\langle H^{(-1)}, \leq, + \rangle$, i.e. a bijection satisfying $A \sqsubseteq B$ iff $A^{(-1)} \leq B^{(-1)}$, $(A \overline{+} B)^{(-1)} = A^{(-1)} + B^{(-1)}$, and $(A^{(-1)})^{(-1)} = A$.[157]

For further properties and the role of these fuzzy real numbers see Höhle 1978a as well as Klement 1985 for their role in the context of integration.

The second approach is closely related and based on the notion of a fuzzy real line (Hutton 1975; Gantner, Steinlage, and Warren 1978) examined on p. 295 above. Recall that the elements of the fuzzy real line $\mathbb{R}([0, 1])$ are certain equivalence classes

---

[155] P. C. Fishburn, *Interval Orders and Interval Graphs* (New York: Wiley, 1985).
[156] This is a particular case of a very general concept of L-fuzzy quantity (Höhle 1981).
[157] $(A^{(-1)})^{(-1)}$ is the pseudoinverse of $A^{(-1)}$.

[λ] of monotone decreasing functions $\lambda : \mathbb{R} \to [0, 1]$ approaching 0 on the left and 1 on the right, and that a real t is represented by a class corresponding to the characteristic functions of $(-\infty, t)$ or, equivalently, $(-\infty, t]$. This approach is thus in a sense symmetric to Höhle's. Rodabaugh (1982) defined an extension of addition of the reals on $\mathbb{R}([0, 1])$ and showed its uniqueness given some continuity assumption. His definition was somewhat involved but Lowen (1983) showed that it may essentially be viewed as if obtained by Zadeh's extension principle. Lowen (1985b) later showed that certain fuzzy order relations extending the ordering of reals are more fundamental for the fuzzy real line than its fuzzy topology.

The third approach (Höhle 1987) is very different from those above. The idea is to apply, in a fuzzy setting, Dedekind's definition of real numbers from rational numbers.[158] According to this definition, reals are essentially pairs $\langle C, D \rangle$ of sets of rational numbers for which C contains all rationals less than or equal to every number in D and D contains all rationals greater than or equal to every number in C. From a more abstract view, the set of reals is just the so-called Dedekind-MacNeille completion of the partially ordered set $\langle \mathbb{Q}, \leq \rangle$. Höhle thus developed the concept of Dedekind-MacNeille completion for fuzzy partial orders and defined fuzzy real numbers as members of this completion. From a more general viewpoint, Höhle's set of fuzzy real numbers is essentially the fuzzy concept lattice $\mathscr{B}(\mathbb{Q}, \mathbb{Q}, \leq)$ built over a suitable complete residuated lattice on $[0, 1]$ as a structure of truth degrees and with a crisp order $\leq$ of rationals as the incidence fuzzy relation (p. 281). Thus, the fuzzy real numbers are pairs $\langle C, D \rangle$ of fuzzy sets of rationals with essentially the same meaning as in the crisp case described above if interpreted according to the principles of fuzzy logic in the narrow sense. Höhle discusses some particular cases. For example, he shows that if the left-continuous t-norm is the minimum t-norm, the fuzzy real numbers are essentially the fuzzy points in the sense of Wong 1974b. Even though Höhle's approach has not been continued, mostly because it is mathematically nontrivial and thus not easily accessible to researchers usually working with fuzzy numbers, it is of great methodological value. It is one of the first attempts to develop within a first-order fuzzy logic a nontrivial mathematical theory.

Mathematical analysis

Mathematical analysis is one of the areas whose generalizations based on replacing reals numbers by fuzzy quantities have been developed in a number of papers.[159] To a large extent, the research and methods in this area closely follow those in set-valued analysis (Aubin and Frankowska 1990) and the usage of the extension principle and the level-cut representation of fuzzy sets.

---

[158] *Stetigkeit und irrationale Zahlen* [Continuity and irrational numbers] (Braunschweig: Vieweg, 1872), *Was sind und was sollen die Zahlen?* [What are numbers and what should they be?] (Braunschweig: Vieweg, 1888).

[159] Other areas, covered below, are probability and statistics, geometry, and various areas of applied mathematics.

Fundamental to mathematical analysis are properties of the topology induced by a given metric on $\mathbb{R}^n$, such as openness and closedness, compactness or completeness, and the same holds true for analysis with fuzzy quantities. In what follows we denote by $\mathscr{F}(\mathbb{R}^n)$ the set of all fuzzy intervals (p. 299), but several results may be obtained without the convexity assumption thus involved. All a-cuts ^aA, $0 < a \leq 1$, as well as the closure $\overline{\text{supp}\,A}$ of the support set are then compact subsets of $\mathbb{R}^n$. Two important metrics on $\mathscr{F}(\mathbb{R}^n)$ are the supremum metric d_∞ and the parameterized metric d_p, $1 \leq p < \infty$, defined for fuzzy quantities $A, B \in \mathscr{F}(\mathbb{R}^n)$ by

$$d_\infty(A,B) = \sup_{a \in [0,1]} d_H(^aA, {}^aB) \text{ and}$$

$$d_p(A,B) = \left(\int_0^1 d_H(^aA, {}^aB)^p \, da \right)^{\frac{1}{p}},$$

where d_H is the Hausdorff distance.[160] The metric d_∞ was first used by Heilpern (1981), and also appears in the work of Puri and Ralescu (1983, 1985, 1986) on fuzzy random variables; the d_p metrics appear in Klement, Puri, and Ralescu 1986, again in the context of fuzzy random variables (section 5.3.5). For other metrics, see e.g. Goetschel and Voxman 1981. A comprehensive treatment is found in the book Diamond and Kloeden 1994. The basic properties of these metrics are shown in the next theorem, in which $\mathscr{F}(K)$ denotes the set of all fuzzy quantities in $K \subseteq \mathbb{R}^n$. The results for d_∞ and d_1 are due to Puri and Ralescu (1985) and Klement, Puri, and Ralescu (1986); the others are due to Diamond and Kloeden (1994).

Theorem 5.30. $\mathscr{F}(\mathbb{R}^n)$ equipped with d_∞ is a complete metric space which is not separable. For each compact $K \subseteq \mathbb{R}^n$, $\mathscr{F}(K)$ equipped with d_p is a complete metric space which is separable, for every $1 \leq p < \infty$.

Further information, particularly about convergence and compactness w.r.t. various metrics, can be found in Diamond and Kloeden 1994.

Let us now turn to the analysis of functions which involve fuzzy quantities and to the problems of differentiation and integration of such functions. It turned out that functions assigning fuzzy quantities to reals are mathematically interesting and feasible generalizations of classical real functions and that, importantly, such functions naturally appear in the area of fuzzy random variables (section 5.3.5), in which context a number of important results have appeared, e.g. Puri and Ralescu 1983, 1985, 1986. Other early works on these problems include Kloeden 1980, 1982, Dubois and Prade 1982b, Goetschel and Voxman 1981, Bobylev 1985a, Kaleva 1987, 1990, Seikkala 1987, and other papers by these authors. It would be more accurate, however, to view such functions as generalizations of set-valued functions, i.e. as general functions from $\mathbb{R}^k$ to the set of subsets of $\mathbb{R}^n$ (Aubin and Frankowska 1990). By a *fuzzy-set-valued mapping* we thus understand functions of the form

---

[160]Denoting by $\|u - v\|$ the Euclidean distance on $\mathbb{R}^n$, d_H is a metric defined for $U, V \subseteq \mathbb{R}^n$ by $d_H(U,V) = \max\{\sup_{u \in U} \inf_{v \in V} \|u - v\|, \sup_{v \in V} \inf_{u \in U} \|u - v\|\}$.

$F : T \to \mathscr{F}(\mathbb{R}^n)$, where T is some subset of $\mathbb{R}^k$. In what follows, we always consider the metric d_∞.

Definitions of some properties are obtained as particular cases of classical notions. For instance, $F : T \to \mathscr{F}(\mathbb{R}^n)$ is called continuous at $t_0 \in T$ if for every $\varepsilon > 0$ there exists $\delta > 0$ such that for each $t \in T$, $\|t - t_0\| < \delta$ implies $d_\infty(F(t), F(t_0)) < \varepsilon$. As an example of a connection to the properties of set-valued functions, note that F is continuous at t_0 if and only if the set-valued mappings ${}^a F$ defined by $({}^a F)(x) = {}^a(F(x))$ are continuous at x_0 uniformly in $a \in [0, 1]$. Uniformity, which is essential here, means that for each $\varepsilon > 0$ there is $\delta > 0$ such that $d_H({}^a F(t), {}^a F(t_0)) < \varepsilon$ holds for all $a \in [0, 1]$ and $t \in T$ satisfying $\|t - t_0\| < \delta$. This view enables us to define the weaker notion of upper semicontinuity, needed for differentiation, which results from the above condition when the respective inequality in the preceding definition is replaced by $d_H^*({}^a F(t), {}^a F(t_0)) < \varepsilon$ where $d_H^*(U, V) = \sup_{u \in U} \inf_{v \in V} \|u - v\|$, and dually for lower semicontinuity. The notion of *strong measurability* of F, which requires that all ${}^a F$ be measurable in the sense of measurability of set-valued functions, is essential for the definition of the integral of a fuzzy-set-valued function. A weaker condition is *measurability* which requires that the preimage of every Borel subset of $\langle \mathscr{F}(\mathbb{R}^n), d_\infty \rangle$ is a Borel subset of $\mathbb{R}^k$. The following theorem shows the basic relationships (Kaleva 1987, Diamond and Kloeden 1994):

Theorem 5.31. For fuzzy-set-valued functions, strong measurability implies measurability and lower or upper semicontinuity at each point implies strong measurability.

Of the several notions of differentiability of fuzzy-set-valued functions, including de Blasi differentiability (Diamond and Kloeden 1994), Hukuhara differentiability (Puri and Ralescu 1983), and Bobylev differentiability (Bobylev 1985a), we briefly mention the first one.[161] A function $F : T \to \mathscr{F}(\mathbb{R}^n)$ is called *de Blasi differentiable* at $t_0 \in T$ if there exists an upper semicontinuous and homogeneous mapping $D_{t_0}^F : \mathbb{R}^k \to \mathscr{F}(\mathbb{R}^n)$, called the *de Blasi differential* of F at t_0, satisfying

$$\lim_{\|\Delta t\| \to 0} \frac{d_\infty(F(t_0 + \Delta t), F(t_0) + D_{t_0}(\Delta t))}{\Delta t} = 0.$$

In this definition, homogeneity of $D_{t_0}^F$ means $D_{t_0}^F(\lambda t) = \lambda \cdot D_{t_0}^F(t)$ for every $\lambda > 0$ and $t \in \mathbb{R}^k$. It follows that if F is de Blasi differentiable at t_0, then each set-valued mapping ${}^a F$, $a \in [0, 1]$, is classically de Blasi differentiable at t_0 and its differential is the a-cut ${}^a D_{t_0}^F$ of the de Blasi differential of F. Moreover, the de Blasi differential is unique if it exists, is linear in that $D_{t_0}^{\lambda F + \varkappa G} = \lambda D_{t_0}^F + \varkappa D_{t_0}^G$, and its existence at t_0 implies continuity of F at t_0. As another example, we present the following generalization of the classical mean value theorem for derivatives. In this theorem,

---

[161] The first two are based on the classical notion developed by F. S. de Blasi, "On the differentiability of multi-functions," *Pacific J Math* 66 (1976): 67–81, and M. Hukuhara, "Intégration des applications measurables dont la valeur est un compact convexe," *Funkcialaj Ekvacioj* 10 (1967): 205–23.

continuous differentiability of F means the existence of the de Blasi differential $D_{t_0}^F$ at every $t_0 \in T$ and the continuity of $D_{t_0}^F(\Delta t)$ as a function of t_0 for each $\Delta t \in \mathbb{R}^k$.

Theorem 5.32. If $T \subseteq \mathbb{R}^k$ is open and convex and $F : T \to \mathscr{F}(\mathbb{R}^n)$ is continuously differentiable, then

$$d_\infty(F(t_1), F(t_2)) \leq \|t_2 - t_1\| \cdot \sup_{t \in \overline{t_1 t_2}} D_t^F((t_2 - t_1)/\|t_2 - t_1\|),$$

where $\overline{t_1 t_2}$ denotes the line segment between t_1 and t_2.

Integration of fuzzy-set-valued functions has been examined for the important case $k = 1$, which appears for instance in considerations regarding fuzzy random variables (section 5.3.5). To illustrate the concept of integration of such functions, we restrict ourselves for simplicity to functions of the form $F : [0,1] \to \mathscr{F}(\mathbb{R}^n)$. First, note that a function F of this form is called *integrably bounded* if there exists an integrable function $h : [0,1] \to \mathbb{R}$ such that $\|F(t)\| \leq h(t)$ for each $t \in [0,1]$ with $\|F(t)\|$ denoting the norm induced by the metric d_∞. It follows that each *a*-cut ^aF of an integrably bounded F is integrably bounded as a set-valued function. If $F : [0,1] \to \mathscr{F}(\mathbb{R}^n)$ is strongly measurable and integrally bounded, then the just-mentioned integrable boundedness of each ^aF ensures that each ^aF is Aumann integrable on $[0,1]$ in the sense of set-valued analysis, i.e. has its Aumann integral $\int_0^1 {}^aF\,dt$.[162] If then there exists a fuzzy set $A \in \mathscr{F}(\mathbb{R}^n)$ such that for each $a \in [0,1]$ we have $^aA = \int_0^1 {}^aF dt$, i.e. the a-cut ^aA equals the classical Aumann integral of ^aF, one calls F *integrable* over $[0,1]$, denotes A by $\int_0^1 F dt$ and calls it the *integral* of F over $[0,1]$. This concept of integral is due to Puri and Ralescu 1986, but appeared earlier in Féron 1979. The following theorem ensures the existence of integrals (Puri and Ralescu 1986); see also Kaleva 1990:

Theorem 5.33. If $F : [0,1] \to \mathscr{F}(\mathbb{R}^n)$ is strongly measurable and integrally bounded, then it is integrable over $[0,1]$.

A continuous function is always integrable and, naturally, the integral has many further properties analogous to those of the Aumann integral, such as $\int_0^1 F + G dt = \int_0^1 F dt + \int_0^1 G dt$, $\int_0^1 \lambda F dt = \lambda \int_0^1 F dt$, or the integral inequality

$$d_\infty(\int_0^1 F dt, \int_0^1 G dt) \leq \int_0^1 d_\infty(F, G) dt.$$

A visible direction in mathematical analysis of functions involving fuzzy quantities is represented by *fuzzy differential equations*. Initial papers on this topic include Kaleva 1987, 1990, Seikkala 1987, and Bobylev 1985b, as well as Hüllermeier

---

[162] R. J. Aumann, "Integrals of set-valued functions," *J Math Anal Appl* 12 (1965): 1–12. A set-valued function G of $[0,1]$ to the set of all convex compact subsets of $\mathbb{R}^n$ is called Aumann integrable if there exists its Aumann integral, $\int_0^1 G(t) dt$, which is defined as $\int_0^1 G(t) dt = \{\int_0^1 g(t) dt \mid g \in \text{sel}(G)\}$, where $\text{sel}(G)$ denotes the set of all integrable selectors of G, i.e. integrable functions $f : [0,1] \to \mathbb{R}^n$ satisfying $g(t) \in G(t)$ for each $t \in [0,1]$.

1999 which examines numerical methods. One version of the problems considered is represented by the initial value problem

$$\frac{dx}{dt}(t) = f(x(t)), \quad x(0) = x_0, \tag{5.52}$$

with some $t_1 < 0 < t_2$, in which $f : \mathscr{F}(\mathbb{R}) \to \mathscr{F}(\mathbb{R})$ is a given function mapping fuzzy intervals to fuzzy intervals and $x_0 \in \mathscr{F}(\mathbb{R})$ is a fuzzy interval representing the initial value. The goal is to find a solution, i.e. a fuzzy-set-valued function $x : (t_1, t_2) \to \mathscr{F}(\mathbb{R})$ satisfying (5.52), in which usually the Hukuhara derivative and the metric d_∞ are used. Note, however, that several other formulations of the initial value problem as well as other problems known from classical differential equations have been proposed; see Diamond and Kloeden 2000, Lakshmikantham and Mohapatra 2003. Since then numerous papers have been published on this topic, mostly in the journal *Fuzzy Sets and Systems* and other journals specializing in fuzzy sets, but a number of papers have also appeared in the journals *Chaos, Solitons & Fractals* and *Nonlinear Analysis*, and occasionally other journals specializing in nonlinear analysis and related fields.

5.3.5 Probability and statistics

We now present significant directions in applying fuzzy sets to extend the capability of probability theory and statistics. Note that here we are neither interested in how probability theory may be utilized in fuzzy logic and its applications nor in the debates on the allegedly competitive character of fuzzy logic and probability theory—these are overviewed in chapter 2. Related to the topics covered here are the developments in logics of uncertainty which are examined in section 4.5.

Probability of fuzzy events

In one of his early papers, Zadeh (1968a) proposed generalizing the notion of event to account for situations in which events are fuzzy rather than crisp. While the classical notion of event, defined as a subset of a given set Ω of outcomes, is appropriate for modeling events such as "temperature is higher than 75 °F," Zadeh conceived of fuzzy events as particular fuzzy sets $A : \Omega \to [0, 1]$, with $A(\omega)$ interpreted as the degree to which A applies to the outcome $\omega \in \Omega$. This way, vaguely specified events such as "high temperature" obtain natural mathematical models. To measure probabilities of fuzzy events, Zadeh assumed that there is a given probability space $\langle \Omega, \mathscr{B}, P \rangle$ with $\Omega = \mathbb{R}^n$, $\mathscr{B}$ the σ-field of Borel sets in $\mathbb{R}^n$, and P a probability measure on $\mathscr{B}$. He defined fuzzy events as Borel-measurable fuzzy sets $A \in [0, 1]^\Omega$, defined the probability of a fuzzy event by the Lebesgue-Stieltjes integral as

$$m(A) = \int_\Omega A \, dP, \tag{5.53}$$

and presented some observations regarding closedness of fuzzy events under fuzzy-set operations, independence, conditional probability, and entropy.

Zadeh's formula (5.53) is natural in that it generalizes the classical one, $P(A) = \int_\Omega \chi_A \, dP$, with χ_A being the characteristic function of a classical event A. On the other hand, it leaves open the question whether reasonable probability measures of fuzzy events have to be induced by classical probability measures as in (5.53). A study of measures on fuzzy sets generally not based on classical measures was initiated by Höhle (1976, 1983). Independently, Klement (1980a,b, 1982b,a) and somewhat later Butnariu (1983, 1986, 1987) started an attempt to generalize notions of classical measure theory, such as that of σ-algebra, measure and probability, from the point of view of fuzzy sets, to provide natural foundations for the ideas propounded by Zadeh (1968a) and, in the case of Butnariu, for considerations regarding fuzzy games. Their work resulted in a joint book (Butnariu and Klement 1993) and in many subsequent developments which we now describe. Note, however, that several other generalizations of the notion of probability space have appeared; see e.g. Khalili 1979, Piasecki 1985, and Smets 1982. In particular, Piasecki's (1985) notion of P-measure has been further developed, e.g. by Dvurečenskij, Mesiar, and Riečan; see e.g. Mesiar 1993b, Dvurečenskij 1996, and the references therein.

A fundamental notion in Butnariu and Klement's approach is that of *tribe* on Ω w.r.t. a given t-norm $\otimes$, or a $\otimes$-tribe, which is a subset $\mathcal{T}$ of $[0,1]^\Omega$ satisfying[163]

$$\emptyset \in \mathcal{T}, \tag{5.54}$$

$$\text{for each } A \in \mathcal{T} \text{ we have } \neg A \in \mathcal{T}, \tag{5.55}$$

$$\text{for each sequence } (A_n)_{n=1}^\infty \in \mathcal{T} \text{ we have } \otimes_{n=1}^\infty A_n \in \mathcal{T}, \tag{5.56}$$

where $\neg$ is the standard negation, $\neg a = 1 - a$, and the operations on fuzzy sets are based on $\neg$ and $\otimes$ and defined componentwise, e.g. $(\neg A)(\omega) = \neg(A(\omega))$. Note that the infinite $\otimes$-based intersection $\otimes_{n=1}^\infty A_n$ always exists and if condition (5.56) is replaced by requiring that $A, B \in \mathcal{T}$ imply $A \otimes B \in \mathcal{T}$, we obtain the notion of *clan*.[164] Due to (5.55) and the duality of t-norms and their t-conorms, $\otimes$ may be replaced by its dual t-norm $\oplus$. Since the operations based on t-norms, t-conorms, and the standard negation generalize the classical intersection, disjunction, and complement, it readily follows that the notion of tribe generalizes σ-algebras, i.e. collections of subsets of Ω containing $\emptyset$ and closed w.r.t. complements and arbitrary unions. Hence, modulo identification of sets with crisp fuzzy sets, the collection of subsets of Ω is a σ-algebra if and only if it is a tribe w.r.t. an arbitrary t-norm.

If $\mathcal{A}$ is an ordinary σ-algebra on Ω and $\otimes$ is Borel-measurable, then the set $\mathcal{A}^\wedge$ of all measurable fuzzy sets in $[0,1]^\Omega$ is a tribe w.r.t. $\otimes$, which is called *generated* (Butnariu and Klement 1993) or *full* (Navara 2005). Interestingly, Zadeh's system of fuzzy events is just a generated tribe. Conversely, the collection $\mathcal{T}^\vee$ of all crisp fuzzy sets of $\mathcal{T}$ forms a tribe which forms a σ-algebra.

---

[163] Variations of this notion have appeared; our presentation is based essentially on Butnariu and Klement 2002.
[164] O. Wyler, "Clans," *Compositio Mathematica* 17 (1966): 172–89.

Naturally, properties of $\otimes$-tribes depend on those of the t-norm $\otimes$. Of particular importance are tribes w.r.t. Frank t-norms (Frank 1979) defined by $a \otimes_\lambda b = \log_\lambda \left(1 + \frac{(\lambda^a - 1)(\lambda^b - 1)}{\lambda - 1}\right)$ for each $\lambda \in [0, \infty]$ except $\otimes_0$, $\otimes_1$, and $\otimes_\infty$, which are the minimum, product, and Łukasiewicz t-norms, respectively. Butnariu and Klement (1993) established the following properties.

Theorem 5.34. Let $\lambda \in (0, \infty]$. Each tribe $\mathscr{T}$ w.r.t. $\otimes_\lambda$ is also a tribe w.r.t. the minimum as well as the Łukasiewicz t-norm, satisfies $\mathscr{T} \subseteq \mathscr{T}^{\vee\wedge}$, and, moreover, is generated iff it contains all fuzzy sets which are constant functions.

According to Butnariu and Klement, for a t-norm $\otimes$ and its t-conorm $\oplus$, a $\otimes$-*measure* on a $\otimes$-tribe is a function $m : \mathscr{T} \to [-\infty, \infty]$ assuming at most one of the values $-\infty$ and ∞ which satisfies for any $A, B, A_n (n \in \mathbb{N}) \in \mathscr{T}$,

$$m(\emptyset) = 0, \tag{5.57}$$

$$m(A \otimes B) + m(A \oplus B) = m(A) + m(B), \tag{5.58}$$

$$\lim_{n \to \infty} m(A_n) = m(C) \text{ whenever } A_n \nearrow C \text{ and } C \in \mathscr{T}. \tag{5.59}$$

Here, $A_n \nearrow C$ means that $(A_n)_{n \in \mathbb{N}}$ is a nondecreasing sequence whose supremum is C. Worth noting is (5.58), which expresses a kind of a valuation property, while (5.59) expresses continuity of m from the left.[165] A measure m is called a *probability* $\otimes$-*measure* if it assumes only values in $[0, 1]$ and $m(\Omega) = 1$. If $\mathscr{T}$ consists of crisp fuzzy sets (and is thus a σ-algebra), then (probability) $\otimes$-measures are just the ordinary (probability) measures modulo identifying crisp fuzzy sets with ordinary sets. An important problem is to provide representations of measures by integrals. For illustration, we provide the following result by Butnariu (1987), improving a previous result by Klement (1982a) and showing that the respective measures admit the representation (5.53) originally suggested by Zadeh (1968a).

Theorem 5.35. If $\mathscr{T}$ is a tribe w.r.t. the Łukasiewicz t-norm $\otimes$ and m a probability $\otimes$-measure on it, then there exists a unique ordinary probability measure P on $\mathscr{T}^\vee$ such that for each A we have $m(A) = \int_\Omega A \, dP$.

Conversely, if P is a probability measure on a σ-algebra $\mathscr{A}$ and $\mathscr{T}$ its generated tribe, then $m_P(A) = \int_\Omega A \, dP$ defines a probability $\otimes$-measure on $\mathscr{T}$, from which the ordinary probability measure P may be recovered by theorem 5.35. Many other properties of tribes and measures, representations such as those by Markov kernels, decompositions, and other properties of $\otimes$-measures have been obtained since the pioneering work by Butnariu and Klement; see e.g. Barbieri, Navara, and Weber 2003; Butnariu 1983, 1986, 1987; Butnariu and Klement 1991; Klement 1980a,b,

---

[165]Later on, Navara slightly changed the definition of a tribe and measure and made it more specific without affecting essentially the results obtained. In particular, he replaced (5.56) by requiring that a tribe be closed under $\otimes$ and under limits of nondecreasing sequences, and considered in addition to (5.59) the dual condition regarding limits of nonincreasing sequences; e.g. Navara 2005.

1982a,b; Mesiar 1993a; Mesiar and Navara 1996; as well as Butnariu and Klement 2002; and Navara 2005, 2012 for further information.

Considerations regarding probabilities may be generalized from the setting of σ-algebras with measures to the abstract setting of Boolean algebras with measures.[166] Replacing Boolean algebras by more general algebras which represent algebras of many-valued sets thus offers considerations regarding probabilities of many-valued sets. Many results in this regard have been obtained for probability measures on MV-algebras. The first such results, obtained in the context of quantum logic, were due to Riečan (1992) and Chovanec (1993).[167] Other early papers include Riečan 2000; Mundici 1995; Di Nola, Georgescu, and Lettieri 1999; Dvurečenskij 2000; Pulmannová 2000; for conditional probabilities see Kroupa 2005; for a generalization beyond MV-algebras see Dvurečenskij and Rachůnek 2006. A *probability*, or a *state*, *on a σ-complete MV-algebra A*, i.e. an MV-algebra in which every countable set has its supremum, is a mapping $m : A \to [0,1]$ which satisfies

$$m(1) = 1,$$
$$\text{if } a \otimes b = 0 \text{ then } m(a \oplus b) = m(a) + m(b),$$
$$\text{if } a_n \nearrow a \text{ then } m(a_n) \nearrow m(a).$$

States on MV-algebras are obviously closely connected to measures on tribes. Namely, due to the properties of the Łukasiewicz t-norm $\otimes$, a $\otimes$-tribe $\mathcal{T}$ is a σ-complete MV-algebra and the notion of probability measure on $\mathcal{T}$ coincides with the notion of state on $\mathcal{T}$ as an MV-algebra. Many results regarding probabilities on MV-algebras have been obtained, particularly in the 1990s. These include results regarding observables, which are the counterparts of the notion of random variable, the central limit theorem, joint observables, conditioning, and the notion of entropy; see Riečan and Mundici 2002 for a nice exposition.

Fuzzy random variables

The concept of random variable—one of the most important in probability theory and statistics—has been studied from the viewpoint of fuzzy sets since the late 1970s, and such studies, along with examinations of related probabilistic and statistical problems, represent ongoing research. An ordinary random variable represents an assignment of real numbers to outcomes of a random experiment, such as the assignment of heights to people selected for a survey. More precisely, given a probability space $\langle \Omega, \mathcal{B}, P \rangle$, a random variable is a mapping $X : \Omega \to \mathbb{R}$ which is measurable w.r.t. the σ-algebra $\mathcal{B}$ and the Borel σ-algebra on $\mathbb{R}$. This condition ensures that we may assess probabilities of statements like "the observed height is in A" where $A \subseteq \mathbb{R}$ is a closed interval or other Borel subset of $\mathbb{R}$: such probability equals $P(\{\omega \in \Omega \mid X(\omega) \in A\})$. Instead of real numbers, a fuzzy random variable assigns

---

[166] E.g. P. R. Halmos, *Lectures on Boolean Algebras* (Heidelberg: Springer, 1974), chap. 15.

[167] Chovanec acknowledges Riečan's priority in introducing the notion of state on MV-algebras.

fuzzy sets of real numbers to the experimental outcomes. Thus, instead of "weight of person ω is 70" we deal with "weight of ω is small" or similar, more expressive assignments.

Of various approaches to fuzzy random variables, two are particularly important. The first uses a modification of the notion of fuzzy random variable proposed by Kwakernaak (1978, 1979). It was developed by Kruse (1982, 1984), later joined by his doctoral student Meyer, and resulted in the influential book Kruse and Meyer 1987, which nicely addresses both the foundational as well as practical questions. In this view, the assigned fuzzy set is regarded as an approximate description of a value of an ill-known ordinary random variable. The other approach is due to Puri and Ralescu 1985, 1986 and was developed in a subsequent series of papers. In this view, the assigned fuzzy set is considered as an exact value on its own. We now describe both approaches and some related developments.

Given a probability space $\langle \Omega, \mathcal{B}, P \rangle$, a *fuzzy random variable* according to Kruse and Meyer is a mapping $\mathscr{X} : \Omega \to F(\mathbb{R})$, where $F(\mathbb{R})$ is the set of all normal fuzzy sets $A : \mathbb{R} \to [0,1]$ with convex and closed a-cuts ${}^aA$ for all $0 < a < 1$, such that for each $a \in (0,1)$ the mappings $\mathscr{X}_a : \Omega \to \mathbb{R}$ and $\mathscr{X}^a : \Omega \to \mathbb{R}$ defined by $\mathscr{X}_a(\omega) = \inf {}^a(\mathscr{X}(\omega))$ and $\mathscr{X}^a(\omega) = \sup {}^a(\mathscr{X}(\omega))$ are real-valued random variables.[168] Of essential importance in this approach are the "original" random variables $X : \Omega \to \mathbb{R}$ of which $\mathscr{X}$ may be thought of as an approximate perception. Naturally, the degree $O_{\mathscr{X}}(X)$ to which such X may be regarded as the original for $\mathscr{X}$ is defined by

$$O_{\mathscr{X}}(X) = \inf_{\omega \in \Omega} [\mathscr{X}(\omega)](X(\omega)).$$

Notice that $O_{\mathscr{X}}(X)$ is naturally interpreted as the truth degree of "for each $\omega \in \Omega$, the actual value $X(\omega)$ is in the perceived $\mathscr{X}(\omega)$," and that $O_{\mathscr{X}}$ is a fuzzy set in the universe $rv(\Omega, \mathcal{B}, P)$ of all ordinary random variables $X : \Omega \to \mathbb{R}$. The way Kruse and Meyer reason further about fuzzy random variables is well-illustrated by the notion of *expected value*, $E\mathscr{X}$, of a fuzzy random variable $\mathscr{X}$, a particular example of the general notion of moment: $E\mathscr{X}$ is the fuzzy set in $\mathbb{R}$ defined as

$$(E\mathscr{X})(t) = \sup\{O_{\mathscr{X}}(X) \mid X \in rv(\Omega, \mathcal{B}, P) \text{ and } EX = t\}.$$

Hence, the degree $(E\mathscr{X})(t)$ is naturally interpreted as the truth degree of "there exists an ordinary random variable X that is an original of $\mathscr{X}$ and has expected value t." Calculations regarding fuzzy random variables are based on the level-cut representation of fuzzy sets and Zadeh's extension principle, which in many cases results in practically usable formulas. One thus obtains formulas such as ${}^a[E\mathscr{X}] = [E\mathscr{X}_a, E\mathscr{X}^a]$, allowing one to compute the a-cut of the expected value of $\mathscr{X}$ from the expected values of the above associated random variables $\mathscr{X}_a$ and $\mathscr{X}^a$.[169] Kruse and Meyer's treatment and later contributions contain several results regarding de-

---

[168] In fact, the definition in Kruse and Meyer 1987 is slightly more general.

[169] This simple formula actually holds for a slightly modified situation involving a probability space richer than $\langle \Omega, \mathcal{B}, P \rangle$ (Kruse and Meyer 1987, chap. 7).

scriptive statistics, limit theorems such as the strong law of large numbers, and estimation and testing of hypotheses, as well as computational considerations.

Now, let $F(\mathbb{R}^n)$ denote the set of all fuzzy sets A in $\mathbb{R}^n$ for which all a-cuts ^aA for $0 < a \leq 1$ and the closure of $\{u \in \mathbb{R}^n \mid A(u) > 0\}$ are compact (i.e. closed and bounded), and denote A_a these respective compact sets. In the approach taken by Puri and Ralescu (1985, 1986), a *fuzzy random variable* is conceived of as a generalization of the notion of random set, thus sometimes called a *fuzzy random set*, and is defined for a given probability space $\langle \Omega, \mathcal{B}, P \rangle$ as a mapping $\mathcal{X} : \Omega \to F(\mathbb{R}^n)$ such that the associated mappings $\mathcal{X}_a$, assigning to each $\omega \in \Omega$ the compact set A_a, are random sets (p. 273) in that they are measurable w.r.t. $\mathcal{B}$ and the Borel σ-algebra generated by the topology of the Hausdorff metric d_{H} on $\mathbb{R}^n$ (p. 306).[170] In an important paper, Klement, Puri, and Ralescu (1986) strengthened this condition by requiring Borel measurability w.r.t. to the sup metric $d_\infty(A, B) = \sup_{a \in [0,1]} d_{\mathrm{H}}(^aA, ^aB)$ on $F(\mathbb{R}^n)$. In this approach, further notions are defined as generalizations of the respective notions for random sets. Thus, for instance, the expected value of integrably bounded $\mathcal{X}$ (p. 308) is defined as the unique fuzzy set $E\mathcal{X} \in F(\mathbb{R}^n)$ such that for each $a \in [0, 1]$, the a-level set $^aE\mathcal{X}$ is the expected value of the random set $\mathcal{X}_a$ defined as the Aumann integral of $\mathcal{X}_a$ (p. 308). Numerous problems involving Puri and Ralescu's notion of fuzzy random variable have been explored, including integration and differentiation in probabilistic settings, weak and strong laws of large numbers, limit theorems, martingales and further stochastic processes, decision models, regression models, or hypotheses testing; see Gil, López-Díaz, and Ralescu 2006, as well as the books Couso, Dubois, and Sanchez 2014 and Nguyen and Wu 2006.

Other developments

The research on probability measures of fuzzy events mentioned above is related to investigations of general uncertainty measures (Höhle 1976, 1983; Höhle and Klement 1984; Höhle and Weber 1997); see Klement and Weber 1999 for a survey. For instance, Höhle and Weber (1997) consider uncertainty measures as functions $m : L \to [0, 1]$ on bounded lattices L which are isotone and map the least and greatest elements to 0 and 1, respectively, and obtain various results for such general as well as more particular measures, especially measures on bounded lattices with involution, which subsume tribes and MV-algebras, as well as other relevant structures such as orthomodular lattices. These general notions subsume as important particular cases various regular measures developed within generalized measure theory (Wang and Klir 2009), which are functions of the form $m : C \to [0, 1]$ for certain $C \subseteq 2^X$, including the possibilistic measures which are examined in section 3.7. They also subsume probability measures on fuzzy events which are of the form $m : C \to [0, 1]$ for certain $C \subseteq [0, 1]^X$. Note also that extensions to fuzzy events of

---

[170] The concept of fuzzy random set was considered for the first time by Féron (1976).

measures other than probability have been examined, in particular the Dempster-Shafer measures (Smets 1981, Yager 1982b, Yen 1990, Grabisch 2009).

Of the many existing approaches combining the notion of fuzzy set with probabilistic notions, Hirota's 1976 notion of probabilistic set is expounded in Czogala and Hirota 1986, which is a function $\mu : U \times \Omega \to [0, 1]$, where $\langle \Omega, \mathcal{B}, P \rangle$ is a given probability space such that for each $u \in U$ the function $\mu(u, \cdot)$ is measurable. Probabilistic sets account for situations in which there is uncertainty associated with elicitation of membership degrees.

Finally, many approaches using fuzzy sets have been taken to generalize the setting of probabilistic and statistical considerations, including the books Viertl 1996, Buckley 2003, and Nguyen et al. 2012; the overview papers Coppi, Gil, and Kiers 2006 and Gebhardt, Gil, and Kruse 1998; as well as other relevant chapters in Słowiński 1998. This is due to the multifaceted character of the phenomenon of uncertainty into which fuzzy logic fits rather naturally.

We consider it important to mention at the end that Zadeh has repeatedly challenged classical probability theory and other classical theories of uncertainty by presenting simple problems which these classical theories are not capable of solving. For example, Zadeh 2006 offers three nice examples: *Balls-in-box problem*: A box contains about 20 balls of various sizes. Most are large. What is the number of small balls? What is the probability that a ball drawn at random is neither small nor large? *Temperature problem*: Usually, the temperature is not very low and not very high. What is the average temperature? *Tall Swedes problem*: Most Swedes are tall. How many are short? What is the average height of Swedes? According to Zadeh, in order to be able to solve these problems, probability theory needs to be extended by the capabilities of fuzzy logic. Zadeh's arguments are simple and convincing. It seems likely that the existing contributions, some of which are presented in this section, represent just the first few steps toward a materialization of Zadeh's proposals.

5.3.6 Geometry

Classical geometric concepts, such as that of point, line, triangle, or incidence, are clear-cut. Thus, an object either is or is not a point, a line, a triangle, or a circle; a point either lies on a line or not, two areas are either overlapping or not, and two lines are either parallel or not. Furthermore, these concepts are idealizations, or absolutely sharpened versions, of natural human geometric concepts. Looking at a photograph, people see approximate lines and other objects rather than lines in the sense of classical geometry and naturally consider two such approximate lines more or less parallel rather than either parallel or not. Approaching geometrical concepts and ideas from the point of view of fuzzy logic thus makes good sense. A closer look reveals that there are in fact several possible approaches to pursue the idea of fuzzifying geometrical concepts. The literature contains some such approaches and these are developed to varying extents.

A brief but interesting consideration is found in part III of Zadeh 1975c, in which Zadeh discusses the intuitive notion of fuzzy theorem as an approximately true statement that may be inferred from axioms by approximate reasoning. On p. 68, Zadeh presents the following example of such a theorem from plane geometry concerning approximate triangles:

> Let AB, BC, and CA be approximate straight lines which form an approximate equilateral triangle with vertices A, B, C. ... Let M_1, M_2, and M_3 be approximate midpoints of the sides BC, CA and AB, respectively. Then the approximate straight lines AM_1, BM_2 and CM_3 form an approximate triangle $T_1 T_2 T_3$ which is more or less (more or less small) in relation to ABC.

Neither is this statement an actual theorem nor is Zadeh's proof of it an actual proof because both are based on intuition and developed by a kind of qualitative reasoning. A closer inspection reveals that the notion of a fuzzy theorem Zadeh proposes may naturally be approached via Pavelka-style fuzzy logic (section 4.3.2). However, up to now these ideas have not been given any significant further development.

The PhD thesis *Fuzzy Geometry* by Poston (1972), despite its title, does not contain any reference to the literature on fuzzy sets even though the topic of Poston's investigation is related to fuzzy sets, in particular to the concept of similarity which is explicitly or implicitly central in many investigations in fuzzy logic. Poston's notion of a fuzzy space is in a sense synonymous with Poincaré's notion of physical continuum or Zeeman's notion of tolerance space and stands for a pair $\langle U, \tau \rangle$ where τ is a tolerance, i.e. a reflexive and symmetric binary relation on a set U.[171] Poston refers to τ as "a fuzzy" and interprets $\langle u, v \rangle \in \tau$ as meaning that u and v are indistinguishable. Central to Poston's work are considerations of notions analogous to those in general but also algebraic topology, a finitistic version of differential geometry, and a variety of further topics, making his thesis basically a theory of tolerance relations with topological and geometrical motivations.

In a series of papers, Rosenfeld has put forward an approach to geometrical notions motivated by computer image processing in which area he was a pioneer (see also chapter 6). This direction started in the late 1970s and many papers have been published in the journal *Pattern Recognition Letters*. Rosenfeld considers fuzzy sets in $\mathbb{R}^n$ and interprets them as nonsharply delineated geometric regions. For instance, a gray-scale picture as well as its parts may be interpreted as fuzzy sets in $\mathbb{R}^2$. For practical purposes, Rosenfeld often uses a two-dimensional grid instead of $\mathbb{R}^2$. Central in this approach are various intuitive geometric notions, from topological properties of connectedness, adjacency, or separation, to metric notions such as area, perimeter, diameter, and distance, as well as various further notions such as those regarding convexity. For example, two points, u and v, are *connected*; see e.g. Rosenfeld 1979, in a fuzzy set A if there exists a (reasonably defined) path π in $\mathbb{R}^n$ from u to v such that

$$A(u) \wedge A(v) \leq \bigwedge\nolimits_{w \in \pi} A(w).$$

---

[171] See pp. 334ff. and n. 215 on p. 334. Zeeman was one of Poston's supervisors.

Thus, with conjunction interpreted by min, u and v are connected if and only if there exists a path π such that the formula "if u is in A and v is in A then each point in π is in A" is true to degree 1, thus generalizing classical connectedness. One may also define related notions and observe natural relationships between them. For instance, the degree $s_A(\pi) = \bigwedge_{w \in \pi} A(w)$ in the previous formula is called the *strength* of π w.r.t. A and may be interpreted as the truth degree of "each point in π belongs to A." The *degree of connectedness* of u and v w.r.t. A is then defined as $c_A(u,v) = \bigvee_{\pi} s_A(\pi)$, where the supremum is taken over all paths π from u to v. Thus, $c_A(u,v)$ may be interpreted as the truth degree of "there exists a path whose points belong all to A." One then easily obtains the following claim (Rosenfeld 1979) and similar intuitive relationships:

Theorem 5.36. $c_A(u,u) = A(u)$; $c_A(u,v) = c_A(v,u)$; u *and* v *are connected in* A *iff* $c_A(u,v) = A(u) \wedge A(v)$.

Metric notions here are defined using the available classical notions; thus, for instance, the area of $A \in [0,1]^{\mathbb{R}^2}$ is defined as an integral of A in $\mathbb{R}^2$. Rosenfeld also fuzzified the classical notions of triangle, rectangle, and other shapes. For instance, a fuzzy set $A \in \mathbb{R}^2$ is called a *fuzzy rectangle* if A is separable and connected. Being *separable* means that there exist fuzzy sets $A_x, A_y \in \mathbb{R}$ such that $A(u,v) = A_x(u) \wedge A_y(v)$ upon a suitable coordinate system. With *convexity* of A meaning that for every point w on the line between u and v we have $A(w) \geq A(u) \wedge A(v)$, one obtains:

Theorem 5.37. A *is a fuzzy rectangle iff it is separable and convex iff* $A(u,v) = A_x(u) \wedge A_y(v)$ *for convex fuzzy sets* $A_x, A_y \in [0,1]^{\mathbb{R}}$.

The level cuts ${}^a A$ of fuzzy rectangles are then just classical rectangles. Many further properties, in a sense quite straightforward generalizations of classical notions, can be found in the literature. For further information, see the overview Rosenfeld 1984, 1998.

A different approach has been initiated in Buckley and Eslami 1997 and in Eslami, Nemat, and Buckley 2001. It represents a relatively straightforward way of generalizing analytic geometry by exploiting level-cuts to solve equations with fuzzy quantities as proposed in Buckley and Eslami 1991. The authors define a *fuzzy point* at $\langle u,v \rangle \in \mathbb{R}^2$ basically as a fuzzy quantity P in $\mathbb{R}^2$, denoted also $P(u,v)$, whose core contains only $\langle u,v \rangle$, i.e. ${}^1 P = \{\langle u,v \rangle\}$. For fuzzy points $P_1(u_1,v_1)$ and $P_2(u_2,v_2)$, they also define *fuzzy distance* as a fuzzy set $D(P_1, P_2)$ in $\mathbb{R}$ defined by its a-cuts as

$$ {}^a D(P_1, P_2) = \{d(u,v) \mid u \in {}^a P_1(u_1,v_1) \text{ and } v \in {}^a P_2(u_2,v_2)\} $$

and prove that $D(P_1, P_2)$ is a fuzzy number. D is shown to be a *fuzzy metric* which is defined as a mapping m assigning fuzzy numbers to pairs of fuzzy points and satisfying certain properties generalizing those of classical metrics. The authors then consider various definitions of the concept of line. They first exclude as too restrictive and thus unsatisfactory the possibility of defining a line as a set of pairs $\langle X, Y \rangle$

of fuzzy numbers satisfying $AX + BY + C = 0$ in which A, B, and C are given fuzzy numbers with operations defined by the extension principle, as well as a possibly less restrictive equation $Y = AX + B$. Instead, they introduce four notions of a line as fuzzy sets L_{12}, L_{21}, L_3, and L_4 in $\mathbb{R}^2$ whose α-cuts are defined as follows for given fuzzy numbers A, B, and C, and fuzzy points P and Q:

$$^\alpha L_{11} = \{\langle x,y \rangle \in \mathbb{R}^2 \mid ax + by = c, a \in {}^\alpha A, b \in {}^\alpha B, c \in {}^\alpha C\},$$
$$^\alpha L_{12} = \{\langle x,y \rangle \in \mathbb{R}^2 \mid y = ax + b, a \in {}^\alpha A, b \in {}^\alpha B\},$$
$$^\alpha L_2 = \{\langle x,y \rangle \in \mathbb{R}^2 \mid y - v = a(x - u), \langle u,v \rangle \in {}^\alpha P, a \in {}^\alpha A\},$$
$$^\alpha L_3 = \{\langle x,y \rangle \in \mathbb{R}^2 \mid \tfrac{y - v_1}{x - u_1} = \tfrac{v_2 - v_1}{u_2 - u_1}, \langle u_1, v_1 \rangle \in {}^\alpha P, \langle u_2, v_2 \rangle \in {}^\alpha Q\}.$$

Call a fuzzy set in $\mathbb{R}^2$ a *fuzzy line*$_*$ if it is of the form L_* as defined above. The authors examine conditions under which, for instance, fuzzy lines$_{11}$ are fuzzy lines$_{12}$ as well as further relationships, and examine the notion of parallel lines and intersection of lines. In part II, Buckley and Eslami (1997) examine the notions of circle and polygon in a similar spirit. They also consider the notions of circumference, area, and perimeter, which they define as fuzzy numbers, again in the spirit sketched above. Even though some theorems are proved, the work by Buckley and Eslami (1997) is mainly a definitional paper. Their results were later extended in Eslami, Nemat, and Buckley 2001 to $\mathbb{R}^3$. Subsequent research includes Ghosh and Chakraborty 2014, which contains further references.

The above two approaches presuppose the classical space $\mathbb{R}^n$ and define fuzzified versions of classical geometric notions by employing fuzzy quantities $\mathbb{R}^n$ and the classical notions, e.g. that of line, as in the definition of a fuzzy line in Buckley and Eslami 1997. A more fundamental viewpoint would, however, require beginning with the notion of incidence, and considering from a fuzzy logic point of view axiomatic approaches to geometry. So far, the work by Gupta and Ray (1993) seems the only exception which adopts this view. The authors attempt to build an axiomatic fuzzy projective geometry. They start with a nonempty set U and define a *fuzzy point* as a pair $\langle u, a \rangle$ where $u \in U$ and $a \in (0,1]$. A *complete set of fuzzy points* is then a set Π of fuzzy points such that for each $u \in U$ there exists a fuzzy point $\langle u, a \rangle \in \Pi$ and a *fuzzy line through* Π is a fuzzy set $l \in [0,1]^U$ such that $l(u) > 0$ implies $\langle u, l(u) \rangle \in \Pi$ for each $u \in U$. A fuzzy point $\langle u, a \rangle$ is *incident* with l if $l(u) = a$. The authors then define a *fuzzy plane projective geometry* as a triplet $\langle \Pi, \Lambda, I \rangle$, for which Π and I are as above, Λ is a set of fuzzy lines through Π, and the following axioms are satisfied:

(a) Given two points $\langle u, a \rangle, \langle v, b \rangle \in \Pi$ with $u \neq v$, there is a unique fuzzy line in Λ with which both points are incident.

(b) Given two distinct fuzzy lines in Λ, there is at least one fuzzy point in Π incident with both lines.

(c) Π contains at least four fuzzy points $\langle u_i, a_i \rangle$ with $u_i \neq u_j$ for $i \neq j$ such that no three of them are incident with the same line in Λ.

These axioms are inspired by the axioms of classical projective geometry.[172] The authors then define three models of their geometry, the so-called straight line model, the model M, and the spherical geodesic model. The first model is defined over $U = \mathbb{R} \cup \{ri \mid 0 \neq r \in \mathbb{R}\} \cup \{\infty\}$ with i denoting the imaginary unit, has the set Π of points of type $\langle u, a \rangle \in \mathbb{R} \times (0,1)$, or $\langle ri, \frac{1}{\pi}\cot^{-1}(r) \rangle$ with $0 \neq r \in \mathbb{R}$, or $\langle \infty, \frac{1}{2} \rangle$, and contains lines of three types, one of them being defined for each real d as a fuzzy line $[d]$ for which $[d](u) = \frac{1}{\pi}\cot^{-1}(d)$ if $u \in \mathbb{R}$, $[d](u) = \frac{1}{2}$ if $u = \infty$, and $[d](u) = 0$ otherwise. The authors derive various theorems from the axioms, examine which of them are true in the respective models, and formulate open problems. These have later been addressed in Kuijken and Van Maldeghem 2003, which is one of the few papers following this work.

Interesting considerations appear in Gerla's work on point-free geometries. A point-free geometry represents an approach to geometry in which the primitive concept is that of a region, rather than a point.[173] Points are viewed as classes of regions, hence are available but as a derived notion. The idea of point-free geometry goes back to Alfred North Whitehead's considerations of how volumes and relationships between them, such as inclusion, may be used to define the notion of point.[174] There exist two basic attempts to describe axiomatically a point-free geometry. One of them is based upon Whitehead's original conception regarding as basic the relation of inclusion. Another, based on a later proposal of Whitehead, regards connectedness as basic.[175] Coppola, Gerla, and Miranda (2010) consider formalizations of these approaches, of which we present a brief sketch. An *inclusion space* is a pair $\langle U, \leq \rangle$ where U is a set and $\leq$ a partial order on U satisfying certain further properties. A *contact space* is a pair $\langle U, C \rangle$ such that C is a binary reflexive and symmetric relation on U satisfying some further properties. The elements of U are interpreted as regions, $u \leq v$ as "u is included in v," and uCv as "u is in contact with v." The authors provide several properties in such formalizations and define four canonical inclusion spaces $\langle U, \subseteq \rangle$ in which U consists of certain subsets $\mathscr{R}$ of $\mathbb{R}^n$ and $\subseteq$ is set-theoretical inclusion. Upon putting uCv iff $u \cap v \neq \emptyset$, one obtains the corresponding canonical contact spaces. Importantly, they prove that such $\langle U, C \rangle$ are indeed contact spaces, that the inclusion relation is first-order definable from C in the canonical contact spaces, and that the converse definability is not possible. This is an important result justifying Whitehead's shift from the inclusion to the contact approach. Interestingly, they also consider fuzzy versions of these notions. In particular, they consider a continuous Archimedean t-norm $\otimes$ and the

[172] A classic text is A. Heyting, *Axiomatic Projective Geometry* (Amsterdam: North-Holland, 1963). The axioms above are derived from those on pp. 24–25 of Heyting's book.

[173] Alternatively, a pointless geometry or geometry without points. For an overview, see G. Gerla, "Pointless geometries," in *Handbook of Incidence Geometry*, ed. F. Buekenhout (Amsterdam: Elsevier, 1995), 1015–31.

[174] A. N. Whitehead, *An Enquiry Concerning the Principles of Natural Knowledge* (Cambridge, UK: Cambridge Univ. Press., 1919); *The Concept of Nature: Tarner Lectures Delivered in Trinity College, November 1919* (Cambridge, UK: Cambridge Univ. Press., 1920).

[175] A. N. Whitehead, *Process and Reality* (New York: Macmillan, 1929).

notion of *graded inclusion space* which involves—as the graded counterpart to the ordinary inclusions relation—a reflexive and $\otimes$-transitive fuzzy relation $\leq$ satisfying certain further properties. They show relationships between such spaces and hemimetric spaces.[176] Natural graded inclusion spaces on $\mathbb{R}$, called canonical spaces, are obtained by the authors by putting

$$u \leq v = f^{(-1)}(e(u,v)),$$

where $e(u,v) = \sup_{p \in u} \inf_{q \in v} d(p,q)$ is the so-called excess measure induced by a metric d and $f^{(-1)}$ is the pseudoinverse of the additive generator of $\otimes$ (p. 275). Interestingly, they prove that unlike the classical case, contact relations are definable in the canonical graded inclusion spaces.

Fractal geometry studies figures that exhibit repeating patterns.[177] Such figures, which are called fractals, are abundant in nature. Fractal geometry thus found many applications. Since images with gray or color levels may be interpreted as fuzzy sets, their structure from the point of view of fractal geometry asks for examination of fuzzy fractals. Such examinations were indeed carried out in some papers which provide another research direction; see e.g. Cabrelli et al. 1992, Pham 2008.

5.3.7 Further developments

In addition to the areas covered in the preceding sections, numerous contributions have been made to other areas of mathematics based on fuzzy logic. These include the various mathematical models of decision making, pioneered by Bellman and Zadeh (1970), which subsume preference modeling and game theory (Branzei, Dimitrov, and Tijs 2005; Butnariu and Klement 1993; Mareš 2001); see also pp. 279 and p. 310; these are also examined in chapter 6. Another area with many contributions is optimization and in particular linear programming, in which many models employing fuzzy quantities were studied; see e.g. Fiedler et al. 2006. Fuzzy quantities are also utilized in the various methods of interpolation and approximation (Diamond and Kloeden 2000), which include the important methods based on fuzzy if-then rules presented in chapter 6, as well as the methods of regression initiated in Tanaka, Uejima, and Asai 1970. Relevant to approximation is fuzzy transform (Perfilieva 2006, Perfilieva and Kreinovich 2011), which is concerned with approximate representation of functions in a similar sense as the classic Fourier or Laplace transforms. Dynamical systems and, in particular, chaos theory have also been approached from the viewpoint of fuzzy logic, see Li, Halang, and Chen 2006 for a collection of survey papers. A number of contributions were made to the theory and applications of matrices with truth degrees in $[0,1]$, which are sometimes called fuzzy matrices, or in more general structures of truth degrees. These matrices gen-

---

[176] Generalized metric spaces possibly violating symmetry and the condition "$d(u,v) = 0$ implies $u = v$."
[177] B. B. Mandelbrot, *The Fractal Geometry of Nature* (New York: Freeman, 1977).

eralize Boolean matrices and are particular cases of matrices over semirings.[178] They may clearly be thought of as representing binary fuzzy relations between finite universes to which they are inherently connected. For instance, compositions of such matrices correspond to relational compositions. The problems studied include the decomposition problems discussed in section 3.5 as well as various problems inspired by the corresponding ones in classical linear algebra. Note also that another area inherently connected to binary fuzzy relations are fuzzy graphs, which are basically graphs with truth degrees, mostly in $[0, 1]$, attached to their edges. Fuzzy graphs correspond to binary fuzzy relations and various problems of a graph theoretic nature have been considered for them, as well as for the more general concept of fuzzy hypergraphs; see e.g. Mordeson and Nair 2000. While many results for matrices over $[0, 1]$ have been published in well-established journals on matrix theory, such as *Linear Algebra and its Applications*, publications regarding fuzzy graphs, while addressing a conceptually interesting topic, contain mostly obvious generalizations of results for classical graphs and have not, by and large, appeared in recognized graph journals.

5.4   Miscellaneous issues

5.4.1   Interpretation of truth degrees

In this section we briefly examine the important and complex question of interpretation and meaning of truth values in many-valued and fuzzy logic, as well as the related question of interpretation of fuzzy sets.

Truth values, significance of their interpretation, and truth degrees

Truth is one of the most basic and delicate eternal themes in philosophy, one which penetrates many fields including linguistics and logic. Philosophers have developed various views of truth, several of which are based on the idea that truth means, in one way or another, a correspondence to reality.[179] This idea is also employed in Tarski's conception of truth in formal languages (Tarski 1936), which is the basis of modern semantic studies in logic. The notion of truth is inherently present in the concept of *truth value*. The term "truth value" is due to Gottob Frege (1848–1925).[180] Frege used the German word "Wahrheitswert," and used "das Wahre" and "das Falsche" for "true" and "false," respectively. He understood truth values as abstract objects

---

[178] K. H. Kim, *Boolean Matrix Theory and Applications* (New York: Dekker, 1982). J. S. Golan, *Semirings and Their Applications* (Dordrecht: Kluwer, 1999).

[179] P. Edwards, ed., *The Encyclopedia of Philosophy* (New York: Macmillan, 1967); R. L. Kirkham, *Theories of Truth: A Critical Introduction* (Cambridge, MA: MIT Press, 1995).

[180] Kneale and Kneale 1962 (p. 413) and Church 1956 (p. 25). Frege was a German mathematician, logician, and philosopher, and is widely regarded as one of the founders of modern logic.

that constitute the *Bedeutung* of sentences and held that there are only two truth values.[181] In Frege 1892, he writes:[182]

> We are therefore driven into accepting the *truth-value* of a sentence as constituting its *Bedeutung*. By the truth-value of a sentence I understand the circumstance that it is true or false. There are no further truth-values. For brevity, I call the one the True, the other the False. Every assertoric sentence concerned with the *Bedeutung* of its words is therefore to be regarded as a proper name, and its *Bedeutung*, if it has one, is either the True or the False. These two objects are recognized, if only implicitly, by everybody who judges something to be true—and so even by a sceptic.

The explicit treatment of truth values as abstract objects appears, however, as early as Peirce 1885 (p. 183), where they are denoted **v** (truth) and **f** (falsity) and called "constant values," and to a certain extent even earlier by Boole (1854, chapter III), who uses the symbols 1 and 0 and interprets them as classes he calls "Universe" and "Nothing."

Notwithstanding the philosophical subtleties and challenges surrounding the notion of truth, the two truth values recognized in classical logic, 1 and 0, are usually considered as having an intuitively clear meaning: they represent truth and falsity, respectively.[183] With the introduction of additional truth values the problem arises of how to actually interpret them and whether and how to reconsider the interpretation of 1 and 0. This problem has two important facets. First, without at least an intuitively clear interpretation of truth values, a given many-valued logic may rightfully be considered a formalistic exercise which can hardly aspire to have the status of a solid logic.[184] Second, a clear interpretation of truth values naturally helps reveal notions and problems which appear in the given logic and which may even be conceptually new compared to classical logic. In this sense, the interpretation helps direct the agenda of the given logic. From this viewpoint, absence of a clear interpretation leaves one with the idea of generalizing the results of classical logic as the only inspiration. Even though such inspiration delivers plenty of interesting research problems, we argue that the interpretation of *truth values as degrees of truth* and the related *graded approach to truth*, which became visible and influential particularly due to Zadeh's writings, represents a significant mark in the development of many-valued logic. In fact, this interpretation has directed the agenda of fuzzy logic in the narrow sense to a significant extent. Fuzzy logic in the narrow sense may correspondingly be characterized as a many-valued logic in which truth values are interpreted as degrees of truth and whose agenda is driven by the aim to model reasoning in natural language and, as such, includes a study of language constructs like linguistic hedges or fuzzy quantifiers not taken into account in other logics.[185] To

---

[181]The German term *Bedeutung* has been variously translated as "reference," "denotation," "meaning," and "significance," among others. For more information, see pp. 36–46 in Beaney 1997.

[182]Engl. transl. on pp. 157–58 in Beaney 1997. Such a view is expressed even earlier in Frege 1891.

[183]From a fuzzy logic point of view, one might critically add that they are considered so because one actually assumes a theory of bivalent truth to which they refer and which is actually presupposed by classical logic.

[184]Such logic may still be technically useful, e.g. in proving independence of axioms as shown in chapter 4.

[185]In our view, such an informal characterization is more appropriate than formal characterizations of fuzzy

substantiate the above claims from a historical perspective is the purpose of the next two sections.

Interpretation of truth values in many-valued logic until the mid-1960s

In chapter 4, we show that some of the pioneers—particularly Łukasiewicz, Kleene, Bochvar, and the inventors of probability logics—had a clear idea of the meaning of truth values. For some—particularly Bernays, to a large extent Post, and Gödel— the meaning of truth values was not relevant and thus was not considered because their truth values had only a technical role or, as with Post (n. 74 on p. 137), the question of meaning was relevant but not developed very far. However, even though Łukasiewicz invented his logic to solve the important philosophical problem of future contingencies and put significant emphasis on it in his first and in some of his later writings such as Łukasiewicz 1953, investigations of Łukasiewicz logics by other researchers were as a rule conducted in a completely detached way from the original motivations or other extramathematical motivations whatsoever. Contributions in which the various many-valued logics were studied as abstract logical systems with no intended interpretation of truth values and in which the possibly existing original motivation had no significance and was not even mentioned became typical in works on many-valued logic.

For Łukasiewicz logic this might also have been due to the following problem with Łukasiewicz's interpretation of $1/2$ as "possible."[186] Let the truth values of φ and ψ both be equal to $1/2$, i.e. both φ and ψ are possible in that they are possibly but not certainly true. If $\varphi = \psi$, then the truth value of $\varphi \rightarrow \psi$ is 1 because $\varphi \rightarrow \varphi$ is certainly true in Łukasiewicz's view. But if φ and ψ are independent, such as "it will rain tomorrow" and "my watch will break tomorrow," then the truth value of $\varphi \rightarrow \psi$ is not known, i.e. $\varphi \rightarrow \psi$ is "possible" and should hence be assigned $1/2$. Therefore, the truth-value of $\varphi \rightarrow \psi$ cannot be determined solely from those of φ and ψ, hence $\rightarrow$ cannot be truth functional, contrary to Łukasiewicz's proposal.

In the introduction to their influential monograph, Rosser and Turquette (1952, 1–2) eloquently describe the situation as follows:

---

logic as a logic of chains (Běhounek and Cintula 2006a) or a logic obeying standard completeness (Metcalfe and Montagna 2007), which nevertheless certainly have good merits in particular when relating fuzzy logics to other logics.

[186] The problem is similar to that of truth functionality in probability and other uncertainty logics (p. 145 and section 4.5), and is described e.g. in Urquhart 1986. An argument against the probabilistic interpretation of truth values in the Łukasiewicz logic appared in Scott 1974, 1976. Even earlier, Moh (1954) pointed out that $1/2 \leftrightarrow \neg 1/2$ should not equal 1 for "we regard a proposition, whether concerning future time or not, as never equivalent to its negation," which hints to the same problem. An early critique of Łukasiewicz pointing out his confusion of truth values and modalities is due to C. A. Baylis, "Are some propositions neither true nor false?" *Philosophy of Science* 3, no. 2 (1936): 156–66.

Worth noting is that as late as 1986, Urquhart maintains in the concluding section that "Łukasiewicz's many-valued systems have remained logical toys or curiosities," a view he connects to the above problem: "This is hardly surprising if we are willing to grant that there seems to be a fundamental error (generalized truth functionality) at the root of Łukasiewicz's system."

> [I]t is our opinion that most interpretations that have so far been proposed cannot be taken too seriously until the precise formal development has been carried to a level of perfection considerably beyond that which is reached even in the present work. Of course we do not wish to deny the possibility of finding interpretations for subsystems of many-valued logic such as the statement calculus or the predicate calculus of first order, but we do consider various recent proposals for interpretations of many-valued logic definitely premature. . . .
>
> As will be seen from the following chapters, the amount of complexity involved in taking our few steps toward the ultimate goal of formalizing many-valued logic is quite considerable, and this might lead the skeptic to question the wisdom of taking such steps, much less any further steps, without considerable assurance of ultimate success in terms of meaning and interpretation. Actually we are willing to take the inevitable risk associated with novel investigations, but in due regard for the skeptic we shall indicate our belief that the gamble has some chance of success. Admittedly, we can not now give conclusive proof of final success, but some favorable evidence can be presented.

Interestingly, Rosser and Turquette mention the theory of numbers as a particular theory to be worked out in many-valued logic, thus vaguely indicating the idea of working out mathematical theories within many-valued logics. The problem of interpretation has in fact been mentioned in several works. Salomaa (1959, 121) put it this way:

> To sum up, we see that the problem of finding an interpretation for the truth-values is still far from a satisfactory solution . . . in the formal development of many-valued logic the semantical meaning of truth-values is quite unessential . . . it is an advantage from the formal point of view that we have no prejudices regarding the possible interpretations.

Rescher's (1969, 102–3, 106, 235) view is different and firmly pragmatic:

> We characterize as "abstract" that approach . . . with assignments of values of some *unspecified* kind, that are not necessarily *truth*-values at all. On such an approach, we overlook wholly the relevance of the "values" at issue to semantical considerations regarding truth and falsity: The assignment of values to propositions is simply viewed as . . . a (possibly very useful) abstract sort of symbolic game. . . .
>
> . . . Without an adequate semantical underpinning such a combinatorial instrumentality . . . does not—it is clear—really qualify as itself constituting a logic.
>
> [A] choice between systems [of many-valued logic] cannot be made . . . on the basis of abstract *theoretical* considerations, but must be made on the basis of *practical* (pragmatic or instrumentalistic) considerations regarding the specific, lower-level purposes at issue.

Scott (1974, 418)—with no awareness of existing works on fuzzy logic—wrote rather critically:[187]

> [F]ew—even the creators of the subject—can understand many-valued truth tables. They were constructed by abstract *analogy* and have never been given, in my opinion, adequate motivation.

In addition to another paper by Scott (1976), which is entitled "Does many-valued logic have any use?" and which we examine shortly, there is also Quine's view presented in his influential book:[188]

> Primarily the motivation of these studies has been abstractly mathematical: the pursuit of analogy and generalization. Studied in this spirit, many-valued logic is logic only analogically speaking; it is uninterpreted theory, abstract algebra.

---

[187] Scott's critique is not quite fair because both Kleene and Bochvar, for example, whom Scott does not mention, gave convincing interpretations for their logics.

[188] W. V. Quine, *Philosophy of Logic* (Cambridge, MA: Harvard Univ. Press, 1970), 84.

These examples help further illustrate that in spite of a few exceptions, most works on many-valued logic have not paid attention to the interpretation of truth values and have aimed instead at generalization of results from classical logic, and that the lack of meaningful interpretations has generally been considered a problem. Consequently, many-valued logic has been portrayed even in influential writings as an enterprise with a rather questionable rationale and applicability.

Emergence of the interpretation of truth values as truth degrees and its significance

The described situation may seem strange to someone familiar with the basics of fuzzy logic because interpreting truth values in $[0,1]$ as degrees of truth of propositions involving predicates such as "small" or "similar" is regarded as a natural and straightforward idea. As we now document, this idea actually appeared a long time before Zadeh's seminal paper on fuzzy sets but it was not until after the appearance of his paper in 1965 that this idea became influential.

Łukasiewicz (1930) spoke of truth values in $[0,1]$ as representing "degrees of possibility" which alludes to the idea that truth values may be ordered by a naturally interpretable order.[189] In most other early works on many-valued logic, however, ordering of truth values is rarely mentioned and no meaning other than a technical one is attached to it.[190] Symptomatic in this regard is Łukasiewicz's (1920) original notation in which falsity and truth are denoted by 0 and 1, whereas the middle truth value is denoted by 2. It was also the habit in many later works to denote truth by 1 and denote the other values by $2, 3, \ldots$, i.e. by larger numbers. Explorations of designated truth values, which as a rule were intervals containing 1 (true) and possibly some other values toward 0 (false), were close to the idea of truth degrees but, in the end, designated truth values represented truth while the others represented falsity. Particularly interesting in this regard is the paper by Rose (1951b), which was inspired by the idea of Birkhoff (1948) to develop logics whose truth values form lattices and may thus be only partially ordered.[191] Rose explored logical properties of such logics but did not consider any particular meaning of the partially ordered truth values. This is different in Rose 1952, where he further develops the lattice-valued logics but also considers a logic with propositions φ regarding geometry to which truth values $\langle a_1, a_2, a_3 \rangle \in \{0,1\}^3$ are assigned in the direct product of three two-element Boolean algebras in such a way that $a_2 = 1$ iff φ is true in Euclidean geometry, and the same for a_1 and a_3 for the two non-Euclidean geometries, the elliptic and hyperbolic ones. Then, e.g. with $D = \{\langle a_1, 1, a_3 \rangle \mid a_1, a_3 \in \{0,1\}\}$ as the set of designated truth values,

---

[189] See the quotation on p. 119 above.

[190] Partial order appears explicitly and plays an important role in Chang's (1958a) algebraic study of Łukasiewicz logic; in his position paper, Rosser (1960) says that "interpretations of such a situation [i.e., Łukasiewicz logic] are necessarily unnatural" but points out that in this logic, implication "can be thought of as embodying the notion of $\leq$" because $P \rightarrow Q$ "can have the ultimate of truth if and only if the value assigned to P is not greater than that assigned to Q," suggesting but not mentioning explicitly gradation of truth.

[191] Birkhoff was inspired by J. M. Keynes, who suggested that modes of probability are only partially ordered.

the logic corresponds to Euclidean geometry, but with $D = \{\langle 1, 1, 1 \rangle, \langle 0, 1, 1 \rangle\}$, it corresponds to the absolute geometry of Bolyai.[192] The idea of nonlinearly ordered sets of truth values and the possibility of taking a direct product of sets of truth values, in which the factors represent truth evaluations corresponding to several views, appeared in a paper by the Czech logician Zich (1938). There, Zich speaks of ordering and scales of truth values, of intermediary truth values, and puts forward almost explicitly the view that truth values in a linearly ordered scale should be interpreted as degrees of truth:

> In such a system it makes no sense to say if this or that form is true or not, but only if it has a "larger [truth] value" than another one which is being compared with it.[193]

Notice how close Zich's view is to the later characterization by Hájek (1998b, 2):

> We shall understand FLn [fuzzy logic in the narrow sense] as a logic with *comparative notion of truth*: sentences may be compared according to their truth values.

Even more clearly, truth degrees are advocated by Waismann (1945–46, 90):

> [W]e might as well introduce a logic with a larger number of truth-gradations, for instance a four-valued logic ("true", "nearly true", "not quite false", "entirely false") . . . Let no one say that these logics are a mere play with symbols. For there are sub-domains in our language in which a logic with graduated truth-values is quite natural. If the question is, *e.g.*, whether a certain comparison is to the point, in many cases instead of saying "Yes" or "No" it is more natural to reply, "Yes, more or less to the point" . . . But in so doing one *accomplishes the transition to a logic with a graduated scale of truth-values.* The same holds good of most statements describing properties capable of gradations.

In spite of the above suggestive interpretations of truth values as truth degrees, as well as several other works examined in chapter 2 employing the idea of gradation of truth or membership which existed until the mid-1960s, the largely skeptical views of many-valued logic described in the previous section were prevailing. In particular, the propositional and predicate Łukasiewicz, Gödel, and other logics already examined by then as well as the corresponding notions of many-valued sets with their basic calculus—thus fuzzy logics and fuzzy sets from today's perspective—were regarded as toys.

The situation began to change only after the emergence of fuzzy sets due to Zadeh. This is documented above in chapter 4 as well as in chapter 3 and in the preceding sections of the present chapter. For one, many examinations were directly inspired by the idea of truth degrees as truth values of propositions involving fuzzy predicates. This includes examinations of fuzzy logic in the narrow sense—starting with Goguen's contributions and continuing with those of many others including Hájek—as well as developments in mathematics based on fuzzy logic. Secondly, the

---

[192] R. L. Faber, *Foundations of Euclidean and Non-Euclidean Geometry* (New York: Dekker, 1983).

[193] "V takovém systému pak nemá smysl říkat, je-li ten a ten útvar pravdivý nebo ne, nýbrž pouze, má-li „větší hodnotu" než jiný, jenž je s ním právě srovnáván" (Zich 1938, 190).

sheer awareness of the usefulness of fuzzy sets served as an impetus for further developments in Łukasiewicz and other logics directly or potentially related to the idea of truth gradation.[194]

Interpretation of truth values, truth degrees, and fuzzy sets: A clarification

Even though the idea of interpreting truth values as truth degrees seems intuitively clear, one may naturally ask for more particular semantics of such degrees and of fuzzy sets, and for operational rules regarding truth degrees, hence for a more comprehensive account of the ramifications of interpreting truth values as truth degrees.

Many early writings on fuzzy sets confusingly interpreted truth-functionally processed truth degrees as probabilities.[195] Indeed, in the early literature on fuzzy sets one finds several meanings attached to membership degrees $A(u)$, such as degrees of similarity of u to some prototype, degrees of preference of u, or degrees of certain kinds of belief involving u.[196] Corresponding to these meanings, various ways to define fuzzy sets have been proposed. For instance, distance has been used to define similarity to a prototype. Dubois and Prade (1997) call the above-mentioned meanings of truth degrees "the three semantics of fuzzy sets." The multitude of meanings led many to criticize fuzzy sets on grounds of the alleged arbitrariness of the meaning of membership degrees.

In our view, such criticism is misconceived.[197] Namely, the same kind of criticism could be applied to ordinary sets because membership of u in set A may also have several meanings—having distance from a given point less than δ, being less than a given number, having probability greater than 0.5 of being chosen, or any number of other potentially infinite number of meanings. One needs to realize that the words "meaning" and "interpretation" in connection with truth values and truth degrees are used in two distinct ways.

The first, as just discussed, is when speaking of meaning or interpretation of membership degrees in fuzzy sets or fuzzy relations. In this case, the multitude of possible meanings is natural and corresponds to the very purpose of mathematics— the notion of fuzzy set, just as that of ordinary set, is an *abstract*, or general, notion and as such, the meaning of degree of membership in a fuzzy set, just as that of membership in a set, varies and depends on the particular situation in which the concept of fuzzy set is applied.

Second, we speak, as in the previous paragraph and the preceding section, of interpreting truth values as truth degrees. In this sense, we view the partially or totally

---

[194] For example, according to Georgescu, Iorgulescu, and Rudeanu (2006, 83), Moisil was thinking of his infinitely-valued algebras for a long time but it was only after the appearance of fuzzy sets that he had a solid motivation to publish his ideas in 1968 (p. 128 above).

[195] Non-truth-functional fuzzy logics in which truth degrees are interpreted as probabilities or in general as degrees of belief are examined in detail in secion 4.5.

[196] Note e.g. the special issue entitled "Interpretation of grades of membership," *Fuzzy Sets Syst* 23, no. 3 (1988).

[197] Unfortunately, several writings defending fuzzy sets are somewhat superficial or even conceptually flawed.

ordered set L bounded by 0 and 1 of elements, which are called truth values, as bearing some relationship to the notion of truth. In doing so, we assume a view of truth, or a theory of truth, according to which truth is not bivalent but rather graded in that it makes sense to speak of *grades of truth* or *truth degrees* with the understanding that there is a largest as well as a smallest truth degree and that between them are several intermediate degrees. The truth values a in L are viewed as representing these truth degrees and are thus called truth degrees themselves. Notice that this view is very much like the classical one in which one also assumes a theory of truth, namely a bivalent one. Such a theory recognizes only truth and falsity and these are denoted by the truth values 1 and 0, respectively, in classical logic. One usually does not speak of the ordering of classical truth values, hence one does not call them truth degrees, but it is implicitly understood that "truth is more than falsity," so to speak. A question then, such as "where do the truth degrees come from?" or "what does it actually mean that a proposition p has truth degree 0.8?," *has the same philosophical bearing* for fuzzy logic as the corresponding questions "where do truth and falsity come from?" and "what does it mean that a proposition p is true?" have for classical logic. One may answer them with a kind of theory of truth or by reference to the intuitive meaning of the notions involved. Admittedly, with intermediary truth degrees these notions seem more complex and a satisfactory theory of truth—which in our view has not been worked out yet—more challenging.[198] Until such a theory is proposed such judgments are, nevertheless, mere speculations. On the other hand, a reference to intuition should not be dismissed as superficial justification too quickly, given the many conflicting views regarding a theory of bivalent truth and thus in fact the lack of an agreed-upon-by-all theory of truth for the classical case.

Be that as it may, the fact remains that many questions regarding interpretation and meaning of truth degrees in fuzzy logic, which have often been raised, have their legitimate counterparts in classical logic. Answering them seems to be—on principle—of the same conceptual difficulty for fuzzy logic as for classical logic. Pointing out the issues involved as problematic and controversial for fuzzy logic and not admitting that they are equally so for classical logic is thus like criticizing the right hand for having no more and no less than five fingers while not looking at the left hand.

The important question of providing semantics and justifications not only for truth values but also for truth functions of logical connectives is of a different nature. While Goguen (1967, 1968–69) argued that a partially ordered set equipped with operations corresponding to logical connectives should form a complete residuated lattice in order to be regarded as a reasonable structure of truth degrees for fuzzy logic (section 4.3.1), various kinds of semantics have later been proposed to justify some particular structures of truth degrees. Several of these are examined in the next section.

---

[198] Certainly relevant are theories of vagueness; see section 5.4.3.

Semantics of truth degrees and logical connectives

We start with Scott (1976) who, in his critical paper mentioned on p. 324 above, proposed interpreting the truth values in an n-valued Łukasiewicz logic as representing *degrees of error* of propositions with 0 and 1 representing $n-1$ (maximum) and 0 errors. His considerations suggest that on this reading the truth table for Łukasiewicz implication seems plausible. In his response to Scott, T. J. Smiley argues that Scott's interpretation suffers basically the same problem as Łukasiewicz's, namely that the logic of error is not truth functional (Scott 1976, 74–88). Note also that a related interpretation was proposed independently by Urquhart.[199] Scott (1976) also presents the later influential interpretation by Giles to which we now turn.

Giles (1974) proposed a particular *game semantics* for the infinitely-valued predicate Łukasiewicz logic.[200] Game semantics were introduced independently by Lorenzen and Hintikka, and have subsequently been thoroughly explored particularly since the 1990s.[201] Giles's semantics was in turn inspired by Lorenzen's logical dialogue games.[202] For simplicity, we restrict discussion here to sentences, i.e. formulas φ with no free variables, and assume that for each element u of a given structure **M** in which formulas are evaluated there is a constant c interpreted by u. The game has two players, referred to as I and *You*. One of them has the role of proponent, P, the other is opponent, O, and during the game their roles may switch. A *state* of the game is given by a pair of multisets $\{\psi_1, \ldots, \psi_m\}$ and $\{\varphi_1, \ldots, \varphi_n\}$, called *your tenet* and *my tenet*, which are thought of as representing formulas asserted by *You* and I.[203] Such a state is denoted $[\psi_1, \ldots, \psi_m \mid \varphi_1, \ldots, \varphi_n]$. A state is *final* if it contains atomic formulas only. At a given nonfinal state, a nonatomic formula χ, called the *current formula*, is picked at random from the ψ_is and φ_js. It may be shown that the power of the players does not depend on this choice. If χ is in my tenet then I acts as P and *You* as O, otherwise I acts as O and *You* as P. The *game for φ under* **M** starts at the state $[\ \mid \varphi]$. At each nonfinal state, the game proceeds by the following rules:

(∧)   if χ is $\theta_1 \wedge \theta_2$ then the game continues in a new state in which the respective occurrence of $\theta_1 \wedge \theta_2$ in Ps tenet is replaced by either θ_1 or θ_2 depending on Os choice;

(∨)   if χ is $\theta_1 \vee \theta_2$ then the game continues in a new state in which the respective

---

[199] "An interpretation of many-valued logic," *Z Math Logik Grundlagen Math* 19 (1973): 111–14. Urquhart attributes priority to Scott who presented his ideas at a symposium in Berkeley, CA, in 1971.

[200] A full version of Giles's paper appeared later in R. Wójcicki and G. Malinowski, eds., *Selected Papers on Łukasiewicz Sentential Calculi* (Wrocław: Ossolineum, 1977), 13–51.

[201] P. Lorenzen, "Logik und Agon," in *Arti del XII Congresso Internazionale di Filosofia* (Venezia, 1958), 187–94; J. Hintikka, "Language-games for quantifiers," in *Studies in Logical Theory*, ed. N. Rescher (Oxford: Blackwell, 1968), 46–72; also K. Lorenz, "Dialogspiele als semantische Grundlage von Logikkalkülen," *Archiv für mathematische Logik und Grundlagenforschung* 11 (1968): 32–55, 73–100.

[202] Our description differs from the original one in Giles 1974. The description follows that of Fermüller and Majer (2013), and has some advantages over the original one w.r.t. current state of the art.

[203] In a multiset, an element may occur several times.

occurrence of $\theta_1 \vee \theta_2$ in Ps tenet is replaced by either θ_1 or θ_2 depending on Ps choice;

($\rightarrow$)　if χ is $\theta_1 \rightarrow \theta_2$ then the game continues in a new state in which the respective occurrence of $\theta_1 \rightarrow \theta_2$ is removed from Ps tenet and O decides whether to proceed at the thus resulting state or to add θ_1 to Os tenet and θ_2 to Ps tenet and to proceed after this addition;

($\forall$)　if χ is $\forall x\theta$ then O choses a constant c and the game continues in a new state in which the occurrence of $\forall x\theta$ in Ps tenet is replaced by $\theta(c)$;

($\exists$)　if χ is $\exists x\theta$ then P choses a constant c and the game continues in a new state in which the occurrence of $\exists x\theta$ in Ps tenet is replaced by $\theta(c)$.

For a final state $[\psi_1,\ldots,\psi_m \mid \varphi_1,\ldots,\varphi_n]$ and a structure $\mathbf{M}$ for the predicate Łukasiewicz logic, the *payoff* for player I is the value

$$m - n + 1 + \sum_{i=1}^{n} \|\varphi_i\|_{\mathbf{M}} - \sum_{j=1}^{m} \|\psi_j\|_{\mathbf{M}},$$

where $\|\theta\|_{\mathbf{M}}$ are the truth degrees of the atomic formulas θ in $\mathbf{M}$. If for every $\varepsilon > 0$ player X, i.e. P or O, has a strategy which guarantees him a payoff of at least $a - \varepsilon$ and Xs opponent has a strategy which guarantees that X has a payoff at most $a + \varepsilon$, then a is called the *value for X of the game for φ under* $\mathbf{M}$. The following theorem is due to Giles:[204]

Theorem 5.38. For the predicate Łukasiewicz logic, the value for player I of the game for φ under $\mathbf{M}$ equals the truth degree $\|\varphi\|_{\mathbf{M}}$ of φ in $\mathbf{M}$.

In his original treatment, Giles assigned certain yes-no experiments to atomic formulas φ with known probabilities $\langle\varphi\rangle$, called risk values, and employed betting money on the results. The risk values are related to the above truth degrees via $\langle\varphi\rangle = 1 - \|\varphi\|_{\mathbf{M}}$. Correspondingly, instead of payoff for player I, Giles worked with a total expected risk and a total expected amount of money I has to pay to player *You*. Note also that Giles originally proposed his game as a foundation for reasoning in physics but later he motivated it by the aim to provide semantics for fuzzy logic and extended it beyond the original proposal; see e.g. Giles 1982. Recently, Giles games and their variants have been examined intensively by Fermüller, Majer, Cintula, Metcalfe, and Roschger; see e.g. Fermüller 2008, Fermüller and Majer 2013, Fermüller and Metcalfe 2009.

A conceptually different game semantics for Łukasiewicz logic was developed by Mundici (1992, 1993), who showed that MV-algebras are naturally interpreted as algebras governing the dynamics of states of knowledge in Ulam games. Basically, the Ulam game generalizes the well-known game of Twenty Questions in which a person thinks of a number x in a finite set S of numbers, say x is between one and one million, and we try to guess this number by asking yes-or-no questions such as

---

[204] Giles outlined the proof; for a complete proof and further considerations see Fermüller and Metcalfe 2009.

"is $x \leq 100$?" or "is x even?" which the person answers truthfully.[205] In the *Ulam game with at most m lies*, the goal is the same but the person is allowed to lie at most m times. Ulam games have broad, interesting connections to searching with error and coding in the presence of noise.[206] Clearly, the questions are conveniently represented by subsets D of S. During the game, our knowledge is naturally represented by a function $\sigma : S \rightarrow \{0, 1, \ldots, m, m+1\}$, where $\sigma(y)$ is the number of answers falsified by y, i.e. the number of questions D containing y that were asked and answered negatively. Alternatively, σ may be represented by the $(m+2)$-valued fuzzy set $\tau : S \rightarrow \{0, 1/m+1, \ldots, m/m+1, 1\}$ defined by $\tau(y) = 1 - \sigma(y)/m+1$. Mundici calls such τ a *state of knowledge* in the game. Intuitively, $\tau(y)$ measures in units of $m+1$ how far y is from falsifying too many answers and may thus be seen as the truth degree of the proposition "y is not falsified too much." Before we start asking, we are in the *initial state*, i.e. such that $\tau(y) = 1$ for each $y \in S$. When we obtain an answer to a given question D, our state of knowledge may naturally change for any given $y \in S$ according to whether y is in D or not and whether the answer to D is positive or negative. For a question D we may formally define the *positive answer* as the fuzzy set D^{yes} given by

$$D^{\text{yes}}(y) = \begin{cases} 1 & \text{if } y \in D, \\ 1 - 1/m+1 & \text{if } y \notin D, \end{cases}$$

and dually for the *negative answer* D^{no}, i.e. $D^{\text{no}} = \overline{D}^{\text{yes}}$. The following observation then shows that the dynamics of states of knowledge in the Ulam game with at most m lies is described by the Łukasiewicz conjunction $\otimes$ in the $(m+2)$-valued MV-algebra $\{0, 1/m+1, \ldots, m/m+1, 1\}$ extended componentwise:

Theorem 5.39. Let τ be the state of knowledge arising from questions $D_1, \ldots, D_t$ and their answers $b_1, \ldots, b_t \in \{\text{yes}, \text{no}\}$. Then $\tau = D_1^{b_1} \otimes \cdots \otimes D_t^{b_t}$.

With more and more questions answered, our state of knowledge gets sharper: we say that τ_1 is sharper than τ_2 (or τ_2 is coarser than τ_1) if $\tau_1(y) \leq \tau_2(y)$ for each $y \in S$. For every state τ, there is the coarsest state $c(\tau)$ *incompatible* with τ in that $\tau \otimes c(\tau) = 0$ and one in fact has $c(\tau) = 1 - \tau$. Denote thus $c(\tau)$ by $\neg \tau$. The operations $\otimes$ and $\neg$ along with the initial state thus form an MV-algebra, the *MV-algebra of states of knowledge*. Moreover, an equation in the language of MV-algebras holds in every MV-algebra if and only if it holds in any MV-algebra of states of knowledge arising from the Ulam game with any finite set S and any number m of lies.

Let us conclude by noting that variants of this semantics have recently been developed for other fuzzy logics, such as Hájek's BL and its two other extensions, the Gödel and product logics (Cicalese and Mundici 2011, Montagna and Corsi 2014).

---

[205] S. M. Ulam, *Adventures of a Mathematician* (New York: Charles Scribner's Sons, 1976), 281. The game is sometimes called the Rényi-Ulam game due to an earlier formulation of this problem by A. Rényi, "On a problem in information theory" [in Hungarian], *Magyar Tudományos Akadémia Matematikai Kutató Intézetének Közleményei* 6 (1961): 505–16.

[206] A. Pelc, "Searching games with errors—fifty years of coping with liars," *Theor Comput Sci* 270 (2002): 71–109.

To present briefly some other contributions, let us mention Paris (1997), who proposed a semantics for the truth functions min, max, and $1-x$ in which the truth degree of a proposition p is viewed as a number of agent's independent arguments for p; Lawry (1998) who—building on the suggestion by Gaines (1976, 1978)—examined semantics in which truth degrees come from populations of voters; and Vetterlein (2008) who provided semantics for Łukasiewicz and Hájek's basic logic.

Psychological significance of truth degrees

It is clearly of great importance for the present considerations to examine the psychological relevance of truth degrees.[207] After the rejection of the classical view in the psychology of concepts due to Eleanor Rosch's pioneering work demonstrating that human concepts are graded rather than clear-cut (Rosch 1973a,b), several experiments have been conducted which actually demonstrate that *truth degrees have psychological significance.* Pioneering experiments in this regard, most of which have appeared in premier psychological journals, were conducted by Kochen and Badre (1974), Hersh and Caramazza (1976), Oden (1977a,b), and McCloskey and Glucksberg (1978). These works confirmed that natural human concepts are fuzzy and indicated that they may reasonably be represented by fuzzy sets, that humans have the capability of consistently processing information involving truth degrees, and that conjunctions and disjunctions of propositions admitting truth degrees are reasonably modeled by the then-used min and product t-norms and their t-conorms. Further experiments appeared somewhat later and include Thole, Zimmermann, and Zysno 1979 and Zimmermann and Zysno 1980, which also demonstrate that truth degrees are psychologically significant. Note also that Parikh (1991), in a two-page report on his experiment in which he did not refer to any of the above studies except McCloskey and Glucksberg 1978, observed inconsistencies in assigning truth degrees to membership in concepts and arrived at the conclusion that employment of truth degrees does not help to "explain our linguistic practices." It must be said, however, that his experiment is much less comprehensive compared to the others mentioned above and that as such it may need a more careful revision, also because of his selection of concepts and exemplars, which themselves may have caused the inconsistencies.

5.4.2 Fuzzy logic and paradoxes

History offers many examples of various kinds of paradoxes.[208] Even though paradoxes may at first sight seem to serve mainly to amuse or perplex us, some have greatly influenced the development of philosophy, mathematics, and science. This is because a paradox often hints at some fundamental problem in current understanding of the issue at stake. Resolving a paradox then amounts to recognizing

---

[207] The role of fuzzy logic in psychology is examined in section 6.7.

[208] N. Rescher, *Paradoxes: Their Roots, Range, and Resolution* (Chicago, IL: Open Court, 2001).

the problem and possibly coming up with a fundamentally new perspective, a new paradigm within which the original paradox is no longer a paradox. The aim of this section is to examine four well-known paradoxes and the role fuzzy logic may play in their possible resolution.[209]

Sorites paradox

The *sorites* (σωρείτης) paradox , also called the *paradox of the heap*, can be described as follows.[210] A single grain of sand does not form a heap; the addition of a single grain does not turn a nonheap into a heap; hence, two grains do not form a heap, three grains do not form a heap, and so on, which leads to the conclusion that no number of grains is sufficient to make a heap. Yet we know that one million grains forms a heap. Hence the paradox.[211]

The term "sorites paradox" sometimes refers to similar kinds of paradoxes involving vague predicates.[212] The paradox is attributed to the Megarian logician Eubulides of Miletus (fl. 4th century BC) and was originally formulated as a puzzle: a single grain does not form a heap; two grains do not either; but certainly, if you keep adding grains, sooner or later a heap appears; so where do you draw the line?

Let us formalize a bit. If nh is a unary predicate symbol representing "nonheap,"[213] the paradox assumes the following form:

$$\begin{array}{ll} \text{assumptions:} & nh(1) \\ & nh(x) \to nh(x+1) \\ \hline \text{conclusion:} & nh(n) \text{ for each positive integer } n \end{array} \qquad (5.60)$$

The conclusion is obtained by repeated application of *modus ponens*: from the first assumption, $nh(1)$, and the instance $nh(1) \to nh(2)$ of the second assumption, the application of *modus ponens* yields $nh(2)$, which together with $nh(2) \to nh(3)$ similarly yields $nh(3)$ and so on up to $nh(n)$.

We now present the essence of a resolution of the sorites paradox using fuzzy logic which is due to Goguen (1968–69). One ingredient is the observation that the properties involved in sorites paradoxes, such as that of not forming a heap, are vague and are thus naturally modeled by fuzzy sets. This observation is essentially due to Black (1937). The second ingredient consists in observing that once we acknowledge that nh represents a fuzzy set, it is intuitively appropriate to say that the truth degree of $nh(x+1)$ is smaller, but only slightly smaller, than that of $nh(x)$, i.e. addition of a single grain may not decrease the degree of not forming a heap too much. This is more appropriate than saying as in the original formulation above

---

[209] Recall in this context Bochvar's three-valued logic which was designed to analyze paradoxes (pp. 142ff.).

[210] The name comes from the Greek word "soros" (σορός) which means heap. There are various alternative formulations of this paradox.

[211] Sometimes, a reverse form is used: one is removing grains from a heap and what remains is still a heap.

[212] E.g. the *falakros* (φαλακρός) paradox, also called the paradox of the bald man: A man with one hair is bald; if a man with n hairs is bald then a man with $n+1$ hairs is bald; hence every man is bald.

[213] If one does not like the negative predicate, one may replace it with "small."

that such an addition does not make a nonheap a heap. Thus modified, this assumption is equivalent to assuming that the formula $nh(x) \to nh(x+1)$ is assigned a high truth degree, say 0.99, which is nevertheless smaller than 1. The first assumption remains fully accepted, i.e. is assumed to degree 1. How does one reason with these assumptions, the second of which is only accepted to degree 0.99? Actually, by Goguen's *modus ponens* (4.58) presented in section 4.3.1, which is tailored for such a scenario. Thus, if for instance the connectives employed are the product conjunction and its residuum, as Goguen (1968–69) assumed, one obtains from $nh(1)$ with degree 1 and $nh(1) \to nh(2)$ with degree 0.99, first $nh(2)$ with degree $1 \cdot 0.99 = 0.99$ from which, employing further $nh(2) \to nh(3)$ with degree 0.99, one gets $nh(3)$ with degree $0.99 \cdot 0.99 = 0.99 \approx 0.98$, and so on, eventually obtaining $nh(n)$ with degree 0.99^{n-1}. Instead of (5.60), we thus get the scheme

assumptions:	$nh(1)$	with degree 1
	$nh(x) \to nh(x+1)$	with degree 0.99
conclusion:	$nh(n)$	with degree 0.99^{n-1}

which no longer represents a paradox. Namely, with increasing n, the degree to which one may assert that n grains do not form a heap, i.e. 0.99^{n-1}, decreases due to repeated application of Goguen's *modus ponens*. Thus, for instance, we obtain $nh(100)$ to degree approximately 0.37 and $nh(1000000)$ to degree approximately 0. Hence, the degree to which we may assert that a collection of one million grains does not form a heap is virtually 0. This certainly represents an appealing account for the sorites paradox. From the present perspective, the classical sorites paradox results from the failure to recognize that the predicates involved, such as heap or bald, are fuzzy rather than crisp and that addition of a single grain or hair actually changes the truth degree even though almost negligibly.

Variations of the presented resolution later appeared along with further considerations e.g. in Hájek and Novák 2003 and in Gerla 2010. However, there also exists a number of other, mutually very different resolutions which have been proposed.[214]

Poincaré paradox

Define for a threshold $\varepsilon \geq 0$ a binary relation T_ε on $\mathbb{R}$ by $\langle u, v \rangle \in T_\varepsilon$ if $|u - v| \leq \varepsilon$. It is nowadays a textbook example motivating the notion of tolerance that while T_0 is just the ordinary equality on $\mathbb{R}$, for $\varepsilon > 0$ the relation T_ε—naturally interpreted as "approximate equality"—is reflexive, symmetric, but not transitive, i.e. T_ε is a non-transitive tolerance relation.[215] The switch from equality to approximate equality is thus accompanied by a loss of transitivity which is certainly counterintuitive because

---

[214] Rescher, op. cit.; R. M. Sainsbury, *Paradoxes*, 3rd. ed. (Cambridge: Cambridge Univ. Press, 2009).

[215] Y. A. Schreider, *Equality, Resemblance, and Order* (Moscow: Mir, 1975). Tolerance relations were introduced in considerations on indistingishability by E. C. Zeeman, "The topology of the brain and visual perception," in *The Topology of 3- Manifolds*, ed. M. K. Fort Jr. (Englewood Clifs, NJ: Prentice Hall, 1962), 240–56, to formalize the notion of resemblance.

one expects that if u is approximately equal to v and so is v to w, then u should be approximately equal to w albeit perhaps not so much as u and v or v and w. This kind of consideration goes back to Henri Poincaré (1854–1912), who on several occasions discussed how the notion of mathematical continuum is derived from that of a physical continuum.[216] He characterized the physical continuum by the formula

$$A = B, \quad B = C, \quad A < C,$$

expressing the experiential statement that on many occasions we are not able to distinguish object A from object B, and B from C, but can distinguish A from C, and said: "whence arises an intolerable contradiction that has been obviated by the introduction of the mathematical continuum" in which $A = B$ and $B = C$ imply $A = C$. The problem thus posed has been termed the *Poincaré paradox* by Menger, who maintained that a closer examination suggests that one should sacrifice more than the transitivity of equality (Menger 1951b, 178):

> We should give up the assumption that equality is a relation. . . . We obtain a more realistic theoretical description of the equality of two elements by associating with A and B a number, namely, the probability of finding A and B indistinguishable. . . . In principle, this idea solves Poincaré's paradox.

As Menger explains, if both the equality of A and B and the equality of B and C are only highly likely, it may well be the case that the equality of A and C is less likely and that this equality is in fact less likely than the inequality of A and C. Denoting the probability that a and b are equal by $E(a, b)$, Menger continues with the postulates for E which demand, in terms of fuzzy logic, that E be a $[0, 1]$-valued fuzzy equality w.r.t. the product t-norm (pp. 277ff. above), i.e. satisfies $E(u, u) = 1$, $E(u, v) = E(v, u)$, and $E(u, v) \cdot E(v, w) \leq E(u, w)$.[217] In a sense, the switch from classical logic which underlies the classical equality to fuzzy logic resolves the paradox. This resolution is similar to that of the sorites paradox in the previous section in that one realizes that our view of the property and relation involved—that of not forming a heap and that of equality—demands that they be fuzzy rather than crisp. The fact that this change of perspective due to the switch from classical to fuzzy logic resolves long-standing paradoxes demonstrates the significance of fuzzy logic.[218]

To realize this seemingly simple point is, nevertheless, not immediate. Note that when discussing similarity and indistinguishability to motivate tolerance relations, Schreider in his well-recognized book actually speaks of degrees and says that "The superlative degree of resemblance is indistinguishability," but he, nevertheless, remains within the realm of crisp relations in which one is left with the choice of either the too restrictive notion of equivalence or the too weak notion of tolerance.[219]

---

[216] *La Science et l'Hypothèse* [Science and hypothesis] (Paris: Flammarion, 1902), chap. II; *La Valeur de la Science* [The value of science] (Paris: Flammarion, 1905), chap. IV.

[217] He in fact gave conditions for fuzzy equivalence but from his further considerations it follows that he assumed fuzzy equality.

[218] For this reason, the paradoxes are used in the introductory chapter of the book Belohlavek 2002c. The Poincaré paradox is also used and examined in many writings of Höhle since the early 1990s.

[219] *Equality, Resemblance, and Order*, op. cit.

It should be clear that fuzzy equalities w.r.t. other nonidempotent t-norms such as the Łukasiewicz t-norm $\otimes$ resolve the paradox as well. Here the point simply is that if u and v are indistinguishable (approximately equal) to a high degree, e.g. $E(u,v) = 0.9$, and the same for v and w, then transitivity in the sense of fuzzy logic ensures that

$$E(u,w) \geq E(u,v) \otimes E(v,w) = 0.9 \otimes 0.9 = 0.8,$$

i.e. u and w are indistinguishable at least to a degree bounded from below by a conjunctive-aggregation of $E(u,v)$ and $E(v,w)$. This approach also ensures another naturally required property, namely that two endpoints of a long chain of highly indistinguishable elements may be mutually well-distinguishable, for if e.g. $E(u_1, u_2) = \cdots = E(u_9, u_{10}) = 0.9$, then transitivity only yields $E(u_1, u_{10}) \geq 0.9 \otimes \cdots \otimes 0.9 = 0$, which is an empty constraint, thus u_1 and u_{10} need not be similar at all. The Łukasiewicz t-norm is employed in considerations of the Poincaré paradox by Höhle and Stout (1991) and Höhle (1996), who examined and proposed solutions to further considerations by Menger.[220] Note also that De Cock and Kerre (2003) point out a possible problem in the above solution—a natural many-valued indistinguishability E need not be separating, i.e. need not satisfy that $E(u,v) = 1$ implies $u = v$, for it is reasonable, they argue, to assume that heights of two people differing by 1 cm are indistinguishable to degree 1. If one accepts this view, a problem remains because then transitivity w.r.t. any t-norm easily yields that any two heights are indistinguishable to degree 1. De Cock and Kerre (2003) propose a solution based on assuming a pseudometric on the underlying universe.[221] A different solution, in a setting which does not refer to the notion of ultrametric and does not leave the first-order framework, is due to Gerla (2008).

Paradoxes of set theory

In section 5.1.2 (p. 235), we examined the famous Russell's paradox and its role in the development of set theory. Russell's paradox in fact is a logical inconsistency in classical logic in that it shows that if the axiom of full comprehension (p. 253) is allowed, i.e. if every property defines a set, then one may derive $(x \in x) \leftrightarrow \neg(x \in x)$, which is a contradictory formula in classical logic and hence every formula is provable from it. Moh (1954) was the first to observe that the situation is radically different in many-valued logics—Russell's contradictory formula $(x \in x) \leftrightarrow \neg(x \in x)$ no longer represents a contradiction in the three-valued Łukaskewicz logic for if $x \in x$ assumes the truth degree $1/2$, $(x \in x) \leftrightarrow \neg(x \in x)$ evaluates to 1.[222]

[220] "Geometry and positivism," in *Selected Papers in Logic and Foundations, Didactics, Economics*, ed. K. Menger (Dordrecht: Reidel, 1979), 225–34; the paper was presented at the 1966 Mach Symposium.

[221] Their paper initiated a discussion in the same issue of *Fuzzy Sets and Systems* involving four additional papers and a rejoinder by De Cock and Kerre; see *Fuzzy Sets Syst* 133, no. 2 (2003), 155–92.

[222] Moh Shaw-kwei (1917–2011) was a Chinese logician. He studied in Bernays's seminar on mathematical logic at the ETH Zürich in 1948. His name is often confused in the literature: Moh is his surname, Shaw-kwei is his given name.

Note that already Łukasiewicz (1920) hinted at the connection between the property $1/2 \leftrightarrow \neg 1/2 = 1$ and the paradoxes by saying: "This accounts for the fact that in the three-valued logic there are no antinomies." Moh in fact examined more general paradoxes which were presented by Curry and which did not contain negation.[223] In addition, he asked whether or not new Russellian paradoxes arise in many-valued logics, in particular whether new contradictions arise in the many-valued logics in which the formulas representing classical paradoxes no longer appear as contradictions. He proved that the answer is affirmative for every finitely-valued Łukasiewicz logic. The question remained open for the infinitely-valued Łukasiewicz logic which led him to pose the problem of whether one can develop set theory with full comprehension in this logic, i.e. whether such a set theory is consistent. Independently, full comprehension in many-valued logics was—essentially for the same reasons—considered by Skolem (1957) and his work initiated an interesting stream of research in many-valued set theory. These developments are presented in detail in section 5.2.3. Recall that even though set theory with full comprehension in the infinitely-valued Łukasiewicz logic has been studied intensively but also quite recently, Moh's (1954) question of its consistency is still open (p. 254 above). Let us conclude by noting that Maydole (1975) provides an overview of various many-valued logics and the respective examinations of consistency of set theories with full comprehension.[224] Of the several logics examined, only three turn out as possible candidates for consistency: the already-mentioned infinitely-valued Łukasiewicz logic, a kind of infinitely-valued Post logic, and a kind of infinitely-valued sequence logic.

Liar paradox

Suppose a person says "I lie." Is the sentence he utters true? If yes then he is lying, hence the uttered sentence is not true; if not, then what he says is a lie, hence the sentence is true. If we denote the sentence by λ then the paradox is represented by $\lambda \leftrightarrow \neg\lambda$. This paradoxical finding is called the *liar paradox*, or *pseudomenos* (ψευδόμενος), and is attributed to Epimenides of Knossos (7th–6th century BC).[225] The liar paradox has a pattern common to many *paradoxes of self-reference*, such as the barber paradox of Russell:[226] define a barber as one who shaves all those, and those only, who do not shave themselves, and ask the question whether a barber shaves himself. Given the observations of Moh (1954) and Skolem (1957) examined in the previous section, it is not difficult to see that the paradox disappears if we

---

[223] H. B. Curry, "The inconsistency of certain formal logics," *J Symb Logic* 7 (1942): 115–17.

[224] Maydole wrote his PhD thesis on this topic: *Many-Valued Logic as a Basis for Set Theory* (Boston, MA: Boston Univ., 1972).

[225] Epimenides, himself a Cretan, is supposed to have said "all Cretans are liars," but it is unsure whether he intended any paradox. According to C. Prantl, *Geschichte der Logik im Abendlande*, vol. I (Leipzig: Hirzel, 1855), 50, pseudomenos had the form of the question "Does one lie when one says that one lies?"

[226] Rescher, *Paradoxes*, op. cit.; B. Russell, "The philosophy of logical atomism and other essays: 1914–1919," in *Collected Papers of Bertrand Russell*, vol. 8, ed. J. G. Slater (London: Allen & Unwin, 1986).

assume that the sentence "I lie" may be assigned the truth degree 1/2 and if we use the connectives of Łukasiewicz's three-valued logic. Because then again, the sentence $\lambda \leftrightarrow \neg\lambda$ is no longer a contradiction and evaluates to 1. Such a resolution actually appears in Varela 1975, with acknowledgment to Varela in Gaines 1976, and within possibility theory in Zadeh 1979c.[227]

For some, however, the assumption that lying can actually be considered a graded predicate may seem strange. Namely, even if one embraces the idea of truth degrees, it is not difficult to see that, still, the view of lying as a *crisp* predicate is legitimate: "when I say that I lie, I mean to assert full truth" goes the explanation of such an understanding of the paradox. If one is not convinced of the legitimacy of the modified paradox—the liar with crisp lying—replace the liar by the barber paradox: assigning to the proposition "person p shaves himself" truth degrees other than 0 and 1 may seem rather strange.[228] The thus-modified paradox may again be formalized in fuzzy logic because fuzzy logic actually can enforce crispness of certain propositions if desired: if we want to ensure that φ is crisp we just add the axiom $\varphi \vee \Delta\varphi$, denoted usually crisp($\varphi$) where Δ is the Baaz delta (p. 204). In such cases, the liar paradox (with crisp lying) is presented by the formulas crisp(λ) and $\lambda \leftrightarrow \neg\lambda$. In classical logic, this is equivalent to the original form. In fuzzy logic, however, this is very different—the above solution does not work because the set with the two above formulas is inconsistent in Łukasiewicz logic and a paradox thus remains. From a semantic viewpoint, the liar paradox in a fuzzy setting thus allows two forms, neither of which may be pronounced the correct one since both make sense.

The above considerations were informal. The paradox appeared because in our natural, informal language we have two essential means of generating self-reference: we may denote propositions and we may speak of their truth and thus say e.g. "the proposition '2 is less than 3' is true." Due to fundamental considerations of Gödel on incompleteness and Tarski on the notion of truth, we know that such means are available even in formal languages which are rich enough to formalize Peano arithmetic. In such languages, we may refer to a formula φ via a term $\ulcorner\varphi\urcorner$, its *name*, and have the *truth predicate*, *Tr*, which is a unary relation symbol such that the intended interpretation of $Tr(\ulcorner\varphi\urcorner)$ is that φ is true.[229] It is well known that the extension of Peano arithmetic by the truth predicate *Tr* and the axiom schema $\varphi \leftrightarrow Tr(\ulcorner\varphi\urcorner)$ is inconsistent in classical first-order logic because Gödel's diagonal argument yields a formula λ—the liar's formula—such that one may prove $\lambda \leftrightarrow \neg Tr(\ulcorner\lambda\urcorner)$ and hence also $\lambda \leftrightarrow \neg\lambda$, which is classically inconsistent. In their interesting paper, Hájek,

[227] B. Skyrms, "Return of the liar: Three-valued logic and the concept of truth," *American Philosophical Quarterly* 7 (1970): 153–61, proposes a different solution based on non-truth-functional logic in which the third value is interpreted as neuter. S. Kripke, "Outline of a theory of truth," *J Philosophy* 72 (1975): 690–716, applies Kleene's three-valued logic to the liar paradox. These are just two examples of attempts to resolve the liar paradox with a many-valued logic different from fuzzy logics.

[228] We do not say that *it is not possible* to assign to this proposition a truth degree other than 0 and 1, but we do say that the primary understanding is that the proposition may only be either true or false.

[229] E.g. P. Hájek and P. Pudlák, *Metamathematics of First-Order Arithmetic* (Berlin: Springer, 1993).

Paris, and Shepherdson (2000b) examine Hájek's (1999) question of what happens if we replace classical logic with the predicate Łukasiewicz logic Ł∀ in the sense that one adds to Ł∀ crisp Peano arithmetic, adding as well a truth predicate Tr which is allowed to be many-valued.[230] They showed that the resulting theory, ŁPATr, is consistent, has nonstandard models (with a crisp arithmetical part) with truth degrees in $[0, 1]$, but the standard model of arithmetic, $\mathbb{N}$, has no many-valued extension to a model of ŁPATr with truth values in $[0, 1]$ or even in an MV-chain.[231] If, however, one adds axioms saying that Tr commutes with connectives, then the resulting theory is inconsistent. This also demonstrates that whether or not a paradox appears when approached in a formal setting with truth degrees is a delicate matter requiring careful examination. Clearly, no less care is needed for assessment of whether this or that formal result offers a plausible resolution of the paradox.

5.4.3 Fuzzy logic and vagueness

In this section we examine the phenomenon of *vagueness*. Since vagueness involves many issues, most of them closely related to fuzzy logic, and since the literature on vagueness is extensive, we need to be selective in our coverage of these issues as well as references. Note also that some of these issues are related to the problem of interpreting truth degrees examined in section 5.4.1. There are several books on vagueness, including the edited collection of influential papers on vagueness by Keefe and Smith (1996), as well as other recent books such as Williamson 1994, Sorensen 2001, Keefe 2000, Shapiro 2006, Smith 2008, and van Deemter 2010, where one may find further information.

Definition of vagueness

The term "vague" has multiple meanings, among them "not clearly expressed," "not having a precise meaning," "not clearly defined or grasped," "not thinking clearly or precisely," "not sharply outlined," etc. In the literature on vagueness, this term has a more specific meaning—we used it in the previous chapters when denoting as vague terms such as "high temperature," "tall man," "bald man," or "heap." In recent literature, vague terms are most often characterized as those which have *unsharp boundaries* (in that there is no clear cut-off separating objects to which the term applies from the others), have *borderline cases* (objects for which it is neither the case that the term applies nor that it does not apply), and are *susceptible to the sorites paradox* (section 5.4.2). Broader meanings that would also subsume generality and ambiguity are usually excluded.[232] Note in this context that instead of vagueness in the above sense, Zadeh speaks of *fuzziness* and characterizes it as unsharpness

---

[230] According to a personal communication from P. Hájek, the authors' interest was purely mathematical—they did not attempt to provide any kind of argument whether or not fuzzy logic reasonably resolves the liar paradox.

[231] That is, contrary to classical logic the liar paradox does not yield a contradiction.

[232] The distinction between generality and ambiguity is due to Russell (1923) and Black (1937).

of boundaries.[233] Attempts to provide more fundamental definitions have also appeared and we return to this issue later.

Emergence and early contributions

Vague predicates appeared in antiquity with Eubulides' sorites paradox which we examined in section 5.4.2. Sorites, which became more widely known due to Diodorus Cronus (died c. 284 BC), has been used as an argument in various debates in antiquity, in particular about the criterion of truth between the Sceptics of the New Academy and the Stoics. The Stoics' theory of knowledge and their concept of cognitive impression—meaning basically a clear-cut and definite idea—defended by Chrysippus, was an obvious target.[234] Subsequently, there was little interest in the sorites paradox and vagueness in general until attempts were made by Lorenzo Valla (1407–1457) to change the logical curriculum toward emphasis on rhetoric within which sorites had been used. Vagueness is the subject of interesting considerations by John Locke (1632–1704), who arrived at it naturally in his empiricist analysis. Locke argued that concepts of species and their boundaries are made by the human mind rather than by the nature of the objects to which the concept is being applied, and he was well aware that the concepts thus formed have borderline cases (Locke 1690):[235]

> [A] continued series of things, that in each remove differ very little one from the other. There are fishes that have wings, and are not strangers to the airy region; and there are some birds that are inhabitants of the water, whose blood is cold as fishes, and their flesh so like in taste, that the scrupulous are allowed them on fish-days. There are animals so near of kin both to birds and beasts, that they are in the middle between both. . . .
>
> . . . I think, I may say, that the certain boundaries of that species are so far from being determined, and the precise number of simple ideas, which make the nominal essence, so far from being settled and perfectly known, that very material doubts may still arise about it. And I imagine, none of the definitions of the word man, which we yet have, nor descriptions of that sort of animal, are so perfect and exact, as to satisfy a considerable inquisitive person; . . .

---

[233] Zadeh in fact repeatedly emphasized the need to distinguish fuzziness from vagueness and criticized the current routine. The following quotation from his e-mail to the BISC group dated January 23, 2013, with the subject "What is fuzzy logic?" is illuminating: "Before fuzzy theory came into existence, the concept of fuzziness was a very infrequent topic of discussion in the literature of logic and philosophy. When it was discussed, the term vague was employed, inaccurately, to describe what should have been called fuzzy. No distinction was made between the concepts of vagueness and fuzziness. Basically, vagueness connotes insufficient specificity, whereas fuzziness connotes unsharpness of class boundaries. I will be back in a few minutes, is fuzzy but not vague. I will be back sometime, is fuzzy and vague. . . . Inappropriate use of the term vague is still a common practice in the literature of philosophy."

[234] It is known that Chrysippus (ca. 279–206 BC) wrote in defense the books with suggestive titles *On soritical arguments against words* and *On the little-by-little argument*, neither of which, however, has survived; see J. Barnes, "Medicine, experience and logic," in *Science and Speculation*, ed. J. Barnes, J. Brunschwig, M. Burnyeat, and M. Schofield (Cambridge: Cambridge Univ. Press, 1982), 41ff.; cf. also n. 19 on p. 113 above.

[235] The quotation is from pp. 416 and 423 of book III and p. 121 of book IV of the 1824 edition published by Valentine Seaman in New York.

In his well-known reply to Locke's book, Leibniz maintained that concept boundaries are fixed by the nature of objects but admits that in some concepts, such as *bald*, opinion plays a role; see G. W. Leibniz, *Nouveaux essais sur l'entendement humain* [New essays on human understanding] (completed 1704, published in Amsterdam and Leipzig: J. Schrender, 1765).

> [T]he extent of these species, with such boundaries, are so unknown and undetermined, that it is impossible with any certainty to affirm, that all men are rational, or that all gold is yellow.

Isaac Watts (1674–1748), who was influenced by Locke, further elaborated on these ideas and spoke of degrees in connection with vague terms in chapter VI of the first part of his influential textbook on logic, which appeared in twenty editions and had been used at Oxford for over 100 years (Watts 1724):[236]

> [T]here are many things which cannot well be defined, either as to the name or the thing, ... Such are ... most of our *simple* ideas, and particularly *sensible* qualities, as *white, blue, red, cold.* ...
>
> The several *species* of beings are seldom precisely limited. . . . The essences of many things do not conflict *in indivisibili*, or in one evident indivisible point, as some have imagined; but by various degrees they approach nearer to, or differ more from, others that are of a kindred nature. So (as I have hinted before) in the very middle of each of the arches of a rainbow, the colours of *green, yellow,* and *red,* are sufficiently distinguished; but near the borders of the several arches they run into one another, so that you hardly know how to limit the colours, nor whether to call it *red* or *yellow, green* or *blue.*

Remarks on the imprecise boundaries of human concepts occasionally appeared in writings on philosophy and logic in the 19th century, but no particular interest was paid to vagueness as such and no clear positions toward vagueness seem to have been taken in these writings. In his book on logic, Alexander Bain (1870, 160–61, 177) mentions the need for scales of numerical degrees as follows:

> The quality, so very decided in the great mass of instances, is found to have degrees, to shade insensibly into the state called liquid, where solidity terminates. . . .
>
> Many couples of qualities, unmistakeably contrasted in the greater number of instances of them, pass into one another by insensible gradations, rendering impossible the drawing of a hard and fast line. . . .
>
> There is but one solution of the riddle. A certain *margin* must be allowed as *indetermined.* . . .
>
> It is essential that we should be able to describe with accuracy all individual facts and observations; consequently names must be devised for all the known qualities of things whether physical or mental, and also modes of signifying differences of degree whenever degree is taken into account. To describe the diamond, we need such names as crystal, refracting power, specific gravity, hardness; and a numerical scale for stating the amount or degree of each property.

A clearly negative view of vagueness was put forward on a number of occasions since the late 1870s by Frege—one of the founders of modern logic. In no less important a work than his monumental *Grundgesetze der Arithmetik* he said:[237]

> A definition of a concept . . . must unambiguously determine, as regards any object, whether or not it falls under the concept. . . . Thus there must not be any object as regards which the definition leaves in doubt whether it falls under the concept;. . . . the concept must have a sharp boundary. . . . To a concept without sharp boundary there would correspond an area that had not a sharp boundary-line all around, but in places just vaguely faded away into the background. . . . and likewise a concept that is not sharply defined is wrongly termed a concept. Such quasiconceptual constructions cannot be recognized as concepts by logic; it is impossible to lay down precise laws for them.

---

[236] The quotation is from pp. 97 and 100–101 of the 1792 edition published by John Cuthell in London.

[237] See n. 182 on p. 322 above; the quotation is from the English translation of Band II (Beaney 1997, 259–65).

Interestingly, S. Puryear, "Frege on vagueness and ordinary language," *Philosophical Quarterly* 63 (2013): 120–40, argues that some of Frege's writings suggest that he thought that unsharp predicates could have meaning, particularly those used in ordinary language.

> [W]e must stick to our point: without complete and final definitions, we have no firm
> ground underfoot, we are not sure about the validity of our theorems, and we cannot confidently
> apply the laws of logic, which certainly presuppose that concepts, and relations too, have sharp
> boundaries.

The end of the 19th century in fact marks a renewal of interest in vagueness. It was Peirce who made vagueness an important subject of scientific inquiry and who—unlike Frege, but in accordance with Locke, Watts, and the later philosophers including Bain—viewed vagueness as a natural feature of language that needed to be taken into account. Peirce was interested in developing a theory of vagueness, as briefly examined in section 2.1. It took some two decades more before the influential paper by Russell (1923) appeared, which identified several issues regarding vagueness, as well as other early writings including Cohen 1927, Black 1937, Aldrich 1937, Copilowish 1939, and Hempel 1939, which are also examined in section 2.1. Early contributions identified some key issues and problems with vagueness and thus set the stage for numerous further writings from which some basic theories of vagueness evolved.[238]

Problems in vagueness

The *sorites paradox* and its variants are usually considered to represent the basic issue regarding vagueness. This problem is known from antiquity and is examined in section 5.4.2. Another one, advanced by Russell (1923), relates to so-called *higher-order vagueness* by which is meant the fact that the notion of a borderline case of a vague predicate is itself vague and needs to be accounted for as such. Thus one may think of borderline borderline cases, borderline borderline borderline cases, and so on. In addition to these problems, which represent tests for various theories of vagueness, there exist a number of further tests that are covered in the literature on vagueness mentioned at the beginning of this section.

Theories of vagueness (other than those based on fuzzy logic)

Epistemicism holds the view, also known as *vagueness as ignorance*, that vague predicates have sharp boundaries but we do not know them. Statements involving vague predicates such as "A collection of 100 grains does not form a heap" are either true or false in this view, but we may not know it. Hence, there is a cut-off separating heaps from nonheaps which is, however, fundamentally unknown to us. The proponents of this view include James Cargile, Richmond Campbell, Roy Sorensen, Timothy Williamson, and Paul Horwich. One of the premises "if n grains do not form a heap then nor do $n + 1$ grains" of the sorites paradox (p. 333) must be false according to this view, because for some n, the antecedent is true but the consequent false. There

---

[238] In the overview below we follow Smith 2008. The theories examined are in fact types of theories since we refrain from further distinguishing details. Similar classification of the theories may be found in other works (Williamson 1994, Keefe and Smith 1996, van Deemter 2010). Here we refrain from presenting pros and cons of one theory from the viewpoint of another.

is hence no paradox. We only erroneously think that there is one because, since we do not know the sharp boundary, we do not know which premise is actually false and so assume that all are true.

Truth gaps may be thought of as situations in which no truth values are assigned to certain sentences, such as "1000 grains do not form a heap." This naturally happens when the characteristic functions of classical or many-valued relations are partial and not total, i.e. if formulas are interpreted in a *partial structure* (also called partial interpretation). This offers the following explanation of vagueness: the characteristic functions of vague predicates are partial—they are not defined for borderline cases. Truth values of compound sentences are determined either according to the truth tables of Kleene's three-valued logic, in which case "not defined" is actually regarded as a third truth value, or according to some other rules.

Supervaluationism represents an influential view which may be regarded as a development of the truth gap approach. Supervaluations were examined by van Fraassen (1966), who also coined the term, but the idea had already appeared in Mehlberg 1958. Supervaluationism as a theory of vagueness has been studied by Fine (1975), Dummett (1975), and Kamp (1975), and later by several other authors. In this view, vague predicates are represented by partial classical relations which are undefined for borderline cases. The collection of such partial relations forms a partial structure **M** for the given logical language, e.g. a first-order language in which one speaks about heaps. One may then consider the collection of all classical *extensions* of **M**, i.e. classical total structures in which the truth gaps are replaced by classical truth values. For instance, if the partial set representing the vague predicate "nonheap" assigns 1 to numbers 1 to 500, is undefined for 501 to 10,000 and assigns 0 to numbers larger than 10,000, its extension assigns *some* truth value to each number from 501 to 10,000 and copies the other truth values which were already assigned. One in fact specifies and is restricted to *admissible* extensions—assigning in the previous example 1 to even numbers and 0 to odd ones in 501 to 10000 is intuitively *not* admissible, for if 0 is assigned to n then 0 needs to be assigned to any number larger than n. A *supervaluation* is the assignment of truth values to formulas φ defined basically as follows: φ is assigned 1 (0) if φ is assigned 1 (0) according to the rules of classical logic in each admissible extension of **M**; otherwise φ is undetermined, i.e. assigned no truth value. The supervaluation thus depends on the partial structure **M**. Some basic, easy-to-observe features of supervaluations are that classical tautologies are assigned 1 while contradictions get 0. Moreover, supervaluations are not truth functional: if e.g. φ and ψ are undetermined then so is $\neg\varphi$ but while $\varphi \lor \psi$ is undetermined, $\varphi \lor \neg\varphi$ gets 1. Sorites is explained as follows. Since in each admissible extension, some $nh(n) \to nh(n+1)$ is violated, the premise of sorites gets assigned 0 by the supervaluation. Since the premise is false, one cannot use it as an argument and the paradox thus disappears.

Plurivaluationism has been identified and named by Smith (2008), who demonstrates that this view, while fundamentally distinct from supervaluationism, has misleadingly been conflated with it. Plurivaluationism is characterized by rejecting a feature common to all of the above views, namely that each discourse has a unique intended interpretation. For instance, from the point of view of fuzzy logic, a vague term is interpreted by a fuzzy set, in epistemicism by a classical set, and in supervaluationism by a partial set. In plurivaluationism, a vague term is interpreted by a collection of classical sets, all of them being thought of as acceptable classical interpretations. Acceptability is determined by the language usage of the respective speech community. From this viewpoint, the classical logic view is a particular case in which there is exactly one acceptable interpretation at a given point.[239] Smith argues that due to the confusion mentioned above, it is difficult to clearly pinpoint proponents of plurivaluationism, but he nevertheless identifies Field (1973) and Przełęcki (1976) as early proponents.

Alternatively, *contextualism* holds that vagueness emerges only when we consider meaning of language over time and thus vagueness is viewed as a diachronic phenomenon. Finally, Hilary Putnam's *intuitionist view* holds that vague predicates need to be treated as undecidable predicates are in intuitionistic logic.

Vagueness and fuzzy logic

It should be clear by now that fuzzy logic offers a basic view of vague predicates as predicates representable by fuzzy sets and fuzzy relations. The view that the application of vague terms is a matter of degree is perhaps the oldest of the various explanations of vagueness. This view is the heart of Zadeh's motivations for fuzzy sets and had clearly been advanced already by Black (1937) and several scholars thereafter, as well as some philosophers as early as the end of the 17th century. Fuzzy logic offers a resolution of the sorites paradox as described in section 5.4.2.

Most proponents of fuzzy logic regard it as obvious that the idea of vague terms, such as "nonheap," "bald," or "tall," applying to degrees is intuitively rather natural, by far more so than views which in one way or another employ only classical truth values. Nevertheless, few of them have considered fuzzy logic as providing a once-and-for-all theory of vague predicates with all their subtleties. Better to say, fuzzy logic has been viewed by its proponents as providing useful conceptual tools which are intuitively appealing for representing vague terms, help naturally explain sorites as well as other puzzling questions relevant to vagueness, and lead to successful applications. The question of whether fuzzy logic represents a full-fledged theory of vagueness immune to possible philosophical objections was by and large not on the agenda of its proponents until the 2000s or so. Such objections, nevertheless, started to appear in the literature on vagueness. The best known is the thesis

---

[239] From a logical point of view, in classical logic a formula is interpreted in a given (single) classical structure at a given point, in plurivaluationism this single structure is replaced by a collection of structures.

that fuzzy logic imposes *artificial precision*: why assign to the proposition "Peter is tall" the truth degree 0.71? Why not 0.7 or 0.714?—such precision is artificial when modeling vague terms. This argument appears in various forms in the literature, e.g. by Haack (1979), Keefe (2000), and Urquhart (1986), who says: "[I]t seems plausible to say that the nature of vague predicates precludes attaching precise numerical values just as much as it precludes attaching precise classical truth values." Such objections have been discussed in many places, including the book by Smith to which we shall turn shortly.[240] On the other hand, philosophers of vagueness have often lacked deeper knowledge of fuzzy logic and thus some of their objections have not been substantiated. Thus for instance, Varzi in his review of the well-known book by Haack (1996)—an updated version of her 1974 book *Deviant Logic*—claims that the updated version no longer has the merits of the 1974 edition, which was well received particularly for its thoughtful defense of classical logic against the then-existing nonclassical logics.[241] Varzi maintains that the updated version is largely out of date, especially in its coverage of new contributions regarding nonclassical logics and other topics criticized by Haack, and writes:

> Even among those forms of deviance that were extensively discussed in the 1974 edition, there are some—such as vagueness—whose account is now seriously defective. . . .
> I have the same complaint about the two essays on fuzzy logic, which are included in the volume precisely to fill in at least one of the gaps of *Deviant Logic*. . . . Haack argues that fuzzy logic is "methodologically extravagant and linguistically incorrect". . . . [T]his argument is disappointing.

An important contribution to this unsettled situation has been provided in the book by Nicholas Smith (2008). Smith embraces the idea of degrees of truth as a natural basic idea for a development of a theory of vagueness. He critically examines the existing views and concludes there is a need for a definition of vagueness that is more fundamental than the characterizations mentioned at the beginning of this section. Such a definition should facilitate a comparison of theories of vagueness and help assess in which respect a given theory is good or bad. For this purpose he proposed the following definition of vagueness as closeness: a predicate P is vague if for any two objects u and v, if u and v are very close in P-relevant aspects then the propositions that P applies to u and that P applies to v are very close with respect to truth. From this viewpoint, he naturally arrives at *fuzzy plurivaluationism*—a new theory of vagueness which combines fuzzy logic and plurivaluationist viewpoints. This theory is then carefully examined, compared to others, and supported by various considerations in Smith's book.

Recent discussions of vagueness by fuzzy logicians include Fermüller 2003, Hájek 1999, 2009, Vetterlein 2012, an interesting conversation by Hájek and Fermüller (2011), and the edited book by Cintula et al. (2011) in which the Hájek-Fermüller

---

[240] See also Belohlavek 2007 and the related discussion in section 3.9.

[241] A. C. Varzi, *Philosophical Review* 107 (1998): 468–71. We include this example because it illustrates well a broader phenomenon, namely criticism of fuzzy logic, often a dismissive one, by philosophers but also psychologists (section 6.7) with a limited understanding of it.

conversation appears. The title of Hájek's conference talk, "Vagueness and fuzzy logic—can logicians learn from philosophers and can philosophers learn from logicians?" to which the conversation refers, is most suggestive in our view—interactions both ways have the potential to improve our understanding of vagueness and provide fuzzy logic with further inspiration and challenges.

Chapter 6

Applications of Fuzzy Logic

6.1 Introduction

OUR PRINCIPAL AIM in this chapter is to characterize the usefulness of fuzzy logic in selected areas of science, engineering, medicine, management, business, and other areas of human affairs. That is, we are concerned in this chapter not only with surveying literature on applications of fuzzy logic in these various areas and describing how they have developed, but also, more importantly, with assessing their usefulness in these respective areas. Assessing the usefulness of fuzzy logic in each area at some particular time is of course far more difficult than just describing the proposed applications.

Each area of human affairs is characterized by certain aims and is therefore associated with people working jointly toward these aims—researchers, practitioners, educators, administrators, and the like—who together form the community of each respective area. We take a position in this chapter that the only genuine assessments of the usefulness of fuzzy logic in each area are those made by members of the respective community. However, assessments by different members or groups of members of the same community are often contradictory. In each such case, we proceed by examining arguments that support the individual contradictory assessments and only then offer our own assessment.

Besides those already mentioned, one additional community—the so-called fuzzy-logic community—plays an especially important role in the development of applications of fuzzy logic. This community can loosely be characterized as a heterogeneous group of researchers, practitioners, educators, etc., who share their common interest in fuzzy logic—an interest motivated primarily by their belief that fuzzy logic has the potential for playing a useful role in their respective areas. The emergence of this community is closely connected with the evolution of a supporting infrastructure for fuzzy logic, within the first two decades after the genesis of fuzzy logic in 1965, as discussed in detail in section 2.5.

When we traced authors of proposed applications of fuzzy logic in various traditional areas, we discovered that almost all of them were members of the communities associated with the respective traditional areas. This was an important and intuitively rather appealing discovery because these are the people who have the proper expertise for proposing significant new directions in their respective areas. However, these authors have often published their proposed applications of fuzzy logic in media (journals, conference proceedings, edited books, etc.) oriented to fuzzy logic rather than to their respective traditional disciplines. We suggest two plausible reasons for this peculiar phenomenon.

The first reason is that it was, naturally, in general easier to publish results in the fuzzy-logic literature. An important factor here was the fact that the community associated with the given area was often antagonistic toward fuzzy logic, making it difficult at that time to publish ideas involving fuzzy logic in the literature representing a traditional discipline. In the early stages of the history of fuzzy logic, this was typical in many traditional areas. While this situation has substantially changed during the 50-year history of fuzzy logic, it still exists, at least to some degree, in some areas, where for various reasons the resistance to fuzzy logic continues to persist.

The second possible reason why authors from traditional academic communities often publish in fuzzy-logic literature is that they are consciously interested in making the fuzzy-logic community aware of their work. There are of course many possible incentives for seeking the awareness of the fuzzy logic community. For example, it is easy to imagine that authors may want their work to be scrutinized and possibly developed further by other members of the fuzzy-logic community, or that they want to establish links with other researchers pursuing similar work that may potentially lead to cooperation.

Our coverage of applications of fuzzy logic in this chapter consists of two parts. The first is devoted to an overview, as comprehensive as possible, of applications developed within each of the primary areas of science, engineering, medicine, etc. This is a very large undertaking due to the virtual explosion of literature dealing with applications of fuzzy logic in recent years. Thus, we have had to be exceedingly selective. According to the spirit of this book, we attempt to capture as best as possible and in sufficient detail the genesis and historical development of various types of applications in each area. However, to capture the current state of the art of fuzzy logic applications in each of the commonly recognized traditional academic disciplines, we have used carefully selected references to highly informative publications, such as major monographs, edited volumes, and in some cases, relevant textbooks. A fair number of publications within this enormous literature describe applications that are either superficial or deficient in some other way. Such applications are usually

proposed by authors who do not have adequate expertise in the respective application area or whose mathematical background is deficient. Thus, in surveying the
literature we have made a conscious effort to "separate the wheat from the chaff."

The second part of our coverage of applications of fuzzy logic deals with the
issue of assessing the usefulness of fuzzy logic in the various application areas. As
previously indicated, to do so we rely on relevant information from members of
the respective communities. Scattered information of this kind is to be found in
various sources, such as interviews, debates, survey papers, forewords to books, and
the like. In several cases, we were able to obtain such information directly from the
members of a given community. Whenever we have encountered information that
was contradictory, we have resorted inevitably to our own informed judgment.

In this chapter, we are also interested in comparing how the initially expected
usefulness of fuzzy logic in each of the traditional areas may differ from the assessed
usefulness at this time—50 years since the genesis of fuzzy logic. Furthermore, we
have also made a conscious effort to explain why the expected usefulness may often
be very different from actual practice.

Note also that the information provided in this chapter is supplemented in section 7.4 with a quantitative assessment of the impact of fuzzy logic in various areas.

6.2 A historical overview

Just one year after the publication of Zadeh's seminal paper, Bellman, Kalaba, and
Zadeh (1966) suggested a sensible application of fuzzy sets to pattern classification.
A few years later, Bellman and Zadeh (1970) described an application of fuzzy sets
to decision making. In particular, they described a multistage decision process involving dynamic programming in which goals and constraints were represented by
fuzzy sets. This paper inspired interest in the role of fuzzy sets in decision making,
as reflected in the impressive number of papers published on various applications
of fuzzy logic to decision making in the 1970s. These papers were critically reviewed
in an important book by Kickert (1978), the very first book fully devoted to applications of fuzzy set theory. Simultaneously with these developments, applications of
fuzzy sets to pattern classification also advanced substantially throughout the 1970s,
as demonstrated by the linguistic approach to pattern classification introduced by
Zadeh (1977) and by the emergence of fuzzy clustering via the work of Bezdek (1973,
1981), Dunn (1973), and others (see section 3.8).

Another important application area of fuzzy sets and fuzzy logic that germi-
nated in the 1970s involved fuzzy controllers. The rationale for fuzzy controllers
based on *fuzzy if-then rules* was discussed in a short paper by Zadeh (1972b). Shortly
after this paper was published, the first experimental fuzzy controller was actually
designed and built by Ebrahim H. Mamdani (1942–2010) and his student Sedrak
Assilian in England at Queen Mary College, University of London, and described

in their paper Mamdani and Assilian 1975. This paper also contains some preliminary experimental results showing that the fuzzy controller, which was designed for controlling a small steam engine, consistently outperformed conventional controllers in the sense that it was substantially faster in reaching the desired value of the controlled variables.

In his recollections, published some twenty years later, Mamdani (1993) describes circumstances that led to his pioneering work on fuzzy controllers as well as difficulties that he encountered in this regard. As he recalls (Mamdani 1993, 340):

> It was Zadeh's paper (Zadeh 1972b) published at that time which persuaded us to use a fuzzy rule-based approach. Between reading and understanding Zadeh's paper and having a working controller took a mere week and it was "surprising" how easy it was to design a rule-based controller . . .
>
> On reflection it becomes clear that our approach to control was based on a different view of control (that based on AI paradigm) than the conventional view of control (that based on the analytical control theory). The latter was at that time a very successful discipline and thus all-powerful. So much so that any discussion on the problem of control could only be expressed in the terminology of that established paradigm. This culture clash was at the very heart of the early criticisms of fuzzy control.

A few applications of fuzzy sets in other areas were also suggested in the 1970s, but the types of applications already mentioned were dominant—pattern classification, decision making, and control. The overall state of the art of fuzzy logic applications around the end of the 1970s is well described in the book by Dubois and Prade (1980a), in which some 100 pages are devoted to applications. In spite of the impressive number of suggested and partially developed applications of fuzzy sets at that time, none of these applications had shown as yet any substantial impact on the application areas involved. Even the fuzzy controller built by Mamdani and Assilian, notwithstanding its promising results, was still a laboratory "toy," not a real application. Hence, the criticism that fuzzy set theory and fuzzy logic had not demonstrated any significant applications was still largely justifiable at that time. However, this situation changed dramatically throughout the 1980s and even more in the 1990s.

In 1980, the first commercial fuzzy controller was permanently installed for controlling a cement kiln owned by F. L. Smidth & Company in Denmark. It is described in a paper written by Holmblad and Østergaard (1982), who designed and built it. Before the fuzzy controller was installed, human operators had controlled the kiln. This was inconvenient and expensive as it took about eight weeks to train a new operator. So the company had already considered in the 1960s the possibility of replacing the human operators with a computer-based controller. In 1973, Lauritz P. Holmblad (1944–2005), a Danish engineer at the Technical University of Denmark, took this challenge and left the university to work on the desired controller. He soon found that the process to be controlled was too complex and unwieldy for a conventional controller, but, fortunately, he came across Mamdani's work on fuzzy control. The idea of fuzzy control immediately appealed to him, due to its focus on modeling an experienced operator rather than the process to be controlled. When he found that the textbook commonly used for training human operators of cement kilns contained 27 control rules described in natural language, which could readily be represented as if-then rules in Mamdani's fuzzy controller, he became convinced that fuzzy control was the right way to deal with this difficult control problem.[1] He recruited his former student, Jens-Jørgen Østergaard, to help him with the design and implementation of a Mamdani-type fuzzy controller for controlling the cement kiln. The final product was quite successful. It not only eliminated the long and expensive process of training human operators, but it even improved slightly their performance and cut fuel consumption.[2]

A much more significant application of fuzzy logic was accomplished in 1987 in Japan, when the subway system in the city of Sendai switched from human-operated trains to fully automatic operation of trains via a sophisticated fuzzy controller involving both feedback and feedforward features. This was the outcome of a large and initially rather risky project that was conceived in 1979 by two researchers at Hitachi Systems Development Laboratory, Seiji Yasunobu and Shoji Miyamoto, and was supported throughout by Hitachi. In the end, the project, described later in Yasunobu and Miyamoto 1985, was a huge success. The fuzzy controller achieved not only substantially higher precision in stopping at any designated point (so very important in Japan during rush hours), but it also made each stop considerably more comfortable by virtually eliminating the usual jerkiness during acceleration and braking. In addition, automating the whole subway system reduced energy consumption by about 10%. This success motivated the Tokyo municipal government to employ similar fuzzy control in newly constructed extensions of the rapidly growing Tokyo subway system in the mid-1990s.

A few days after the Sendai subway system switched from manual to fully automated operation based on its fuzzy controller, Tokyo hosted the Second IFSA World Congress. This was a unique opportunity for foreign participants at the congress to see the subway and get firsthand experience of its wonderful performance. It is not surprising that fuzzy control was the subject of many spirited discussions at coffee breaks and other informal gatherings at the congress. The congress was of course the prime place for presenting the best outcomes of fuzzy-set research in many countries at that time, and a large part of that research was application-oriented, with Japanese

---

[1] K. E. Peray and J. J. Wadell, *The Rotary Cement Kiln* (New York: Chemical Publ., 1972).

[2] The significance of this pioneering work was recognized in a specialized textbook on fuzzy control by Jantzen (2013), which was dedicated to the memory of Mamdani and Holmblad. His colleague, Jens-Jørgen Østergaard, wrote a foreword to the book that contains some interesting historical information: "The first experiments using a real cement kiln were carried out at the beginning of 1978 at an FL Smidth cement plant in Denmark. At this stage of the development work, the attitude of the management was skeptical, partly because of the strange name 'fuzzy'. Other names were suggested, but eventually, with an increased understanding by the management of the concept, it was decided to stay with the word fuzzy, a decision that has never been regretted since. ... In 1980, FL Smidth launched the first commercial computer system for automatic kiln control based on fuzzy logic. To date, hundreds of kilns, mills, and other processes have been equipped with high-level fuzzy control by FL Smidth and other suppliers of similar systems."

researchers being on the cutting edge. This all contributed to the development of highly positive attitudes toward fuzzy logic in Japan, which was clearly expressed when Lotfi Zadeh, in July 1989, received the prestigious Honda Prize, whose aim is "to honor technology that helps to develop a humane civilization."

These rapidly changing attitudes toward fuzzy logic in Japan motivated many Japanese industries to explore the potential of fuzzy-logic applications, which resulted in a surprising variety of innovative and sometimes unexpected applications that turned out to be highly successful commercially.[3] Most of them were applications of fuzzy control, such as control of industrial plants, elevators in high-rise buildings, traffic in large cities and other complex transportation situations, as well as various functionalities in automobiles (transmissions, brakes, cruise control, suspension, traction, and others). The most visible, as well as the most commercially successful, was the introduction of fuzzy control into consumer products of amazing variety, such as intelligent washing machines based on fuzzy-set representation of knowledge of an experienced user, camcorders with digital image stabilizers, rice cookers, refrigerators, air conditioners, and many other products. These products are the main factor responsible for the fuzzy boom in Japan mentioned below. One of the most sophisticated fuzzy controllers was designed, implemented, and successfully tested by Michio Sugeno for controlling the flight of a helicopter without a pilot by instructions expressed in natural language and sent to the controller via wireless communication from land. The controller and its development, which took about ten years, are described in some detail in the book *Selected Papers by M. Sugeno* edited by Nguyen and Prasad (1999, 13–43).

In parallel with developments of these many applications of fuzzy logic, software tools for such applications were also developed and a fair number of them gradually became commercially available, ranging from fuzzy expert shells and various other tools for specific applications to fully general tools, such as the Fuzzy Logic Toolbox developed for use with MATLAB (Sivanandam, Sumathi, and Deepa 2007).[4] It goes without saying that these tools greatly encouraged and enhanced further developments of fuzzy-logic applications.

All these highly visible and successful applications of fuzzy logic led eventually to a partnership of major Japanese industries, including initially 49 companies, with the Japanese government in supporting a large-scale research center dedicated to organized research on a wide range of engineering applications of fuzzy logic. The center opened in Yokohama in 1989 under the name "Laboratory for International Fuzzy Engineering" (LIFE), with the following stated purpose:[5]

---

[3] Important applications of fuzzy control that were developed in Japan in the early 1980s, including the subway system automatically operated by fuzzy control in Sendai, are described in the book edited by Sugeno (1985). This book also contains an annotated bibliography of works on fuzzy control (pp. 249–69), which represented the state of the art in the area of fuzzy control at that time. For commercial success of fuzzy logic, see also p. 441.

[4] MATLAB is a programming language developed by MathWorks, a privately held multinational corporation that specializes in mathematical computing software.

[5] From an eight-page brochure, *Introduction to the Laboratory for International Fuzzy Engineering Research*,

> In an attempt to realize a highly sophisticated information-intensive society, this laboratory is intended to vitalize fuzzy theory basic study, research on its efficient utilization by strengthening ties between industrial and academic circles, and to promote international technological exchange.

The overall spirit and results produced at LIFE during the initial three-year period (1989–91) are documented in a book edited by Anca Ralescu (1994), who was a visiting scholar at LIFE at that time.

Another research center was founded under the name "Fuzzy Logic Systems Institute" (FLSI) in Iizuka on Kyushu Island in southwest Japan in 1990, exactly one year after LIFE, primarily through the efforts and vision of Takeshi Yamakawa. The Center was initially supported by 13 companies and was loosely connected with the Kyushu Institute of Technology, where Yamakawa was a professor. A special feature of this institute was its focus on research to develop neuro-fuzzy systems and other hybrid systems and especially hardware for supporting these systems.

The extraordinary period in the history of fuzzy logic from about the mid-1980s to the mid-1990s, during which the broad applicability of fuzzy logic was firmly established in Japan, is often referred to as the "fuzzy boom." It is described in detail in a well-researched book by McNeill and Freiberger (1993) and also summarized in a paper by Schwartz et al. (1994). Most of the industrial applications developed and implemented during this period are described in the books edited by Hirota (1993) and Hirota and Sugeno (1995).

As one would expect, the fuzzy boom has affected in a positive way attitudes toward fuzzy logic in many other countries. In the United States, for example, the paper by Bernard (1988) demonstrated convincingly these changing attitudes. It was the first publication in the United States in which fuzzy control was taken seriously by the community of classical control theorists and compared fairly with conventional control. The author describes in some detail a fairly extensive study conducted at MIT. The purpose of the study was to compare fuzzy controllers with conventional controllers in controlling the MIT research reactor under both steady-state and transient conditions.[6] The following quotation captures the overall conclusions from the study (Bernard 1988, 10–11):

> It is evident from the successful applications of the fuzzy, rule-based approach to the control of cement kilns and other ill-defined systems that the technology is practical and beneficial for situations in which there is no accurate plant model. The question that the research being conducted at MIT seeks to address is whether or not the fuzzy, rule-based approach is also of value if the process is well characterized. In the course of the MIT work, several disadvantages to the rule-based methodology became apparent. First, there are few guidelines for the implementation of this technology. Second, there is the nagging question of the completeness of the rule base. Third, there is the difficulty of calibrating and updating rule-based controllers. Despite these disadvantages, there do appear to be at least two roles for which the fuzzy, rule-based approach is eminently suited. These are:

---

prepared by LIFE in 1989. The brochure describes the mission and organizational structure of LIFE.

[6] The author, John A. Bernard, received his doctoral degree from MIT in 1984 for research work on classical control theory and was principal engineer at the MIT Research Reactor Laboratory responsible for control studies, such as the one described in this paper.

1. Rule-based controllers are generally more robust than analytic ones. Hence, they should be used in a support role, with the objective of bringing a plant to a safe condition should the analytic controller fail for some reason, such as loss of a sensor. . . .
2. The rule-based approach could be used to improve the man-machine interface. One of the disadvantages of the analytic approach is that human operators do not think in terms of formalized control laws and, therefore, may not understand the behavior of an automated system. . . .

In summary, the rule-based and analytic approaches should not be viewed as competing technologies—each has certain weaknesses and certain strengths. These two methodologies should be used in tandem to create truly robust control systems.

We should add that virtually all the disadvantages of fuzzy controllers, which were correctly identified in Bernard's paper, were later alleviated by further research. The author was in fact aware of this possibility: "The fuzzy, rule-based approach is at a disadvantage because, being an emerging technology, standardized methods for its application have not yet been developed" (Bernard 1988, 5).

One positive outcome of the fuzzy boom was that industries as well as governments in some countries, not only in Japan, became more receptive to supporting research on fuzzy logic, especially application-oriented research. Moreover, some major professional societies, such as the IEEE, embarked on various activities (conferences, publications, etc.) intended to encourage research on applications of fuzzy logic. However, in spite of all these developments, fuzzy logic was still occasionally criticized for lack of useful applications, as is illustrated by the following quotation from a book by the well-known mathematician, Saunders Mac Lane, which was published during the middle of the fuzzy boom (Mac Lane 1986, 439–40):

Not all outside influences [on mathematics] are really fruitful. For example, one engineer came up with the notion of a *fuzzy set*—a set X where a statement $x \in X$ of membership may be neither true nor false but lies somewhere in between, say between 0 and 1. It was hoped that this ingenious notion would lead to all sorts of fruitful applications, to fuzzy automata, fuzzy decision theory and elsewhere. However, as yet most of the intended applications turn out to be just extensive exercises, not actually applicable; there has been a spate of such exercises. After all, if all Mathematics can be built from sets, then the whole lot of variant (or, should we say, deviant) Mathematics can be built by fuzzifying these sets.

This statement, which seems to express honestly Mac Lane's beliefs about fuzzy logic, would have been perfectly reasonable in the 1960s, or even in the 1970s, but certainly not in 1986, when the reality regarding applicability of fuzzy logic was already fundamentally different from its portrayal in Mac Lane's statement.

During the 1990s, it was increasingly recognized that combining fuzzy logic with artificial neural networks could enhance applications of fuzzy logic. On the one hand, neural networks added some useful capabilities to fuzzy systems, such as adaptability, learning of membership functions and operations from examples, and solving efficiently some inverse problems emerging from various applications of fuzzy logic. On the other hand, fuzzifying neural networks resulted in their more flexible and robust counterparts. These hybrid systems, which became known as *neuro-fuzzy systems*, were initially introduced and examined in a pioneering book by Kosko (1992a). Three years later, von Altrock (1995) described over 30 practical case studies of neuro-fuzzy systems and provided simulation software to experiment with such systems.

In the late 1990s, it was also recognized by numerous researchers (Sanchez, Shibata, and Zadeh 1997) that combining fuzzy logic with genetic algorithms and related computational methods, later subsumed under the name *evolutionary computation*, could enhance applications of fuzzy logic as well. The book by Cordón et al. (2001) covers this subject quite comprehensively, including systems in which fuzzy logic is combined with both neural networks and evolutionary computation, both of which have the potential to further enhance some applications of fuzzy logic.

Observing how these combinations of fuzzy logic with other methodological tools—so-called hybrid methodologies—enhance the applicability of fuzzy logic, Zadeh increasingly felt that a single generic name was needed for capturing the essence of these hybrid methodologies, and he eventually coined for them the name *soft computing*. He first suggested this name early in the 1990s at various conference presentations, but made it more visible in 1994 by publishing two papers fully devoted to discussing the idea of soft computing (Zadeh 1994a,b). He described it as follows (Zadeh 1994a, 77–78):

> In traditional—hard—computing, the prime desiderata are precision, certainty, and rigor. By contrast, the point of departure in soft computing is the thesis that precision and certainty carry a cost and that computation, reasoning, and decision making should exploit—wherever possible—the tolerance for imprecision and uncertainty.
>
> A case in point is the problem of parking an automobile. Most people are able to park an automobile quite easily because the final position of the vehicle and its orientation are not specified precisely. If they were, the difficulty of parking would grow geometrically with the increase in precision and eventually would become unmanageable for humans. What is important to observe is that the problem of parking is easy for humans when it is formulated imprecisely and difficult to solve by traditional methods because such methods do not exploit the tolerance for imprecision.
>
> The exploitation of the tolerance for imprecision and uncertainty underlies the remarkable human ability to understand distorted speech, decipher sloppy handwriting, comprehend nuances of natural language, summarize text, recognize and classify images, drive a vehicle in dense traffic and, more generally, make rational decisions in an environment of uncertainty and imprecision. In effect, in raising the banner of "Exploit the tolerance for imprecision and uncertainty," soft computing uses the human mind as a role model and, at the same time, aims at formalization of the cognitive processes humans employ so effectively in the performance of daily tasks. . . .
>
> At this juncture, the principal constituents of soft computing are fuzzy logic (FL), neural network theory (NN), and probabilistic reasoning (PR), with the latter subsuming belief networks, genetic algorithms, parts of learning theory, and chaotic systems. In this triumvirate, FL is primarily concerned with imprecision, NN with learning, and PR with uncertainty. . . . FL, NN, and PR are complementary rather than competitive. For this reason, it is frequently advantageous to employ FL, NN, and PR in combination rather than exclusively.

Research communities in the areas subsumed under soft computing have by and large endorsed this new concept. In fact, soft computing is now viewed as a separate research area, which has been growing rapidly since the mid-1990s in terms of virtually every relevant measure. Here are a few examples: A journal dedicated to soft computing—actually entitled *Soft Computing*—has been published by Springer since 1997; a book series entitled *Studies in Fuzziness and Soft Computing* has been published by Physica-Verlag since 1996; the European Centre for Soft Computing was founded in Mieres in northwest Spain in 2006 and, among its activities, it began publishing a journal entitled *Archives for the Philosophy and History of Soft Computing* in 2013; and EUSFLAT has published *Mathware & Soft Computing* magazine since 1994.

We should also mention that soft computing is a methodological base for designing and building systems known as *intelligent systems*. In fact, the research area of intelligent systems emerged around the mid-1990s in close connection with the genesis of soft computing. Intelligent systems are typically defined as human-made systems that are capable of achieving highly complex tasks in *human-like* intelligent ways. By taking the qualifier "human-like" seriously and employing soft computing, the area of intelligent systems is clearly distinguished from the mainstream in the broader area of artificial intelligence, which has been largely committed to traditional computing and has not paid so much attention to implementing various intelligent tasks in a *human-like way*. Two representative early publications in the area of intelligent systems are the books by Albus and Meystel (2001) and Meystel and Albus (2002).

In the following nine sections of this chapter, we describe representative examples of fuzzy-logic applications in nine broad areas of human affairs and for each area supplement them with appropriate references.

6.3 Engineering

Engineering was certainly not among the areas in which the need for fuzzy logic was initially anticipated. Paradoxically, it was the first area in which the utility of fuzzy logic was established for the first time, primarily due to the idea of fuzzy control—an idea radically different from that of classical control. While classical control is based on a mathematical model of the process to be controlled, fuzzy control is based on a mathematical representation of knowledge of experienced human operators—domain experts—expressed by statements in a natural language whose meaning is expressed in turn by fuzzy if-then rules via precisiation (Zadeh 1984).

6.3.1 Fuzzy control

In order to illustrate basic ideas of fuzzy control, we first describe in some detail a very simple fuzzy controller of the Mamdani type mentioned in section 6.2. Although other types of fuzzy controllers were developed later, most notably the one introduced by Takagi and Sugeno (1985), Mamdani type of fuzzy controllers is not only historically significant, but it also continues to be highly popular, primarily due to its simplicity and good performance.[7]

The fuzzy controller we have chosen to describe is the one whose purpose is to stabilize an inverted pendulum. This control problem may be described as follows. A movable pole is attached to a vehicle through a pivot. It is assumed that the pole and the vehicle can move only to the right or left. The control problem is to keep the pole (inverted pendulum) in the vertical position, which is clearly a highly unstable state. The controller is capable of moving the vehicle appropriately to counter any disturbances with the aim of keeping the pole in a vertical position. Two input variables and one output variable are involved in this control problem. One of the input variables is the angle (positive or negative) between the actual position of the pole and its desirable vertical position, i.e. an error, measured by an appropriate angle sensor. We denote this variable by x. The second input variable is the angular speed (rate of change of variable x), which we denote by y. The output variable, denoted here by z, represents actions taken by the controller. Each value of this variable represents a particular velocity of the vehicle (positive or negative), which is needed to correct the error x. It must be converted to a suitable physical quantity (typically an electric current for a motor) that governs the desired movement of the vehicle. For the sake of simplicity, we do not address this and other implementation details.

We chose the problem of stabilizing an inverted pendulum to illustrate basic ideas of fuzzy controllers because this problem is easy to understand and yet, it is a challenging problem under severe disturbances or various unfavorable physical characteristics of the inverted pendulum (e.g. very small length of the pole). For these reasons, its pedagogical value was recognized a long time ago. Controllers of inverted pendulums have routinely been included in the inventory of laboratory tools for teaching classical control theory. Their fuzzy counterparts have also been popular for demonstrating and teaching basic principles of fuzzy control since the first fuzzy controller of an inverted pendulum, designed and implemented by Yamakawa (1989), was publicly demonstrated at the Second IFSA Congress in Tokyo in 1987. It was a great success, as it showed the remarkable robustness of fuzzy control under severe environmental disturbances. Using just seven if-then rules, the controller worked perfectly even after Yamakawa placed a flower atop the controlled pole. Later, he demonstrated that his fuzzy controller could handle even greater challenges. For example, he attached a small platform to the top of the pole and placed a wine glass on it containing a liquid or even a live mouse. The fuzzy controller perfectly countered the turbulence of the liquid as well as the erratic movements of the mouse.

---

[7] When Ebrahim H. Mamdani unexpectedly passed away in January 2010, the fuzzy-logic community chose to pay homage to him for his groundbreaking work on fuzzy control by publishing a book edited by Trillas et al. (2012). This is a significant book from the historical point of view as it captures a great deal of personal information (recollections, correspondence, and the like) from Mamdani's contemporaries about his critical role in fuzzy logic during the 1970s, a period that was not particularly hospitable to fuzzy logic.

Classical controllers typically employ mathematical models of systems to be controlled, from which the desired control actions are determined. A model for an inverted pendulum may be expressed as a set of nonlinear differential equations, such as

$$I\ddot{x} = VL\sin x - HL\cos x,$$
$$V - mg = -mL(\ddot{x}\sin x + \dot{x}^2\cos x),$$
$$H = m\ddot{p} + mL(\ddot{x}\cos x - \dot{x}^2\sin x),$$
$$U - H = M\ddot{p},$$

where the symbols denote the following: $x, \dot{x}, \ddot{x}$—error and its first and second derivatives; $2L$—length of the pendulum; $p, \ddot{p}$—position of the vehicle and its second derivative; m—mass of the pendulum; M—mass of the vehicle; H—horizontal force at the pivot; V—vertical force at the pivot; U—driving force given to the vehicle; $I = \frac{1}{3}mL^2$—moment of inertia (Yamakawa 1989).

Fuzzy controllers, on the contrary, do not require mathematical models of systems to be controlled. Instead, they operate in terms of fuzzy if-then rules expressed in natural language and formulated in terms of states of appropriate linguistic variables defined for input and output variables of the controller. The fuzzy rules are virtually independent of the various physical parameters, such as the mass or length of the pendulum, which play essential roles in the differential equations. Whenever some of them change, corresponding parameters in the differential equations must be properly adjusted, but the if-then rules remain basically the same. Fuzzy controllers are thus far more robust than the classical ones. In fact, fuzzy controllers are capable of working even under severe control circumstances, such as the control of a very short pendulum under heavy and chaotic disturbances, provided that their operational speed is adequate. This indicates the importance of specialized fuzzy hardware (section 6.3.2).

In our particular example, input variables of the fuzzy controller are x and y, and the only output variable is z. Values of each of these variables range over some interval of real numbers of the form $[-v, v]$, where the actual value of v is in general distinct for the three variables. An example of a possible linguistic variable defined for each of the three variables by appropriate values of v, say values $\hat{x}$, $\hat{y}$, and $\hat{z}$, is shown in figure 6.1. That is, the ranges of the three variables are $X = [-\hat{x}, \hat{x}]$, $Y = [-\hat{y}, \hat{y}]$, $Z = [-\hat{z}, \hat{z}]$, and their joint space is $U = X \times Y \times Z$. The seven triangular-shaped fuzzy sets represent for each of the three variables the following linguistic terms: AZ—Approximately Zero; NS—Negative Small; NM—Negative Medium; NL—Negative Large; PS—Positive Small; PM—Positive Medium; and PL—Positive Large.

With seven linguistic terms for each of the two input variables, the total number of possible fuzzy if-then rules is 49 (seven times seven). The matrix in table 6.1 conveniently defines these rules. Normally, these rules are elicited from experienced

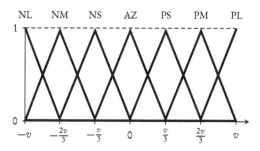

Figure 6.1: A linguistic variable employed in the discussed fuzzy controller.

Table 6.1: Fuzzy if-then rules for a fuzzy controller of an inverted pendulum.

	z	NL	NM	NS	AZ	PS	PM	PL
	NL	NL	NL	NL	NL	NM	NS	AZ
	NM	NL	NL	NL	NM	NS	AZ	PS
	NS	NL	NL	NM	NS	AZ	PS	PM
x	AZ	NL	NM	NS	AZ	PS	PM	PL
	PS	NM	NS	AZ	PS	PM	PL	PL
	PM	NS	AZ	PS	PM	PL	PL	PL
	PL	AZ	PS	PM	PL	PL	PL	PL

human operators or constructed from data obtained by observing their actions. In our example, as was already mentioned, the rules are highly intuitive and can be fairly easily determined by common sense. Consider, for example, all rules in which the two input variables are approximately equal in terms of levels (low, medium, high), but one of them is positive and the other negative. They clearly approximately correct one another, so no positive or negative action is needed; hence the inferred approximate state of the output variable should be AZ. Similarly, when both input variables are AZ, the inferred approximate state of the output variable should also be AZ. Consider now, for example, that one of the input variables is NL. Then, clearly, the largest possible move of the vehicle to the left is needed unless the second input variable is positive at some level; hence, the output variable should be NL if the second variable is NL, NM, NS, or AZ. Similarly, the remaining entries in table 6.1 can be determined by common sense.

Each of the possible 49 if-then rules in a fuzzy controller of the inverted pendulum has the following canonical form:

$$\text{Rule } k: \text{if } x \text{ is } A_k \text{ and } y \text{ is } B_k, \text{ then } z \text{ is } C_k, \tag{6.1}$$

where A_k, B_k, and C_k ($k \in \mathbb{N}_{49}$) are fuzzy sets defined on sets X, Y, and Z, respectively, which represent the meaning of the linguistic terms NL, NM, NS, AZ, PS, PM, and PL, and their combinations correspond to table 6.1.

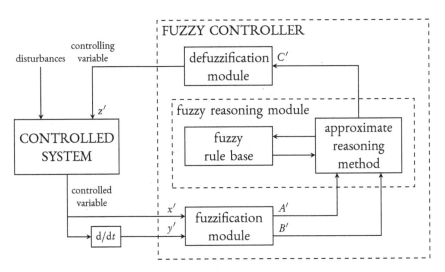

Figure 6.2: A block diagram for the discussed fuzzy controller.

Fuzzy controllers for control problems similar to the one of an inverted pendulum consist of three modules: a fuzzification module, a fuzzy reasoning module, and a defuzzification module. Interconnections between these modules and the controlled system are shown in figure 6.2. The controller operates by repeating the same cycle consisting of the following three steps:

Step 1. Measurements of input variables x and y in each particular cycle, which we denote by x' and y', are processed by the fuzzification module. The purpose of fuzzification, which is optional, is to represent each measured real number, say x', more realistically by taking into account measurement uncertainties. This is particularly important when sensors employed in the controller are of low quality (not very precise) or are subject to deterioration under strenuous conditions. Via fuzzification, the real number x' is replaced with a fuzzy number A' that is supposed to characterize the concept of *approximately x'*, as is exemplified by a triangular-shaped fuzzy number whose level-cut representation has the form

$$^{\alpha}A' = [x_j + \varepsilon(\alpha - 1), x_j + \varepsilon(1 - \alpha)],$$

where $\varepsilon \geq 0$ is a constant chosen by the designer of the controller. This fuzzy number expresses in this example how much uncertainty is introduced by the chosen fuzzification. The uncertainty increases with the increasing value of ε. When $\varepsilon = 0$, clearly, $^{\alpha}A' = [x_j, x_j] = \{x_j\}$ for all $\alpha \in [0, 1]$, so in this case no fuzzification is employed. Fuzzy numbers A' defined in this way thus also include the option when no fuzzification is chosen. In general, a properly designed fuzzification makes the controller more robust.

Step 2. The two outcomes of the fuzzification module in each cycle, fuzzy numbers A' and B', are applied to the fuzzy reasoning module, which consists of two components: (a) *fuzzy rule base*—a set of chosen fuzzy if-then rules of the form

(6.1); (b) *approximate reasoning method* (or *inference method*) of some particular type chosen by the designer of the controller. In our example, we chose to use the Mamdani interpolative reasoning method. The first step of this method in each cycle is to determine for each rule (6.1) the degree to which the fuzzy sets A' and B', which represent the input values x' and y', match the antecedent of the rule. For rule k, this degree, m_k, is computed as the minimum of the degrees a_k and b_k to which A' matches A_k and to which B' matches B_k, respectively. These degrees are defined by

$$a_k = \sup_{x^* \in X} \min(A'(x^*), A_k(x^*)) \text{ and } b_k = \sup_{y^* \in Y} \min(A'(y^*), A_k(y^*)).$$

Hence, a_k may be interpreted as the truth degree of the proposition "there exists value $x^* \in X$ to which both A' and A_k apply," and similarly for b_k. Since

$$m_k = \min(a_k, b_k), \tag{6.2}$$

m_k is indeed interpretable as the truth degree of the proposition "the representations A' and B' of the input values x' and y' match A_k and B_k." Note that m_k is also called the degree of applicability of rule k or the degree to which rule k fires. For each k, m_k is then used to compute the actual output C'_k of rule k that corresponds to the input values by cutting off the consequent C_k of the rule at the level m_k, i.e.

$$C'_k(z) = \min(m_k, C_k(z))$$

for each $z \in Z$. The fuzzy sets C_k resulting this way represent the contributions of the respective rules to the overall conclusion of the inference in the given cycle. These contributions must then be properly aggregated to obtain a fuzzy set C'— the overall conclusion of the inference in this cycle. In Mamdani's method, C' is obtained as the standard union of the contributions C'_k of the particular rules, i.e.

$$C'(z) = \max_k C'_k(z)$$

for each $z \in Z$.

It is worth mentioning that Mamdani's method may alternatively be described in terms of the compositional rule of inference (section 3.6) as follows: C' is exactly the result of the sup-min composition of $A' \times B'$ and the fuzzy relation R, where $A' \times B'$ is the Cartesian product of A' and B' given by $(A' \times B')(x, y) = \min(A'(x), B'(y))$, and R is the standard union of fuzzy relations R_k on $X \times Y \times Z$ defined by $R_k(x, y, z) = \min(A_k(x), B_k(y), C_k(z))$ for every $x \in X$, $y \in Y$, and $z \in Z$. Hence, R is defined by $R(x, y, z) = \max_k R_k(x, y, z)$ and may be regarded as a fuzzy relation representing the whole fuzzy rule base. Moreover, the inference method may be represented graphically. This fact is important because the graphical representation makes it possible to understand the method easily in intuitive terms by engineers and other researchers who employ fuzzy-rule-based systems. This factor contributed to the rapid development of various products based on fuzzy logic during the boom in Japan described in section 6.2. Figure 6.3 provides a graphical representation of Mamdani's inference based on two of the 49 rules of the form (6.1) depicted in table 6.1, given that the fuzzy sets involved are defined as in figure 6.1 and that the input fuzzy sets A' and B' are just singletons representing the values x' and y', respectively. In the first rule, A_1, B_1, and C_1 represent the meaning of the linguistic terms PS, NM, and NS, while in the second, A_2, B_2, and C_2 represent PS, NS, and AZ, respectively.

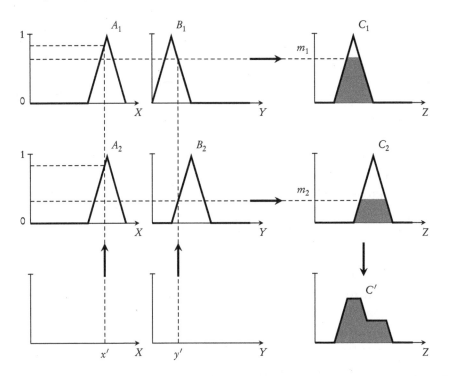

Figure 6.3: Mamdani's inference method. For rule 1, represented in the uppermost part, the inputs x' and y' determine degrees of match with the rule premises A_1 and B_1, the smaller of which, m_1, is used to cut off the consequent C_1. The same procedure applies to cutting off C_2 in rule 2. The cut-off fuzzy sets, represented by the gray subareas of C_1 and C_2, are then combined to yield the output fuzzy set C'.

Step 3. Fuzzy set C' is processed by the defuzzification module. Defuzzification is not optional since the fuzzy set C' that results from the reasoning process does not describe a particular action to be performed by the fuzzy controller. The role of defuzzification is to convert this fuzzy set to a unique action, z', which in some sense best represents the fuzzy set C'. The selection depends of course on the chosen criterion of goodness and, hence, it is not unique.[8] In his interpolation reasoning method, Mamdani employed one of the earliest and still widely used defuzzification

---

[8] The problem of defuzzification was introduced and discussed (not under this name) for the first time by Zadeh (1968b) in the broad context of fuzzy algorithms. It reappeared in the mid-1970s, under the name "output selection process," in the narrower context of fuzzy controllers. The term "defuzzification" seems to have been used for the first time by Larkin (1985). Numerous defuzzification methods have been suggested and investigated over the years, including some parameterized classes of methods. A comprehensive survey describing the history and development of these methods is covered in Martin and Klir 2007.

methods, which is called *center of area* or *centroid* method, in which the defuzzified value, $d(C')$, of a fuzzy set C' on $\mathbb{R}$ is defined as

$$d(C') = \frac{\int_{S(C')} C'(z)zdz}{\int_{S(C')} C'(z)dz},$$

where $S(C')$ denotes the support of fuzzy set C'. This definition is of course applicable under the assumptions that function C' is integrable and that $\int_{S(C')} C'(x)dx > 0$, which are virtually always valid in fuzzy controllers. The value $d(C')$ may be interpreted as the value for which the area under the graph of function C' is divided into two equal subareas or, alternatively, as the expected value of the function C'. In implementations on digital computers, the following discrete variant of the formula is usually used: $d(C') = \frac{\sum_{i=1}^{p} C'(z_i)z_i}{\sum_{i=1}^{p} C'(z_i)}$, where $Z = \{z_1, \ldots, z_p\}$ is the discretized universe. The action of the described Mamdani-type fuzzy controller of an inverted pendulum in the given cycle, z', is thus defined as $z' = d(C')$.

Our example of the fuzzy control of an inverted pendulum is rather rudimentary, expressing an early stage in the development of fuzzy controllers. For example, it does not include the requirement that the position of the vehicle also be stabilized. Two additional input variables were later added to the controller to take care of this requirement: a variable expressing the distance (positive or negative) between the actual position of the vehicle and its desired position, and a variable expressing the velocity of the vehicle (positive or negative). Moreover, in the 1990s it was shown in numerous ways that fuzzy controllers are also capable of controlling double and triple inverted pendulums, as, for example, in Cheng et al. 1996 and Zhang et al. 1993.[9] Later, Zhang, Wang, and Li (2012) showed that they are even capable of controlling quadruple inverted pendulums.

After the very successful applications of fuzzy control in Japan in the 1980s and 1990s, which are described in section 6.2, research on fuzzy control, both theoretical and applied, virtually exploded. The following are some of the advances made over the years by theoretical research.

Numerous other types of fuzzy controllers, including some variants of Mamdani's type, have been suggested and explored (Foulloy and Galichet 1995). Among them, the most visible are controllers based on fuzzy logic inference, such as the compositional rule of inference described in section 3.6, and another type of interpolative fuzzy controllers introduced by Takagi and Sugeno (1985). The latter fuzzy controllers are referred to as Takagi-Sugeno, or TS, controllers. Considering again the control of an inverted pendulum as an example, in order to see clearly the similarities and differences between the two types of interpolative fuzzy controllers, the

---

[9] A double inverted pendulum is essentially an inverted pendulum that consists of two poles connected to each other via a pivot, one of which is also connected to a vehicle in the same way as in a single inverted pendulum. The aim of the controller is now to keep the connected poles in the vertical position, which is of course a much harder task. When inverted pendulums consist of three or four connected poles, they are referred to as triple or quadruple inverted pendulums, respectively.

if-then rules in the TS controller have the following form in this example:

Rule k: if x is A_k and y is B_k, then z is $f_k(x,y)$,

where f_k is usually a linear function. The action z of the TS controller in the given cycle for measured input values x' and y' is expressed directly by the formula

$$z = \frac{\sum_k m_k f_k(x',y')}{\sum_k m_k},$$

where m_k is defined by (6.2) as in Mamdani's approach.

The amazingly successful applications of fuzzy controllers in the 1980s were often explained on various intuitive grounds. However, as was pointed out by some skeptics, no underlying theoretical explanation was available for this success. Some researchers working in the area of fuzzy systems and fuzzy control began to investigate this issue mathematically in the early 1990s. Through this mathematical research, it was eventually established that fuzzy controllers have the capability to approximate functions defined on compact subsets of the n-dimensional Euclidean space $\mathbb{R}^n$ to any desired degree of accuracy. That is, fuzzy controllers or, more generally, fuzzy systems equipped with defuzzification were established as *universal approximators*. Some initial results regarding this theoretical problem, which appeared in the literature almost simultaneously (Kosko 1992b, Wang 1992, Buckley 1992), showed only that fuzzy controllers of some rather special types are universal approximators. Later, the universal approximation was established for virtually all known types of fuzzy controllers: for example, Ying (1994) and Castro (1995) proved it for Mamdani-type controllers, Ying (1998) proved it for Takagi-Sugeno controllers with linear functions as rule consequents, and Klawonn and Novák (1996) proved it for controllers based on fuzzy logic inference with Łukasiewicz implication. It was also established that fuzzy systems with defuzzification and feedforward neural networks could approximate each other with arbitrary precision.[10] A fairly comprehensive survey of issues, results, and literature regarding universal approximation by fuzzy controllers is given on pp. 135–95 in chapter 4 of Nguyen and Sugeno 1998.

After the surprising success of fuzzy controllers in the 1980s and 1990s, research on fuzzy controllers, both theoretical and experimental, became more focused. It has been primarily concerned with various theoretical or methodological issues that have gradually emerged from attempts to apply fuzzy controllers to increasingly more challenging practical control problems. Among these issues are, for example, the issue of how to deal with the complexity of control problems that involve so many variables that the number of possible if-then rules is too large to be directly handled, how to analyze stability of fuzzy controllers, how performance of fuzzy controllers can be optimized, how to design fuzzy systems with adaptive capabilities, and the like. As is quite typical, each of these issues has been approached in several fundamentally different ways, and various methods have been proposed and

---

[10] See Buckley and Hayashi 1993 for an overview.

developed within each of these approaches. This captures in a nutshell the rapidly growing research on fuzzy controllers over the last twenty years or so. The body of literature in this rather special area of applied fuzzy logic is now so extensive that it is possible to mention here only a few carefully selected books that best capture the current state of the art in this area.

The monograph by Yager and Filev (1994) was the first comprehensive book on fuzzy control and it remains one of the best, even though it is no longer fully up-to-date. Among more recent monographs are the books by Margaliot and Langholz (2000), Piegat (2001), and Michels et al. (2006). Many edited books on fuzzy control have also been published, each of which attempts to survey some of the advancements in this rapidly growing area. Together, these books characterize the current state of the art in the area of fuzzy control quite well. However, recently, no single, comprehensive survey book of these advancements has been published. The last comprehensive book on fuzzy control was the one edited by Nguyen and Sugeno (1998), but that, unfortunately, is now out of date. An interesting phenomenon is the growing number of books that specialize on various subareas of fuzzy control, such as the book by Wang (1994), which was the first book specializing on adaptive fuzzy control, the more recent book by Angelov (2002) that specializes on flexible fuzzy adaptive systems, the book edited by Aracil and Gordillo (2000) that focuses on stability issues of fuzzy controllers, and the book by Zhang, Phillis, and Kouikoglou (2005) that specializes on control of queuing systems. This seems to be an indication of the growing scope as well as the depth of the area of fuzzy control. A more recent book by Siddique (2014), which comprehensively covers hybrid approaches to intelligent control systems based on fuzzy logic, neural networks and genetic algorithms, should also be mentioned. Finally, a few good examples of textbooks on fuzzy control, chosen from a fairly large list, are those by Passino and Yurkovich (1998), Nguyen et al. (2003), Lilly (2010), and Jantzen (2013). We should add that methods developed for fuzzy control can also be employed (with or without defuzzification) for qualitative systems modeling in the sense discussed by Sugeno and Yasukawa (1993). This is a universal methodological tool with broad applicability in many areas.

6.3.2 Other engineering applications

Fuzzy control has been undoubtedly the major engineering application of fuzzy logic, but it is by no means the only one. In this subsection, we survey some other significant applications. We begin with one that is closely connected with fuzzy control—specialized computer hardware for fuzzy logic. Such engineering products, *fuzzy hardware* for short, in their many forms have made substantial contributions to the success of fuzzy controllers, primarily because hardware implementations of fuzzy controllers are always substantially faster than their software counterparts. The processing speed of fuzzy controllers, which is usually measured in terms of the number of fuzzy logic inferences per second (FLIPS), is critically important for their performance. Increasing the number of FLIPS makes fuzzy controllers more robust and, in addition, increases their scope of applicability. Consider, for example, once more the problem of controlling an inverted pendulum. We generally know from our common-sense experience that a short pole is more difficult to balance than a long pole and that the reason for this difference is that balancing a short pole requires reacting faster than balancing a long one. This common-sense experience can be directly interpreted in terms of fuzzy controllers: the shorter the pole, the faster the controller must be to control it. Similarly, the stronger the disturbances, the faster the controller must be to counter them. In his demonstration of the first fuzzy controller for an inverted pendulum at the Second IFSA World Congress in Tokyo, Yamakawa was able to show the incredible robustness of fuzzy controllers, to the complete amazement of the audience, primarily because his controller was implemented in analog hardware (of Yamakawa's own design) that made the controller very fast.

The idea of building specialized computer hardware for fuzzy logic is almost as old as the idea of fuzzy logic itself. In 1966, just one year after publication of Zadeh's first paper on fuzzy sets, Peter N. Marinos suggested this idea in some detail in a memorandum at Bell Telephone Laboratories in Holmdel, New Jersey (Marinos 1966). He also published it in an abridged version three years later in Marinos 1969, when he was at Duke University. However, neither Bell Telephone Laboratories nor Marinos have pursued this idea further—it seems that it had emerged too early. However, the idea reappeared some twenty years later in the narrower context of fuzzy controllers during the fuzzy boom in Japan.

The pioneers of fuzzy hardware were Takeshi Yamakawa and Tsutomu Miki in Japan, and Masaki Togai and Hiroyuki Watanabe in the United States. They developed their versions of fuzzy hardware almost simultaneously around the mid-1980s and described them in Yamakawa and Miki 1986 and Togai and Watanabe 1986a,b. The former has been manufactured by Omron Corporation, the latter by Togai InfraStructure, a California-based company founded by Masaki Togai in 1987. After this pioneering work, a great number of hardware products to support fuzzy control and other applications of fuzzy logic have been developed and manufactured, reflecting emerging needs (e.g. to support neuro-fuzzy systems or fuzzy systems employing type-2 fuzzy sets and the like), as well as advances in technological processes for manufacturing integrated circuits. Fuzzy hardware is surveyed in books edited by Kandel and Langholz (1998) and Teodorescu, Jain, and Kandel (2001); an overview of more recent developments can be found in a paper by Zavala and Nieto (2012).

As already mentioned, the primary reason for using fuzzy hardware is to increase the speed of fuzzy controllers or other processors employing fuzzy logic. This is possible primarily due to the capability of fuzzy hardware to perform operations on fuzzy if-then rules in an analog way. Analog hardware for fuzzy logic employs basically a modeling relation between a mathematical system—a mathematical description of a chosen fuzzy reasoning process employing if-then rules—and a physical system—a fuzzy hardware component designed to model the mathematical system. The term "analog" is borrowed from the area of analog computers, which are based on a similar modeling relation, in which the mathematical systems are typically differential equations. As is well known, analog computers are considerably faster in solving differential equations than digital computers. However, they have two shortcomings in comparison with digital computers. One is that their precision is ultimately limited by the unavoidable limit of physical measurement. Digital computers do not have, at least in principle, such limitations. The second shortcoming of analog computers is their seriously restricted flexibility. Although analog fuzzy processors are not totally rigid, they are programmable only within a restricted application domain (say a particular class of fuzzy controllers) in the same way as analog computers are programmable within their limited scope of applications (a class of differential equations). In applying analog computations to fuzzy reasoning, fortunately, measurement limitations are in many cases not of major concern. However, their limited flexibility is a serious shortcoming even in this application domain. These considerations led to the development of the following two fundamentally different types of processors employing fuzzy hardware:

1. Fuzzy hardware is employed in a standard digital computer to add to its instruction set suitable clusters of operations on fuzzy sets. In this arrangement, the overall operation of the computer remains basically the same, but its speed in dealing with fuzzy-set applications is moderately increased.
2. Fuzzy hardware is employed in an external coprocessor, connected to the main computer, which executes all operations involving fuzzy sets. In this arrangement, the speeds of fuzzy-set applications is substantially increased due to hardware-supported parallel processing in the coprocessor, but the coordination between the two processors is a limiting factor.

Whenever the predominant aim is to build a high-speed fuzzy processor, the most appropriate and also most natural way is to implement it fully via analog fuzzy hardware. Fuzzy processors based on analog hardware have been the trademark of Yamakawa, Miki and the Omron Corporation. As early as in 1993, they described analog fuzzy processors that achieved the speed of one mega-FLIPS (Miki et al. 1993). The whole process of designing, implementing, and testing analog fuzzy controllers is described in detail in the book by Dualibe, Verleysen, and Jespers (2003).

We turn now to the established role of fuzzy logic in *engineering design*, which may be broadly described as follows. Experienced designers know that early stages in engineering design are far more important than later stages. If good alternatives in a design problem are eliminated early in the design process, they cannot be recovered in later stages. Moreover, if early decisions are required to be precise, then it is virtually impossible to determine how good or bad each decision may be in terms of the subsequent design steps and, eventually, in terms of the final design. Experienced designers are aware of this difficulty and begin the design process with complete but highly imprecise descriptions, expressed often in natural language, of the desired artifact to be designed. As the design process advances from the formative stages to more detailed designs, the descriptive imprecision at each stage allows the designer to evaluate the consequences of possible decisions at previous stages and eliminate those leading to undesirable consequences. That is, imprecision is gradually reduced. At the end of the whole design cycle, all imprecision is virtually eliminated except for unavoidable tolerances resulting from imperfections in the manufacturing process.

In the late 1980s, it was recognized, primarily in the context of mechanical engineering design at the California Institute of Technology (Caltech), that fuzzy set theory was eminently suited for representing mathematically the above-described way of dealing with engineering design problems by experienced designers.[11] The actual work on developing this idea started at Caltech in the early 1990s (Wood, Otto, and Antonsson 1992). In spite of initially being greeted with skepticism, this work has turned out to be successful and the usefulness of fuzzy set theory in engineering design is now generally accepted (Sebastian and Antonsson 1996, Antonsson and Sebastian 1999).

Among all engineering disciplines, *civil engineering* was the first in which some applications of fuzzy logic were developed. The relevance of fuzzy logic to civil engineering was recognized quite early (Blockley 1977, 1979), and the civil engineering community has been surprisingly active since the early 1980s in exploring how to utilize it for dealing with numerous civil engineering problems.[12] This can be explained by the fact that civil engineering is fundamentally different from other engineering disciplines in that each project is by and large unique. Hence, available theories are never fully applicable to these projects and, moreover, there is virtually never a chance to test a prototype. As a consequence, the uncertainty in applying scientific knowledge to civil engineering projects is very high and the uses of subjective judgment and context-sensitive, common-sense reasoning based on designers' experiences are unavoidable. This seems to explain why many civil engineers demonstrated so early in the history of fuzzy logic their strong affinity with it.

One class of problems in civil engineering for which fuzzy logic has been proven useful consists of assessing or evaluating existing constructions. Typical examples of such problems are the assessment of fatigue in metal structures, assessment of the quality of highway pavements, evaluation of damage in buildings after earthquakes or severe storms, and the like. To illustrate the use of fuzzy logic in these problems,

---

[11] The doctoral dissertation by Kristin L. Wood at Caltech, *A Method for Representing and Manipulating Uncertainties in Preliminary Engineering Design* (1989), likely played some role in this recognition.

[12] The journal *Civil Engineering Systems*, which began publication in 1984, became a primary journal at that time for fuzzy-logic applications in civil engineering. Many more papers in this area were published there than we can refer to in this book.

we describe the problem of assessing the physical conditions of bridges, as described by Tee, Bowman, and Sinha (1988).

All accumulated experience regarding the assessment of physical conditions of bridges is summarized in the *Bridge Inspector's Training Manual*, published by the US Department of Transportation. This manual provides bridge inspectors with basic guidelines and procedures for performing their tasks. It specifies three components for each bridge, its deck, superstructure, and substructure, which are further divided into 13, 16, and 20 subcomponents, respectively. The inspector is required to assess the condition of each subcomponent individually, to assess its relative structural importance in the context of each particular bridge inspection, and to aggregate these individual assessments to obtain an overall assessment of the bridge.

Since each bridge inspection involves highly imprecise information, it is based to a large extent on the experience, intuition, and subjective judgments of the inspector. Human subjectivity and imprecision can be taken into account by employing an appropriate fuzzy number to express the condition of each inspected component and by employing another fuzzy number to express the structural importance of the component. The latter fuzzy sets are obtained by surveying a group of engineers and inspectors regarding the dependence of structural importance on rating the condition of each component and averaging their responses.

Once the two kinds of fuzzy numbers are determined for all components (see the original paper for more details), the overall evaluation is obtained by computing the weighted average of all the fuzzy numbers representing conditions of the evaluated components, each weighted by the fuzzy number representing its structural importance. In order to compute this weighted average correctly, clearly, the constrained fuzzy arithmetic (introduced in section 3.4) must be employed.

Other examples of using fuzzy logic for assessing or evaluating existing constructions include highway pavement evaluation (Gunaratne, Chameau, and Altschaeffl 1988), risk assessment for gas pipelines (Ikejima and Frangopol 1987), and assessment of earthquake intensity based on building damage records (Juang and Elton 1986). A fair number of other civil engineering applications related to earthquakes are also included in Feng and Liu 1985, 1986.

There are of course many other applications of fuzzy logic in civil engineering, including a fair number of applications in transportation (Teodorević and Vukadinović 1998). The overview by Wong, Chou, and Yao (1999), which is accompanied by appropriate references, is a useful resource regarding these applications.

Another engineering application in which the utility of fuzzy logic has already been established is the theory of *reliability* pertaining to engineering products. In this context, reliability is a measure of the expected capability of an engineering product to operate under specified conditions without failures for a given period of time. Classical reliability theory, which has been a subject of research since the early 1940s, is based on two assumptions:

(a) *Probability assumption*—the behavior of the engineering product with respect to reliability can adequately be expressed in terms of probability theory.

(b) *Assumption of binary states*—at any given time, the engineering product is either in a functioning state or in a failed state.

These assumptions, which in classical reliability theory are taken for granted, are under closer scrutiny questionable in many practical situations. Assumption (a) makes sense only when probabilities of the recognized states, in whatever way obtained, are sole indicators of the underlying physical process that affect reliability, and when adequate statistical data that are relevant to the engineering product are available. These stringent conditions are not fulfilled if the engineering product operates under unique circumstances, such as most structural constructions in civil engineering, and available statistical data are not fully relevant at best. They are certainly not fulfilled when human operators are involved or when factors that affect failures are not of a statistical nature, such as bugs in computer programs and the like. Assumption (b), for example, is too rigid for studying the reliability of all systems known as degradable systems (Cai, Wen, and Zhang 1991a).

These and many other circumstances pertaining to the reliability of engineering products under which assumptions (a) and (b) of classical reliability theory are questionable or even not valid have been increasingly recognized since at least the mid-1980s and led eventually to suggestions of alternative reliability theories, rooted in fuzzy-set interpretation of possibility theory. These alternative reliability theories are based on fundamentally different assumptions:

(a′) *Possibility assumption*—the behavior of the engineering product with respect to reliability can adequately be expressed in terms of possibility theory.

(b′) *Assumption of fuzzy states*—at any given time, the engineering product is in functioning state to some degree and in failed state to another degree.

By accepting either assumption (a) or (a′) and, similarly, either (b) or (b′), we obtain four possible reliability theories:

(a-b)   This is a reliability theory based on probability theory and binary states, denoted in the literature usually by the acronym probist (probability theory and binary states). In probist, the term "probability theory" is usually not understood in the strict sense of classical probability theory, but in a broader sense, covering also fuzzy probabilities and other kinds of imprecise probabilities, as well as probabilities of fuzzy events.

(a′-b) This is a reliability theory based on combining probability theory (viewed again in the broader sense) with fuzzy states, whose acronym is profust.

(a-b′) This reliability theory is based on a fuzzy-set interpretation of possibility theory and binary states; it is known in the literature as posbist reliability theory.

(a'-b') This reliability theory is formalized in terms of possibility theory and fuzzy states, so its acronym is posfust.

These four reliability theories are complementary in the sense that each is suitable for dealing with reliability in certain application domains. Together, they substantially enlarge the application domain of classical reliability theory. They have been formally investigated since the early 1990s (Cai, Wen, and Zhang 1991a,b). The literature covering these theories and their applications is now quite extensive. Two excellent references include a special issue of *Fuzzy Sets and Systems* edited by Cai (1996a) and his comprehensive monograph Cai 1996b on the four reliability theories. A more recent book by Verma, Srividya, and Prabhu Gaonkar (2007), which includes further advances in fuzzy reliability theory in engineering, is also suitable as a textbook.

Fuzzy logic has been applied as well in the area of industrial engineering. These applications are well described in the books Karwowski and Mital 1986 and Evans, Karwowski, and Wilhelm 1989, as well as in the chapter "Quality Control and Maintenance" in Zimmermann 1999 (pp. 162–84).

6.4 Decision making

The idea of employing fuzzy logic in decision making (fuzzy decision making, or FDM, for short) was introduced and thoroughly discussed for the first time in the classic paper by Bellman and Zadeh (1970), even though some hints of it appeared in earlier publications by Zadeh (1968b) and Chang (1969). This paper aroused considerable interest within the fuzzy-logic community in this promising application area. This stimulated research on reformulating various types of classical decision-making problems and the associated classical optimization problems (especially linear programming) to their fuzzy counterparts. Although the fuzzy-logic community was still very small at that time, a surprisingly large number of papers were published on the subject within just a few years. These early contributions are surveyed in an organized way in a historically important book by Kickert (1978)—the first book fully devoted to applications of fuzzy logic to decision making.

Among the papers on FDM published in the 1970s, Zadeh 1976b, which is not listed among references in Kickert's book, deserves special attention. Zadeh explains in this paper clearly the principal rationale for using fuzzy logic in decision making—to allow decision makers to express objectives and constraints of their decision problems in an ordinary, common-sense way—by statements in natural language—and deal with them via their appropriate fuzzy-set representations. In many decision problems involving a single decision maker, it is reasonable to assume that the decision maker understands the meaning of linguistic descriptions and is able to define fuzzy sets that reasonably well represent these meanings. If this is not possible, the process of meaning precisiation developed by Zadeh (1971a, 1978a, 1984) can be employed.

It is quite amazing how much the very small group of researchers, motivated by their common conviction that fuzzy logic has the potential to make decision making more realistic and practical, accomplished within just a few years in spite of the largely hostile academic environment at that time. In broad terms, they established that the use of fuzzy logic in decision making is feasible and worthy of further investigation. In the narrow sense, they developed various fuzzy counterparts of the classical linear programming problem, which has played an important role in classical decision-making problems, whose decision alternatives are real numbers or vectors of real numbers within some designated intervals.[13] A general fuzzy counterpart of classical linear programming, which subsumes the various special cases, is formulated as follows: Maximize $\sum_{j=1}^{n} C_j X_j$ subject to $\sum_{j=1}^{n} A_{ij} X_j \leq B_i$ ($i \in \mathbb{N}_m$) and $X_j \geq 0$, where coefficients A_{ij}, B_i, C_j are fuzzy intervals, X_j are fuzzy intervals representing states of n linguistic variables, and symbols $\leq$ (or $\geq$) denote ordering of fuzzy intervals.

The strong early research on FDM, as captured in the book by Kickert (1978), intensified in the 1980s. Results obtained during this decade are documented by numerous journal articles on this subject (more than 70 articles in *Fuzzy Sets and Systems* alone), by many papers in conference proceedings, and by chapters in several edited books on fuzzy decision making. Three larger works deserve special recognition: a highly original book by Kacprzyk (1983), which substantially expanded the ideas presented in the classic paper by Bellman and Zadeh (1970); an edited book on fuzzy decision making by Zimmermann, Zadeh, and Gaines (1984); and, above all, the first textbook fully devoted to fuzzy decision making by Zimmermann (1987).

Since 1990, the literature on FDM and the associated fuzzy optimization has been growing extremely rapidly, second only to the rapid growth of literature on fuzzy control. Moreover, this growing literature has some distinctive characteristics that indicate that FDM is already well developed and that the ongoing research in virtually all other areas is growing. The literature contains many monographs that focus on various special areas of FDM or on associated fuzzy optimization; most cover the respective areas quite comprehensively. It also contains textbooks on some of these special areas, indicating that these areas have reached a high level of maturity. A surprisingly large number of special issues on various themes pertaining to FDM and fuzzy optimization have been organized. For example, at least 20 such special issues have been published prior to 2014 in *Fuzzy Sets and Systems* alone. A specialized journal in this area, *Fuzzy Optimization and Decision Making*, was founded in 2002. In order to assess briefly the overall accomplishments of research on FDM,

---

[13] A particular fuzzy counterpart of the classical linear programming problem was proposed and examined for the first time in an application context by Zimmermann (1976).

we first introduce a rough classification of this large area into special subareas. We then examine relevant literature from the standpoint of this classification and with the aim of assessing the level of development in each of the recognized subareas of FDM.

The whole area of decision making, classical as well as fuzzy, is usually classified into two broad classes: single-person decision making and multiperson decision making. The former, which we examine first, is commonly divided into three subareas that are usually referred to as multiattribute, multiobjective, and multicriteria decision making. In multiattribute FDM, a single decision objective is expressed via several attributes. This type of FDM is the simplest of the three types. Multiobjective FDM involves two or more distinct decision objectives, each expressed via its own attributes. This type of FDM is considerably more complex since the objectives are often not commensurable and may conflict with one another. In multicriteria decision-making problems, the decision maker is interested in exploring a given decision-making problem (of either of the previous two types) from several alternative points of view based on distinct criteria (pragmatic, ethical, aesthetic, etc.). This adds some additional methodological issues to the process of decision making, such as the issue of appropriate aggregation of outcomes obtained for different criteria.[14]

Literature dealing with the three subareas of single-person FDM is extensive. It is thus sensible to identify in each subarea only the most representative monographs. Multiattribute FDM and the associated varieties of fuzzy linear programming problems are covered quite comprehensively in the relatively early books by Chen and Hwang (1992) and Lai and Hwang (1992), respectively. By examining these books, our judgment is that this subarea of FDM is now well developed. Multiobjective FDM is well represented by three books of Sakawa (1993, 2000, 2002) and two large monographs by Xu and Zhou (2011) and Xu and Zeng (2014). Not only has this subarea of FDM reached a high level of sophistication from philosophical, mathematical, computational, as well as pragmatic points of view, but considerable research is still being devoted to it. For multicriteria FDM, a relatively early book by Fodor and Roubens (1994) is important due to its rigorous mathematical analysis of fuzzy preference ordering in the context of multicriteria FDM. A more recent work by Pedrycz, Ekel, and Parreiras (2011) is currently the most comprehensive book dealing with multicriteria FDM.

As for the area of multiperson FDM, this concerns situations in which several individuals interact with one another in dealing with a given decision problem. Depending on the kind of interaction, two main subareas of multiperson decision making (classsical as well fuzzy) have been recognized:[15]

---

[14] As an interesting historical remark, the three types of one-person FDM were already discussed at length in Zimmermann, Zadeh, and Gaines 1984 (pp. 241–320); a full chapter was devoted to multicriteria FDM in the early book by Kickert (1978, 61–84).

[15] Another type of multiperson decision making, referred to as *team decision making*, is occasionally mentioned in the fuzzy-set literature. The classical theory of team decision making was developed in a book, *Economic Theory*

Group decision making. Given a group of individuals, a set of possible group actions, a shared belief by all members of the group that some group action is needed, and their individual preference orderings on the set of possible actions, the decision problem is then to find an action (or a set of feasible actions) that in some sense best reflects the individual preferences of the group. There are of course many ways of dealing with this problem depending on how the term "best reflects" is interpreted. Observe that fuzzy group decision making is closely connected with multicriteria FDM. In the former, distinct preference orderings result from criteria employed by group members; in the latter, they result from distinct criteria chosen by a single decision maker. Due to this connection, fuzzy group decision making is covered, jointly with multicriteria FDM, in the book by Pedrycz, Ekel, and Parreiras (2011).

Gaming. The term "gaming" refers to any multiperson decision making that always involves some form of competition, expressed by conflicting objectives among the individual decision makers, but may also involve various forms of cooperation. The common theoretical base for all varieties of this type of multiperson decision making is the mathematical theory of games that was formulated in the book by von Neumann and Morgenstern.[16] A game is formally described by a set of rules that individual decision makers in a group (usually called players) are obliged to follow. It is assumed that in each given situation, the set of possible outcomes is known to all players and that each player has a particular preference ordering among them. Each player is required to make a particular choice from a given set of possible alternatives without any knowledge about choices made by the other players. However, it is assumed that all players know the outcome for each combination of possible choices made by all players. Each player in the group deals with the same problem: what choice should he or she make to influence the outcome in the most favorable way according to his or her preference ordering? Game theory allows the players to deal with this problem mathematically. As von Neumann and Morgenstern recognized in their now classic book, players of any given group with three or more players may decide to form coalitions within which they partially cooperate and game theory should be able to handle this possibility. This leads to two classes of games: (a) *noncooperative games*, which are strictly competitive and do not allow for any cooperation; and (b) *cooperative games*, in which cooperation is allowed as part of the competition.

A few early proposals for fuzzifying classical game theory were made from the mid-1970s to the mid-1980s by Aubin (1976, 1981, 1984), Ragade (1976), and Butnariu (1978, 1979). These publications were oriented primarily to cooperative games with fuzzy coalitions. Interest in fuzzy games has increased considerably since the mid-1980s and the literature on this subject has grown rapidly in various directions.

---

of Teams, by Jacob Marschak and Roy Radner (New Haven, CT: Yale Univ. Press, 1972). Its fuzzy counterpart was suggested by H. Nojiri (1979), but has not shown any significant follow-up.

[16]J. von Neumann and O. Morgenstern, *Theory of Games and Economic Behavior* (Princeton, NJ: Princeton Univ. Press, 1944).

For our purposes in this book, we recommend the following four representative monographs on fuzzy games, which together capture the state of the art in this area: Butnariu and Klement 1993; Mareš 2001; Nishizaki and Sakawa 2001; and Branzei, Dimitrov, and Tijs 2005.

6.5 Natural sciences

Compared with engineering and decision making, the process of recognizing that fuzzy logic might be useful in the natural sciences has been much slower. Surprisingly, the utility of fuzzy logic was first recognized in chemistry.

6.5.1 Chemistry

Among the earliest suggestions of potentially useful applications of fuzzy logic in chemistry were those made in the early 1990s by Otto (1990), Otto and Yager (1992), and Singer and Singer (1993). At that time, a few researchers in theoretical chemistry began to recognize that fuzzy logic could play an essential role for dealing with some unresolved problems in this area. These problems emanated from the traditional way of viewing concepts such as symmetry or chirality as bivalent—either true or false in each of their applications in chemistry. Researchers in theoretical chemistry increasingly felt that viewing these concepts as fuzzy could alleviate certain problems, as their truth values would be in each application a matter of degree. Zabrodsky, Peleg, and Avnir (1992, 1993), as well as Zabrodsky and Avnir (1995a,b), analyzed this issue in great detail.[17] It is worth quoting from their first paper, in which they deal with the concept of symmetry as follows (Zabrodsky, Peleg, and Avnir 1992, 7843):

> One of the most deeply-rooted paradigms of scientific thought is that Nature is governed in many of its manifestations by strict symmetry laws. The continuing justification of that paradigm lies within the very achievements of human knowledge it has created over the centuries. Yet we argue that the treatment of natural phenomena in terms of "either/or," when it comes to a symmetry characteristic property, may become restrictive to the extent that some of the fine details of phenomenological observations and of their theoretical interpretation may be lost. Atkins writes in his widely-used text on physical chemistry:[18] "Some objects are more symmetrical than others," signaling that a scale, quantifying this most basic property, may be in order. The view we wish to defend in this report is that symmetry can be and, in many instances, should be treated as a continuous "gray" property, and not necessarily as a "black or white" property which exists or does not exist. Why is such continuous symmetry measure important? In short, replacing a "yes or no" information processing filter, which acts as a threshold decision-making barrier which differentiates between two states, with a filter allowing a full range of "maybe's," enriches, in principle, the information content available for analysis.

---

[17] A short, common-sense discussion of this issue, exemplified by the concept of symmetry, may even be found in *Nature* (Fowler 1992).

[18] P. W. Atkins, *Physical Chemistry*, 3rd ed. (Oxford: Oxford Univ. Press, 1986), 406.

These developments led to the decision to devote one of the annual Mathematical Chemistry Conferences entirely to the role of fuzzy logic in chemistry.[19] The conference, "Are the Concepts of Chemistry All Fuzzy?," was held in Pitlochry, Scotland, in 1996. An important outcome of this conference was a carefully edited book by Rouvray (1997), which was the first major publication on applications of fuzzy logic in the natural sciences. The principal contribution of this book was that it convincingly demonstrated that fuzzy logic was useful not only for realistically representing some concepts in theoretical chemistry, such as symmetry, chirality, molecular structure, or molecular shape and size, but also for dealing properly and effectively with some methodological problems in chemistry, such as the problems of molecular recognition, hierarchical classification, or computer-aided elucidation of molecular structures.

Although no additional books on fuzzy logic in chemistry have as yet been published, research in this domain has been ongoing. This can be seen from an overview of applications of fuzzy logic in chemistry and chemical engineering given in chapter 9 of Zimmermann 1999 and from the ongoing stream of publications after this overview appeared, not only on this subject, but on various problems in other areas of chemistry as well. For example, the paper by Sârbu (1999) describes the use of hierarchical fuzzy clustering for classification of chemical elements based on various groups of physical characteristics of the elements. For readers with a strong background in physical organic chemistry, we recommend a historical and philosophical paper by Akeroyd (2000).

6.5.2 Biology

The fact that the initial applications of fuzzy logic in the natural sciences emerged in chemistry and not in biology is very surprising. It was biology that initially motivated Zadeh to embark on his groundbreaking work on fuzzy set theory, as he expressed it very clearly on p. 857 in Zadeh 1962:

> For coping with the analysis of biological systems and to deal effectively with such systems, which are generally orders of magnitude more complex than man-made systems, we need a radically different kind of mathematics, the mathematics of fuzzy or cloudy quantities that are not describable in terms of probability distributions.

For a long time, unfortunately, biologists showed virtually no interest in fuzzy logic except for some occasional and isolated explorations, such as those described in Luo, Ji, and Li 1995 and in von Sternberg and Klir 1998, which the biological community by and large ignored. The useful role of fuzzy logic in biology was eventually recognized, but not until about the mid-2000s and only in the rapidly growing new interdisciplinary area of *bioinformatics*.[20] This rather narrow, but highly important

---

[19] These annual conferences were organized under the auspices of the International Society for Mathematical Chemistry.

[20] The term *bioinformatics* was coined in the mid-1980s. The *Merriam-Webster Dictionary* defines bioinformatics as "the collection, classification, storage, and analysis of biochemical and biological information using computers especially as applied to molecular genetics and genomics."

and visible area is closely connected with the Human Genome Project, which was implemented during the period 1990–2003.[21] The outcome of this project was a very large database containing structures of all human genes. This database and a growing number of other biomolecular databases provide researchers in molecular genetics with a huge amount of information. The challenge is to utilize this information for advancing biological knowledge by answering many profound biological questions, such as those regarding functions of individual genes, processes leading to the three-dimensional structures of proteins, functions of these structures, and the like. It is this analytical part of bioinformatics, where the usefulness of fuzzy logic is suggestive and, in fact, has already been demonstrated in numerous ways.

Fuzzy logic entered into bioinformatics only at the very beginning of the new millennium, some fifteen years after bioinformatics emerged. In spite of its late entry, the visibility of fuzzy logic as a powerful resource for bioinformatics has slowly but steadily grown after a few successful early applications of fuzzy logic to bioinformatics were published. Among them, the earliest are the papers by Sadegh-Zadeh (2000a), in which the author introduced a broad framework for dealing with genetic information via ordered fuzzy sets represented by points in an n-dimensional unit cube, and Woolf and Wang (2000), in which the first successful application of fuzzy logic in bioinformatics was described and evaluated.[22] Looking just at the first decade of the new millennium, the growth of a wide variety of applications of fuzzy logic and soft computing in bioinformatics is best documented in three large edited volumes devoted to these applications—Pelta and Krasnogor 2005, Corne et al. 2006, and Jin and Wang 2009, which together contain 775 pages, along with the monograph by Xu et al. (2008) that specializes in applications of fuzzy logic in bioinformatics. According to this book, there are three main types of problems in bioinformatics for which fuzzy logic has proven especially useful. The first type involves general applications of various fuzzy similarity measures (defined on objects represented in terms of gene ontology) together with fuzzy clustering methods based on these similarity measures for data mining in bioinformatics.[23] The second type consists of problems known in bioinformatics as structural problems—the problems of characterization, comparison, and prediction of three-dimensional structures of

---

[21] The term "Human Genome Project" (HGP) refers to an international, collaborative research program whose objective was to determine the structures—sequences of deoxyribonucleic (DNA) molecules—of all genes of human beings.

[22] This application is designed for analyzing massive gene expression data, obtained by so-called expression profiling techniques, involving regulations performed by biological cells of their genes in response to environmental changes. These regulations are normally aimed at protecting the cells in face of environmental changes. However, errors in these regulations may lead to serious diseases. The work by Woolf and Wang (2000) was clearly motivated by these medical implications. This work was later improved by Ressom, Reynolds, and Varghese (2003).

[23] The term "ontology" is not used here in the philosophical sense, but in the sense commonly recognized in computer science. In this latter sense, an ontology consists of a set of concepts regarding some area of interest together with the relationship among the concepts. Gene ontology focuses on the role of genes in biological organisms. It has thus a narrower scope than the ontology of the entire realm of molecular biology.

proteins. The third type consists of all problems that involve functioning of genes in biological cells including their regulation by cells.

Outside of bioinformatics, the potential utility of fuzzy logic in biology has remained by and large unappreciated. However, we should mention two exceptions. The first is the paper by Center and Verma (1998), in which the authors argue that systems based on linguistic variables are more realistic for systems modeling in biology and agriculture than their conventional counterparts with numerical variables. The second is the book by Sandler and Tsitolovsky (2008), in which the authors explore the use of fuzzy differential equations for modeling the dynamic behavior of neural cells.

6.5.3 Physics

Contrary to biology, physics—one of the most advanced and successful areas of science—was not among the areas of human affairs that motivated the emergence of fuzzy logic. It is perhaps due to the enormous success of physics that the developed measurement routines in physics have rarely been questioned by physicists. A rare exception was the outstanding American physicist Percy Williams Bridgman (1882–1961).[24] He expresses his rather unorthodox views on measurement in physics as follows (Bridgman 1959, 226–28, italics added):

> There is no question but that we do talk about aspects of experience in yes-no terms, and in so far as any field of experience has such yes-no sharpness it has it in virtue of the fact that it is a verbal activity. One may well question, however, whether we have any right to ascribe such yes-no properties to any verbal activity. What are these words anyhow? They are not static things, but are themselves a form of activity which varies in some way with every so-called repetition of the word. The word as we use it is part of a terribly complicated system, involving both present structure in the brain and the past experience of the brain, most of which we cannot possibly be conscious of. . . .
>
> The physics of measurement and of the laboratory does not have the yes-no sharpness of mathematics, but nevertheless employs conventional mathematics as an indispensable tool. Every physicist combines in his own person, to greater or lesser degree, the experimental physicist who makes measurements in the laboratory, and the theoretical physicist who represents the results of the measurements by the numbers of mathematics. These numbers are things he says or writes on paper. The jump by which he passes from the operations of the laboratory to what he mathematically says about the operations is a jump which may not be bridged logically, and is furthermore a jump which ignores certain essential features of the physical situation. For the mathematics which the physicist uses does not exactly correspond to what happens to him. *In the laboratory every measurement is fuzzy because of error.* As far as reproducing what happens to him is concerned, the mathematics of the physicist might equally well be the mathematics of the rational numbers, in which such irrationals as $\sqrt{2}$ or pi do not occur. Now one would certainly be going out of one's way to attempt to force theoretical physics into a straightjacket of the mathematics of the rational numbers as distinguished from the mathematics of all real numbers, but *by forcing it into the straightjacket of any kind of mathematics at all, with its yes-no sharpness, one is discarding an essential aspect of physical experience and to that extent renouncing the*

---

[24] Bridgman, who was a professor of physics at Harvard University during most of his academic career, was an experimental physicist par excellence. In 1946, he won the Nobel Prize in physics for his groundbreaking work on the physics of high pressures. He also wrote extensively on measurement in physics and on various aspects of philosophy of science.

Another dissenter was Duhem; see p. 16 above for his idea of "mathematics of approximation."

> *possibility of exactly reproducing that experience.* In this sense, the commitment of physics to the use of mathematics itself constitutes, paradoxically, a renunciation of the possibility of rigor. . . .
>
> Now, it appears to me, the linkage of error in every sort of physical measurement must be regarded as inevitable when it is considered that the knowledge of the measurement, which is all we can be concerned with, is a result of the coupling of the external situations with a human brain. Even if we had adequate knowledge of the details of this coupling we admittedly could not yet use this knowledge in formulating in detail how the unavoidable fuzziness should be incorporated in our description of the world nor how we should modify our present use of mathematics. About the only thing we can do at present is to continue in our present use of mathematics, but with the addition of a caveat to every equation, warning that things are not quite as they seem.

It is certainly interesting that Bridgman used the terms *fuzzy* and *fuzziness* in 1959, six years before they were introduced by Zadeh (1965a), and his use of these terms was quite compatible with Zadeh's use (to describe situations in which yes-no sharpness was not realistic). In fact, these terms even appear in some of his earlier writings.

Bridgman's critical comments about measurement procedures in physics did not have any visible impact on physics during his life. However, since the early 1990s some physicists specializing on measurement have occasionally referred to Bridgman's criticism (Mari 1996, 2000, 2003) as well as to the possibility of employing fuzzy sets in measurement (Mari 2000, Zingales and Mari 2000). This is encouraging since at least some physicists now seem to be aware of fuzzy set theory and its potential role in physical measurement. However, very little of this potential has been explored so far. As yet, so far as we know, only two such explorations have been made, each with a rather modest aim. The first one involves the use of fuzzy equivalence and compatibility relations for defining fuzzy compatibility on a set of classical intervals of real numbers, suggested by Mari (1991). The second is a method based on fuzzy set theory for detecting systematic errors in a measurement process that involves a small number of measurements and an unknown distribution of the measured data, which is described in Gao et al. 2002. Although these beginnings are encouraging, we have not identified as yet any serious effort in the physical sciences literature to develop a more expressive theory of measurement based on fuzzy logic. Perhaps the new formalization of measurement proposed by Mari (2000), which seems more realistic than traditional representational formalization, will prove to be a more suitable base for developing such a theory. While in the representational formalization homomorphism from the empirical relations to the numerical relations is defined mathematically, assuming that the empirical relations are perceivable, this assumption is not needed in Mari's new formalization since the homomorphism is embedded in the construction of measuring systems and their use.

A rare application of fuzzy logic in astronomy, one of the oldest areas of physics, was developed and tested by Shamir and Nemiroff (2005, 2008). This deals with the problem of automatic identification of astronomical objects. They describe the essence of their application in the abstract of Shamir and Nemiroff 2005 as follows:

> Accurate automatic identification of astronomical objects in an imperfect world of nonlinear wide-angle optics, imperfect optics, inaccurately pointed telescopes, and defect-ridden cameras

is not always a trivial first step. In the past few years, this problem has been exacerbated by the rise of digital imaging, providing vast digital streams of astronomical images and data. In the modern age of increasing bandwidth, human identifications are many times impracticably slow. In order to perform an automatic computer-based analysis of astronomical frames, a quick and accurate identification of astronomical objects is required. Such identification must follow a rigorous transformation from topocentric celestial coordinates into image coordinates on a CCD frame.[25] This paper presents a fuzzy logic based algorithm that estimates needed coordinate transformations in a practical setting. Using a training set of reference stars, the algorithm statically builds a fuzzy logic model. At runtime, the algorithm uses this model to associate stellar objects visible in the frames to known-catalogued objects, and generates files that contain photometry information of objects visible in the frame. Use of this algorithm facilitates real-time monitoring of stars and bright transients, allowing identification and alerts to be issued more reliably. The algorithm is being implemented by the Night Sky Live all-sky monitoring global network and has shown itself significantly more reliable than the previously used nonfuzzy logic algorithm.

This interesting and successful application of fuzzy logic in astronomy is mentioned here only because it is currently such a rarity in physics.

Let us turn now to the connection of fuzzy logic (and many-valued logics) with quantum mechanics. Since many-valued logics emerged almost simultaneously with quantum mechanics, the connection between the two areas has a long and rich history. Details of the early history (prior to about the mid-1970s) can be found in chapter 8 of Jammer 1974.

The very first connection between quantum mechanics and multivalued logic was made by Zawirski (1931a,b), as mentioned at the end of section 4.2.3.[26] He argued that the complementarity of particle and wave theories in quantum mechanics cannot be reconciled within classical two-valued logic, but can be reconciled within at least three-valued logic. These papers passed almost unnoticed prior to the publication of Rescher 1969.

One year after these publications by Zawirski, an interesting letter to the editor of the *Physical Review* by Fritz Zwicky—an astrophysicist at the California Institute of Technology—was published (Zwicky 1933). In this letter Zwicky argues that "every scientific statement referring to observations should possess a certain minimum degree of flexibility" and, therefore, "[f]ormulations of scientific truths intrinsically must be *many-valued*." He elaborates on these broad ideas by illustrating their meaning with a few examples from physics, mostly from quantum mechanics. Reaction to the paper was quick and mixed—Eric T. Bell (1933), a historian of mathematics, was favorable but Henry Margenau (1934), a physicist and philosopher of science, was critical.

An important step in developing an appropriate logic for quantum mechanics is closely associated with John von Neumann's work in the early 1930s on the mathematical foundations of quantum mechanics. When he wrote his well-known

---

[25] A charge-coupled devise (CCD), which was invented in 1969 at Bell Laboratories by William Boyle and George E. Smith, plays an important role in digital imaging.

[26] Zygmunt Zawirski (1882–1948) was an important figure in the famous Lvov-Warsaw philosophical school founded by Kazimierz Twardowski; see I. Szumilewicz-Lachman, *Zygmunt Zawirski: His Life and Work* (Dordrecht: Kluwer, 1994).

treatise on this subject (von Neumann 1932), he suspected that the logic of quantum mechanics should differ in certain aspects from the usual, classical logic employed in classical mechanics, and he hit upon the idea of how this unorthodox logic could be formulated. He discussed this "half-baked" idea with Garrett Birkhoff, a mathematician at Harvard University, and they set out to work together to identify the structure of quantum logic and express it axiomatically. The result of their joint work was the classic paper Birkhoff and von Neumann 1936 on the formal properties of quantum logic (p. 147 above). Their primary result was a demonstration that the structure of quantum logic, when compared with the Boolean algebra structure of classical logic, is weaker since it cannot satisfy the distributivity properties of Boolean algebra. Moreover, their arguments showed that "the physically significant statements in quantum mechanics actually constitute a sort of projective geometry, while the physically significant statements concerning a given system in classical dynamics constitute a Boolean algebra." The paper stimulated considerable mathematical research resulting in numerous publications, including several monographs. Most of the publications in this domain are devoted to further investigations of mathematical structures that emerge from Birkhoff and von Neumann's paper. Publications that investigate structures of quantum logic from the point of view of fuzzy logic are relatively rare. In the rest of this section, we mention several such works.

From the historical point of view, two papers by Prugovečki (1974, 1975) are significant. These are the very first publications on applications of fuzzy logic (or rather fuzzy sets) in quantum mechanics. In both papers, which are connected, the author employs the concepts of fuzzy events and fuzzy measurements and develops a generalization of conventional Kolmogorov probability theory with the aim of formalizing measurements in quantum mechanics in a more realistic way.

Among the various monographs devoted to quantum logics, Dalla Chiara, Giuntini, and Greechie 2004 is the only one that covers both mathematical structures that conform to the principle of bivalence and those that do not. These are called, respectively, *sharp quantum logics* and *unsharp quantum logics*. In particular, the authors seem to be well aware of the importance of unsharp quantum logics (Dalla Chiara, Giuntini, and Greechie 2004, 5):

> Strangely enough, from the historical point of view, the abstract researches on fuzzy structures and on quantum structures have undergone quite independent developments for many decades during the 20th century.
>
> Only after the Eighties, there emerged an interesting convergence between the investigations about fuzzy and quantum structures, in the framework of the so-called *unsharp approach to quantum theory*. In this connection a significant conjecture has been proposed: perhaps some apparent mysteries of the quantum world should be described as special cases of some more general *fuzzy phenomena*, whose behavior has not yet been fully understood.

Moreover, on p. 37 of their book, the authors describe a specific example in quantum theory, in which the principle of bivalence fails.

The book by Dalla Chiara, Giuntini, and Greechie (2004) has an extensive bibliography on quantum logics and a large part of it contains references to papers on unsharp quantum logics. It is thus reasonable to conclude this section by giving references to a few important papers devoted to fuzzy logic and quantum logics that are not included in this bibliography. The most important is perhaps a series of papers by Pykacz (1992, 1993, 1994, 1999), as well as papers by Pykacz and D'Hooghe (2001), Dvurečenskij and Riečan (1991), and Riečan (2007).

6.6   Earth sciences

In this section, we review applications of fuzzy logic across a broad spectrum of sciences subsumed under the Earth sciences, such as geology, hydrology, meteorology, soil science, oceanography, seismology, physical geography, paleontology and others, each with its various subareas. These sciences were certainly not among those in which the need for a new mathematics based on fuzzy sets was initially perceived. It is thus not surprising that the applicability of fuzzy logic in Earth sciences was recognized rather late, some three decades after the emergence of fuzzy set theory itself. However, once it was recognized, it quickly became one of the most successful applications areas of fuzzy logic.

The first significant publication on applications of fuzzy logic to the Earth sciences was a large, two-volume book edited by Feng and Liu (1985, 1986), based on contributions at the International Symposium on Fuzzy Mathematics in Earthquake Researches held in Beijing in 1985. The symposium was intended to cover areas of seismology, engineering seismology, and earthquake engineering, so a fair number of participants were civil engineers. Problems covered in the book include earthquake prediction, seismic microzonation, seismic hazard analysis, earthquake-induced disaster estimation, seismic design of engineering structures, and decision making for earthquake preparedness. This book undoubtedly had a strong influence, especially in China, on using fuzzy logic in subsequent earthquake research, as is clear from an overview prepared about ten years later by Huang (2004).

Notwithstanding the early recognition of the utility of fuzzy logic in seismology and related areas, the development of wider applications of fuzzy logic in Earth sciences began, by and large, during the 1990s and then expanded rapidly in the first decade of the 21st century. In the early book by Bárdossy and Duckstein (1995), the authors attempted to encourage researchers in nonengineering areas to explore uses of fuzzy logic, so successful in engineering, in their own areas. For this purpose, they described four prospective applications—weather classification, forecasting of water demand, soil water movement, and sustainable reservoir operation.

After the publication of this book, literature devoted to applications of fuzzy logic in many areas of the Earth sciences visibly increased. It is significant that most applications were developed by specialists in the four areas just mentioned and published in relevant specialized journals or proceedings of specialized conferences. This seems to indicate that the utility of fuzzy logic has been by and large recognized within the Earth sciences.

However, the dispersion of literature on applications of fuzzy logic in the Earth sciences makes it difficult to capture the essence of these applications. Fortunately, the book edited by Demicco and Klir (2004) includes an overview of basic ideas regarding the use of fuzzy logic in the Earth sciences, as well as an annotated bibliography listing the main publications devoted to applications of fuzzy logic in some specific areas of Earth sciences: (1) surface hydrology; (2) subsurface hydrology; (3) groundwater risk assessment; (4) geotechnical engineering; (5) hydrocarbon exploration; (6) seismology; (7) soil science and landscape development; (8) sedimentology; and (9) miscellaneous other areas. In addition, the book contains an extensive survey of fuzzy logic applications in sedimentology, hydrology and water resources, paleontology, and seismology, as well as two case studies in which original new approaches are employed to deal with the reef growth problem and the problem of estimating ancient sea levels. Thanks to its comprehensive coverage, the book characterizes the impressive state of the art of applications of fuzzy logic in the Earth sciences during the first decade of the 21st century.

As for the classical branch of Earth sciences—geology—we address the following question: How can the broad endorsement of fuzzy logic in geology be explained when there was virtually no interest in fuzzy logic in classical physics? This considerable difference may be explained primarily by the substantial differences in the nature of problems studied in these scientific areas, and also, at least to some degree, by their histories. While classical physics deals by and large with problems regarding relationships among numerical variables, natural language seems better able to deal with problems in geology (as well as other branches of the Earth sciences) than any of the precise mathematical languages. There are also substantial historical differences between these two areas of science. Physics is one of the oldest natural sciences, under which all other natural sciences and the Earth sciences were initially subsumed.[27] Geology emerged as a separate science of its own only in the 17th century. Its initial preoccupation with the description and classification of geological objects was gradually replaced throughout the 19th century with attempts to understand how the surface of the Earth had developed and changed. As a result, a substantial amount of knowledge pertaining to this issue was produced by the work of many geologists via their extensive, systematic, and painstaking observations, combined with commonsense reasoning. This knowledge, which was described in natural language with virtually no use of mathematics, is preserved in the scholarly geological literature, especially books, published mostly within the first half of the 20th century. After the emergence of computer technology, this knowledge was gradually discounted

---

[27] The term "physics" is derived from the ancient Greek word "physis," which means nature.

as useless, as it could not be represented mathematically in a computer-acceptable language. This led to the development of mathematical geology.

The older geological knowledge expressed in natural language regained its value with the emergence of fuzzy logic, but it took some three decades for geologists to recognize it. Most of the geological knowledge described verbally in the older geological books, often directly in the form of if-then statements, can now be represented mathematically by translating it into fuzzy if-then rules. The apparatus of fuzzy logic handles such knowledge in terms of linguistic variables and procedures of approximate reasoning. It is fortunate that this possibility was recognized and utilized in geology and other areas of the Earth sciences beginning in the mid-1990s. A typical example is the early use of fuzzy logic in stratigraphic modeling described by Nordlund (1996), who also developed a specialized software tool (FUZZIM) to facilitate this kind of modeling (Nordlund 1999). See also Demicco and Klir 2001 and Warren, Demicco, and Bartek 2007.

Since the publication of Demicco and Klir 2004, the literature dealing with applications of fuzzy logic in the Earth sciences has expanded substantially, indicating that the utility of fuzzy logic in the Earth sciences has at least been established. Three important monographs, Bárdossy and Fodor 2004, Shepard 2005, Şen 2010, all deal with problem areas where the use of fuzzy logic is crucial.

Finally, there is the role of fuzzy logic in ecology, an interdisciplinary area at the border of Earth sciences and biology. As in the other Earth sciences, interest in applications of fuzzy logic in ecology has become visible since the early 1990s. However, two earlier, pioneering papers should be mentioned since they likely stimulated this interest. The first one is the paper by Bosserman and Ragade (1982), in which fuzzy set theory was introduced as a useful mathematical tool for developing an alternative way of analyzing ecological systems. The second paper by Roberts (1989) was more specific. It described the useful role of fuzzy graphs for analyzing the problem of forest succession. Both papers appeared in the journal *Ecological Modeling*, where most papers on fuzzy logic in ecology in the 1990s were published, including two special issues edited by Salski, Fränzle, and Kandzia (1996) and Li and Rykiel (1996). An overview paper by Salski (1999) describes the level of development of applications of fuzzy logic in ecology around the end of the 20th century. After this overview paper, fuzzy logic has increasingly been combined with other tools for dealing with ecological problems and, hence, the word "fuzzy" now appears very rarely in titles of ecological publications. This makes it rather difficult to trace ecological publications in which fuzzy logic plays some role. Fortunately, Salski, Holsten, and Trepel (2009) have prepared a more recent overview of fuzzy-logic applications in ecology. Other relevant books include the monographs by Shepard (2005) and Chockalingam (2012), and the book edited by Recknabel (2006).

6.7　Psychology

The history of connections between psychology and fuzzy logic is quite extraordinary. One of the most preeminent figures in mathematical psychology, Robert Duncan Luce (1925–2012), commented as follows in a work that appeared one year before the publication of Zadeh's seminal paper on fuzzy sets (Luce 1964, 376–77):

> It should be pointed out that no fundamentally new mathematical ideas have yet arisen from work on psychological problems. To date, we are parasites on mathematics. I doubt that this will always be so, but the time span is likely to be such that our contributions to mathematics will have little or no impact on anyone doing research in mathematics today. The reason that I believe an influence is bound to occur ultimately is that psychological problems exist that, on the one hand, seem to be meaningful, if not yet precisely stated, and about which we can experiment, if not yet incisively, and on the other hand, that do not seem to fit comfortably into the existing mathematical language. In some cases this may simply reflect a lack of ingenuity, but in others I suspect that the appropriate mathematics does not yet exist. . . .
>
> 　　The language of sets does not always seem adequate to formulate psychological problems. . . . The boundaries of many of my "sets" and of ones that my subjects ordinarily deal with are a good deal fuzzier than those of mathematics. . . . It is quite difficult to pin down just what elements are and are not members of that set, and I am not sure that it is possible in principle . . . Even supposing that it is impossible, I do not believe that this means that attempts to understand behavior are ridiculous, although ultimately it may be deemed inappropriate to try to cast theories of behavior in current mathematical language. . . .
>
> 　　Even assuming that there are profound troubles, mathematical psychologists are not about to call a moratorium until the troubles are resolved. Some, indeed most of us, will skirt around those aspects of behavior for which the difficulties are especially pronounced. Others will try to tackle them more directly and to the extent of their success, they will enrich mathematics as well as psychology.

We have tried to find out if Luce became aware of Zadeh's seminal paper on fuzzy sets, which was published one year after he made the above quoted remarks. After careful research, it seems that he was never aware of Zadeh's paper and subsequent developments, which he almost anticipated. However, it is clear from the following conclusions in his extensive autobiographical article (Luce 1989) that his doubts about the relevance of classical mathematics to psychology were still on his mind:[28]

> I suspect that much of our problem in using mathematics effectively arises from the state of conceptualization in psychology rather than from the appropriateness of mathematics in formulating psychological theory. But there does remain the haunting fear that the existing mathematics is not, in fact, particularly suited to the problems of psychology. Consider, for example, the representation of uncertainty in decision making. I can never get over the feeling that to cast it into probabilistic terms is misguided; intuitively, I sense that however human beings handle uncertainty, their calculus is different from probability. Or take memory and learning: can it be that the troubles we have had have to do with the fact that memories seem to be diffusely represented in the brain and so may not be very amenable to our usual set theoretic formulations? . . .
>
> 　　Perhaps only rarely—psychophysics may be the prime example—is the existing mathematics well suited to the phenomenon; in other areas we may have to become involved in the creation of new sorts of mathematics. If, as I believe, this is the case, our time perspective had better be a long one, for we await a latterday Newton.

　　Zadeh's main motivation for introducing fuzzy sets was to represent and deal with classes of objects that do not have sharp boundaries. This was the main theme

---

[28] This article appears in several slightly different versions. Its initial draft was apparently circulated in the 1970s to a few colleagues who "figured large in it" and was revised on the basis of some of their comments.

of Zadeh 1965a as well as Bellman, Kalaba, and Zadeh 1966. Since most of the concepts employed in human thinking and communication stand for classes of this kind, it was tacitly assumed that fuzzy sets would play a useful role in psychology, especially in the psychology of concepts. This connection of fuzzy sets to psychology was made more explicit in the paper by Goguen (1968–69, 325):

> The 'hard' sciences, such as physics and chemistry, construct exact mathematical models of empirical phenomena, and then use these models to make predictions. Certain aspects of reality always escape such models, and we look hopefully to future refinements. But sometimes there is an elusive fuzziness, a readjustment to context, or an effect of observer upon observed. These phenomena are particularly indigenous to natural language, and are common in the 'soft' sciences, such as biology and psychology. . . .
>
> 'Exact concepts' are the sort envisaged in pure mathematics, while 'inexact concepts' are rampant in everyday life. . . . Ordinary logic is much used in mathematics, but applications to everyday life have been criticized because our normal language habits seem so different. Various modifications of orthodox logic have been suggested as remedies. . . .
>
> Without a semantic representation for inexact concepts it is hard to see that one modification of traditional logic really provides a more satisfactory syntactic theory of inexact concepts than another. However, such a representation is now available (Zadeh 1965a).

In general, a concept is a mental representation of a class of real or abstract entities, usually referred to as a *concept category*. In the psychology of concepts, a concept is usually viewed as a body of knowledge regarding the entities in the associated concept category that is stored in long-term memory (also referred to as semantic memory) and employed by default in most of the cognitive processes underlying higher cognitive competencies.

Prior to the 1970s, it had tacitly been assumed in the psychology of concepts that each concept category is a classical set defined by a collection of attributes that are separately necessary and jointly sufficient. The view of concepts under this assumption, referred to as the *classical view*, was seriously challenged in the early 1970s by results from a series of psychological experiments designed and performed by Eleanor Rosch (1973a,b).[29] These experiments were designed to answer the question of whether people perceive membership in concept categories as a yes-or-no matter or a matter of degree. The results obtained consistently demonstrated that membership in concept categories—and not only in perceptual ones but also in various nonperceptual semantic categories—is not a yes-or-no matter, as had been assumed by the classical view, but rather is a matter of degree. This led by and large to almost universal rejection of the classical view.

Rosch's experiments also revealed that some individuals in concept categories more typically exemplify the respective concepts than others. That is, each concept category is associated with an ordering relation reflecting the typicality of individuals in the category as examples of the associated concept. The most typical individual(s) can be viewed as natural prototype(s) of the category. The ordering relation in the category can then be defined via a suitable similarity measure, as thoroughly

---

[29] These two papers are chosen here as representative of Rosch's numerous contributions published in the 1970s. Prior to 1973, she had published a few papers, which describe her initial experiments, under the name "Heider"—her married name at that time.

investigated by Tversky (1977). This, in a nutshell, is the essence of a prototype view of concepts that emerged from Rosch's experiments as a natural successor to the classical view of concepts.

Although the results obtained by Rosch and the emerging prototype view of concepts were suggestive of possible uses of fuzzy sets in the psychology of concepts, Rosch herself did not recognize at that time any connection between these two areas.[30] However, her results have stimulated other psychologists to examine the potential role of fuzzy sets in psychology. This has led to a lively discussion of this matter in the psychological literature throughout the 1970s. Perhaps the first researcher who recognized Rosch's early experiments and responded to them by exploring the potential role of fuzzy sets in psychology as well as in natural language was George Lakoff, an American cognitive linguist and, interestingly, a colleague of both Rosch and Zadeh at the University of California at Berkeley. As Lakoff explains (Lakoff 1973, 458, 491):

> Logicians have, by and large, engaged in the convenient fiction that sentences of natural languages (at least declarative sentences) are either true or false or, at worst, lack a truth value, or have a third value often interpreted as 'nonsense.' And most contemporary linguists who have thought seriously about semantics, especially formal semantics, have largely shared this fiction, primarily for lack of a sensible alternative.
>
> Yet students of language, especially psychologists and linguistic philosophers, have long been attuned to the fact that natural language concepts have vague boundaries and fuzzy edges and that, consequently, natural language sentences will very often be neither true, nor false, nor nonsensical, but rather true to a certain extent and false to a certain extent, true in certain respects and false in other respects. . . .
>
> Fuzzy concepts have had a bad press among logicians, especially in this century when the formal analysis of axiomatic and semantic systems reached a high degree of sophistication. It has been generally assumed that such concepts were not amenable to serious formal study. I believe that the development of fuzzy set theory by Zadeh . . . makes such serious study possible.

Several psychological experiments, which were inspired by the discussion about fuzzy sets and concepts in Lakoff's paper and by the work of Labov (1973), are described in a paper by Hersh and Caramazza (1976). The outcomes of these experiments led the authors to propose "that natural language concepts be considered as inherently vague and, specifically, as fuzzy sets of meaning components [and] . . . that language operators—negative markers, adverbs, and adjectives—be considered as operators on fuzzy sets." They conclude the paper by expressing their belief (resulting from the experimental outcomes) that "the formal treatment of vagueness is an important and necessary step toward a more comprehensive handling of natural language phenomena and the communication of vague information."

In two papers published by psychologist Gregg C. Oden in 1977, the author discusses the connection between concepts and fuzzy sets, which he clearly recognizes as potentially fruitful. In one of them, Oden 1977a, he refers to previous psychological studies that, in his opinion, have shown that many subjective categories are fuzzy sets. He then describes psychological experiments designed to address the question

---

[30] In fact, according to our correspondence with Rosch, she was not even aware of fuzzy sets at that time.

of whether people are capable of processing fuzzy information (involving degrees of truth or membership) in a consistent manner. He concludes that results of these experiments indicate that, indeed, people possess this capability. In the other paper, Oden 1977b, he describes experiments aimed at comparing human judgments regarding conjunctions and disjunctions of fuzzy statements with their counterparts obtained formally by either one of two operations of conjunction that were at that time discussed in the fuzzy-set literature—the minimum and product operations—and the corresponding operations of disjunction obtained from the two conjunctions and the standard operation of negation via the De Morgan law—the maximum and the algebraic sum. He found that the operations of product and algebraic sum better represented human judgment in his experiments than the minimum and maximum operations. This result led to the important insight that "it is not unreasonable for different rules to be used for conjunction under various situations" (Oden 1977b, 572).

The four above-mentioned papers exemplify the fact that an appreciable number of psychologists, especially within the psychology of concepts, became genuinely interested throughout the 1970s in exploring the role and potential utility of fuzzy logic in psychology. This spirit is captured nicely in the last section, entitled "Importance of Studying Fuzzy Psycho-Logic," in Oden 1977b (p. 574):

> It has long been recognized (Luchins and Luchins 1965)[31] that standard formal logic provides a poor description of the processes of actual human thought and that this is in large part because traditional logic is completely discrete and adheres strictly to the principle of the excluded middle: A statement must either be true or false.[32] Clearly, this principle does not characterize much of our subjective experience and knowledge. Given that humans must interact with a world in which the boundaries between things are often not distinct, it is probably a very important and necessary human asset that we are competent at processing fuzzy information.
>
> This fuzziness competency undoubtedly has important implications for the nature of nearly all semantic information processing. . . . Clearly, however, before such theories of fuzzy semantic information processing may be formulated in any detail, it will be necessary to have a much better understanding of the nature of the basic cognitive processes used in dealing with fuzzy information. The present article may provide a beginning for the development of such a model of fuzzy psycho-logic.

These very positive attitudes toward fuzzy logic of many psychologists in the 1970s began to decline visibly in the early 1980s. This change was especially striking in the psychology of concepts, where fuzzy logic started to be portrayed as completely useless. It is now well established, as we describe later, that a paper written by two highly influential cognitive psychologists, Osherson and Smith (1981), triggered this drastic change. According to Gregory Murphy (2002, 24), this paper rep-

---

[31]This book explores connections between foundational issues in mathematics and psychology as well as the role of mathematics in psychology and vice versa. Written jointly by Abraham S. Luchins (1914–2005), one of the most important American Gestalt psychologists, and his wife, Edith Hirsch Luchins (1921–2002), a Polish-American mathematician, the book is highly sophisticated from the standpoint of both mathematics and psychology, as well as their historical development.

[32]Observe that Oden incorrectly interprets the principle of excluded middle, a subject of severe criticism from the standpoint of psychology in Luchins and Luchins 1965, as the principle of bivalence. It seems that the latter was actually on his mind and he just overlooked the subtle distinction between the two principles.

resents one of several attempts made by some psychologists in the late 1970s and early 1980s to revise the classical view of concepts in order to overcome the considerable evidence against it emerging from the above-mentioned experiments by Rosch. In order to understand the specific aims of this paper, consider what the authors themselves have to say (Osherson and Smith 1981, 36):

> The organization of this paper is as follows. We first present one version of prototype theory. We then show how it might be extended to account for conceptual combination by means of principles derived from *fuzzy-set theory* (e.g. Zadeh 1965a). This extension is demonstrated to be fraught with difficulties. We then move on to the issue of truth conditions for thoughts, again using fuzzy-set theory as a means of implementing the prototype approach, and again demonstrating that this implementation won't work. In a final section, we establish that our analysis holds for virtually any version of prototype theory, and consider ways of reconciling previous evidence for this theory with the wisdom of the older kind of theory of concepts.

A short response to Osherson and Smith's paper was published one year later by Zadeh (1982a). After explaining briefly how some of the negative claims about fuzzy logic made by Osherson and Smith could be resolved, he criticized Oherson and Smith's definition of prototype theory, and suggested a more adequate definition. Unfortunately, Zadeh's paper was virtually ignored.

The second response to Osherson and Smith 1981, on a a much larger scale, was presented in a series of five closely connected papers by Fuhrmann (1988a,b, 1989, 1990, 1991).[33] These address not only numerous problems with Osherson and Smith's arguments, but also examine in detail how fuzzy set theory and prototype theory of concepts have developed totally independently of one another since the 1960s. Fuhrmann considers this unfortunate and argues that an appropriate cross-fertilization between the two areas should have been beneficial to both of them. He discusses in great detail how fuzzy set theory should be extended and how the prototype theory of concepts should be reformulated in order to achieve their integration. To this end, he introduces a new type of fuzzy sets, called modified fuzzy sets (or M-fuzzy sets), which would make it possible to achieve this integration.

In spite of the remarkably rich proposal for a feasible connection between fuzzy set theory and the psychology of concepts developed in Fuhrmann's five papers, published within a period of four years (1988–91), his ideas have had virtually no influence on either of the two communities. We can think of several reasons for this unfortunate outcome: (1) The papers were published in journals devoted neither to fuzzy set theory nor to cognitive psychology. (2) After publishing his five papers, Fuhrmann virtually disappeared and, as far as we know, has not published anything since.[34] In any case, his five papers on the relationship between fuzzy set theory and

---

[33] Fuhrmann 1989 is a reprinted version with some technical corrections of Fuhrmann's earlier paper in Zétényi 1988 (pp. 155–202).

[34] Geörgy Fuhrmann is a somewhat mysterious figure. In spite of extensive efforts, only very little seems to be known about him. He received his doctoral degree in Biological Sciences with a specialization in Theoretical Psychophysics from the Hungarian Academy of Sciences in 1987; he worked at the Computer Science Laboratory of the Technical University of Budapest in 1988 and 1989; he was a visiting scholar at the Institute of Cognitive Studies of the University of California at Berkeley in 1990; and he returned to the Technical University of Budapest in 1991. We have not been able to establish any communication with him and we do not even know if he is still alive.

the psychology of concepts indicate his solid background in both these areas. They might eventually provide an operational basis for promoting a fruitful cooperation between the two communities, which is long overdue.

The third and most comprehensive, specific response to Osherson and Smith's paper appeared more than 20 years after its publication. It is expressed in two papers by Belohlavek et al. (2002, 2009) and in the edited book Belohlavek and Klir 2011. The genesis of this response and why it is spread over almost ten years is explained in the preface to Belohlavek and Klir 2011. In a nutshell, a careful analysis of Osherson and Smith 1981 performed by the four authors of Belohlavek et al. 2002, 2009 revealed, surprisingly, that virtually all of Osherson and Smith's conclusions about fuzzy logic were erroneous. In some cases, their mistakes were based on obvious mathematical errors, in other cases on various oversights, misunderstandings, and misconceptions. While Belohlavek et al. 2002 focuses primarily on identifying specific errors, Belohlavek et al. 2009 documents, in addition, how the erroneous conclusions were by and large uncritically accepted within the psychology of concepts and inhibited cooperation with the fuzzy-logic community for some 30 years. The publication of Belohlavek and Klir 2011 by the MIT Press — a well-recognized publisher in the area of psychology—was an effort to stimulate a renewal of this potentially fruitful cooperation.

6.8   Social sciences

6.8.1   A historical overview

Although social sciences are not mentioned in the seminal paper by Zadeh (1965a) nor in any of his other papers published in the 1960s, the connection between fuzzy logic and the social sciences was apparently on his mind, as revealed by this short remark in Zadeh 1971c (pp. 469–470):

> What we still lack, and lack rather acutely, are methods for dealing with systems which are too complex or too ill-defined to admit of precise analysis. Such systems pervade life sciences, social sciences, philosophy, economics, psychology and many other "soft" fields. . . .
>
> Perhaps the major reason for the ineffectiveness of classical mathematical techniques in dealing with systems of high order of complexity lies in their failure to come to grips with the issue of fuzziness, that is, with imprecision that stems not from randomness but from a lack of sharp transition from a membership in a class to nonmembership in it.

He eventually devoted an entire article to this issue, in which he observes (Zadeh 1982b, 25–26):

> During the past decade the focus of systems research in systems theory has shifted increasingly toward the analysis of large-scale systems in which human judgment, perception, and emotions play an important role. Such *humanistic systems* are typified by socioeconomic systems, transportation systems, environmental control systems, food production systems, education systems, health-care delivery systems, criminal justice systems, information dissemination systems, and the like. . . .
>
> The systems theory of the future—the systems theory that will be applicable to the analysis of humanistic systems—is certain to be quite different in spirit as well as in substance from

systems theory as we know it today. I will take the liberty of referring to it as *fuzzy systems theory* because I believe that its distinguishing characteristic will be a conceptual framework for dealing with a key aspect of humanistic systems—namely, the pervasive fuzziness of almost all phenomena that are associated with their external as well as internal behavior.

In spite of these reasonable expectations, the interest of most social scientists in fuzzy logic has been lukewarm at best. A rare exception has been Michael Smithson, an Australian behavioral and social scientist, who has recognized since the early 1980s the importance of fuzzy set theory for behavioral and social sciences, and wrote the first book on this subject (Smithson 1987). When this book was published, he was invited to write a survey paper on the status of fuzzy set theory in the social sciences for *Fuzzy Sets and Systems* (Smithson 1988). His paper contains some interesting observations (pp. 1–2):

> The human sciences have been rather slow to utilize or even evaluate fuzzy set theory and its spinoffs, and the social sciences particularly so. . . .
>
> Yet there is little doubt that the human sciences require formal and even mathematical frameworks for handling graded categories with blurred boundaries. Not only are the phenomena of human thought and behavior inherently fuzzy, but researchers in these fields must use concepts and theoretical schema which themselves are fuzzy (even in a nonpejorative sense). . . .
>
> Why have so few researchers in these fields utilized fuzzy set theory, and why has the dialog between them and fuzzy set theorists been so underdeveloped? At least two reasons immediately spring to mind. First, the human sciences tend to be methodologically conservative when mathematically sophisticated, and mathematically ignorant when methodologically innovative. There are exceptions, of course, but these are general trends in several fields. . . .
>
> A related problem stems from the fact that fuzzy set theory itself is not 'readymade' for applications in the human sciences. It lacks a clear basis in measurement, and many of the connections between fuzzy set concepts and mainstream research concerns in these areas have not been clearly drawn. Crudely put, fuzzy set theory has not been presented in a user-friendly manner.

The classical book by Smithson (1987) has attracted some attention to the useful role of fuzzy sets in the social sciences, but for many years this happened almost exclusively within the fuzzy-logic community. Fortunately, it also attracted the attention of Charles Ragin, a social scientist, who became interested in exploring the use of fuzzy sets for bridging the gap between quantitative and qualitative methods in social-science research, which led eventually to the publication of his book (Ragin 2000), the second major book on the role of fuzzy sets in the social sciences written by a social scientist. While some social scientists praised the book, such as the well-known sociologists Howard S. Becker and John W. Mayer, most remained skeptical (Ragin 2000, 3):

> Social scientists generally stay away from anything labeled "fuzzy" because their work is so often described this way by others, especially by scholars in the "hard" sciences. My initial title for this book, *Fuzzy Social Science*, made so many of my colleagues cringe that I felt compelled to change it so that the adjective "fuzzy" applied to sets, not to social science.

Eight years later, Ragin published another book about using fuzzy sets in social sciences (Ragin 2008), in which he challenges the conventional approach to social science research and proposes an alternative approach based on fuzzy set theory that transcends the various limitations of conventional quantitative as well as qualitative social-science research. As a consequence, as he argues and demonstrates experimentally, this new approach has the capability of narrowing the gap between knowledge obtained by qualitative social scientists and knowledge obtained by quantitative social scientists.

The two books by Ragin make an important statement about the utility of applying fuzzy set theory in the social sciences. Unfortunately, the use of fuzzy-set methods in social sciences has not yet been widely accepted by social scientists, especially those with a quantitative orientation. A simple, human friendly introduction to fuzzy set theory for social scientists written by Smithson and Verkuilen (2006) may possibly help in this regard.

Perhaps the most important and highly original contribution to the use of fuzzy logic in the social sciences at large is at this time the book by Badredine Arfi (2010), a Finnish political scientist. He basically employs the idea of computing with words, first suggested by Zadeh (1996b), and applies this to a wide range of problems in social sciences. Moreover, he allows both membership grades and truth values to be linguistic variables. The book contains forewords by Ragin and Zadeh, who both praise it very highly. According to its preface (Arfi 2010, xi):

> Human beings cannot live and communicate without using natural languages, yet the latter are inherently vague and equivocal. Addressing the problem of linguistic vagueness is no easy task because most social sciences variables are often difficult to precisely operationalize, a problem that confronts equally those inclined to use either qualitative or quantitative methods of analysis. Constantly seeking more conceptual and operational precision and crispness is hence commonly believed to be a golden rule to the production of high quality research work in social sciences. Contra this conventional wisdom, this book raises the question: *What if instead of seeking to get rid of language-caused vagueness we make its preservation an essential requirement for analyzing social science phenomena, both theoretically and empirically?* This book develops a novel approach, termed as linguistic fuzzy-logic methodology, which in essence addresses the problem of language-related vagueness in analyzing social sciences phenomena. Exploring the relatively new field of fuzzy-logic as applied to social and political phenomena takes us into a yet to be explored rich territory of research questions, theoretical arguments, and policy insights. Every theory and every empirical analysis of social phenomena is underpinned in a given logic of reasoning for judging coherence, consistency and the possibility of falsification. Much of what counts as social science today is based on a Boolean logic with two truth values, 0 and 1. The task set out for this book is to go beyond this assumption by showing that linguistic fuzzy logic can be very useful in analyzing empirical data, studying causation, and strategic reasoning in social science theories.

This concludes our historical overview of principal developments pertaining to applications of fuzzy logic in the social sciences at large. In the rest of this section, we cover in more detail some specific developments in two particular social sciences—economics and political science.

6.8.2 Economics

Economics, a social science whose subject is the study of how societies deal with problems that result from relative scarcity, is generally considered to be the most advanced of the social sciences. This assessment is primarily based on the extensive role that mathematics has played in economics since the late 19th century. As is well documented, it was the British economist Alfred Marshall (1842–1924) who first introduced mathematics into economics in his monumental book, *Principles of Economics*, published first in 1890 by Macmillan in London.[35] It is interesting that all mathematics introduced in Marshall's book is presented concisely in a rather short Mathematical Appendix (pp. 838–58) divided into 24 sections labeled Notes I–XXIV, which consists of less than 2.5% of the entire book. The rest of the book is devoted to discussing problems of economics strictly verbally with no mathematics at all. This was in fact an expression of Marshall's hesitancy to overly emphasize the role of mathematics in economics, which he also expressed more explicitly in occasional remarks elsewhere. In *Principles of Economics*, for example, he makes the following remark on p. 780:

> Economics has made greater advances than any other branch of the social sciences, because it is more definite and exact than any other. But every widening of its scope involves some loss of this scientific precision; and the question of whether the loss is greater than the gain resulting from its greater breath of outlook, is not to be decided by any hard and fast rule.

At another place in *Principles of Economics* (p. 368), after discussing his mathematical model of economic equilibria,[36] which involves a relationship among the demand for final products, the supply of final products, the demand for factors of production, and the supply of factors of production, and after specifying the conditions that would exist in an economy in long-run equilibrium, he remarks:

> Nothing of this is true in the world in which we live. Here every economic force is constantly changing its action, under the influence of other forces which are acting around it. Here changes in the volume of production, in its method and its cost are ever mutually modifying one another; they are always affecting and being affected by the character and the extent of demand. Further all these mutual influences take time to work themselves out, and, as a rule, no two influences move at an equal pace. In this world therefore every plain and simple doctrine as to the relations between costs of production, demand and value is necessarily false: and the greater the appearance of lucidity which is given to it by skillful exposition, the more mischievous it is. A man is likely to be a better economist if he trusts to his common senses, and practical instincts, than if he professes to study the theory of value and is resolved to find it easy.

Finally, it is worth mentioning that he made the following two remarks in letters to one of his best students, Arthur L. Bowley, who eventually became a professor of economics and statistics at the London School of Economics:

> In my view every economic fact, whether or not it is of such a nature as to be expressed in numbers, stands in relation as cause and effect to many other facts: and since it *never* happens that all of them can be expressed in numbers, the application of exact mathematical methods to those which can is nearly always a waste of time, while in the large majority of cases it is positively misleading; and the world would have been further on its way forward if the work had never been done at all. . . .
>
> I know I had a growing feeling in the latter years of my work at the subject that a good mathematical theorem dealing with economic hypotheses was very unlikely to be good economics: and I went more and more on the rules—(1) Use mathematics as a shorthand language, rather

---

[35] For our purposes we use the ninth (variorum) edition, with annotations by C. W. Guillebaud, published in 1961 by Macmillan in London and New York, to which also the page numbers given below refer.

[36] The model is formulated in one page as Note XXI in the Mathematical Appendix (p. 855).

than as an engine of inquiry. (2) Keep to them till you have done. (3) Translate to English. (4) Then illustrate by examples that are important in real life. (5) Burn the mathematics. (6) If you can't succeed in 4, burn 3. This last I did often.[37]

Marshall's concerns—and in some sense warnings—about the unguarded use of mathematics in economics have unfortunately been ignored by many economists, who felt that the right way to make progress in economics is to improve Marshall's mathematical treatment and develop it further. Mathematization of economics has been the hallmark of mainstream economics throughout the 20th century. However, the increasing mathematical sophistication of economic theories has not been matched by their predictive capabilities.[38] Arguably, one reason for this discrepancy is the unrealistic precision of classical mathematics for dealing with economic reality. While predictions obtained from economic theories based on classical mathematics are always precise, they are virtually never fully accurate and, more importantly, they are often highly inaccurate and misleading.

It is well known that experienced economists often express fairly accurate economic predictions in linguistic terms, such as "The price of oil is *not likely to increase substantially* in the *near future*." Such predictions are determined by common-sense reasoning, employing the economist's knowledge and relevant information, which is often expressed in linguistic terms as well. Fuzzy logic has the capability of representing such linguistic statements mathematically.

The earliest attempts by some economists to explore the potential of fuzzy set theory for reformulating the texture of economic theory began in the 1970s. These were primarily French economists, who learned about fuzzy sets from the series of four French books on fuzzy sets by Kaufmann (n. 26 on p. 29). Their early explorations, all published in French, are surveyed on p. 360 in Dubois and Prade 1980a. Among these French economists who were early to recognize the utility of fuzzy set theory in economics, Claude Ponsard (1927–1990) seems to be by far the most influential. In order to recognize his leading role in demonstrating the importance of fuzzy set theory in economics, a special issue of *Fuzzy Sets and Systems* edited by Antoine Billot (1992a), an accomplished French economist, was posthumously dedicated to Ponsard. This special issue contains an overview by Billot and Thisse (1992) of Ponsard's two major contributions to economics—the important advances he made in the previously somewhat neglected spatial economic theory (Ponsard 1983) between 1953 and 1974, and his pioneering work that introduced fuzzy set theory to economics, especially to spatial economics, during the period from 1975 until his untimely death in 1990.

After his initial explorations of using fuzzy sets in economics in the late 1970s, Ponsard's primary contributions in this area were published in the 1980s, and most of them were in English. In one of these early papers, Ponsard (1981) shows, for ex-

---

[37] A. C. Pigou, ed., *Memorials of Alfred Marshall* (London: Macmillan, 1925), 422, 427.

[38] This has been observed by historians of economics; see, for example, H. Landreth, *History of Economic Theory* (Boston: Houghton Mifflin, 1976).

ample, how fuzzy sets can be used to reformulate the classical theory of consumer behavior in mathematical economics by discarding its two unrealistic assumptions: the
assumption that the consumer can perfectly discriminate between different goods,
and the assumption that the goods satisfying customer's needs are all supplied at
a unique point in space. The result is a considerably more realistic theory.

This was followed by a series of papers by Ponsard (1982a,b, 1985a,b), in which
he fuzzified other areas of classical economics, and which culminated in a book edited
by Ponsard and Fustier (1986). This book contains ten important papers on the use
of fuzzy logic in economics, including two contributions by Ponsard, one on a theory of spatial general equilibrium in a fuzzy economics, and one on viewing spatial
oligopoly as a fuzzy game.[39] All of the papers in this book were written by members of the Institute for Mathematical Economics at the University of Dijon, which
at that time was directed by Ponsard. Later on, Ponsard (1987a) generalized the famous Nash equilibrium concept by showing that an n-person noncooperative fuzzy
game with mixed strategies has at least one equilibrium point.[40]

In Ponsard 1988, which provided an excellent survey of established fuzzy mathematical models in economics at that time, Ponsard begins with a question raised
by Zadeh in his foreword to the classical book by Zimmermann (1985): "Are there,
in fact, any significant problem areas in which the use of the theory of fuzzy sets
leads to results which could not be obtained by classical methods?" He ends the paper by answering this question: "In economics, the answer is positive. The use of
fuzzy subset theory leads to results which could not be obtained by classical methods." This is a significant statement, since at that time no one was as erudite in fuzzy
economics as Ponsard.

In spite of Ponsard's impressive accomplishments, the community of mainstream economists showed virtually no interest in his work. The only recognition
he received was an invitation to write an article on fuzzy sets for the first edition of
The New Palgrave: A Dictionary of Economics (Ponsard 1987b).

In 1988, Ponsard began working on a major book with the tentative title *Fuzzy
Economic Space: An Axiomatic Approach*. When he died unexpectedly on March
24, 1990, his manuscript was not yet fully completed. Fortunately, his main ideas
are preserved and further developed in an important book by Billot (1992b), who
was a doctoral student of Ponsard's in the late 1980s. The significance of this book
is well described in a fairly extensive preface by Zimmermann (Billot 1992b, IV-X):

> There seems to be one major difference between Dr. Billot's contribution and the other pub-
> lications in which fuzzy set theory is applied to either game theory, group decision making or
> economic theory: Some authors use fuzzy set theory in order to gain a mathematically more at-
> tractive or efficient model for their problems of concern, others use a specific problem area just
> in order to demonstrate the capabilities of fuzzy set theory without a real justification of whether
> this "fuzzification" of a traditional theory makes sense or is justified by better results. For Dr.
> Billot, not surprisingly similar to Professor Ponsard, the economic thinking seems to be the pre-

---

[39] The term "fuzzy economics" was coined by Ponsard for mathematical economics based on fuzzy set theory.
[40] J. F. Nash, "Equilibrium points in n-person games," *Proc Natl Acad Sci USA* 36 (1950): 48–49.

dominant factor and fuzzy set theory is only a tool to improve an economic model or theory where it seems possible or appropriate. Hence, his book should not primarily be considered as a contribution in fuzzy set theory, but as one in economic theory, which merges microeconomic considerations with macroeconomic observations.

It is clear that Ponsard and the group of French economists influenced by him, including Billot, represent the most important school of thought in fuzzy economics, and thus their work suffices to exemplify the utility of fuzzy logic in economics. Among other developments pertaining to fuzzy economics, there is also a specialized journal on fuzzy economics, *Fuzzy Economic Review*, sponsored by the International Association for Fuzzy-Set Management and Economy, which commenced its biannual publication in 1996.

6.8.3 Political science

Among the uses of fuzzy logic in political science, many results have emerged from fairly recent collaborative efforts involving a group of political scientists and a group of mathematicians specializing on the mathematics of uncertainty at Creighton University in Omaha, Nebraska. The outcome of these efforts is covered comprehensively in a book by Clark et al. (2008a). Around the time their work on the project described in the book began, a historical overview of the use of fuzzy set theory in political science was published by Nurmi and Kacprzyk (2007), who conclude (p. 297):

The impact of fuzzy sets on the mainstream political science has thus far been marginal. Despite the intuitive applicability of fuzzy sets to the analysis of political discourse and other forms of behavior, the existing fuzzy sets applications focus mainly on individual and group decisions and comparative politics. The most important empirical results have been achieved in comparative politics and—to a lesser extent—in international relations. Yet, we believe the potential of fuzzy logic is far from exhausted by these applications. We are particularly optimistic about the possibilities opened by advances in computing with words.

Here the last sentence refers to a paper by Arfi (2005), which introduced computing with words into political science, and which Nurmi and Kacprzyk discuss in their overview in great detail.[41] They also mention that the most important results from fuzzy-set applications in political science have been obtained in two areas: international relations and comparative politics. Fuzzy-set applications in international relations are well characterized in a paper by Cioffi-Revilla (1981); those in comparative politics are closely associated with the above-mentioned collaboration between political scientists and mathematicians at Creighton University, which is our main topic in this section. The genesis of this collaboration is described in detail in Clark et al. 2008a. The following is a brief summary.

Interest in theoretical approaches to comparative politics greatly increased in political science since the abrupt end of the Cold War in the early 1990s, when many countries became involved in the transformation from Communist Party rule to

---

[41]Arfi is a political scientist who later published an important book on methods of linguistic fuzzy logic in the broader context of the social sciences (Arfi 2010), which is discussed in section 6.8.1.

functioning democracies. There was virtually no experience with political and so-cial transformations of this kind and that greatly motivated theoretical research in comparative politics—the study and comparison of the world's democratic political systems.

Terry Clark received his PhD in political science from the University of Illinois at Urbana-Champaign in 1992, shortly after the end of the Cold War, and immedi-ately became interested in studying the new democratic political systems in eastern and east-central Europe, especially those that had already been consolidated. After embarking on his academic career in political science at Creighton University in 1993, he devoted considerable attention to studying these emerging political systems. Ten years later, he published a book on his studies whose preface begins as follows:[42]

> It has been my good fortune that my academic career began on the eve of the collapse of com-munism in eastern and east-central Europe. I, along with my colleagues, have had the privilege of studying the profound social, economic, and political challenges that have attended that col-lapse. In contrast to those who labored before us, we have had access to rich repositories of data of all types and descriptions, as a consequence of which we have been able to pursue a greater universe of questions. Nonetheless, despite both the energy and excitement that all of this has generated, I must confess to one frustration. Almost ten years after the collapse, we still do not have adequate theories capable of guiding research in many major areas of concern.

In his studies, Clark followed the common assumption in comparative politics that individuals in democratic political systems prefer in a rational way some political outcomes to others and he employed so-called spatial modeling, commonly used in comparative politics for deriving and testing hypotheses on political outcomes resulting from institutional designs in democratic systems of government.[43] In spa-tial modeling, some elementary concepts of classical (Euclidean) geometry, such as the Euclidean distance from an individual's preferred alternative to other available alternatives, all viewed as spatial points, are utilized for analyzing political systems.

Clark's frustration about the lack of adequate theories guiding research in the area of comparative politics, mentioned in the last sentence of the above quotation, resulted from his dissatisfaction with spatial modeling. He increasingly recognized that it is highly unrealistic for applications in comparative politics, primarily due to the required precision of classical geometry. In the late spring of 2005, in the Department of Mathematics at Creighton University, he happened to meet with John Mordeson, a mathematician specializing on the mathematics of uncertainty and especially on mathematics based on fuzzy logic. At this meeting Clark described the various shortcomings of classical spatial modeling in comparative politics and Mordeson thus became interested in exploring fuzzy geometry to overcome some of these difficulties. This led to their cooperation, which eventually included a few ad-ditional researchers from their respective teams. The first results of this cooperation

---

[42] T. D. Clark, *Beyond Post-Communist Studies: Political Science and the New Democracies of Europe* (Armonk, NY: Sharpe, 2002).

[43] See, for example, the book by D. Austen-Smith and J. S. Banks, *Positive Political Theory I: Collective Preference* (Ann Arbor, MI: Univ. of Michigan Press, 1999).

are covered in Clark et al. 2008a. After discussing the motivation for applying fuzzy set theory in comparative politics in chapter 1, the basics of fuzzy set theory, written primarily for political scientists, are presented in chapter 2. This is followed by introducing in chapter 3 a version of fuzzy plane geometry developed by Buckley and Eslami (1997). The principal results obtained by applying one- and two-dimensional fuzzy spatial models to comparative politics are covered, respectively, in chapter 4 and chapter 5. Both of these, but especially the two-dimensional models, are shown to be superior to their classical counterparts. The remaining two chapters of the book, one on introducing the broad range of aggregation operations that qualify for aggregating fuzzy preferences in spatial analysis and one on the notorious problem of cycling in spatial models, both classical and fuzzy, are oriented more toward preparing the ground for future research.

Shortly after publishing the book (Clark et al. 2008a), the two collaborating groups published a paper (Clark et al. 2008b) in which they present a fuzzy reformulation of the so-called portfolio allocation model.[44] This reformulation is extended to surplus majority governments and illustrated by a case study of the 1996 surplus majority coalition government in Lithuania. The study shows that the fuzzy version of the portfolio allocation model is clearly superior to its original version. It has the potential, for example, of enlarging the number of realistic governments that might be predicted to emerge from the cabinet-making process.

Five years after publishing their first book (Clark et al. 2008a), another book, coauthored this time by Mordeson et al. (2013), emerged from the interdisciplinary collaboration at Creighton University. This book, however, deals with very different issues than the first one. It explores new mathematical techniques involving fuzzy set theory and other types of uncertainty for modeling global issues in the areas of comparative politics, international relations and public policy, such as political stability, economic freedom, quality of life, and nuclear deterrence. The aim of these techniques is to determine linear equations between a dependent variable and one or more independent variables in cases where conventional techniques such as linear regression are of little use due to the nonrepeatability of experiments, availability of only a small number of data points, or dependence on data derived from experts' opinions. A large part of the book is devoted to case studies in which the various proposed techniques are compared.

---

[44] This is a formal model regarding cabinet-making processes in parliamentary systems that was introduced in a book by M. Laver and K. A. Shepsle, *Making and Breaking Governments: Cabinets and Legislatures in Parliamentary Governments* (Cambridge: Cambridge Univ. Press, 1996).

6.9 Computer science

Computer science is an area in which fuzzy logic has been applied widely.[45] Computer science encompasses many diverse areas, including theoretical, experimental, and engineering fields.[46] Since it would be virtually impossible to characterize the many areas in computer science in which attempts to apply fuzzy logic have been made, we instead concentrate on three which are significant, illustrate a typical reason for employing fuzzy logic in computer science, namely its ability to model inexact concepts used by humans to describe data, and illustrate a reliance of certain applications on fuzzy logic in the narrow sense.

Databases and information retrieval

Databases, i.e. computer systems which allow efficient storage, querying, and retrieval of data, represent an area in which fuzzy logic has a great potential for significant applications.[47] The first contributions in applying fuzzy logic to databases appeared in the late 1970s and include Tahani 1977; Mizumoto, Umano, and Tanaka 1977; and Giardina 1979. These were followed in the early 1980s by several other papers as well the first book by Zemankova-Leech and Kandel (1984). Among them is the conceptually important paper by Buckles and Petry (1982), who used the term "fuzzy relational database" and introduced two basic ideas. The first is to store in a table entry corresponding to a row and column, a whole *set* of values of the corresponding domain instead of a *single* value as in ordinary relational databases. With domains $D_1, \ldots, D_n$, they thus consider the table as a classical relation $R \subseteq 2^{D_1} \times \cdots \times 2^{D_n}$ instead of $R \subseteq D_1 \times \cdots \times D_n$. The second is to equip each *domain with a similarity relation* which is a fuzzy equivalence. Later, this idea proved to be of great importance because it allows similarity querying of the database, i.e. queries such as "show houses which cost approximately $200,000 and are near Ithaca, NY." The authors, however, consider only crisp queries with similarity thresholds such as "show houses with price similar to $200,000 to degree at least 0.8," the result of which is a classical relation with tuples containing possibly sets of domain values. The natural next step to consider—in addition to similarities on domains—also *degrees attached to tuples* is due to Raju and Majumdar (1988) who hence consider a database table as a fuzzy relation $R : D_1 \times D_n \to [0,1]$, such as[48]

---

[45] This is also apparent from the quantitative assessment in section 7.4.

[46] National Research Council, *Computer Science: Reflections on the Field, Reflections from the Field* (Washington, DC: National Academies Press, 2004).

[47] Strictly speaking, a database is a collection of data. The term "database" is also used to denote what should properly be called a database management system and we use it in this meaning. Each database management system is based on a certain database model which represents a conceptual foundation for the system, i.e. its theoretical principles and concepts. The most important is Codd's relational model of data and the so-called relational databases, due to E. F. Codd, "A relational model of data for large shared data banks," *Commun ACM* 13 (1970): 377–87.

[48] Attaching degrees to tuples also appeared in the early works of the late 1970s.

	ID	PRICE	BEDROOMS	LOCATION
0.98	62	195,000	2	Ithaca, NY
0.85	14	230,000	2	Trumansburg, NY
0.73	23	145,000	1	Ithaca, NY
0.40	59	320,000	4	Candor, NY

Nevertheless, their interpretation of the degree $R(t)$ to which a tuple t belongs to R, such as the above 0.98, 0.85, 0.73, or 0.40, as "a possibility measure or a measure of association of the items of a tuple" is somewhat unclear, and is also reflected in their not fully exploiting such degrees. Still, their paper is one of the most influential ones on extending the classical relational model from a fuzzy logic point of view. In their later studies, Belohlavek and Vychodil (2006c, 2010) argued that $R(t)$ is naturally interpreted as the degree to which t matches a similarity query. With appropriately defined similarity relations, the above table can be seen as a result of the similarity query "show houses which cost approximately $200,000 and are near Ithaca, NY." Thus, house ID 62 matches the query to degree 0.98 and so on. In this view, stored data in which all the degrees $R(t)$ equal 1 correspond to the empty query. The concept of a data table as a fuzzy relation has then the same interpretation and role as the concept of a table as a relation in the classical Codd's model. In another early paper, Prade and Testemale (1984) proposed a generalization of the relational model in which the values in tables are known imprecisely and—more generally than Buckles and Petry—suggested that the values be represented by possibility distributions, that possibility distributions may even appear in queries, and developed the basics of the thus extended relational algebra. Later on, the above ideas as well as modifications and some new ideas were examined in a number of papers. These focused on various issues including the important query languages (Buckles and Petry 1985; Kacprzyk, Zadrożny, and Ziolkowski 1989; Bosc and Pivert 1995). For overviews of these developments, see Bosc and Kacprzyk 1995; Petry 1996; Galindo, Urrutia, and Piattini 2006; and the comprehensive handbook Galindo 2008.

    The potential of using fuzzy logic in databases derives basically from two natural needs reflected in the above contributions. One is the need to retrieve data using queries including the above similarity queries and possibly also other human-like queries whose answers require *approximate match*, i.e. a match to a certain degree, rather than an exact, yes-or-no match. The other is the need to *represent and store inexact data* and, in general, information which is in various ways deficient. Evidence that these problems have for long been perceived as important in mainstream database research is well documented by the Lowell database assessment of 25 leading experts.[49] This concludes that "current DBMS have no facilities for either approximate data or imprecise queries" and lists the management of uncertainty in data among its six of the most important research directions. In light of this commonly recognized need, it may be puzzling that with some few exceptions, such as Raju and Majumdar 1988, the numerous contributions on fuzzy logic to databases

---

[49] S. Abiteboul et al., "The Lowell database research self-assessment," *Commun ACM* 48 (2005): 111–18.

have not appeared in major database journals or conferences, but, instead, have only appeared in journals and conferences devoted to fuzzy logic. As a result, these contributions have not generally been recognized or taken seriously in mainstream database research, which is in sharp contrast to the appearance of numerous contributions in major database forums on probabilistic databases which also address the challenges mentioned above but from a conceptually different perspective.[50] This phenomenon is partly examined by Belohlavek and Vychodil (2011), who comment upon the large number of papers on fuzzy logic approaches to Codd's relational model. They recall that the key factors of success of the ordinary Codd's model of data is its foundation and clear rooting in predicate logic. Thus, one of the major figures in the field, Christopher J. Date, says: "The relational approach really is rock solid, owing (once again) to its basis in mathematics and predicate logic."[51] They further argue that as a rule, the existing fuzzy logic extensions of Codd's model are ad hoc in that a clear link to a corresponding logic framework is missing or not handled appropriately. Clearly, such a link is extremely important for these extensions because with intermediary grades of truth, things become more technically involved and, consequently, the need for a coherent bundle of logical concepts, principles, and results is apparent. Belohlavek and Vychodil contend that the lack of solid logical foundations is the main reason why most of the existing contributions to fuzzy logic extensions of Codd's model are, by and large, definitional papers. While Codd's model represents a complex body of mutually interrelated notions, such as those regarding redundancy, data dependencies, query systems, or domain and tuple calculi,[52] the existing fuzzy logic contributions typically focus on particular issues only, and several problems—fundamentally important for design, feasibility, efficiency and conceptual clarity—are left with little consideration. They further argue that with the recent developments of fuzzy logic in the narrow sense, solid foundations for fuzzy logic extensions of the relational model may be worked out and demonstrate their claim by their own developments, e.g. of syntactico-semantically complete calculi for fuzzy functional dependencies or complete domain and tuple calculi and other issues regarding query systems of similarity databases (Belohlavek and Vychodil 2006c, 2010).

From this standpoint, the view of database researchers that "fuzzy databases" are not well developed is justified.[53] On the other hand, similarity queries have been intensively studied by the database community but their studies assume the classi-

---

[50] E.g. N. Dalvi and D. Suciu, "Management of probabilistic data: Foundations and challenges," in *Proc 26th ACM Symp on Principles of Database Systems* (New York: ACM, 2007), 1–12; N. Dalvi and D. Suciu, "Probabilistic databases: Diamonds in the dirt," *Commun ACM* 52 (2009): 86–94.

[51] C. J. Date, *Database Relational Model: A Retrospective Review and Analysis* (Reading, MA: Addison-Wesley, 2000), 138.

[52] For a clear exposition, see the classic book by D. Maier, *The Theory of Relational Databases* (Rockville: Computer Science Press, 1983).

[53] Very often it is simply conceptually confusing and garbled writing on "fuzzy databases" that makes a bad impression.

cal Codd's model with a similarity module involving metric-based similarity on top of it.[54] The fuzzy logic extension based on adding similarity relations to domains may from this perspective be regarded as conceptually clean and offering a more correct way of looking at similarity queries—instead of adding metric structures atop the classical model, the change from classical to fuzzy logic, and the ensuing change from equality to similarity, provides an appropriate framework for similarity queries which is purely logically based and hence conceptually of the same kind as the classical Codd's model. To illustrate using a highly visible example that fuzzy logic ideas are indeed of fundamental relevance for databases, we now turn to interesting research on aggregation in queries involving fuzzy predicates.

In the 1996 Symposium on Principles of Database Systems, a leading database conference, Ronald Fagin, a computer scientist and logician at the IBM Almaden Research Center, presented a paper entitled "Combining fuzzy information from multiple systems." This paper, of which an extended version was later published as Fagin 1999, as well as its follow-ups (Fagin 1998; Fagin, Lotem, and Naor 2003; Fagin, Kumar, and Sivakumar 2003), contain new results important for database theory and practice as well as for analysis of algorithms and computational complexity theory. For their 2003 paper, Fagin, Lotem, and Naor were awarded the prestigious 2014 Gödel Prize. In this paper, they "produced a groundbreaking result . . . by introducing the powerful 'threshold algorithm.' The combination of its elegant mathematical property and its simplicity contributed to making the threshold algorithm a foundation for much follow-on research, and it is now widely used in numerous applications and systems."[55] Because of their importance and simplicity, the details are worth describing.

The problem addressed by Fagin results in multimedia database systems which store images, video, and other complex data. To access such data, the systems must be able to handle user queries naturally formulated in terms which are inherently fuzzy. A user may want to get images with objects which are red and round. A degree to which an object x satisfies such a query is naturally obtained by an aggregation, using min or another fuzzy logic conjunction or aggregation function t, of degrees to which x is red and to which x is round. Fagin assumes a scenario in which the N objects in the database are accessed in terms of m fuzzy attributes, such as "red" or "round." He assumes that for each attribute $i = 1, \ldots, m$, the middleware may obtain a list L_i of pairs $\langle x, x_i \rangle$ sorted in a descending order by x_i where $x_i \in [0,1]$ is the grade to which attribute i applies to the object x.[56] The middleware is in fact assumed to be capable of *sorted* (or sequential) *access*, i.e. browsing L_i from its first

---

[54] E.g. C. Li, K. C.-C. Chang, I. F. Ilyas, and S. Song, "RankSQL: Query algebra and optimization for relational top-k queries," in *Proc 2005 ACM Conf on Management of Data* (New York: ACM, 2005), 131–42.

[55] "2014 Gödel prize," *ACM SIGACT News* 45, no. 2 (2014): 5. The prize is given annually for outstanding papers in theoretical computer science awarded jointly by the Association for Computing Machinery Special Interest Group on Algorithms and Computational Theory (ACM SIGACT) and the European Association for Theoretical Computer Science (EATCS).

[56] Middleware is a software system on top of its various subsystems used to access data in the subsystems.

pair to the second to the third and so on. If $\langle x, x_i \rangle$ is on place l in L_i, one needs l sorted accesses to retrieve x_i. It may also do a *random access*, in which the grade x_i is retrieved with one access. If s and r sorted and random accesses are performed, the (total) *middleware cost* is $c_s \cdot s + c_r \cdot r$. Now, the problem is to find an algorithm which computes the *top k* objects best satisfying the query, i.e. k objects x with top aggregate degrees $t(x_1, \ldots, x_m)$, with a low middleware cost and with ties broken arbitrarily. The result may thus be seen as a fuzzy set $\{t^{(x_1, \ldots, x_m)}/x \mid x$ is one of the top k objects$\}$. A naive algorithm would need to browse all m lists. The first efficient algorithm, described in Fagin 1999 and nowadays called *Fagin's algorithm*, is optimal with high probability in the worst case if the atomic queries are independent in a certain precisely defined sense. Nevertheless, it is not optimal in all cases. The *threshold algorithm* (Fagin, Lotem, and Naor 2003), which is inspired by Fagin's algorithm and which appeared independently elsewhere, improves this substantially.[57] The algorithm is simple enough to be presented here:

1. Set $j = 1$.
2. For every $i = 1, \ldots, m$, do a sorted access to the jth pair $\langle x, x_i \rangle$ in L_i. If x has not been previously seen, do random accesses in the other lists, $L_p \neq L_i$, to obtain its grades $x_1, \ldots, x_m$. If $t(x_1, \ldots, x_m)$ is one of the k highest so far, then remember x and its $t(x_1, \ldots, x_m)$.
3. Take the grades $a_1, \ldots, a_m$ of the jth objects in lists $L_1, \ldots, L_m$ and compute the threshold $\tau = t(a_1, \ldots, a_m)$. If at least k of the remembered objects x satisfy $t(x_1, \ldots, x_m) \geq \tau$ then output the k best objects x with their $t(x_1, \ldots, x_m)$ and stop; otherwise increase j by 1 and continue on line 2.

The threshold algorithm is not only correct but, importantly, it is *instance optimal*, meaning that in a sense it is optimal for every instance, i.e. in a very strong sense, stronger than being optimal in the worst or average case. The notion of instance optimality introduced in the paper was itself new and proved to be useful in computational complexity theory. Note also that even though the threshold algorithm is rather short, its various complexity analyses in Fagin, Lotem, and Naor 2003 are highly nontrivial.

Fuzzy logic has also been used in the related area of *information retrieval*, where the basic problem is to retrieve documents from a large collection that are relevant with respect to keywords specified by a user. The first works in this area include Radecki 1979, 1981; Nakamura and Iwai 1982; Kraft and Buell 1983; and Miyamoto, Miyake, and Nakayama 1983. The principal idea with variations that have appeared in numerous subsequent papers, consists in modeling the keyword-document relationship by a fuzzy relation representing the degree of relevance of keywords to

---

[57] In S. Nepal and M. V. Ramakrishna, "Query processing issues in image (multimedia) databases," in *Proc 15th Int. Conf. on Data Engineering* (IEEE, 1999), 22–29; and in U. Güntzer, W. T. Balke, and W. Kießling, "Optimizing multi-feature queries in image databases," in *Proc 26th Very Large Databases* (San Francisco, CA: Morgan Kaufmann, 2000), 419–28. An extended abstract of Fagin et al.'s paper is in *Proc 20th ACM Symp on Principles of Database Systems*, (New York: ACM, 2001), 102–13.

documents and possibly also assigning truth degrees to search keywords to model their importance in the search. A separate section of the widely recognized book on information retrieval by Baeza-Yates and Ribeiro-Neto is devoted to the so-called fuzzy set model of retrieval.[58]

Machine learning and data mining

Machine learning and data mining are two closely related areas in which the usefulness of fuzzy logic has been explored in many contributions. These include a number of contributions in mainstream journals and conferences. Historically, the first successful and nowadays widely recognized machine learning method was Bezdek's *fuzzy c-means clustering*. Many contributions on several aspects of this method have been published in leading journals such as *Pattern Recognition, Pattern Recognition Letters,* or *IEEE Trans. on Pattern Analysis and Machine Intelligence,* and the method is now included routinely in standard books on clustering. This method is presented in the context of the general idea of fuzzy clustering in section 3.8; see also the recent paper Khalilia et al. 2014 for further information.

The reason fuzzy logic is useful in clustering is in principle the same as with the other methods of machine learning and data mining—fuzzy logic provides a means of representing concepts that people may learn from data as well as concepts that people use to describe knowledge hidden in complex data. These concepts, which are central to machine learning and data mining, are inherently fuzzy. We now survey some contributions which clearly demonstrate the role of fuzzy logic in machine learning and data mining, and which have appeared in leading journals and conferences.

First, there is the interesting work by Kruse and his coworkers, mainly former doctoral students, on learning various kinds of *graphical models* from data. These models include Bayesian networks but also networks representing dependencies in relational data and data with possibilistic uncertainty. The methods were used in many projects of Kruse's group, e.g. in telecommunications or the automotive industry. See in particular the book by Borgelt and Kruse (2002) for more information, as well as the recent book Kruse et al. 2013 on the broader topic of computational intelligence.

In a series of papers, various authors including Ada Fu, Man-hon Wong, Ben Yahia, Gyenesei, Guoqing Chen, Hüllermeier, Dubois, Prade, Bosc, Pivert, and Sudkamp, have explored *fuzzy association rules* which are dependencies of the form $A \Rightarrow B$ where A and B are sets of attributes. The attributes are fuzzy in that they apply to particular objects to degrees. The aim is to extract from a large data table, in which rows and columns represent degrees to which the attributes apply to objects, a concise set of informative rules such as {obesity, little sport} ⇒ {high blood

---

[58]R. Baeza-Yates and B. Ribeiro-Neto, *Modern Information Retrieval* (Harlow, England: Addison-Wesley, 1999), 2nd edition published in 2011.

pressure}. An informative rule needs to have high confidence in that many objects satisfying A satisfy B and have high support in that a good portion of objects satisfies both A and B.[59] Such rules extend the classical association rules which are dependencies in data with Boolean rather than fuzzy attributes, and which are among the most widely used methods of modern data mining. For further information including existing approaches and numerous references, see Dubois, Hüllermeier, and Prade 2006.

Data representing a fuzzy relation between objects and attributes are the subject of several papers on *formal concept analysis* of data with fuzzy attributes, initiated independently by Burusco and Fuentes-González in 1994, Pollandt in 1997, and Belohlavek in 1998.[60] Its foundations were worked out in the late 1990s, particularly by Belohlavek and in the early 2000s also by his colleagues and other researchers, including Krajči and Vychodil. Further information including references may be found in Belohlavek 2002c, 2004; Ben Yahia and Jaoua 2001; Burusco and Fuentes-González 1994; Krajči 2005; and Pollandt 1997. Here the aim is to extract and further use so-called formal concepts which are certain natural representations of human-like concepts with fuzzy boundaries characterized by their extents and intents. The method has strong logico-algebraic foundations and also involves rules $A \Rightarrow B$ related to the ones mentioned in the previous paragraph. An interesting utilization is in factor analysis of data with both Boolean and fuzzy attributes for which formal concepts, which are the fixpoints of certain operators associated with the input data table, were proved to be optimal factors. The method, initially developed by Belohlavek and Vychodil, enables the extraction of naturally interpretable factors in relational data, including ordinal data (Belohlavek and Krmelova 2013), and has been used for analysis of numerous datasets such as sports data and other performance data.

A series of contributions to machine learning and data mining employing fuzzy logic, including the above-mentioned work on association rules, is due to Hüllermeier. For instance, Hüllermeier (2003), building on his previous joint work with Dubois and Prade, explored fuzzy sets and possibility theory as an alternative to the existing probabilistic approaches in the case of *instance-based learning* which is a particular case of supervised learning. In particular, he proposed a similarity-based principle of extrapolation which implements David Hume's well-known general principle underlying much of instance-based learning, according to which similar causes lead to similar effects. Binary fuzzy relations capturing the properties of graded preference relationships are employed in preference learning tasks in various

---

[59] Confidence and support are technical terms in this area. In the context of data mining, association rules were introduced by R. Agrawal, T. Imielinski, and A. Swami, "Mining association rules between sets of items in large databases," in *Proc 1993 ACM Conf on Management of Data* (New York: ACM, 1993), 207–16. Much more general rules were considered in the GUHA method introduced in the late 1960s; see Hájek and Havránek 1978.

[60] Classical formal concept analysis was invented by Rudolf Wille in the early 1980s. The basic text is by B. Ganter and R. Wille, *Formal Concept Analysis: Mathematical Foundations* (Berlin: Springer, 1999).

ranking problems; see e.g.   Hüllermeier et al. 2008 on so-called label ranking or Pahikkala et al. 2013 on conditioned ranking of relational data. Hüllermeier's further contributions along with others to machine learning and data mining which employ fuzzy logic are surveyed in Hüllermeier 2011.

Logic programming and automated proving

Examining logic programming and, more generally, resolution and other automated theorem proving methods from the point of view of fuzzy logic is clearly a natural move. The first examinations of resolution in the context of fuzzy logic was due to Lee (1972), whose work is considered above (p. 181). Lee used min, max, and $1 - x$ for conjunction, disjunction, and negation, and formulated a resolution principle in this setting. Several subsequent papers, some coauthored by M. Mukaidono—see e.g. Mukaidono, Shen, and Ding 1989—consider cases in which only the S-implication is considered. A survey of early approaches to fuzzy logic programming is provided in Dubois, Lang, and Prade 1991. Fuzzy logic programming with Łukasiewicz connectives, for which S-implication coincides with the residuated one, was later studied by Klawonn and Kruse (1994) and Thiele and Lehmke (1994). While the above resolution methods worked with formulas in the form of disjunction of atomic formulas and their negations, later on, they were also developed for logics with residuated implications, such as in Mukaidono and Kikuchi 1993 for Gödel logic and in Vojtáš 2001 for a more general setting. The main results obtained are various forms of completeness for the resolutions examined, including Pavelka-style completeness by Vojtáš (2001). Some of the approaches have also been implemented as computer programs. Various automated theorem proving methods, including resolution, have also been considered in the context of general many-valued logics in a number of works. In particular, Hähnle 1994 was the first book on this topic; later Baaz and Fermüller (1995) and also O'Hearn and Stachniak (1992) provided general frameworks for first-order many-valued logics. For more information, see chapter 7 in Gottwald 2001.

Other developments

Numerous other contributions to various areas of computer science have been made as well, including studies of various kinds of *automata*. The problem of general fuzzy automata, which is connected to the very notion of computability, is examined in section 4.6.2. This section illustrates that computability notions examined from a fuzzy logic viewpoint lead to interesting fundamental questions. Related to these are investigations on certain particular and other types of automata. For works on finite fuzzy automata, see the early advanced study Wechler 1978, the monograph Mordeson and Malik 2002, and also Ćirić et al. 2010—one of many papers that have recently appeared in the journals *Fuzzy Sets and Systems* and *Information Sciences*. These papers employ residuated lattices as structures of truth degrees and, in this regard, may be viewed as continuing the direction started by Wechsler (1978). On the other hand, such automata are special cases of automata over semirings and are related to other kinds of weighted automata.[61] Other sorts of automata have been studied as well, such as fuzzy cellular automata (Cattaneo et al. 1997). Another area particularly worth mentioning is image processing—covered to some extent in section 6.12. For further information about fuzzy logic in computer science, see the chapter by Belohlavek, Kruse, and Moewes (2011) in a book on the current state of computer science from a historical perspective edited by Edward Blum and Alfred Aho, leading experts in the field.

6.10 Medicine

Medicine is one field in which the utility of fuzzy set theory was appreciated quite early, by about the mid-1970s. The need for fuzzy logic in medicine was initially rec-
· ognized in the area of medical diagnosis. This is not surprising since the most useful descriptors of disease features are usually linguistic terms that cannot be adequately represented in mathematics based on classical set theory. For example, the following statement is a typical description of hepatitis: "Total proteins are *usually normal*, albumin is *decreased*, α-globulins are *slightly decreased*, β-globulins are *slightly decreased*, and γ-globulins are *increased*." Clearly, the italicized terms are inherently fuzzy. They can be adequately represented by appropriate fuzzy sets, but not by any classical sets.

The possibility of using fuzzy set theory in medical diagnosis was first suggested and discussed in a doctoral dissertation by Merle Anne Albin (1975), and shortly after that in two early papers by Elie Sanchez (1977, 1979).[62] Sanchez character-izes medical diagnosis in his papers in terms of fuzzy relational equations, which he introduced in his earlier paper Sanchez 1976. In his model, a physician's med-ical knowledge in some medical specialization is represented as a fuzzy relation K defined on $S \times D$ where S and D denote sets of symptoms and diseases, respectively. For each pair, $\langle s, d \rangle \in S \times D$, the value $K(s, d)$ represents the degree of association between symptom s and disease d. Then, given the relation K and a fuzzy set, S_p, which describes a physician's assessment of the degrees of severity of all symptoms in S for a particular patient p, the sought fuzzy set, D_p, which describes the degrees of possibility that patient p suffers from individual diseases in set D, can be obtained via the fuzzy relational equation $D_p = S_p \circ K$ (section 3.5). Of course, when Sanchez

---

[61] M. Droste, W. Kuich, and H. Vogler (eds.), *Handbook of Weighted Automata* (Berlin: Springer, 2009). Semi-ring automata were studied by M. P. Schützenberger, "On a theorem of E. Jungen," *Proc AMS* 13 (1962): 885–90.

[62] Albin's dissertation was done in mathematics under the supervision of Hans Bremermann, but Lotfi Zadeh was also a member of her doctoral committee. In her dissertation, Albin focuses mostly on the role of fuzzy cluster-ing and on the method of deformable prototypes developed by Bremermann (1976). After completing her studies in mathematics, she studied medicine at the Medical School of the University of California at San Diego. After graduating, she never returned to her historically significant work on fuzzy logic in medicine.

published his early papers on medical diagnoses, he employed max-min composition, which was the only one considered at that time. That is, the above relational equation represents in fact a set of equations, one for each $d \in D$ of the form

$$D_p(d) = \max_{s \in S}\{\min[S_p(s), K(s,d)]\}.$$

Sanchez also showed in his second paper (Sanchez 1979) that fuzzy relational equations could be used, at least in principle, to construct medical knowledge (fuzzy relation K) from databases containing diagnostic records of a set of patients, P. This is based on the assumption that the diagnostic record for each patient $p \in P$ consists of two pieces of information: a fuzzy set S_p, describing the severity of relevant symptoms of patient p, and a fuzzy set D_p, describing the physician's final judgment regarding the possibility degrees of associated diseases of patient p. From these records, two fuzzy relations can be formed: a relation between patients and symptoms, R_s, and a relation between patients and diseases, R_d. These two fuzzy relations can then be employed in the fuzzy relational equation $R_d = R_s \circ K$, where K denotes the medical knowledge (a fuzzy relation between symptoms and diseases) constructed from the database. Clearly, K can be obtained by solving this equation. Among the solutions, it is prudent to take the unique maximum solution or, if no solution exists, take the best approximate solution (section 3.5).

Among other early papers discussing the utility of fuzzy set theory in medicine, notable is a pair of papers by Esogbue and Elder (1979, 1980), in which the authors examine in detail various stages in the overall process of making a diagnosis by a physician for a particular patient and discuss how the use of fuzzy set theory in some of these stages can make the whole process more realistic.

Since the early 1980s, the literature on medical applications of fuzzy set theory has been steadily growing. A large part of this literature is oriented to the development of prospective fuzzy-set-based systems for computer-assisted medical diagnosis. Research into this new territory has been made largely under the leadership of Klaus-Peter Adlassnig at the Department of Medical Computer Science of the University of Vienna Medical School in Austria. He conceptualized such systems in his doctoral dissertation Adlassnig 1983.[63]

While still working on his dissertation, Adlassnig published three early papers. In Adlassnig 1980, he described his conception of a computer-assisted medical diagnosis based on fuzzy set theory; in Adlassnig 1982, he surveyed relevant literature in the area; and in the joint paper Adlassnig and Kolarz 1982, the authors described in some detail an actual fuzzy-set-based system for computer-assisted medical diagnosis, called CADIAG-2. This system was a successor of a similar system, CADIAG-1, which was based on three-valued logic and in whose design Adlassnig previously

---

[63] Shortly after completing his doctoral dissertation in 1983, Adlassnig chose to begin his career in 1984–85 as a postdoctoral fellow with Lotfi Zadeh at UC Berkeley. After returning to Austria, he joined the Department of Medical Computer Sciences at the Vienna Medical School, where he embarked on research fully dedicated to medical diagnosis based on fuzzy set theory. He became professor of medical informatics in 1992.

participated. According to Adlassnig and Kolarz 1982, CADIAG-2 had already been designed and implemented and was under testing in the area of rheumatology when the paper was published. Adlassnig played a major role in all subsequent developments of CADIAG-2, as well as in developments of other medical support systems based on fuzzy set theory at the University of Vienna Medical School (Adlassnig et al. 1985), and he concisely describes these developments over a period of some twenty years in a survey paper Adlassnig 2001. This paper captures the state of the art of these systems except for one, quite significant feature that was added to the systems fairly recently. This feature is a fuzzy programming language, called Fuzzy Arden Syntax, designed for use in clinical decision support systems.[64] Basic features of this fuzzy programming language, as well as its history and its role in clinical decision support systems, are described in detail in a paper by Vetterlein, Mandl, and Adlassnig (2010).

In spite of further developments in the area of medical diagnosis based on fuzzy set theory, there is only one relatively small monograph devoted to this subject, which is currently on the market. It is written by Rakus-Anderson (2007) in a textbook-like style. The author describes the process of medical diagnosis and discusses also the use of fuzzy decision making for dealing with the associated optimization process of selecting appropriate medication.

In addition to its use in the area of medical diagnosis, fuzzy set theory has been utilized in medicine in numerous other ways, too many in fact to be adequately covered in this section. For example, fuzzy controllers have been widely applied in medicine, but descriptions of their varied applications are scattered over the extensive medical literature. As a consequence, it is virtually impossible to characterize them comprehensively in simple terms. However, we can reasonably exemplify this wide variety of medical applications of fuzzy control by one, quite visible example— the use of fuzzy controllers in anesthesiology (Adamus and Belohlavek 2007). Such fuzzy controllers are often combined with the monitoring of a patient's vital parameters (e.g. blood pressure, ECG, EEG,[65] etc.), especially during highly invasive surgery. This enables identification of critical situations in which case an alarm is produced for the anesthetist (Jungk, Thull, and Rau 2002). It is now well established that the combination of fuzzy controllers and monitoring systems has substantially improved patients' safety during highly invasive surgical procedures.[66]

Medicine is one of many areas in which digital images, such as mammograms, three-dimensional images constructed from two-dimensional slices, and the like, play an important role. It has recently been recognized that methods based on fuzzy

---

[64] The name *clinical decision support systems* is used for a broader class of computer-based systems that include not only systems for computer-aided medical diagnosis, but also systems with various other functionalities that are needed for clinical decision making.

[65] ECG and EEG are common abbreviations of "electrocardiogram" and "electroencephalogram," respectively.

[66] See R. K. Webb et al., "Which monitor? An analysis of 2000 incident reports," *Anaesthesia and Intensive Care* 21, no. 5 (1993): 529–42.

set theory for analyzing, processing, and interpreting digital images are particularly suitable for medical applications. Since the variety of these applications is rather large and their descriptions are scattered throughout the extensive and often highly specialized medical literature, it is exceedingly difficult to characterize them in simple overall terms, especially without the use of elaborate medical jargon. However, chapter 12 in Zimmermann 1999 (pp. 363–416) provides a useful overview of such applications, and a large section on image processing and interpretation may be found on pp. 281–447 in Szczepaniak, Lisboa, and Kacprzyk 2000, which contains detailed descriptions of some interesting case studies of these applications. The carefully edited book by Barro and Marín (2002) is another useful resource, covering some of the most representative applications of fuzzy set theory in medicine, including those involving digital images.

Finally, the scholarly work of Kazem Sadegh-Zadeh in the area of the analytic philosophy of medicine features fuzzy logic in a major role.[67] In many of his writings during the last twenty years or so, as exemplified by Sadegh-Zadeh (2000a,b, 2001, 2007, 2008), he has consistently maintained on philosophical grounds that fuzzy logic is the only fully adequate logic for medical practice. His arguments are thoroughly covered and substantially expanded in his *Handbook of Analytic Philosophy of Medicine* (Sadegh-Zadeh 2012), which is a culmination of his lifelong work. This large monograph, consisting of 1,133 pages, covers comprehensively and in considerable detail the principal philosophical issues associated with medicine. Among the most interesting are the various logical issues involved in clinical reasoning, to which some 40% of the material in this monograph (excluding references and indexes) is devoted.

In discussing these issues, Sadegh-Zadeh argues that classical first-order predicate logic is hopelessly inadequate in medicine as it is capable of representing only a very small fraction of language employed in medicine. He then examines the various modal extensions of classical logic and shows that classical logic together with these extensions is still far from being sufficiently expressive to represent medical language. Further, he examines various nonclassical logics, such as paraconsistent, intuitionistic, and many-valued logics, and argues that their use in medicine to supplement classical logic and its various modal extensions could help to overcome some unavoidable difficulties, such as dealing with contradictory medical data. Finally, he examines fuzzy logic in detail and claims that it has all of the ingredients needed in medicine. He then concludes that fuzzy logic is the only one among all currently recognized logics that is adequate to represent clinical reasoning in medicine. After reaching this conclusion, he refers to the emergence of fuzzy logic as a "fuzzy revolution" (p. 1000); in fact, he coined this term in his earlier paper Sadegh-Zadeh 2001.

---

[67] Kazem Sadegh-Zadeh was born in Tabriz, Iran, in 1942. In the 1960s and 1970s, he studied medicine and philosophy at the German universities of Münster, Berlin, and Göttingen. He is professor emeritus of philosophy of medicine at the University of Münster, where he worked on clinical logic and methodology, and on the analytic philosophy of medicine.

Of course, this short summary of the monograph by Sadegh-Zadeh (2012) captures only the thrust of his extensive argumentation supplemented with numerous examples from medical practice. His monograph stimulated publication of a companion volume edited by Seising and Tabacchi (2013), consisting of 27 chapters on various medical applications of fuzzy logic.

6.11   Management and business

Of the great variety of applications of fuzzy logic that have developed since the early 1970s, some of them are relevant to the areas of management and business. Most notable are applications of fuzzy logic to decision making, covered in section 6.4. Among other early publications of applications of fuzzy logic in management and business are those involving optimization (Zimmermann 1976), planning (Chanas and Kamburowski 1981), personnel management (Ollero and Freire 1981), resource allocation (Mjelde 1986), finance (Li Calzi 1990), insurance (Lemaire 1990), scheduling (Türkşen, Ulguray, and Wang 1992), and trading (Deboeck 1994). However, these authors and those of similar publications did not by and large associate them explicitly with management or business. This began to change in the mid-1990s. In the English translation of a Japanese book edited by Asai (1995), the role of fuzzy logic in management was explicitly recognized for the first time. Two years later, the role of fuzzy logic in management as well as in business was systematically discussed in a textbook written by George and Maria Bojadziev (1997). That same year, a very different book on the role of fuzzy logic in management was published by Keith Grint (1997). In this thought-provoking book, the author argues on strictly intuitive grounds, but quite convincingly, that fuzzy logic has a great potential to bridge the existing wide gap between conventional management theories and management practices.[68]

In a large survey book on practical applications edited by Zimmermann (1999), part III (consisting of chapters 13–16), which amounts to more than 20% of the whole book, is devoted to applications in management. Chapters 13–15 deal, respectively, with the three basic levels of planning—strategic planning, planning of research and development, and production planning, which also covers production scheduling. The focus is on describing in detail how various methodological tools based on fuzzy logic, such as fuzzy modeling, decision making, optimization, clustering, and the like, may be combined to deal with the complex issues in planning.

Chapter 16, written by Derrig and Ostaszewski, is substantially different; it deals with the use of fuzzy logic in actuarial science.[69] The authors describe how the roles

---

[68] Keith Grint is a professor of public leadership at Warwick University and associate fellow at Green Templeton College, University of Oxford.

[69] In general, actuarial science is concerned with the study of all aspects of financial security systems. These are economic systems designed to transfer, in an economically feasible way, economic risk from the individual to a collective of individuals. Financial security systems are based, by and large, on the idea of pooling unavoidable

of actuaries vary significantly, depending on the nature of governmental regulations in different countries, degrees of economic stability, and other factors, between the following two extremes. At one extreme, the role of an actuary is limited to a professional calculation of premium payments for individual insurance cases by a methodology prescribed by a legislature and using official values of relevant economic parameters. At the other extreme, actuaries are completely free to create, in accordance with the usual standards of professional practice, a methodology based on their own assumptions and goals. Circumstances close to the former extreme are certainly not conducive to introducing innovations, such as the use of fuzzy logic, into actuarial practice.

The need for employing fuzzy sets in actuarial practice was suggested for the first time in a paper by the distinguished Dutch scholar in the area of actuarial science, Wilem De Wit (1982). In particular, he shows in this paper that the process of underwriting (involving the selection and evaluation of risks to be insured) cannot be adequately handled by probability theory alone—the traditional tool for dealing with uncertainty in actuarial science. Jean Lemaire (1990)[70] expanded on these early ideas of De Wit by proposing an overall methodology for insurance underwriting based on fuzzy logic, showing how insurance premiums and reserves can also be calculated using fuzzy arithmetic. Further developments regarding the use of fuzzy logic in actuarial science in the 1990s are summarized in chapter 16 of Zimmermann 1999. Among them, the book by Ostaszewski (1993)—the first and as yet only book on fuzzy logic in actuarial science—should be highlighted.

The fact that more than 20% of the space in Zimmermann 1999—the prime book surveying applications of fuzzy logic around the end of the 20th century—was dedicated to applications in management is a clear indication that fuzzy logic was already recognized at that time as eminently suitable for dealing with the complex and often ill-defined problems in the area of management. At the same time, many ideas regarding the role of fuzzy logic and soft computing in various financial applications have also been explored, as exemplified by a large book edited by Ribeiro et al. (1999) and a book by Peray (1999), in which a particular approach to investment based on fuzzy logic is discussed.

In the new millennium, publications devoted to applications of fuzzy logic or soft computing in management and business virtually exploded. Considering the aims of this book, it is sensible to identify from this very extensive literature several key monographs and some representative examples of the best edited volumes.

Among the key monographs dealing with the role of fuzzy logic in management is Carlsson, Fedrizzi, and Fullér 2004. In order to capture the spirit of this book, the following from the short introduction to the book is indicative (pp. xiii–xiv):

---

economic risk, resulting in a small loss to many rather than a large loss to the unfortunate few. Prime examples of financial security systems are various insurance systems, such as life insurance, health insurance, property insurance, etc. Experts in actuarial science are usually referred to as actuaries.

[70] Professor of insurance and actuarial science at the Wharton School of the University of Pennsylvania.

> Some of the best insights in the field were summarized by Steen Hildbrand and Erik Johnsen:[71] "We know more (about management) than we believe, but we use less of what we know than we should." They use Danish management challenges and principles as their context and bring out the observations that successful companies should (i) master effectiveness and quality in both the details and the whole, (ii) build on and work with flexibility, and should (iii) support continuous learning in both the organizational and the individual level. Understanding and mastering all these components in a dynamic and sometimes drastically changing environment is often classified as complicated, sometimes as impossible.
>
> In this monograph we want to show that the complexity of management can be reduced and that the changes in the environment can be more easily handled by bringing fuzzy logic into the management models and into the practice of management.

The authors show, in some detail and on the basis of their own experience, how the use of fuzzy logic results in significant improvements in various areas of management, such as group decision support systems, management in supply chains, or management of knowledge. In the context of management support systems, for example, they explain how fuzzy logic makes it possible to overcome the classical Arrow's impossibility theorem and, as a consequence, to develop sound processes for achieving a meaningful consensus in decision making involving groups, which is very important in management.[72] They also show, for example, how the notorious problem of information distortion in supply chains (also known as the bullwhip effect)[73] can be alleviated by using a simple Mamdani-type fuzzy controller. In general, this is a well-written book that clearly demonstrates the great utility of fuzzy logic in management.

As far as business is concerned, of two early monographs, one, written by Aliev, Fazlollahi, and Aliev (2004), focuses on the role of soft computing in various areas of business, such as business and economic forecasting, business decision making, marketing, operations management, and finance.

The second early monograph, written by Gil-Lafuente (2005), is a thorough, up-to-date, and well-argued exposition of the growing role of fuzzy logic in the financial activity of contemporary businesses. Throughout the book, the author systematically explains how fuzzy logic helps to deal with uncertainty in various business activities.

Among the numerous more recent books discussing the expanding role of fuzzy logic in management and business, the following two carefully edited books, both focusing on marketing and customer relationships, are representative. The book edited by Casillas and Martínez-López (2010) deals with the use of soft computing in management support systems with a particular focus on marketing and customer relationship. The aim of the other, edited by Meier and Donzé (2012), is to serve as a reference source, as comprehensive as possible, of methods based on fuzzy logic

---

[71] S. Hildebrand and E. Johnsen, *Ledelse nu* [Management now] (Copenhagen: Borsen Boger, 1994).

[72] K. J. Arrow, *Social Choice and Individual Values* (New Haven, CT: Yale Univ. Press, 1951). Arrow's theorem, also known as Arrow's paradox, states that certain appealing criteria for ranking methods are mutually incompatible. This implies a serious difficulty in formalizing collective social choice.

[73] See, for example, the paper by H. L. Lee, V. Padmanabhan, and S. Whang, "Information distortion in a supply chain: The bullwhip effect," *Management Science* 43 (1997): 546–58.

that are applicable to customer relationship management and marketing. The book is well organized and contains an extensive compilation of relevant references.

6.12   Other applications

The purpose of this section is twofold. First, we want to describe several additional types of applications of fuzzy logic that are well developed but are for various reasons not included in the previous sections of this chapter. Second, we want to present an overview of various other applications of fuzzy logic, actual or prospective, and explain why we have not covered them in this book.

Image analysis

Among the well-developed applications of fuzzy logic, some of the most important are undoubtedly its various applications within the field known as image analysis, which plays important roles in several of the previously discussed areas, especially medicine, Earth sciences, and biology.[74]

As is generally recognized, Azriel Rosenfeld (1931–2004) was not only considered the founder of image analysis, but was also a leading researcher in this field for almost fifty years. He published more than 30 books and well over 600 research articles, virtually all on computer image analysis, which influenced in one way or another many subsequent contributions to this field. His lifelong work on computer image analysis commenced in the late 1950s, primarily in the context of computer vision. Around the mid-1970s, he recognized that fuzzy set theory might play a useful role in computer image analysis, and began to develop some fuzzified areas of mathematics, such as *fuzzy graphs, fuzzy geometry, fuzzy digital topology*, and others, some of which are examined in chapter 5. He used them readily for advancing his own research, but also urged from time to time other researchers to be open to the ideas emerging from fuzzy set theory. His many contributions, especially to problems of image understanding, are well characterized in books edited by Bowyer and Ahuja (1996) and Davis (2001), which are fully dedicated to his accomplishments.[75] The latter contains a chapter in which Rosenfeld himself describes the genesis and evolution of his lifelong work in the field of image analysis. It also contains a chapter by John Mordeson on Rosenfeld's contributions to mathematics based on fuzzy set theory.

Computer image analysis, including as well computer image processing and understanding, is now a huge field in which fuzzy logic has found interesting applications. An interesting early monograph was written by Pienkowski (1989), in which

---

[74] The term *image processing* is often used instead of image analysis. We use both these terms as appropriate.

[75] The term *image understanding* refers, roughly speaking, to the higher-level processes of interpreting the various components recognized in an image, such as objects or regions of various shapes and colors, etc., as well as their spatial relationship, within a particular application context pertaining to the image.

the author investigates in considerable detail the use of fuzzy logic for machine perception of color in image analysis. Another monograph by Chi, Yan, and Pham (1996) offers a survey of fuzzy algorithms for image processing and pattern recognition compiled at that time from the literature. A fairly comprehensive overview of fuzzy methods developed for image processing is included in one of the handbooks of *Fuzzy Sets Series*, edited by Bezdek et al. (1999). More recently, two books on fuzzy logic in image processing have appeared, one edited by Kerre and Nachtegael (2000), the other by Nachtegael et al. (2003). There is also the special issue of *Fuzzy Sets and Systems* edited by Sobrevilla and Montseny (2007), and a monograph by Chaira and Ray (2010).

Spatial information

The applications of fuzzy logic in image analysis just mentioned are connected to its more recent role in the relatively new area of research that is concerned with all aspects of processing and dealing with spatial information. This area, sometimes referred to as *spatial analysis* (Leung 1988, 1997), emerged in the 1980s from geography, primarily in the context of geographic information systems (Cobb, Petry, and Robinson 2000; Lodwig 2008).[76] An excellent overview of the use of fuzzy logic in this area around the end of the 20th century, written by Yee Leung, may be found in chapter 8 of Zimmermann 1999 (pp. 267–300). In order to describe how this area has been understood more recently, the following quotation from the introduction to the book edited by Jeansoulin et al. (2010)—currently the major source covering this relatively new area—is informative:

> The term *spatial information* refers to pieces of information that are associated with locations, which typically refer to points or regions in some two- or three-dimensional space. Many applications deal with geographic information, in which case the space under consideration is the surface of the Earth. Other applications, however, deal with spatial information of a quite different kind, ranging from medical images (e.g. MRI scans) to industrial product specifications (e.g. computer aided design and manufacturing), or layouts of buildings or campuses. In addition to describing aspects of the real world, spatial information may also describe virtual environments. Beyond virtual environments, we may even consider space in a metaphorical way for describing or reasoning about meaning of concepts, viewed as regions in a multi-dimensional space (e.g. the conceptual spaces of Gärdenfors (2000)).

After examining the various methods for dealing with spatial information that are discussed in Jeansoulin et al. 2010 as well as in earlier books by Leung (1988, 1997), it is clear that fuzzy logic plays an important role in the relatively new and currently very active research area of spatial analysis.

---

[76] Applications of fuzzy logic in geography are not discussed in previous sections of this chapter because geography is not fully subsumed under any of the areas covered in these sections: it is partially subsumed under Earth sciences (so-called *physical geography*) and partially under social sciences (so-called *human geography*). However, applications of fuzzy logic in both physical and human geography fit well into the new and broader area of spatial analysis.

Robotics

Robotics is another area that has not been discussed in previous sections of chapter 6, primarily due to its multidisciplinary nature. Fuzzy logic has been utilized in robotics, often in the context of soft computing and intelligent systems, since the early 1990s. This is documented, for example, by relevant papers reprinted in chapter 3 of the book edited by Marks (1994, 115–73). However, information about the actual applications of fuzzy logic in robotics is widely scattered throughout the huge literature dealing with robotics. Specialized books focusing on fuzzy-logic applications in robotics are very scarce. Nevertheless, to illustrate the utility of fuzzy logic in robotics, three monographs are noteworthy, especially those written by Jacak (1999), Katic and Vukobratovic (2003), and Cuesta and Ollero (2005), as well as two major collections of papers, those edited by Jain and Fukuda (1998) and Driankov and Saffiotti (2001).

Risk analysis

Risk analysis is one of the areas in which the utility of fuzzy logic was recognized quite early. The usefulness of fuzzy logic for this area was demonstrated for the first time in the context of computer security systems in a doctoral dissertation by Clements (1977) at the University of California at Berkeley.[77] Recognizing the absence of sufficient data for statistical analysis in this application domain, Clements developed a method based on linguistic variables in which subjective experts' assessments could be adequately utilized, together with any other (objective) data available, to assess security risks in various computer security systems. His dissertation stimulated further work on the use of fuzzy logic for risk analysis. An early paper by Hoffman and Neitzel (1981) and a small but important monograph by Schmucker (1984) exemplify quite well this work. The principal aim of Schmucker's monograph was to describe in fair detail an automated risk analysis utility, called *Fuzzy Risk Analyzer*, which had been developed and implemented by the Computer Security Research Group at George Washington University in Washington, DC. In addition, the monograph contains an excellent annotated bibliography, very valuable at that time, of literature dealing with the use of fuzzy set theory for modeling expressions in natural language, especially in the context of risk analysis.

Three additional monographs on fuzzy risk analysis have been published after the pioneering work by Schmucker. One monograph, written by Gheorghe and Mock (1999), deals with a broader area that covers not only analysis, but also management of various health and environmental risks. It is significant that fuzzy logic plays a prominent role in this monograph—almost 40% of its text material is allocated to methods based on fuzzy logic, compared with less than 25% allocated to the use of conventional methods.

---

[77] It is interesting, although not surprising, that Lotfi Zadeh was a member of the dissertation committee. The chairman of the committee was Lance Hoffman, a pioneer and top expert in the area of computer security systems.

The second monograph, written by Huang and Shi (2002), consists of two parts. Part I is devoted to the so-called *principle of information diffusion*, which was initially introduced in a paper by Huang (1997). The basic idea of this principle is to replace traditional numerical input-output data that are incomplete or ill justified with their appropriately fuzzified counterparts in order to improve the identified input-output relationship. The difficult issue of how to determine in each particular application an appropriate fuzzification that is in some sense optimal or close to optimal is thoroughly investigated. Part II deals with applications of the principle of information diffusion. These concern almost exclusively applications to fuzzy risk analysis, even though the principle has a much broader applicability.

In the third and most recent monograph, Irina Georgescu (2012) deals with risk analysis in the context of economic and business activities in terms of the mathematical theory of graded possibilities, predominantly based on its fuzzy-set interpretation. Throughout the book, she compares the various features of the possibilistic formalization with their traditional probabilistic counterparts. In particular, she gives a good deal of attention to a possibilistic formalization of risk aversion, which is an important feature of risk analysis in the areas of economics and business; for more information on risk aversion, see a relevant paper by Pratt and a small book of lectures on this topic by Arrow.[78] Georgescu also includes one chapter on credibility theory and its applications to risk analysis. Credibility theory is a mathematical theory based on the theory of graded possibilities in the following way: for any given pair of dual possibility and necessity measures, Pos and Nec, respectively, a credibility measure, Cr, is defined by the arithmetic average of Pos and Nec for each set on which the possibility and necessity measures are defined.[79]

Fuzzy cognitive maps

Another group of applications of fuzzy logic that is not covered in the previous sections are those employing *fuzzy cognitive maps*. The idea of a cognitive map was first suggested and applied in 1948 in a paper by the cognitive psychologist Edward Tolman, entitled "Cognitive maps in rats and men," *Psychological Review* 55, no. 4: 189–208. In a nutshell, a cognitive map is a directed graph whose nodes are variables representing some concepts and whose connections (edges) indicate causal influences between the variables. For each pair of nodes, say A and B, a connection from A to B in the graph indicates that the variable assigned to node B is causally influenced by the one assigned to node A. The connections are usually labeled by either + or − depending upon whether the influence is either positive or negative.

Fuzzy cognitive maps, which were introduced by Bart Kosko (1986a), are extensions of classical cognitive maps in which some features of the latter are fuzzi-

---

[78] J. W. Pratt, "Risk aversion in the small and in the large," *Econometrics* 32, no. 1–2 (1964): 122–36. K. J. Arrow, *Aspects of the Theory of Risk-Bearing* (Helsinki: Jahnssonin Säätiö, 1965).

[79] Credibility theory was introduced and developed by Liu (2004).

fied. The most obvious feature of classical cognitive maps to fuzzify seems to be the concept of causality. It is certainly in the spirit of fuzzy logic to view causality as a matter of degree. However, when employing this view, it is crucial to recognize that causality is distinct from logical implication. In his paper, Kosko carefully explains this distinction and shows how each negative causal influence in a classical cognitive map can be conveniently replaced with a positive causal influence. Fuzzy cognitive maps with fuzzified causality are thus directed graphs in which for each pair of nodes, say A and B, a connection from A to B is labeled with the degree of causal influence of A on B. This degree is normally expressed by a real number from the unit interval $[0, 1]$ or, alternatively, from the interval $[-1, 1]$. However, it can also be expressed qualitatively by fuzzy numbers (states, granules) of a chosen linguistic variable. In a similar way, variables associated with nodes of fuzzy cognitive maps can be expressed qualitatively via appropriate linguistic variables.

Fuzzy cognitive maps are constructed, by and large, on the basis of experts' experience, although various machine-learning capabilities have lately been added to the construction process. In general, each fuzzy cognitive map is a representation of knowledge regarding the behavior of a particular dynamic system. This behavior can be made explicit via an iterative process. In each iteration, the present states of all variables are changed to their next states on the basis of causal influences, both direct and indirect, between the variables. The indirect influences between a pair of variables are associated with all paths in the directed graph that begin at the node of the influencing variable and end at the node of the influenced one. All causal influences on each variable must of course be properly aggregated. Specific aggregation procedures depend not only on which features in the given fuzzy cognitive map are fuzzified and how, but also on the choice of proper aggregation operations in the context of each application.

Research on the theory and applications of fuzzy cognitive maps has been particularly active since the beginning of this century. Papageorgiou and Salmeron (2013) review the impressive research in this area during the first decade. A large book edited by Glykas (2010) is another important resource, which covers in fair detail the main advances in the area of fuzzy cognitive maps during this decade. A more recent book edited by Papageorgiou (2014) covers the very recent advances in this area.

Music

Before bringing this chapter to a close, we should discuss one additional application area, where fuzzy logic turns out to play a key role—music. At present, this is a rare exception among the arts. While applications of fuzzy logic have occasionally been suggested in various other areas of the arts (painting, sculpture, architecture, poetry, etc.), no significant interest has resulted from these rather isolated suggestions to pursue any of them further.

The fact that the use of fuzzy logic in music is quite natural is made clear in the preface to one of the prime monographs on the mathematical theory of tone systems by Ján Haluška (2004):

> The sculptor working in marble has set his limit by the choice of this material to the exclusion of all other materials. Analogously, the musician has to select the tone system he wants to use. This strong emphasis on the necessity of limitation reflects not a subjective prejudice but it is a fundamental artistic law. There is no art without limitation.
>
> There are four important and mutually interacting attributes that we can manipulate to create or describe any sound. And we can work with these attributes in two different ways: we can measure them and we can hear them. If we measure them, they are physical attributes; if we hear them, they are perceptual attributes. The four physical attributes are: frequency, amplitude, waveform, and duration. Their perceptual counterparts are: pitch, loudness, timbre and (psychological) time. There is similarity between hearing and measuring these attributes; however, it is a complex correlation. The two are not exactly parallel.

As Haluška explains, the basic elements of music—musical tones—can be viewed and studied either as physical entities produced by various musical instruments or as perceptions of the physical products of these entities by the human senses. Clearly, this fundamental dichotomy applies to various systems of musical tones as well. When these systems consist of tones viewed as physical entities, classical mathematics based on bivalent logic is perfectly adequate to deal with them. However, when the tones are viewed as human perceptions, the use of mathematics based on fuzzy logic is useful for dealing with tone systems viewed in this way.

The two most important characteristics of each tonal system, when tones are viewed as physical entities, are the frequencies of all tones recognized in the system and their differences for all pairs of the recognized tones. The latter, called musical intervals, are described physically as ratios of their frequencies. A particular interval whose frequency ratio is 2 is called an octave. Two tones whose distance is equal to one or more octaves are viewed as equivalent. Most frequently, especially in Western classical music, twelve tones within each octave are chosen according to some rules that govern, in general, the intervals between consecutive tones in each octave.

It is well established that human auditory perceptive capabilities are remarkably tolerant to small deviations from the ideal (physical) frequencies representing individual notes. That is, tones whose actual frequencies are sufficiently close to the ideal frequency defining a particular tone in a given tone system are perceived as the same pitch. The concept of "being sufficiently close to a number expressing the ideal frequency" can be approximated in a natural way by an appropriate fuzzy number (granule) constructed on the basis of available knowledge regarding characteristics of human auditory perception. Similarly, musical intervals are perceived as approximate ratios. This fuzzy approximation plays an especially important role in the so-called well-tempered tuning within a given tonal system. The aim of this tuning is to make small deviations (tolerated by human perception) from perfect tuning of individual notes in a particular key in order to achieve a perceptually acceptable tuning in all keys. This allows instruments such as the piano, once tuned in a well-tempered way, to play compositions written in any key and they are all perceived as well tuned.[80] Although the tone intervals cannot be exactly the same in all keys under a well-tempered tuning, this is generally viewed as a musical advantage, as it gives a slightly distinctive character to compositions written in different keys.

As for works dealing with fuzzy logic in music, except for one rather early paper by Goguen (1977), virtually all relevant literature is associated with the 21st century. Among the most informative publications related to the above discussion are those by Haluška (1997, 2000, 2002, 2004), Liern (2005), and León and Liern (2012).

Miscellaneous additional applications

We could certainly go on and describe actual or potential applications in many other areas; see, for example, applications covered in the book edited by Dadios (2012). But we do not do so for several reasons. First, it is not the aim of this book to cover everything that concerns fuzzy logic. Second, there are some additional areas in which fuzzy logic has been extensively and successfully applied, but these are by and large applications in areas that have already been covered in previous sections of this chapter, such as fuzzy control, fuzzy decision making, fuzzy image processing, fuzzy clustering, and the like. This characterizes typical applications of fuzzy logic in areas such as, for example, forensic science, archeology, and paleontology, among others. Third, in some areas, such as the legal profession, the potential utility of fuzzy logic is highly suggestive, but its actual utility has not been realized as yet due to various virtually insurmountable barriers (political, ethical, religious, and others). Finally, there are many other areas in which applications of fuzzy logic have occasionally been suggested, but these suggestions have not been sufficiently developed as yet, so to discuss them here would be premature.

---

[80] The famous forty-eight compositions by Johann Sebastian Bach, "Das Wohltemperierte Klavier" (The Well-Tempered Clavier), are written in all 12 major and 12 minor keys. If they are all played on a piano and each is perceived as well tuned, then the piano tuning is said to be well-tempered.

Chapter 7

Significance of Fuzzy Logic

7.1 Introduction

IN THIS CHAPTER, our aim is to make a sound appraisal of the significance of fuzzy logic based on the historical analysis we have presented in this book, as well as on our vision of prospective future developments in fuzzy logic. We thus begin with a retrospective overview which we present in section 7.2. We then employ the well-known concepts of paradigm and paradigm shift as useful metaphors. These are the subject of section 7.3. Our assessment of the significance of fuzzy logic, on which the entire book has been focused, is the content of section 7.4. In section 7.5, we conclude with our prospective views concerning the future of fuzzy logic.

7.2 A retrospective overview of fuzzy logic

The aim of this section is to summarize concisely the evolution of fuzzy logic since its genesis in the mid-1960s. To this end, we draw largely on the material presented in previous chapters of this book, focusing on the most prominent characteristics of fuzzy logic that have emerged in the course of its 50-year history.

We adopt three approaches to examine the evolution of fuzzy logic. The first focuses on relevant theoretical developments. These include not only those in fuzzy logic in the broad as well as the narrow sense, but also those in mathematics based on fuzzy logic. The second approach evaluates applications of fuzzy logic in various areas of science, engineering, and other areas, such as medicine, management, and business. The third approach assesses developments of the organizational infrastructure that has played an important role in supporting research, education, and other activities crucial for the advancement of fuzzy logic and its applications.

Theoretical developments

The dichotomy in theoretical research between fuzzy logic in the broad sense (FLb) and fuzzy logic in the narrow sense (FLn) was already tacitly recognized in the seminal paper by Zadeh (1965a), although not under these names.

The agenda of FLb has its roots in Zadeh's motivations for introducing the concept of a fuzzy set and his vision of its potential utility. Initially, the agenda of FLb focused primarily on utilizing the expressive power of intuitive fuzzy set theory, as conceived by Zadeh, for emulating various unique capabilities of human beings, such as working with classes that do not have sharp boundaries, reasoning and decision making with statements expressed in natural language, acting on the basis of perceptions, and the like. Gradually, the agenda has become more specific by introducing and investigating relevant new concepts (e.g. linguistic variables, fuzzy intervals or fuzzy clusters), principles (e.g. extension principles or rules of inference), and problems and methods (e.g. computing with fuzzy intervals, precisiation of meanings of expressions in natural language by fuzzy sets, or fuzzy clustering). All these aspects of FLb are covered from a historical perspective in chapter 3.

The agenda of FLn is closely connected with research on many-valued logics which were already substantially developed before the 1960s. Zadeh's idea of a fuzzy set convincingly suggested a new way to interpret truth values, namely as degrees of truth. This idea, along with Zadeh's other investigations regarding reasoning in natural language, provided impetus for examining FLn as well as conceptually new problems on its agenda. This is best documented in the early contributions by Goguen and the subsequent development of Pavelka-style logics in which not only truth but also the concept of entailment and other metalogical concepts are a matter of degree. The resulting framework is quite general and encompasses both truth-functional as well as non-truth-functional logics. An important turning point in the development of FLn is marked with the appearance of Hájek's work in the early 1990s. Under his leadership, FLn gradually became an established area of mathematical logic. As a result, various logical calculi were subsequently developed, including propositional and predicate logics and higher-order logics, as well as various kinds of other logics such as modal logics and logics of uncertainty. The problems studied involved mostly traditional problems in logic such as questions of axiomatization and computational complexity. Attention has also been paid to the relationship between FLn and FLb, particularly to examination of problems of FLb from the viewpoint of FLn, but the extent of such investigations has been limited and we return to it in section 7.5. All of these investigations are examined in detail in chapter 4.

Developments of various areas of mathematics based on fuzzy logic, which are thoroughly covered in chapter 5, began as early as the late 1960s. Although related developments—such as those in axiomatic theories of many-valued sets—occasionally appeared before the publication of Zadeh's 1965 paper, the natural motivation and rationale provided by Zadeh encouraged examinations of various parts of mathematics in which it is naturally desirable to replace sets by fuzzy sets and in general to examine situations involving degrees of truth. This impulse and the circumstances described in detail in chapter 5 eventually led to fuzzifications in virtually all areas of mathematics including set theory, algebra, topology, geometry, analysis, probability and statistics, as well as many areas of applied mathematics. Although the respective developments vary significantly in their extent, sophistication, and the motivations which led to these developments, several areas have now been considerably advanced and provide novel concepts and methods which have profoundly enhanced ordinary mathematical methods. Foundational studies in mathematics based on fuzzy logic have also been examined on a larger scale since the 1990s. Even though significant progress has been made, these studies are best regarded as still relatively recent and call for further exploration. They include category-theoretic approaches, which are closely connected to the theory of topoi, axiomatic set-theoretic approaches, which are related to axiomatic set theories developed in other contexts, as well as higher-order logic approaches. Foundational aspects regarding mathematics based on fuzzy logic open interesting questions which are further examined in section 7.5.

Applications

Developments of applications of fuzzy logic in a wide variety of human affairs during the first fifty years of fuzzy logic are described in detail in chapter 6. Although examples of prospective applications of fuzzy logic have been proposed in virtually all areas, they have inspired interest for further developments only in some areas and to varying degrees. We now summarize the main features of these developments.

In his seminal paper, Zadeh argued that fuzzy sets provide a natural tool for "dealing with problems in which the source of imprecision is the absence of sharply defined criteria of class membership rather than the presence of random variables" (Zadeh 1965a, 339). Within the next five years, he published several papers in which suggestive problems were illustrated by those of optimization under ill-defined constraints, pattern classification in which unsharp boundaries are allowed, and decision making in which goals and/or constraints constitute classes of alternatives whose boundaries are not sharply defined. These papers stimulated early research that led to some significant results in the 1970s. This research subsequently intensified and, as described in chapter 6, resulted in numerous publications on a broad spectrum of successful applications of fuzzy logic.

In other areas, however, the development of successful applications was quite different. This is particularly true in biology and psychology. As is explained in section 6.5.2, the anticipated need for mathematics based on fuzzy logic in biology was one of the primary motivations for introducing fuzzy sets by Zadeh. Yet, biologists at first showed virtually no interest in exploring this new mathematics for almost 50 years. Only very recently has some interest in the use of fuzzy logic in biology been shown, but only within a narrow yet rapidly growing new area—bioinformatics. In psychology, as in biology, expected applications of fuzzy logic have not materialized as yet, but for very different reasons, which are described in detail in section 6.7.

In sharp contrast to the situation in biology and psychology, significant applications of fuzzy logic have been made and to a large extent accepted in a number of areas, even though they were not initially expected or at least were not among the original motivations for introducing fuzzy sets. Among these are engineering, computer science, chemistry, geology, management, business, medicine, and music (see the respective sections in chapter 6).

Physics and various social sciences (especially economics and political science) are areas in which significant applications of fuzzy logic have also been developed, but the acceptance of fuzzy logic by their respective communities has been somewhat lukewarm for reasons that are specific to each area, as described in chapter 6.

The overall significance of the many applications of fuzzy logic during its first fifty years can be summarized as follows. The significance varies substantially among areas if measured by the degrees to which they have been accepted. In some areas (engineering, computer science, chemistry, geology, medicine, or music), applications of fuzzy logic have largely been accepted. On the contrary, surprisingly, biology and psychology are areas in which little interest has been observed so far. In most areas, however, such as management, business, economics, political science, and others, the applications have been accepted to various degrees, and it seems that support in these fields is slowly growing.

Supporting infrastructure

Early developments in infrastructure supporting fuzzy logic, roughly during the first two decades after the publication of Zadeh's prophetic paper on fuzzy logic, are described in section 2.5. The following are the most important events during this critical early stage:

1. The first academic journal specializing on all aspects of fuzzy logic—*Fuzzy Sets and Systems*—was launched in 1978.
2. Annual International Seminars on Fuzzy Set Theory (the so-called Linz Seminars), which have played a major role in advancing theoretical aspects of fuzzy logic, were initiated in 1979.
3. A quarterly bulletin known under the acronym *BUSEFAL* (n. 33 on p. 40) was initiated in 1980 as a medium for effective communication and discussion of preliminary ideas pertaining to fuzzy logic.
4. The first professional society supporting fuzzy logic—the North American Fuzzy Information Processing Society (NAFIPS)—was established in 1981.
5. In 1982, NAFIPS began to organize annual conferences on fuzzy logic.

6. The first international organization supporting fuzzy logic—the International Fuzzy Systems Association (IFSA)—was established in 1984.
7. In 1985, IFSA organized the first of its biennial international conferences, known as the IFSA World Congresses.
8. Textbooks on fuzzy logic began to appear in the 1980s.

These eight milestones were crucial for the emerging new area of fuzzy logic. They gave researchers working in this area a sense of community and, more importantly, provided them with adequate organizational support. Many later events further expanded the supporting infrastructure, most notably the endorsement of fuzzy logic by IEEE in the early 1990s and the emergence of various national and regional organizations supporting fuzzy logic, the majority of which later became institutional members of IFSA. However, each of the eight early events listed above were critically important for the initial advances in the theory as well as applications of fuzzy logic.

7.3 Paradigm shifts in science, mathematics, engineering, and other areas

For our discussion regarding the significance of fuzzy logic, two concepts central to Thomas Kuhn's highly influential book, *The Structure of Scientific Revolutions* (Kuhn 1962), are especially useful.[1] The first is the concept of a scientific paradigm, the second is the related concept of a paradigm shift, both of which serve here as useful metaphors. The aim of this section is to describe briefly how these concepts were understood by Kuhn, and then to explain how we broaden them to suit our purposes here.

Paradigms and paradigm shifts

What Kuhn proposed regarding scientific revolutions was a new view wherein the concept of a scientific paradigm plays a central role. According to Kuhn's view of scientific development, work in any area of science is normally pursued under the constraints of a particular paradigm. Basically, a *scientific paradigm* is taken to be a collection of concepts, presuppositions, principles, beliefs, theories, methods, and habits of mind, along with any particular commitments that are shared and taken for granted by members of a scientific community in any given area. Paradigms define implicitly legitimate problems and methods, as well as specific rules and standards for scientific practice. Moreover, education in the given area is also based upon the prevailing paradigm.

---

[1]Thomas S. Kuhn (1922–1996) was an American physicist who received his PhD from Harvard University in 1949. Thereafter, during three years as a Harvard Junior Fellow, his interests changed to history and philosophy of science, fields to which he made his major scholarly contributions.

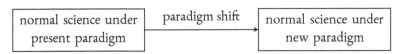

Figure 7.1: The structure of scientific revolutions.

Each period during which science in a given area is practiced under a particular paradigm is called by Kuhn a period of *normal science*. As long as no serious difficulties are encountered in dealing with the recognized problems in a given area of science, the paradigm is not challenged. However, when it becomes increasingly difficult or even impossible to deal with some problems due to, for example, persistent paradoxes or observed anomalies, the time is ripe for the emergence of a new paradigm, one via which the difficulties can be overcome. As a rule, the new paradigm is incommensurable with the old one. This means in effect that the new paradigm cannot be built upon the old one, but must supplant it. Consequently, it usually takes considerable time before the new paradigm is generally accepted by the scientific community at large. The process by which a new paradigm comes to be generally accepted by the scientific community is called by Kuhn a *paradigm shift*. When the new paradigm is eventually by and large accepted, Kuhn views the paradigm shift as having constituted a revolution in the respective area.[2] Scientific development in each particular area of science is thus characterized as a sequence of periods of normal science that are connected by paradigm shifts, as depicted schematically in figure 7.1.

Kuhn's book has stimulated many spirited discussions among historians and philosophers of science.[3] Although most participants in these discussions, including Kuhn himself in his later writings, disagreed with various details in the book, Kuhn's overall characterization of scientific progress—as depicted in figure 7.1—has been generally accepted by working scientists. However interesting these discussions may be considered, they are not directly relevant to our purpose—the assessment of the significance of fuzzy logic 50 years after its genesis—because our paradigm shift involves primarily logic and the associated mathematics, which entails special considerations apart from those concerning the nature of scientific revolutions in the sciences generally.

The issue of paradigm shifts and revolutions in mathematics has been discussed by historians of mathematics since the 1970s. Some have argued that revolutions in

---

[2] The completion of a paradigm shift from one paradigm to another in a given area of science is viewed by Kuhn as a *scientific revolution* in that area. It is often this final conclusion, the result of the process, that is meant when the term "paradigm shift" is used colloquially—simply the (often radical) change of worldview—without reference to the specificities of Kuhn's historical argument.

[3] A particularly good overview of typical discussions of Kuhn's ideas and his reflection on these discussions is presented in the book *Criticism and the Growth of Knowledge*, ed. I. Lakatos and A. Musgrave (London: Cambridge Univ. Press, 1970). For a general overview of Kuhn 1962, see the studies by I. Bernard Cohen, *Revolutions in Science* (Cambridge, MA: Harvard Univ. Press, 1985), 22–23, and H. Floris Cohen, *The Scientific Revolution: A Historiographical Inquiry* (Chicago: Univ. of Chicago Press, 1994).

mathematics are not possible because, contrary to the sciences, emerging new theories in mathematics do not invalidate older, established theories. The main proponent of this view has been Michael J. Crowe of the University of Notre Dame, who actually initiated these discussions by his stimulating paper Crowe 1975, which is also reprinted in Gillies 1992. In this paper, Crowe asserted ten "laws" concerning patterns of change in the history of mathematics, one of which (law 10) was expressed by a simple categorical statement: "Revolutions never occur in mathematics." In his later paper (Crowe 1992), he described the evolution of his own views on the nature of developments in mathematics and explained in particular how he had revised them regarding the impossibility of revolutions in mathematics. The paper ends with the following summary (Crowe 1992, 316):

> In concluding this paper, I wish to note that reading the fine essays in this volume [i.e. Gillies 1992] has left me with three chief convictions:
>
> 1. The question of whether revolutions occur in mathematics is in substantial measure definitional.
> 2. It is more evident than ever (as these essays show) that the new historiography of science can be usefully applied to the history of mathematics.
> 3. A revolution is underway in the historiography of mathematics, a revolution that is enabling a discipline that dates back to Eudemus of Rhodes to attain new and unprecedented levels of insight and interest.[4]

Other historians of mathematics have argued, contrary to the strong initial claims made by Crowe, that revolutions in mathematics are not only possible, but inevitable. The main proponent of this view has been Joseph W. Dauben. On the nature of revolutions in mathematics, he argues as follows (Dauben 1992, 81):

> Each generation, every age sets its own boundaries, limits, blinders to what is possible, to what is acceptable. Revolutions in mathematics take the next generation beyond what has been established to entirely new possibilities, usually inconceivable from the previous generation's point of view. The truly revolutionary insights have opened the mind to new connections and possibilities, to new elements, diverse methods, and greater levels of abstraction and generality. Revolutions obviously do occur *within* mathematics. Were this not the case, we would still be counting on our fingers.

This may be further elaborated in terms of paradigms (Dauben 1996, 143–44):

> Revolutions in mathematics may be identified with changes in the paradigms in terms of which mathematicians think about mathematics. Although such revolutions may not be frequent, they are surely decisive in enlarging the domain of mathematics in ways that do not bar the monsters that challenge conventions and in turn provide new ideas—like incommensurable magnitudes, infinitesimals or transfinite numbers. Instead, the revolutionaries find that indeed, the monsters cannot be avoided if true progress is to be made. In fact, confronting the monsters, in finding ways to tame them, the beast may become the beauty, an essential addition to the body of mathematics that represents its most fundamental sort of progress—*El Progreso Matemático*—in its truest sense.

Dauben also illustrates in his papers some actual revolutions via exemplars from the history of mathematics. These include the Pythagorean discovery of incommensurable quantities in ancient Greece (Dauben 1984, 1996), the emergence of calculus

---

[4] Eudemus of Rhodes was an ancient Greek philosopher who is generally considered the first historian of science and mathematics. He lived approximately from 370 BC to 300 BC.

in the 17th century (Dauben 1992), Georg Cantor's development of transfinite set theory (Dauben 1984), and Abraham Robinson's nonstandard analysis (Dauben 1992). The last two cases are also covered in detail in his extensive monographs.[5] There are of course additional examples of revolutions in the history of mathematics that have been discussed by other historians of mathematics. For example: the emergence of non-Euclidean geometries in the 19th century, the revolution created by Frege in logic at the end of the century, and in the 20th century the establishment by Gödel's incompleteness theorems of the severe limitations of Hilbert's program (paradigm) concerning the foundations of mathematics.[6]

Before proceeding further, we should clarify our understanding of the notion of a new paradigm that challenges one that is established and initiates a paradigm shift in any area of science, mathematics, or some other area of human affairs. According to all discussions about paradigm shifts in the literature, new paradigms challenge existing ones in a fundamental way, sufficiently powerful to overcome the recognized difficulties or limitations of the existing paradigm. Regardless of how the new paradigm shift may be characterized as *highly significant* or *sufficiently large*, or in some other way, the characterization is always described in some appropriate terms of natural language, which are genuinely vague. Consequently, the very notion of a new paradigm must be amenable to description and analysis in terms of fuzzy logic. That is to say, the concept of a new paradigm is genuinely vague as to when it emerges and when its acceptance may be regarded as successful or complete, or even the extent to which a new paradigm may be described as an "earthquake" or, indeed, as a "major revolution," or perhaps only a minor one. That is, whether some challenge to the existing paradigm can be viewed as a new paradigm, which initiates a paradigm shift, is in our view always a matter of degree.

Fuzzy logic as a new paradigm

Let us now turn to our primary concern in this book—fuzzy logic and its significance. Classical logic may certainly be viewed as an established paradigm in the area of logic and fuzzy logic as a challenging paradigm. The challenge of this new paradigm is the rejection of one principle upon which logic has been based for millennia—the principle of bivalence. A motivation for this challenge—a recognition

---

[5] J. W. Dauben, *Georg Cantor: His Mathematics and Philosophy of the Infinite* (Cambridge, MA: Harvard Univ. Press, 1979; reprint Princeton, NJ: Princeton Univ. Press, 1990); *Abraham Robinson: The Creation of Nonstandard Analysis, A Personal and Mathematical Odyssey* (Princeton, NJ: Princeton Univ. Press, 1995).

[6] Yuxin Zheng discusses "Non-Euclidean geometry and revolutions in mathematics," in Gillies 1992, 169–82, and more recently Richard J. Trudeau has devoted a book to *The Non-Euclidean Revolution* (Basel: Birkhäuser, 2001; reprint 2008). As for the revolution instigated by Frege, see Donald Gillies, "The Fregean revolution in logic," in Gillies 1992, 265–305. Curtis Franks has described the "revolutionary insight that led [Hilbert] to forge his new science (meta-mathematics)" in his study, *The Autonomy of Mathematical Knowledge: Hilbert's Program Revisited* (Cambridge: Cambridge Univ. Press, 2009), 2. Likewise, Jaako Hintikka and Gabriel Sandu have characterized Gödel's first incompleteness theorem as the "major earthquake of twentieth-century logic," and analyze what they call the "revolutionary consequences of the incompleteness of IF (independence friendly) first-order logic" in their article, "A revolution in logic?" *Nordic J Philosophical Logic* 1, no. 2 (1996), 169–83, esp. p. 177.

that bivalence severely restricts the expressiveness of logic and hence its domain of applicability—was explained with great clarity in the seminal paper by Zadeh (1965a) and the somewhat later paper by Goguen (1968–69). In this sense, the paradigm shift from classical logic to fuzzy logic began in the mid-1960s. At the present time, 50 years since the genesis of fuzzy logic, the paradigm shift is still ongoing, but there is no doubt that it has considerably advanced.

Now, this paradigm shift in logic induces an associated paradigm shift in mathematics—a shift from mathematics based on classical logic to mathematics based on fuzzy logic. Although the latter paradigm shift has been guided by some general principles discussed in chapter 5, it has manifested itself somewhat differently in the various subareas of mathematics. It is thus reasonable to distinguish a general paradigm shift in mathematics from specialized paradigm shifts in different subareas of mathematics. All these paradigm shifts are of course at this time still ongoing.

Next, the paradigm shifts in logic and mathematics just mentioned induce possible paradigm shifts in all areas of human affairs that employ in some way logic or the associated mathematics. This includes not only the various areas of science, but also other areas, such as engineering, medicine, management, business, decision-making, risk analysis, and many others. The impact of this particular paradigm shift in logic thus extends far beyond logic. It is clearly a paradigm shift on a very large scale, which may justifiably be called a *grand paradigm shift*.

According to Kuhn, his concepts of scientific communities and scientific paradigms were inspired, at least to some degree, by similar concepts—*thought collectives* and *thought styles*—that were introduced in 1935 by Ludwig Fleck (1896–1961) — a Polish microbiologist of Jewish origin, known for his significant research in several areas of medicine.[7] Fleck's concepts are broader than Kuhn's corresponding concepts as they are not restricted to the sciences and, hence, they are more fitting for our discussion. A "thought collective" is roughly viewed by Fleck as any "community of persons mutually exchanging ideas or maintaining intellectual interaction." He argues that the individuals of any given thought collective are linked together by a particular thought style they share. His concept of a "thought style" is almost the same as Kuhn's concept of a paradigm except that, again, it is not restricted to scientific communities. Fleck's concept of a thought style is thus strictly speaking more relevant to our discussion than Kuhn's concept of a paradigm. However, we prefer to use the established term "paradigm" with the understanding that it is not in our case restricted only to scientific communities.

The paradigm shift transcending the limitations of classical logic to allow the radical notions of fuzzy logic with its implications for mathematics and many other

---

[7] Fleck introduced these concepts in his book *Entstehung und Entwicklung einer wissenschaftlichen Tatsache: Einführung in die Lehre vom Denkstil und Denkkollektiv* (Basel, Switzerland: Benno Schwabe, 1935). For an English translation of the book (with a foreword by Kuhn, as well as a preface, a biographical sketch of Fleck, and a commentary by one of the book's translators, Thaddeus Trenn), see L. Fleck, *Genesis and Development of a Scientific Fact* (Chicago: Univ. of Chicago Press, 1979).

areas of human affairs has been ongoing for fifty years and is still far from being completed. To provide an assessment of the significance of fuzzy logic as a new paradigm from a variety of relevant viewpoints is the goal of the next section.

7.4 Assessment of the significance of fuzzy logic

The various particular achievements regarding foundations and applications of fuzzy logic, but also the development of its supporting infrastructure, thoroughly examined in chapters 2–6, are summarized and assessed in section 7.2. In the present section, the culmination of our assessment, we focus on major aspects regarding the importance of fuzzy logic which are crucial for a proper understanding of its significance.

Fuzzy logic as a response to the prevailing neglect of a fundamental phenomenon

Throughout the history of science and philosophy since the ancient Greeks, determinacy, definiteness, exactness, and precision, have been considered as indisputable ideals—as definitional characteristics and maxims of methodological rigor. Descriptions of reality, one's observations, scientific laws, as well as philosophical views have routinely been formulated in natural language and it was taken for granted that reasoning in natural language obeyed the laws of classical logic. Among these laws was the *principle of bivalence*, which was explicitly formulated by Greek philosophers and according to which every statement is either true or false. Since this includes statements about classes of objects and relations among them, all classes and relations were considered as clear-cut. Mathematics, sometimes seen as the queen of the exact sciences, naturally accepted the dictum of bivalence with the prospect that the clear-cut nature of mathematical concepts would lead to our ability to obtain precise models of the physical world. This perspective was reinforced by the development of modern logic toward the end of the 19th century. For example, Frege—one of the founders of modern logic—explicitly argued that classes and relationships that are not clear-cut should be eliminated from the realm of logic and science (p. 341 above). This view—which may aptly be called the *classical view*—gradually became the backbone of virtually all fields of science, engineering, and other areas of human affairs, and led to highly successful applications.

What was virtually overlooked, however, was that a key ingredient of the classical view—*the assumption that all classes and relations employed in reasoning about the physical world are clear-cut and hence also the principle of bivalence itself—both represent a gross oversimplification*. Aristotle—who is considered the founder of classical logic and a paragon of precision—recognized that certain human concepts are not clear-cut, admit borderline cases, and are naturally viewed as admitting of degrees (p. 6 above). Such concepts, which are nowadays called vague, are the crux of the long-unresolved sorites paradox which has, since antiquity, been regarded as a logical mystery (p. 333 above). The phenomenon of vague concepts started to appear regularly in writings on philosophy and logic only since the end of the 17th century, but except for its recognition and occasional remarks on the graded nature of vague concepts, no particular attention was paid to them. This started to change at the beginning of the 20th century in the writings of Peirce, Russell, and Black, who recognized that vagueness and *the lack of sharp boundaries is not a peripheral issue but rather a characteristic feature of all empirical concepts*, and that the bivalent character of logic and mathematics is in sharp contrast to this fact (sections 2.1 and 2.2). In the first half of the 20th century, when modern mathematical methods began to be routinely employed in various domains, several distinguished scholars in different fields expressed, in one way or another, *the inadequacy of classical mathematics and the need for a new mathematics which was capable of dealing with concepts that are not clear-cut.* Such calls came from fields as diverse as physics, biology, psychology, and linguistics, but also from philosophy of science. In response to such concerns, several new formalisms emerged, including Black's consistency profiles and Kaplan and Schott's calculus of empirical classes (section 2.2), as well as the view that the *intermediate degrees of application associated with vague concepts may be conceived of as truth values* in the sense of many-valued logic.[8] These explorations, which were typically conducted independently, culminated in the groundbreaking paper by Zadeh (1965a), who introduced the concept of a *fuzzy set as a simple and appealing formal notion representing the intuitive idea of a set with unsharp boundaries.* This paper represents a clear milestone and marks the true beginning of fuzzy logic for the reasons explained in section 2.2 and other parts of this book.

The emergence of fuzzy logic may thus be seen as a response to a prevailing neglect of—but often also negative attitudes toward—a phenomenon which is ubiquitous in human descriptions of the physical world and human reasoning about it, namely the lack of sharp boundaries of most human concepts and, in this sense, their inexactness.

Efficiency and inevitability of inexact concepts

Fuzzy logic approaches such phenomena by what has been aptly termed the *graded approach to truth*—in addition to *true* and *false*, it recognizes intermediate truth degrees assigned to propositions. These include propositions regarding membership of objects in categories representing concepts with unsharp boundaries. Such an approach is simple and appealing, as best demonstrated by the elegant *resolution of a long-standing paradox*—the sorites paradox (section 5.4.2).

The ability of a new approach to resolve a long-standing paradox usually indicates a deeper significance, and fuzzy logic is a case in point. Its practical significance in fact reaches far beyond the resolution of paradoxes. It derives from the observation that inexact concepts are not only conveniently employed in human descrip-

---

[8] Such a view was clearly formulated by Waismann (1945–46), see p. 25 above.

tions of the physical world, but are highly efficient and inevitable in such descriptions. The issue at stake has lucidly been described by Zadeh (1973) in terms of his *principle of incompatibility*:

> Stated informally, the essence of this principle is that as the complexity of a system increases, our ability to make precise and yet significant statements about its behavior diminishes until a threshold is reached beyond which precision and significance (or relevance) become almost mutually exclusive characteristics.

In complex situations, statements involving inexact concepts may still be accurate—even though not precise in the traditional sense—and yet significant. Thus, for instance, while a classical mathematical description of a control strategy that involves exact quantities may turn out to be infeasible, an experienced human operator may still be able to describe a successfully employed control strategy in a natural language using terms such as "high temperature" or "small change."

The fuzzy logic gambit

From a methodological point of view, the expressiveness of fuzzy logic to represent faithfully inexact concepts and emulate reasoning with such concepts is a critical factor in its successful applications. This ability, repeatedly demonstrated throughout this book, is due to the admission of truth degrees by fuzzy logic. Fuzzy logic represents inexact concepts by fuzzy sets which are exact mathematical objects. It is in this sense that fuzzy logic employs *precisiation of meaning*—one moves from inexact concepts involved in natural language descriptions to exact mathematical objects which, nevertheless, represent the inexact concepts faithfully enough. In a similar vein, one moves from natural language constructs such as logical connectives ("and," "or," etc.), linguistic hedges ("very," "rather," etc.), or quantifiers ("for all," "for many," etc.) to their exact mathematical representations.

But there is yet another, no less important ingredient which is rarely mentioned explicitly and thus often not properly appreciated or even misunderstood. It may be called a *purposeful employment of imprecision*: even though precise values are available, it may still be preferable to employ inexact concepts with the prospect that the resulting model will be feasible and yet sufficiently accurate.[9] Thus, even though one could employ temperature as a classical variable with its numerical values, one might prefer to use corresponding linguistic variables having as values terms such as "small" or "very high."

The exploitation of these ingredients—the purposeful employment of imprecision followed by precisiation of meaning—is essentially what Zadeh calls the *fuzzy logic gambit*.[10] The fuzzy logic gambit represents a methodologically new approach which is clearly made possible by a shift from the paradigm of classical logic to the new paradigm of fuzzy logic.

---

[9] In this context, one often speaks of tolerance toward imprecision. We find the term "purposeful employment of imprecision," which expresses a stronger attitude than tolerance, as describing the situation better.

[10] E.g. in his e-mail message with the subject "What is fuzzy logic?" to the BISC group from January 26, 2013.

Why is fuzzy logic more powerful than classical logic?

A simple answer is that since fuzzy logic generalizes classical logic, it is by definition at least as powerful as classical logic. This answer is correct but trivial. We are interested in a different aspect of this question. Namely, why does the admission of intermediate truth degrees by fuzzy logic and the replacement of classical sets and relations by fuzzy sets and relations lead to models which are substantially more powerful than those of classical logic? This question is clearly related to the expressive power of inexact concepts in describing the physical world.

In a sense, classical logic has only one facet—the symbolic facet. It enables us to express information by statements such as "*x* is related to *y*." Such information is *flat* in a sense. Fuzzy logic, on the other hand, has two facets, the *symbolic facet* and the *numerical facet*. The symbolic, or qualitative, facet retains the capability of classical logic to describe the physical world using a simple symbolic language. This ensures conceptual clarity because the language enables us to form statements referring to classes of objects and relationships among them and combine such statements using logical connectives just as natural language does. The numerical, or quantitative, facet makes it possible to differentiate—we not only have the possibility of uttering a statement but may also attach a degree to it and thus express, for instance, that "*x* is related to *y* to degree *a*." Since *a* may range over a whole scale of truth degrees, the numerical facet greatly enhances the expressive power of the symbolic one. The combination of the symbolic and numerical facets thus makes it possible to express information which is no longer flat.

The numerical facet also involves truth functions of logical connectives which additionally enhance the capability of representing complex notions and relationships among them and which are also employed in making inferences. To mention just one example, recall theorem 5.9 according to which fuzzy equalities with respect to certain t-norms are in a natural correspondence with metrics. Thus in a sense, having a fuzzy equality on a given universe U is the same as having a metric on U. Therefore, whenever one uses a first-order fuzzy logic with equality, one actually has the considerably more powerful concept of a metric space at one's disposal. Compare this to the classical paradigm: the same scenario yields instead of the concept of metric on a set the rather trivial concept of ordinary equality on a set. This clearly shows that the numerical facet provides a great conceptual enhancement.

It is also appropriate to mention at this point the considerable variety of truth functions of logical connectives, such as the parameterized classes of t-norms. Employing any such t-norm as a truth function of conjunction does not in any way impair conceptual clarity of the resulting model because its symbolic part—the logical formulas describing it—remains the same. However, the possibility of choosing an appropriate t-norm makes the model more flexible. Note as well that fuzzy logic offers the possibility of employing connectives that do not appear in classical logic, such as linguistic modifiers, averaging functions, and various other aggregation functions.

Is mathematics based on fuzzy logic a new kind of mathematics?

Several early writings in the first half of the 1950s, examined in chapter 2 and other parts of this book, expressed a need for a new kind of mathematics—one which could deal with sets and relations that do not have sharp boundaries. Since the mathematics based on fuzzy logic may naturally be regarded as a response to such needs, it is proper to address the question of whether and in what sense mathematics based on fuzzy logic may be regarded as a new kind of mathematics. This question has several dimensions.

Let us first clarify an issue which has been a source of some early misunderstandings, one already addressed by Goguen (1967).[11] Mathematics based on fuzzy logic is not a kind of mathematics whose definitions, proofs, and theorems are inexact, imprecise, or vague in any way, and thus would lack precise meaning. Rather, it is a rigorous mathematics whose definitions and theorems have precise meanings and whose proofs are conducted according to precisely specified rules. Thus they obey the commonly accepted standards of mathematics in general. Yet, it is a mathematics concerned with inexact classes and relations. In this respect, the early misunderstandings failed to see that inexact classes and relations are modeled by the precisely defined concepts of fuzzy sets and fuzzy relations.

The next aspect of our question is whether and in what sense may the concept of fuzzy set be considered as fundamental, just as the concept of set is considered a fundamental concept of classical mathematics. This question naturally comes to the fore when one observes the practice of a mathematician who works with fuzzy sets. Such a mathematician works with fuzzy sets as functions $f : U \to L$, where U is a given universe and L a set of truth degrees. This, in principle, corresponds to the common practice of an ordinary mathematician who works with sets which are subsets of a given universe. But while for the ordinary mathematician, this basic concept—that of a set—is conceived of as a primitive concept and is approached in a so-called intuitive manner, the situation is different for the mathematician working with fuzzy sets. Namely, the concept of a fuzzy set is now conceived of as a classical function, f, from a classical set, U, to another classical set, L. Since classical functions are classical sets, the fuzzy sets with which this mathematician works are actually classical sets and, therefore, the primitive notion which this mathematician uses is still the notion of a set.[12]

Therefore, a mathematician working with fuzzy sets as just described actually works within classical set theory in the sense just explained. To be more precise,

---

[11] See e.g. the critical remarks by Kahan quoted on p. 31 above, which display this kind of misunderstanding; for Goguen's remark, see the quotation on p. 244 above.

[12] A classical function is a classical set of ordered pairs.

this mathematician works within an intuitive set theory—the same framework in which classical mathematicians work. Mathematics based on fuzzy logic may hence be put on a solid ground in exactly the same way as classical mathematics—all notions, proofs, and theorems may be developed formally within a classical axiomatic set theory. Fuzzy sets may therefore be thought of as constituting a universe within classical axiomatic set theory—in a sense the world in which the common mathematician working with fuzzy sets acts. The construction and exploration of such a universe is exactly the point of the early works by Klaua examined along with subsequent developments in section 5.2.3. Since such a framework is based upon classical axiomatic set theory, i.e. a particular first-order theory, it is based on classical logic. This feature is manifested in that when a mathematician proves theorems *about fuzzy sets*, the reasoning is based on classical logic.

This still leaves open the question of whether it is possible to develop an axiomatic theory of fuzzy sets based on a fuzzy logic rather than on classical logic, and thus whether the concept of fuzzy set may truly be regarded as a primitive concept. The answer to this question is positive and the corresponding developments are covered in section 5.2.3. Even though the related developments are still relatively recent, they reveal the following picture. One may develop within the framework of predicate fuzzy logic an axiomatic theory of fuzzy sets, FST, which has the following properties. First, FST is—in terms of mathematical logic—interpretable in classical Zermelo-Fraenkel axiomatic set theory, ZF. This means that one may develop a universe of fuzzy sets within the axiomatic theory of classical sets using classical logic. Conceptually, this option is the same as the one described in the previous paragraph—one starts with classical sets and obtains fuzzy sets as a derived concept. Second, ZF is interpretable in FST. Importantly, this means that one may start with fuzzy sets as a primitive concept and provide an axiomatic theory for them within which one works according to the rules of fuzzy logic and within which one has at one's disposal classical sets which obey the laws of classical logic. In principle, one may develop mathematics based on fuzzy logic within FST. In this sense, *mathematics based on fuzzy logic may be developed axiomatically with a primitive notion of a fuzzy set that is not derived from the notion of a classical set.* Such mathematics then includes classical mathematics based on sets.

So far, the question of whether mathematics based on fuzzy logic may be viewed as a new kind of mathematics has been considered from a formal point of view. The importance of this view should not, however, preclude us from taking into account yet another aspect of this question, which is perhaps even more important. It depends upon realizing that the concept of a fuzzy set—whether formally considered as an object of axiomatic set theory within classical logic or an axiomatic theory of fuzzy sets developed within a formal fuzzy logic—allows for a new way of thinking about the physical world, namely in terms of fuzzy sets rather than classical sets.

Fuzzy logic as a new way of thinking

When speaking of a new way of thinking in the preceding paragraph, we mean that a mathematician or any researcher for that matter who employs the language of mathematics to formulate new observations and hypotheses, express laws and rules, describe decision strategies and algorithms, and reason about natural phenomena, may now do so using mathematics based on fuzzy logic. The importance of this new option derives from the fact that it provides a means of *employing directly*—via the concept of a fuzzy set—*inexact concepts* which are an inherent part of common-sense reasoning about the phenomena. To use the metaphor from the quotation on p. 427 above, inexact concepts—the monsters from the viewpoint of the classical paradigm—thus become tamed and very useful in the new paradigm which is based on fuzzy logic.[13] That inexact concepts are employed directly in mathematical models based on fuzzy logic, rather than being distorted or even circumvented and thus virtually ignored—as they are in the classical paradigm—is of crucial importance. Namely, researchers not only describe the phenomena they inspect in terms of the formal notions involved in their mathematical models, they eventually even think about them in terms of these notions. From this viewpoint, fuzzy logic makes it possible to *correlate formal notions naturally with one's common-sense concepts* about the physical world.

Fuzzy if-then rules and in particular automatic control using these rules—the most visible and commercially highly successful application of fuzzy logic examined in section 6.3.1—make it crystal clear that such opportunities may materialize and that this may indeed happen on a grand scale. In such cases, traditional control methods are replaced by fundamentally different ones—those which are based on control strategies formulated in natural language and which thus employ inexact concepts such as "high temperature" and mimic human reasoning with such concepts. The potential for such applications is not limited to any particular field, and the many examples of applications of fuzzy logic reviewed throughout this book document this convincingly. The question of to what extent fuzzy logic methods will become generally recognized and widely used in the great variety of prospective application fields is, nevertheless, impossible to answer at this point. Nevertheless, we can describe the current situation as follows. In several fields, fuzzy logic methods are routinely used and have led to considerable commercial success. In other fields, these methods are being explored and appear promising. There are also fields where these methods have not been utilized as yet or where there seems to be resistance

---

[13] An analogy with Robinson's *nonstandard analysis* seems particularly suggestive. Its crucial concept, that of an infinitely small number or an *infinitesimal*, is naturally involved in mathematicians' reasoning about calculus. Yet it resisted a rigorous mathematical treatment until Robinson developed nonstandard analysis in the 1960s. According to this idea, infinitesimals are conceived as elements of the so-called hyperreal field—a particular ultra-product of the field of reals—and are thus part of classical mathematics. See A. Robinson, *Non-standard Analysis* (Princeton, NJ: Princeton Univ. Press, 1974).

toward them. Why the situation has developed in this way is another interesting question to which we return in section 7.5.

Quantitative assessment

We now provide various quantitative data which document the overall impact of fuzzy logic. We start by examining the influence of Zadeh's 1965 paper, "Fuzzy sets," in terms of the number of citations this paper has received. As of July 2015, these numbers according to three well-recognized databases are as follows:[14]

Scopus	Web of Science	Google Scholar
26,535	20,935	57,345

These numbers are considerably higher than those for any other of Zadeh's highly cited papers. For example, the second and third most cited papers received the following numbers of citations:

	Scopus	Web of Science	Google Scholar
Zadeh 1973	3,558	3,303	17,399
Zadeh 1975c, part I	4,341	4,708	10,937

Even though the numbers for these two papers are much higher than those for the papers written by other contributors to fuzzy logic, the extraordinarily high numbers for Zadeh's seminal paper clearly demonstrate its groundbreaking character. It also is worth noting that many papers utilizing the concept of a fuzzy set no longer cite Zadeh's 1965 paper and, since this is now considered to be commonly known, they refer instead to suitable textbooks, which certainly reduces the citation counts.

The influence of Zadeh's 1965 paper in terms of citation counts is also apparent from a comparison with other highly cited papers. An article of October 2014 in *Nature* analyzed the 100 most-cited research papers of all time.[15] The vast majority of these are papers in biology, chemistry, and physics. In these fields, citation counts are generally much higher than in mathematics, computer science, or engineering. Still, among the top 100 papers, 11 are classified as contributions to mathematics and statistics. Zadeh's paper, "Fuzzy sets" of 1965, ranks third among these and 46th overall.[16] This remarkable status of Zadeh's paper clearly points to its exceptional nature. It is, however, not only the high number of citations of the 1965 paper that

---

[14] Scopus is a bibliographic database owned by Elsevier covering titles from over 5,000 publishers in various areas. Web of Science (WoS) is an online scientific citation indexing service provided by Thomson Reuters. Google Scholar (GS) is a freely accessible web search engine indexing scholarly literature of various formats in numerous areas. Scopus, WoS, and GS differ in their coverage of particular areas as well as in the kinds of materials included.

[15] R. Van Noorden, B. Maher, and R. Nuzzo, "The top 100 papers: *Nature* explores the most-cited research of all time," *Nature* 514 (2014): 550–53, with analysis at http://www.nature.com/news/the-top-100-papers-1.16224. The citation counts were made using the Web of Science.

[16] The first two are E. L. Kaplan and P. Meier, "Nonparametric estimation from incomplete observations," *J American Statistical Association* 53 (1958): 457–81, and D. R. Cox, "Regression models and life-tables," *J Royal Statistical Society* 34 (1972): 187–220. It should be noted, however, that the list does not include some highly cited

Table 7.1: Citations of Zadeh 1965a in Scopus by subject areas as of July 2015.

Computer Science	14,562	Biochem., Genetics, Mol. Biol.	785
Engineering	11,555	Chem. Eng., Chemistry	780
Mathematics	8,620	Agricultural and Biological Sci.	672
Decision Sciences	3,364	Medicine	622
Environmental Science	1,471	Energy	588
Social Sciences	1,364	Materials Science	493
Business, Mgmt., and Acct.	1,166	Economics, Ecmtr., and Finance	411
Earth and Planetary Sciences	1,092	Psychology	182
Physics and Astronomy	808	Arts and Humanities	175

is impressive, but also their diversity in terms of the areas in which the work is cited. This is demonstrated clearly by table 7.1.[17]

The impact of the inception of fuzzy logic on different research areas, and on mathematics in particular, is documented well by the numbers of papers indexed in Scopus as an "article" or "conference paper" whose title, abstract, or keywords contain the terms "fuzzy," "fuzzy sets," or "fuzzy logic." These are shown in table 7.2. The clusters by five-year periods in table 7.2 as well as the classification according to subject areas, which is provided by table 7.3, shows that the influence of fuzzy logic is steadily growing and that it spans across all fields although to varying extents. While these numbers illustrate well the influence of fuzzy logic, they do not take into account growth over the years in numbers of publications in general. This factor is taken into account in table 7.4, which summarizes the share of papers on fuzzy logic among all papers in mathematics and among papers in all areas according to Scopus. In mathematics, the share steadily increased until around 2000 from which date on it seems to have begun to stabilize.

The dramatic appearance of a growing body of research drawing on fuzzy logic methods within little more than a decade of Zadeh's seminal paper was soon apparent to the editors of *Mathematical Reviews* and the *Zentralblatt für Mathematik*. Consequently, "Fuzzy Logic" and "Fuzzy Set Theory" were introduced as new subclasses in the Mathematics Subject Classification (MSC) scheme—and this turns out to have been done surprising early—in 1980.[18] The current version, MSC2010, contains 21 classifications containing the term "fuzzy."

---

papers, such as C. E. Shannon, "A mathematical theory of communication," *Bell System Technical J* 27 (1948): 379–423, which has almost 75,000 citations according to Google Scholar. Nor may the list be understood as containing the most *important* papers because several papers considered as revolutionary, such as Albert Einstein's paper on the general theory of relativity, do not appear in it. This is the result of the chosen method and the Web of Science coverage.

[17] The abbreviated subject areas here, respectively, stand for "Business, Management, and Accounting"; "Biochemistry, Genetics, and Molecular Biology"; "Chemical Engineering"; "Agricultural and Biological Sciences"; and "Economics, Econometrics, and Finance."

[18] With thanks to Norman Reichert, managing editor, *Mathematical Reviews.*

Table 7.2: Numbers of papers in Scopus and papers in Scopus's subject area "Mathematics" as of July 2015.

Period	Papers in Scopus			Scopus/Mathematics		
	fuzzy	fuzzy set	fuzzy logic	fuzzy	fuzzy set	fuzzy logic
1965–69	26	9	0	5	2	0
1970–74	81	24	6	25	14	2
1975–79	258	108	23	109	62	13
1980–84	1,049	550	104	418	212	32
1985–89	2,417	1,549	385	850	488	103
1990–94	7,013	5,425	2,222	1,740	1,048	311
1995–99	16,844	11,737	5,457	2,744	1,765	604
2000–04	25,939	16,804	7,041	3,752	2,443	729
2005–09	60,099	26,004	18,825	13,856	6,995	3,817
2010–14	82,010	25,624	20,348	20,216	7,876	4,055

Table 7.3: Numbers of papers in Scopus whose title, abstract, or keywords contain "fuzzy" by subject areas as of July 2015.

Engineering	78,034	Earth and Planetary Sciences	4,116
Computer Science	72,240	Environmental Science	3,943
Mathematics	31,165	Biochem., Genetics, Mol. Biol.	3,871
Decision Sciences	9,773	Business, Mgmt. and Acct.	3,509
Chem. Eng., Chemistry	7,062	Medicine	3,168
Physics and Astronomy	6,436	Agricultural and Biological Sci.	2,564
Energy	4,973	Economics, Ecmtr. and Finance	819
Materials Science	4,284	Psychology	468
Social Sciences	4,257	Arts and Humanities	453

Table 7.4: Share in % of papers whose title, abstract, or keywords contain "fuzzy" in papers in English indexed in Scopus in subject area "Mathematics" as of July 2015. The numbers in parentheses represent the share with respect to all subject areas.

1965–69	1970–74	1975–79	1980–84	1985–89
0.15 (0.003)	0.27 (0.008)	0.65 (0.02)	1.91 (0.06)	2.85 (0.098)
1990–94	1995–99	2000–2004	2005–9	2010–14
6.21 (0.23)	8.57 (0.38)	10.49 (0.51)	12.07 (0.81)	10.99 (0.84)

Another quantitative measure of the growing significance of fuzzy logic as a legitimate and important subject for study is reflected in funding for research concerned with fuzzy logic and its applications. Here we only offer one such indicator, but this in itself is indicative of how quickly fuzzy logic was able to establish itself as a legitimate subject, despite the skepticism of some and the open hostility of others. In this connection we note what Maria Zemankova, currently program director for Information Integration & Informatics (III) of the Information & Intelligent Systems Division (IIS), which is part of the Computer & Information Science & Engineering Directorate (CISE) at the U.S. National Science Foundation (NSF), has said about NSF funding for fuzzy logic research: "The academic community has been very antagonistic to fuzzy logic" (McNeill and Freiberger 1993, 76).

At NSF, this has meant that some program officers went out of their way, particularly in the early years, to prevent funding for any research related to fuzzy logic, whereas other program officers, like Zemankova who was more sympathetic, treated research applications involving fuzzy logic fairly. As she recalls, on one occasion: "Someone came into my office and said, 'Take this proposal. I don't fund fuzzy logic.' I said, 'This is in your area scientifically and just because it uses fuzzy logic to solve problems in your area, you can't say, "I don't fund fuzzy logic".' ... He had no choice but to take it back, but I know what'll happen. He'll send it to reviewers who will kill it" (McNeill and Freiberger 1993, 258–59).

In some areas of NSF, program officers were willing to fund appropriate and promising research, and the actual numbers, in five-year increments beginning with 1970, both in terms of the number of grants awarded and the actual dollar amounts of research funded confirm the publication trends we have already compiled; see table 7.5.[19]

Research on fuzzy logic has by no means been solely theoretical or limited to kinds of results usually reported in journals and conference papers. A considerable amount of research has been devoted to applications that have actually led to patents. The numbers regarding patents as of 2014 are as follows:[20]

Patents related to fuzzy logic			
United States	21,878 (issued)	WIPO	50,999
United States	22,272 (applied)	EPO	3,268
Japan	7,149	Australia	2,350
China	25,454	Canada	556

---

[19] Data based on searches for "fuzzy" in proposal titles and abstracts for NSF grants, as retrieved from the database of NSF Award Abstracts (http://www.nsf.gov/awardsearch/, accessed in November 2015) of all funded projects with award dates between January 1, 1970 and December 31, 2014.

[20] Based on data as of July 2015 on the webpage at http://www.cs.berkeley.edu/ zadeh/stimfl.html. "WIPO" is the acronym for "World Intellectual Property Organization," "EPO" for "European Patent Office." The number for EPO does not include patents in the individual countries in Europe. The number for China does not include all years and does not include Taiwan and Hong Kong.

Table 7.5: Funding of research on fuzzy logic by NSF.

Period	NSF grants	NSF grants in US$
1970–74	2	85,000
1975–79	2	253,690
1980–84	11	751,509
1985–89	25	1,924,520
1990–94	69	7,941,173
1995–99	84	12,937,526
2000–04	52	12,143,272
2005–09	38	8,519,173
2010–14	37	8,731,404
total	320	53,287,267

The remarkable extent of the impact of fuzzy logic in terms of real-world applications and commercial success is best illustrated by a few salient facts regarding the fuzzy boom in Japan. This began with the implementation in 1987 of a new subway system in Sendai (section 6.2), which soon became an international sensation thanks to its use of fuzzy logic to control the trains. The system, developed by Hitachi, uses fuzzy logic to control the speed and thereby achieve relatively smoother rides compared with other trains; it has also proven to be 10% more energy efficient than human-controlled acceleration. The entire system has recently received a major upgrade that was completed in 2013, and includes a safety system that is also based on fuzzy logic.

Subsequently, according to one estimation provided by the Japanese Ministry of International Trade and Industry (MITI), fuzzy-logic based products—70% of which are to be found in consumer electronics—accounted for about $1.5 billion in revenue in 1990 alone, and over $2 billion in 1991. Estimating the global market for computer services, software, and hardware to be about $200 billion in 1990, this means that fuzzy products then accounted for roughly 1% of the global computer market (Kosko 1993, 71). A more recent example of the successful applications of fuzzy logic by Japanese companies is provided by the cumulative sales of the Omron home-use blood pressure monitor, which is based on fuzzy logic. Since the first such model—the Omron Digital Automatic Blood Pressure Monitor, Model HEM-706 Fuzzy—was introduced in 1991, total global sales have surpassed 120 million units.[21]

---

[21] *Unbridled Spirit: The Story Behind Fuzzy Blood Pressure Monitor Development,* booklet by Omron Healthcare Co., Ltd., 2013.

7.5 A prospective view on the 50th anniversary of fuzzy logic

We close this chapter by providing our assessment of the future prospects of fuzzy logic which we regard as an important part of our overall assessment of its significance. Given the 50 years that have passed since its inception, the picture clearly indicates that even though considerable progress has been made both in terms of theory and applications, fuzzy logic has been developed in many directions but to different extents and a number of foundational questions still remain open. Our aims in this section are naturally modest—we do not aspire to predict the future of fuzzy logic. Rather, we outline certain broader issues which require further attention and whose examination may bring further advances in the foreseeable future.

We start with the problem of *general awareness of fuzzy logic* by researchers as its potential users. It is to be expected that with further advancement, new research will increasingly appear in venues that do not specialize in fuzzy logic but rather in journals on logic, mathematics, and the various areas in which fuzzy logic is potentially useful. This is a natural process that will simultaneously raise general awareness of fuzzy logic among researchers in various areas. Such a process is also healthy in the sense that it provides valuable feedback. Nevertheless, there is one particular aspect worth emphasizing, namely awareness of fuzzy logic as a mathematical tool among students of various fields in which mathematics is part of the standard curricula.

Availability of *general purpose textbooks* is obviously essential for this purpose. Although some such textbooks exist (section 2.5), they are mostly intended for students in mathematics, computer science, and engineering. It is important to realize that students in natural sciences and even more so in social sciences, have very different skills and attitudes toward mathematics and the existing textbooks are generally too difficult for them to comprehend. Just as there exist textbooks on statistics for the social sciences, it would be very helpful if there were similar kinds of introductory texts on fuzzy logic or if such texts were parts of introductory textbooks on mathematics for students in fields where fuzzy logic would be potentially useful.

The educational prospects mentioned above are clearly connected to further development of *applications of fuzzy logic*. In this respect, we find it particularly important to reflect on the existing discrepancy between the initial expectations regarding applications of fuzzy logic on the one hand and their actual materialization on the other hand in some soft sciences such as psychology but in other fields as well, such as biology. Initial expectations were naturally high because inexact concepts are abundant in these fields for which renowned experts had expressed the need for mathematics capable of dealing with inexact concepts, as we have seen in chapters 2 and 6. The reasons why currently available applications of fuzzy logic have not been made in some fields, despite expectations, is certainly an interesting phenomenon which apparently has no simple explanation. We believe that a large part of the explanation lies in low awareness and various kinds of misconceptions regarding fuzzy logic. This may convincingly be shown in psychology in terms of the historical circumstances analyzed in section 6.7. Another possible factor derives from the fact that the use of fuzzy logic methods requires a kind of mathematical sophistication—the inexact concepts which play a crucial role but have previously been approached rather informally, should now be dealt with using mathematics based on fuzzy logic. But at first this may well prove challenging for researchers in the soft sciences who usually have modest mathematical skills and a generally ambivalent attitude toward using mathematical methods.

The potential of fuzzy logic for future applications can best be enhanced by establishing its *foundations*. Foundations not only represent the backbone for applications but, importantly, also help secure the overall status of fuzzy logic as a coherent body of principles, concepts, and methods.

In this respect, we believe that various fundamental philosophical problems pertaining to fuzzy logic should be explored further. In particular, questions regarding the meaning of truth degrees discussed in section 5.4.1 and which have to some extent been investigated in studies of vagueness (section 5.4.3), need further study. It is especially important to work out a satisfactory theory of truth congruent with the basic idea of fuzzy logic, namely the graded approach to truth.

Another fundamental issue concerns interactions with psychology. We do not mean here the possible applications of fuzzy logic in psychology discussed in section 6.7. Rather, we mean problems regarding the psychological status of truth degrees and operations with them, which include the question of the psychological status of truth functions of logical connectives. Such questions have been partly explored, particularly in the early 1970s (section 5.4.1). Their further examination is likely to shed new light on the problem of the choice of fuzzy logic connectives but also on other problems as well. Especially appealing in this regard seems the idea of approaching such problems via the theory of measurement—a mathematically advanced field of psychology which studies scales of measurement, operations on scales, the notion of meaningfulness of such operations, and other concepts. While some contributions have been made along these lines, this direction is greatly underdeveloped. While attention to psychological issues in logic books and vice versa is apparent from earlier texts on logic, the end of the 19th century brought a separation of logic and psychology and in particular, elimination of psychological issues from logical investigations.[22] It is interesting that natural questions in fuzzy logic now again ask for psychological investigations.

Concerning further developments in the various areas of mathematics based on fuzzy logic, we view this as a continuing process which should primarily be directed by the needs of applications and proper motivations rather than by a desire to fuzzify

---

[22] The elimination of psychology was basically a consequence of the successful effort to expel so-called psychologism from logic, i.e. confusion of psychological with logical entities.

the whole of classical mathematics. On the one hand, the various concepts and theories developed should become parts of traditional mathematical areas, such as topology and probability, and extend their application capabilities. On the other hand, one may also consider the structure of mathematics based on fuzzy logic itself. This structure is likely to be different from the structure of classical mathematics in that counterparts of some large areas of classical mathematics may appear much smaller in mathematics based on fuzzy logic due to a lack of natural needs to develop them. We believe that such a pattern is indeed emerging and will continue to emerge.

As regards *mathematical foundations of fuzzy logic*, considerable work has yet to be done. This includes further progress in relatively recent developments, such as axiomatic theories of fuzzy sets and category-theoretic foundations (sections 5.2.3 and 5.2.4), but also conceptually novel approaches. A promising and highly compelling approach is the *alternative set theory* of Petr Vopěnka (1935–2015), a pioneer in developing Boolean models and applying them in independence proofs in classical set theory (p. 256 above). Later on, he developed alternative set theories motivated by the aim to model unsharply delineated classes of objects. It is a coincidence that he developed the first such theory, the *theory of semisets*, together with Hájek (Vopěnka and Hájek 1972). More sophisticated from a philosophical viewpoint is his alternative set theory (AST).[23] AST, which is a bivalent theory, approaches vagueness from a radically different viewpoint than fuzzy logic, namely via a notion of horizon in one's view of the universe. AST has been carefully explored as regards its foundations. Moreover, certain parts of mathematics have been developed within AST. Some relationships between AST and fuzzy logic have been considered by Novák (1992) and by Zlatoš (1997). Thorough examinations of relationships between AST and fuzzy logic represent a highly interesting project. In addition, the book by Rescher and Grim (2011) provides an account of philosophical foundations for modeling collections of objects whose membership is not clear-cut for a variety of possible reasons, including vagueness of the description of such collections, and deserves attention.

Foundational considerations regarding fuzzy logic admit interesting nonconventional questions. To illustrate, there is a thought-provoking issue already alluded to in Post 1921 and Rosser 1941. In developing various fuzzy logics and their associated mathematics, we eventually need—at the metalevel—a two-valued logic and our reasoning at the metalevel is governed by the laws of classical logic. For instance, a given formula of a predicate fuzzy logic is eventually regarded as provable or not. A given theorem regarding certain properties of fuzzy relations is either provable or not. Even in Pavelka-style fuzzy logic, where provability is a matter of degree, the metalevel is two-valued: a given formula in the end either is or is not

---

[23] An early account is provided in Vopěnka 1979 which is written in English but is rather incomplete in providing a full account, particularly with respect to philosophical views which are an essential ingredient. Such an account is given in Vopěnka 1989a which, unfortunately, has not been translated into English. There are several introductory texts on AST written in English, such as Vopěnka 1989b, 1991. A rich source of information about Vopěnka and his work is the special issue of *Ann Pure Appl Logic* dedicated to Petr Vopěnka, vol. 109, no. 1–2 (2001), pp. 1–138, edited by B. Balcar and P. Simon.

provable to a given degree. In this sense, two-valued logic seems to have a methodological priority which is a consequence of the rules of conduct of the present-day
methodology of logic and mathematics. Whether this must be so is an intriguing
question that seems to have been virtually ignored. This question obviously relates
to a compelling philosophical problem already mentioned by Goguen (1967, 147),
who wrote:

> It is somewhat unsatisfying philosophically to ground a logic of fuzziness in a logic of exactness;
> it would seem to ask for an independent postulational formulation. On the other hand, our
> method shows that if mathematics, as we use it, is consistent, so is fuzziness, as we formulate it.

Returning to *fuzzy logic in the narrow sense* as presently practiced, we have
argued in section 7.2 that it has advanced considerably, especially during the last
twenty years. Attention has been paid particularly to axiomatizations of various
kinds of propositional and predicate logics, but also to other questions traditionally
studied in logic. Model theory for fuzzy logic, however, is underdeveloped. This
area of logic is of particular interest in a fuzzy setting because several notions regarding general fuzzy relational structures—whose study belongs to this area—are
of practical importance. The notion of isomorphism of relational structures is but
one simple example. In a fuzzy setting, one may regard degrees of isomorphism,
consider them as expressing the appealing concept of similarity of fuzzy relational
structures, and examine traditional model-theoretic results from this perspective.
Note too that a highly interesting relevant approach is represented by the somewhat
overlooked continuous model theory (p. 154). Pavelka-style fuzzy logic with its degrees of consequence, degrees of provability, and further metalogical notions also
deserves further exploration. In particular, a thorough study of logical problems
from the point of view of graded consequence as a primary concept may offer a new,
unifying perspective on various kinds of logics, including the non-truth-functional
ones. Here Gerla's pioneering results and the tradition of the Polish school may
serve as inspiration. Another possibly interesting direction would be to examine
from the viewpoint of fuzzy logic various logics extending classical logic. Such examinations have already been carried out to some extent (chapter 4), such as those
of intuitionistic logic, modal logics, or paraconsistent logics, which are sometimes
characterized as inconsistency-tolerant logics and thus seem naturally suited for such
examinations because the very concept of inconsistency may arguably be viewed as
admitting degrees of truth.

Throughout its history until the late 19th century, logic's main theme has been
human reasoning. Its study has included both deductive and inductive reasoning
along with numerous issues extending to philosophy and psychology. These matters have been conceived rather broadly but reasoning in natural language has always
been the main subject. Since the mid-19th century, logic has increasingly been more
considered as a science of deductive reasoning, leaving the subject of inductive reasoning to probability and statistics. A radical move came in the late 19th and early
20th centuries which mark the beginnings of modern formal logic as we know it

today. Since then, logic has primarily been seen as concerned with mathematical reasoning rather than reasoning in natural language. This has brought considerable simplification but also restrictions of its applicability. From this perspective, Zadeh's concept of fuzzy logic and his contributions may justly be seen as an *attempt to revive the study of reasoning in natural language as a grand theme of logic*.[24] Even though considerable progress has been made during the past decades, many important problems still remain unresolved. We believe that some such problems are of fundamental importance but are largely being ignored. Part of the reason is that the perspective outlined above has not been properly recognized. More concretely, we believe that there are two basic issues at stake.

The first concerns fuzzy logic as formal logic. So far, it has been developed as fuzzy logic in the narrow sense, also referred to as mathematical fuzzy logic. In the prevalent view, it is conceived of and approached as a generalization of classical logic in which truth functions of classical connectives are replaced by their many-valued counterparts satisfying certain constraints which are viewed as congruent with the idea of degrees of truth. We regard this conception as too restrictive and believe that to reach its full potential, fuzzy logic in the narrow sense needs to extend beyond it. An obvious way to proceed toward such an extended conception would be to look for inspiration to the challenging problems considered by Zadeh in the context of fuzzy logic in the broad sense. These problems involve reasoning with natural language statements which include various kinds of language constructs which are foreign to classical logic. Recall in this context that some effort has already been made in this direction and that we examined relevant contributions in chapter 4. In particular, Hájek has repeatedly emphasized the need for interaction between fuzzy logic in the narrow and the broad sense, to which he contributed with his comprehensive book Hájek 1998b.[25] We believe that more work in this direction may bring further advances in the study of reasoning in natural language—the long-standing grand theme of logic.

The second issue is concerned with the separation of deductive and inductive reasoning and inclusion of the latter within probability and statistics. This separation has also meant the exclusion of natural language from considerations of probability. That such exclusion presents a serious limitation to the applicability of classical probability theory was recognized by Zadeh in the late 1960s. In his writings, he began to study probabilities of fuzzy events, such as the event that a randomly picked man will be tall, which are out of the conceptual reach of classical probability but frequently occur in natural language statements. In his later writings, such

---

[24] Some directions in logic such as modal logic or intensional logic, which are inspired by reasoning in natural language, may also be looked at from this perspective. Nevertheless, we consider these efforts much narrower in their aims.

[25] The distinguishing roles of the adjectives "narrow" and "broad" used in connection with the term "fuzzy logic" are certainly useful and reasonably justified. However, the adjectives also indicate—and justly so—a kind of separation regarding the subject of interest. The above perspective and its future development may hopefully overcome this somewhat unnatural situation.

as Zadeh 1996b, 2005, 2015, he presented further challenging problems—some of which are summarized on p. 315 above—whose resolution is beyond the capability of classical probability theory. Probabilities of fuzzy events have led to numerous investigations, which naturally call for further developments toward general measures of uncertainty (section 5.3.5). Nevertheless, little attention has been paid to the problems from Zadeh's above-mentioned later writings. In view of the above remarks, such problems naturally fit the original motivation of pursuing logic as a broadly conceived study of reasoning in natural language and their exploration presents another grand problem which deserves to be examined thoroughly.

We conclude with the following quotation from the close of the paper entitled "Fuzzy logic—a personal perspective" (Zadeh 2015), which was written by Zadeh on the occasion of the 50th anniversary of his first paper on fuzzy sets:

> In large measure, science is based on the classical, Aristotelian, bivalent logic. In science, binarization is the norm rather than exception. In human cognition, the opposite is true. One of the principal contributions of fuzzy logic is that of providing a basis for a far-reaching shift—in almost all fields of science—from binarism to pluralism, from black and white to shades of gray. In coming years, this move is likely to accelerate, and the impact of fuzzy logic is likely to become more visible and more substantive. Eventually, what is likely is that in science—as in fuzzy logic—almost everything will be, or will be allowed to be, a matter of degree. This is what I see in my crystal ball.

It will be interesting to reflect on Zadeh's prediction and to see how accurately his crystal ball may indeed have successfully foretold the future of fuzzy logic fifty years from now.

Appendix A

The Enigma of Cox's Proof

THE CLAIM in some of the debates described in section 2.4 that the only way to deal with uncertainty is to use the rules of probability theory is often justified by referring to a proof by Cox (1946), which is also presented without any substantial change in Cox 1961. According to Cox, the aim of the proof is "to show that by employing the algebra of symbolic logic it is possible to derive the rules of probability from two quite primitive notions." Although his derivation is actually not presented as a proof of a theorem, but rather as a sequence of intertwined formal and intuitive arguments, it has been routinely referred to as "Cox's proof" or "Cox's theorem." In order to describe this derivation, we consider it appropriate to adhere to Cox's original notation: Letters **a**, **b**, **c**, ... denote propositions; $\sim$ **a** denotes the negation of proposition **a**; **ab** and **a**$\lor$**b** denote, respectively, the conjunction and disjunction of propositions **a** and **b**; and **b**|**a** denotes "some measure of the reasonable credibility of the proposition **b** when the proposition **a** is known to be true."

Using this notation and employing the Boolean algebra of classical propositional logic, Cox aims at proving that the only sensible way to combine reasonable credibilities is to use the rules of probability theory. He begins with the assumption (axiom) that

$$\mathbf{cb}|\mathbf{a} = F(\mathbf{c}|\mathbf{ba}, \mathbf{b}|\mathbf{a}),$$

where F is some function of two variables to be determined. Employing the associative law of conjunctions of propositions, he derives the equation

$$F(F(\mathbf{d}|\mathbf{cba}, \mathbf{c}|\mathbf{ba}), \mathbf{b}|\mathbf{a}) = F(\mathbf{d}|\mathbf{cba}, F(\mathbf{c}|\mathbf{ba}, \mathbf{b}|\mathbf{a})), \tag{A.1}$$

where **a**, **b**, **c**, **d** are any propositions. Letting $d|cba = x$, $c|ba = y$, and $b|a = z$, equation (A.1) becomes

$$F(F(x,y),z) = F(x,F(y,z)). \tag{A.2}$$

Function F must satisfy this functional equation, known as the associativity equation (Aczél 1966, 253), for arbitrary values x, y, and z.

Once Cox converted his original problem into this purely mathematical problem, his challenge was to solve this equation. In a long and tedious derivation, described in detail in a large appendix to his paper, Cox managed to show the following: If F has continuous second-order derivatives, then

$$Cf(F(p,q)) = f(p)f(q)$$

is the solution of (A.2), where, as stated by Cox, "f is an arbitrary function of a single variable, and C is an arbitrary constant." This derivation is correct and it was a significant contribution to the theory of functional equations when Cox's paper was published.

In order to determine a relation between $b|a$ and $\sim b \,|\, a$, Cox furthermore assumes (employs as an axiom) that

$$\sim b \,|\, a = S(b|a),$$

where S is some function to be determined. Employing the law of double negation and one of the De Morgan's laws of classical logic, he derives the functional equation

$$xS(S(y)/x) = yS(S(x)/y), \tag{A.3}$$

where $x = c|a$ and $y = S(cd|a)$. He then shows (again by a tedious but correct derivation in the appendix of his paper) the following: If S is twice differentiable, then

$$S(p) = (1 - p^m)^{\frac{1}{m}}$$

is the solution of (A.3), where m is an arbitrary constant. Cox considers the value of m purely conventional and chooses $m = 1$ to obtain the desired formula of probability theory.

The results obtained by Cox became a subject of controversy in some of the debates described in section 2.4. The controversy was triggered by two closely related claims: a rather extreme claim that "the only satisfactory description of uncertainty is probability" (Lindley 1987, 17) and an associated claim that "the strongest argument for the use of standard probability theory is a proof by Cox" (Cheeseman 1988, 60). Since Cox's proof was published more than 40 years before the debates and in a journal specializing on physics, it is understandable that many people participating in the debates were initially not aware of it. However, some of them recognized in their commentaries that the proof was contingent upon the assumption that function F in the associativity functional equation (A.2) has a continuous second derivative. This, they argued, excludes possibility and necessity measures since they are based on max and min functions, which clearly do not satisfy this assumption. Next, some advocates of the original meaning of Cox's proof responded by defending it via reference to another method for solving equation (A.2), which was developed by Aczél (1966). For example, Smith and Erickson (1989, 38) wrote:

> By assuming that F is twice differentiable in both variables, Cox derived from Eq. (A.3) a differential equation, which he then solved. Some fuzzy set advocates have pounced upon this assumption as invalidating Cox's theory, in evident ignorance of the work of (Aczél 1966, 1987), who derived the same general solution without assuming differentiability.[1]

This was more than a decade later still echoed by Jaynes (2003, 668):

> The issue of nondifferentiable functions arises from time to time in probability theory. In particular, when one solves a functional equation such as those studied in Chapter 2,[2] to assume differentiability is to have a horde of compulsive mathematical nitpickers descend upon one, with claims that we are excluding a large class of potentially important solutions. However, we noted that this is not the case; Aczél demonstrated that Cox's functional equations can all be solved without assuming differentiability (at the cost of much longer derivation) and with just the same solution that we found above.

It is correct that Aczél was able to solve (A.2) without the assumption that function F has a continuous second derivative, but assuming instead that F is reducible on both sides, which he defines (Aczél 1966, 255) as "$F(t, u) = F(t, v)$ or $F(u, w) = F(v, w)$ only if $u = v$." The authors of the above statements seem to tacitly assume that the requirement that F be reducible on both sides is weaker than the requirement that F has a continuous second derivative and, hence, the controversy regarding functions max and min is resolved. However, such an assumption is wrong. Functions max and min do not have continuous second derivatives, but they are nonreducible on both sides as well. The two requirements are actually not comparable. In addition to the product function, which clearly satisfies both of them, and the max and min functions, which satisfy neither of them, there also exist associative functions that satisfy only one of them. For example, function $F(x, y) = \sqrt{x^2 + y^2}$ is associative and twice differentiable, but nor reducible. On the other hand, function $F(x, y) = f^{-1}(f(x) \cdot f(y))$ where $f(x) = \frac{x}{2}$ for $x \in [0, 0.5]$ and $f(x) = 1.5x - 0.5$ for $x \in (0.5, 1]$ (and $f(y)$ is defined in the same way), is clearly associative and not differentiable, but it is reducible from both sides. To show its reducibility, let $F(x, y) = F(x, z)$. Then, $f^{-1}(f(x) \cdot f(y)) = f^{-1}(f(x) \cdot f(z))$, hence $f(x) \cdot f(y) = f(x) \cdot f(z)$, and thus $f(y) = f(z)$. Since f is a bijective function, we obtain that $y = z$, so F is reducible from one side. Moreover, since F is a symmetric function, it is also reducible from the other side.

In this appendix, we set out to examine, strictly on mathematical grounds, why Cox's proof does not justify the extreme claims made by some advocates of probability theory (Lindley, Cheeseman, and others) that *probability is the only sensible description of uncertainty*. However, Cox's proof, as a justification for such extreme

---

[1] "Eq. (A.3)" is equation (29) in Smith and Erickson 1989; Aczél's work of 1987 refers to J. Aczél, *A Short Course on Functional Equations* (Dordrecht: Reidel, 1987).

[2] In chapter 2 of his book, Jaynes basically outlines Cox's proof and only refers to the alternative, much longer proof by Aczél.

claims, is also vulnerable on philosophical grounds. This is beyond the aim of this appendix, but we consider it worth referring to a paper by Colyvan (2004), where these aspects are thoroughly discussed.

Appendix B

Overview of Classical Logic

The scope of logic and classical logic

L OGIC studies reasoning. Such study has many facets but the central one is consequence. That is to say, the question of whether a given proposition is a consequence of some other propositions may be regarded as the very central subject of logic.

In pursuing its goal, logic studies correct modes of reasoning. As an example, from the propositions "It rains" and "If it rains then streets are wet" we may infer the proposition "Streets are wet." This is an example of correct reasoning. On the other hand, deducing "It rains" from the propositions "Streets are wet" and "If it rains then streets are wet" is not correct, because streets may be wet for a reason other than rain. Modern logic is not concerned with the content of the propositions involved in reasoning. Rather, it is concerned with their *form*. The inference described above has the same form as the following one: from "I have money" and "If I have money then I can buy goods" infer "I can buy goods." This form may succinctly be represented by the rule:

$$\frac{P, \text{ if } P \text{ then } Q}{Q},$$

which is commonly known as the rule of *modus ponens*, also called the rule of detachment. The emphasis on form rather than content along with the fact that modern logic deals with propositions in formalized languages, i.e. languages built of abstract symbols, are the reasons why modern logic is often called *formal logic* or *symbolic logic*. In referring to modern logic, the term *mathematical logic* is also used to emphasize the fact that it utilizes methods that are common in mathematics and that research in modern logic has many features in common with mathematical research.

Reasoning is a complex subject. In their endeavors, logicians impose various restrictions to make the subject more tractable and the study easier. The form of propositions involved in reasoning is one example of such restrictions. For example, consider propositions such as "If John is home after 8 am then it is probable that John goes to work by car." This proposition results by applying the logical connective of implication, expressed here by "if . . . then . . . ," to two simpler propositions, namely "John is home after 8 am" and "It is probable that John goes to work by car." The second proposition consists of the qualifier expressed by "It is probable that . . ." and the proposition "John goes to work by car" to which the qualifier is applied. The original proposition is therefore a compound proposition of the form

$$\text{if } P \text{ then it is probable that } Q.$$

While the connective of implication is relatively simple, the qualifier "It is probable that . . ." is considerably more involved. Namely, this qualifier may be regarded at the level of syntax, i.e. when it comes to the formal rules for handling the symbols of a given language, as a unary connective, i.e. a connective that is applied to the proposition "John goes to work by car." In a similar manner, we consider negation, which is expressed by "It is not true that . . ." or simply by "Not . . . ," another unary connective. However, at the level of semantics, i.e. when it comes to meaning, negation is much simpler than the connective "It is probable that . . ." Namely, according to ordinary usage, "Not P" is considered true if "P" is false and is considered false if "P" is true. In this sense, negation is a truth-functional logical connective because for every proposition "P," the truth value of "Not P" is a function of the truth value of "P." On the other hand, "It is probable that . . ." is not truth-functional since the truth value of "It is probable that P" does not depend only on the truth value of "P." For example, the proposition "It is probable that it is now raining in Seattle" may be considered true due to available meteorological evidence, no matter whether it is actually raining or not in the city of Seattle, i.e. irrespective of the truth value of the proposition "It is now raining in Seattle." In order to avoid the perplexities of probability and statements involving probabilities, one may want to restrict oneself to propositions that do not involve connectives such as "It is probable that . . ."

Clearly, any such restriction may be considered a drawback because it implies a loss in expressivity. On the other hand, even propositions restricted this way may still be useful and applicable in a wide range of situations. Since the restricted propositions are simpler and hence easier to analyze, it may be possible to obtain deeper results regarding reasoning with them. Moreover, the study of reasoning with restricted forms of propositions makes clear the cost of not restricting, or equivalently, the gain that may be realized by restricting propositions. Some natural logical questions have different answers for different restrictions. For example, the problem of whether a given proposition is entailed by some other given propositions may be efficiently decidable for some kinds of propositions but only inefficiently decidable or even undecidable algorithmically for other kinds of propositions. These reasons make the study of logical systems which deal with propositions restricted in various ways an important endeavor.

The best known and most widely employed logical system is represented by *classical logic*. Classical logic is characterized by its two most important features: *bivalence* and *truth functionality*. Bivalence means that one recognizes only two truth values, i.e. values that may be assigned to propositions, commonly called *truth* and *falsity* and often denoted, respectively, by 1 and 0, sometimes by other symbols such as t and f, $\top$ and $\bot$, among others.[1] A proposition to which we assign 1 is considered a true proposition, a proposition assigned 0 is considered false. Hence, bivalence may succinctly be characterized as follows:

> Every proposition is either true or false.

Truth functionality (also called compositionality) means:

> The truth value of a compound proposition is uniquely determined by the truth values of its constituent propositions.

We explained above what truth functionality means for negation. Similarly, if a compound proposition is the conjunction of two simpler propositions, i.e. "P and Q," truth functionality means that the truth value of "P and Q" is solely determined by the truth value of P and the truth value of Q. In particular, "P and Q" is true if both P and Q are true and false otherwise.

What follows provides a brief overview of some basic notions and results of classical logic. We begin with classical propositional logic followed by classical predicate logic. Separating the treatment of propositional logic as a simpler part of predicate logic is a usual practice in logic, which is convenient because of our presentation of the development of fuzzy logic in the narrow sense. In classical propositional logic, propositions are typically built from elementary propositions by connectives of negation, conjunction, disjunction, implication, equivalence, and possibly other truth-functional connectives. Elementary propositions are regarded as atomic statements whose possible inner structure is unknown. From the point of view of propositional logic, the proposition

> "John is male and everybody likes Mary"

is of the form

$$P \text{ and } Q.$$

Predicate logic, on the other hand, adds to propositional logic the possibility of analyzing the structure of propositions in greater detail. Namely, it makes it possible to formalize certain statements about individuals involving relations among individuals, functions over individuals, and universal and existential quantification over individual variables, i.e. expressions such as "for every x: ..." and "for some x: ..."

---

[1] The term "truth value" and its understanding as an abstract object is due to Frege (1891) who used the German word "Wahrheitswert." The explicit use of two symbols representing truth and falsity also appears on p. 183 in Peirce 1885. See Church 1956 (p. 25), Kneale and Kneale 1962 (p. 413), and p. 321 above for additional information.

In predicate logic, the above proposition "John is male and everybody likes Mary" is considered to be of the form

$$\text{male(John) and } (\forall x)\, \text{likes}(x, \text{Mary}).$$

Thus, propositions are expressed in greater detail in predicate logic than in propositional logic.

There exist many books on classical logic, both introductory and advanced.[2] The history of logic is covered in two classic books, Bocheński 1961 and Kneale and Kneale 1962.[3]

Classical propositional logic

The language of propositional logic consists of *propositional symbols, symbols of logical connectives*, and auxiliary symbols.[4] In addition, symbols of truth values are sometimes included in the language as well. Propositional symbols are usually letters such as p, q, and r, possibly with indices, like p_1 and p_3, and one usually assumes that there is an infinite number of them. For logical connectives of negation, conjunction, disjunction, implication, and equivalence, we use symbols $\neg$, $\wedge$, $\vee$, $\rightarrow$, and $\leftrightarrow$, respectively.[5] Various kinds of parentheses and brackets are used as auxiliary symbols.

These symbols are used to build *formulas* of propositional logic. Formulas are certain sequences of symbols and are inductively defined as follows: every propositional symbol is a formula, so-called atomic formula; if φ and ψ are formulas, then $\neg\varphi$, $(\varphi \wedge \psi)$, $(\varphi \vee \psi)$, $(\varphi \rightarrow \psi)$, and $(\varphi \leftrightarrow \psi)$ are formulas, so-called compound formulas; there are no other formulas.[6] Formulas represent simple forms of natural language propositions. For instance, the sequence $((p \wedge q) \rightarrow r)$ is a formula which represents the form of the proposition "If weather is nice and we have time then we go outside" as well as the form of "If $2+2=4$ and $4+1=5$ then $2+2+1=5$."

The various symbols along with the rules to form formulas constitute the *syntax* of propositional logic. As purely syntactic objects, formulas have no truth values by

---

[2] E.g. Bell and Machover 1977, Church 1956, Kleene 1967, Mendelson 1964, Quine 1982, and Shoenfield 1967.

[3] For recent works dealing with the history of modern logic, see T. Drucker (ed.), *Perspectives on the History of Mathematical Logic* (Basel: Birkhäuser, 2008); L. Haaparanta (ed.), *The History of Modern Logic* (New York and Oxford: Oxford Univ. Press, 2009); and J. van Benthem and G. Heinzmann (eds.), *The Age of Alternative Logics: Assessing Philosophy of Logic and Mathematics Today* (Dordrecht: Springer, 2009). There is also a journal devoted to the history of logic, *History and Philosophy of Logic*, founded by Ivor Grattan-Guinness in 1980 and originally published by Abacus Press; the journal now appears quarterly and is published by Taylor and Francis.

[4] Other common terms for this logic are propositional calculus, sentential logic, and sentential calculus.

[5] Various other symbols of connectives are in use. For instance, one also uses $\sim$, -, non for negation; & and et for conjunction, vel for disjunction, $\Rightarrow$ and $\supset$ for implication, and $\Leftrightarrow$ and $\equiv$ for equivalence.

[6] Sometimes, "well-formed formula" is used instead of "formula," in which case "formula" refers to any sequence of the symbols of the language. One usually adopts conventions to facilitate the reading of formulas. For instance, the outer parentheses may be omitted. This allows us to write $p \rightarrow (q \vee r)$ instead of $(p \rightarrow (q \vee r))$ and the like. Further omission of parentheses is possible if we adopt conventions regarding priority of connectives. The sequence $\neg$, $\wedge$, $\vee$, $\rightarrow$, $\leftrightarrow$ indicates the priority from the highest to the lowest. Therefore, $p \rightarrow q \wedge \neg r_1 \vee t_2$ denotes the formula $(p \rightarrow ((q \wedge \neg r_1) \vee t_2))$.

themselves. To determine their truth values, one first needs to assign truth values to propositional symbols, i.e. to specify a mapping of the set of propositional symbols to the set $\{0,1\}$ of truth values. Such a mapping, called a (*truth*) *evaluation*, represents a particular situation about which we know whether the actual proposition referred to by the propositional symbol is true or false. If p refers to "It rains" then an evaluation e for which $e(p) = 1$ represents a situation in which it rains while an evaluation for which $e(p) = 0$ represents a situation in which it does not rain. Given an evaluation e, the truth value of any formula is determined according to the following tables, called *truth tables*:[7]

φ	$\neg\varphi$
0	1
1	0

φ	ψ	$\varphi \wedge \psi$	$\varphi \vee \psi$	$\varphi \rightarrow \psi$	$\varphi \leftrightarrow \psi$
0	0	0	0	1	1
0	1	0	1	1	0
1	0	0	1	0	0
1	1	1	1	1	1

The tables describe how the truth values of a compound formula are obtained from those of the constituent formulas. For instance, the first table says that if the truth value of a formula φ is 0, the truth value of its negation $\neg\varphi$ is 1 and if the truth value of φ is 1, the truth value of $\neg\varphi$ is 0. In other words, if a formula is false, its negation is true, and if the formula is true, its negation is false. Clearly, such a meaning of negation concurs with its use in natural language. Similarly, the second table describes the meaning of other logical connectives. In particular, a conjunction of two formulas is true if and only if both formulas are true. A disjunction is true if and only if at least one of the constituent formulas is true.[8] An implication $\varphi \rightarrow \psi$ is false if φ, called the *antecedent* or *premise*, is true and ψ, called the *consequent* or *conclusion*, is false, and is true otherwise. An equivalence is true if and only if both constituent formulas have equal truth values, justifying the term "equivalence." These are the basic rules of semantics of propositional logic. The distinction between syntax and semantics is an important aspect in logic.

For negation, conjunction, and (nonexclusive) disjunction, the tables describe the only possible ways of defining the meaning of these connectives compatible with their use in natural language. For implication, the situation is more involved. The implication with its meaning given by the above table is called *material implication* and there have been many disputes about the adequacy of such an understanding

---

[7] Truth tables were used by both Frege (1879) and Peirce (1885), and became generally popular due to L. Wittgenstein's "Logisch-philosophiche Abhandlung," *Annalen der Naturphilosophie* 14 (1921): 185–262, translated into English by C. K. Ogden and F. P. Ramsey as *Tractatus logico-philosophicus* (London: Kegan Paul, 1922), and due to their use by Łukasiewicz and Post. The term "truth table" was used in Post 1920a, 1921 and in Wittgenstein's *Tractatus*, where the German word "Wahrheitstafel" appears as "truth table." For a study of the history of truth tables, see I. H. Anellis, "The genesis of the truth-table device," *Russell: The Journal of Bertrand Russell Studies* 24 (2004): 55–70.

[8] Such disjunction is nonexclusive. Different is the exclusive disjunction, described by "either φ or ψ but not both." Note that while English is ambiguous in that it uses one word for both of these connectives, other languages are more precise; Latin, for example, uses "vel" for the nonexclusive disjunction and "aut" for the exclusive one.

of implication.[9] One issue is connected to the reading of $\varphi \rightarrow \psi$ as "φ implies ψ." If φ and ψ represent "Elephants are green" and "$1 > 2$," respectively, then "φ implies ψ" is true according to the table. This may be regarded as highly counterintuitive because φ bears no connection to the meaning of ψ. From this viewpoint, a reading of $\varphi \rightarrow \psi$ as "if φ then ψ" is more appropriate.[10] Another issue is connected to so-called *counterfactual conditionals*. These are statements with verbs in the subjunctive mood such as "If Peter had not come, I would have finished my homework." By asserting such a conditional, one indicates the falsity of the antecedent, i.e. one indicates that Peter actually came. The conditional is understood as an assertion about a supposed situation that is similar to but different from the actual situation in that the antecedent holds in the supposed situation. The truth value of a counterfactual conditional is not determined by the truth values of its antecedent and consequent, i.e. the corresponding logical connective is not truth functional. Namely, some conditionals with a false antecedent and a false consequent are true and some are false. However interesting the study of various types of conditional statements may be, it is clear that for implications other than material implication, one needs recourse to a different way of definition than the tabular one. Namely, one may argue that material implication is the only reasonable implication definable by a table like the one above. Indeed, the ordinary use of implication in natural language requires that $\varphi \rightarrow \psi$ be true if both φ and ψ are true (fourth, i.e. the last row) and that $\varphi \rightarrow \psi$ be false if φ is true and ψ is false (third row). If we changed the definition for the second row only, i.e. switched 1 to 0, we would obtain the definition of equivalence. If we changed the definition for the first row only, we would lose the property that $\varphi \rightarrow \psi$ has the same truth value as $\neg\psi \rightarrow \neg\varphi$ which is regarded natural when bivalence is assumed. If we changed both the first and the second row, we would obtain the definition of conjunction. From this point of view, the material implication is the only reasonable truth-functional implication. It should be noted that material implication is a natural form of statement. One uses these statements, such as "If Peter comes, I finish my homework" in case we do not know whether the antecedent is true or not. Such statements may be thought of as predictions in the following sense.[11] If Peter comes, my prediction will turn out to be true or false depending on whether I will finish the homework or not. If Peter does not come, my prediction is not falsified and since we assume bivalence of propositions, the prediction will be considered true, although not interesting. Similar considerations as those for implication apply to equivalence, sometimes called *material equivalence*.

---

[9] According to Kneale and Kneale (1962), the problem of implicational statements was debated for the first time by Diodorus Cronus (unknown–c. 284 BC) and his pupil Philo of Megara (fl. 300 BC). See section 3 of chapter III in Kneale and Kneale 1962 for detailed information about these debates. For modern debates, see C. I. Lewis, "Implication and the algebra of logic," *Mind* 21 (1912): 522–31; C. I. Lewis, "The issues concerning material implication," *J Philosophy, Psychology and Scientific Method* 14 (1917): 350–56; N. Goodman, "The problem of counterfactual conditionals," *J Philosophy* 44 (1947): 113–28; chapter 1 in Kleene 1967, and chapter 3 in Quine 1982.

[10] Some authors, including Quine, speak of "conditional" instead of "implication"; see Quine 1982.

[11] This view is presented in Kleene 1967.

The above tables may equivalently be regarded as functions. Such functions are called the *truth functions of logical connectives*. For instance, the function corresponding to $\neg$ is a unary function $\neg^2$ in $\{0, 1\}$ defined by

$$\neg^2 0 = 1 \quad \text{and} \quad \neg^2 1 = 0,$$

the function corresponding to $\rightarrow$ is a binary function $\rightarrow^2$ in $\{0, 1\}$ defined by

$$0 \rightarrow^2 0 = 1, \quad 0 \rightarrow^2 1 = 1, \quad 1 \rightarrow^2 0 = 0, \quad \text{and} \quad 1 \rightarrow^2 1 = 1,$$

and similarly for the other connectives.[12] Using these functions, the truth value $\|\varphi\|_e$ of formula φ in the evaluation e may thus inductively be defined as follows:

- atomic formulas: for each propositional symbol p,
$$\|p\|_e = e(p);$$
- compound formulas: for all formulas φ and ψ,
$$\|\neg\varphi\|_e = \neg^2 \|\varphi\|_e,$$
$$\|\varphi \wedge \psi\|_e = \|\varphi\|_e \wedge^2 \|\psi\|_e,$$
$$\|\varphi \vee \psi\|_e = \|\varphi\|_e \vee^2 \|\psi\|_e,$$
$$\|\varphi \rightarrow \psi\|_e = \|\varphi\|_e \rightarrow^2 \|\psi\|_e,$$
$$\|\varphi \leftrightarrow \psi\|_e = \|\varphi\|_e \leftrightarrow^2 \|\psi\|_e.$$

Such a definition makes explicit the truth functionality of logical connectives. Note also that the set $\{0, 1\}$ equipped with the truth functions of logical connectives forms a two-element Boolean algebra—the algebraic counterpart of classical logic.[13]

As mentioned above, one of the crucial notions in logic is *consequence*, or *entailment*. In logic, it coincides with its intuitive meaning: a proposition P follows from (is a consequence of) propositions $P_1, \ldots, P_n$ if P is true in every situation in which all $P_1, \ldots, P_n$ are true. In propositional logic, consequence is formalized in the following manner. A formula φ *follows* from a set T of formulas if φ is true in each truth evaluation that makes true all the formulas in T. In this case, we write

$$T \models \varphi.$$

---

[12] We use the superscript ^{2} to distinguish symbols of connectives and their truth functions. We adopt the convention that if a function corresponding to a symbol f is a part of a structure $\mathbf{M}$, the function is denoted by $f^{\mathbf{M}}$. In this case, the structure **2** is the two-element Boolean algebra (see n. 13 below).

[13] Boolean algebras are named after George Boole (1815–1864), who initiated their study in his famous Boole 1847, 1854. In the second half of the 19th century, Boolean algebras were further studied by Augustus De Morgan (1806–1871), W. Stanley Jevons (1835–1882), Charles Sanders Peirce (1839–1914), and Ernst Schröder (1841–1902). The modern study of Boolean algebras as an algebraic discipline goes back to E. V. Huntington, "Sets of independent postulates for the algebra of logic," *Trans AMS* 5 (1904): 288–309. For a modern treatment of Boolean algebras, see S. Givant and P. Halmos, *Introduction to Boolean Algebras* (New York: Springer, 2009).

Boolean algebras may be defined in several equivalent ways, e.g. as complemented distributive lattices, i.e. structures $\langle L, \wedge, \vee, ', o, i \rangle$ such that $\wedge$ and $\vee$ are idempotent, commutative, and associative binary operations in L satisfying the absorption law, $x \wedge (x \vee y) = x = x \vee (x \wedge y)$, and the complementation law, $x \wedge x' = o$ and $x \vee x' = i$ with o and i being the least and the greatest elements in L in that $o = o \wedge x$ and $i = i \vee x$ for every x. A Boolean algebra is obtained when we take $\{0, 1\}$ for L, $\neg^2$ for $'$, $\wedge^2$ for $\wedge$, $\vee^2$ for $\vee$, and 0 and 1 for o and i.

A formula that is always true, i.e. true in each truth evaluation, is called a *tautology*.[14] Clearly, φ is a tautology if and only if $T \models \varphi$ for T being the empty set, in which case we simply write $\models \varphi$.[15] For example, one easily verifies that $p \rightarrow (q \rightarrow p)$ is a tautology. There exist an infinite number of tautologies because if φ is a tautology, then $\psi \rightarrow \varphi$ is a tautology for every formula ψ. Tautologies have always attracted the attention of philosophers and logicians, and many tautologies have their own names.[16]

The above notion of entailment is naturally defined with recourse to semantics, namely to the notion of truth evaluation. This is why one also uses the term "semantic entailment" to refer to entailment as defined above. It is, however, of considerable interest to know whether entailment can be defined on a purely syntactic basis. In particular, is it possible to derive the semantic consequences of T by a simple syntactic manipulation, i.e. without recourse to semantic notions? This question and its various forms regarding the broad variety of particular systems of logic is one of the central questions of concern to modern logicians. The answer to this question is positive and is based on the notion of a proof.

Informally, a proof from a set T of formulas is a sequence of formulas which may be obtained from certain predefined formulas, called *axioms*, and the formulas in T by some predefined *deduction rules*. The role of axioms is to postulate essential properties of logical connectives. The set T is usually called a *theory* for its formulas are viewed as representing our assumptions which may be used in proofs. The deduction rules, sometimes called inference rules, represent correct elementary inference steps.

Before we proceed with the formal notion of a proof, let us explain a convenient conceptual simplification which is often utilized by logicians. Since the language of propositional logic, as defined above, contains the symbols of five connectives, namely negation, conjunction, disjunction, implication, and equivalence, we would need a relatively large number of axioms to postulate their properties. Alternatively, one may use the fact that some connectives may be defined in terms of the others. In particular, since for any truth values a and b we have

$$a \wedge^2 b = \neg^2(a \rightarrow^2 (\neg^2 b)),$$

conjunction may be defined in terms of negation and implication. The same is true of disjunction and equivalence since

$$a \vee^2 b = (\neg^2 a) \rightarrow^2 b \quad \text{and} \quad a \leftrightarrow^2 b = \neg^2((a \rightarrow^2 b) \rightarrow^2 \neg^2(b \rightarrow^2 a)).$$

This makes it possible to simplify the language so that it contains only two symbols

---

[14] The use of the term "tautology" in logic goes back to Wittgenstein's *Tractatus*; see Church 1956 (p. 162) and Kleene 1967 (p. 12).

[15] The use of "$\models$" goes back to S. C. Kleene, *Sets, Logic, and Mathematical Foundations*, notes by H. W. Oliver on lectures at an NSF Summer Institute for Teachers of Secondary and College Mathematics, Williams College, Williamstown, MA, mimeographed, v+169 pp.; see Kleene 1967 (p. 36).

[16] See e.g. Kneale and Kneale 1962.

of connectives, namely $\neg$ and $\rightarrow$. Its formulas are just the formulas of the original language containing $\neg$ and $\rightarrow$ only. The other formulas of the original language, e.g. $p \wedge (q \rightarrow r)$, are then looked at as shorthands:

$$\varphi \wedge \psi \quad \text{for} \quad \neg(\varphi \rightarrow \neg\psi),$$
$$\varphi \vee \psi \quad \text{for} \quad \neg\varphi \rightarrow \psi, \text{ and}$$
$$\varphi \leftrightarrow \psi \quad \text{for} \quad \neg((\varphi \rightarrow \psi) \rightarrow \neg(\psi \rightarrow \varphi)).$$

The new language is therefore simpler and the axioms only need to postulate the properties of two connectives instead of five. Still, we can work with all of the formulas of the original language—some of them are proper formulas of the new language, some of them are shorthands for proper formulas. In what follows, we shall make use of this simplification and assume that the language contains only two symbols of connectives, namely $\neg$ and $\rightarrow$.[17]

The most basic deduction rule, present in various forms in various systems of logic, is the rule of *modus ponens* mentioned above. This rule may be represented by the scheme

$$\frac{\varphi, \varphi \rightarrow \psi}{\psi} \tag{B.1}$$

and allows us to infer the formula φ whenever we have inferred φ and $\varphi \rightarrow \psi$. Assume that we are given a set Ax of axioms, such as[18]

$$\varphi \rightarrow (\psi \rightarrow \varphi) \tag{B.2}$$

$$(\varphi \rightarrow (\psi \rightarrow \chi)) \rightarrow ((\varphi \rightarrow \psi) \rightarrow (\varphi \rightarrow \chi)) \tag{B.3}$$

$$(\neg\psi \rightarrow \neg\varphi) \rightarrow (\varphi \rightarrow \psi) \tag{B.4}$$

A *proof* of a formula φ from a theory T and axioms Ax is a finite sequence $\varphi_1, \ldots, \varphi_n$

---

[17] The choice of $\neg$ and $\rightarrow$ is not the only one possible. Instead, one could take $\neg$ and $\wedge$, because $\vee$, $\rightarrow$, and $\leftrightarrow$ are definable in terms of negation and conjunction; or $\neg$ and $\vee$, for the same reason. However, none of the five connectives alone is sufficient to define the remaining four. Interestingly, there exist two binary connectives, each of which alone is sufficient to define the five connectives (in fact, sufficient to define all propositional connectives). The first is the so-called Sheffer stroke, or NAND, denoted by $|$ and defined so that $\varphi | \psi$ is true iff at least one of φ and ψ is false. The other is called the Peirce arrow, or NOR, and is denoted by $\downarrow$. $\varphi \downarrow \psi$ is true iff neither φ nor ψ are true, i.e. both φ and ψ are false. Each of $\neg$, $\wedge$, $\vee$, $\rightarrow$, and $\leftrightarrow$ is definable in terms of $|$, and the same holds for $\downarrow$. For instance, we can define $\neg\varphi$ as $\varphi | \varphi$ and $\varphi \wedge \psi$ as $(\varphi|\psi)|(\varphi|\psi)$; $\neg\varphi$ as $\varphi \downarrow \varphi$ and $\varphi \wedge \psi$ as $(\varphi|\varphi)|(\psi|\psi)$. That a single binary connective is sufficient to define other connectives was discovered by C. S. Peirce; see his *The Simplest Mathematics*, vol. 4, *Collected Papers of Charles Sanders Peirce*, ed. C. Hartshorne and P. Weiss (Cambridge, MA: Harvard Univ. Press, 1933), 12–20, 215–16, and rediscovered by H. M. Sheffer, "A set of five independent postulates for Boolean algebras, with application to logical constants," *Trans AMS* 14 (1913): 481–88. The symbol "|" was introduced in J. G. P. Nicod, "A reduction in the number of primitive propositions of logic," *Proc Cambridge Philosophical Society* 19 (1917): 32–41.

[18] In fact, (B.2)–(B.4) are axiom schemes. Each scheme represents an infinite number of axioms, i.e. formulas, namely those resulting by replacement of the symbols "φ," "ψ," and "χ" by formulas. Alternatively, one may consider (B.2)–(B.4) as three axioms with "φ," "ψ," and "χ" being three distinct propositional symbols and an additional deduction rule, the *rule of substitution*, which allows us to infer from a given formula φ any formula that results by replacing any propositional variable in φ by an arbitrary formula, thus e.g. infer $p \rightarrow ((\neg r \wedge p) \rightarrow p)$ from $p \rightarrow (q \rightarrow p)$. Clearly, the three axiom schemes allow us to infer the same formulas as the three axioms along with the rule of substitution.

The axioms (B.2), (B.3), and (B.4) are called the law of affirmation of the consequent, the self-distributive law of implication, and the converse law of contraposition, respectively.

such that $\varphi_n = \varphi$ and for each $i = 1,\ldots,n$, some of the following conditions are satisfied:

- φ_i is an axiom in Ax,
- φ_i is in T,
- φ_i may be obtained from some of the preceding formulas by *modus ponens*, i.e. for some $j, k < i$, the formula φ_k is $\varphi_j \to \varphi_i$.

We say that φ is *provable* from T if there is a proof of φ from T.[19] In this case we write

$$T \vdash \varphi,$$

or $T \vdash_{Ax} \varphi$ to emphasize that Ax is the given set of axioms.[20] The notion of a proof is a simple formalization of the informal concept of proof as a sequence of claims that makes use of evident truths, assumptions, and elementary deduction steps.

There are many useful properties regarding the formal notion of a proof, one of the most important among them being the *deduction theorem* which says that for every set T of formulas and any formulas φ and ψ,

$$T \cup \{\varphi\} \vdash \psi \quad \text{if and only if} \quad T \vdash \varphi \to \psi.\text{[21]} \qquad (B.5)$$

We can now proceed with stating the so-called *completeness theorem* for the axiomatic system described above. This theorem says that for any set T of formulas and any formula φ,

$$T \models \varphi \quad \text{if and only if} \quad T \vdash \varphi, \qquad (B.6)$$

i.e. φ follows from T if and only if φ is provable from T.[22] In other words, semantic consequence represented by $\models$ is fully captured by syntactic consequence represented by $\vdash$. The part of the completeness theorem asserting that if φ is provable from T then φ follows from T is called *soundness* and is easy to obtain. Namely, it follows easily from the fact that every axiom is a tautology and that *modus ponens* preserves truth in that if φ and $\varphi \to \psi$ are true in a truth evaluation e, then ψ is true in e, for every e. Soundness says that our system is reasonable in that it does not allow us to prove untrue formulas. The other part, namely that if φ follows from T then

---

[19] The formal system presented here is an example of a Hilbert-type system, named after David Hilbert (1862–1943). The systems of natural deduction and the closely related sequent calculi, also called Gentzen-type systems after Gerhard Gentzen (1909–1945), are different. See § 77 in Kleene 1952 and pp. 126–27, 305–6 in Kleene 1967.

[20] Using "⊢" to denote provability goes back to Frege (1879). The present use is due to J. B. Rosser, "A mathematical logic without variables. I," *Ann Math* 36 (1935): 127–50, and to S. C. Kleene, "Proof by cases in formal logic," *Ann Math* 35 (1934): 529–44. See also p. 30 in Kleene 1967.

[21] The deduction theorem holds true in various forms for various systems of logic. It was published for the first time by Jacques Herbrand in 1928 and proved in his dissertation in 1930. See p. 164 in Church 1956.

[22] The system consisting of (B.2)–(B.4) is one of many which have been proposed; see chapter II in Church 1956 as well as Łukasiewicz and Tarski 1930 for details. The first, modern, Hilbert-style axiom system with *modus ponens* and the (not explicitly stated) rule of substitution, was presented in Frege 1879. Other early formulations are due to Bertrand Russell (1872–1970) in his *The Principles of Mathematics* (W. W. Norton, 1903) and, with Alfred North Whitehead (1861–1947), their Whitehead and Russell 1910–13. Our system is due to Jan Łukasiewicz (1878–1956), who derived it by simplification from that of Frege (1879) in Łukasiewicz 1929a; see theorem 6 and n. 11 in Łukasiewicz and Tarski 1930. The first published proofs of the completeness theorem are due to Bernays (1918) and Post (1921); see also Zach 1999.

φ is provable from T, sometimes called completeness, is the substantial claim of the theorem. It says that the axioms and the deduction rule involved are complete in that the inference from T based on them exhausts all semantic consequences of T. If T is empty, the completeness theorem says that the formulas provable from the axioms without further assumptions are just the tautologies. For T being finite, the corresponding theorem is also called *weak completeness*, while *strong completeness* then refers to what we called completeness, i.e. with T possibly infinite.[23]

The completeness theorem may be equivalently phrased in terms of two important logical concepts, namely (a semantic concept of) *satisfiability* and (a syntactic concept of) *consistency*. A set T of formulas is called satisfiable if there exists a truth evaluation e such that every formula in T is true in e. T is called consistent if there exists a formula that is not provable from T, i.e. it is not the case that all formulas are provable from T. One may demonstrate that the completeness theorem is equivalent to the claim that for every T,

$$T \text{ is satisfiable if and only if } T \text{ is consistent.} \tag{B.7}$$

Classical predicate logic

The language of predicate logic consists of symbols of *relations* (also called symbols of predicates or just predicates), symbols of *functions*, symbols of (*individual*) *variables*, symbols of logical *connectives*, symbols of *quantifiers*, and auxiliary symbols.[24] Symbols of connectives and auxiliary symbols are the same as in the language of propositional logic. Symbols of relations, functions, and variables are usually the letters $r, s, \ldots, f, g, \ldots$, and $x, y, \ldots$, respectively, possibly with indices. One assumes that at least one symbol of relation and an infinite number of variables are present in the language. With every symbol S of relation or function there is associated a nonnegative integer $\sigma(S)$, called the *arity* of S, specifying the number of arguments for S; S is then called $\sigma(S)$-ary. As usual, for $n = 1, 2, 3$, one uses "unary," "binary," and "ternary" instead of "n-ary." 0-ary function symbols are thought of as symbols of constants and 0-ary relation symbols may be thought of as propositional symbols. Consequently, propositional logic may be thought of as a particular case of predicate logic in which one only has 0-ary relation symbols. There are two quantifiers in predicate logic, the *universal quantifier* denoted by ∀ and the *existen-*

---

[23] The term "completeness" has different meanings in logic. Unlike the present completeness, which is a property of the axiom system relative to the semantics, a system of axioms is called (absolutely) complete if adding any nonprovable formula to the axioms results in an inconsistent system, i.e. every formula becomes provable in such a system. See Church 1956 for more information.

[24] Other common terms are "first-order logic," "predicate calculus," and (formerly also) "functional calculus." The rationale of the first term is that it allows for quantification over elements of the universe of discourse. Second-order logics allow for quantification over sets of elements of the universe of discourse, and more generally, over relations and functions. Still more general quantification is allowed in higher-order logics. From this point of view, the propositional logic may be considered a zeroth-order logic.

tial quantifier denoted by $\exists$.[25] A language of a predicate logic is hence determined by the triplet $\langle R, F, \sigma \rangle$, called the *type* of the language, where R and F are the sets of relation and function symbols (other symbols are common to all languages).

The basic notions of syntax of predicate logic are terms and formulas. *Terms* are expressions, such as $3 \cdot x$, whose values are elements of a given set in a suitable interpretation of symbols. Formally, terms are defined as follows: every symbol of a variable is a term; if $t_1, \ldots, t_n$ are terms and f is an n-ary function symbol, then $f(t_1, \ldots, t_n)$ is a term; there are no other terms. *Formulas* are expressions such as $\sin(x) \leq x^2 + 1$ whose values are truth values in suitable interpretations and are defined as follows: if $t_1, \ldots, t_n$ are terms and r is an n-ary relation symbol, then $r(t_1, \ldots, t_n)$ is a formula (atomic formula); if φ and ψ are formulas then $\neg\varphi, (\varphi \wedge \psi)$, $(\varphi \vee \psi), (\varphi \rightarrow \psi), (\varphi \leftrightarrow \psi)$ are formulas; if φ is a formula and x a symbol of a variable, then $(\forall x)\varphi$ and $(\exists x)\varphi$ are formulas. The formulas $(\forall x)\varphi$ and $(\exists x)\varphi$ read "for all x, φ" and "there exists x such that φ." For convenience, one uses the infix notation, i.e. $x + y$ and $x \leq y$ instead of $+(x, y)$ and $\leq (x, y)$, respectively. For instance, $0 + x$ and $x \cdot (y + 2 \cdot z)$ are terms while $0 \leq x + x$, $x + y = y + x$, and $\neg x = 0 \rightarrow (\forall y)(\forall z)(x \cdot y = x \cdot z \rightarrow y = z)$ are formulas of a language containing a nullary function symbol 0, binary function symbols $+$ and $\cdot$, and binary relation symbols $\leq$ and $=$.

Terms and formulas have no values by themselves. In order to assign values to them, it is necessary to interpret the symbols of functions and relations and the variables involved. The symbols of a given language are interpreted in structures for this language. A *structure* **M** for a given language consists of a nonempty set M, called the *universe of discourse*, a set of relations containing for each n-ary relation symbol r an n-ary relation $r^{\mathbf{M}}$ in M, and a set of functions containing for each n-ary function symbol f an n-ary function $f^{\mathbf{M}}$ in M.[26] Formally, a structure for a language of type $\langle R, F, \sigma \rangle$ is a triplet $\mathbf{M} = \langle M, R^{\mathbf{M}}, F^{\mathbf{M}} \rangle$ where $R^{\mathbf{M}}$ and $F^{\mathbf{M}}$ are the sets of relations and functions, i.e.

$$R^{\mathbf{M}} = \{r^{\mathbf{M}} \subseteq M^{\sigma(r)} \mid r \in R\} \text{ and } F^{\mathbf{M}} = \{f^{\mathbf{M}} : M^{\sigma(f)} \rightarrow M \mid f \in F\}.$$

A *valuation* of variables is any mapping v assigning elements in M to variables. The

---

[25] The use of quantifiers in logic originates with Frege 1879. Independently, quantifiers appeared in the work of O. H. Mitchell, a student of Peirce, "On a new algebra of logic," in *Studies in Logic: By Members of the Johns Hopkins University*, ed. C. S. Peirce, (Boston, MA: Little & Brown, 1883), 72–106, to whom Peirce refers in Peirce 1885, where the quantifiers are further explored. Quantifiers later appeared in the works of Schröder, Giuseppe Peano (1858–1932), and Russell.

The term "quantifier" was introduced by Peirce. Peirce used Π_x and Σ_x for $\forall x$ and $\exists x$, which notation was also used by many logicians later on, including Schröder and Kurt Gödel (1906–1978). Peano invented the notation (x) for $\forall x$ and invented $\exists x$—a notation adopted in Russell and Whitehead's *Principia Mathematica* and also by other logicians including Alonzo Church (1903–1995) and Willard Van Orman Quine (1908–2000). The symbol $\forall$ was invented by Gentzen (inverted "A" from the German word "All-Zeichen," an analogy to $\exists$ derived from "E"). Later on, $\wedge$ and $\vee$ for $\forall$ and $\exists$ were used by Polish logicians.

For discussions of the history of quantification and the roles of Frege and Peirce in particular, see H. Putnam, "Peirce the logician," *Historia Mathematica* 9 (1982): 290–301; and W. D. Goldfarb, "Logic in the twenties: The nature of the quantifier," *J Symb Logic* 44, no. 3 (1979): 351–68.

[26] The term "universe of discourse" was introduced by Augustus De Morgan.

element assigned to the variable x is denoted $v(x)$. Given a structure $\mathbf{M}$ and a valuation v for a given language, the value $\|t\|_{\mathbf{M},v}$ of a term t is defined as follows. If t is a variable, its value equals $v(x)$. If t is a compound term of the form $t = f(t_1, \ldots, t_n)$, the value of t is obtained by applying the function $f^{\mathbf{M}}$ to the values assigned to $t_1, \ldots, t_n$. Formally, $\|t\|_{\mathbf{M},v}$ is defined inductively as follows:

- variables: for each symbol of variable x, $\|x\|_{\mathbf{M},v} = v(x)$;
- compound terms: for term $f(t_1, \ldots, t_n)$, $\|t\|_{\mathbf{M},v} = f^{\mathbf{M}}(\|t_1\|_{\mathbf{M},v}, \ldots, \|t_1\|_{\mathbf{M},v})$.

Given a structure $\mathbf{M}$ and a valuation v, the truth value $\|\varphi\|_{\mathbf{M},v}$ of a formula φ is defined as follows. If φ is an atomic formula of the form $r(t_1, \ldots, t_n)$, its truth value is 1 or 0 depending on whether the elements obtained by interpreting the terms $t_1, \ldots, t_n$ are in the relation $r^{\mathbf{M}}$ or not. The truth values of a compound formula of the form $\neg\varphi$, $(\varphi \wedge \psi)$, $(\varphi \vee \psi)$, $(\varphi \rightarrow \psi)$, and $(\varphi \leftrightarrow \psi)$ are defined as in propositional logic. A formula of the form $(\forall x)\varphi$ is true if φ is true in $\mathbf{M}$ for every valuation v' that differs from v only in what value it assigns to x (which fact is denoted by $v' =_x v$). Similarly, $(\exists x)\varphi$ is true if φ is true in $\mathbf{M}$ for at least one valuation $v' =_x v$. These definitions may concisely be described in an inductive manner as follows:

- for atomic formulas,

$$\|r(t_1, \ldots, t_n)\|_{\mathbf{M},v} = \begin{cases} 1 & \text{if } \langle \|t_1\|_{\mathbf{M},v}, \ldots, \|t_n\|_{\mathbf{M},v} \rangle \in r^{\mathbf{M}}, \\ 0 & \text{if } \langle \|t_1\|_{\mathbf{M},v}, \ldots, \|t_n\|_{\mathbf{M},v} \rangle \notin r^{\mathbf{M}}; \end{cases}$$

- for formulas φ and ψ,

$$\|\varphi \wedge \psi\|_{\mathbf{M},v} = \|\varphi\|_{\mathbf{M},v} \wedge^2 \|\psi\|_{\mathbf{M},v},$$
$$\|\varphi \vee \psi\|_{\mathbf{M},v} = \|\varphi\|_{\mathbf{M},v} \vee^2 \|\psi\|_{\mathbf{M},v},$$
$$\|\varphi \rightarrow \psi\|_{\mathbf{M},v} = \|\varphi\|_{\mathbf{M},v} \rightarrow^2 \|\psi\|_{\mathbf{M},v},$$
$$\|\varphi \leftrightarrow \psi\|_{\mathbf{M},v} = \|\varphi\|_{\mathbf{M},v} \leftrightarrow^2 \|\psi\|_{\mathbf{M},v};$$

- for a formula φ and a variable x,

$$\|(\forall x)\varphi\|_{\mathbf{M},v} = \min\{\|\varphi\|_{\mathbf{M},v'} \mid v' =_x v\},$$
$$\|(\exists x)\varphi\|_{\mathbf{M},v} = \max\{\|\varphi\|_{\mathbf{M},v'} \mid v' =_x v\}.$$

Therefore, structures and valuations in predicate logic have a similar role to that of truth evaluations in propositional logic, namely that of the basic semantic structures in which the syntactic objects, namely terms and formulas, are interpreted. The notions of entailment and tautology are defined essentially as in propositional logic. In particular, a structure $\mathbf{M}$ is a *model* of a formula φ (or φ is true in $\mathbf{M}$) if $\|\varphi\|_{\mathbf{M},v} = 1$ for every valuation v. A formula φ *follows* from a set T of formulas (called a *theory*), in symbols $T \models \varphi$, if φ is true in every structure that is a model of all formulas in T. A formula is called a *tautology* if it is true in every structure.

When axiomatizing predicate logic, one may utilize a similar convention as the one explained above for propositional logic. Namely, one considers $\neg$ and $\rightarrow$ as the primitive symbols (i.e. contained in the language) and considers the expression containing $\land$, $\lor$, and $\leftrightarrow$ as shorthands. In addition, we regard only $\forall$ as a symbol of the language and regard $\exists$ as derived. In particular, we consider expressions of the form $(\exists x)\varphi$ as shorthands for $\neg(\forall x)\neg\varphi$. This is because the existential quantifier may be defined in terms of the universal quantifier and negation in that

$$\|(\exists x)\varphi\|_{M,v} = \|\neg(\forall x)\neg\varphi\|_{M,v}.$$

As in the case of propositional logic, such a convention simplifies the language while retaining its expressive power.

We now consider a Hilbert-style system for predicate logic extending the one described above for propositional logic.[27] Its deduction rules are the rule of *modus ponens* and the rule of *generalization*. The former rule has again the form (B.1) except that the formulas involved are the formulas of predicate logic. The rule of generalization has the form

$$\frac{\varphi}{(\forall x)\varphi},$$

i.e. it makes it possible to infer the formula $(\forall x)\varphi$ from any formula φ of predicate logic. The axioms are the three axioms (B.2)–(B.4) and two new ones,[28]

$$(\forall x)\varphi \to \varphi(x/t) \tag{B.8}$$

where t is a term substitutable for the variable x, and

$$(\forall x)(\varphi \to \psi) \to (\varphi \to (\forall x)\psi) \tag{B.9}$$

if x does not have a free occurrence in φ.

The axioms (B.8) and (B.9) include important syntactic restrictions. The restrictions ensure that (B.8) and (B.9), as well as the first three axioms, are tautologies. The restrictions refer to further syntactic notions which have the following meaning. An occurrence of x in φ is called *free* if this is not an occurrence of x in some subformula $(\forall x)\psi$ (i.e., a substring) of φ; otherwise, the occurrence is called *bound*; the free (bound) variables of φ are just those having a free (bound) occurrence in φ. In (B.8), $\varphi(x/t)$ denotes the formula resulting from φ by substituting t for every free occurrence of x in φ. A term t is substitutable for x in φ if no occurrence of any variable in t becomes bound in the resulting formula $\varphi(x/t)$.

The notion of proof for this system is analogous to that for propositional logic. In particular, a proof from a theory T is again a finite sequence of formulas satisfying the constraints that a formula may appear in the sequence if it is an axiom, or if it is in T, or if it may be inferred from the previous formulas in the proof by a deduction rule. The difference is that the notion concerns predicate logic formulas

---

[27] Predicate logic as an independent logical system appeared for the first time in Hilbert and Ackermann 1928; see Church 1956 (p. 289). As in the case of propositional logic, many axiomatic systems for prediate logic have been proposed. The system presented here appears in Church 1956 (his system F^1) and results by some modifications from the system in B. Russell, "Mathematical logic as based on the theory of types," *American J Math* 30 (1908): 222–62.

[28] (B.8) and (B.9) are called the axiom of substitution and the axiom of distributivity.

and involves the five axioms and the two deduction rules of predicate logic described above. Interestingly, for this extension of the axiomatic system of the propositional logic, many theorems of propositional logic, possibly with appropriate modifications, still hold. In particular, the completeness theorem (B.6) and its equivalent formulation (B.7), in which satisfiability means having a model, hold for any set T of formulas and any formula φ, while the deduction theorem (B.5) holds for any set T of formulas and any closed, i.e. without free variables, formula φ.[29] On the other hand, there exist important differences between propositional and predicate logic, amounting to the fact that propositional logic is significantly simpler. For example, while propositional logic is decidable, i.e. there exists an algorithm deciding whether a given formula is the consequence of a given set of axioms, predicate logic is undecidable.[30]

---

[29] The completeness theorem for the predicate logic is due to Gödel (1930); several other proofs appeared later; see e.g. Church 1956, Mendelson 1964.

[30] The undecidability of predicate logic was established independently by A. Church in "An unsolvable problem of elementary number theory," *American J Math* 58 (1936): 345–63, and Alan M. Turing (1912–1954) in "On computable numbers, with an application to the Entscheidungsproblem," *Proc London Math Soc* 42 (1936–37): 230–65.

Appendix C

Photographs

THE PRIMARY AIM of this appendix is to present portraits of several of the most consequential figures in the history of fuzzy logic. Included as well is a photograph illustrating the site of one of the debates over fuzzy logic and probability theory, held during the Eighth Maximum Entropy Workshop at St. John's College, Cambridge University, United Kingdom, August 1–5, 1988.

Lotfi A. Zadeh (b. 1921). He is widely recognized as the founder of fuzzy logic due to his seminal paper Zadeh 1965a. He is also recognized for conceiving the agenda of fuzzy logic in the broad sense and for his leadership in pursuing this agenda (chapter 3). Basic biographical information about Zadeh is given in section 2.2; the influence of his work on mathematics and other areas of human affairs is assessed in section 7.4.

Jan Łukasiewicz (1878–1956). This famous Polish logician is generally recognized as one of the most important pioneers of many-valued logics, which played a significant role in preparing for the emergence of fuzzy logic. Łukasiewicz's work is mentioned briefly in section 2.1, and is discussed in detail in section 4.2.1, where basic biographical information about him is also presented.

Joseph A. Goguen (1941–2006). He is noteworthy for his important early contributions (Goguen 1967, 1968–69), which greatly influenced subsequent developments in fuzzy logic and mathematics based on it, as is discussed in detail in section 4.3.1 and in various parts of chapter 5. See also the obituary for him by Zadeh (2007).

Michio Sugeno (b. 1940). He is widely recognized for his significant early contributions to fuzzy modeling and control, as documented by a book of his selected works edited by Nguyen and Prasad (1999). This book also contains relevant biographical information about Sugeno and a foreword by Lotfi Zadeh, expressing his personal assessment of Sugeno's profound contributions to fuzzy logic. For biographical information about Sugeno, see pp. 144–48 in McNeill and Freiberger 1993.

Jan Pavelka (1948–2007). He is widely recognized for his series of three important early papers on fuzzy logic (Pavelka 1979), which was inspired by ideas of Goguen. The content of these papers and their significance for the history of fuzzy logic is discussed in detail in section 4.3.2, which also contains basic biographical information about Pavelka.

Takeshi Yamakawa (b. 1946). He is recognized for his many innovative applications of fuzzy logic, but above all, for his pioneering work on analog-like hardware for fuzzy control and other applications of fuzzy logic, which are discussed in section 6.3.2. Relevant biographical information about Yamakawa is covered, for example, on pp. 152–54 in McNeill and Freiberger 1993 and on pp. 197–200 in Kosko 1993.

Petr Hájek (b. 1940; photograph courtesy of Petr Cintula). He is recognized for his pioneering work on fuzzy logic in the narrow sense that culminated in the publication of his groundbreaking book Hájek 1998b. His work is discussed in detail, especially in section 4.4.3, which also contains basic biographical information about him; for additional information see especially Montagna 2015.

Participants in the Eighth Maximum Entropy Workshop at St. John's College, Cambridge, in 1988, where the Cheeseman-Klir debate described in detail in section 2.4 took place. Peter Cheeseman and George Klir are seated in the front row, sixth and seventh from the right, respectively.[a]

[a]Permission to use this photograph in this book, given to us by the copyright holder—Lafayette Photography in Cambridge, UK (www.lafayettephotography.com)—is gratefully acknowledged.

1. Robert Papoular 2. Alastair Livesey 3. Tom Marsh 4. David Morrison 5. Paul Jowitt
6. Richard Bryan 7. F.Fröhner 8. Chris Gilmore 9. Keith Henderson 10. R.B.Hicks
11. Louis Roemer 12. Colin Bannister 13. Robert Collins 14. Juana Sanchez.
15. Carlos Rodriguez 16. Stephen Wilkins 17. Andrew Templeman 18. Stig Steenstrup
19. Randall Barron 20. Martin Barth 21. Alan M.Thompson 22. C.J.Shelton
23. D.van Ormondt 24. Monika Sebold 25. John Close 26. John Richardson
27. David Drabold 28. E.R.Podolyak 29. Ernest Laue.
30. John Karkheck 31. Myron Tribus 32. Ken Packer 33. Ofer Lahav 34. Marc A.Delsuc
35. Birgit Meyer 36. H.A.Mayer-Hasselwander 37. Nailong Wu 38. E.L.Kosarev
39. David Hestenes 40. J.K.Elder 41. Michael K.-S.Tso 42. David Larner 43. Odet Pols
44. Dave Wilkinson 45. Mark Charter 46. Dursun Ustundag 47. John Burg.
48. John Dougherty 49. Chris Burrows 50. Ali Mohammad-Djafari 51. Sibusiso Sibisi
52. Andrew Strong 53. Jacob Bekenstein 54. Ron Canterna 55. Do Kester 56. A.Baruya
57. Romke Bontekoe 58. A.D.McLachlan 59. Colin Fox 60. R.T.Constable 61. N.A.Farrow
62. R.D.Levine 63. Y.Tikochinsky 64. G.J.Daniell 65. R.Huis 66. John Pendrell
67. S.Hildebrand.
68. Keith Horne 69. Steve Luttrell 70. Lee Schick 71. Gary Erickson
72. Larry Bretthorst 73. Tom Grandy 74. Ray Smith 75. Steve Gull 76. Martin Skilling
77. Jennifer Skilling 78 John Skilling 79. Edwin Jaynes 80. Rabinder N.Madan
81. Paul F.Fougere 82. George Klir 83. Peter Cheeseman 84. Anthony Garrett
85. Andrew Gelman 86. Matthew Self 87. Roger Balian 88. Lucien Preuss

Names of participants shown in the above photograph.

Glossary of Symbols

Numbers

$\mathbb{N}$	set of all positive integers
$\mathbb{N}_n$	set $\{1,\ldots,n\}$
$\mathbb{Z}$	set of all integers
$\mathbb{Q}$	set of all rational numbers
$\mathbb{R}$	set of all real numbers
$[a,b]$	closed interval of real numbers between a and b
$(a,b], [a,b)$	half-open intervals of real numbers between a and b
(a,b)	open interval of real numbers between a and b

Classical sets, relations, and functions

$A,B,X,Y,U,V,\ldots$	sets
$x,y,u,v,\ldots$	elements of sets
$x \in A$	x belongs to set A
$\{x,y,\ldots\}$	set consisting of elements $x,y,\ldots$
$\{x \in U \mid \varphi(x)\}$	set of elements in universe U that satisfy property φ
$\emptyset$	empty set
2^U	set of all subsets of set U
$A \subseteq B$	set A is included in set B
$A \cap B$	intersection of sets A and B
$A \cup B$	union of sets A and B
$A - B$	set of all elements in set A that are not in set B
$\overline{A}$	complement of set A
$\langle u_1,\ldots,u_n \rangle$	tuple of elements $u_1,\ldots,u_n$
$U_1 \times \cdots \times U_n$	Cartesian product of sets $U_1,\ldots,U_n$
U^2	Cartesian product $U \times U$
$R \subseteq U_1 \times \cdots \times U_n$	R is relation among sets $U_1,\ldots,U_n$
$[u]_R$	class of relation R determined by element u
U/R	factor set of set U modulo equivalence R
$f : U \to V$	f is a function (mapping) of set U to set V
f^{-1}	inverse function to f

473

$f \circ g$	composition of functions f and g
$\inf A$ or $\bigwedge A$	infimum of A
$\sup A$ or $\bigvee A$	supremum of A
$\mathscr{A}, \mathscr{B}, \ldots$	families of sets

Fuzzy sets and logic

L	set of truth degrees (e.g., $L = [0,1]$)
$a, b, c \ldots$	truth degrees
$\mathbf{L}$	structure of truth degrees (set L equipped with operations)
$A, B, C, \ldots$	fuzzy sets
$P, Q, R, \ldots$	fuzzy relations
L^U or $\mathbf{L}^U$	set of all fuzzy sets in universe U with truth degrees in set L
$A(x)$	degree of membership of element x in fuzzy set A
$A \subseteq B$	fuzzy set A is (fully) included in fuzzy set B
$S(A,B)$	degree of inclusion of fuzzy set A in fuzzy set B
$A \approx B$	degree of equality of fuzzy sets A and B
$A \cap B$ or $i(A,B)$	intersection of fuzzy sets A and B
$A \cup B$ or $u(A,B)$	union of fuzzy sets A and B
${}^a A$	a-cut of fuzzy set A
$p, q, r, \ldots$	propositional symbols
$\varphi, \psi, \chi, \ldots$	logical formulas
$\neg$ and $\neg^{\mathbf{L}}$ or $\neg^{\cdot}$	symbol of negation and truth function of negation
$\otimes, \wedge$ and $\otimes^{\mathbf{L}}, \wedge^{\mathbf{L}}$ or $\otimes^{\cdot}, \wedge^{\cdot}$	symbols of conjunction and truth functions of conjunction
$\oplus, \vee$ and $\oplus^{\mathbf{L}}, \vee^{\mathbf{L}}$ or $\oplus^{\cdot}, \vee^{\cdot}$	symbols of disjunction and truth functions of disjunction
$\rightarrow$ and $\rightarrow^{\mathbf{L}}$ or $\rightarrow^{\cdot}$	symbol of implication and truth function of implication
$\leftrightarrow$ and $\leftrightarrow^{\mathbf{L}}$ or $\leftrightarrow^{\cdot}$	symbol of equivalence and truth function of equivalence
$\lVert \varphi \rVert_e$	truth degree of formula φ in truth evaluation e
$\mathbf{M}$	first-order structure
$x, y, z, \ldots$	object variables
$\forall$ and $\exists$	universal and existential quantifier
$r, s, \ldots$ and $r^{\mathbf{M}}, s^{\mathbf{M}}, \ldots$	symbols of relations and corresponding relations in $\mathbf{M}$
$f, g, \ldots$ and $f^{\mathbf{M}}, g^{\mathbf{M}}, \ldots$	symbols of functions and corresponding functions in $\mathbf{M}$
$\lVert \varphi \rVert_{\mathbf{M},v}$	truth degree of formula φ in structure $\mathbf{M}$ and valuation v
$\mathrm{Mod}(T)$	set of all models of theory T
$T \models \varphi$	formula φ semantically follows from theory T
$T \vdash \varphi$	formula φ is provable from theory T
$\lVert \varphi \rVert_T$	degree to which φ semantically follows from T
$\lvert \varphi \rvert_T$	degree to which φ is provable from T

References

Ackermann, R. 1967. *An Introduction to Many-Valued Logics.* London: Routledge & Kegan Paul.

Aczél, J. 1966. *Lectures on Functional Equations and Their Applications.* New York: Academic Press.

Adams, E. W. 1998. *A Primer of Probability Logic.* Stanford, CA: Center for the Study of Language and Information.

Adams, M. M., and N. Kretzmann, eds. 1983. *William Ockham: Predestination, God's Foreknowledge, and Future Contingents.* 2nd ed. Indianapolis, IN: Hackett.

Adamus, M., and R. Belohlavek. 2007. Fuzzy control of neuromuscular block during general anesthesia—system design, development and implementation. *Int J Gen Syst* 36 (6): 733–43.

Adlassnig, K.-P. 1980. A fuzzy logical model of computer-assisted medical diagnosis. *Methods of Information in Medicine* 19 (3): 141–48.

———1982. A survey of medical diagnosis and fuzzy subsets. In *Approximate Reasoning in Decision Analysis*, ed. M. M. Gupta and E. Sanchez, 203–17. Amsterdam: North-Holland.

———1983. *Ein Computerunterstütztes Medizinisches Diagnosesystem unter Verwendung von Fuzzy Teilmengen* [A computer-aided medical diagnosis system using fuzzy subsets]. PhD diss., Technical Univ. of Vienna, Austria.

———2001. The section on medical expert and knowledge-based systems at the Department of Medical Computer Sciences of the University of Vienna Medical School. *Artificial Intelligence in Medicine* 21 (1): 139–46.

Adlassnig, K.-P., and G. Kolarz. 1982. CADIAG-2: Computer-assisted medical diagnosis using fuzzy subsets. In *Approximate Reasoning in Decision Analysis*, ed. M. M. Gupta and E. Sanchez, 219–47. Amsterdam: North-Holland.

Adlassnig, K.-P., G. Kolarz, W. Scheithauer, H. Effenberger, and G. Grabner. 1985. CADIAG: Approaches to computer-assisted medical diagnosis. *Computers in Biology and Medicine* 15 (5): 315–35.

Aguzzoli, S., S. Bova, and B. Gerla. 2011. Free algebras and functional representation for fuzzy logics. In Cintula, Hájek, and Noguera 2011, 713–91.

Aguzzoli, S., and B. Gerla. 2002. Finite-valued reductions of infinite-valued logics. *Arch Math Logic* 41 (4): 361–99.

Akeroyd, F. M. 2000. Why was a fuzzy model so successful in physical organic chemistry? *Int J for Philosophy of Chemistry* 6:161–73.

Albin, M. A. 1975. *Fuzzy Sets and Their Applications to Medical Diagnosis and Pattern Recognition.* PhD diss., Univ. of California, Berkeley.

Albus, J. S., and A. M. Meystel. 2001. *Engineering of Mind: The Science of Intelligent Systems.* New York: Wiley.

Aldrich, V. C. 1937. Some meanings of 'vague.' *Analysis* 4 (6): 89–95.

Alefeld, G., and J. Herzberger. 1983. *Introduction to Interval Computations.* New York: Academic Press.

Aliev, R. A., B. Fazlollahi, and R. R. Aliev, eds. 2004. *Soft Computing and Its Applications in Business and Economics.* Berlin: Springer.

Alsina, C., M. J. Frank, and B. Schweizer. 2006. *Associative Functions: Triangular Norms and Copulas.* Singapore: World Scientific.

Alsina, C., E. Trillas, and L. Valverde. 1980. On non-distributive logical connectives for fuzzy sets theory. *BUSEFAL* 3:18–29.

———1983. On some logical connectives for fuzzy sets theory. *J Math Anal Appl* 93:15–26.

Angelov, P. P. 2002. *Evolving Rule-Based Models: A Tool for Design of Flexible Adaptive Systems.* Heidelberg: Physica-Verlag.

Anthony, J. M., and H. Sherwood. 1979. Fuzzy groups redefined. *J Math Anal Appl* 69:124–30.

———1982. A characterization of fuzzy subgroups. *Fuzzy Sets Syst* 7 (3): 297–305.

Antonsson, E. K., and H.-J. Sebastian. 1999. Fuzzy sets in engineering design. In Zimmermann 1999, 57–117.

Aracil, J., and F. Gordillo, eds. 2000. *Stability Issues in Fuzzy Control*. Heidelberg: Physica-Verlag.

Arfi, B. 2005. Fuzzy decision making in politics: a linguistic fuzzy-set approach. *Political Analysis* 13:23–56.

———— 2010. *Linguistic Fuzzy Logic Methods in Social Science*. Berlin: Springer.

Asai, K., ed. 1995. *Fuzzy Systems for Management*. Tokyo: Ohmsha.

Atanassov, K. T. 1986. Intuitionistic fuzzy sets. *Fuzzy Sets Syst* 20 (1): 87–96.

———— 1999. *Intuitionistic Fuzzy Sets: Theory and Applications*. Heidelberg: Physica-Verlag.

———— 2005. Answer to D. Dubois, S. Gottwald, P. Hájek, J. Kacprzyk, and H. Prade's paper "Terminological difficulties in fuzzy set theory—the case of 'Intuitionistic Fuzzy Sets.'" *Fuzzy Sets Syst* 156 (3): 496–99.

Atanassov, K. T., and S. Stoeva. 1983. Intuitionistic fuzzy sets. In *Proc Polish Symp on Interval and Fuzzy Mathematics*, 23–26. Poznan.

Aubin, J.-P. 1976. Fuzzy core and equilibria of games defined in strategic form. In *Directions in Large-Scale Systems*, ed. Y. C. Ho and S. K. Mitter, 371–88. New York: Plenum Press.

———— 1981. Cooperative fuzzy games. *Mathematics of Operations Research* 6:1–13.

———— 1984. Cooperative fuzzy games: The static and dynamic points of view. In Zimmermann, Zadeh, and Gaines 1984, 407–28.

Aubin, J.-P., and H. Frankowska. 1990. *Set-Valued Analysis*. Basel: Birkhäuser.

Avron, A. 1991. Hypersequents, logical consequence and intermediate logics for concurrency. *Ann Mathematics and Artificial Intelligence* 4 (3–4): 225–48.

Baaz, M. 1996. Infinite-valued Gödel logics with 0-1 projections and relativizations. *Lect Notes Logic* 6:23–33.

Baaz, M., A. Ciabattoni, and F. Montagna. 2004. Analytic calculi for monoidal t-norm based logic. *Fund Inform* 59 (4): 315–32.

Baaz, M., and C. G. Fermüller. 1995. Resolution-based theorem proving for many-valued logics. *J Symbolic Computation* 19:353–91.

Baaz, M., P. Hájek, F. Montagna, and H. Veith. 2001. Complexity of t-tautologies. *Ann Pure Appl Logic* 113 (1–3): 3–11.

Baaz, M., P. Hájek, D. Švejda, and J. Krajíček. 1998. Embedding logics into product logic. *Studia Logica* 61:35–47.

Baaz, M., A. Leitsch, and R. Zach. 1996. Incompleteness of a first-order Gödel logic and some temporal logics of programs. *Lect Notes Comput Sci* 1092:1–15.

Baaz, M., and R. Zach. 1998. Compact propositional Gödel logics. In *Proc 28th IEEE Int Symp on Multiple-Valued Logic*, 108–13. Los Alamitos, CA: IEEE Computer Society Press.

Baczyński, M., and B. Jayaram. 2008. *Fuzzy Implications*. Berlin: Springer.

Bain, A. 1870. *Logic: Part Second. Induction*. London: Longmans, Green, Reader & Dyer.

Baldwin, J. F., and B. W. Pilsworth. 1980. Axiomatic approach to implication for approximate reasoning with fuzzy logic. *Fuzzy Sets Syst* 3:193–219

Bandler, W., and L. J. Kohout. 1978. *Fuzzy Relational Products and Fuzzy Implication Operators*. Report no. FRP-1, Dept. of Mathematics, Univ. of Essex, Colchester. Presented at the Workshop on Fuzzy Reasoning—Theory and Applications. London: Queen Mary College.

———— 1980a. Fuzzy power sets and fuzzy implication operators. *Fuzzy Sets Syst* 4:13–30.

———— 1980b. Fuzzy relational products as a tool for analysis and synthesis of the behavior of complex natural and artificial systems. In *Fuzzy Sets: Theory and Applications to Policy Analysis and Information Systems*, ed. P. P. Wang and S. K. Chang, 341–67. New York: Plenum Press.

———— 1980c. Semantics of implication operators and fuzzy relational products. *Int J Man Mach Stud* 12:89–116.

———— 1993. Cuts commute with closures. In *Fuzzy Logic*, ed. R. Lowen and M. Roubens, 161–67. Dordrecht: Kluwer.

Barbieri, G., M. Navara, and H. Weber. 2003. Characterization of T-measures. *Soft Computing* 8:44–50.

Bárdossy, G., and L. Duckstein. 1995. *Fuzzy Rule-Based Modeling with Applications to Geophysical, Biological, and Engineering Systems*. Boca Raton, FL: CRC Press.

Bárdossy, G., and J. Fodor. 2004. *Evaluation of Uncertainties and Risks in Geology: New Mathematical Approaches for Their Handling*. Berlin: Springer.

Barro, S., and R. Marín, eds. 2002. *Fuzzy Logic in Medicine*. Heidelberg: Physica-Verlag.

Bartl, E., and R. Belohlavek. 2015. Hardness of solving relational equations. *IEEE Trans Fuzzy Syst* 23 (6): 2435–38.

Baudry, L., ed. 1989. *The Quarrel over Future Contingents (Louvain 1465–1475)*. Dordrecht: Kluwer.

Beaney, M., ed. 1997. *The Frege Reader*. Oxford, UK: Blackwell.

Běhounek, L., U. Bodenhofer, and P. Cintula. 2008. Relations in fuzzy class theory: Initial steps. *Fuzzy Sets Syst* 159:1729–72.

Běhounek, L., and P. Cintula. 2005. Fuzzy class theory. *Fuzzy Sets Syst* 154:34–55.

———— 2006a. Fuzzy logics as the logics of chains. *Fuzzy Sets Syst* 157 (5): 604–10.

———— 2006b. From fuzzy logic to fuzzy mathematics: A methodological manifesto. *Fuzzy Sets Syst* 157:642–46.

Běhounek, L., and M. Daňková. 2009. Relational compositions in fuzzy class theory. *Fuzzy Sets Syst* 160:1005–36.

Bell, E. T. 1933. Remarks on the preceding note on many-valued truths. *Physical Review* 43 (12): 1033.

Bell, J. L. 2005. *Set Theory: Boolean-Valued Models and Independence Proofs*. 3rd ed. Oxford, UK: Oxford Univ. Press.

Bell, J. L., and M. Machover. 1977. *A Course in Mathematical Logic*. Amsterdam: North-Holland.

Bellman, R. E., and M. Giertz. 1973. On the analytic formalism of the theory of fuzzy sets. *Inf Sci* 5:149–56.

Bellman, R. E., R. Kalaba, and L. A. Zadeh. 1966. Abstraction and pattern classification *J Math Anal Appl* 13 (1): 1–7.

Bellman, R. E., and L. A. Zadeh. 1970. Decision-making in a fuzzy environment. *Management Science* 17 (4): B-141–B-164.

———1977. Local and fuzzy logics. In Dunn and Epstein 1977, 103–65.

Belluce, L. P. 1960. Some remarks on the completeness of infinite valued predicate logic. Abstract. *Notices AMS* 7:633.

———1964. Further results on infinite valued predicate logic. *J Symb Logic* 29 (2): 69–78.

———1986. Semisimple algebras of infinite valued logic and bold fuzzy set theory. *Canadian J Math* 38 (6): 1356–79.

———1992. Semi-simple and complete MV-algebras. *Algebra Universalis* 29:1–9.

Belluce, L. P., and C. C. Chang. 1960. A weak completeness theorem for infinite valued predicate logic. Abstract. *Notices AMS* 7:632–33.

———1963. A weak completeness theorem for infinite valued first-order logic. *J Symb Logic* 28 (1): 43–50.

Belohlavek, R. 1999. Fuzzy Galois connections. *Math Logic Quart* 45:497–504.

———2001a. Fuzzy closure operators. *J Math Anal Appl* 262:473–89.

———2001b. Lattice type fuzzy order and closure operators in fuzzy ordered sets. In *Proc Joint 9th IFSA World Congress and 20th NAFIPS Int Conf*, 2281–86. Vancouver: IEEE Press.

———2002a. Fuzzy closure operators II: Induced relations, representation, and examples. *Soft Computing* 7:53–64.

———2002b. Fuzzy equational logic. *Arch Math Logic* 41:83–90.

———2002c. *Fuzzy Relational Systems: Foundations and Principles*. New York: Kluwer.

———2003a. Birkhoff variety theorem and fuzzy logic. *Arch Math Logic* 42:781–90.

———2003b. Cutlike semantics for fuzzy logic and its applications. *Int J Gen Syst* 32 (4): 305–19.

———2004. Concept lattices and order in fuzzy logic. *Ann Pure Appl Logic* 128:277–98.

———2007. Do exact shapes of fuzzy sets matter? *Int J Gen Syst* 36 (5): 513–25.

Belohlavek, R., and T. Funioková. 2004. Similarity and fuzzy tolerance spaces. *J Logic Comput* 14:827–55.

Belohlavek, R., and G. J. Klir. 2007. On Elkan's theorems: clarifying their meaning via simple proofs. *Int J Intell Syst* 22 (2): 203–7.

———eds. 2011. *Concepts and Fuzzy Logic*. Cambridge, MA: MIT Press.

Belohlavek, R., G. J. Klir, H. W. Lewis III, and E. Way. 2002. On the capability of fuzzy set theory to represent concepts. *Int J Gen Syst* 31 (6): 569–85.

———2009. Concepts and fuzzy sets: Misunderstandings, misconceptions, and oversights. *Int J Approx Reason* 51 (1): 23–34.

Belohlavek, R., and M. Krmelova. 2013. Beyond Boolean matrix decompositions: toward factor analysis and dimensionality reduction of ordinal data. In *Proc IEEE ICDM*, 961–66.

Belohlavek, R., R. Kruse, and C. Moewes. 2011. Fuzzy logic in computer science. In *Computer Science: The Hardware, Software and Heart of It*, ed. E. K. Blum and A. V. Aho, 385–419. New York: Springer.

Belohlavek, R., and V. Vychodil. 2005. *Fuzzy Equational Logic*. Berlin: Springer.

———2006a. Algebras with fuzzy equalities. *Fuzzy Sets Syst* 157 (2): 161–201.

———2006b. Fuzzy Horn logic I: Proof theory; II: Implicationally defined classes. *Arch Math Logic* 45 (1): 3–51; 45 (2): 149–77.

———2006c. Data tables with similarity relations: Functional dependencies, complete rules and non-redundant bases. *Lect Notes Comput Sci* 3882:644–58.

———2010. Query systems in similarity-based databases: Logical foundations, expressive power, and completeness. In *Proc ACM SAC*, 1648–55. New York: ACM Press.

———2011. Codd's relational model from the point of view of fuzzy logic. *J Logic Comput* 21:851–62.

———2012. Formal concept analysis and linguistic hedges. *Int J Gen Syst* 41 (5): 503–32.

———2015. A logic of graded attributes. *Arch Math Logic* 54:785–802.

Ben Yahia, S., and A. Jaoua. 2001. Discovering knowledge from fuzzy concept lattice. In *Data Mining and Computational Intelligence*, ed. A. Kandel, M. Last, and H. Bunke, 167–90. Heidelberg: Physica-Verlag.

Bendová, K., and P. Hájek. 1993. Possibilistic logic as tense logic. In *Qualitative Reasoning and Decision Technologies*, ed. P. Carrete and M. Singh, 441–50. Barcelona: CIMNE Barcelona.

Bergmann, M. 2008. *An Introduction to Many-Valued and Fuzzy Logic*. New York: Cambridge Univ. Press.

Bernard, J. A. 1988. Use of a rule-based system for process control. *IEEE Control Systems Magazine* 8 (5): 3–13.

Bernays, P. 1918. Beiträge zur axiomatischen Behandlung des Logik-Kalküls [Contributions to the axiomatic treatment of logical calculus]. Habilitationsschrift, Universität Göttingen.

——— 1926. Axiomatische Untersuchung des Aussagen-Kalküls der „Principia Mathematica" [Axiomatic investigation of the propositional calculus of „Principia Mathematica"]. *Mathematische Zeitschrift* 25:305–20.

Bezdek, J. C. 1973. *Fuzzy Mathematics in Pattern Classification*. PhD diss., Cornell Univ., Ithaca, NY.

——— 1980. A convergence theorem for the fuzzy ISODATA clustering algorithm. *IEEE Trans on Pattern Analysis and Machine Intelligence* 2 (1): 1–8.

——— 1981. *Pattern Recognition with Fuzzy Objective Function Algorithms*. New York: Plenum Press.

Bezdek, J. C., D. Dubois, and H. Prade, eds. 1999. *Fuzzy Sets in Approximate Reasoning and Information Systems*. Boston: Kluwer.

Bezdek, J. C., and J. D. Harris. 1978. Fuzzy partitions and relations: An axiomatic basis for clustering. *Fuzzy Sets Syst* 1:111–27.

Bezdek, J. C., J. Keller, R. Krisnapuram, and N. R. Pal. 1999. *Fuzzy Models and Algorithms for Pattern Recognition and Image Processing*. Boston: Kluwer.

Bezdek, J. C., and S. K. Pal, eds. 1992. *Fuzzy Models for Pattern Recognition: Methods That Search for Structures in Data*. Piscataway, NJ: IEEE Press.

Biacino, L., and G. Gerla. 1984. Closure sytems and *L*-subalgebras. *Inf Sci* 33:181–95.

——— 1987. Recursively enumerable L-sets. *Z Math Logik Grundlagen Math* 33:107–13.

——— 1989. Decidability, recursive enumerability and Kleene hierarchy for L-subsets. *Z Math Logik Grundlagen Math* 35:49–62.

——— 1996. An extension principle for closure operators. *J Math Anal Appl* 198:1–24.

——— 2002. Fuzzy logic, continuity and effectiveness. *Arch Math Logic* 41:643–67.

Biacino, L., G. Gerla, and M. Ying. 2000. Approximate reasoning based on similarity. *Math Logic Quart* 46:77–86.

Biacino, L., and A. Lettieri. 1989. Equations with fuzzy numbers. *Inf Sci* 47:63–76.

Bilgiç, T., and I. B. Türkşen. 2000. Measurement of membership functions: theoretical and empirical work. In Dubois and Prade 2000, 195–227.

Billot, A., ed. 1992a. Special issue dedicated to Professor Claude Ponsard. *Fuzzy Sets Syst* 49 (1): 1–90.

——— 1992b. *Economic Theory of Fuzzy Equilibria: An Axiomatic Analysis*. Berlin: Springer.

Billot, A., and J.-F. Thisse. 1992. Claude Ponsard (1927–1990): A biographical essay. *Fuzzy Sets Syst* 49 (1): 3–8.

Birkhoff, G. 1948. *Lattice Theory*. 2nd ed. Providence, RI: American Mathematical Society.

Birkhoff, G., and J. von Neumann. 1936. The logic of quantum mechanics. *Ann Math* 37 (4): 823–43.

Black, M. 1937. Vagueness: An exercise in logical analysis. *Philos Sci* 4 (4): 427–55.

——— 1949. *Language and Philosophy*. Ithaca, NY: Cornell Univ. Press.

——— 1963. Reasoning with loose concepts. *Dialogue* 2:1–12.

Blanshard, B. 1939. *The Nature of Thought*. 2 vols. London: George Allen & Unwin.

Blizard, W. D. 1989. Real-valued multisets and fuzzy sets. *Fuzzy Sets Syst* 33:77–97.

Blockley, D. I. 1977. Analysis of structural failures. *Proc Inst of Civil Engineers* 62:51–74.

——— 1979. The role of fuzzy sets in civil engineering. *Fuzzy Sets Syst* 2 (4): 267–78.

Blok, W. J., and D. Pigozzi. 1989. *Algebraizable Logics*. Providence, RI: American Mathematical Society.

Blok, W. J., and C. J. van Alten. 2002. The finite embeddability property for residuated lattices, pocrims and BCK-algebras. *Algebra Universalis* 48 (3): 253–71.

Bobylev, V. N. 1985a. Опорная функция нечёткого множества и её характеристические свойства [Support function of a fuzzy set and its characteristic properties]. *Matematicheskie Zametki* 37 (4): 507–13.

——— 1985b. Cauchy problem under fuzzy control. *BUSEFAL* 21:117–26.

Bocheński, I. M. 1961. *A History of Formal Logic*. Notre Dame, IN: Univ. of Notre Dame Press.

Bochvar, D. A. 1938. Об одном трехзначном исчислении и его применении к анализу парадоксов классического расширенного функционального исчисления. *Matematicheskii Sbornik* 4:287–308. Trans. M. Bergmann 1981. On a three-valued logical calculus and its application to the analysis of the paradoxes of the classical extended functional calculus. *History and Philosophy of Logic* 2:87–112.

——— 1943. К вопросу о непротиворечивости одного трехзначного исчисления [On the consistency of a three-valued calculus]. *Matematicheskii Sbornik* 12 (54): 353–69.

Bodenhofer, U. 1998. *A Similarity-Based Generalization of Fuzzy Orderings*. PhD diss., Johannes Kepler Universität Linz.

——— 2000. A similarity-based generalization of fuzzy orderings preserving the classical axioms. *Int J UFKBS* 8:593–610.

Boehner, P., ed. 1945. *The Tractatus de Praedestinatione et de Praescientia Dei et de Futuris Contingentibus of*

William Ockham, Edited with a Study on the Mediaeval Problem of a Three-Valued Logic by Philotheus Boehner, O.F.M. New York: The Franciscan Institute of St. Bonaventure College.

Boicescu, V., A. Filipoiu, G. Georgescu, and S. Rudeanu. 1991. *Lukasiewicz-Moisil Algebras.* Amsterdam: North-Holland.

Bojadziev, G., and M. Bojadziev. 1995. *Fuzzy Sets, Fuzzy Logic, Applications.* Singapore: World Scientific.

——— 1997. *Fuzzy Logic for Business, Finance, and Management.* 2nd ed. Singapore: World Scientific.

Bolc, L., and P. Borowik. 1992. *Many-Valued Logics 1: Theoretical Foundations.* Berlin: Springer.

Boole, G. 1847. *The Mathematical Analysis of Logic.* Cambridge, UK: Macmillan, Barclay & Macmillan.

——— 1854. *An Investigation of the Laws of Thought, on Which Are Founded the Mathematical Theory of Logic and Probabilities.* London: Walton and Maberly. Repr., New York: Dover, 1951.

Borgelt, C., and R. Kruse. 2002. *Graphical Models: Methods for Data Analysis and Mining.* Chichester, UK: Wiley.

Borkowski, L., ed. 1970. *Jan Łukasiewicz: Selected Works.* Amsterdam: North-Holland.

Bosbach, B. 1981. Concerning bricks. *Acta Mathematica Academiae Scientiarum Hungarica* 38:89–104.

Bosc, P., and J. Kacprzyk, eds. 1995. *Fuzziness in Database Management Systems.* Heidelberg: Physica-Verlag.

Bosc, P., and O. Pivert. 1995. SQLf: A relational database language for fuzzy querying. *IEEE Trans Fuzzy Syst* 3:1–17.

Bosserman, R. W., and R. K. Ragade. 1982. Ecosystem analysis using fuzzy set theory. *Ecological Modeling* 16 (2–4): 191–208.

Bouchon-Meunier, B., D. Dubois, L. Godo, and H. Prade. 1999. Fuzzy sets and possibility theory in approximate and plausible reasoning. In *Fuzzy Sets in Approximate Reasoning and Information Systems,* ed. J. C. Bezdek, D. Dubois, and H. Prade, 15–190. Dordrecht: Kluwer.

Bouchon-Meunier, B., M. Rifqi, and S. Bothorel. 1996. Towards general measures of comparison of objects. *Fuzzy Sets Syst* 84 (2): 143–53.

Bova, S., and F. Montagna. 2008. Proof search in Hájek's basic logic. *ACM Trans on Computational Logic* 9 (3): article 21.

Bowyer, K. W., and N. Ahuja, eds. 1996. *Advances in Image Understanding: A Festschrift for Azriel Rosenfeld.* Los Alamitos, CA: Wiley-IEEE Computer Society.

Bradley, F. H. 1893. *Appearance and Reality: A Metaphysical Essay.* London: Swan Sonnenschein.

Branzei, R., D. Dimitrov, and S. Tijs. 2005. *Models in Cooperative Game Theory: Crisp, Fuzzy, and Multi-Choice Games.* Berlin: Springer.

Bremermann, H. 1976. Pattern recognition. In *Systems Science in the Social Sciences,* ed. H. Bossel, S. Klaczko, and N. Müller, 116–59. Basel and Stuttgart: Birkhäuser.

Bridgman, P. W. 1959. How much rigor is possible in physics? In *The Axiomatic Method,* ed. L. Henkin, P. Suppes, and A. Tarski, 225–37. Amsterdam: North-Holland.

Buckles, B. P., and F. E. Petry. 1982. A fuzzy representation of data for relational databases. *Fuzzy Sets Syst* 7:213–26.

——— 1985. Query languages for fuzzy databases. In *Management Decision Support Systems Using Fuzzy Sets and Possibility Theory,* ed. J. Kacprzyk and R. R. Yager, 241–52. Cologne: Verlag TÜV Rheinland.

Buckley, J. J. 1992. Universal fuzzy controllers. *Automatica* 28 (6): 1245–48.

——— 2003. *Fuzzy Probabilities.* Heidelberg: Springer.

Buckley, J. J., and E. Eslami. 1997. Fuzzy plane geometry I: Points and lines; II: Circles and polygons. *Fuzzy Sets Syst* 86:179–87; 87:79–85.

Buckley, J. J., and Y. Hayashi. 1993. Numerical relationships between neural networks, continuous functions, and fuzzy systems. *Fuzzy Sets Syst* 60 (1): 1–8.

Buckley, J. J., and Y. Qu. 1991. Solving fuzzy equations: A new solution concept. *Fuzzy Sets Syst* 39:291–301.

Budimirović, B., V. Budimirović, B. Šešelja, and A. Tepavčević. 2014. Fuzzy identities with application to fuzzy semigroups. *Inf Sci* 266:148–59.

Buff, H. W. 1985. Decidable and undecidable MV-algebras. *Algebra Universalis* 21 (2–3): 234–49

Burusco, A., and R. Fuentes-González. 1994. The study of the L-fuzzy concept lattice. *Mathware & Soft Computing* 3:209–18.

Busaniche, M., and F. Montagna. 2011. Hájek's logic BL and BL-algebras. In Cintula, Hájek, and Noguera 2011, 355–447.

Butnariu, D. 1978. Fuzzy games: A description of a concept. *Fuzzy Sets Syst* 1 (3): 181–92.

——— 1979. Solution concepts for *n*-persons fuzzy games. In *Advances in Fuzzy Set Theory and Applications,* ed. M. M. Gupta, R. K. Ragade, and R. R. Yager, 339–59. Amsterdam: North-Holland.

——— 1983. Additive fuzzy measures and integrals I. *J Math Anal Appl* 93:436–52.

——— 1986. Fuzzy measurability and integrability. *J Math Anal Appl* 117:385–410.

——— 1987. Values and cores of fuzzy games with infinitely many players. *J Game Theory* 16:43–68.

Butnariu, D., and E. P. Klement. 1991. Triangular norm-based measures and their Markov kernel representation. *J Math Anal Appl* 162:111–43.

————— 1993. *Triangular Norm-Based Measures and Games with Fuzzy Coalitions.* Dordrecht: Kluwer.

————— 2002. Triangular norm-based measures. In *Handbook of Measure Theory*, ed. E. Pap, 947–1010. Amsterdam: Elsevier.

Butnariu, D., E. P. Klement, and S. Zafrany. 1995. On triangular norm-based propositional fuzzy logics. *Fuzzy Sets Syst* 69:241–55.

Cabrelli, C. A., B. Forte, U. M. Molter, and E. R. Vrscay. 1992. Iterated fuzzy set systems: A new approach to the inverse for fractals and other sets. *J Math Anal Appl* 171:79–100.

Cai, K. Y., ed. 1996a. Fuzzy methodology in system failure engineering. Special issue, *Fuzzy Sets Syst* 83 (2): 111–290.

————— 1996b. *Introduction to Fuzzy Reliability.* Dordrecht: Kluwer.

Cai, K. Y., C. Y. Wen, and M. L. Zhang. 1991a. Fuzzy reliability modeling of gracefully degradable systems. *Reliability Engineering and System Safety* 33 (1): 141–57.

————— 1991b. Fuzzy variables as basis for a theory of fuzzy reliability in the possibility context. *Fuzzy Sets Syst* 42 (2): 145–72.

Carlsson, C., M. Fedrizzi, and R. Fullér. 2004. *Fuzzy Logic in Management.* Boston: Kluwer.

Carnielli, W. A. 1987. Systematization of finite many-valued logics through the method of tableaux. *J Symb Logic* 52:473–93.

Casillas, J., and F. J. Martínez-López, eds. 2010. *Marketing Intelligent Systems Using Soft Computing: Managerial and Research Applications.* Berlin: Springer.

Castro, J. L. 1994. Fuzzy logics as families of bivaluated logics. *Fuzzy Sets Syst* 64:321–32.

————— 1995. Fuzzy logic controllers are universal approximators. *IEEE Trans Syst Man Cyb* 25 (4): 629–35.

Castro, J. L., and E. Trillas. 1991. Tarski's fuzzy consequences. In *Proc Int Fuzzy Engineering Symp*, vol. 1, 70–81.

Castro, J. L., E. Trillas, and S. Cubillo. 1994. On consequence in approximate reasoning. *J Applied Non-Classical Logics* 4:91–103.

Cattaneo, G., and D. Ciucci. 2003a. Generalized negations and intuitionistic fuzzy sets. A criticism to a widely used terminology. In *Proc 3rd Conf of EUSFLAT*, 147–52. Zittau, Germany: Univ. of Applied Sciences.

————— 2003b. Intuitionistic fuzzy sets or orthopair fuzzy sets? In *Proc 3rd Conf of EUSFLAT*, 153–58. Zittau, Germany: Univ. of Applied Sciences.

————— 2006. Basic intuitionistic principles in fuzzy set theories and its extensions (A terminological debate on Atanassov IFS). *Fuzzy Sets Syst* 157 (24): 3198–3219.

Cattaneo, G., P. Flocchini, G. Mauri, C. Q. Vogliotti, and N. Santoro. 1997. Cellular automata in fuzzy background. *Physica D: Nonlinear Phenomena* 105 (1–3): 105–20.

Cattaneo, G., and G. Nisticò. 1989. Brouwer-Zadeh posets and three-valued Łukasiewicz posets. *Fuzzy Sets Syst* 33 (2): 165–90.

Cavallo, R. E., and G. J. Klir. 1982. Reconstruction of possibilistic behavior systems. *Fuzzy Sets Syst* 8 (2): 175–97.

Center, B., and B. P. Verma. 1998. Fuzzy logic for biological and agricultural systems. *Artif Intell Rev* 12:213–25.

Chaira, T., and A. K. Ray. 2010. *Fuzzy Image Processing and Applications with MATLAB.* Boca Raton, FL: CRC Press.

Chakraborty, M. K. 1988. Use of fuzzy set theory in introducing graded consequence in multiple valued logic. In *Fuzzy Logic in Knowledge-Based Systems, Decision, and Control*, ed. M. M. Gupta and T. Yamakawa, 247–57. Amsterdam: North-Holland.

————— 1995. Graded consequence: Further studies. *J Applied Non-Classical Logics* 5:227–38.

Chanas, S., and J. Kamburowski. 1981. The use of fuzzy variables in PERT. *Fuzzy Sets Syst* 5 (1): 11–19.

Chang, C. C. 1958a. Algebraic analysis of many-valued logics. *Trans AMS* 88:476–90.

————— 1958b. Proof of an axiom of Lukasiewicz. *Trans AMS* 87:55–56.

————— 1959. A new proof of the completeness of the Lukasiewicz axioms. *Trans AMS* 93:74–80.

————— 1961. Theory of models of infinite valued logic. I–IV. Abstract. *Notices AMS* 8:68–69.

————— 1963. The axiom of comprehension in infinite valued logic. *Mathematica Scandinavica* 13:9–30.

————— 1998. The writing of the MV-algebras. *Studia Logica* 61:3–6.

Chang, C. C., and H. J. Keisler. 1962. Model theories with truth values in a uniform space. *Bull AMS* 68 (2): 107–9.

————— 1966. *Continuous Model Theory.* Princeton, NJ: Princeton Univ. Press.

————— 1990. *Model Theory.* 3rd ed. Amsterdam: North-Holland.

Chang, C. L. 1967. *Fuzzy Sets and Pattern Recognition.* PhD diss., Dept. of Electrical Engineering, Univ. of California, Berkeley.

————— 1968. Fuzzy topological spaces. *J Math Anal Appl* 24 (1): 182–90.

Chang, S. S. L. 1969. Fuzzy dynamic programming and decision making process. In *Proc 3rd Princeton Conf on Information Sciences and Systems*, 200–203.

Chapin Jr., E. W. 1974. Set-valued set theory: Part one. *Notre Dame J Formal Logic* 15:619–34.

———— 1975. Set-valued set theory: Part two. *Notre Dame J Formal Logic* 16:255–67.

Cheeseman, P. 1986. Probabilistic versus fuzzy reasoning. In *Uncertainty in Artificial Intelligence*, ed. L. N. Kanal and J. F. Lemmer, 85–102. Amsterdam: North-Holland.

———— 1988a. An inquiry into computer understanding. *Comput Intell (Canadian)* 4:58–66.

———— 1988b. In defense of *An inquiry into computer understanding*. *Comput Intell (Canadian)* 4:129–42.

Chen, S. J., and C. L. Hwang. 1992. *Fuzzy Multiple Attribute Decision Making*. Berlin: Springer.

Cheng, F., G. Zhong, Y. Li, and Z. Xu. 1996. Fuzzy control of a double-inverted pendulum. *Fuzzy Sets Syst* 79 (3): 315–21.

Chi, Z., H. Yan, and T. D. Pham. 1996. *Fuzzy Algorithms: With Applications to Image Processing and Pattern Recognition*. Singapore: World Scientific.

Chockalingam, J. 2012. *Quantifying Sustainability of Forest Management Using Fuzzy Logic: Focussing toward Forest Certification*. Saarbrücken, Germany: Lambert Acad. Publ.

Chovanec, F. 1993. States and observables on MV algebras. *Tatra Mt Math Publ* 3:55–65.

Church, A. 1956 *An Introduction to Mathematical Logic*. Princeton, NJ: Princeton Univ. Press.

Chvalovský, K. 2012. On the independence of axioms in BL and MTL. *Fuzzy Sets Syst* 197:123–29.

Ciabattoni, A., F. Esteva, and L. Godo. 2002. T-norm based logics with n-contraction. *Neural Network World* 12 (5): 441–52.

Cicalese, F., and D. Mundici. 2011. Recent developments of feedback coding and its relations with many-valued logic. In *Proof, Computation and Agency*, ed. J. van Benthem, R. Parikh, and A. Gupta, 115–31. Dordrecht: Springer.

Cignoli, R. 1982. Proper n-valued Łukasiewicz algebras as S-algebras of Łukasiewicz n-valued propositional calculi. *Studia Logica* 41:3–16.

———— 1993. Free lattice-ordered abelian groups and varieties of MV-algebras. *Notas de lógica matemática* 38:113–18.

Cignoli, R. L. O., I. M. L. D'Ottaviano, and D. Mundici. 2000. *Algebraic Foundations of Many-Valued Reasoning*. Dordrecht: Kluwer.

Cignoli, R., F. Esteva, L. Godo, and A. Torrens. 2000. Basic fuzzy logic is the logic of continuous t-norms and their residua. *Soft Computing* 4 (2): 106–12.

Cignoli, R., and D. Mundici. 1997. An elementary proof of Chang's completeness theorem for the infinite-valued calculus of Łukasiewicz. *Studia Logica* 58:79–97.

Cintula, P. 2001a. About axiomatic systems of product fuzzy logic. *Soft Computing* 5:243–44.

———— 2001b. An alternative approach to the ŁΠ logic. *Neural Network World* 124:561–71.

———— 2001c. The ŁΠ and ŁΠ$\frac{1}{2}$ propositional and predicate logics. *Fuzzy Sets Syst* 124 (3): 289–302.

———— 2003. Advances in the ŁΠ and ŁΠ$\frac{1}{2}$ logics. *Arch Math Logic* 42 (5): 449–68.

———— 2005a. A note to the definition of the ŁΠ-algebras. *Soft Computing* 9 (8): 575–78.

———— 2005b. Short note: On the redundancy of axiom (A3) in BL and MTL. *Soft Computing* 9 (12): 942.

———— 2006. Weakly implicative (fuzzy) logics I: Basic properties. *Arch Math Logic* 45 (6): 673–704.

Cintula, P., F. Esteva, J. Gispert, L. Godo, F. Montagna, and C. Noguera. 2009. Distinguished algebraic semantics for t-norm based fuzzy logics: Methods and algebraic equivalencies. *Ann Pure Appl Logic* 160:53–81.

Cintula, P., C. G. Fermüller, L. Godo, and P. Hájek, eds. 2011. *Understanding Vagueness: Logical, Philosophical and Linguistic Perspectives*. London: College Publications.

Cintula, P., and B. Gerla. 2004. Semi-normal forms and functional representation of product fuzzy logic. *Fuzzy Sets Syst* 143 (1): 89–110.

Cintula, P., and P. Hájek. 2010. Triangular norm based predicate fuzzy logics. *Fuzzy Sets Syst* 161:311–46.

Cintula, P., P. Hájek, and R. Horčík. 2007. Formal systems of fuzzy logic and their fragments. *Ann Pure Appl Logic* 150 (1–3): 40–65.

Cintula, P., P. Hájek, and C. Noguera, eds. 2011. *Handbook of Mathematical Fuzzy Logic*. 2 vols. London: College Publications.

Cintula, P., and C. Noguera. 2010. Implicational (semilinear) logics I: A new hierarchy. *Arch Math Logic* 49 (4): 417–46.

Cioffi-Revilla, C. A. 1981. Fuzzy sets and models of international relations. *American J Political Science* 25 (1): 129–59.

Ćirić, M., M. Droste, J. Ignjatović, and H. Vogler. 2010. Determinization of weighted finite automata over strong bimonoids. *Inf Sci* 180:3497–3520.

Clark, T. D., J. M. Larson, J. N. Mordeson, J. D. Potter, and M. J. Wierman. 2008a. *Applying Fuzzy Mathematics to Formal Models in Comparative Politics*. Berlin: Springer.

Clark, T. D., J. M. Larson, J. N. Mordeson, and M. J. Wierman. 2008b. Extension of the portfolio allocation model to surplus majority governments: a fuzzy approach. *Public Choice* 134 (3–4): 179–99.

Clements, D. P. 1977. *Fuzzy Ratings for Computer Security Evaluation*. PhD diss., Univ. of California, Berkeley.

Cobb, M., F. Petry, and V. Robinson, eds. 2000. Uncertainty in Geographic Information Systems and Spatial Data. Special issue, *Fuzzy Sets Syst* 113 (1): 1–159.

Cohen, L. J. 1962. *The Diversity of Meaning*. London: Methuen.

———1970. *The Implications of Induction*. London: Methuen.

———1973. A note on inductive logic. *J Philosophy* 70 (2): 27–40.

———1977. *The Probable and the Provable*. Oxford, UK: Oxford Univ. Press.

———1989. *An Introduction to the Philosophy of Induction and Probability*. Oxford, UK: Oxford Univ. Press.

Cohen, M. R. 1927. Concepts and twilight zones. *J Philosophy* 24:673–83.

Colyvan, M. 2004. The philosophical significance of Cox's theorem. *Int J Approx Reason* 37 (1): 71–85.

Copilowish, I. M. 1939. Border-line cases, vagueness, and ambiguity. *Philosophy of Science* 6:181–95.

Coppi, R., M. A. Gil, and H. A. L. Kiers. 2006. The fuzzy approach to statistical analysis. *Computational Statistics & Data Analysis* 51:1–14.

Coppola, C., G. Gerla, and A. Miranda. 2010. Point-free foundation of geometry and multi-valued logic. *Notre Dame J Formal Logic* 51:383–405.

Cordón, O., F. Herrera, F. Hoffmann, and L. Magdalena. 2001. *Genetic Fuzzy Systems: Evolutionary Tuning and Learning of Fuzzy Knowledge Bases*. Singapore: World Scientific.

Corne, D., G. Fogel, J. C. Rajapakse, and L. Wang, eds. 2006. Soft Computing for Bioinformatics and Medical Informatics. Special issue, *Soft Computing* 10 (4): 285–403.

Corsi, G. 1992. Completeness theorem for Dummett's LC quantified and some of its extensions. *Studia Logica* 51 (2): 317–35.

Couso, I., D. Dubois, and L. Sanchez. 2014. *Random Sets and Random Fuzzy Sets as Ill-Perceived Random Variables*. Heidelberg: Springer.

Cox, R. T. 1946. Probability, frequency and reasonable expectation. *American J Phys* 14 (1): 1–13.

———1961. *The Algebra of Probable Inference*. Baltimore: Johns Hopkins Press.

Cross, V., and T. Sudkamp. 2002. *Similarity and Compatibility in Fuzzy Set Theory: Assessment and Applications*. Heidelberg: Physica-Verlag.

Crowe, M. J. 1975. Ten "laws" concerning patterns of change in the history of mathematics. *Historia Mathematica* 2:161–66. Reprinted in Gillies 1992, 15–20.

———1992. Afterword (1992): A revolution in the historiography of mathematics? In Gillies 1992, 306–16.

Cuesta, F., and A. Ollero. 2005. *Intelligent Mobile Robot Navigation*. Berlin: Springer.

Czogala, E., J. Drewniak, and W. Pedrycz. 1982. Fuzzy relation equations on a finite set. *Fuzzy Sets Syst* 7:89–101.

Czogala, E., and K. Hirota. 1986. *Probabilistic Sets: Fuzzy and Stochastic Approach to Decision, Control, and Recognition Processes*. Cologne: Verlag TÜV Rheinland.

Dadios, E. P., ed. 2012. *Fuzzy Logic: Emerging Technologies and Applications*. Rijeka, Croatia: InTech.

Dalla Chiara, M., R. Giuntini, and R. Greechie. 2004. *Reasoning in Quantum Theory: Sharp and Unsharp Quantum Logics*. Dordrecht: Kluwer.

Das, P. S. 1981. Fuzzy groups and level subgroups. *J Math Anal Appl* 84:264–69

Dauben, J. W. 1984. Conceptual revolutions and the history of mathematics: Two studies in the growth of knowledge. In *Transformation and Tradition in the Sciences: Essays in Honor of I. Bernard Cohen*, ed. E. Mendelsohn, 81–103. Cambridge, UK: Cambridge Univ. Press. Reprinted in Gillies 1992, 49–71.

———1992. Appendix (1992): Revolutions revisited. In Gillies 1992, 72–82.

———1996. Paradigms and proofs: How revolutions transform mathematics. In *Paradigms and Mathematics*, ed. E. Ausejo and M. Hormigón, 117–48. Madrid: Siglo de España XXI Editores.

Davis, L. S., ed. 2001. *Foundations of Image Understanding*. Boston: Kluwer.

De Baets, B. 2000. Analytical solution methods for fuzzy relational equations. In Dubois and Prade 2000, 291–340.

De Baets, B., and H. De Meyer. 2003a. On the existence and construction of T-transitive closures. *Inf Sci* 152:167–79.

———2003b. Transitive approximation of fuzzy relations by alternating closures and openings. *Soft Computing* 7:210–19.

———2005. Transitivity-preserving fuzzification schemes for cardinality-based similarity measures. *Europ J Operational Research* 160 (3): 726–40.

De Baets, B., and E. Kerre. 1994. The cutting of compositions. *Fuzzy Sets Syst* 62:295–309.

De Baets, B., and R. Mesiar. 1998. 𝒯-partitions. *Fuzzy Sets Syst* 97 (2): 211–23.

———2002. Metrics and 𝒯-equalities. *J Math Anal Appl* 267:531–47.

De Cock, M., and E. Kerre. 2003. On (un)suitable fuzzy relations to model approximate equality. *Fuzzy Sets Syst* 133:137–53.

De Cooman, G. 1997. Possibility theory I: The measure- and integral-theoretic groundwork; II: Conditional possibility; III: Possibilistic independence. *Int J Gen Syst* 25 (4): 291–323; 325–51; 353–71.

De Cooman, G., and E. E. Kerre. 1993. Ample fields. *Bull Belgian Mathematical Society—Simon Stevin* 67 (3–4): 235–44.

De Wit, G. W. 1982. Underwriting and uncertainty. *Insurance: Mathematics and Economics* 1 (4): 277–85.

Deboeck, G. J., ed. 1994. *Trading on The Edge: Neural, Genetic, and Fuzzy Systems for Chaotic Financial Markets.* New York: Wiley.

Delgado, M., M. D. Ruiz, D. Sánchez, and M. A. Vila. 2014. Fuzzy quantification: A state of the art. *Fuzzy Sets Syst* 242:1–30.

Demicco, R. V., and G. J. Klir. 2001. Stratigraphic simulation using fuzzy logic to model sediment dispersal. *J Petroleum Science and Engineering* 31 (2–4): 135–55.

———— eds. 2004. *Fuzzy Logic in Geology.* San Diego: Academic Press.

Demirci, M. 1999. Fuzzy functions and their fundamental properties. *Fuzzy Sets Syst* 106:239–46.

———— 2003. Foundations of fuzzy functions and vague algebra based on many-valued equivalence relations, part II: Vague algebraic notions. *Int J Gen Syst* 32:157–75.

Demirci, M., and D. Çoker. 1993. On the axiomatic theory of fuzzy sets. *Fuzzy Sets Syst* 60:181–98.

Di Nola, A. 1993. MV-algebras in the treatment of uncertainty. In *Fuzzy Logic,* ed. R. Lowen and M. Roubens, 123–31. Dordrecht: Kluwer.

Di Nola, A., G. Georgescu, and A. Lettieri. 1999. Extending probabilities to states of MV-algebras. *Ann Kurt Gödel Society* 3:31–50.

Di Nola, A., and G. Gerla. 1986. Fuzzy models of first order languages. *Z Math Logik Grundlagen Math* 32:331–40.

———— 1987. Lattice valued algebras. *Stochastica* 11:137–50.

Di Nola, A., and I. Leuştean. 2011. Łukasiewicz logic and MV-algebras. In Cintula, Hájek, and Noguera 2011, 469–583.

Di Nola, A., S. Sessa, W. Pedrycz, and E. Sanchez. 1989. *Fuzzy Relation Equations and Their Applications to Knowledge Engineering.* Dordrecht: Kluwer.

Diamond, P., and P. Kloeden. 1994. *Metric Spaces of Fuzzy Sets: Theory and Applications.* Singapore: World Scientific.

———— 2000. Metric topology of fuzzy numbers and fuzzy analysis. In Dubois and Prade 2000, 583–641.

Dienes, Z. P. 1949. On an implication function in many-valued systems of logic. *J Symb Logic* 14 (2):95–97.

Dilworth, R. P. 1939. Non-commutative residuated lattices. *Trans AMS* 46:426–44.

Dombi, J. 1982. A general class of fuzzy operators, the De Morgan class of fuzzy operators and fuzziness measures induced by fuzzy operators. *Fuzzy Sets Syst* 8:149–63.

Driankov, D., and A. Saffiotti, eds. 2001. *Fuzzy Logic Techniques for Autonomous Vehicle Navigation.* Heidelberg: Physica-Verlag.

Dualibe, C., M. Verleysen, and P. G. A. Jespers. 2003. *Design of Analog Fuzzy Logic Controlleres in CMOS Technologies: Implementation, Test and Application.* Dordrecht: Kluwer.

Dubois, D., F. Esteva, P. Garcia, L. Godo, and H. Prade. 1997. A logical approach to interpolation based on similarity relations. *Int J Approx Reason* 17 (1): 1–36.

Dubois, D., F. Esteva, L. Godo, and H. Prade. 2007. Fuzzy-set based logics—an history-oriented presentation of their main developments. In *Handbook of the History of Logic,* vol 8., ed. D. M. Gabbay and J. Woods, 325–449. Amsterdam: North-Holland.

Dubois, D., S. Gottwald, P. Hájek, J. Kacprzyk, and H. Prade. 2005. Terminological difficulties in fuzzy set theory: The case of "intuitionistic fuzzy sets." *Fuzzy Sets Syst* 156 (3): 485–91.

Dubois, D., E. Hüllermeier, and H. Prade. 2006. A systematic approach to the assessment of fuzzy association rules. *Data Mining and Knowledge Discovery* 13:167–92.

Dubois, D., E. Kerre, R. Mesiar, and H. Prade. 2000. Fuzzy interval analysis. In Dubois and Prade 2000, 483–581.

Dubois, D., J. Lang, and H. Prade. 1991. Fuzzy sets in approximate reasoning, part 2: Logical approaches. *Fuzzy Sets Syst* 40:203–44.

———— 1994. Possibilistic logic. In *Handbook of Logic in Artificial Intelligence and Logic Programming,* vol. 3, ed. D. M. Gabbay, C. J. Hogger, and J. A. Robinson, 439–513. Oxford, UK: Oxford Univ. Press.

Dubois, D., and H. Prade. 1978. Operations on fuzzy numbers. *Int J Systems Science* 9:613–26.

———— 1980a. *Fuzzy Sets and Systems: Theory and Applications.* New York: Academic Press.

———— 1980b. New results about properties and semantics of fuzzy set-theoretic operations. In *Fuzzy Sets Theory and Applications to Policy Analysis and Information Systems,* ed. P. P. Wang and S. K. Chang, 59–75. New York: Plenum Press.

———— 1981. Addition of interactive fuzzy numbers. *IEEE Trans Automat Contr* 26 (4): 926–36.

———— 1982a. A unifying view of comparison indices in a fuzzy set-theoretic framework. In *Fuzzy Set and Possibility Theory: Recent Developments,* ed. R. R. Yager, 3–13. Oxford, UK: Pergamon Press.

———— 1982b. Towards fuzzy differential calculus. 1: Integration of fuzzy mappings; 2: Integration on fuzzy intervals; 3: Differentiation. *Fuzzy Sets Syst* 8:1–17, 105–16, 225–33.

——— 1984. A theorem on implication functions defined from triangular norms. *Stochastica* 8:267–79.

——— 1987a. An alternative approach to the handling of subnormal possibility distributions. *Fuzzy Sets Syst* 24 (1): 123–26.

——— 1987b. Fuzzy numbers: An overview. In *Mathematics and Logic*, vol. 1 of *Analysis of Fuzzy Information*, ed. J. C. Bezdek, 3–39. Boca Raton, FL: CRC Press.

——— 1988. *Possibility Theory: An Approach to Computerized Processing of Uncertainty*. New York: Plenum Press.

——— 1990. Rough fuzzy sets and fuzzy rough sets. *Int J Gen Syst* 17 (2–3): 191–209.

——— 1991. Fuzzy sets in approximate reasoning, part 1: Inference with possibility distributions. *Fuzzy Sets Syst* 40 (1): 143–202.

——— 1992. Putting rough sets and fuzzy sets together. In *Intelligent Decision Support: Handbook of Applications and Advances in the Rough Set Theory*, ed. R. Słowiński, 203–32. Dordrecht: Kluwer.

——— 1997. The three semantics of fuzzy sets. *Fuzzy Sets Syst* 90:141–50.

——— eds. 2000. *Fundamentals of Fuzzy Sets*. Boston: Kluwer.

——— 2004. Possibilistic logic: a retrospective and prospective view. *Fuzzy Sets Syst* 144 (1): 3–23.

Duhem, P. 1906. *La Théorie Physique: Son Objet, Sa Structure* [Physical theory: Its aim and structure]. Paris: Chavalier & Rivière.

——— 1954. *The Aim and Structure of Physical Theory*. Princeton, NJ: Princeton Univ. Press. Trans. of the 2nd ed. of Duhem 1906.

Dummett, M. 1959. A propositional calculus with denumerable matrix. *J Symb Logic* 24 (2): 97–106.

——— 1975. Wang's paradox. *Synthese* 30:301–24.

Dunn, J. C. 1973. A fuzzy relative of the ISODATA process and its use in detecting compact well-separated clusters. *J Cybernetics* 3 (3): 32–57.

Dunn, J. M., and G. Epstein, eds. 1977. *Modern Uses of Multiple-Valued Logic*. Dordrecht: Reidel.

Dvurečenskij, A. 1996. On a representation of observables in fuzzy measurable spaces. *J Math Anal Appl* 197 (2): 579–85.

——— 2000. Loomis-Sikorski theorem for σ-complete MV-algebras and ℓ-groups. *J Australasian Mathematical Society* 68 (2): 261–77.

Dvurečenskij, A., and J. Rachůnek. 2006. Probabilistic averaging in bounded Rl-monoids. *Semigroup Forum* 72:190–206.

Dvurečenskij, A., and B. Riečan. 1991. Fuzzy quantum models. *Int J Gen Syst* 20 (1): 39–54.

Dwinger, P. 1977. A survey of the theory of Post algebras and their generalizations. In Dunn and Epstein 1977 51–75.

Elkan, C. 1993. The paradoxical success of fuzzy logic. In *Proc 11th National Conf on Artificial Intelligence*, 698–703. Menlo Park, CA: MIT Press.

Elorza, J., and P. Burillo. 2003. Connecting fuzzy preorders, fuzzy consequence operators and fuzzy closure and co-closure systems. *Fuzzy Sets Syst* 139:601–13.

Epstein, G. 1960. The lattice theory of Post algebras. *Trans AMS* 95 (2): 300–317.

Eslami, E., F. Nemat, and J. J. Buckley. 2001. Fuzzy space geometry I: Points, lines, planes; II: Subpoints and sublines. *J Fuzzy Math* 9:659–75; 693–700.

Esogbue, A. O., and R. C. Elder. 1979. Fuzzy sets and the modelling of physician decision processes, part I: The initial interview-information gathering session. *Fuzzy Sets Syst* 2 (4): 279–91.

——— 1980. Fuzzy sets and the modelling of physician decision processes, part II: Fuzzy diagnosis decision models. *Fuzzy Sets Syst* 3 (1): 1–9.

Esteva, F., and X. Domingo. 1980. Sobre funciones de negación en $[0,1]$ [On functions of negation on $[0,1]$]. *Stochastica* IV:141–66.

Esteva, F., P. Garcia, L. Godo, and R. Rodríguez. 1997. A modal account of similarity-based reasoning. *Int J Approx Reason* 16:235–60.

Esteva, F., P. Garcia-Calvés, and L. Godo. 1994. Relating and extending semantical approaches to possibilistic reasoning. *Int J Approx Reason* 10 (4): 311–44.

Esteva, F., J. Gispert, L. Godo, and F. Montagna. 2002. On the standard and rational completeness of some axiomatic extensions of the monoidal t-norm logic. *Studia Logica* 71:199–226.

Esteva, F., J. Gispert, L. Godo, and C. Noguera. 2007. Adding truth-constants to logics of continuous t-norms: Axiomatization and completeness results. *Fuzzy Sets Syst* 158:597–618.

Esteva, F., and L. Godo. 1999. Putting together Łukasiewicz and product logics. *Mathware & Soft Computing* 6:219–34.

——— 2001. Monoidal t-norm based logic: Towards a logic for left-continuous t-norms. *Fuzzy Sets Syst* 123 (3): 271–88.

Esteva, F., L. Godo, P. Hájek, and F. Montagna. 2003. Hoops and fuzzy logic. *J Logic Comput* 13:532–55.

Esteva, F., L. Godo, P. Hájek, and M. Navara. 2000. Residuated fuzzy logics with an involutive negation. *Arch Math Logic* 39:103–24.

Esteva, F., L. Godo, and F. Montagna. 2001. The ŁΠ and ŁΠ$\frac{1}{2}$ logics: Two complete fuzzy systems joining Łukasiewicz and product logics. *Arch Math Logic* 40 (1): 39–67.

————— 2004. Equational characterization of the subvarieties of BL generated by t-norm algebras. *Studia Logica* 76:161–200.

Esteva, F., L. Godo, and C. Noguera. 2006. On rational weak nilpotent minimum logics. *J Multiple-Valued Logic & Soft Computing* 12 (1–2): 9–32.

————— 2010. Expanding the propositional logic of a t-norm with truth-constants: Completeness results for rational semantics. *Soft Computing* 14 (3): 273–84.

————— 2013. A logical approach to fuzzy truth hedges. *Inf Sci* 232:366–85.

Evans, G., W. Karwowski, and M. Wilhelm, eds. 1989. *Applications of Fuzzy Methodologies in Industrial Engineering*. New York: Elsevier.

Evans, T., and P. B. Schwartz. 1958. On Słupecki T-functions. *J Symb Logic* 23 (3): 267–70

Eytan, M. 1981. Fuzzy sets: A topos-logical point of view. *Fuzzy Sets Syst* 5:47–67.

Fagin, R. 1998. Fuzzy queries in multimedia database systems. In *Proc 17th ACM Symp on Principles of Database Systems*, 1–10.

————— 1999. Combining fuzzy information from multiple systems. *J Comput Syst Sci* 58:83–99.

Fagin, R., R. Kumar, and D. Sivakumar. 2003. Efficient similarity search and classification via rank aggregation. In *Proc 2003 ACM Conf on Management of Data*, 301–12.

Fagin, R., A. Lotem, and M. Naor. 2003. Optimal aggregation algorithms for middleware. *J Comput Syst Sci* 66:614–56.

Feng D., and X. Liu, eds. 1985. *Fuzzy Mathematics in Earthquake Researches*, vol. 1. Beijing: Seismological Press.

————— eds. 1986. *Fuzzy Mathematics in Earthquake Researches*, vol. 2. Beijing: Seismological Press.

Fenstad, J. E. 1964. On the consistency of the axiom of comprehension in the Łukasiewicz infinite valued logic. *Mathematica Scandinavica* 14:65–74.

Fermüller, C. G. 2003. Theories of vagueness versus fuzzy logic: Can logicians learn from philosophers? *Neural Network World* 13:455–65.

————— 2008. Dialogue games for many-valued logics—an overview. *Studia Logica* 90:43–68.

Fermüller, C. G., and O. Majer. 2013. On semantic games for Łukasiewicz logic, preprint.

Fermüller, C. G., and G. Metcalfe. 2009. Giles's game and the proof theory of Łukasiewicz logic. *Studia Logica* 92:27–61.

Féron, R. 1976. Ensembles aléatoires flous [Random fuzzy sets]. *Comptes Rendus de l'Académie des Sciences* 282:903–6.

————— 1979. Ensembles aléatoires flous dont la fonction d'appartenance prend ses valeurs dans un treillis distributif fermé [Random fuzzy sets whose membership function takes its values in a closed distributive lattice]. *Publications Économétriques* 12:81–118.

Fiedler, M., J. Nedoma, J. Ramík, J. Rohn, and K. Zimmermann. 2006. *Linear Optimization Problems with Inexact Data*. New York: Springer.

Field, H. 1973. Theory change and the indeterminacy of reference. *J Philosophy* 70:462–81.

Fine, K. 1975. Vagueness, truth and logic. *Synthese* 30:265–300.

Fisch, M., and A. Turquette. 1966. Peirce's triadic logic. *Trans Charles S. Peirce Society* 2 (2): 71–85.

Fitting, M. 1991. Many-valued modal logics. *Fund Inform* 15 (3–4): 235–54.

————— 1992. Many-valued modal logics II. *Fund Inform* 17:55–73.

Flondor, P., G. Georgescu, and A. Iorgulescu. 2001. Pseudo t-norms and pseudo-BL-algebras. *Soft Computing* 5 (5): 355–71.

Fodor, J. 1991. On fuzzy implication operators. *Fuzzy Sets Syst* 42:293–300.

————— 1995. Contrapositive symmetry of fuzzy implications. *Fuzzy Sets Syst* 69:141–56.

Fodor, J., and M. Roubens. 1994. *Fuzzy Preference Modelling and Multicriteria Decision Support*. Dordrecht: Kluwer.

Fodor, J., and R. R. Yager. 2000. Fuzzy set-theoretic operators and quantifiers. In Dubois and Prade 2000, 125–93.

Fonck, P., J. Fodor, and M. Roubens. 1998. An application of aggregation procedures to the definition of measures of similarity between fuzzy sets. *Fuzzy Sets Syst* 97:67–74.

Font, J. M., and R. Jansana. 1996. *A General Algebraic Semantics for Sentential Logics*. Berlin: Springer.

Font, J. M., A. J. Rodríguez, and A. Torrens. 1984. Wajsberg algebras. *Stochastica* 8:5–31.

Formato, F., G. Gerla, and L. Scarpati. 1999. Fuzzy subgroups and similarities. *Soft Computing* 3:1–6.

Fortet, R., and M. Kambouzia. 1976. Ensembles aléatoires et ensembles flous [Random sets and fuzzy sets]. *Publications Économétriques* 9 (1): 1–23.

Foster, D. H. 1979. Fuzzy topological groups. *J Math Anal Appl* 67:549–64.

Foulloy, L., and S. Galichet. 1995. Typology of fuzzy controllers. In *Theoretical Aspects of Fuzzy Control*, ed. H. T. Nguyen, M. Sugeno, R. Tong, and R. R. Yager, 65–90. New York: Wiley.

Fourman, M. P., and D. S. Scott. 1979. Sheaves and logic. *Lect Notes Math* 753: 302–401.

Fowler, P. W. 1992. Vocabulary for fuzzy symmetry. *Nature* 360 (6405): 626.

Frank, M. J. 1979. On the simultaneous associativity of $F(x,y)$ and $x + y − F(x,y)$. *Aeq Math* 19:194–226.

Frege, G. 1879. *Begriffsschrift, eine der arithmetischen nachgebildete Formelsprache des reinen Denkens.* Halle: Nebert. Trans. S. Bauer-Mengelberg 1967. *Begriffsschrift*, a formula language, modeled upon that of arithmetic, for pure thought. In van Heijenoort 1967, 1–82.

——— 1891. *Funktion und Begriff.* Jena: Hermann Pohle. Trans. P. Geach 1997. Function and concept. In Beaney 1997, 130–148.

——— 1892. Über Sinn und Bedeutung. *Zeitschrift für Philosophie und philosophische Kritik, Neue Folge* 100 (1): 25–50. Trans. M. Black 1997. On *Sinn* and *Bedeutung*. In Beaney 1997, 151–71.

Friedman, H. 1973. Some applications of Kleene's methods for intuitionistic systems. *Lect Notes Math* 337:113–70.

Frink Jr., O. 1938. New algebras of logic. *American Mathematical Monthly* 45 (4): 210–19.

Fuhrmann, G. 1988a. Prototypes and "fuzziness" in the logic of concepts. *Synthese* 75:317–47.

——— 1988b. Fuzziness of concepts and concepts of fuzziness. *Synthese* 75:349–72.

——— 1989. M-fuzziness in brain/mind modelling. *Cybernetica* 32:355–90.

——— 1990. Note on the generality of fuzzy sets. *Inf Sci* 51:143–52.

——— 1991. Note on the integration of prototype theory and fuzzy-set theory. *Synthese* 86:1–27.

Fullér, R., and T. Keresztfalvi. 1991. On generalization of Nguyen's theorem. *Fuzzy Sets Syst* 41:371–74.

Fung, L. W., and K. S. Fu. 1975. An axiomatic approach to rational decision making in a fuzzy environment. In Zadeh et al. 1975, 227–56.

Gabbay, D. M. 1999. *Fibring Logics*. Oxford, UK: Clarendon Press.

Gaines, B. R. 1976. Foundations of fuzzy reasoning. *Int J Man Mach Stud* 8:623–68.

——— 1978. Fuzzy and probability uncertainty logics. *Inf Control* 38:154–69.

Galatos, N., P. Jipsen, T. Kowalski, and H. Ono. 2007. *Residuated Lattices: An Algebraic Glimpse at Substructural Logics*. Amsterdam: Elsevier.

Galindo, J., ed. 2008. *Handbook of Research on Fuzzy Information Processing in Databases*. Hershey, PA: Information Science Reference.

Galindo, J., A. Urrutia, and M. Piattini. 2006. *Fuzzy Databases: Modeling, Design and Implementation*. Hersey, PA: Idea Group Publ.

Gantner, T. E., R. C. Steinlage, and R. H. Warren. 1978. Compactness in fuzzy topological spaces. *J Math Anal Appl* 62:547–62.

Gao, Y., X. Xia, Z. Wang, and Z. Tao. 2002. Detection of systematic errors in a measurement process using fuzzy set theory. *Review of Scientific Instruments* 73 (4): 1786–94.

Gärdenfors, P. 1975. Qualitative probability as an intensional logic. *J Philosophical Logic* 4:171–85.

——— 2000. *Conceptual Spaces: The Geometry of Thought*. Cambridge, MA: MIT Press.

Garmendia, L., R. González del Campo, V. López, and J. Recasens. 2009. An algorithm to compute the transitive closure, a transitive approximation and a transitive opening of a fuzzy proximity. *Mathware & Soft Computing* 16:175–91.

Garmendia, L., and J. Recasens. 2009. How to make T-transitive a proximity relation. *IEEE Trans Fuzzy Syst* 17:200–207.

Gebhardt, J., M. A. Gil, and R. Kruse. 1998. Fuzzy set-theoretic methods in statistics. In Słowiński 1998, 311–47.

Gentilhomme, Y. 1968. Les ensembles flous en linguistique [Fuzzy sets in linguistics]. *Cahiers de Linguistique Théorique et Appliquée* 5:47–63.

Georgescu, G., A. Iorgulescu, and S. Rudeanu. 2006. Grigore C. Moisil (1906–1973) and his school in algebraic logic. *Int J Computers, Communications & Control* 1 (1), 81–99.

Georgescu, G., and A. Popescu. 2004. Non-dual fuzzy connections. *Arch Math Logic* 43:1009–39.

Georgescu, I. 2012. *Possibility Theory and the Risk*. Berlin: Springer.

Gerla, G. 1982. Sharpness relation and decidable fuzzy sets. *IEEE Trans Automat Contr* 27 (5): 1113.

——— 1985. Pavelka's fuzzy logic and free L-subsemigroups. *Z Math Logik Grundlagen Math* 31:123–29.

——— 1987. Decidability, partial decidability and sharpness relation for *L*-subsets. *Studia Logica* 46:227–38.

——— 1989. Turing *L*-machines and recursive computability for *L*-maps. *Studia Logica* 48:179–92.

——— 1994a. An extension principle for fuzzy logics. *Math Logic Quart* 40:357–80.

——— 1994b. Comparing fuzzy and crisp deduction systems. *Fuzzy Sets Syst* 67:317–28.

——— 1994c. Inferences in probability logic. *Artificial Intelligence* 70:33–52.

——— 1996. Graded consequence relations and fuzzy closure operator. *J Applied Non-Classical Logics* 6:369–79.

——— 1997. Probability-like functionals and fuzzy logic. *J Math Anal Appl* 216:438–65.

——— 2000. A note on functions associated with Gödel formulas. *Soft Computing* 4 (4): 206–9.

———— 2001. *Fuzzy Logic: Mathematical Tools for Approximate Reasoning*. Dordrecht: Kluwer.

———— 2006. Effectiveness and multivalued logics. *J Symb Logic* 71(1):137–62.

———— 2008. Approximate similarities and Poincaré paradox. *Notre Dame J Formal Logic* 49:203–26.

———— 2010. Why I have an extra-terrestrial ancestor. In *Percorsi incrociati (in ricordo di Vittorio Cafagna)*, 179–91. Università degli Studi di Salerno: Collana Scientifica di Salerno.

Gheorghe, A. V., and R. Mock. 1999. *Risk Engineering: Bringing Risk Analysis with Stakeholders Values*. Dordrecht: Kluwer.

Ghosh, D., and D. Chakraborty. 2014. Analytical fuzzy plane geometry II. *Fuzzy Sets Syst* 243:84–109.

Giardina, C. 1979. *Fuzzy Databases and Fuzzy Relational Associative Processors*. Technical report. Hoboken, NJ: Stevens Inst. Technology.

Gil, M. A., M. López-Díaz, and D. A. Ralescu. 2006. Overview on the development of fuzzy random variables. *Fuzzy Sets Syst* 157:2546–57.

Gil-Lafuente, A. M. 2005. *Fuzzy Logic in Financial Analysis*. Berlin: Springer.

Giles, R. 1974. A non-classical logic for physics. *Studia Logica* 33:397–415.

———— 1976. Łukasiewicz logic and fuzzy set theory. *Int J Man Mach Stud* 8:313–27.

———— 1982. Semantics for fuzzy reasoning. *Int J Man Mach Stud* 17:401–15.

Gillies, D., ed. 1992. *Revolutions in Mathematics*. New York: Oxford Univ. Press.

Girard, J.-Y. 1987. Linear logic. *Theor Comput Sci* 50 (1): 1–102.

Gispert, J., and A. Torrens. 2005. Axiomatic extensions of IMT₃ logic. *Studia Logica* 81 (3): 311–24.

Glöckner, I. 2006. *Fuzzy Quantifiers: A Computational Theory*. Berlin: Springer.

Glykas, M., ed. 2010. *Fuzzy Cognitive Maps: Advances in Theory, Methodologies, Tools and Applications*. Berlin: Springer.

Gödel, K. 1930. Die Vollständigkeit der Axiome des logischen Funktionenkalküls. *Monatshefte für Mathematik und Physik* 37:349–60. Trans. S. Bauer-Mengelberg 1967. The completeness of the axioms of the functional calculus of logic. In van Heijenoort 1967, 582–91.

———— 1932. Zum intuitionistischen Aussagenkalkül [On the intuitionistic propositional calculus]. *Anzeiger der Akademie der Wissenschaften in Wien, Mathematisch-naturwissenschaftliche Klasse* 69:65–66.

Godo, L., and P. Hájek. 1999. Fuzzy inference as deduction. *J Applied Non-Classical Logics* 9 (1): 37–60.

Godo, L., and R. O. Rodríguez. 2002. Graded similarity-based semantics for nonmonotonic inferences. *Ann Mathematics and Artificial Intelligence* 34:89–105.

Goetschel Jr., R., and W. Voxman. 1981. A pseudometric for fuzzy sets and certain related results. *J Math Anal Appl* 81:507–23.

Goguen, J. A. 1967. L-fuzzy sets. *J Math Anal Appl* 18:145–74.

———— 1968. *Categories of Fuzzy Sets: Applications of Non-Cantorian Set Theory*. PhD diss., Univ. California, Berkeley.

———— 1968–69. The logic of inexact concepts. *Synthese* 19 (3–4): 325–73.

———— 1969. Categories of V-sets. *Bull AMS* 75:622–24.

———— 1973. The fuzzy Tychonoff theorem. *J Math Anal Appl* 43:734–42.

———— 1974. Concept respresentation in natural and artificial languages: Axioms, extensions and applications for fuzzy sets. *Int J Man Mach Stud* 6:513–61.

———— 1977. Complexity of hierarchically organized systems and the structure of musical experiences. *Int J Gen Syst* 3 (4): 233–51.

Goodman, I. R. 1980. Identification of fuzzy sets with a class of canonically induced random sets. In *Proc 19th IEEE Conf on Decision and Control*, 352–57. Albuquerque, NM: IEEE Press.

———— 1998. Random sets and fuzzy sets: A special connection. In *Proc FUSION 98*, 93–100.

Gottwald, S. 1976. A cumulative system of fuzzy sets. *Lect Notes Math* 537:109–19.

———— 1976–77. Untersuchungen zur mehrwertigen Mengenlehre. I, II, III [Investigations in many-valued set theory. I, II, III]. *Mathematische Nachrichten* 72:297–303; 74:329–36; 79:207–17.

———— 1979. Set theory for fuzzy sets of higher levels. *Fuzzy Sets Syst* 2 (2): 125–51.

———— 1986. Characterizations of the solvability of fuzzy equations. *Elektronische Informationsverarbeitung Kybernetik* 22:67–91.

———— 1989. *Mehrwertige Logik: Eine Einführung in die Theorie und Anwendungen* [Many-valued logic: An introduction to theory and applications]. Berlin: Akademie-Verlag.

———— 1993. *Fuzzy Sets and Fuzzy Logic: Foundations of Applications—From a Mathematical Point of View*. Wiesbaden: Vieweg.

———— 1994. Approximately solving fuzzy relation equations: Some mathematical results and some heuristic proposals. *Fuzzy Sets Syst* 66 (2): 175–93.

———— 1995. Approximate solutions of fuzzy relational equations and a characterization of t-norms that define metrics for fuzzy sets. *Fuzzy Sets Syst* 75 (2): 189–201.

————1999. Axiomatizing t-norm based logic. In *Proc EUSFLAT-ESTYLF Joint Conf*, 303–6.

————2001. *A Treatise on Many-Valued Logics*. Baldock, UK: Research Studies Press.

————2006. Universes of fuzzy sets and axiomatization of fuzzy set theory, part I: Model-based and axiomatic approaches. *Studia Logica* 82:211–44.

————2008. Calculi of information granules: Fuzzy relational equations. In *Handbook of Granular Computing*, ed. W. Pedrycz, A. Skowron, and V. Kreinovich, 225–48. Hoboken, NJ: Wiley.

Gottwald, S., and S. Jenei. 2001. A new axiomatization for involutive monoidal t-norm-based logic. *Fuzzy Sets Syst* 124:303–7.

Grabisch, M. 2009. Belief functions on lattices. *Int J Intell Syst* 24 (1): 76–95.

Grabisch, M., J.-L. Marichal, R. Mesiar, and E. Pap. 2009. *Aggregation Functions*. Cambridge, UK: Cambridge Univ. Press.

Grattan-Guinness, I. 1976. Fuzzy membership mapped onto intervals and many-valued quantities. *Z Math Logik Grundlagen Math* 22:149–60.

Grayson, R. J. 1979. Heyting-valued models for intuitionistic set theory. *Lect Notes Math* 753:402–14.

Grigolia, R. S. 1973. Алгебраический анализ *n*-значных логических систем Лукасевича–Тарского [Algebraic analysis of Łukasiewicz-Tarski *n*-valued logical systems]. *Trudy Tbilisskogo Gosudarstvennogo Universiteta* A6-7:121–32.

————1977. Algebraic analysis of Łukasiewicz-Tarski's *n*-valued logical systems. In *Selected Papers on Łukasiewicz Sentential Calculi*, ed. R. Wójcicki and G. Malinowski, 81–92. Wrocław: Ossolineum.

Grint, K. 1997. *Fuzzy Management: Contemporary Ideas and Practices at Work*. New York: Oxford Univ. Press.

Grzegorzewski, P., and E. Mrówka. 2005. Some notes on (Atanassov's) intuitionistic fuzzy sets. *Fuzzy Sets Syst* 156 (3): 492–95.

Gunaratne, M., J.-L. Chameau, and A. G. Altschaeffl. 1988. A successive fuzzification technique and its application to pavement evaluation. *Civil Eng Syst* 5 (2): 77–80.

Gupta, K. C., and S. Ray. 1993. Fuzzy plane projective geometry. *Fuzzy Sets Syst* 54:191–206.

Haack, S. 1979. Do we need fuzzy logic? *Int J Man Mach Stud* 11:437–45.

————1996. *Deviant Logic, Fuzzy Logic: Beyond the Formalism*. Chicago: Univ. Chicago Press.

Hähnle, R. 1994. *Automated Deduction in Multiple-Valued Logics*. Oxford, UK: Oxford Univ. Press.

Hailperin, T. 1996. *Sentential Probability Logic: Origins, Development, Current Status, and Technical Applications*. Bethlehem, PA: Lehigh Univ. Press.

Hájek, P. 1994. On logics of approximate reasoning. *Lect Notes Comput Sci* 808:17–29.

————1995a. Fuzzy logic and arithmetical hierarchy. *Fuzzy Sets Syst* 73:359–63.

————1995b. Possibilistic logic as interpretability logic. *Lect Notes Comput Sci* 945:273–80.

————1997. Fuzzy logic and arithmetical hierarchy II. *Studia Logica* 58:129–41.

————1998a. Basic fuzzy logic and BL-algebras. *Soft Computing* 2 (3): 124–28.

————1998b. *Metamathematics of Fuzzy Logic*. Dordrecht: Kluwer.

————1999. Ten questions and one problem on fuzzy logic. *Ann Pure Appl Logic* 96:157–65.

————2000. Function symbols in fuzzy predicate logic. In *Proc East-West Fuzzy Logic Days*, 2–8. Zittau–Görlitz: IPM.

————2001a. Fuzzy logic and arithmetical hierarchy III. *Studia Logica* 68 (1): 129–42.

————2001b. On very true. *Fuzzy Sets Syst* 124:329–33.

————2002. Observations on the monoidal t-norm logic. *Fuzzy Sets Syst* 132 (1): 107–12.

————2003a. Fuzzy logics with noncommutative conjunctions. *J Logic Comput* 13 (4): 469–79.

————2003b. Observations on non-commutative fuzzy logic. *Soft Computing* 8 (1): 38–43.

————2004a. A true unprovable formula of fuzzy predicate logic. *Lect Notes Comput Sci* 3075:1–5.

————2004b. Fuzzy logic and arithmetical hierarchy IV. In *First-Order Logic Revised*, ed. V. F. Hendricks, F. Neuhaus, S. A. Pedersen, U. Scheffler, and H. Wansing, 107–15. Berlin: Logos Verlag.

————2005. On arithmetic in the Cantor-Łukasiewicz fuzzy set theory. *Arch Math Logic* 44:763–82.

————2006a. Mathematical fuzzy logic—what can it learn from Mostowski and Rasiowa. *Studia Logica* 84:51–62.

————2006b. What is mathematical fuzzy logic. *Fuzzy Sets Syst* 157:597–603.

————2007a. On witnessed models in fuzzy logic. *Math Logic Quart* 53:66–77.

————2007b. On witnessed models in fuzzy logic II. *Math Logic Quart* 53:610–15.

————2009. On vagueness, truth values and fuzzy logics. *Studia Logica* 91:367–82

————2010. On witnessed models in fuzzy logic III—witnessed Gödel logics. *Math Logic Quart* 56:171–74.

————2013a. On equality and natural numbers in Cantor-Łukasiewicz set theory. *Logic J IGPL* 21:91–100.

————2013b. Some remarks on Cantor-Łukasiewicz fuzzy set theory. *Logic J IGPL* 21:183–86.

————2013c. Towards metamathematics of weak arithmetics over fuzzy logic. *Logic J IGPL* 19:467–75.

Hájek, P., and P. Cintula. 2006. On theories and models in fuzzy predicate logics. *J Symb Logic* 71 (3): 863–80.

Hájek, P., and C. G. Fermüller. 2011. A conversation about fuzzy logic and vagueness. In Cintula et al. 2011, 405–416.

Hájek, P., L. Godo, and F. Esteva. 1995. Fuzzy logic and probability. In *Proc 11th Conf on Uncertainty in Artificial Intelligence*, 237–44. San Francisco, CA: Morgan Kaufmann.

———— 1996. A complete many-valued logic with product conjunction. *Arch Math Logic* 35 (3): 191–208.

Hájek, P., and Z. Haniková. 2001. A set theory within fuzzy logic. In *Proc 31st IEEE Int Symp on Multiple-Valued Logic*, 319–23. Los Alamitos, CA: IEEE Computer Society Press.

———— 2003. A development of set theory in fuzzy logic. In *Beyond Two: Theory and Applications of Multiple-Valued Logic*, ed. M. Fitting and E. Orłowska, 273–85. Heidelberg: Physica-Verlag.

———— 2013. Interpreting lattice-valued set theory in fuzzy set theory. *Logic J IGPL* 21:77–90.

Hájek, P., and D. Harmancová. 1996. A many-valued modal logic. In *Proc 6th Conf on Information Processing and Management of Uncertainty in Knowledge-Based Systems*, 1021–24.

Hájek, P., D. Harmancová, F. Esteva, P. Garcia, and L. Godo. 1994. On modal logics for qualitative possibility in a fuzzy setting. In *Proc 10th Conf on Uncertainty in Artificial Intelligence*, 278–85. San Francisco, CA: Morgan Kaufmann.

Hájek, P., D. Harmancová, and R. Verbrugge. 1995. A qualitative fuzzy possibilistic logic. *Int J Approx Reason* 12:1–19.

Hájek, P., and T. Havránek. 1978. *Mechanizing Hypotheses Formation: Mathematical Foundations for a General Theory*. Berlin: Springer.

Hájek, P., F. Montagna, and C. Noguera. 2011. Arithmetical complexity of first-order fuzzy logics. In Cintula, Hájek, and Noguera 2011, 853–908.

Hájek, P., and V. Novák. 2003. The sorites paradox and fuzzy logic. *Int J Gen Syst* 32:373–83.

Hájek, P., and J. Paris. 1997. A dialogue on fuzzy logic. *Soft Computing* 1:3–5.

Hájek, P., J. Paris, and J. Shepherdson. 2000a. Rational Pavelka predicate logic is a conservative extension of Łukasiewicz predicate logic. *J Symb Logic* 65 (2): 669–82.

———— 2000b. The liar paradox and fuzzy logic. *J Symb Logic* 65 (1): 339–46.

Halpern, J. 2003. *Reasoning about Uncertainty*. Cambridge, MA: MIT Press.

Haluška, J. 1997. Uncertanty and tuning in music. *Tatra Mt Math Publ* 12:113–29.

———— 2000. Equal temperament and Pythagorean tuning: A geometrical interpretation in the plane. *Fuzzy Sets Syst* 114 (2): 261–69.

———— 2002. Uncertainty measures of well-tempered systems. *Int J Gen Syst* 31 (1): 73–96.

———— 2004. *The Mathematical Theory of Tone Systems*. New York: Marcel Dekker.

Hamacher, H. 1978. Über logische Verknüpfungen unscharfer Aussagen und deren zugehörige Bewertungsfunktionen [On logical connections of unsharp propositions and their evaluation functions]. In *Progress in Cybernetics and Systems Research*, vol. 3, ed. R. Trappl, G. J. Klir, and L. Ricciardi, 276–88. Washington, DC: Hemisphere.

Hamblin, C. L. 1959. The modal "probably." *Mind* 68 (270): 234–40.

Haniková, Z. 2011. Computational complexity of propositional fuzzy logics. In Cintula, Hájek, and Noguera 2011, 793–851.

———— 2014. Varieties generated by standard BL-algebras. *Order* 31:15–33.

Hansen, E. 1992. *Global Optimization Using Interval Analysis*. New York: Marcel Dekker.

Hanss, M. 2005. *Applied Fuzzy Arithmetic: An Introduction with Engineering Applications*. Berlin: Springer.

Harkleroad, L. 1984. Fuzzy recursion, RET's and isols. *Z Math Logik Grundlagen Math* 30:425–36.

Harmanec, D., and P. Hájek. 1994. A qualitative belief logic. *Int J UFKBS* 2:227–36.

Hartley, R. V. L. 1928. Transmission of information. *Bell System Technical J* 7 (3): 535–63.

Hay, L. S. 1959. *An Axiomatization of the Infinitely Many-Valued Predicate Calculus*. MS thesis, Cornell Univ., Ithaca, NY.

———— 1963. Axiomatization of the infinite-valued predicate calculus. *J Symb Logic* 28 (1): 77–86.

Head, T. 1995. A metatheorem for deriving fuzzy theorems from crisp versions. *Fuzzy Sets Syst* 73:349–58.

Heilpern, S. 1981. Fuzzy mappings and fixed point theorem. *J Math Anal Appl* 83:566–69.

Hekrdla, J., E. P. Klement, and M. Navara. 2003. Two approaches to fuzzy propositional logics. *J Multiple-Valued Logic & Soft Computing* 9:343–60.

Hempel, C. G. 1939. Vagueness and logic. *Philos Sci* 6 (2): 163–80.

Henkin, L. 1949. The completeness of the first-order functional calculus. *J Symb Logic* 14:159–66.

Hersh, H. M., and A. Caramazza. 1976. A fuzzy set approach to modifiers and vagueness in natural language. *J Experimental Psychology: General* 105 (3): 254–76.

Heyting, A. 1930. Die formalen Regeln der intuitionistischen Logik. *Sitzungsberichte der Preussischen Akademie der Wissenschaften. Physikalisch-mathematische Klasse* 42–56. Trans. The formal rules of intuitionistic logic. In Mancosu 1998, 311–27.

———— 1956. *Intuitionism: An Introduction*. Amsterdam: North-Holland.

Higashi, M., and G. J. Klir. 1984. Resolution of finite fuzzy relational equations. *Fuzzy Sets Syst* 13 (1): 65–82.

Hilbert, D., and W. Ackermann. 1928. *Grundzüge der theoretischen Logik*. Berlin: Springer. Trans. M. Hammond
    M., G. G. Leckie, and F. Steinhardt 1950. *Principles of Mathematical Logic*. Providence, RI: American
    Mathematical Society.

Hirota, K., ed. 1993. *Industrial Applications of Fuzzy Technology*. Tokyo: Springer.

Hirota, K., and M. Sugeno, eds. 1995. *Industrial Applications of Fuzzy Technology in the World*. Singapore: World
    Scientific.

Hisdal, E. 1981. The IF THEN ELSE statement and interval-valued fuzzy sets of higher type. *Int J Man Mach Stud*
    15:385–455.

Hoffman, L. J., and L. A. Neitzel. 1981. Inexact analysis of risk. *Computer Security J* 1 (1): 61–72.

Höhle, U. 1976. Maße auf unscharfen Mengen [Measures on unsharp sets]. *Zeitschrift für Wahrscheinlichkeitsthe-
    orie und verwandte Gebiete* 36:179–88.

———1978a. Probabilistische Metriken auf der Menge der nicht negativen Verteilungsfunktionen [Probabilistic
    metrics on the set of non-negative distributive functions]. *Aeq Math* 18:345–56.

———1978b. Probabilistische Topologien [Probabilistic topologies]. *Manuscripta Mathematica* 26:223–45.

———1981. Representation theorems for *L*-fuzzy quantities. *Fuzzy Sets Syst* 5:83–107.

———1983. Fuzzy measures as extensions of stochastic measures. *J Math Anal Appl* 92:372–80.

———1987. Fuzzy real numbers as Dedekind cuts with respect to a multiple-valued logic. *Fuzzy Sets Syst* 24 (3):
    263–78.

———1991. Monoidal closed categories, weak topoi and generalized logics. *Fuzzy Sets Syst* 42:15–35.

———1992. M-valued sets and sheaves over integral commutative CL-monoids. In *Applications of Category The-
    ory to Fuzzy Subsets*, ed. S. E. Rodabaugh, E. P. Klement, and U. Höhle, 33–72. Dordrecht: Kluwer.

———1994. Monoidal logics. In *Fuzzy Systems in Computer Science*, ed. R. Kruse and J., Gebhardt and Palm R.,
    233–43. Braunschweig: Vieweg.

———1995a. Commutative, residuated l-monoids. In Höhle and Klement 1995, 53–106.

———1995b. Presheaves over GL-monoids. In Höhle and Klement 1995, 127–57.

———1996. On the fundamentals of fuzzy set theory. *J Math Anal Appl* 201:786–826.

———1998. Many-valued equalities, singletons and fuzzy partitions. *Soft Computing* 2:134–40.

———2000. Classification of subsheaves over GL-algebras. *Lect Notes Logic* 13: 238–61.

———2001. *Many Valued Topology and Its Applications*. Boston: Kluwer.

———2003. Many valued topologies and Borel probability measures. In Rodabaugh and Klement 2003, 115–35.

———2005. Many-valued equalities and their representations. In Klement and Mesiar 2005, 301–19.

———2007. Fuzzy sets and sheaves. Part I: Basic concepts; part II: Sheaf-theoretic foundations of fuzzy set theory
    with applications to algebra and topology. *Fuzzy Sets Syst* 158:1143–74; 1175–1212.

Höhle, U., and N. Blanchard. 1985. Partial ordering in *L*-undeterminate sets. *Inf Sci* 35:133–44.

Höhle, U., and E. P. Klement. 1984. Plausibility measures: A general framework for possibility and fuzzy proba-
    bility measures. In Skala, Termini, and Trillas 1984, 31–50.

———1995. *Non-classical Logics and Their Applications to Fuzzy Subsets*. Dordrecht: Kluwer.

Höhle, U., and S. E. Rodabaugh, eds. 1999. *Mathematics of Fuzzy Sets*. Boston: Kluwer.

Höhle, U., and A. P. Šostak. 1999. Axiomatic foundations of fixed-basis fuzzy topology. In Höhle and Rodabaugh
    1999, 123–272.

Höhle, U., and L. N. Stout. 1991. Foundations of fuzzy sets. *Fuzzy Sets Syst* 40:257–96.

Höhle, U., and S. Weber. 1997. Uncertainty measures, realizations, and entropies. In *Random Sets: Theory and
    Applications*, ed. J. Goutsias, R. P. S. Mahler, and H. T. Nguyen, 259–95. New York: Springer.

Holmblad, L. P., and J.-J. Østergaard. 1982. Control of a cement kiln by fuzzy logic. In *Fuzzy Information and
    Decision Processes*, ed. M. M. Gupta and E. Sanchez, 389–99. Amsterdam: North-Holland,

Höppner, F., F. Klawonn, R. Kruse, and T. Runkler. 1999. *Fuzzy Cluster Analysis: Methods for Classification,
    Data Analysis and Image Recognition*. Chichester, UK: Wiley.

Horčík, R. 2005. Standard completeness theorem for ΠMTL. *Arch Math Logic* 44 (4): 413–24.

———2006. Decidability of cancellative extension of monoidal t-norm based logic. *Logic J IGPL* 14 (6): 827–43.

———2007. On the failure of standard completeness in ΠMTL for infinite theories. *Fuzzy Sets Syst* 158 (6):
    619–24.

Horčík, R., and P. Cintula. 2004. Product Łukasiewicz logic. *Arch Math Logic* 43 (4): 477–503.

Horčík, R., and K. Terui. 2011. Disjunction property and complexity of substructural logics. *Theor Comput Sci* 412
    (31): 3992–4006.

Horn, A. 1969. Logic with truth values in a linearly ordered Heyting algebra. *J Symb Logic* 34 (3): 395–408.

Huang, C. 1997. Principle of information diffusion. *Fuzzy Sets Syst* 91 (1): 69–90.

———2004. Fuzzy logic and earthquake research. In *Fuzzy Logic in Geology*, ed. R. V. Demicco and G. J. Klir,
    239–74. San Diego: Academic Press.

Huang, C., and Y. Shi. 2002. *Toward Efficient Fuzzy Information Processing: Using the Principle of Information Diffusion*. Heidelberg: Physica-Verlag.

Hughes, G. E., and M. J. Cresswell. 1996. *A New Introduction to Modal Logic*. New York: Routledge.

Hüllermeier, E. 1999. Numerical methods for fuzzy initial value problems. *Int J UFKBS* 7 (5): 439–61.

——— 2003. Possibilistic instance-based learning. *Artificial Intelligence* 148:335–83.

——— 2011. Fuzzy sets in machine learning and data mining. *Applied Soft Computing* 11:1493–1505.

Hüllermeier, E., J. Fürnkranz, W. Cheng, and K. Brinker. 2008. Label ranking by learning pairwise preferences. *Artificial Intelligence* 172:1897–1916.

Hutton, B. 1975. Normality in fuzzy topological spaces. *J Math Anal Appl* 50:74–79.

Ikejima, K., and M. Frangopol. 1987. Risk assessment for gas pipeline using fuzzy sets. *Civil Eng Syst* 4 (3): 147–52.

Iséki, K., and S. Tanaka. 1978. An introduction to the theory of BCK-algebras. *Mathematica Japonica* 23:1–26.

Jablonski, S. V. 1958. Функциональные построения в к-значной логике [Functional constructions in k-valued logic]. *Trudy Matematicheskogo Instituta imeni V. A. Steklova* 124:5–142.

Jacak, W. 1999. *Intelligent Robotic Systems: Design, Planning, and Control*. Dordrecht: Kluwer.

Jain, L. C., and T. Fukuda, eds. 1998. *Soft Computing for Intelligent Robotic Systems*. Heidelberg: Physica-Verlag.

Jain, R. 1976a. Outline of an approach for the analysis of fuzzy systems. *Int J Control* 23:627–40.

——— 1976b. Tolerance analysis using fuzzy sets. *Int J Systems Science* 7 (12): 1393–1401.

Jammer, M. 1974. *The Philosophy of Quantum Mechanics: The Interpretation of Quantum Mechanics in Historical Perspective*. New York: Wiley.

Jantzen, J. 2013. *Foundations of Fuzzy Control*. Substantially enlarged 2nd ed. Chichester, UK: Wiley.

Jaśkowski, S. 1936. Recherches sur le système de la logique intuitioniste. *Actualités scientifiques et industrielles* 393:58–61. Trans. S. McCall 1967. Investigations into the system of intuitionist logic. In McCall 1967, 259–63.

Jayaram, B., and R. Mesiar. 2009. *I*-fuzzy equivalence relations and *I*-fuzzy partitions. *Inf Sci* 179:1278–97.

Jaynes, E. T. 2003. *Probability Theory: The Logic of Science*. Cambridge, UK: Cambridge Univ. Press.

Jeansoulin, R., O. Papini, H. Prade, and S. Schockaert, eds. 2010. *Methods for Handling Imperfect Spatial Information*. Berlin: Springer.

Jech, T. 2003. *Set Theory: The Third Millennium Edition, Revised and Expanded*. Berlin: Springer.

Jenei, S., and F. Montagna. 2002. A proof of standard completeness for Esteva and Godo's logic MTL. *Studia Logica* 70 (2): 183–92.

——— 2003. A proof of standard completeness for non-commutative monoidal t-norm logic. *Neural Network World* 13 (5): 481–89.

Jin, Y., and L. Wang, eds. 2009. *Fuzzy Systems in Bioinformatics and Computational Biology*. Berlin: Springer.

John, R. 1998. Type 2 fuzzy sets: An appraisal of theory and applications. *Int J UFKBS* 6 (6): 563–76.

Juang, C. H., and D. J. Elton. 1986. Fuzzy logic for estimation of earthquake intensity based on building damage records. *Civil Eng Syst* 3 (4): 187–91.

Jungk, A., B. Thull, and G. Rau. 2002. Intelligent alarms for anesthesia monitoring based on fuzzy logic approach. In *Fuzzy Logic in Medicine*, ed. S. Barro and R. Marín, 113–38. Heidelberg: Physica-Verlag.

Kacprzyk, J. 1983. *Multistage Decision-Making under Fuzziness*. Cologne: Verlag TÜV Rheinland.

Kacprzyk, J., S. Zadrożny, and A. Ziolkowski. 1989. FQUERY III+: A 'human-consistent' database querying system based on fuzzy logic with linguistic quantifiers. *Information Systems* 14 (6): 443–53.

Kaleva, O. 1987. Fuzzy differential equations. *Fuzzy Sets Syst* 24:301–17.

——— 1990. The Cauchy problem for fuzzy differential equations. *Fuzzy Sets Syst* 35:389–96.

Kamp, J. A. W. 1975. Two theories about adjectives. In *Formal Semantics of Natural Language*, ed. E. L. Keenan, 123–55. Cambridge, UK: Cambridge Univ. Press.

Kandel, A., and G. Langholz, eds. 1998. *Fuzzy Hardware: Architectures and Applications*. Dordrecht: Kluwer.

Kandel, A., and S. C. Lee. 1979. *Fuzzy Switching and Automata: Theory and Applications*. New York: Crane, Russak.

Kaplan, A., and H. F. Schott. 1951. A calculus for empirical classes. *Methodos* 3:165–90.

Karnik, N. N., J. M. Mendel, and Q. Liang. 1999. Type-2 fuzzy logic systems. *IEEE Trans Fuzzy Syst* 7 (6): 643–58.

Karwowski, W., and A. Mital, eds. 1986. *Applications of Fuzzy Set Theory in Human Factors*. Amsterdam: Elsevier.

Katic, D., and M. Vukobratovic. 2003. *Intelligent Control of Robotic Systems*. Dordrecht: Kluwer.

Katsaras, A. K., and D. B. Liu. 1977. Fuzzy vector spaces and fuzzy topological vector spaces. *J Math Anal Appl* 58:135–46.

Kaufmann, A. 1975. *Fundamental Theoretical Elements*. Vol. 1 of *Introduction to the Theory of Fuzzy Subsets*. New York: Academic Press.

Kaufmann, A., and M. M. Gupta. 1985. *Introduction to Fuzzy Arithmetic: Theory and Applications*. New York: Van Nostrand Reinhold.

Keefe, R. 2000. *Theories of Vagueness*. Cambridge, UK: Cambridge Univ. Press.

Keefe, R., and P. Smith, eds. 1996. *Vagueness: A Reader*. Cambridge, MA: MIT Press.

Kerre, E. E., and M. De Cock. 1999. Linguistic modifiers: An overview. In *Fuzzy Logic and Soft Computing*, ed. G. Chen, M. Ying, and K. Y. Cai, 69–85. Dordrecht: Kluwer.

Kerre, E. E., and M. Nachtegael, eds. 2000. *Fuzzy Techniques in Image Processing*. Heidelberg: Physica-Verlag.

Khalili, S. 1979. Fuzzy measures and mappings. *J Math Anal Appl* 68:92–99.

Khalilia, M., J. C. Bezdek, M. Popescu, and J. M. Keller. 2014. Improvements to the relational fuzzy c-means clustering algorithm. *Pattern Recognition* 47:3920–30.

Kickert, W. J. M. 1978. *Fuzzy Theories on Decision-Making: A Critical Review*. Leiden: Martinus Nijhoff.

Kitainik, L. 1993. *Fuzzy Decision Procedures with Binary Relations: Towards a Unified Theory*. Dordrecht: Kluwer.

Klaua, D. 1965. Über einen Ansatz zur mehrwertigen Mengenlehre [On an approach to many-valued set theory]. *Monatsberichte der Deutschen Akademie der Wissenschaften zu Berlin* 7:859–67.

——— 1966. Über einen zweiten Ansatz zur mehrwertigen Mengenlehre [On a second approach to many-valued set theory]. *Monatsberichte der Deutschen Akademie der Wissenschaften zu Berlin* 8:161–77.

——— 1970. Stetige Gleichmächtigkeiten kontinuierlich-wertiger Mengen [Continuous equicardinalities of continuum-valued sets]. *Monatsberichte der Deutschen Akademie der Wissenschaften zu Berlin* 12:749–58.

Klawonn, F., and J. L. Castro. 1995. Similarity in fuzzy reasoning. *Mathware & Soft Computing* 2:197–228.

Klawonn, F., and R. Kruse. 1994. A Łukasiewicz logic based Prolog. *Mathware & Soft Computing* 1:5–29.

Klawonn, F., and V. Novák. 1996. The relation between inference and interpolation in the framework of fuzzy systems. *Fuzzy Sets Syst* 81 (3): 331–54.

Kleene, S. C. 1938. On a notation for ordinal numbers. *J Symb Logic* 3:150–55.

——— 1952. *Introduction to Metamathematics*. Amsterdam: North-Holland.

——— 1967. *Mathematical Logic*. New York: Wiley.

Klement, E. P. 1980a. Fuzzy σ-algebras and fuzzy measurable functions. *Fuzzy Sets Syst* 4:83–93.

——— 1980b. Characterization of finite fuzzy measures using Markoff-kernels. *J Math Anal Appl* 75:330–39.

——— 1981. Operations on fuzzy sets and fuzzy numbers related to triangular norms. In *Proc 11th IEEE Int Symp on Multiple-Valued Logic*, 218–25.

——— 1982a. Characterization of fuzzy measures constructed by means of triangular norms. *J Math Anal Appl* 86:345–58.

——— 1982b. Construction of fuzzy σ-algebras using triangular norms. *J Math Anal Appl* 85:543–65.

——— 1982c. Operations on fuzzy sets—an axiomatic approach. *Inf Sci* 27:221–32.

——— 1985. Integration of fuzzy-valued functions. *Revue Roumaine de Mathématique Pures et Appliquées* 30:375–84.

Klement, E. P., and R. Mesiar, eds. 2005. *Logical, Algebraic, Analytic, and Probabilistic Aspects of Triangular Norms*. Amsterdam: Elsevier.

Klement, E. P., R. Mesiar, and E. Pap. 2000. *Triangular Norms*. Dordrecht: Kluwer.

Klement, E. P., M. L. Puri, and D. A. Ralescu. 1986. Limit theorems for fuzzy random variables. *Proc Royal Society of London A* 407:171–82.

Klement, E. P., and S. Weber. 1999. Fundamentals of a generalized measure theory. In Höhle and Rodabaugh 1999, 633–51.

Klir, G. J. 1988. Methodological principles of uncertainty in inductive modelling: A new perspective. In *Maximum-Entropy and Bayesian Methods in Science and Engineering*, vol. 1, *Foundations*, ed. G. J. Erickson and C. R. Smith, 295–304. Dordrecht: Kluwer.

——— 1989. Is there more to uncertainty than some probability theorists might have us believe? *Int J Gen Syst* 15 (4): 347–78.

——— 1997a. Fuzzy arithmetic with requisite constraints. *Fuzzy Sets Syst* 91 (2): 165–75.

——— 1997b. The role of constrained fuzzy arithmetic in engineering. In *Uncertainty Analysis in Engineering and Sciences: Fuzzy Logic, Statistics, and Neural Network Approach*, ed. B. L. Ayyub and M. M. Gupta, 1–19. Boston: Kluwer.

——— 1999. On fuzzy-set interpretation of possibility theory. *Fuzzy Sets Syst* 108 (3): 263–73.

——— 2006. *Uncertainty and Information: Foundations of Generalized Information Theory*. Hoboken, NJ: Wiley.

Klir, G. J., and J. A. Cooper. 1996. On constrained fuzzy arithmetic. In *Proc 5th IEEE Int Conf on Fuzzy Systems*, 1285–90. Piscataway, NJ: IEEE Press.

Klir, G. J., and T. A. Folger. 1988. *Fuzzy Sets, Uncertainty, and Information*. Englewood Cliffs, NJ: Prentice Hall.

Klir, G. J., and D. Harmanec. 1994. On modal logic interpretation of possibility theory. *Int J UFKBS* 2(2): 237–45.

Klir, G. J., and Y. Pan. 1998. Constrained fuzzy arithmetic: Basic questions and some answers. *Soft Computing* 2 (2): 100–108.

Klir, G. J., and B. Yuan. 1994. Approximate solutions of systems of fuzzy relation equations. In *Proc 3rd IEEE Int Conf on Fuzzy Systems*, 1452–57. Orlando, FL: IEEE Press.

——— 1995. *Fuzzy Sets and Fuzzy Logic: Theory and Applications*. Upper Saddle River, NJ: Prentice Hall.

——— eds. 1996. *Fuzzy Sets, Fuzzy Logic, and Fuzzy Systems: Selected Papers by Lotfi A. Zadeh*. Singapore: World Scientific.

Kloeden, P. 1980. Compact supported endographs and fuzzy sets. *Fuzzy Sets Syst* 4:193–201.

——— 1982. Fuzzy dynamical systems. *Fuzzy Sets Syst* 7:275–96.

Kneale, W., and M. Kneale. 1962 *The Development of Logic*. Oxford, UK: Clarendon Press.

Kochen, M., and A. N. Badre. 1974. On the precision of adjectives which denote fuzzy sets. *J Cybernetics* 4:49–59.

Kolařík, M. 2013. Independence of the axiomatic system for MV-algebras. *Mathematica Slovaca* 63 (1): 1–4.

Komori, Y. 1981. Super Łukasiewicz propositional logics. *Nagoya Mathematical J* 84:119–33.

Körner, S. 1951. Ostensive predicates. *Mind* 60:80–89.

——— 1966. *Experience and Theory*. London: Routledge & Kegan Paul.

Kosko, B. 1986a. Fuzzy cognitive maps. *Int J Man Mach Stud* 24 (1): 65–75.

——— 1986b. Fuzzy entropy and conditioning. *Inf Sci* 40:165–74.

——— 1992a. *Neural Networks and Fuzzy Systems: A Dynamial Systems Approach to Machine Intelligence*. Englewood Cliffs, NJ: Prentice Hall.

——— 1992b. Fuzzy systems as universal approximators. In *Proc 1st IEEE Int Conf on Fuzzy Systems*, 1153–62. San Diego, CA: IEEE Press.

——— 1993. *Fuzzy Thinking*. New York: Hyperion.

Kraft, D. H., and D. A. Buell. 1983. Fuzzy sets and generalized Boolean retrieval systems. *Int J Man Mach Stud* 19:45–56.

Krajči, S. 2005. A generalized concept lattice. *Logic J IGPL* 13 (5): 543–50.

Kreinovich, V. 2005. Optimal finite characterization of linear problems with inexact data. *Reliable Computing* 11 (6): 479–89.

Kreinovich, V., M. Chiangpradit, and W. Panichkitkosolkul. 2012. Efficient algorithms for heavy-tail analysis under interval uncertainty. *Ann Operations Research* 195 (1): 73–96.

Kreinovich, V., A. Lakeyev, J. Rohn, and P. Kahl. 1997. *Computational Complexity and Feasibility of Data Processing and Interval Computations*. Dordrecht: Kluwer.

Kroupa, T. 2005. Conditional probability on MV-algebras. *Fuzzy Sets Syst* 149:369–81.

Krupka, M. 2010. An alternative version of the main theorem of fuzzy concept lattices. In *Cybernetics and Systems*, ed. R. Trappl, 9–14. Vienna: Austrian Society for Cybernetic Studies.

Kruse, R. 1982. The strong law of large numbers for fuzzy random variables. *Inf Sci* 28:233–41.

——— 1984. Statistical estimation with linguistic data. *Inf Sci* 33:197–207.

Kruse, R., C. Borgelt, F. Klawonn, C. Moewes, M. Steinbrecher, and P. Held. 2013. *Computational Intelligence: A Methodological Introduction*. London: Springer.

Kruse, R., J. Gebhardt, and F. Klawonn. 1994. *Foundations of Fuzzy Systems*. Chichester, UK: Wiley.

Kruse, R., and K. D. Meyer. 1987. *Statistics with Vague Data*. Dordrecht: Reidel.

Kubin, W. 1979. Eine Axiomatisierung der Mehrwertigen Logiken von Gödel [An axiomatization of many-valued logics of Gödel]. *Z Math Logik Grundlagen Math* 25:549–58.

Kubiński, T. 1958. Nazwy nieostre [Unsharp names]. *Studia Logica* 7:115–79.

——— 1960. An attempt to bring logic nearer to colloquial language. *Studia Logica* 10:61–75.

Kuhn, T. S. 1962. *The Structure of Scientific Revolutions*. Chicago: Univ. of Chicago Press.

Kühr, J. 2003. Pseudo-BL algebras and PRl-monoids. *Mathematica Bohemica* 128:199–208.

Kuijken, L., and H. Van Maldeghem. 2003. On the definition and some conjectures of fuzzy projective planes by Gupta and Ray, and a new definition of fuzzy building geometries. *Fuzzy Sets Syst* 138:667–85.

Kwakernaak, H. 1978. Fuzzy random variables—I: Definitions and theorems. *Inf Sci* 15:1–29.

——— 1979. Fuzzy random variables. Part II: Algorithms and examples for the discrete case. *Inf Sci* 17:253–78.

Labov, W. 1973. The boundaries of words and their meanings. In *New Ways of Analyzing Variation in English*, ed. C.-J. N. Bailey and R. W. Shuy, 340–73. Washington, D.C.: Georgetown Univ. Press.

Lacava, F. 1979. Alcune proprietà delle L-algebre e delle L-algebre esistenzialmente chiuse [Some properties of L-algebras and existentially closed L-algebras]. *Bolletino Unione Matematica Italiana* 16 (2): 360–66.

Lai, Y. J., and C. L. Hwang. 1992. *Fuzzy Mathematical Programming*. Berlin: Springer Verlag.

Lake, J. 1976. Sets, fuzzy sets, multisets and functions. *J London Mathematical Society*, 2nd ser., 12:323–26.

Lakoff, G. 1973. Hedges: A study in meaning criteria and the logic of fuzzy concepts. *J Philosophical Logic* 2 (4): 458–508.

Lakshmikantham, V., and R. N. Mohapatra. 2003. *Theory of Fuzzy Differential Equations and Inclusions*. London: Taylor and Francis.

Lang, J. 1991. *Logique possibiliste: Aspects formels, déduction automatique et applications* [Possibilistic logic: Formal aspects, automatic deduction, and applications]. PhD diss., Université Paul Sabatier, Toulouse, France.

———— 2000. Possibilistic logic: Complexity and algorithms. In *Handbook of Defeasible Reasoning and Uncertainty Management Systems*, vol. 5, ed. D. M. Gabbay and P. Smets, 179–220. Dordrecht: Kluwer.

Larkin, L. I. 1985. Fuzzy logic controller for aircraft flight control. In *Industrial Applications of Fuzzy Control*, ed. M. Sugeno, 87–103. Amsterdam: North-Holland.

Laviolette, M., and J. W. Seaman Jr. 1992. Evaluating fuzzy representations of uncertainty. *Mathematical Scientist* 17:26–41.

———— 1994. The efficacy of fuzzy representation of uncertainty. *IEEE Trans Fuzzy Syst* 2 (1): 4–15.

Laviolette, M., J. W. Seaman Jr., J. D. Barrett, and W. H. Woodall. 1995. Probabilistic and statistical view of fuzzy methods. *Technometrics* 37 (3): 249–61.

Lawry, J. 1998. A voting mechanism for fuzzy logic. *Int J Approx Reason* 19:315–33.

Lee, E. T., and L. A. Zadeh. 1969. Note on fuzzy languages. *Inf Sci* 1 (4): 421–34.

Lee, R. C. T. 1972. Fuzzy logic and the resolution principle. *J ACM* 19 (1): 109–19.

Lee, R. C. T., and C. L. Chang. 1971. Some properties of fuzzy logic. *Inf Control* 19 (5): 417–31.

Lehmke, S. 2004. Fun with automated proof search in basic propositional fuzzy logic. In *Abstracts of the Seventh International Conference FSTA*, ed. P. E. Klement, R. Mesiar, E. Drobná, and F. Chovanec, 78–80.

Lemaire, J. 1990. Fuzzy insurance. *ASTIN Bulletin* 20 (1): 33–55.

León, T., and V. Liern. 2012. Mathematics and soft computing in music. In *Soft Computing in Humanities and Social Sciences*, ed. R. Seising and V. Sanz, 451–65. Berlin: Springer.

Leung, Y. 1988. *Spatial Analysis and Planning under Imprecission*. New York: North-Holland.

———— 1997. *Intelligent Spatial Decision Support Systems*. Berlin: Springer.

Lewis, C. I., and C. H. Langford. 1932. *Symbolic Logic*. New York: Century Company.

Li, B. L., and E. J. Rykiel Jr., eds. 1996. Fuzzy Modeling in Ecology. Special issue, *Ecological Modeling* 90 (2): 109–85.

Li, Y. 2009. Lattice-valued fuzzy Turing machines: Computing power, universality and efficiency. *Fuzzy Sets Syst* 160:3453–74.

Li, Z., W. A. Halang, and G. Chen. 2006. *Integration of Fuzzy Logic and Chaos Theory*. Berlin: Springer.

Li Calzi, M. 1990. Towards a general setting for the fuzzy mathematics of finance. *Fuzzy Sets Syst* 35 (3): 265–80.

Liern, V. 2005. Fuzzy tuning systems: the mathematics of musicians. *Fuzzy Sets Syst* 150 (1): 35–52.

Lilly, J. H. 2010. *Fuzzy Control and Identification*. Hoboken, NJ: Wiley.

Lindley, D. V. 1987. The probability approach to the treatment of uncertainty in artificial inteligence and expert systems. *Statistical Science* 2 (1): 17–24.

Ling, C. H. 1965. Representation of associative functions. *Publicationes Mathematicae Debrecen* 12:189–212.

Liu, B. 2004. *Uncertainty Theory*. Berlin: Springer.

Liu Y. M., and M. K. Luo. 1997. *Fuzzy Topology*. Singapore: World Scientific.

Locke, J. 1690. *An Essay Concerning Human Understanding*. London: Basset.

Lodwig, W., ed. 2008. *Fuzzy Surfaces in GIS: Theory, Analytical Methods, Algorithms, and Applications*. Boca Raton, FL: CRC Press.

Lowen, R. 1976. Fuzzy topological spaces and fuzzy compactness. *J Math Anal Appl* 56:621–33.

———— 1977. Initial and final fuzzy topologies and the fuzzy Tychonoff theorem. *J Math Anal Appl* 58:11–21.

———— 1978. A comparison of different compactness notions in fuzzy topological spaces. *J Math Anal Appl* 64:446–54.

———— 1983. On $(\mathbb{R}(L), \oplus)$. *Fuzzy Sets Syst* 10:203–9.

———— 1985a. Metric spaces viewed as fuzzy topological spaces induced by Lipschitz functions. *Mathematische Nachrichten* 120:249–65.

———— 1985b. The order aspect of the fuzzy real line. *Manuscripta Mathematica* 49:293–309.

Luce, R. D. 1964. The mathematics used in mathematical psychology. *American Mathematical Monthly* 71 (4): 364–78.

———— 1989. R. Duncan Luce. In *A History of Psychology in Autobiography*, vol. 8, ed. G. Lindzey, 244–89. Stanford, CA: Stanford Univ. Press.

Luchins, A. S., and E. H. Luchins. 1965. *Logical Foundations of Mathematics for Behavioral Scientists*. New York: Holt, Rinehart, and Winston.

Łukasiewicz, J. 1906. Analiza i konstrukcja pojęcia przyczyny [Analysis and construction of the concept of cause]. *Przegląd Filozoficzny* 9:105–79.

———— 1910. *O zasadzie sprzeczności u Arystotelesa: studium krytyczne*. Kraków: Polska Akademia Umiejętności. Trans. V. Wedin 1971. On the principle of contradiction in Aristotle. *Review of Metaphysics* 24 (3):485–509. Originally published in a shorter version 1910. Über den Satz des Widerspruchs bei Aristoteles. *Bulletin international de l'Académie des sciences de Cracovie, Cl. d'Histoire et de Philosophie*.

——— 1913. *Die logische Grundlagen der Wahrscheinlichkeitsrechnung.* Krakow: Polska Akademia Umiejętności. Trans. O. Wojtasiewicz 1970. Logical foundations of probability. In Borkowski 1970, 16–63.

——— 1918. Treść wykładu pożegnalnego prof. Jana Łukasiewicza, wygłoszonego w auli Uniwersytetu Warszawskiego dnia 7-go marca 1918. *Pro Arte et Studio* 3:3–4. Trans. O. Wojtasiewicz 1970. Farewell lecture by Professor Jan Łukasiewicz, delivered in the Warsaw University Lecture Hall on March 7, 1918. In Borkowski 1970, 84–86.

——— 1920. O logice trójwartościowej. *Ruch Filozoficzny* 5:170–71. Trans. O. Wojtasiewicz 1970. On three-valued logic. In Borkowski 1970, 87–88.

——— 1923. Interpretacja liczbowa teorii zdań. *Ruch Filozoficzny* 7:92–93. Trans. O. Wojtasiewicz 1970. A numerical interpretation of the theory of propositions. In Borkowski 1970, 129–130.

——— 1929a. *Elementy Logiki Matematycznej.* Trans. O. Wojtasiewicz 1963. *Elements of Mathematical Logic.* Oxford, UK: Pergamon Press.

——— 1929b. O znaczeniu i potrzebach logiki matematycznej w Polsce [On the importance and needs of mathemtical logic in Poland]. *Nauka Polska* 10:604–20.

——— 1930. Philosophische Bemerkungen zu mehrwertigen Systemen des Aussagenkalkuüls. *Comptes rendus des Séances de la Societé des Sciences et des Lettres de Varsovie, Cl. iii* 23:51–77. Trans. H. Weber 1970. Philosphical remarks on many-valued systems of propositional logic. In Borkowski 1970, 153–78.

——— 1953. A system of modal logic. *J Computing Systems* 1:111–49.

——— 1961. O determinizmie. In *Z zagadnień logiki i filozofii,* ed. J. Słupecki, Warsaw. Trans. Z. Jordan 1970. On deteminism. In Borkowski 1970, 110–28.

——— 1994. Curriculum vitae of Jan Lukasiewicz. *Metalogicon* 7 (2): 133–37.

Łukasiewicz, J., and A. Tarski. 1930. Untersuchungen über den Aussagenkalkül. *Comptes rendus de Séances de la Société des Sciences et des Lettres de Varsovie, Cl. iii* 23:39–50. Trans. J. H. Woodger 1970. Investigations into the sentential calculus. In Borkowski 1970, 131–52.

Luo, L., F. Ji, and H. Li. 1995. Fuzzy classification of nucleotide sequences and bacterial evolution. *Bull Mathematical Biology* 57 (4): 527–37.

Mac Lane, S. 1986. *Mathematics: Form and Function.* New York: Springer.

MacColl, H. 1877–98. The calculus of equivalent statements. *Proc London Math Soc* 9 (1877–78): 177–86; 10 (1878–79): 16–28; 11 (1879–80): 113–21; 28 (1896–97): 156–83, 555–79; 29 (1897–98): 98–109.

——— 1906. *Symbolic Logic and Its Applications.* London: Longmans, Green.

Malinowski, G. 1993. *Many-Valued Logics.* Oxford, UK: Clarendon Press.

Mamdani, E. H. 1993. Twenty years of fuzzy control: Experiences gained and lessons learnt. In *Proc 2nd IEEE Int Conf on Fuzzy Systems,* 339–44. Piscataway, NJ: IEEE Press. Reprinted in Marks 1994, 19–24.

Mamdani, E. H., and S. Assilian. 1975. An experiment in linguistic synthesis with a fuzzy logic controller. *Int J Man Mach Stud* 7 (1): 1–13.

Mancosu, P., ed. 1998. *From Brouwer to Hilbert: The Debate on the Foundations of Mathematics in the 1920s.* New York: Oxford Univ. Press.

Mangani, P. 1973. Su certe algebre connesse con logiche a più valori [On some algebras associated with multiple logical values]. *Bolletino dell'Unione Mathematica Italiana* 8 (4): 68–78.

Mao, J., and A. K. Jain. 1996. A self-organizing network for hyperellipsoidal clustering (HEC). *IEEE Trans on Neural Networks* 7 (1): 16–29.

Mareš, M. 1977. How to handle fuzzy-quantities? *Kybernetika* 13:23–40.

——— 1989. Addition of fuzzy quantities: Disjunction-conjunction approach. *Kybernetika* 25:104–16.

——— 1994. *Computation over Fuzzy Quantities.* Boca Raton, FL: CRC Press.

——— 2001. *Fuzzy Cooperative Games: Cooperation with Vague Expectations.* Heidelberg: Physica-Verlag.

Margaliot, M., and G. Langholz. 2000. *New Approaches to Fuzzy Modeling and Control: Design and Analysis.* Singapore: World Scientific.

Margenau, H. 1934. On the application of many-valued systems of logic in physics. *Philos Sci* 1 (1): 118–21.

Mari, L. 1991. On the comparison of inexact measurement results. *Measurement* 9 (4): 157–62.

——— 1996. The meaning of "quantity" in measurement. *Measurement* 17 (2): 127–38.

——— 2000. Beyond the representational viewpoint: A new formalization of measurement. *Measurement* 27 (2): 71–84.

——— 2003. Epistemology of measurement. *Measurement* 34 (1): 17–30.

Marinos, P. N. 1966. *Fuzzy Logic.* Tech Memo 66-3344-1, Holmdel, NJ: Bell Telephone Labs.

——— 1969. Fuzzy logic and its application to switching systems. *IEEE Trans on Computers* 18 (4): 343–48.

Markovskii, A. V. 2005. On the relation between equations with max-product composition and the covering problem. *Fuzzy Sets Syst* 153:261–73.

Marks II, R. J., ed. 1994. *Fuzzy Logic Technology and Applications.* New York: IEEE Press.

Martin, O., and G. J. Klir. 2007. Defuzzification as a special way of dealing with retranslation. *Int J Gen Syst* 36 (6): 683–701.

Martinek, P. 2011. Completely lattice L-ordered sets with and without L-equality. *Fuzzy Sets Syst* 166:44–55.

Mashhour, A. S., and M. H. Ghanim. 1985. Fuzzy closure spaces. *J Math Anal Appl* 106:154–70.

Maydole, R. E. 1975. Paradoxes and many-valued set theory. *J Philosophical Logic* 4:269–91.

Mazurkiewicz, S. 1932. Zur Axiomatik der Wahrscheinlichkeitsrechnung [On axiomatics of probability calculus]. *Comptes rendus des Séances de la Societé des Sciences et des Lettres de Varsovie, Cl. iii* 25:1–4.

—— 1934. Über die Grundlagen der Wahrscheinlichkeitsrechnung I [On the foundations of probability calculus I]. *Monatshefte für Mathematik und Physik* 41:343–52.

McCall, S., ed. 1967. *Polish Logic 1920–1939.* London: Oxford Univ. Press.

McCloskey, M. E., and S. Glucksberg. 1978. Natural categories: Well defined or fuzzy sets? *Memory & Cognition* 6:462–72.

McKeon, R., ed. 1941. *The Basic Works of Aristotle.* New York: Random House.

McLeish, M., ed. 1988. Forum: An inquiry into computer understanding. Special issue, *Comput Intell (Canadian)* 4:55–142.

McNaughton, R. 1951. A theorem about infinite-valued sentential logic. *J Symb Logic* 16 (1): 1–13.

McNeill, D., and P. Freiberger. 1993. *Fuzzy Logic: The Discovery of a Revolutionary Computer Technology and How It Is Changing Our World.* New York: Simon & Schuster.

Mehlberg, H. 1958. *The Reach of Science.* Toronto: Univ. of Toronto Press.

Meier, A., and L. Donzé, eds. 2012. *Fuzzy Methods for Customer Relationship Management and Marketing: Applications and Classifications.* Hershey, PA: Business Science Reference.

Mendel, J. M. 2001. *Uncertain Rule-Based Fuzzy Logic Systems: Introduction and New Directions.* Upper Saddle River, NJ: Prentice Hall.

Mendelson, E. 1964. *Introduction to Mathematical Logic.* Princeton, NJ: Van Nostrand.

Menger, K. 1942. Statistical metrics. *Proc Natl Acad Sci USA* 28 (12): 535–37.

—— 1951a. Ensembles flous et fonctions aléatoires [Fuzzy sets and random functions]. *Comptes Rendus de l'Académie des Sciences* 232:2001–3.

—— 1951b. Probabilistic theories of relations. *Proc Natl Acad Sci USA* 37:178–80.

Menu, J., and J. Pavelka. 1976. A note on tensor products on the unit interval. *Commentationes Mathematicae Universitatis Carolinae* 17 (1): 71–83.

Meredith, C. A. 1958. The dependence of an axiom of Łukasiewicz. *Trans AMS* 87:54.

Mesiar, R. 1993a. Fundamental triangular norm based tribes and measures. *J Math Anal Appl* 177:633–40.

—— 1993b. Fuzzy observables. *J Math Anal Appl* 174:178–93.

Mesiar, R., and M. Navara. 1996. T_s-tribes and T_s-measures. *J Math Anal Appl* 201:91–102.

Mesiar, R., B. Reusch, and H. Thiele. 2006. Fuzzy equivalence relations and fuzzy partitions. *J Multiple-Valued Logic & Soft Computing* 12:167–81.

Metcalfe, G. 2011. Proof theory for mathematical fuzzy logic. In Cintula, Hájek, and Noguera 2011, 209–82.

Metcalfe, G., and F. Montagna. 2007. Substructural fuzzy logics. *J Symb Logic* 72 (3): 834–64.

Metcalfe, G., N. Olivetti, and D. Gabbay. 2004. Analytic proof calculi for product logics. *Arch Math Logic* 43 (7): 859–89.

—— 2005. Sequent and hypersequent calculi for Abelian and Łukasiewicz logics. *ACM Trans on Computational Logic* 6 (3): 578–613.

—— 2008. *Proof Theory for Fuzzy Logics.* Berlin: Springer.

Meystel, A. M., and J. S. Albus. 2002. *Intelligent Systems: Architecture, Design, and Control.* New York: Wiley.

Michálek, J. 1975. Fuzzy topologies. *Kybernetika* 11:345–54.

Michels, K., F. Klawonn, R. Kruse, and A. Nürnberger. 2006. *Fuzzy Control: Fundamentals, Stability and Design of Fuzzy Controllers.* Berlin: Springer.

Miki, T., H. Matsumoto, K. Ohto, and T. Yamakawa. 1993. Silicon implementation for a novel high-speed fuzzy inference engine: Mega-FLIPS analog fuzzy processor. *J Intelligent and Fuzzy Systems* 1 (1): 27–42.

Miyakoshi, M., and M. Shimbo. 1985. Solutions of composite fuzzy relational operations with triangular norms. *Fuzzy Sets Syst* 16:53–63.

Miyamoto, S., T. Miyake, and K. Nakayama. 1983. Generation of a pseudothesaurus for information retrieval based on cooccurrences and fuzzy set operations. *IEEE Trans Syst Man Cyb* 13:62–70.

Mizumoto, M., and K. Tanaka. 1976a. Algebraic properties of fuzzy numbers. In *Proc Int Conf on Cybernetics and Society,* 559–63. Washington, DC: IEEE Press.

—— 1976b. Some properties of fuzzy sets of type 2. *Inf Control* 31:312–40.

—— 1981. Fuzzy sets of type 2 under algebraic product and algebraic sum. *Fuzzy Sets Syst* 5 (3): 277–90.

Mizumoto, M., M. Umano, and K. Tanaka. 1977. Implementation of a fuzzy-set theoretic data structure system. In *Proc Int Conf on Very Large Databases,* 59–69.

Mjelde, K. M. 1986. Fuzzy resource allocation. *Fuzzy Sets Syst* 19 (3): 239–50.

Močkoř, J. 1999. Fuzzy and non-deterministic automata. *Soft Computing* 3:221–26.

Moh, S.-K. 1954. Logical paradoxes for many-valued systems. *J Symb Logic* 19:37–40.

Moisil, G. C. 1940 Recherches sur les logiques non-chrysippiennes [Investigations of non-Chrysippean logics]. *Annales Scientifiques de l'Université de Jassy* 26:431–66.

———— 1941. Notes sur les logiques non-chrysippiennes [Notes on non-Chrysippean logics]. *Annales Scientifiques de l'Université de Jassy* 27:86–98.

Montagna, F. 2000. An algebraic approach to propositional fuzzy logic. *J Language, Logic and Information* 9:91–124.

———— 2001. Three complexity problems in quantified fuzzy logic. *Studia Logica* 68 (1): 143–52.

———— 2005. On the predicate logics of continuous t-norm BL-algebras. *Arch Math Logic* 44 (1): 97–114.

———— (ed.) 2015. *Petr Hájek on Mathematical Fuzzy Logic*. Cham, Switzerland: Springer.

Montagna, F., and E. Corsi. 2014. Variants of Ulam game and game semantics for many-valued logics. In *Beyond True and False: Logic, Algebra and Topology*, 26–27. Florence, Italy.

Montagna, F., and H. Ono. 2002. Kripke semantics, undecidability and standard completeness for Esteva and Godo's logic MTL∀. *Studia Logica* 71 (2): 227–45.

Montagna, F., and L. Sacchetti. 2003. Kripke-style semantics for many-valued logic. *Math Logic Quart* 49:629–41.

———— 2004. Corrigendum to "Kripke-style semantics for many-valued logic." *Math Logic Quart* 50:104–7.

Moore, R. E. 1966. *Interval Analysis*. Englewood Cliffs, NJ: Prentice-Hall.

———— 1979. *Methods and Applications of Interval Analysis*. Philadelphia: Society for Industrial & Applied Mathematics.

Mordeson, J. N., and D. S. Malik. 1998. *Fuzzy Commutative Algebra*. Singapore: World Scientific.

———— 2002. *Fuzzy Automata and Languages: Theory and Applications*. Boca Raton, FL: Chapman and Hall/CRC.

Mordeson, J. N., and P. S. Nair. 2000. *Fuzzy Graphs and Fuzzy Hypergraphs*. Heidelberg: Physica-Verlag.

Mordeson, J. N., M. J. Wierman, T. D. Clark, A. Pham, and M. A. Redmond. 2013. *Linear Models in the Mathematics of Uncertainty*. Berlin: Springer.

Mostert, P. S., and A. L. Shields. 1957. On the structure of semigroups on a compact manifold with boundary. *Ann Math* 65:117–43.

Mostowski, A. 1957. On a generalization of quantifiers. *Fund Math* 44:12–36.

———— 1961a. Axiomatizability of some many valued predicate calculi. *Fund Math* 50:165–90.

———— 1961b. An example of a non-axiomatizable many valued logic. *Z Math Logik Grundlagen Math* 7:72–76.

Mukaidono, M., and H. Kikuchi. 1993. Foundations of fuzzy logic programming, In *Between Mind and Computer: Fuzzy Science and Engineering*, ed. P.-Z. Wang and K.-F. Loe, 225–44. Singapore: World Scientific.

Mukaidono, M., Z. Shen, and L. Ding. 1989. Fundamentals of fuzzy Prolog. *Int J Approx Reason* 3:179–93.

Mundici, D. 1986a. Interpretation of AF C*-algebras in Łukasiewicz sentential calculus. *J Functional Analysis* 65:15–63.

———— 1986b. Mapping Abelian *l*-groups with strong unit one-one into MV-algebras. *J Algebra* 98:76–81.

———— 1987. Satisfiability in many-valued sentential logic is NP-complete. *Theor Comput Sci* 52:145–53.

———— 1992. The logic of Ulam's game with lies. In *Knowledge, Belief and Strategic Interaction*, ed. C. Bicchieri and M. L. Dalla Chiara, 275–84. Cambridge, UK: Cambridge Univ. Press.

———— 1993. Ulam games, Łukasiewicz logic, and AF C*-algebras. *Fund Inform* 18:151–61.

———— 1994. A constructive proof of McNaughton's theorem in infinite-valued logic. *J Symb Logic* 59 (2): 596–602.

———— 1995. Averaging the truth value in Łukasiewicz sentential logic. *Studia Logica* 55:113–27.

———— 2011. *Advanced Łukasiewicz calculus and MV-algebras*. Dordrecht: Springer.

Murali, V. 1991. Lattice of fuzzy subalgebras and closure systems in I^X. *Fuzzy Sets Syst* 41:101–11.

Murphy, G. L. 2002. *The Big Book of Concepts*. Cambridge, MA: MIT Press.

Myhill, J. 1973. Some properties of intuitionistic Zermelo-Fraenkel set theory. *Lect Notes Math* 337:206–31.

Nachtegael, M., D. Van der Weken, D. Van De Ville, and E. E. Kerre, eds. 2003. *Fuzzy Filters for Image Processing*. Berlin: Springer.

Nahmias, S. 1978. Fuzzy variables. *Fuzzy Sets Syst* 1 (2): 97–110.

Nakamura, K., and S. Iwai. 1982. Topological fuzzy sets as a quantitative description of analogical inference and its application to question-answering systems for information retrieval. *IEEE Trans Syst Man Cyb* 12:193–204.

Navara, M. 2005. Triangular norms and measures of fuzzy sets. In Klement and Mesiar 2005, 345–90.

———— 2012. An algebraic generalization of the notion of tribe. *Fuzzy Sets Syst* 192:123–33.

Negoiţă, C. V., and D. A. Ralescu. 1975a. *Applications of Fuzzy Sets to Systems Analysis*. Basel: Birkhäuser.

———— 1975b. Representation theorems for fuzzy concepts. *Kybernetes* 4:169–74.

Neumaier, A. 1990. *Interval Methods for Systems of Equations*. Cambridge, UK: Cambridge Univ. Press.

Nguyen, H. T. 1978. A note on the extension principle for fuzzy sets. *J Math Anal Appl* 64 (2): 369–80.

Nguyen, H. T., O. M. Kosheleva, and V. Kreinovich. 1996. Is the success of fuzzy logic really paradoxical?: Toward the actual logic behind expert systems. *Int J Intell Syst* 11 (5): 295–326.

Nguyen, H. T., V. Kreinovich, B. Wu, and G. Xiang. 2012. *Computing Statistics under Interval and Fuzzy Uncertainty*. Berlin: Springer.

Nguyen, H. T., and N. R. Prasad, eds. 1999. *Fuzzy Modeling and Control: Selected Works of M. Sugeno*. Boca Raton, FL: CRC Press.

Nguyen, H. T., N. R. Prasad, C. L. Walker, and E. A. Walker. 2003. *A First Course in Fuzzy and Neural Control*. Boca Raton, FL: CRC Press.

Nguyen, H. T., and M. Sugeno, eds. 1998. *Fuzzy Systems*. Boston: Kluwer.

Nguyen, H. T., and E. A. Walker. 1997. *A First Course in Fuzzy Logic*. Boca Raton, FL: CRC Press.

Nguyen, H. T., and B. Wu. 2006. *Fundamentals of Statistics with Fuzzy Data*. Berlin: Springer.

Nieminen, J. 1977. On the algebraic structure of fuzzy sets of type 2. *Kybernetika* 13 (4): 261–73.

Nishizaki, I., and M. Sakawa. 2001. *Fuzzy and Multiobjective Games for Conflict Resolution*. Heidelberg: Physica-Verlag.

Nojiri, H. 1979. A model of fuzzy team decision. *Fuzzy Sets Syst* 2 (3): 201–12.

Nordlund, U. 1996. Formalizing geological knowledge—with an example of modeling stratigraphy using fuzzy logic. *J Sedimentary Research* 66 (4): 689–98.

———1999. FUZZIM: Forward stratigraphic modeling made simple. *Computers & Geoscienceces* 25 (4): 449–56.

Novák, V. 1980. An attempt at Gödel-Bernays-like axiomatization of fuzzy sets. *Fuzzy Sets Syst* 3:323–25.

———1987. First-order fuzzy logic. *Studia Logica* 46:87–109.

———1989. *Fuzzy Sets and Their Applications*. Bristol: Adam Hilger.

———1990. On the syntactico-semantical completeness of first-order fuzzy logic. Part I: Syntax and semantics; part II: Main results. *Kybernetika* 26:47–66; 134–54.

———1992. *The Alternative Mathematical Model of Linguistic Semantics and Pragmatics*. New York: Plenum Press.

———1995a. A new proof of completeness of fuzzy logic and some conclusions for approximate reasoning. In *Proc 4th IEEE Int Conf on Fuzzy Systems*, 1461–68. Piscataway, NJ: IEEE Press.

———1995b. Linguistically oriented fuzzy logic control and its design. *Int J Approx Reason* 12 (3–4): 263–77.

———1996. Paradigm, formal properties and limits of fuzzy logic. *Int J Gen Syst* 24 (4): 377–405.

———2005. On fuzzy type theory. *Fuzzy Sets Syst* 149 (2): 235–73.

———2012. Elements of model theory in higher-order fuzzy logic. *Fuzzy Sets Syst* 205:101–15.

———2012. Reasoning about mathematical fuzzy logic and its future. *Fuzzy Sets Syst* 192:25–44.

Novák, V., and A. Dvořák. 2011. Formalization of commonsense reasoning in fuzzy logic in broader sense. *Applied and Computational Mathematics* 10:106–21.

Novák, V., I. Perfilieva, and J. Močkoř. 1999. *Mathematical Principles of Fuzzy Logic*. Boston: Kluwer.

Nurmi, H., and J. Kacprzyk. 2007. Fuzzy sets in political science: An overview. *New Mathematics and Natural Computation* 3 (3): 281–99.

Ockham, W. 1321–24. *Tractatus de praedestinatione et de praescientia Dei et de futuris contingentibus*. Trans. M. M. Adams and N. Kretzmann 1983. *William Ockham: Predestination, God's Foreknowledge, and Future Contingents*. In Adams and Kretzmann 1983, 34–79.

Oden, G. C. 1977a. Fuzziness in semantic memory: Choosing exemplars of subjective categories. *Memory & Cognition* 5:198–204.

———1977b. Integration of fuzzy logical information. *J Experimental Psychology: Human Perception and Performance* 3 (4): 565–75.

O'Hearn, P. W., and Z. Stachniak. 1992. A resolution framework for finitely-valued first-order logics. *J Symbolic Computation* 13:235–254.

Ollero, A., and E. Freire. 1981. The structure of relations in personnel management. *Fuzzy Sets Syst* 5 (2): 115–25.

Orlov, A. I. 1978. Нечеткие и случайные множества [Fuzzy and random sets]. In *Prikladnoj Mnogomernij Statisticheskij Analiz*, 262–80. Moskva: Nauka.

Orlovsky, S. A. 1978. Decision-making with a fuzzy preference relation. *Fuzzy Sets Syst* 1:155–67.

Osherson, D. N., and E. E. Smith. 1981. On the adequacy of prototype theory as a theory of concepts. *Cognition* 9:35–58.

Ostaszewski, K. 1993. *An Investigation into Possible Applications of Fuzzy Methods in Actuarial Science*. Schaumburg, IL: Society of Actuaries.

Otto, M. 1990. Fuzzy theory: A promising tool for computerized chemistry. *Analytica Chimica Acta* 235:169–75.

Otto, M., and R. R. Yager. 1992. An application of approximate reasoning to chemical knowledge. *Revue Internationale de Systémique* 6:465–81.

Ovchinnikov, S. 2000. An introduction to fuzzy relations. In Dubois and Prade 2000, 233–59.

Pacheco, R., A. Martins, and A. Kandel. 1996. On the power of fuzzy logic. *Int J Intell Syst* 11 (10): 779–89.

Pahikkala, T., A. Airola, M. Stock, B. De Baets, and W. Waegeman. 2013. Efficient regularized least-squares algorithms for conditional ranking on relational data. *Machine Learning* 93:321–56.

Panti, G. 1995. A geometric proof of the completeness of the Łukasiewicz calculus. *J Symb Logic* 60 (2): 563–78.

Papageorgiou, E. I., ed. 2014. *Fuzzy Cognitive Maps for Applied Sciences and Engineering*. Berlin: Springer.

Papageorgiou, E. I., and J. L. Salmeron. 2013. A review of fuzzy cognitive maps research during the last decade. *IEEE Trans Fuzzy Syst* 21 (1): 66–79.

Parikh, R. 1991. A test for fuzzy logic. *ACM SIGACT News* 22:49–50.

Paris, J. B. 1994. *The Uncertain Reasoner's Companion: A Mathematical Perspective*. Cambridge, UK: Cambridge Univ. Press.

———1997. A semantics for fuzzy logic. *Soft Computing* 1:143–47.

Passino, K. M., and S. Yurkovich. 1998. *Fuzzy Control*. Reading, MA: Addison-Wesley.

Pavelka, J. 1979. On fuzzy logic I: Many-valued rules of inference; II: Enriched residuated lattices and semantics of propositional calculi; III: Semantical completeness of some many-valued propositional calculi. *Z Math Logik Grundlagen Math* 25:45–52; 119–34; 447–64.

Pawlak, Z. 1991. *Rough Sets: Theoretical Aspects of Reasoning about Data*. Dordrecht: Kluwer.

Pedrycz, W. 1982. *Fuzzy Control and Fuzzy Systems*. Report 82 14, Dept. of Mathematics, Delft Univ. Technology.

———1983. Fuzzy relational equations with generalized connectives and their applications. *Fuzzy Sets Syst* 10 (2): 185–201.

———2005. *Knowledge-Based Clustering: From Data to Information Granules*. Hoboken, NJ: Wiley.

Pedrycz, W., P. Ekel, and R. Parreiras. 2011. *Models and Methods of Fuzzy Multicriteria Decision-Making and Their Applications*. Chichester, UK: Wiley.

Pedrycz, W., and F. Gomide. 1998. *An Introduction to Fuzzy Sets: Analysis and Design*. Cambridge, MA: MIT Press.

Peeva, K., and Y. Kyosev. 2004. *Fuzzy Relational Calculus: Theory, Applications and Software*. Singapore: World Scientific.

Peirce, C. S. 1885. On the algebra of logic: A contribution to the philosophy of notation. *American J Math* 7 (2): 180–196.

Pelta, D., and N. Krasnogor, eds. 2005. Fuzzy Sets in Bioinformatics. Special issue, *Fuzzy Sets Syst* 152 (1): 1–158.

Peray, K. 1999. *Investing in Mutual Funds Using Fuzzy Logic*. Boca Raton, FL: St. Lucie Press.

Perfilieva, I. 2006. Fuzzy transforms: theory and applications. *Fuzzy Sets Syst* 157:993–1023.

Perfilieva, I., and S. Gottwald. 2003. Solvability and approximate solvability of fuzzy relation equations. *Int J Gen Syst* 32 (4): 361–72.

Perfilieva, I., and V. Kreinovich. 2011. Editorial: Fuzzy transform as a new paradigm in fuzzy modeling. *Fuzzy Sets Syst* 180:1–2.

Peters, S., and D. Westerståhl. 2006. *Quantifiers in Language and Logic*. Oxford, UK: Clarendon Press.

Petry, F. E. 1996. *Fuzzy Databases: Principles and Applications*. Dordrecht: Kluwer.

Pham, T. D. 2008. Fuzzy fractal analysis of molecular imaging data. *Proc IEEE* 96:1332–47.

Piasecki, K. 1985. Probability of fuzzy events defined as denumerable additivity measure. *Fuzzy Sets Syst* 17:271–84.

Piegat, A. 2001. *Fuzzy Modeling and Control*. Heidelberg: Physica-Verlag.

Pienkowski, A. E. K. 1989. *Artificial Colour Perception Using Fuzzy Techniques in Digital Image Processing*. Cologne: Verlag TÜV Rheinland.

Pollandt, S. 1997. *Fuzzy-Begriffe: Formale Begriffsanalyse unscharfer Daten* [Fuzzy concepts: Formal concept analysis of unsharp data]. Berlin: Springer.

Ponsard, C. 1981. An application of fuzzy subsets theory to the analysis of the consumer's spatial preferences. *Fuzzy Sets Syst* 5 (3): 235–44.

———1982a. Producer's spatial equilibrium with a fuzzy constraint. *Europ J Operational Research* 10 (3): 302–13.

———1982b. Partial spatial equilibria with fuzzy constraints. *J Regional Science* 22 (2): 159–75.

———1983. *History of Spatial Economic Theory*. Berlin: Springer.

———1985a. Fuzzy data analysis in a spatial context. In *Measuring the Unmeasurable*, ed. P. Nijkamp, H. Leitner, and N. Wrigley, 487–508. Dordrecht: Martinus Nijhoff.

———1985b. Fuzzy sets in economics: Foundations of soft decision theory. In *Management Decision Support Systems Using Fuzzy Sets and Possibility Theory*, ed. J. Kacprzyk and R. R. Yager, 25–37. Cologne: Verlag TÜV Rheinland

———1987a. Nash fuzzy equilibrium: Theory and applications to spatial duopoly. *Europ J Operational Research* 31:376–84.

———1987b. Fuzzy sets. In *The New Palgrave: A Dictionary of Economics*, vol. 2, ed. J. Eatwell, M. Milgate, and P. Newman, 449–52.

———1988. Fuzzy mathematical models in economics. *Fuzzy Sets Syst* 28 (3): 273–83.

Ponsard, C., and B. Fustier, eds. 1986. *Fuzzy Economics and Spatial Analysis*. Dijon, France: Institut de Mathématiques Economiques et Librairie de l'Université.

Post, E. L. 1920a. Determination of all closed systems of truth tables. Abstract. *Bull AMS* 26:437.

———1920b. Introduction to a general theory of elementary propositions. Abstract. *Bull AMS* 26:437.

———1921. Introduction to a general theory of elementary propositions. *American J Math* 43 (3): 163–85.

Poston, T. 1972. *Fuzzy Geometry*. PhD diss., Univ. of Warwick, England.

Powell, W. C. 1975. Extending Gödel's negative interpretation to ZF. *J Symb Logic* 40:221–29.

Prade, H., and C. Testemale. 1984. Generalizing database relational algebra for the treatment of incomplete or uncertain information and vague queries. *Inf Sci* 34:115–43.

Prati, N. 1988. An axiomatization of fuzzy classes. *Stochastica* 12:65–78.

Preparata, F. P., and R. T. Yeh. 1972. Continuously valued logic. *J Comput Syst Sci* 6:397–418.

Prugovečki, E. 1974. Fuzzy sets in the theory of measurement of incompatible observables. *Foundations of Physics* 4 (1): 9–18.

———1975. Measurement in quantum mechanics as a stochastic process on spaces of fuzzy events. *Foundations of Physics* 5 (4): 557–71.

Przełęcki, M. 1976. Fuzziness as multiplicity. *Erkenntnis* 10:371–80.

Pulmannová, S. 2000. A note on observables on MV-algebras. *Soft Computing* 4:45–48.

Pultr, A. 1976. Closed categories of L-fuzzy sets. In *Vorträge aus dem Problemseminar Automaten- und Algorithmentheorie*, 60–68. Dresden: Techn. Univ. Dresden.

———1982. Fuzziness and fuzzy equality. *Commentationes Mathematicae Universitatis Carolinae* 23:249–67.

Puri, M. L., and D. A. Ralescu. 1983. Differentials of fuzzy functions. *J Math Anal Appl* 91:552–58.

———1985. The concept of normality for fuzzy random variables. *Ann Probability* 13:1373–79.

———1986. Fuzzy random variables *J Math Anal Appl* 114:409–22.

Pykacz, J. 1992. Fuzzy set ideas in quantum logics. *Int J Theoretical Physics* 31 (9): 1767–83.

———1993. Fuzzy quantum logic. *Int J Theoretical Physics* 32 (10): 1691–1707.

———1994. Fuzzy quantum logic and infinite-valued Łukasiewicz logic. *Int J Theoretical Physics* 33 (7): 1403–16.

———1999. Non-classical logics, non-classical sets, and non-classical physics. In *Quantum Structures and the Nature of Reality*, ed. D. Aerts and J. Pykacz, 67–101. Dordrecht: Kluwer.

Pykacz, J., and B. D'Hooghe. 2001. Bell-type inequalities in fuzzy probability calculus. *Int J UFKBS* 9 (2): 263–75.

Quine, W. V. 1982. *Methods of Logic*. 4th ed. Cambridge, MA: Harvard Univ. Press.

Radecki, T. 1979. Fuzzy set theoretical approach to document retrieval. *Information Processing and Management* 15:247–59.

———1981. Outline of a fuzzy logic approach to information retrieval. *Int J Man Mach Stud* 14:169–78.

Ragade, R. K. 1976. Fuzzy games in the analysis of options. *J Cybernetics* 6:213–21.

Ragaz, M. 1981. Arithmetische Klassifikation von Formelmengen der unendlichwertigen Logik [Arithmetic classification of sets of formulas of infinitely-valued logic]. PhD diss., ETH Zürich.

———1983. Die Unentscheidbarkeit der einstelligen unendlichwertigen Prädikatenlogik [The undecibility of monadic infinitely-valued predicate logic] *Archiv für mathematische Logik und Grundlagenforschung* 23 (1): 129–39.

Ragin, C. C. 2000. *Fuzzy-Set Social Science*. Chicago: Univ. of Chicago Press.

———2008. *Redesigning Social Inquiry: Fuzzy Sets and Beyond*. Chicago: Univ. of Chicago Press.

Raju, K. V. S. V. N., and A. K. Majumdar. 1988. Fuzzy functional dependencies and lossless join decomposition of fuzzy relational database systems. *ACM Trans Database Systems* 13:129–66.

Rakus-Anderson, E. 2007. *Fuzzy and Rough Techniques in Medical Diagnosis and Medication*. Berlin: Springer.

Ralescu, A. L., ed. 1994. *Applied Research in Fuzzy Technology*. Dordrecht: Kluwer.

Ralescu, D. 1978. Fuzzy subobjects in a category and the theory of 𝒞-sets. *Fuzzy Sets Syst* 1:193–202.

Rasiowa, H. 1964. A generalization of a formalized theory of fields of sets on non-classical logics. *Rozprawy Matematyczne* 42:3–29.

———1974. *An Algebraic Approach to Non-classical Logics*. Amsterdam: North-Holland.

Rasiowa, H., and R. Sikorski. 1963. *The Mathematics of Metamathematics*. Warsaw: PWN—Polish Scientific Publishers.

———1950. A proof of the completness theorem of Gödel. *Fund Math* 37:193–200.

Recasens, J. 2010. *Indistinguishability Operators: Modelling Fuzzy Equalities and Fuzzy Equivalence Relations*. Berlin: Springer.

Recknabel, F., ed. 2006. *Ecological Informatics: Scope, Techniques, and Applications*. Heidelberg: Springer.

Reichenbach, H. 1932. Wahrscheinlichkeitslogik [Probabilistic logic]. *Sitzungsberichte der Preussischen Akademie der Wissenschaften, Physikalisch-mathematische Klasse* 476–88.

———1935. *Wahrscheinlichkeitslehre. Eine Untersuchung über die logischen und mathematischen Grundlagen der Wahrscheinlichkeitsrechnung*. Trans. E. H. Hutten and M. Reichenbach 1971. *The Theory of Probability: An Inquiry into the Logical and Mathematical Foundations of the Calculus of Probability*. Berkeley, CA: Univ. of California Press.

———— 1935–36. Wahrscheinlichkeitslogik und Alternativlogik [Probabilistic logic and alternative logic]. *Erkenntnis* 5:177–78.

———— 1944. *Philosophic Foundations of Quantum Mechanics*. Berkeley and Los Angeles: Univ. of California Press.

Rényi, A. 1970. *Probability Theory*. Amsterdam: North-Holland.

Rescher, N. 1969. *Many-Valued Logic*. New York: McGraw-Hill.

Rescher, N., and P. Grim. 2011. *Beyond Sets: A Venture in Collection-Theoretic Revisionism*. Frankfurt: Ontos.

Ressom, H., R. Reynolds, and R. S. Varghese. 2003. Increasing the efficiency of fuzzy logic-based gene expression data analysis. *Physiological Genomics* 13 (2): 107–17.

Ribeiro, R. A., H.-J. Zimmermann, R. R. Yager, and J. Kacprzyk, eds. 1999. *Soft Computing in Financial Engineering*. Heidelberg: Physica-Verlag.

Riečan, B. 1992. Fuzzy connectives and quantum models. In *Cybernetics and System Research*, vol. 1, ed. R. Trappl, 335–38. Singapore: World Scientific. In *Cybernetics and Systems*, ed. R. Trappl, 9–14. Vienna: Austrian Society for Cybernetic Studies.

———— 2000. On the probability theory on MV algebras. *Soft Computing* 4 (1): 49–57.

———— 2007. On some contributions to quantum structures by fuzzy sets. *Kybernetika* 43 (4): 481–90.

Riečan, B., and D. Mundici. 2002. Probability on MV-algebras. In *Handbook of Measure Theory*, ed. E. Pap, 869–909. Amsterdam: Elsevier.

Rine, D. C., ed. 1984. *Computer Science and Multiple-Valued Logic: Theory and Applications*. Amsterdam: North-Holland.

Roberts, D. W. 1989. Analysis of forest succession with fuzzy graph theory. *Ecological Modeling* 45 (4): 261–74.

Rodabaugh, S. E. 1982. Fuzzy addition in the *L*-fuzzy real line. *Fuzzy Sets Syst* 8:39–52.

———— 1999. Categorical foundations of variable-basis fuzzy topology. In Höhle and Rodabaugh 1999, 273–388.

Rodabaugh, S. E., and E. P. Klement, eds. 2003. *Topological and Algebraic Structures in Fuzzy Sets*. Dordrecht: Kluwer.

Rodríguez, R. O., F. Esteva, P. Garcia, and L. Godo. 2003. On implicative closure operators in approximate reasoning. *Int J Approx Reason* 33:159–84.

Rosch, E. H. 1973a. Natural categories. *Cognitive Psychology* 4 (3): 328–50.

———— 1973b. On the internal structure of perceptual and semantic categories. In *Cognitive Development and the Acquisition of Language*, ed. T. M. Moore, 111–44. New York: Academic Press.

Rose, A. 1951a. A lattice-theoretic characterisation of the $\aleph_0$-valued propositional calculus. *Mathematische Annalen* 123 (1): 285–87.

———— 1951b. Systems of logic whose truth-values form lattices. *Mathematische Annalen* 123 (1): 152–65.

———— 1952a. Eight-valued geometry. *Proc London Mathematical Society* 3 (2): 30–44.

———— 1952b. The degree of completeness of the *M*-valued Łukasiewicz propositional calculus. *J London Mathematical Society* s1-27:92–102.

———— 1953. The degree of completeness of the $\aleph_0$-valued Łukasiewicz propositional calculus. *J London Mathematical Society* s1-28:176–84.

———— 1969. The degree of completeness of the *m*-valued Łukasiewicz propositional calculus. Correction and addendum. *J London Mathematical Society* s1-44 (1): 587–91.

Rose, A., and J. B. Rosser. 1958. Fragments of many valued statement calculi. *Trans AMS* 87:1–53.

Rosenberg, I. 1970. Über die funktionale Vollstandigkeit in den mehrwertigen Logiken [On functional completeness in many-valued logics]. *Rozpravy Československé akademia věd, Řada matematických a přírodních věd* 80:3–93.

Rosenbloom, P. C. 1942. Post algebras. I. Postulates and general theory. *American J Math* 64:167–88.

Rosenfeld, A. 1971. Fuzzy groups. *J Math Anal Appl* 35:512–17.

———— 1979. Fuzzy digital topology. *Inf Control* 40:76–87.

———— 1984. The fuzzy geometry of image subsets. *Pattern Recognition Letters* 2:311–17.

———— 1998. Fuzzy geometry: An updated overview. *Inf Sci* 110:127–33.

Rosental, C. 2008. *Weaving Self-Evidence: A Sociology of Logic*. Princeton, NJ: Princeton Univ. Press.

Ross, T. J. 1995. *Fuzzy Logic with Engineering Applications*. New York: McGraw-Hill.

Rosser, B. 1941. On the many-valued logics. *American J Phys* 9:207–12.

Rosser, J. B. 1960. Axiomatization of infinite-valued logics. *Logique et Analyse* 3:137–53.

Rosser, J. B., and A. R. Turquette. 1945. Axiom schemes for *M*-valued propositional calculi. *J Symb Logic* 10:61–82.

———— 1948. Axiom schemes for *M*-valued functional calculi of first order: Part I. Definition of axiom schemes and proof of plausibility. *J Symb Logic* 13 (4): 177–92.

———— 1950. A note on the deductive completeness of M-valued propositional calculi. *J Symb Logic* 14 (4): 219–25.

———— 1951. Axiom schemes for M-valued functional calculi of first order. Part II. Deductive completeness. *J Symb Logic* 16 (1): 22–34.

———— 1952. *Many-Valued Logics.* Amsterdam: North-Holland.

Rouvray, D. H., ed. 1997. *Fuzzy Logic in Chemistry.* San Diego: Academic Press.

Ruspini, E. H. 1969. A new approach to clustering. *Inf Control* 15:22–32.

———— 1982. Recent developments in fuzzy clustering. In *Fuzzy Sets and Possibility Theory: Recent Developments,* ed. R. R. Yager, 133–47. New York: Pergamon Press.

———— 1991. On the semantics of fuzzy logic. *Int J Approx Reason* 5:45–88.

Ruspini, E. H., P. P. Bonissone, and W. Pedrycz, eds. 1998. *Handbook of Fuzzy Computation.* Bristol and Philadelphia: Institute of Physics Publishing.

Russell, B. 1903. *The Principles of Mathematics.* Cambridge, UK: Cambridge Univ. Press.

———— 1923. Vagueness. *Australasian J Psychology and Philosophy* 1 (2): 84–92.

Rutkowski, L. 2004. *Flexible Neuro-Fuzzy Systems: Structures, Learning and Performance Evaluation.* Boston: Kluwer.

———— 2008. *Computational Intelligence: Methods and Techniques.* Berlin: Springer.

Rutledge, J. D. 1959. *A Preliminary Investigation of the Infinitely Many-Valued Predicate Calculus.* PhD diss., Cornell Univ., Ithaca, NY.

———— 1960. On the definition of an infinitely-many-valued predicate calculus. *J Symb Logic* 25 (3): 212–16.

Saaty, T. L. 1974. Measuring the fuzziness of sets. *J Cybernetics* 4 (4): 53–61.

———— 1980. *The Analytic Hierarchy Process.* New York: McGraw-Hill.

———— 1986. Scaling the membership function. *Europ J Operational Research* 25 (3): 320–29.

Sadegh-Zadeh, K. 2000a. Fuzzy genomes. *Artificial Intelligence in Medicine* 18:1–28.

———— 2000b. Fuzzy health, illness, and disease. *J Medicine and Philosophy* 25 (5): 605–38.

———— 2001. The fuzzy revolution: Goodbye to the Aristotelian Weltanschauung. *Artificial Intelligence in Medicine* 21 (1): 1–25.

———— 2007. The fuzzy polynucleotide space revisited. *Artificial Intelligence in Medicine* 41 (1): 69–80.

———— 2008. The prototype resemblance theory of disease. *J Medicine and Philosophy* 33 (2): 106–39.

———— 2012. *Handbook of Analytic Philosophy of Medicine.* Dordrecht: Springer.

Sadeghian, A., J. M. Mendel, and H. Tahayori, eds. 2013. *Advances in Type-2 Fuzzy Sets and Systems: Theory and Applications.* New York: Springer.

Saffiotti, A. 1987. An AI view of the treatment of uncertainty. *Knowledge Engineering Review* 2 (2): 75–97.

Sakawa, M. 1993. *Fuzzy Sets and Interactive Multiobjective Optimization.* New York: Plenum Press.

———— 2000. *Large Scale Interactive Fuzzy Multiobjective Programming: Decomposition Approaches.* Heidelberg: Physica-Verlag.

———— 2002. *Genetic Algorithms and Fuzzy Multiobjective Optimization.* Dordrecht: Kluwer.

Salij, V. N. 1965. Бинарные £-отношения [Binary £-relations]. *Izvestija Vysshich Uchebnych Zavedenij, Matematika* 1:133–45.

Salomaa, A. 1959. On many-valued systems of logic. *Ajatus* 22:115–59.

Salski, A. 1999. Ecological modeling and data analysis. In Zimmermann 1999, 247–66.

Salski, A., O. Fränzle, and P. Kandzia, eds. 1996. Fuzzy Logic in Ecological Modeling. Special issue, *Ecological Modeling* 85 (1): 1–98.

Salski, A., B. Holsten, and M. Trepel. 2009. A fuzzy approach to ecological modeling and data analysis. In *Handbook of Ecological Modeling and Informatics,* ed. S. E. Jøgensen, T.-S. Chon, and F. Rechnagel, 125–39. Ashurst, UK: WIT Press.

Sanchez, E. 1976. Resolution of composite fuzzy relation equations. *Inf Control* 30:38–48.

———— 1977. Solutions in composite fuzzy relation equations: applications to medical diagnosis in Brouwerian logic. In *Fuzzy Automata in Decision Processes,* ed. M. M. Gupta, G. N. Saridis, and B. R. Gaines, 221–34. New York: North-Holland.

———— 1979. Medical diagnosis and composite fuzzy relations. In *Advances in Fuzzy Set Theory and Applications,* ed. M. M. Gupta, R. K. Ragade, and R. R. Yager, 437–44. Amsterdam: North-Holland.

———— 1984. Solution of fuzzy equations with extended operations. *Fuzzy Sets Syst* 12:237–48.

Sanchez, E., T. Shibata, and L. A. Zadeh, eds. 1997. *Genetic Algorithms and Fuzzy Logic Systems: Soft Computing Perspectives.* Singapore: World Scientific.

Sandler, U., and L. Tsitolovsky. 2008. *Neural Cell Behavior and Fuzzy Logic.* New York: Springer.

Santos, E. S. 1970. Fuzzy algorithms. *Inf Control* 17:326–39.

———— 1976. Fuzzy and probabilistic programs. *Inf Sci* 10:331–45.

Sârbu, C. 1999. Fuzzy classification of the chemical elements. Vol. 65, supplement 28 of *Encyclopedia of Library and Information Science,* ed. A. Kent, 112–38. New York: Marcel Dekker.

Savický, P., R. Cignoli, F. Esteva, L. Godo, and C. Noguera. 2006. On product logic with truth constants. *J Logic Comput* 16 (2): 205–25.

Scarpellini, B. 1962. Die Nichtaxiomatisierbarkeit des unendlichwertigen Prädikatenkalküls von Łukasiewicz [The non-axiomatizability of the infinitely-valued predicate calculus of Łukasiewicz]. *J Symb Logic* 27 (2): 159–70.

Schmechel, N. 1996. On the isomorphic lattices of fuzzy equivalence relations and fuzzy partitions. *Multiple Valued Logic* 2:1–46.

Schmucker, K. J. 1984. *Fuzzy Sets, Natural Language, and Risk Analysis*. Rockville, MD: Computer Science Press.

Schröter, K. 1955. Methoden zur Axiomatisierung beliebiger Aussagen- und Prädikatenkalkülle [Methods for axiomatization of arbitrary propositional and predicate calculi]. *Z Math Logik Grundlagen Math* 1:241–51.

Schwartz, D. 1972. Mengenlehre über vorgegebenen algebraischen Systemen [Set theory over given algebraic systems]. *Mathematische Nachrichten* 53:365–70.

———— 1977. Theorie der polyadischen MV-Algebren endlicher Ordnung [Theory of polyadic MV-algebras of finite order]. *Mathematische Nachrichten* 78:131–38.

———— 1980. Polyadic MV-algebras. *Z Math Logik Grundlagen Math* 26:561–64.

Schwartz, D. G., G. J. Klir, H. W. Lewis III, and Y. Ezawa. 1994. Applications of fuzzy sets and approximate reasoning. *Proc IEEE* 82 (4): 482–98.

Schweizer, B. 1975. Multiplications on the space of probability distribution functions. *Aeq Math* 12:156–83.

———— 2005. Triangular norms, looking back—triangle functions, looking ahead. In Klement and Mesiar 2005, 3–15.

Schweizer, B., and A. Sklar. 1983. *Probabilistic Metric Spaces*. Amsterdam: North-Holland.

Scott, D. S. 1967. A proof of independence of the continuum hypothesis. *Mathematical Systems Theory* 1:89–111.

———— 1974. Completeness and axiomatizability in many-valued logic. In *Proc Tarski Symp*, vol. 25, ed. L. Henkin, 411–36. Providence, RI: American Mathematical Society.

———— 1976. Does many-valued logic have any use? In *Philosophy of Logic*, ed. S. Körner, 64–95. Berkeley, CA: Univ. of California Press.

———— 1979. Identity and existence in intuitionistic logic. *Lect Notes Math* 753:660–96.

Sebastian, H.-J., and E. K. Antonsson, eds. 1996. *Fuzzy Sets in Engineering Design and Configuration*. Dordrecht: Kluwer.

Segerberg, K. 1971. Qualitative probability in a modal setting. In *Proc 2nd Scand Log Symp*, ed. E. Fenstad, 341–52. Amsterdam: North-Holland.

Seikkala, S. 1987. On the fuzzy initial value problem. *Fuzzy Sets Syst* 24:319–30.

Seising, R., and M. E. Tabacchi, eds. 2013. *Fuzziness and Medicine: Philosophical Reflections and Application Systems in Health Care*. Berlin: Springer.

Şen, Z. 2010. *Fuzzy Logic and Hydrological Modeling*. Boca Raton, FL: CRC Press.

Shackle, G. L. S. 1949. *Expectation in Economics*. Cambridge, UK: Cambridge Univ. Press.

———— 1955. *Uncertainty in Economics and Other Reflections*. Cambridge, UK: Cambridge Univ. Press.

———— 1958. *Time in Economics*. Amsterdam: North-Holland.

———— 1961. *Decision, Order and Time in Human Affairs*. Cambridge, UK: Cambridge Univ. Press.

———— 1979. *Imagination and the Nature of Choice*. Edinburgh: Edinburgh Univ. Press.

Shafer, G. 1976. *A Mathematical Theory of Evidence*. Princeton, NJ: Princeton Univ. Press.

Shamir, L., and R. J. Nemiroff. 2005. A fuzzy logic based algorithm for finding astronomical objects in wide-angle frames. *Publications of the Astronomical Society of Australia* 22 (2): 111–17.

———— 2008. Astronomical pipeline processing using fuzzy logic. *Applied Soft Computing* 8 (1): 79–87.

Shapiro, S. 2006. *Vagueness in Context*. New York: Oxford Univ. Press.

Shepard, R. B. 2005. *Quantifying Environmental Impact Assessment Using Fuzzy Logic*. New York: Springer.

Shiina, K. 1988. A fuzzy-set-theoretic feature model and its application to asymmetric similarity data analysis. *Japanese Psychological Research* 30:95–104.

Shirai, T. 1937. On the pseudo-set. *Memoirs of the College of Science, Kyoto Imperial University, ser. A*, 20:153–56.

Shoenfield, J. R. 1967. *Mathematical Logic*. Reading, MA: Addison-Wesley.

Siddique, N. 2014. *Intelligent Control: A Hybrid Approach Based on Fuzzy Logic, Neural Networks and Genetic Algorithms*. Heidelberg: Springer.

Singer, D., and P. G. Singer. 1993. Fuzzy chemical kinetics: An algorithmic approach. *Int J Systems Science* 24 (7): 1363–76.

Sinha, D., and E. R. Dougherty. 1993. Fuzzification of set inclusion: Theory and applications. *Fuzzy Sets Syst* 55:15–42.

Sivanandam, S. N., S. Sumathi, and S. N. Deepa. 2007. *Introduction to Fuzzy Logic Using MATLAB*. Berlin: Springer.

Skala, H. J., S. Termini, and E. Trillas, eds. 1984. *Aspects of Vagueness*, Dordrecht: Reidel.

Skolem, T. 1957. Bemerkungen zum Komprehensionsaxiom [Remarks on the axiom of comprehension]. *Z Math Logik Grundlagen Math* 3:1–17.

Słowiński, R., ed. 1998. *Fuzzy Sets in Decision Analysis, Operations Research and Statistics*. Boston: Kluwer.

Słupecki, J. 1936. Der volle dreiwertige Aussagenkalkül. *Comptes Rendus des Séances de la Société des Sciences et des Lettres de Varsovie, Cl. iii* 29:9–11. Trans. S. McCall 1967. The full three-valued propositional calculus. In McCall 1967, 335–37.

———— 1939a. Dowód aksjomatyzowalności pełnych systemów wielowartościowych rachunku zdań. *Comptes Rendus des Séances de la Société des Sciences et des Lettres de Varsovie, Cl. iii* 32:110–28. Trans. J. Słupecki 1971. Proof of axiomatizability of full many-valued systems of calculus of propositions. *Studia Logica* 29 (1): 155–68.

———— 1939b. Kryterium pełności wielowartościowych systemów logiki zdań. *Comptes Rendus des Séances de la Sociétdes Sciences et des Lettres de Varsovie, Cl. iii* 3:102–9. Trans. J. Słupecki 1972. A criterion of fullness of many-valued systems of propositional logic. *Studia Logica* 30 (1): 153–57.

———— 1946. The complete three-valued propositional calculus. *Annales Universitatis Mariae Curie-Sklodowska* 1:193–209.

Smets, P. 1981. The degree of belief in a fuzzy event. *Inf Sci* 25:1–19.

———— 1982. Probability of a fuzzy event: An axiomatic approach. *Fuzzy Sets Syst* 7:153–64.

Smets, P., and P. Magrez. 1987. Implication in fuzzy logic. *Int J Approx Reason* 1:327–47.

Smith, C. R., and G. J. Erickson. 1989. From rationality and consistency to Bayesian probability. In *Maximum Entropy and Bayesian Methods*, ed. J. Skilling, 29–44. Dordrecht: Kluwer.

Smith, N. J. J. 2008. *Vagueness and Degrees of Truth*. New York: Oxford Univ. Press.

Smithson, M. 1987. *Fuzzy Set Analysis for Behavioral and Social Sciences*. New York: Springer.

———— 1988. Fuzzy set theory and the social sciences: The scope for applications. *Fuzzy Sets Syst* 26 (1): 1–21.

Smithson, M., and J. Verkuilen. 2006. *Fuzzy Set Theory: Applications in the Social Sciences*. Thousand Oaks, CA: Sage.

Smuts, J. C. 1926. *Holism and Evolution*. New York: Macmillan.

Sobociński, B. 1936. Aksjomatyzacja pewnych wielowartościowych systemów teorji dedukcji [Axiomatization of certain many-valued systems of the theory of deduction]. *Roczniki prac naukowych zrzeszenia asystentów Uniwersytetu Józefa Pilsudskiego w Warszawie* 1:399–419.

Sobrevilla, P., and E. Montseny, eds. 2007. Image Processing. Special issue, *Fuzzy Sets Syst* 158 (3): 213–347.

Sorensen, R. 2001. *Vagueness and Contradiction*. New York: Oxford Univ. Press.

Šostak, A. P. 1985. On a fuzzy topological structure. *Supplemento ai Rendiconti del Circolo Matematico di Palermo*, 2nd ser., no. 11: 89–103.

———— 1989. Два десятилетия нечеткой топологии: Основные идеи, понятия и результаты [Two decades of fuzzy topology: Basic ideas, notions, and results]. *Russian Mathematical Surveys* 44:125–86.

———— 1996. Basic structures of fuzzy topology. *J Mathematical Sciences* 78:662–701.

Stone, M. H. 1936. The theory of representations for Boolean algebras. *Trans AMS* 40:37–111.

Stout, L. N. 1991. A survey of fuzzy set and topos theory. *Fuzzy Sets Syst* 42:3–14.

———— 1995. Categories of fuzzy sets with values in a quantale or projectale. In Höhle and Klement 1995, 219–34.

Sugeno, M. 1977. Fuzzy measures and fuzzy integrals: A survey. In *Fuzzy Automata and Decision Processes*, ed. M. Gupta, G. N. Saridis, and B. R. Gaines, 89–102. New York: North-Holland.

———— ed. 1985. *Industrial Applications of Fuzzy Control*. Amsterdam: North-Holland.

Sugeno, M., and T. Yasukawa. 1993. A fuzzy-logic-based approach to qualitative modeling. *IEEE Trans Fuzzy Syst* 1 (1): 7–31.

Syropoulos, A. 2014. *Theory of Fuzzy Computation*. New York: Springer.

Szczepaniak, P. S., P. J. G. Lisboa, and J. Kacprzyk, eds. 2000. *Fuzzy Systems in Medicine*. Heidelberg: Physica-Verlag.

Szpilrajn, E. 1936. Sur l'equivalence des suites d'ensembles et l'equivalence des fonctions [On the equivalence of sequences of sets and equivalence of functions]. *Fund Math* 26:302–26.

Tahani, V. 1977. A conceptual framework for fuzzy query processing: A step toward very intelligent database systems. *Information Processing and Management* 13:289–303.

Takagi, H., and I. Hayashi. 1991. NN-driven fuzzy reasoning. *Int J Approx Reason* 5 (3): 191–212.

Takagi, T., and M. Sugeno. 1985. Fuzzy identification of systems and its application to modeling and control. *IEEE Trans Syst Man Cyb* 15 (1): 116–32.

Takano, M. 1987. Another proof of strong completeness of the intuitionistic fuzzy logic. *Tsukuba J Math* 11 (1): 101–5.

Takeuti, G., and S. Titani. 1984. Intuitionistic fuzzy logic and intuitionistic fuzzy set theory. *J Symb Logic* 49 (3): 851–66.

———— 1992. Fuzzy logic and fuzzy set theory. *Arch Math Logic* 32:1–32.

Tanaka, H., S. Uejima, and K. Asai. 1982. Linear regression analysis with fuzzy model. *IEEE Trans Syst Man Cyb* 12:903–7.

Tarski, A. 1930a. Über einige fundamentale Begriffe der Metamathematik. *Comptes Rendus des Séances de la Société des Sciences et des Lettres de Varsovie, Cl. iii* 23:22–29. Trans. J. H. Woodger 1956. On some fundamental concepts of metamathematics. In Woodger 1956, 30–37.

——— 1930b. Fundamentale Begriffe der Methodologie der deduktiven Wissenschaften. I. *Monatshefte für Mathematik und Physik* 37 (1): 361–404. Trans. J. H. Woodger 1956. Fundamental concepts of the methodology of the deductive sciences. In Woodger 1956, 60–109.

——— 1935. Wahrscheinlichkeitslehre und mehrwertige Logik [The probability calculus and many-valued logic]. *Erkenntnis* 5:174–75.

——— 1935–36. Grundzüge des Systemenkalküls. Erster Teil. Zweiter Teil. *Fund Math* 25:503–26, 26:283–301. Trans. J. H. Woodger 1956. Foundations of the calculus of systems. In Woodger 1956, 342–83.

——— 1936. Der Wahrheitsbegriff in den formalisierten Sprachen. *Studia Philosophica* 1:261–405. Trans. J. H. Woodger 1956. The concept of truth in formalized languages. In Woodger 1956, 152–278.

——— 1938. Der Aussagenkalkül und die Topologie. *Fund Math* 31: 103–34. Trans. J. H. Woodger 1956. Sentential calculus and topology. In Woodger 1956, 421–54.

Tee, A. B., M. D. Bowman, and K. C. Sinha. 1988. A fuzzy mathematical approach for bridge condition evaluation. *Civil Eng Syst* 5 (1): 17–24.

Teodorescu, H.-N., L. C. Jain, and A. Kandel, eds. 2001. *Hardware Implementation of Intelligent Systems*. Heidelberg: Physica-Verlag.

Teodorević, D., and K. Vukadinović. 1998. *Traffic Control and Transport Planning: A Fuzzy Sets and Neural Networks Approach*. Dordrecht: Kluwer.

Thiele, H. 1958. Theorie der endlichwertigen Łukasiewiczschen Prädikatenkalküle der ersten Stuffe [Theory of finite-valued Łukasiewicz predicate calculi of first order]. *Z Math Logik Grundlagen Math* 4:108–42.

Thiele, H., and S. Lehmke. 1994. On "bold" resolution theory. In *Proc 3rd IEEE Int Conf on Fuzzy Systems*, 1945–50. Piscataway, NJ: IEEE Press.

Thiele, H., and N. Schmechel. 1995. On the mutual definability of fuzzy equivalence relations and fuzzy partitions. In *Proc Int Joint Conf 4th IEEE Int Conf and 2nd Int Fuzzy Engineering Symp*, 1383–90. Piscataway, NJ: IEEE Press.

Thole, U., H.-J. Zimmermann, and P. Zysno. 1979. On the suitability of minimum and product operators for the intersection of fuzzy sets. *Fuzzy Sets Syst* 2:167–80.

Titani, S. 1999. A lattice-valued set theory. *Arch Math Logic* 38:395–421.

Togai, M., and H. Watanabe. 1986a. A VLSI implementation of a fuzzy inference engine: Toward an expert system on a chip. *Inf Sci* 38:147–63.

——— 1986b. Expert system on a chip: An engine for real-time approximate reasoning. *IEEE Expert* 1 (3): 55–62.

Tokarz, M. 1974. A method of axiomatization of Łukasiewicz logics. *Studia Logica* 33 (4): 333–38.

——— 1974. Invariant systems of Łukasiewicz calculi. *Z Math Logik Grundlagen Math* 20:221–28.

Toth, H. 1993. Axiomatic f-set theory. I: Basic concepts. *J Fuzzy Math* 1:109–35.

Trillas, E. 1979. Sobre funciones de negación en la teoría de conjuntos difusos [On negation functions in fuzzy set theory]. *Stochastica* 3:47–59.

——— 1982. Assaig sobre les relacions d'indistingibilitat [An essay on indistinguishability relations]. In *Actes del Primer Congrés Català de Lògica Matemàtica*, 51–59.

Trillas, E., and C. Alsina. 1993. Logic: Going farther from Tarski? *Fuzzy Sets Syst* 53:1–13.

——— 2001. Elkan's theoretical argument, reconsidered. *Int J Approx Reason* 26 (2): 145–52.

Trillas, E., P. P. Bonissone, L. Magdalena, and J. Kacprzyk, eds. 2012. *Combining Experimentation and Theory: A Homage to Abe Mamdani*. Berlin: Springer.

Trillas, E., and L. Valverde. 1981. On some functionally expressable implications for fuzzy set theory. In *Proc 3rd Int Seminar on Fuzzy Set Theory*, 173–90. Linz, Austria: Johannes Kepler Univ.

——— 1984. An inquiry into indistinguishability operators. In Skala, Termini, and Trillas 1984, 231–56.

——— 1985a. On implication and indistinguishability in the setting of fuzzy logic. In *Management Decision Support Systems Using Fuzzy Sets and Possibility Theory*, ed. J. Kacprzyk and R. R. Yager, 198–212. Cologne: Verlag TÜV Rheinland.

——— 1985b. On mode and implication in approximate reasoning. In *Approximate Reasoning in Expert Systems*, ed. M. M. Gupta, A. Kandel, W. Bandler, and J. B. Kiszka, 157–66. Amsterdam: North-Holland.

Türkşen, I. B. 1991. Measurement of membership functions and their acquisition. *Fuzzy Sets Syst* 40 (1): 5–38.

Türkşen, I. B., D. Ulguray, and Q. Wang. 1992. Hierarchical scheduling based on approximate resoning—a comparison with ISIS. *Fuzzy Sets Syst* 46 (3): 349–71.

Turquette, A. R. 1963. Independent axioms for infinite-valued logic. *J Symb Logic* 28 (3): 217–21.

Turunen, E. 1995. Well-defined fuzzy sentential logic. *Math Logic Quart* 41:236–48.

Turunen, E., M. Öztürk, and A. Tsoukiás. 2010. Paraconsistent semantics for Pavelka style fuzzy sentential logic. *Fuzzy Sets Syst* 161:1926–40.

Tversky, A. 1977. Features of similarity. *Psychological Review* 84 (4): 327–52.

Urquhart, A. 1973. An interpretation of many-valued logic. *Z Math Logik Grundlagen Math* 19:111–14.

——— 1986. Many-valued logic. In *Alternatives to Classical Logic*, vol. 2 of *Handbook of Philosophical Logic*, ed. D. M. Gabbay and F. Guenthner, 71–116. Dordrecht: Reidel.

Valverde, L. 1985. On the structure of F-indistinguishability operators. *Fuzzy Sets Syst* 17:313–28.

van Deemter, K. 2010. *Not Exactly: In Praise of Vagueness*. New York: Oxford Univ. Press.

van Fraassen, B. C. 1966. Singular terms, truth-value gaps, and free logic. *J Philosophy* 63:481–95.

van Heijenoort, J., ed. 1967. *From Frege to Gödel: A Source Book in Mathematical Logic, 1879–1931*. Cambridge, MA: Harvard Univ. Press.

Varela, F. J. 1975. A calculus for self-reference. *Int J Gen Syst* 2:5–24.

Vasil'ev, N. A. 1910. О частных суждениях, о треугольнике противоположностей, о законе исключенного четвертого [On partial judgments, on the triangle of oppositions, on the law of excluded fourth, Russian]. *Scientific Notes of Kazan University* 77:1–47.

Verkuilen, J. 2005. Assigning membership in a fuzzy set analysis. *Sociological Methods & Research* 33 (4): 462–96.

Verkuilen, J., R. A. Kievit, and A. Zand Scholten. 2011. Representing concepts by fuzzy sets. In Belohlavek and Klir 2011, 149–76.

Verma, A. K., A. Srividya, and R. S. Prabhu Gaonkar. 2007. *Fuzzy Reliability Engineering: Concepts and Applications*. New Delhi: Narosa.

Vetterlein, T. 2008. A way to interpret Łukasiewicz logic and basic logic. *Studia Logica* 90:407–23.

——— 2012. Vagueness: Where degree-based approaches are useful, and where we can do without. *Soft Computing* 16:1833–44.

Vetterlein, T., H. Mandl, and K.-P. Adlassnig. 2010. Fuzzy Arden syntax: A fuzzy programming language for medicine. *Artificial Intelligence in Medicine* 49 (1): 1–10.

Viertl, R. 1996. *Statistical Methods for Non-Precise Data*. Boca Raton, FL: CRC Press.

Vojtáš, P. 2001. Fuzzy logic programming. *Fuzzy Sets Syst* 124:361–70.

von Altrock, C. 1995. *Fuzzy Logic & NeuroFuzzy Applications Explained*. Englewood Cliffs, NJ: Prentice Hall.

von Neumann, J. 1932. *Mathematische Grundlagen der Quantenmechanik*. Berlin: Springer. Trans. R. T. Beyer 1955. *Mathematical Foundations of Quantum Mechanics*. Princeton, NJ: Princeton Univ. Press.

von Sternberg, R., and G. J. Klir. 1998. Generative archetypes and taxa: A fuzzy set formalization. *Rivista di Biologia (Biology Forum)* 91:403–24.

Vopěnka, P. 1979. *Mathematics in the Alternative Set Theory*. Leipzig: Teubner.

——— 1989a. *Úvod do matematiky v alternatívnej teórii množín* [Introduction to mathematics in the alternative set theory]. Bratislava: Alfa.

——— 1989b. Alternative set theory—all about. In *Proc 1st Symp Mathematics in the Alternative Set Theory*, ed. J. Mlček, M. Benešová, and B. Vojtášková, 28–40. Bratislava: Association of Slovak Mathematicians and Physicists.

——— 1991. The philosophical foundations of alternative set theory. *Int J Gen Syst* 20 (1): 115–26.

Vopěnka, P., and P. Hájek. 1972. *The Theory of Semisets*. Prague: Academia. Amsterdam: North-Holland.

Vychodil, V. 2006. Truth-depressing hedges and BL-logic. *Fuzzy Sets Syst* 157:2074–90.

——— 2007. Direct limits and reduced products of algebras with fuzzy equalities. *J Multiple-Valued Logic & Soft Computing* 13:1–28.

Wade, L. I. 1945. Post algebras and rings. *Duke Mathematical J* 12:389–95.

Waismann, F. 1945–46. Are there alternative logics? *Proc Aristotelian Society* 46:77–104.

Wajsberg, M. 1931. Aksjomatyzacja trójwartościowego rachunku zdań. *Comptes Rendus des Séances de la Société des Sciences et des Lettres de Varsovie, Cl. iii* 24:126–45. Trans. B. Gruchman and S. McCall 1967. Axiomatization of the three-valued propositional calculus. In McCall 1967, 264–84.

——— 1935. Beiträge zum Metaaussagenkalkül I [Contributions to propositional metacalculus I]. *Monatshefte für Mathematik und Physik* 42:221–42.

Wang, G. J. 1999. On the logic foundation of fuzzy reasoning. *Inf Sci* 117:47–88.

Wang, L. X. 1992. Fuzzy systems are universal approximators. In *Proc 1st IEEE Int Conf on Fuzzy Systems*, 1163–70. San Diego, CA: IEEE Press.

——— 1994. *Adaptive Fuzzy Systems and Control: Design and Stability Analysis*. Englewood Cliffs, NJ: Prentice Hall.

Wang, P. Z., and E. Sanchez. 1982. Treating a fuzzy subset as a projectable random subset. In *Fuzzy Information and Decision Processes*, ed. M. M. Gupta and E. Sanchez, 213–20. Amsterdam: North-Holland.

Wang, Z., and G. J. Klir. 2009. *Generalized Measure Theory*. New York: Springer.

Ward, M., and R. P. Dilworth. 1939. Residuated lattices. *Trans AMS* 45:335–54.

Warren, J. D., R. V. Demicco, and L. R. Bartek. 2007. Fuzzy predictive earth analysis constrained by heuristics applied to stratigraphic modeling. In *Fuzzy Logic: A Spectrum of Theoretical & Practical Issues*, ed. P. P. Wang, D. Ruan, and E. E. Kerre, 381–430. Berlin: Springer.

Watts, I. 1724. *Logick: Or, the Right Use of Reason in the Inquiry after Truth, with a Variety of Rules To Guard against Error in the Affairs of Religion and Human Life, As Well As in the Sciences*. London. New edition in 1781 by Charles Elliot.

Weber, S. 1983. A general concept of fuzzy connectives, negations and implications based on t-norms and t-conorms. *Fuzzy Sets Syst* 11:115–34.

Wechler, W. 1978. *The Concept of Fuzziness in Automata and Language Theory*. Berlin: Akademie-Verlag.

Wee, W. G., and K. S. Fu. 1969. A formulation of fuzzy automata and its application as a model of learning systems. *IEEE Trans Syst Sci Cyb* 5 (3): 215–23.

Weidner, A. J. 1974. *An Axiomatization of Fuzzy Set Theory*. PhD diss., Univ. Notre Dame.

————1981. Fuzzy sets and Boolean-valued universes. *Fuzzy Sets Syst* 6:61–72.

Weiss, M. D. 1975. Fixed points, separation, and induced topologies for fuzzy sets. *J Math Anal Appl* 50 (1): 142–50.

Weyl, H. 1940. The ghost of modality. In *Philosophical Essays in Memory of Edmund Husserl*, ed. M. Farber, 278–303. Cambridge, MA: Harvard Univ. Press.

White, R. B. 1979. The consistency of the axiom of comprehension in the infinite-valued predicate logic of Łukasiewicz. *J Philosophical Logic* 8:509–34.

Whitehead, A. N., and B. Russell. 1910–13. *Principia Mathematica*. 3 vols. Cambridge, UK: Cambridge Univ. Press.

Wiedermann, J. 2002. Fuzzy Turing machines revised. *Computing and Informatics* 21:1–13.

Williamson, T. 1994. *Vagueness*. London: Routledge.

Willmott, R. 1986. On the transitivity of containment and equivalence in fuzzy power set theory. *J Math Anal Appl* 120:384–96.

Wójcicki, R. 1973. On matrix representations of consequence operations of Łukasiewicz's sentential calculi. *Z Math Logik Grundlagen Math* 19:239—247.

Woleński, J. 2004. Polish logic. *Logic J IGPL* 12 (5): 399–428.

Wolf, R. G. 1977. A survey of many-valued logic (1966–1974). In Dunn and Epstein 1977, 167–323.

Wong, C. K. 1973. Covering properties of fuzzy topological spaces. *J Math Anal Appl* 43:697–704.

————1974a. Fuzzy topology: Product and quotient theorems. *J Math Anal Appl* 45:512–21

————1974b. Fuzzy points and local properties of fuzzy topology. *J Math Anal Appl* 46:316–28.

Wong, F., K. Chou, and J. Yao. 1999. Civil engineering including earthquake engineering. In Zimmermann 1999, 207–45.

Wood, K. L., K. N. Otto, and E. K. Antonsson. 1992. Engineering design calculations with fuzzy parameters. *Fuzzy Sets Syst* 52 (1): 1–20.

Woodger, J. H., ed. 1956. *Logic, Semantics, Metamathematics: Papers from 1923 to 1938 by Alfred Tarski*. Oxford, UK: Clarendon Press. 2nd ed.: Corcoran J., ed. 1983. Indianapolis, IN: Hackett.

Woolf, P. J., and Y. Wang. 2000. A fuzzy logic approach to analyzing gene expression data. *Physiological Genomics* 3 (1): 9–15.

Wu, W. 1986. Fuzzy reasoning and fuzzy relational equations. *Fuzzy Sets Syst* 20 (1): 67–78.

Wygralak, M. 1983. Fuzzy inclusion and fuzzy equality of two fuzzy subsets, fuzzy operations for fuzzy subsets. *Fuzzy Sets Syst* 10:157–68.

————2003. *Cardinalities of Fuzzy Sets*. Berlin: Springer.

Wyler, O. 1995. Fuzzy logic and categories of fuzzy sets. In Höhle and Klement 1995, 235–68.

Xiang, G., and V. Kreinovich. 2013. Towards fast and accurate algorithms for processing fuzzy data: Interval computations revisited. *Int J Gen Syst* 42:197–223.

Xu, D., J. M. Keller, M. Popescu, and R. Bondugula. 2008. *Applications of Fuzzy Logic in Bioinformatics*. London: Imperial College Press.

Xu, J., and Z. Zeng. 2014. *Fuzzy-Like Multiple Objective Multistage Decision Making*. Cham, Switzerland: Springer.

Xu, J., and X. Zhou. 2011. *Fuzzy-Like Multiple Objective Decision Making*. Berlin: Springer.

Xu, R., and D. C. Wunsch II. 2009. *Clustering*. Hoboken, NJ: Wiley.

Yager, R. R. 1980a. On a general class of fuzzy connectives. *Fuzzy Sets Syst* 4 (3): 235–42.

————1980b. Quantified propositions in a linguistic logic. In *Proc 2nd Int Seminar on Fuzzy Set Theory*, ed. E. P. Klement, 69–124. Linz, Austria: Johannes Kepler Univ.

————1982a. A new approach to the summarization of data. *Inf Sci* 28:69–86.

————1982b. Generalized probabilities of fuzzy events from fuzzy belief structures. *Inf Sci* 28:45–62.

————1983. Quantified propositions in a linguistic logic. *Int J Man Mach Stud* 19 (2): 195–227.

————1985. Reasoning with fuzzy quantified statements: Part I. *Kybernetes* 14:233–40.

————1986a. Reasoning with fuzzy quantified statements: Part II. *Kybernetes* 15:111–20.

———— 1986b. A modification of the certainty measure to handle subnormal distributions. *Fuzzy Sets Syst* 20 (3): 317–24.

———— 1988. On ordered weighted averaging aggregation operators in multicriteria decision making. *IEEE Trans Syst Man Cyb* 18 (1): 183–90.

Yager, R. R., and D. P. Filev. 1994. *Essentials of Fuzzy Modeling and Control*. New York: Wiley.

Yager, R. R., S. Ovchinnikov, R. M. Tong, and H. T. Nguyen, eds. 1987. *Fuzzy Sets and Applications: Selected Papers by L. A. Zadeh*. New York: Wiley.

Yager, R. R., and A. Rybalov. 1996. Uninorm aggregation operators. *Fuzzy Sets Syst* 80 (1): 111–20.

Yamakawa, T. 1989. Stabilization of an inverted pendulum by a high-speed fuzzy logic controller hardware system. *Fuzzy Sets Syst* 32 (2): 161–80.

Yamakawa, T., and T. Miki. 1986. The current mode fuzzy logic integrated circuits fabricated by the standard CMOS process. *IEEE Trans on Computers* 35 (2): 161–67.

Yao, W., and L.-X. Lu. 2009. Fuzzy Galois connections on fuzzy posets. *Math Logic Quart* 55:105–12.

Yasunobu, S., and S. Miyamoto. 1985. Automatic train operation system by predictive fuzzy control. In *Industrial Applications of Fuzzy Control*, ed. M. Sugeno, 1–18. Amsterdam: North-Holland.

Yatabe, S. 2007. Distinguishing non-standard natural numbers in a set theory within Łukasiewicz logic. *Arch Math Logic* 46:281–87.

Yen, J. 1990. Generalizing the Dempster-Shafer theory to fuzzy sets. *IEEE Trans Syst Man Cyb* 20 (3): 559–69.

Ying, H. 1994. Sufficient conditions on general fuzzy systems as function approximators. *Automatica* 30 (3): 521–25.

———— 1998. General Takagi-Sugeno fuzzy systems with simplified linear rule consequent are universal controllers, models and filters. *Inf Sci* 108:91–107.

Ying, M. 1991. Deduction theorem for many-valued inference. *Z Math Logik Grundlagen Math* 37:533–37.

———— 1991–93. A new approach to fuzzy topology (I), (II), (III). *Fuzzy Sets Syst* 39 (1991): 303–21; 47 (1992): 221–32; 55 (1993): 193–207.

———— 1992a. The fundamental theorem of ultraproduct in Pavelka's logic. *Z Math Logik Grundlagen Math* 38:197–201.

———— 1992b. Compactness, the Löwenheim-Skolem property and the direct-product of lattices of truth values. *Z Math Logik Grundlagen Math* 38:521–24.

———— 1993. Fuzzifying topology based on complete residuated lattice-valued logic (I). *Fuzzy Sets Syst* 56:337–73.

———— 1994. A logic for approximate reasoning. *J Symb Logic* 59:830–37.

Young, R. C. 1931. The algebra of many-valued quantities. *Mathematische Annalen* 104 (1): 260–90.

Zabrodsky, H., and D. Avnir. 1995a. Continuous symmetry measures. 4. Chirality. *J American Chem Soc* 117 (1): 462–73.

———— 1995b. Measuring symmetry in structural chemistry. In *Advances in Molecular Structure Research*, vol. 1, ed. M. Hargittai and I. Hargittai, 1–31. London: JAI Press.

Zabrodsky, H., S. Peleg, and D. Avnir. 1992. Continuous symmetry measures. *J American Chem Soc* 114 (20): 7843–51.

———— 1993. Continuous symmetry measures. 2. Symmetry groups and the tetrahedron. *J American Chem Soc* 115 (24): 8278–89.

Zach, R. 1999. Completeness before Post: Bernays, Hilbert, and the development of propositional logic. *Bull Symbolic Logic* 5 (3): 331–66.

Zadeh, L. A. 1962. From circuit theory to system theory. *Proc IRE* 50 (5): 856–65.

———— 1965a. Fuzzy sets. *Inf Control* 8 (3): 338–53.

———— 1965b. Fuzzy sets and systems. In *Proc Symp on System Theory*, 29–37. Brooklyn, NY: Polytechnic Press.

———— 1966. Тени нечетких множеств. *Problemy Peredachi Informatsii* 2 (1): 37–44. Trans. L. A. Zadeh 1996. Shadows of fuzzy sets. In Klir and Yuan 1996, 51–59.

———— 1968a. Probability measures and fuzzy events. *J Math Anal Appl* 23 (2): 421–27.

———— 1968b. Fuzzy algorithms. *Inf Control* 12 (2): 94–102.

———— 1969. Biological applications of the theory of fuzzy sets and systems. In *Proc Int Symp on the Biocybernetics of the Central Nervous System*, ed. L. D. Proctor, 199–206. Boston: Little, Brown.

———— 1971a. Quantitative fuzzy semantics. *Inf Sci* 3 (2): 159–76.

———— 1971b. Similarity relations and fuzzy orderings. *Inf Sci* 3 (2): 177–200.

———— 1971c. Towards a theory of fuzzy systems. In *Aspects of Network and System Theory*, ed. R. E. Kalman and R. N. De Claris, 469–90. New York: Holt, Rinehart & Winston.

———— 1972a. A fuzzy-set-theoretic interpretation of linguistic hedges. *J Cybernetics* 2 (3): 4–34.

———— 1972b. A rationale for fuzzy control. *J Dynamic Systems, Measurement, and Control* 94 (1): 3–4.

———— 1973. Outline of a new approach to the analysis of complex systems and decision processes. *IEEE Trans Syst Man Cyb* 3 (1): 28–44.

————— 1974a. Fuzzy logic and its application to approximate reasoning. In *Information Processing 74: Proc IFIP Congress 74*, ed. J. L. Rosenfeld, 591–94. Amsterdam: North-Holland.

————— 1974b. A new approach to systems analysis. In *Proc 2nd Int Conf on Man and Computer*, ed. M. Marois, 55–94. Amsterdam: North-Holland.

————— 1975a. Calculus of fuzzy restrictions. In Zadeh et al. 1975, 1–39.

————— 1975b. Fuzzy logic and approximate reasoning. *Synthese* 30:407–28.

————— 1975c. The concept of linguistic variable and its application to approximate reasoning—I, II, III. *Inf Sci* 8:199–249; 8:301–57; 9:43–80.

————— 1976a. A fuzzy-algorithmic approach to the definition of complex or imprecise concepts. *Int J Man Mach Stud* 8 (3): 249–91.

————— 1976b. The linguistic approach and its application to decision analysis. In *Directions in Large-Scale Systems*, ed. Y. C. Ho and S. K. Mitter, 339–70. New York: Plenum Press.

————— 1977. Fuzzy sets and their application to pattern classification and clustering analysis. In *Classification and Clustering*, ed. J. Van Ryzin, 251–99. New York: Academic Press.

————— 1978a. PRUF—a meaning representation language for natural languages. *Int J Man Mach Stud* 10 (4): 395–460.

————— 1978b. Fuzzy sets as a basis for a theory of possibility. *Fuzzy Sets Syst* 1 (1): 3–28.

————— 1979a. A theory of approximate reasoning. In *Machine Intelligence*, ed. J. E. Hayes, D. Michie, and L. I. Mikulich, 149–94. New York: Halstead Press.

————— 1979b. Fuzzy sets and information granularity. In *Advances in Fuzzy Set Theory and Applications*, ed. M. M. Gupta, R. K. Ragade, and R. R. Yager, 3–18. Amsterdam: North-Holland.

————— 1979c. Liar's paradox and truth-qualification principle. ERL Memorandum M79/34, Univ. of California, Berkeley. Reprinted in Klir and Yuan 1996, 449–63.

————— 1981a. Possibility theory and soft data analysis. In *Mathematical Frontiers of the Social and Policy Sciences*, ed. L. Cobb and R. M. Thrall, 69–129. Boulder, CO: Westview Press.

————— 1981b. Test-score semantics for natural languages and meaning representation via PRUF. In *Empirical Semantics*, ed. B. Rieger, 281–349. Bochum, Germany: Studienverlag Brockmeyer.

————— 1982a. A note on prototype theory and fuzzy sets. *Cognition* 12:291–97.

————— 1982b. Fuzzy systems theory: A framework for the analysis of humanistic systems. In *Systems Methodology in Social Science Research: Recent Developments*, ed. R. Cavallo, 25–41. Boston: Kluwer.

————— 1983a. A computational approach to fuzzy quantifiers in natural languages. *Computers and Mathematics with Applications* 9 (1): 149–84.

————— 1983b. The role of fuzzy logic in the management of uncertainty in expert systems. *Fuzzy Sets Syst* 11 (3): 199–227.

————— 1984. Precisiation of meaning via translation into PRUF. In *Cognitive Constraints and Communication*, ed. L. Vaina and J. Hintikka, 373–401. Boston: Reidel.

————— 1986. Outline of a theory of usuality based on fuzzy logic. In *Fuzzy Sets Theory and Applications*, ed. A. Jones, A. Kaufmann, and H.-J. Zimmermann, 79–97. Dordrecht: Reidel.

————— 1987. A computational theory of dispositions. *Int J Intell Syst* 2 (1): 39–63.

————— 1994a. Fuzzy logic, neural networks, and soft computing. *Commun ACM* 37 (3): 77–84.

————— 1994b. Soft computing and fuzzy logic. *IEEE Software* 11 (6): 48–56.

————— 1996a. Fuzzy logic and the calculi of fuzzy rules and fuzzy graphs: A précis. *J Multiple-Valued Logic* 1 (1): 1–38.

————— 1996b. Fuzzy logic = computing with words. *IEEE Trans Fuzzy Syst* 4 (2): 103–11.

————— 1999a. The birth and evolution of fuzzy logic—a personal perspective. *Journal of Japan Society for Fuzzy Theory and Systems* 11 (6): 891–905.

————— 1999b. From computing with numbers to computing with words: From manipulation of measurements to manipulation of perceptions. *IEEE Trans on Circuits and Systems—I: Fundamental Theory and Applications* 45 (1): 105–19.

————— 2005. Toward a generalized theory of uncertainty (GTU)—an outline. *Inf Sci* 172 (1–2): 1–40.

————— 2006. Generalized theory of uncertainty (GTU): Principal concepts and ideas. *Computational Statistics and Data Analysis* 51 (1): 15–46.

————— 2007. Joseph Amadee Goguen (1941–2006)—a personal tribute. *Fuzzy Sets Syst* 158 (3): 809–10.

————— 2015. Fuzzy logic—a personal perspective. *Fuzzy Sets Syst* 281:4–20.

Zadeh, L. A., K. S. Fu, K. Tanaka, and M. Shimura, eds. 1975. *Fuzzy Sets and Their Applications to Cognitive and Decision Processes*. New York: Academic Press.

Zavala, A. H., and O. C. Nieto. 2012. Fuzzy hardware: A retrospective and analysis. *IEEE Trans Fuzzy Syst* 20 (4): 623–35.

Zawirski, Z. 1931a. *W sprawie indeterminizmu fizyki kwantowej* [On the indeterminism of quantum physics]. Lvov: Książnica-Atlas.

———— 1931b. Logika trójwartościowa Jana Łukasiewicza. O logice L. E. J. Brouwera. Próby stosoqania logiki wielowartościowej do współczesnego przyrodoznawstwa [Jan Łukasiewicz's three-valued logic. On the logic of L. E. J. Brouwer. Attempts at applications of many-valued logic to contemporary natural science]. *Sprawozdania Poznańskiego Towatzystwa Przyjaciol Nauk* 2:35–42.

———— 1934a. Znaczenie logiki wielowartościowej dla poznania i związek jej z rachunkiem prawdopodobieństwa [Significance of many-valued logic for cognition and its connection with the calculus of probability]. *Przegląd Filozoficzny* 37:393–98.

———— 1934b. Stosunek logiki wielowartościowej do rachunku prawdopodobieństwa [The relation between many-valued logic and probability calculus]. *Prace Komisji Filozoficznej Poznańskiego Towarystwa Przyjaciól Nauk* 4:155–240. Short English summary available in *Zygmunt Zawirski: His Life and Work, with Selected Writings on Time, Logic & the Methodology of Science*, ed. I. Szumilewicz-Lachman, 112–16. Dordrecht: Kluwer.

———— 1935. Über das Verhältnis mehrwertiger Logik zur Wahrscheinlichkeitsrechnung [On the relationalship of many-valued logic to the calculus of probability]. *Studia Philosophica* 1:407–42

Zemankova-Leech, M., and A. Kandel. 1984. *Fuzzy Relational Databases: A Key to Expert Systems*. Cologne: Verlag TÜV Rheinland.

Zétényi, T., ed. 1988. *Fuzzy Sets in Psychology*. Amsterdam: North-Holland.

Zhang, D. 2010. Implication structures, fuzzy subsets, and enriched categories. *Fuzzy Sets Syst* 161:1205–23.

Zhang, H., X. Ma, W. Xu, and P. Z. Wang. 1993. Design fuzzy controllers for complex systems with an application to 3-stage inverted pendulums. *Inf Sci* 72 (3): 271–84.

Zhang, J. 1983. Fuzzy set structure with strong implication. In *Advances in Fuzzy Sets, Possibility Theory, and Applications*, ed. P. P. Wang, 107–36. New York: Plenum Press.

Zhang, R., Y. A. Phillis, and V. S. Kouikoglou. 2005. *Fuzzy Control of Queuing Systems*. London: Springer.

Zhang, Y., J. Wang, and H. Li. 2012. Stabilization of the quadruple inverted pendulum by variable universe adaptive fuzzy controller based on variable gain H_{∞} regulator. *J Systems Science and Complexity* 25:856–72.

Zhao, H., and D. Zhang. 2008. Many valued lattices and their representations. *Fuzzy Sets Syst* 159:81–94.

Zich, O. V. 1938. Výrokový počet s komplexními hodnotami [Propositional calculus with complex values]. *Česká mysl* 34:189–96.

Zimmermann, H.-J. 1976. Description and optimization of fuzzy systems. *Int J Gen Syst* 2 (4): 209–15.

———— 1985. *Fuzzy Set Theory—and Its Applications*. 4th ed. Boston: Kluwer.

———— 1987. *Fuzzy Sets, Decision Making, and Expert Systems*. Boston: Kluwer.

———— ed. 1999. *Practical Applications of Fuzzy Technologies*. Boston: Kluwer.

Zimmermann, H.-J., L. A. Zadeh, and B. R. Gaines, eds. 1984. *Fuzzy Sets and Decision Analysis*. Amsterdam: North-Holland.

Zimmermann, H.-J., and P. Zysno. 1980. Latent connectives in human decision making. *Fuzzy Sets Syst* 4 (1): 37–51.

Zingales, G., and L. Mari. 2000. Uncertainty in the measurement science. In *Measurement Science: A Discussion*, ed. K. Kariya and L. Finkelstein, 135–47. Tokyo: Ohmsha.

Zinov'ev, A. A. 1963. *Philosophical Problems of Many-Valued Logic*. Dordrecht: Reidel.

Zlatoš, P. 1997. The alternative set theory and fuzzy sets. In *Proc IFSA 1997*, 36–47. Prague: Academia.

Zwick, R., E. Carlstein, and D. V. Budescu. 1987. Measures of similarity among fuzzy concepts: A comparative analysis. *Int J Approx Reason* 1:221–42.

Zwicky, F. 1933. On a new type of reasoning and some of its possible consequences. *Physical Review* 43 (12): 1031–33.

Name Index

Subject Index

Printed in the USA/Agawam, MA
January 25, 2018

668460.001